S0-AHS-545

Handbook of MINERALOGY

VOLUME I

Elements, Sulfides, Sulfosalts

John W. Anthony
Richard A. Bideaux
Kenneth W. Bladh
Monte C. Nichols

MINERAL DATA PUBLISHING
Tucson, Arizona

Credit: Crystal drawing of epitaxial overgrowths of tennantite on octahedral pyrite from Quiruvilca, Peru. See J.W. Anthony and W.J. McLean (1973) *Mineral. Record,* 4, 159–163.

IBM-PC is a trademark of International Business Machines.

TeX is a trademark of the American Mathematical Society.

Library of Congress Cataloging-in-Publication Data

Handbook of mineralogy / by John W. Anthony ... [et al.].
p. cm.
Includes bibliographical references.
Contents: v. 1. Elements, sulfides, sulfosalts.
ISBN 0-9622097-0-8 (v. 1) :
1. Mineralogy–Handbooks, manuals, etc. I. Anthony, John W. (John Williams), 1920–
QE366.8.H36 1990
549–dc20 89-13673
CIP

9 8 7 6 5 4

Note: It is not practical for a book to provide complete cross-referencing between mineral names and elemental constituents. Accordingly, a computer program, **SEARCH,** for searching the chemical formulae of all mineral species, with its associated up-to-date file of mineral formulae and names, has been written. This program runs on any IBM-PC or compatible computer having 640K memory and MS-DOS 2.2 or above. It is supplied on a $5\frac{1}{4}$ inch, 360K floppy disk. Further information is available from Mineral Data Publishing at the address below.

Mineral Data Publishing
P.O. Box 37072
Tucson, Arizona 85740

ACKNOWLEDGEMENTS

We have benefited from the encouragement, constructive criticism, and cheerful skepticism of many of our colleagues and associates and happily acknowledge their helpful interest in this project. We are particularly indebted to Richard Erd for his critical review of the overall style of this work, and to John White, Jr., William Pinch, and Hans-Jürgen Wilke for their knowledgeable checking of locality data. Of course, remaining errors are ours.

Our colleagues Steven Steinke, William Wilkerson, and Michael Wood have done a great deal of the culling of mineral data from the literature and we are pleased to acknowledge both their help and good mineralogical judgement.

Much of the spectral reflectance data in the first volume has been taken from the compilation prepared within the Department of Geology and Geophysics of the University of Missouri at Rolla.

The *Mineral Powder Diffraction File Data Book* of the Joint Committee on Powder Diffraction Standards — International Centre for Powder Diffraction Standards was consulted for comparison to our original literature sources, and to insure the completeness of our data. In those cases where a mineral's pattern was available only from the JCPDS, they have generously granted us permission for reproduction.

The relative importance of localities for inclusion in the Distribution section was judged in part through examination of the following extensive collections, by courtesy of their curatorial staffs: the National Museum of Natural History, Washington, D.C., through John White, Jr.; Harvard University, Cambridge, Massachusetts, through Carl Francis; in London, the British Museum (Natural History), through Peter Embrey and Robert Symes; and in Paris, at the School of Mines, Claude Guillemin and Jean-François Poullen; at the Sorbonne, Pierre Bariand; and at the Natural History Museum, Henri-Jean Schubnel.

Peter Embrey, Jeanette Bideaux, and John Trelawney were of great assistance on certain linguistic and grammatical matters. Barbara Nichols spent long hours carefully checking various drafts for accuracy and consistency.

Martha Scott and Karen Schwartz patiently deciphered and entered the information from our hand-written forms into the initial computer files. Curtis Schuh several times has converted this ever-increasing mass of data from one computer system to another.

Typesetting for draft and final book pages of this volume was accomplished using the TeX computerized typesetting system developed at Stanford University by Donald Knuth. Use of this system greatly eased our burden of producing typographically elegant printed pages directly from a database.

We have been able to examine original literature in nearly all cases, through the Science Library of the University of Arizona, the Orton Memorial Library of Geology at the Ohio State University, the Branner Geology Library at Stanford University, and the USGS libraries at Menlo Park and Denver. Access to obscure references was facilitated by use of the abstracts provided by the *American Mineralogist*, *Mineralogical Abstracts*, and *Chemical Abstracts*. These were especially useful for their coverage and translation of the Russian and Chinese literature.

For this first volume, the major international mineralogical journals were examined through 1988. Numerous less accessible references were called to our attention through use of the USGS Ford-Fleischer File, the latest edition of which encompasses the mineralogical literature through 1987. Still, we recognize that our coverage of the literature is perforce less than perfect, and we solicit pertinent information for inclusion of corrections and additions in future volumes and re-issues.

John W. Anthony
Richard A. Bideaux
Kenneth W. Bladh
Monte C. Nichols

January, 1990

INTRODUCTION

The four decades following the appearance of the first two volumes of the seventh edition of *Dana's System of Mineralogy* have seen the number of known mineral species more than double — from about 1500 species at mid-century to something over 3300 at this writing.

This proliferation stems from several factors: the continuing development of a firm base of physical and chemical theory for the crystalline state, the appearance of a panoply of remarkably sophisticated instrumental methods nicely suited to the study of minerals, and the timely blossoming of computer science, principal progeny of the transistor.

The digital computer and its associated software have not only eased the burden of that computationally most arduous of crystallographic tasks, the determination of atomic sites within the unit cell, they have markedly accelerated the collection and interpretation of all manner of mineralogical data. Indeed, their presence is now central to the operation of essentially all analytical instrumentation having any degree of complexity.

This *Handbook of Mineralogy* series was conceived in order to gather in convenient form the data crucial to identification of all mineral species and to provide relatively up-to-date references containing information central to the definition of each species. Our intent is to provide data sufficient to distinguish a species from all others. If additional information seems desirable, primary or secondary references are given for each mineral in which the species is further discussed.

In undertaking this compilation, there is no intention of emulating the *System* or comparable scholarly resources. Rather, our interest is in creating a useful stopgap until a modern work comparable in scope and stature to the *System* becomes available.

A cursory glance will reveal what may appear to be a somewhat unconventional organizational scheme. The well-known chemical-structural classification of the *System* has been adopted for the pattern of the overall work because, from a geological point of view, this classification has the advantage of tending to cluster together minerals of similar provenance. However, within each volume of this work, the minerals have been arranged alphabetically. We have selected this scheme to facilitate the location of individual minerals.

The distribution of minerals among the five volumes is indicated in Table 1. Major categories are listed in bold type in the table and on the spines of the corresponding volumes as a quick guide to the contents. Minor categories are listed in alphabetical order following the major categories.

We have distilled the pertinent facts defining each mineral to one page. Although this concentration may result in seemingly callous winnowing of sacred classical descriptive matter, such scholarly facts, peripheral to our present objectives, can always be consulted in the literature referenced for each species. The hallowed angle table and crystal drawings were early victims of this parsimony.

For species names and chemical formulae, principal guides are the *American Mineralogist*, the recommendations of the International Mineralogical Association's Commission on New Minerals and Mineral Names, and Michael Fleischer's *Glossary of Mineral Species*. Where the philosophies of these are in conflict, we attempt to steer a prudent course.

We have tended to incorporate data from the naturally occurring mineral, preferably type material where available, rather than from its synthetic equivalent. Where data are missing and should be determined, the designation "n.d." (not determined) appears, so highlighting species for which essential observations are lacking.

Our order of presentation of mineral properties should prove an expeditious alternative to conventional formats. In our departure from those conventions we have tried to establish logical groupings of related properties. While we believe these arrangements to be consistent and self-explanatory, we present an explanatory

Table 1. Distribution of mineral categories among the *Handbook of Mineralogy* volumes. Major categories are listed in bold type followed by minor categories in alphabetical order.

VOLUME I	VOLUME II	VOLUME III	VOLUME IV	VOLUME V
Elements	**Silica**	**Halides**	**Arsenates**	**Borates**
Sulfides	**Silicates**	**Hydroxides**	**Phosphates**	**Carbonates**
Sulfosalts		**Oxides**	**Uranates**	**Sulfates**
Alloys		Antimonates	**Vanadates**	Chromates
Antimonides		Antimonites		Germanates
Arsenides		Arsenites		Iodates
Bismuthinides		Carbides		Molybdates
Intermetallics		Nitrides		Nitrates
Selenides		Phosphides		Organics
Sulfhalides		Silicides		Selenates
Sulfoxides		V-oxysalts		Selenites
Tellurides				Sulfites
				Tellurates
				Tellurites
				Tungstates

page at the end of this introduction as a further guide to both the order and content of the mineral entries presented in this volume.

X-ray powder diffraction lines and their intensities are listed for all those minerals for which they were available, preference again being given to patterns from natural material whenever possible. The origin of the X-ray powder data is included in the references, and the locality or source of the sample precedes each pattern listing.

Up to four chemical analyses have been selected from the literature to illustrate the range of substitution in a mineral's structure, usually for major elements only. We often include an idealized analysis for comparison, calculated especially for this work using the atomic weights given in Table 2.

Localities given under "Distribution" are limited to about a dozen worldwide. This seems not unduly restrictive since about half of the minerals are known from only a single locality, and a further quarter from no more than six. For the less common and rare species we have attempted to provide comprehensive locality information.

For species having numerous localities, criteria for inclusion of a locality are the occurrence of especially well crystallized, pure, or abundant material. We have tended to drop occurrences stated as "reported from" in the older literature; these often require authentication by modern methods. We have also tried to avoid incompletely given locality data unless they are the only data available for a species. Locality information is not referenced.

The whereabouts of type material is noted usually only if mentioned in one of the references, or if the material is in the National Museum of Natural History or Harvard University. If unknown to us, this is noted as "n.d." only if the mineral was described after 1900. We specify neither the nature of type material, whether holotype, cotype, etc., nor the type locality.

In our reviews of the literature we have tried to minimize distortions of the facts we have selected. In general, we have not listed those references from which we have extracted no data, the principal exception being papers on crystal structures which we mention whether or not we actually quote from them. We always provide the full reference, even for those important references which occur repeatedly. Abbreviations of mineralogical terms largely conform to the practice of the *American Mineralogist*.

Table 2. Listing of element names, element symbols, atomic numbers and atomic weights used for calculations of density, Z, and idealized chemical analyses made especially for these volumes. The values for the atomic weights are those given in: Commission on Atomic Weights and Isotopic Abundances (1986) Atomic weights of the elements 1985. *Pure and Applied Chemistry,* 58, 1677–1692. (© 1986 IUPAC).

Element	Sym.	At. No.	At. Wt.	Element	Sym.	At. No.	At. Wt.
Actinium†	Ac	89	227.028	Mendelevium†	Md	101	258.10
Aluminum	Al	13	26.982	Mercury	Hg	80	200.59
Americium†	Am	95	243.061	Molybdenum	Mo	42	95.94
Antimony	Sb	51	121.75	Neodymium	Nd	60	144.24
Argon	Ar	18	39.948	Neon	Ne	10	20.180
Arsenic	As	33	74.922	Neptunium†	Np	93	237.048
Astatine†	At	85	209.987	Nickel	Ni	28	58.69
Barium	Ba	56	137.327	Niobium	Nb	41	92.906
Berkelium†	Bk	97	249.075	Nitrogen	N	7	14.007
Beryllium	Be	4	9.012	Nobelium†	No	102	259.101
Bismuth	Bi	83	208.980	Osmium	Os	76	190.2
Boron	B	5	10.811	Oxygen	O	8	15.999
Bromine	Br	35	79.904	Palladium	Pd	46	106.42
Cadmium	Cd	48	112.411	Phosphorus	P	15	30.974
Calcium	Ca	20	40.078	Platinum	Pt	78	195.08
Californium†	Cf	98	242.059	Plutonium†	Pu	94	244.064
Carbon	C	6	12.011	Polonium†	Po	84	208.982
Cerium	Ce	58	140.115	Potassium	K	19	39.098
Cesium	Cs	55	132.905	Praseodymium	Pr	59	140.908
Chlorine	Cl	17	35.453	Promethium†	Pm	61	144.913
Chromium	Cr	24	51.996	Protactinium†	Pa	91	231.036
Cobalt	Co	27	58.933	Radium†	Ra	88	226.025
Copper	Cu	29	63.546	Radon†	Rn	86	222.018
Curium†	Cm	96	247.070	Rhenium	Re	75	186.207
Dysprosium	Dy	66	162.50	Rhodium	Rh	45	102.906
Einsteinium†	Es	99	252.083	Rubidium	Rb	37	85.468
Erbium	Er	68	167.26	Ruthenium	Ru	44	101.07
Europium	Eu	63	151.965	Samarium	Sm	62	150.36
Fermium†	Fm	100	257.095	Scandium	Sc	21	44.956
Fluorine	F	9	18.998	Selenium	Se	34	78.96
Francium†	Fr	87	223.020	Silicon	Si	14	28.086
Gadolinium	Gd	64	157.25	Silver	Ag	47	107.868
Gallium	Ga	31	69.723	Sodium	Na	11	22.990
Germanium	Ge	32	72.61	Strontium	Sr	38	87.62
Gold	Au	79	196.966	Sulfur	S	16	32.066
Hafnium	Hf	72	178.49	Tantalum	Ta	73	180.948
Helium	He	2	4.003	Technetium†	Tc	43	97.907
Holmium	Ho	67	164.930	Tellurium	Te	52	127.60
Hydrogen	H	1	1.008	Terbium	Tb	65	158.925
Indium	In	49	114.82	Thallium	Tl	81	204.383
Iodine	I	53	126.904	Thorium†	Th	90	232.038
Iridium	Ir	77	192.22	Thulium	Tm	69	168.934
Iron	Fe	26	55.847	Tin	Sn	50	118.710
Krypton	Kr	36	83.80	Titanium	Ti	22	47.88
Kurchatovium	Ku	104	261.11	Tungsten	W	74	183.85
Lanthanum	La	57	138.906	Uranium†	U	92	238.029
Lawrencium†	Lr	103	260.105	Vanadium	V	23	50.942
Lead	Pb	82	207.2	Xenon	Xe	54	131.29
Lithium	Li	3	6.941	Ytterbium	Yb	70	173.04
Lutetium	Lu	71	174.967	Yttrium	Y	39	88.906
Magnesium	Mg	12	24.305	Zinc	Zn	30	65.39
Manganese	Mn	25	54.938	Zirconium	Zr	40	91.224

† These elements have no stable isotopes; the atomic weight of the most stable isotope is listed.

Mineral Name **Idealized Chemical Formula**

Crystal Data: Crystal system and point group in Hermann-Mauguin symbols. Description of the visual appearance of single crystals and aggregates. Common crystal forms and twin laws are indicated.

Physical Properties: Megascopic and simple measurable properties of the pure mineral, especially those depending on cohesion. Density is in g/cm^3.

Optical Properties: Characteristics depending on the interaction with white light, first with a hand sample (transparency, color, and streak); secondly in polished or thin section, viewed through a reflecting or polarizing light microscope (color, pleochroism, and anisotropism). Spectral reflectance data (R or R_1–R_2) are given as available, with each wavelength (in nanometers) listed in parentheses followed by the corresponding reflectance value in percent.

Cell Data: Space group symbol and unit cell parameters in Ångstrom units.

X-ray Powder Pattern: Source of material, followed by up to seven most intense maxima, presented as d-spacing in Å followed by relative intensity in parentheses. Maxima known to be unresolved doublets have a "d" appended to their respective intensity values. A "b" is appended to the intensity of those broadened for other reasons.

Chemistry: Determinations of major and minor elements in the natural mineral, reported as weight percentages. Non-idealized empirical formulae are from the references; calculated analyses are based on idealized chemical formulae. Sources of the materials analyzed are given.

Polymorphism & Series: Noted as appropriate. Series are isomorphous with complete substitution and continuous variation of properties with composition.

Occurrence: A description of the geological processes and physicochemical environments inferred for the species' origin.

Association: Minerals in close spatial and inferred genetic relationship with the species. In order of closest association (equilibrium) to farthest, or most abundant to least, where known. Multiple lists indicate species with multiple parageneses, and their sources.

Distribution: The most important geographic localities for occurrences of the species, for up to a dozen localities.

Name: Significance and origin of the name.

Type Material: Museum holdings of type material.

References: Citations used to define the data summary for the species. We are the source for [calc. dens.] or [Z] values in square brackets and for all idealized chemical analyses.

VOLUME I
Elements, Sulfides, Sulfosalts

Crystal Data: Monoclinic, pseudo-orthorhombic. *Point Group:* $2/m$. Crystals rare, prismatic to long prismatic along [001], up to 2.5 cm. Commonly paramorphic after the cubic high-temperature phase ("argentite"), of pseudocubic or pseudo-octahedral habit, to 8 cm. *Twinning:* Polysynthetic on $\{\bar{1}11\}$; contact on $(\bar{1}01)$.

Physical Properties: *Cleavage:* Indistinct. *Fracture:* Uneven. *Tenacity:* Sectile. Hardness = 2.0–2.5 VHN = 21–25 (50 g load). D(meas.) = 7.22 D(calc.) = 7.24

Optical Properties: Opaque. *Color:* Iron-black. *Streak:* Black. *Luster:* Metallic. *Anisotropism:* Weak.

R: (400) 32.8, (420) 32.9, (440) 33.0, (460) 33.1, (480) 33.0, (500) 32.7, (520) 32.0, (540) 31.2, (560) 30.5, (580) 29.9, (600) 29.2, (620) 28.7, (640) 28.2, (660) 27.6, (680) 27.0, (700) 26.4

Cell Data: *Space Group:* $P2_1/c$. $a = 4.231$ $b = 6.930$ $c = 9.526$ $\beta = 125°29'$ Z = 4

X-ray Powder Pattern: Synthetic.
2.606 (100), 2.440 (80), 2.383 (75), 2.836 (70), 2.583 (70), 2.456 (70), 3.080 (60)

Chemistry:

	(1)	(2)
Ag	86.4	87.06
Se	1.6	
S	12.0	12.94
Total	100.0	100.00

(1) Guanajuato, Mexico; by electron microprobe. (2) Ag_2S.

Polymorphism & Series: The high-temperature cubic form ("argentite") inverts to acanthite at 173°C; below this temperature, acanthite is the stable phase and forms directly.

Occurrence: In moderately low-temperature sulfide veins, and in zones of secondary enrichment.

Association: Silver, pyrargyrite, proustite, polybasite, stephanite, aguilarite, galena, chalcopyrite, sphalerite, calcite, quartz.

Distribution: Widespread in silver deposits. Localities for fine primary and paramorphic crystals include, in Germany, Freiberg, Schneeberg, Annaberg, and Marienberg, Saxony; and St. Andreasberg, Harz. At Jáchymov (Joachimsthal), Czechoslovakia. In Mexico, paramorphs from Arizpe, Sonora; the Rayas mine and others in Guanajuato; and many mines in Zacatecas, Chihuahua, etc. In the USA, at Butte, Silver Bow Co., Montana; Tonopah, Esmeralda Co., and the Comstock Lode, Virginia City, Storey Co., Nevada. From various mines at Cobalt, Ontario, Canada. At Chañarcillo, Atacama, Chile.

Name: From the Greek for *thorn*, in allusion to the shape of the crystals.

Type Material: National Museum of Natural History, Washington, D.C., USA, 105328.

References: (1) Palache, C., H. Berman, and C. Frondel (1944) Dana's system of mineralogy, (7th edition), v. I, 191–192 (acanthite), 176–178 ("argentite"). (2) Frueh, A.J., Jr. (1958) The crystallography of silver sulfide, Ag_2S. Zeits. Krist., 110, 136–144. (3) Sadanaga, R. and S. Sueno (1967) X-ray study on the α–β transition of Ag_2S. Mineral. J. (Japan), 5, 124–143. (4) Petruk, W., D.R. Owens, J.M. Stewart, and E.J. Murray (1974) Observations on acanthite, aguilarite and naumannite. Can. Mineral., 12, 365–369. (5) (1960) NBS Circ. 539, 10, 51.

Crystal Data: Orthorhombic. *Point Group:* n.d. In skeletal pseudododecahedral crystals, often elongate in the direction of a pseudocubic or pseudo-octahedral edge, to 3 cm; also massive and in intergrowths with acanthite or naumannite.

Physical Properties: *Fracture:* Hackly. *Tenacity:* Sectile. Hardness = 2.5 VHN = n.d. D(meas.) = 7.40–7.53 D(calc.) = [7.65]

Optical Properties: Opaque. *Color:* Bright lead-gray on fresh surfaces; dull iron-black on exposed surfaces. *Luster:* Metallic. *Anisotropism:* Very weak or isotropic.
R: (400) 33.3, (420) 33.4, (440) 33.5, (460) 33.7, (480) 33.8, (500) 33.4, (520) 32.8, (540) 31.9, (560) 31.0, (580) 30.3, (600) 29.8, (620) 29.5, (640) 29.3, (660) 29.2, (680) 29.2, (700) 29.1

Cell Data: *Space Group:* n.d. $a = 4.33$ $b = 7.09$ $c = 7.76$ Z = 2

X-ray Powder Pattern: Guanajuato, Mexico.
2.43 (100), 2.88 (50), 1.48 (40), 2.23 (30), 1.73 (30), 2.67 (20), 2.59 (20)

Chemistry:

	(1)	(2)	(3)	(4)
Ag	79.6	80.7	79.41	79.50
Cu			0.50	
Se	15.2	13.4	13.96	14.59
S	6.4	7.1	5.93	5.91
Total	101.2	101.2	99.80	100.00

(1) Silver City, Idaho, USA. (2–3) Guanajuato, Mexico. (4) Ag_4SeS.

Polymorphism & Series: Inverted from a higher temperature cubic form.

Occurrence: As a relatively low-temperature mineral in hydrothermal deposits rich in silver and selenium, but manifestly deficient in sulfur.

Association: Acanthite, naumannite, silver, stephanite, proustite, pearceite, calcite, quartz.

Distribution: At the San Carlos and other mines, Guanajuato; and the Chontalpan mine, Taxco, Guerrero, Mexico. In the USA, in Idaho, from Silver City, Owyhee Co., and the 4th of July mine, Yankee Fork, Custer Co.; from the Comstock Lode, Virginia City, Storey Co., Nevada; in Washington, at the L-D mine, Wenatchee, Chelan Co. and the Knob Hill mine, Republic, Ferry Co.; and from Cuchillo, Winston district, Sierra Co., New Mexico. At the Maritoto mine, northeast of Paeroa, New Zealand. From Woluma goldfield, New South Wales, Australia. In the Sanru mine, Hokkaido, Japan.

Name: After P. Aguilar, Superintendent of the San Carlos mine, Guanajuato, Mexico, where the material was originally found.

References: (1) Palache, C., H. Berman, and C. Frondel (1944) Dana's system of mineralogy, (7th edition), v. I, 178. (2) Earley, J.W. (1950) Description and synthesis of the selenide minerals. Amer. Mineral., 35, 337–364. (3) Petruk, W., D.R. Owens, J.M. Stewart, and E.J. Murray (1974) Observations on acanthite, aguilarite and naumannite. Can. Mineral., 12, 365–369.

Crystal Data: Orthorhombic. *Point Group:* $2/m$ $2/m$ $2/m$. Crystals prismatic to acicular, to 3 cm, striated || [001], sometimes radiating; also massive.

Physical Properties: *Cleavage:* Indistinct on {010}. *Fracture:* Uneven. Hardness = 2.0–2.5 VHN = n.d. D(meas.) = 7.07 D(calc.) = 7.255

Optical Properties: Opaque. *Color:* Blackish lead-gray, tarnishes brown or copper-red, sometimes with a yellowish green coating; in polished section, cream-white. *Streak:* Grayish black. *Luster:* Metallic. *Pleochroism:* Cream-white to pure white or light brown. *Anisotropism:* Distinct.

R_1–R_2: (400) 44.7–45.7, (420) 44.9–45.8, (440) 45.1–45.9, (460) 45.4–46.2, (480) 45.8–46.8, (500) 46.3–47.3, (520) 46.4–47.5, (540) 46.3–47.3, (560) 46.0–47.0, (580) 45.5–46.7, (600) 45.0–46.3, (620) 44.6–45.9, (640) 44.2–45.4, (660) 43.7–45.0, (680) 43.3–44.5, (700) 43.0–44.0

Cell Data: *Space Group:* *Pnma*. $a = 11.638(3)$ $b = 4.039(1)$ $c = 11.319(2)$ Z = 4

X-ray Powder Pattern: Beresovsk, USSR.
3.68 (100), 3.19 (80), 2.88 (70), 3.59 (60), 2.59 (50), 1.989 (40), 4.08 (30)

Chemistry:

	(1)	(2)
Pb	35.15	35.98
Cu	11.11	11.03
Bi	36.25	36.29
S	16.56	16.70
Total	99.07	100.00

(1) Beresovsk, USSR. (2) $PbCuBiS_3$.

Occurrence: Found in hydrothermal veins.

Association: Gold, pyrite, galena, tennantite, enargite, chalcopyrite, quartz.

Distribution: There are numerous minor localities. In the USA, in Utah, at the Sells mine, Alta, Salt Lake Co. In Nevada, near Cucomungo Spring and near the Sylvania Mountains, Esmeralda Co. At the Old Lout mine, Poughkeepsie Gulch, near Ouray, San Juan Co., and the Sunnyside mine, Eureka Co., Colorado. At the Chantilly quarry, near Arcola, Loudoun Co., Virginia. From Dobšiná (Dobschau), Czechoslovakia. In the Beresovsk district, near Sverdlovsk (Ekaterinburg), Ural Mountains, USSR, exceptional crystal sprays. In France, at the Gardette mine near Bourg d'Oisans, Isère. At Tasco and Huitzuco, Guerrero, Mexico.

Name: For Dr. Arthur Aikin (1773–1854), a founder and long-time Secretary of the Geological Society of London, England.

References: (1) Palache, C., H. Berman, and C. Frondel (1944) Dana's system of mineralogy, (7th edition), v. I, 412–413. (2) Berry, L.G. and R.M. Thompson (1962) X-ray powder data for the ore minerals. Geol. Soc. Amer. Mem. 85, 135. (3) Moore, P.B. (1967) A classification of sulfosalt structures derived from the structure of aikinite. Amer. Mineral., 52, 1874–1876. (4) Kohotsu, I. and B.J. Wuensch (1971) The crystal structure of aikinite, $PbCuBiS_3$. Acta Cryst., B27, 1245–1252. (5) Mumme, W.G., E. Welin, and B.J. Wuensch (1976) Crystal chemistry and proposed nomenclature for sulfosalts intermediate in the system bismuthinite–aikinite (Bi_2S_3—$CuPbBiS_3$). Amer. Mineral., 61, 15–20. (6) Harris, D.G. and T.T. Chen (1976) Crystal chemistry and re-examination of nomenclature of sulfosalts in the aikinite–bismuthinite series. Can. Mineral., 14, 194–205. (7) Ohmasa, M. and W. Nowacki (1970) A redetermination of the crystal structure of aikinite $[BiS_2|S|Cu^{IV}Pb^{VII}]$. Zeits. Krist., 132, 71–86.

Crystal Data: Hexagonal, pseudocubic. *Point Group:* 3. Rarely in crystals resembling trigonal pyramids, to 0.2 mm; as xenomorphic grains and granular aggregates.

Physical Properties: *Cleavage:* Irregular to conchoidal. *Tenacity:* Brittle. Hardness = ~3.5 VHN = 300–346, 313 average (50 g load). D(meas.) = 5.5 D(calc.) = 5.72

Optical Properties: Opaque. *Color:* Gray-black; white in reflected light. *Streak:* Black. *Luster:* Metallic. *Anisotropism:* Weak, in shades of blue.
R_1–R_2: n.d.

Cell Data: *Space Group:* $R3$. $a = 13.730(3)$ $c = 9.329(1)$ Z = 3

X-ray Powder Pattern: Gal-Khaya deposit, USSR.
3.10 (100), 1.903 (100), 1.621 (100), 2.69 (70), 1.236 (40), 1.345 (30), 4.04 (20)

Chemistry:

	(1)	(2)	(3)
Cu	23.38	23.2	22.86
Zn		0.15	
Hg	32.54	35.4	36.09
As	18.20	18.9	17.97
Sb	2.55	0.41	
S	23.80	23.6	23.08
Total	100.47	101.66	100.00

(1) Aktash deposit, USSR; by electron microprobe, average of two analyses. (2) Gal-Khaya deposit, USSR; by electron microprobe. (3) $Cu_6Hg_3As_4S_{12}$.

Polymorphism & Series: Forms a series with gruzdevite.

Occurrence: Of hydrothermal origin.

Association: Stibnite, chalcostibite, mercurian tetrahedrite, tennantite, luzonite, enargite, cinnabar, chalcopyrite, pyrite, sphalerite, realgar, orpiment, dickite, quartz, calcite.

Distribution: In the USSR, at the Aktash mercury deposit, Altai Mountains; Chauvai, Central Asia; and the Gal-Khaya deposit, Yakutia. From Moctezuma, Sonora, Mexico. In the Hemlo gold deposit, Thunder Bay district, Ontario, Canada. From the Jas Roux deposit, Hautes-Alpes, France.

Name: For its occurrence at the Aktash deposit, USSR.

Type Material: n.d.

References: (1) Vasil'ev, V.I. (1968) New ore minerals of the mercury deposits of Gornyi Altai and their parageneses. In: Problems of the metallogeny of mercury. Izdat. "Nauka" Moscow, 111–129 (in Russian). (2) (1971) Amer. Mineral., 56, 358 (abs. ref. 1). (3) Gruzdev, V.S., N.M. Chernitsova, and N.G. Shumakove (1972) Aktashite, $Cu_6Hg_3As_5S_{12}$, new data. Doklady Acad. Nauk SSSR, 206, 694–697 (in Russian). (4) (1973) Amer. Mineral., 58, 562 (abs. ref. 3). (5) Kaplunnik, L.N., E.A. Pobedimskaya, and N.V. Belov (1980) Crystal structure of aktashite ($Cu_6Hg_3As_4S_{12}$). Doklady Acad. Nauk SSSR, 251, 96–98 (in Russian). (6) (1980) Chem. Abs., 92, 224689 (abs. ref. 5).

Crystal Data: Cubic. *Point Group:* $4/m\ \overline{3}\ 2/m$. Crystals cubic or octahedral. Usually massive, granular. *Twinning:* Lamellar $\parallel$ {111}.

Physical Properties: *Cleavage:* Perfect on {001}. *Fracture:* Uneven. *Tenacity:* Brittle. Hardness = 3.5–4 VHN = 164–174, 167 average (100 g load). D(meas.) = 3.95–4.04 D(calc.) = 4.053

Optical Properties: Opaque, but translucent in very thin splinters. *Color:* Iron-black, tarnishing brown; in polished section, gray-white; deep green to brown and red in thin slivers. *Luster:* Submetallic.

Optical Class: Isotropic. $n = 2.70$ (Li)

R: (400) 30.9, (420) 28.8, (440) 26.7, (460) 25.4, (480) 24.5, (500) 23.8, (520) 23.2, (540) 22.8, (560) 22.4, (580) 22.0, (600) 21.8, (620) 21.6, (640) 21.5, (660) 21.3, (680) 21.2, (700) 21.1

Cell Data: *Space Group:* $Fm3m$. $a = 5.2236$ Z = 4

X-ray Powder Pattern: Synthetic.
2.612 (100), 1.847 (50), 1.509 (20), 1.1682 (16), 1.0662 (16), 3.015 (14), 1.306 (8)

Chemistry:

	(1)	(2)	(3)
Mn	63.03	57.09	63.15
Fe		6.86	
Co		0.03	
S	36.91	36.83	36.85
Total	99.94	100.81	100.00

(1) Arizona, USA. (2) Sotkamo, Finland. (3) MnS.

Occurrence: In epithermal sulfide vein deposits. A rare constituent of a number of meteorites.

Association: Galena, chalcopyrite, sphalerite, pyrite, acanthite, tellurium, rhodochrosite, calcite, rhodonite, quartz.

Distribution: Selected localities include: Alabanda, Caria, southwestern Turkey. From Baia-de-Arieş (Offenbanya), Săcărâmb (Nagyág), and Cavnic (Kapnik), Romania. At Gersdorf, Saxony, Germany. From Adervielle, Hautes-Pyrénées, France. In Peru, at Morococha, Junin. At the Yakumo and Inakuraishi mines, Hokkaido, the Taisei mine, Akita Prefecture, and many other localities in Japan. In the Preciosa mine, Puebla, Mexico. In the USA, in the Lucky Cuss mine, Tombstone, Cochise Co., Arizona; the Queen of the West mine, Summit Co., Colorado; and from Schellbourne, White Pine Co., Montana.

Name: From the locality at Alabanda in Turkey.

References: (1) Palache, C., H. Berman, and C. Frondel (1944) Dana's system of mineralogy, (7th edition), v. I, 207–208. (2) (1955) NBS Circ. 539, 4, 11. (3) Törnroos, R. (1982) Properties of alabandite; alabandite from Finland. Neues Jahrb. Mineral., Abh., 144, 107–123.

Crystal Data: Monoclinic *Point Group:* $2/m$. As flattened and prismatic grains, to 0.5 mm, pinacoidal, prismatic, and flattened [100]; striated || [001] on {100}, other faces dull or tarnished.

Physical Properties: *Cleavage:* Imperfect on {100}. *Fracture:* Conchoidal. *Tenacity:* Very brittle. Hardness = 1.5 VHN = 69 (20 g load). D(meas.) = 3.43(3) D(calc.) = 3.43

Optical Properties: Transparent. *Color:* Light gray in reflected light, with rose-yellow internal reflections; orange-yellow in transmitted light. *Streak:* Orange-yellow. *Luster:* Adamantine, greasy.
Optical Class: Biaxial, positive (+). $\alpha = 2.39(1)$ $\gamma = 2.52(2)$
R_1–R_2: (400) 13.0–14.0, (425) 13.2–14.6, (450) 13.3–14.8, (475) 13.4–14.8, (500) 13.3–14.5, (525) 13.1–14.3, (550) 13.2–14.5, (575) 13.4–14.7, (600) 13.5–14.8, (625) 13.6–14.9, (650) 13.7–15.0, (675) 13.8–15.0, (700) 13.9–15.1

Cell Data: *Space Group:* $P2/c$. $a = 9.89(2)$ $b = 9.73(2)$ $c = 9.13(1)$ $\beta = 101.84(5)°$
Z = 2

X-ray Powder Pattern: Uson caldera, USSR.
3.064 (100), 5.91 (90), 2.950 (90), 5.11 (80), 4.05 (70), 3.291 (50)

Chemistry:

	(1)
As	67.35
S	32.61
Total	99.96

(1) Uson caldera, USSR; by electron microprobe, corresponding to $As_{7.98}S_{9.02}$.

Occurrence: As cement in a sandy gravel (Uson caldera, USSR); in hydrothermal As-S veins (Alacran deposit, Chile).

Association: Realgar, usonite (Uson caldera, USSR); realgar, orpiment, arsenic, sulfur, stibnite, pyrite, greigite, arsenopyrite, arsenolamprite, sphalerite, acanthite, barite, quartz, calcite (Alacran deposit, Chile).

Distribution: In the Uson caldera, Kamchatka, USSR. From the Alacran deposit, Pampa Larga, Chile.

Name: For the occurrence in the Alacran deposit, Chile.

Type Material: Il'menskii Preserve Museum, Miass; A.E. Fersman Mineralogical Museum, Academy of Sciences, Moscow, USSR.

References: (1) Popova, V.I., V.A. Popov, A. Clark, V.O. Polyakov, and S.E. Borisovskii (1986) Alacranite—A new mineral. Zap. Vses. Mineral. Obshch., 115, 360–368 (in Russian). (2) (1988) Amer. Mineral., 73, 189 (abs. ref. 1).

Crystal Data: Hexagonal. *Point Group:* n.d. Platy grains up to 1 mm.

Physical Properties: *Cleavage:* Perfect on {0001}. Hardness = n.d. VHN = 40–65, 51 average (20 g load). D(meas.) = n.d. D(calc.) = 7.80

Optical Properties: Opaque. *Color:* In polished section, light pale gray with slight greenish tint. *Anisotropism:* Weak.
R: (400) — , (420) — , (440) — , (460) 51.4, (480) 51.8, (500) 52.7, (520) 52.8, (540) 53.2, (560) 53.1, (580) 53.4, (600) 53.9, (620) 53.9, (640) 54.2, (660) 54.5, (680) 54.4, (700) 54.9

Cell Data: *Space Group:* n.d. $a = 4.238(1)$ $c = 79.76(2)$ Z = 6

X-ray Powder Pattern: Alekseev mine, USSR.
3.09 (100), 2.12 (60), 2.25 (40), 1.348 (40), 1.307 (40), 3.63 (30), 1.974 (30)

Chemistry:

	(1)	(2)	(3)
Pb	20.3	20.5	21.94
Bi	46.0	45.5	44.25
Te	27.3	27.3	27.02
S	6.3	6.3	6.79
Total	99.9	99.6	100.00

(1) Alekseev mine, USSR; by electron microprobe, leading to $Pb_{0.94}Bi_{2.11}Te_{2.06}S_{1.89}$. (2) Do.; leading to $Pb_{0.95}Bi_{2.10}Te_{2.06}S_{1.89}$. (3) $PbBi_2Te_2S_2$.

Occurrence: Of hydrothermal origin.

Association: Galena, gold, altaite, tetradymite, rucklidgeite, quartz.

Distribution: In the Alekseev mine, Sutemskii region, Stanovoi Range, USSR. From near Tybo, Nye Co., Nevada. In the Barringer mine, Timmins, Ontario, Canada.

Name: For the Alekseev mine, USSR.

Type Material: Gosudarst University, Moscow, USSR.

References: (1) Lipovetskii, A.G., Y.S. Borodaev, and E.N. Zav'yalov (1978) Aleksite, $PbBi_2Te_2S_2$, a new mineral. Zap. Vses. Mineral. Obshch., 107, 315–321. (2) (1979) Amer. Mineral., 64, 652 (abs. ref. 1).

Crystal Data: Hexagonal. *Point Group:* $6/m\ 2/m\ 2/m$. As incrustation of minute crystals; commonly massive and granular.

Physical Properties: *Fracture:* Subconchoidal. Hardness = 4 VHN = 245–302 (100 g load). D(meas.) = 8.38 D(calc.) = 8.72

Optical Properties: Opaque. *Color:* Steel-gray to silver-white, tarnishes dull on exposure; in polished section, bright cream-white. *Luster:* Bright metallic. *Anisotropism:* Weak.
R: (400) 54.2, (420) 55.5, (440) 56.8, (460) 58.2, (480) 59.4, (500) 60.6, (520) 61.8, (540) 62.8, (560) 63.8, (580) 64.7, (600) 65.5, (620) 66.2, (640) 66.8, (660) 67.3, (680) 67.7, (700) 68.0

Cell Data: *Space Group:* $P6_3/mmc$. $a = 2.600$ $c = 4.228$ Z = 2

X-ray Powder Pattern: Mohawk mine, Michigan, USA.
1.989 (100), 2.11 (40), 2.25 (20), 1.299 (20), 1.194 (20), 1.110 (20), 0.837 (20)

Chemistry:

	(1)	(2)	(3)
Cu	83.11	83.53	83.58
Ag	trace		
As	16.44	16.55	16.42
Total	99.55	100.08	100.00

(1) Rancagua, Chile. (2) Champion mine, Michigan, USA. (3) Cu_6As.

Occurrence: In hydrothermal deposits, intimately associated with other copper arsenides.

Association: Copper (usually arsenian), silver, domeykite.

Distribution: In the USA, in Michigan, from Keweenaw Co. at the Mohawk, Pewabic, Seneca, Ahmeek, and Champion mines, and at Painesdale, Houghton Co.; also from Baraga Co. In Chile, at the Algodones silver mine near Coquimbo, Atacama; Cerro de las Seguas, Rancagua, Cachapoal; and at Coro-Coro, La Paz. In the Kokito II mine, Neuquén Province, Argentina. At Långban, Värmland, Sweden. In France, from Daluis, Alpes-Maritimes. From Tsumeb, Namibia.

Name: For the Algodones mine, Chile.

References: (1) Palache, C., H. Berman, and C. Frondel (1944) Dana's system of mineralogy, (7th edition), v. I, 171. (2) Williams, S.A. (1963) Crystals of rammelsbergite and algodonite. Amer. Mineral., 48, 421–422. (3) Berry, L.G. and R.M. Thompson (1962) X-ray powder data for the ore minerals. Geol. Soc. Amer. Mem. 85, 14. (4) Ramdohr, P. (1969) The ore minerals and their intergrowths, (3rd edition), 393–398.

Crystal Data: Hexagonal. *Point Group:* n.d. In complex intergrowths with silver, and as very small grains.

Physical Properties: Hardness = n.d. VHN = 172–203, 189 average. D(meas.) = 10.0 (synthetic). D(calc.) = 10.12

Optical Properties: Opaque. *Color:* Silver. *Luster:* Metallic. *Anisotropism:* Weak.
R_1–R_2: (400) 67.0–68.8, (420) 66.5–68.2, (440) 66.0–67.6, (460) 65.9–67.3, (480) 67.1–68.6, (500) 67.4–68.7, (520) 68.9–70.4, (540) 69.5–70.9, (560) 70.2–71.6, (580) 70.5–71.8, (600) 71.0–72.3, (620) 71.7–73.0, (640) 72.1–73.4, (660) 72.6–73.9, (680) 73.1–74.3, (700) 73.8–74.9.

Cell Data: *Space Group:* n.d. $a = 2.952$ $c = 4.773$ Z = 2

X-ray Powder Pattern: Cobalt, Canada.
2.370 (100), 2.252 (60), 2.548 (40), 1.353 (40), 1.756 (30), 0.943 (30), 1.473 (20)

Chemistry:

	(1)	(2)	(3)
Ag	84.3	83.09	84.48
Hg	0.3		
Ni		0.05	
Sb	15.3	15.82	15.52
As		0.08	
Total	99.9	99.04	100.00

(1) Cobalt, Canada; by electron microprobe, average of 15 analyses, corresponding to $(Ag,Hg)_{6.46}Sb$; the formula taken as $Ag_{1-x}Sb_x$, x = 0.09–0.16. (2) Wasserfall, France; by electron microprobe. (3) $Ag_{0.86}Sb_{0.14}$.

Occurrence: In high-grade Ag-Sb ores.

Association: Silver containing antimony and mercury, breithauptite (Cobalt, Canada); antimonian silver, dyscrasite (Broken Hill, Australia); dyscrasite, kutinaite (Wasserfall, France).

Distribution: In Ontario, Canada, abundant in many mines of the Cobalt area; also in the Red Lake area. From the Consols and Junction mines, Broken Hill, New South Wales, and North Arm, Queensland, Australia. At Wasserfall, southern Vosges, France. From Rejská, near Kutná Hora, Czechoslovakia. In Germany, at Hartenstein, Saxony. At Hällefors, Bergslagen, Sweden.

Name: From the Greek for *another* and the Latin *argentum*, silver.

Type Material: National Museum of Canada, Ottawa, Canada; National Museum of Natural History, Washington, D.C., USA, 135409.

References: (1) Petruk, W.L., L.J. Cabri, D.C. Harris, J.M. Stewart, and L.A. Clark (1970) Allargentum, redefined. Can. Mineral., 10, 163–172. (2) (1971) Amer. Mineral., 56, 638 (abs. ref. 1). (3) Picot, P. and F. Ruhlmann (1978) Présence d'arséniures de cuivre de haute température dans le granite des Ballons (Vosges méridionales). Bull. Minéral., 101, 563–569 (in French with English abs.).

Alloclasite $(Co, Fe)AsS$

Crystal Data: Monoclinic. *Point Group:* 2. Prismatic ‖ [010], commonly in columnar to radiating aggregates.

Physical Properties: *Cleavage:* Perfect on {101}, distinct on {010}. *Fracture:* Uneven to subconchoidal. *Tenacity:* Brittle. Hardness = 5 VHN = 818–940 (100 g load). D(meas.) = 5.95 D(calc.) = 6.188

Optical Properties: Opaque. *Color:* Steel-gray to silver. *Streak:* Nearly black. *Luster:* Bright metallic.

R_1–R_2: (400) 47.2–49.2, (420) 47.5–49.6, (440) 47.8–50.0, (460) 48.2–50.4, (480) 48.7–50.9, (500) 49.1–51.3, (520) 49.4–51.5, (540) 49.3–51.5, (560) 49.1–51.3, (580) 48.7–51.0, (600) 48.1–50.6, (620) 47.6–50.1, (640) 47.2–49.8, (660) 46.9–49.5, (680) 46.7–49.2, (700) 46.6–49.0

Cell Data: *Space Group:* $P2_1$. $a = 4.661$ $b = 5.602$ $c = 3.411$ $\beta = 90°2(5)'$ Z = 2

X-ray Powder Pattern: Oraviţa, Romania.
2.750 (100), 2.469 (90), 1.817 (70), 2.401 (50), 3.58 (30), 2.802 (30), 1.707 (30)

Chemistry:

	(1)	(2)	(3)	(4)
Co	25.1	26.5	23.6	23.3
Fe	6.7	6.0	9.1	7.7
Ni	1.0	trace		
As	47.5	49.0	45.5	47.3
S	19.9	17.0	20.9	20.7
Total	100.2	98.5	99.1	99.0

(1) Elizabeth mine, Romania; by electron microprobe. (2) Silverfields mine, Cobalt, Canada; by electron microprobe. (3) North Rhine-Westphalia, Germany; by electron microprobe. (4) Dogatani mine, Japan; by electron microprobe.

Polymorphism & Series: Dimorphous with glaucodot.

Occurrence: In calcite or quartz veins of apparently low temperature, late stage hydrothermal origin. Also in silicified, recrystallized metamorphic rock.

Association: Gold, silver, glaucodot, cobaltite, cobaltian arsenopyrite, sphalerite, calcite, quartz.

Distribution: In the Elizabeth mine, Oraviţa (Oravicza), Romania. From the Silverfields mine, Cobalt, and the Siscoe Metals mine, Miller Lake, Gowganda, Ontario, Canada. From Bou-Azzer, Morocco. At the Scar Crag mine, Braithwaite, Keswick, Cumbria, England. In the Lautaret Pass, Hautes-Alpes, France. From the Dogatani mine, Japan. At Mount Isa, Queensland, Australia. From the USA, in the Kibblehouse quarry, Perkiomenville, Montgomery Co., Pennsylvania.

Name: From the Greek for *other*, and *to break*, because its cleavage was believed to be different from marcasite, which it resembles.

Type Material: n.d.

References: (1) Kingston, P.W. (1971) On alloclasite, a Co–Fe sulpharsenide. Can. Mineral., 10, 838–846. (2) (1972) Amer. Mineral., 57, 1561 (abs. ref. 1). (3) Scott, J.D. and W. Nowacki (1976) The crystal structure of alloclasite, CoAsS, and the alloclasite–cobaltite transformation. Can. Mineral., 14, 561–566. (4) Petruk, W., D.C. Harris, and J.M. Stewart (1971) Characteristics of the arsenides, sulpharsenides and antimonides. Can. Mineral., 11, 149–186.

Crystal Data: Cubic. *Point Group:* $4/m\ \overline{3}\ 2/m$. Usually massive; rarely in cubes and octahedra.

Physical Properties: *Cleavage:* {001} perfect. *Fracture:* Subconchoidal. *Tenacity:* Sectile. Hardness = 3 VHN = 47–53, 51 average. D(meas.) = 8.19 D(calc.) = 8.27

Optical Properties: Opaque. *Color:* Tin-white with yellowish tint, tarnishes bronze; in polished section, white with a delicate greenish hue. *Luster:* Metallic.
R: (400) 65.1, (420) 66.4, (440) 67.7, (460) 68.8, (480) 69.6, (500) 70.0, (520) 70.0, (540) 69.7, (560) 68.8, (580) 67.5, (600) 66.0, (620) 64.6, (640) 63.3, (660) 62.0, (680) 61.0, (700) 60.2

Cell Data: *Space Group:* $Fm3m$. $a = 6.439$ Z = 4

X-ray Powder Pattern: Kirkland Lake, Ontario, Canada.
3.23 (100), 2.29 (80), 1.443 (50), 1.314 (40), 1.859 (30), 1.610 (20), 1.074 (20)

Chemistry:

	(1)	(2)	(3)	(4)
Pb	61.26	61.52	61.33	61.91
Ag			0.43	
Au			0.02	
Cu	0.20		0.01	
Fe	0.64		0.13	
Te	36.84	38.48	38.43	38.09
Se			0.08	
S	0.29			
rem.	0.46			
Total	99.69	100.00	100.43	100.00

(1) Lake Shore mine, Canada. (2) Red Cloud mine, Colorado, USA. (3) Kalgoorlie, Western Australia. (4) PbTe.

Occurrence: In hydrothermal vein gold deposits.

Association: Gold, silver, antimony, tellurium, tellurantimony, galena, pyrite, hessite, nagyagite, tetrahedrite, sylvanite, petzite, arsenopyrite, sphalerite, chalcopyrite, jamesonite, boulangerite, bournonite, aguilarite, pyrrhotite, siderite, cerussite, quartz.

Distribution: Occurrences are too numerous to fully list. Selected localities are: in the Savodinskii mine, Altai Mountains, USSR. From Kalgoorlie, Western Australia. At several mines in the Kirkland Lake area, Ontario, Canada. In the USA, in the Kings Mountain mine, Gaston Co., North Carolina; the Red Cloud mine, Gold Hill, Boulder Co., Colorado; the Hilltop mine, near Las Cruces, Dona Ana Co., New Mexico; and the Stanislaus mine, Carson Hill, Calaveras Co., California.

Name: After the locality in the Altai Mountains, USSR.

References: (1) Palache, C., H. Berman, and C. Frondel (1944) Dana's system of mineralogy, (7th edition), v. I, 205–207. (2) Berry, L.G. and R.M. Thompson (1962) X-ray powder data for the ore minerals. Geol. Soc. Amer. Mem. 85, 48. (3) Sindeeva, N.D. (1964) Mineralogy and types of deposits of selenium and tellurium, 121–125.

Crystal Data: Orthorhombic. *Point Group:* $2/m$ $2/m$ $2/m$. Crystals stout prismatic, up to 3 cm; also thick and thin tabular on {100}; striated ∥ [001]. Massive. *Twinning:* Reported on {110}.

Physical Properties: *Fracture:* Smooth conchoidal. *Tenacity:* Brittle. Hardness = 3–3.5 VHN = n.d. D(meas.) = 5.35 D(calc.) = [5.40]

Optical Properties: Opaque. *Color:* Dark steel-gray, sometimes tarnishing yellow or iridescent; in polished section, white. *Streak:* Black. *Luster:* Metallic.
R_1–R_2: (400) 39.2–42.3, (420) 38.8–42.0, (440) 38.4–41.7, (460) 38.0–41.4, (480) 37.5–41.1, (500) 37.0–40.9, (520) 36.7–40.6, (540) 36.3–40.4, (560) 36.0–40.0, (580) 35.5–39.6, (600) 35.2–39.0, (620) 34.8–38.5, (640) 34.5–38.0, (660) 34.1–37.5, (680) 33.7–37.0, (700) 33.3–36.6

Cell Data: *Space Group: Pmma.* $a = 13.01$ $b = 19.19$ $c = 4.27$ Z = [4]

X-ray Powder Pattern: Takla Lake, Canada.
3.30 (100), 2.90 (80), 3.45 (40), 2.76 (40), 3.74 (30), 2.06 (30), 1.888 (30)

Chemistry:

	(1)	(2)	(3)
Pb	24.10	22.25	23.75
Ag	10.94	10.90	12.36
Cu	0.68	0.96	
Zn		0.31	
Fe	0.30	0.75	
Sb	41.31	40.75	41.87
S	22.06	24.26	22.02
Total	99.39	100.18	100.00

(1) Oruro, Bolivia. (2) Baia Sprie, Romania. (3) $PbAgSb_3S_6$.

Occurrence: Of hydrothermal origin.

Association: Stibnite, sphalerite, barite, fluorite, siderite, quartz (Baia Sprie, Romania); cassiterite, arsenopyrite, stannite, zinkenite, tetrahedrite, pyrite, alunite, quartz (Itos mine, Bolivia); pyrargyrite, stephanite, sphalerite, rhodochrosite, quartz (Morey, Nevada, USA).

Distribution: In the USA, at the Keyser mine, Morey district, Nye Co., Nevada; in the Thompson mine, Darwin district, Inyo Co., California; and at Bear Basin, King Co., Washington. From near Takla Lake, British Columbia, and near Nansen Creek, Yukon Territory, Canada. In Romania, at Baia Sprie (Felsőbánya). From the Třebsko deposit, near Příbram, Czechoslovakia. In the Itos and San José mines, Oruro; and the Tatasi mine, Potosí, Bolivia. In the Les Farges mine, near Ussel, Corrèze; and Bournac, Montagne Noire, France. In Australia, from the Meerschaum mine, north of Omeo, Victoria.

Name: For Andor von Semsey (1833–1923), Hungarian nobleman, who was also an amateur mineralogist.

References: (1) Palache, C., H. Berman, and C. Frondel (1944) Dana's system of mineralogy, (7th edition), v. I, 457–459. (2) Berry, L.G. and R.M. Thompson (1962) X-ray powder data for the ore minerals. Geol. Soc. Amer. Mem. 85, 155–156. (3) Donnay, J.D.H. and G. Donnay (1954) Syntaxic intergrowths in the andorite series. Amer. Mineral., 39, 161–171.

Crystal Data: Orthorhombic. *Point Group:* $2/m\ 2/m\ 2/m$ or $mm2$. As massive grains or granular aggregates (60–100 μm).

Physical Properties: *Cleavage:* At least one noted. *Tenacity:* Brittle. Hardness = n.d. VHN = 1077.9 (50 g load). D(meas.) = n.d. D(calc.) = 8.692

Optical Properties: Opaque. *Color:* Lead-gray; in polished section, white with a pink tint. *Streak:* Grayish black. *Luster:* Dull metallic. *Pleochroism:* Noticeable in air, from pinkish grayish white to pinkish white; distinct in oil. *Anisotropism:* Marked in air from brownish yellow to grayish green, pale red to pale green; pronounced in oil, from reddish yellow, greenish yellow to purplish gray.
R_1–R_2: n.d.

Cell Data: *Space Group:* *Pnnm* or *Pnn*2. $a = 5.41$ $b = 6.206$ $c = 3.01$ Z = 2

X-ray Powder Pattern: Anduo, Tibet, China.
1.920 (100), 1.501 (90), 1.095 (90), 1.133 (80), 1.210 (70), 1.187 (70), 2.000 (50)

Chemistry:

	(1)
Ru	33.61
Os	6.96
Ir	2.44
Cu	0.61
As	56.51
Sb	0.04
S	0.19
Total	100.36

(1) Anduo, Tibet, China; by electron microprobe, corresponding to $(Ru_{0.850}Os_{0.093}Ir_{0.033}Cu_{0.024})_{\Sigma=1.000}(As_{1.928}S_{0.015}Sb_{0.001})_{\Sigma=1.944}$.

Polymorphism & Series: Forms a series with omeiite.

Occurrence: In a chromium deposit related to augite peridotite and dunite.

Association: Chromian spinel, pyrite, pyrrhotite, marcasite, magnetite, chalcopyrite, molybdenite, galena, millerite, violarite. Other platinum group minerals present are ruthenarsenite, sperrylite, ruthenian iridarsenite, irarsenite, osarsite, osmiridium, ruthenian osmiridium, laurite, rutheniridosmine.

Distribution: From Anduo, Tibet, China.

Name: For the Chinese locality at Anduo.

Type Material: Chinese Geological Museum (City ?).

References: (1) Yu, T. and Chou, H. (1979) Anduoite, a new ruthenium arsenide. Kexue Tongbao, 15, 704–708 (in Chinese with English abs.). (2) (1980) Amer. Mineral., 65, 808–809 (abs. ref. 1). (3) Cabri, L.J., Ed. (1981) Platinum group elements: mineralogy, geology, recovery. Can. Inst. Min. & Met., 95–96.

Crystal Data: Orthorhombic. *Point Group:* $2/m\ 2/m\ 2/m$. As prismatic or platy crystals, up to 5 mm. *Twinning:* Present in all crystals examined. The twin operations are symmetry operations that act on a pseudocubic subcell, producing "neighboring twins" in which individuals are difficult to distinguish.

Physical Properties: *Tenacity:* Sectile. Hardness = ~3 VHN = n.d. D(meas.) = n.d. D(calc.) = 5.68

Optical Properties: Opaque. *Color:* Bluish gray. *Streak:* Black. *Luster:* Metallic. R_1–R_2: n.d.

Cell Data: *Space Group:* *Pnma.* $a = 7.89$ $b = 7.84$ $c = 11.01$ Z = 4

X-ray Powder Pattern: Ani mine, Japan. Many of the reported associations of digenite and djurleite, identified by powder diffraction, could be anilite and djurleite, as anilite transforms to digenite during grinding.
1.956 (100), 2.77 (65), 3.20 (57), 2.16 (39), 1.677 (35), 2.54 (31), 2.59 (29)

Chemistry:

	(1)	(2)	(3)
Cu	79.2	80.1	77.6
S	21.7	22.7	22.4
Total	100.9	102.8	100.0

(1–2) Ani mine, Japan; by electron microprobe. All specimens contained some djurleite.
(3) Cu_7S_4.

Occurrence: In a drusy quartz vein (Ani mine, Japan).

Association: Djurleite, covellite (Ani mine, Japan); yarrowite, spionkopite, djurleite, wittichenite, tennantite, chalcopyrite, bornite (Yarrow Creek, Canada).

Distribution: In the Ani mine, Akita Prefecture, Japan. At Neudorf, Harz, Germany. From the Lubin mine, Lower Silesia, Poland. In the Estrella mine, Atacama Province, Chile. From Yarrow Creek, southwestern Alberta, Canada. At Wallaroo, South Australia. From near Nizhni Tagil, Ural Mountains, USSR. At Bor, Yugoslavia.

Name: For the Ani mine, Japan.

Type Material: Sakurai Museum, Tokyo, Japan.

References: (1) Morimoto, N., K. Koto, and Y. Shimazaki (1969) Anilite, Cu_7S_4, a new mineral. Amer. Mineral., 54, 1256–1268. (2) Koto, K. and N. Morimoto (1970) The crystal structure of anilite. Acta Cryst., B26, 915–924.

Crystal Data: Monoclinic. *Point Group:* $2/m$. Crystals are thin, tabular, pseudohexagonal, up to 2 cm across. In subparallel and rosette-like groups. *Twinning:* On {110}.

Physical Properties: *Fracture:* Irregular to conchoidal. *Tenacity:* Brittle. Hardness = 3 VHN = n.d. D(meas.) = 6.33–6.35 D(calc.) = 6.40

Optical Properties: Opaque. *Color:* Black. *Streak:* Black. *Luster:* Submetallic. R_1–R_2: n.d.

Cell Data: *Space Group:* $C2/m$. $a = 12.81$ $b = 7.41$ $c = 11.91$ $\beta = 90°$ Z = 2

X-ray Powder Pattern: Synthetic. (JCPDS 36-392).
2.83 (100), 2.96 (90), 3.11 (70), 2.49 (60), 2.38 (50), 1.86 (50), 2.31 (40)

Chemistry:

	(1)	(2)
Ag	62.54	68.39
Cu	8.90	5.13
Fe	0.05	
As	1.43	0.50
Sb	9.65	10.64
S	17.62	15.43
Total	100.19	100.09

(1) Sonora, Mexico. (2) Guanajuato, Mexico.

Occurrence: In epithermal precious metal veins.

Association: Acanthite, amethyst.

Distribution: In Mexico, from an unspecified locality in Sonora, and from Guanajuato. At Taltal, Chile. From the Clara mine, near Wolfrath, Black Forest, Germany. At Serra S'Ilixi, Sarrabus, Sardinia, Italy. In the Seikoshi mine, Shizuoka Prefecture, Japan. In the USA, at the North Lily mine, Tintic, Juab Co., Utah.

Name: For its compositional relationship to pearceite.

Type Material: Harvard University, Cambridge, Massachusetts, USA, 93159.

References: (1) Frondel, C. (1963) Isodimorphism of the polybasite and pearceite series. Amer. Mineral., 48, 565–572. (2) Harris, D.C., E.W. Nuffield, and M.H. Frohberg (1965) Studies of polybasite, pearceite, antimonpearceite, and arsenpolybasite. Can. Mineral., 8, 172–184. (3) (1965) Amer. Mineral., 50, 1507 (abs. ref. 2).

Crystal Data: Hexagonal. *Point Group:* $\overline{3}\ 2/m$. Prominent $\{01\overline{1}2\}$ yields pseudocubic crystals to 1 cm; also rounded and hoppered; usually massive or cleavable lamellar, also radiated; sometimes botryoidal or reniform with granular texture. *Twinning:* On $\{01\overline{1}4\}$, commonly forming complex groups, fourlings, sixlings; also polysynthetic twins.

Physical Properties: *Cleavage:* Perfect on {0001}, $\{10\overline{1}1\}$ distinct, $\{10\overline{1}4\}$ imperfect, $\{11\overline{2}0\}$ indistinct. *Fracture:* Uneven. *Tenacity:* Brittle. Hardness = 3–3.5 VHN = 50–69 (100 g load). D(meas.) = 6.61–6.72 D(calc.) = 6.697

Optical Properties: Opaque. *Color:* Tin-white. *Streak:* Gray. *Luster:* Metallic. *Pleochroism:* Very feeble. *Anisotropism:* Weak in air, lively in oil.

R_1–R_2: (400) 71.2–73.4, (420) 71.4–73.2, (440) 71.5–73.0, (460) 71.4–72.9, (480) 71.2–72.6, (500) 70.9–72.4, (520) 70.5–72.0, (540) 70.0–71.6, (560) 69.4–71.2, (580) 69.0–70.8, (600) 68.6–70.6, (620) 68.4–70.3, (640) 68.3–70.2, (660) 68.2–70.1, (680) 68.3–70.2, (700) 68.4–70.3

Cell Data: *Space Group:* $R\overline{3}m$. $a = 4.307$ $c = 11.273$ Z = 6

X-ray Powder Pattern: Synthetic.
3.109 (100), 2.248 (70), 1.368 (70), 1.416 (60), 2.152 (60), 1.261 (40), 1.878 (40)

Chemistry:

	(1)
Sb	99.2
As	0.2
Total	99.4

(1) Gottes Segnen mine, St. Andreasberg, Germany; by electron microprobe.

Occurrence: In hydrothermal veins.

Association: Silver, stibnite, allemontite, sphalerite, pyrite, galena, quartz.

Distribution: Numerous localities. In the USA, from South Riverside, Riverside Co.; in kg masses from Erskine Creek, Kern Co., California. From Příbram, Czechoslovakia. At St. Andreasberg, Harz Mountains, Germany. In France, from Allemont, Isère. At Sarrabus, Sardinia, Italy. From Sala, Västmanland, Sweden. At Kalliolampi, Nurmo, Finland. In Sarawak Province, Borneo. At Broken Hill, New South Wales, Australia. From Huasco, Atacama, Chile. In Canada, at South Ham, Wolfe Co., Quebec. At Arechuybo, Chihuahua, Mexico.

Name: From the Latin *antimonium*; possibly of Arabic origin; the chemical symbol from the Latin *stibium, mark.*

References: (1) Palache, C., H. Berman, and C. Frondel (1944) Dana's system of mineralogy, (7th edition), v. I, 132–133. (2) (1954) NBS Circ. 539, 3, 14. (3) Criddle, A.J. and C.J. Stanley, Eds. (1986) The quantitative data file for ore minerals. British Museum (Natural History), London, England, 11.

Crystal Data: Triclinic. *Point Group:* $\overline{1}$. As thin plates parallel to (010) showing striations along {100} and {001}. *Twinning:* About [$\overline{1}$01] with the composition plane near {$\overline{1}$01}.

Physical Properties: *Cleavage:* Perfect on {010}, {100} fair, {001} poor. *Tenacity:* Sectile, pliable but not elastic. Hardness = 2.5 VHN = n.d. D(meas.) = 5.602 D(calc.) = 5.624

Optical Properties: Nearly opaque. *Color:* Iron-black, with thin edges showing deep blood-red; white in reflected light. *Streak:* Black to red-brown. *Luster:* Brilliant metallic. *Pleochroism:* Distinct in oil. *Anisotropism:* Very high in oil.

R_1–R_2: (400) 40.5–45.9, (420) 40.5–45.6, (440) 40.5–45.3, (460) 40.4–44.6, (480) 40.1–43.8, (500) 39.8–43.2, (520) 39.6–42.6, (540) 39.0–42.2, (560) 38.2–42.0, (580) 37.4–41.6, (600) 36.3–41.0, (620) 35.5–40.5, (640) 34.7–39.9, (660) 34.0–39.2, (680) 33.5–38.6, (700) 33.0–38.2

Cell Data: *Space Group:* $P\overline{1}$. $a = 7.72$ $b = 8.82$ $c = 8.30$ $\alpha = 100°22.5'$ $\beta = 90°$ $\gamma = 103°54'$ Z = 6

X-ray Powder Pattern: Animas mine, Bolivia.
2.806 (100), 3.21 (40), 1.940 (30), 3.43 (20), 2.048 (20), 1.705 (20), 1.402 (20)

Chemistry:

	(1)
Ag	34.74
Cu	0.53
Fe	trace
Sb	29.95
Bi	13.75
S	20.87
Total	99.84

(1) Animas mine, Bolivia.

Occurrence: In vein-controlled hydrothermal tin and silver ores.

Association: Tetrahedrite, miargyrite, stannite, pyrite, quartz.

Distribution: In the Animas mine, Chocaya, Sud-Chichas, Potosí, Bolivia. From Herminia, Julcani district; and in the San Genaro mine, Huancavelica, Peru. In the El Indio mine, east of Coquimbo, Chile. From Niarada, Flathead Co., Montana, USA. At Monts de Blond and Le Bourneix, Hautes-Vienne, and Sainte-Marie-aux-Mines, Haut-Rhin, France.

Name: Named for Felix Avelino Aramayo, former Managing Director of the Compagnie Aramayo de Mines en Bolivie.

Type Material: British Museum (Natural History), London, England; National Museum of Natural History, Washington, D.C., USA, 95553.

References: (1) Palache, C., H. Berman, and C. Frondel (1944) Dana's system of mineralogy, (7th edition), v. I, 427–428. (2) Graham, A.R. (1951) Matildite, aramayoite, miargyrite. Amer. Mineral., 36, 436–449. (3) Mullen, D.J.E. and W. Nowacki (1974) The crystal structure of aramayoite, $Ag(Sb,Bi)S_2$. Zeits. Krist., 139, 54–69.

Crystal Data: n.d. *Point Group:* n.d. In anhedral grains of 0.05 mm average diameter.

Physical Properties: Hardness = n.d. VHN = n.d. D(meas.) = n.d. D(calc.) = n.d.

Optical Properties: Opaque. *Color:* In polished section, light gray with faint olive-brown tint. *Luster:* Metallic. *Anisotropism:* Distinct in air, strong in oil, cream to bluish green.

R_1–R_2: (400) — , (420) — , (440) 33.6–33.4, (460) 33.5–33.8, (480) 33.1–33.6, (500) 32.5–33.9, (520) 32.1–33.2, (540) 32.1–33.7, (560) 31.8–33.7, (580) 31.4–33.9, (600) 31.2–33.7, (620) 31.0–34.1, (640) 31.0–33.5, (660) 30.9–33.4, (680) — , (700) —

Cell Data: *Space Group:* n.d. Z = n.d.

X-ray Powder Pattern: n.d.

Chemistry:

	(1)	(2)	(3)	(4)
Ag	64.7	64.0	61.0	59.1
Cu	6.2	5.4	4.3	6.8
Pb	0.8	1.4	1.3	3.1
Fe	1.1	0.8	1.1	0.6
Bi	16.9	16.8	16.0	17.0
Te	1.5	1.8	1.9	2.3
S	11.6	12.6	11.8	13.7
Total	102.8	102.8	97.4	102.6

(1–4) Ivigtut, Greenland; by electron microprobe, the average leading to $Ag_{6.01}Cu_{0.95}Bi_{0.84}Pb_{0.07}Fe_{0.13}Te_{0.18}S_{4.01}$.

Occurrence: In a cryolite deposit.

Association: Galena, hessite, aikinite, matildite, berryite, tellurian canfieldite, acanthite, bismuth, quartz, fluorite, cryolite.

Distribution: From Ivigtut, southern Greenland.

Name: For the major metallic elements in the mineral: ARgentum, *silver*; CUprum, *copper*; and BISmuth.

Type Material: Mineralogical Museum, University of Copenhagen, Denmark, 1973,114.

References: (1) Karup-Møller, S. (1976) Arcubisite and mineral B—two new minerals from the cryolite deposit at Ivigtut, South Greenland. Lithos, 9, 253–257. (2) (1978) Amer. Mineral., 63, 424 (abs. ref. 1).

Crystal Data: Monoclinic. *Point Group:* n.d. Occurs as aggregates (up to 50 μm) of acicular crystals (1–2 μm), as isolated inclusions in galena, sphalerite, and sulfosalts.

Physical Properties: Hardness = n.d. VHN = n.d. D(meas.) = n.d. D(calc.) = [6.44]

Optical Properties: Opaque. *Color:* Grayish green or bluish green. *Luster:* Metallic. *Pleochroism:* Distinct. *Anisotropism:* Distinct.

R_1–R_2: (440) 31.3–33.2, (480) 32.1–34.4, (520) 32.3–35.1, (580) 31.7–34.7, (620) 31.1–33.9, (660) 30.6–32.8, (700) 30.3–31.8

Cell Data: *Space Group:* n.d. $a = 21.09$ $b = 22.11$ $c = 8.05$ $\beta = 103.02°$ Z = 2

X-ray Powder Pattern: Synthetic.
3.43 (100), 2.83 (80), 3.13 (40), 1.90 (30), 4.24 (20), 3.90 (20), 2.36 (20)

Chemistry:

	(1)	(2)	(3)
Pb	56.50	57.94	57.14
Ag	0.04	0.00	
Fe	0.00	0.31	
Sb	22.48	21.44	22.97
S	15.56	15.44	16.29
Cl	3.78	4.39	3.60
Total	98.36	99.52	100.00

(1) Madjarovo, Bulgaria; by electron microprobe. (2) Gruvåsen, Sweden; by electron microprobe. (3) $Pb_{19}Sb_{13}S_{35}Cl_7$.

Occurrence: In polymetallic ore deposits.

Association: Galena, nadorite, chlorian robinsonite, chlorian semseyite, pyrostilpnite, argentian tetrahedrite, anglesite (Madjarovo, Bulgaria); sphalerite, pyrrhotite, chalcopyrite, magnetite, scheelite, stannite, pyrargyrite, silver, antimony, arsenopyrite, nisbite, graphite (Gruvåsen, Sweden).

Distribution: From Madjarovo, Bulgaria. In the Dressfall mine, Gruvåsen, Bergslagen, Sweden.

Name: For the Arda River, Bulgaria.

Type Material: n.d.

References: (1) Breskovska, V.V., N.N. Mozgova, N.S. Bortnikov, A.I. Gorshkov, and A.I. Tsepin (1982) Ardaite—a new lead–antimony chlorosulphosalt. Mineral. Mag., 46, 357–61. (2) (1983) Amer. Mineral., 68, 642 (abs. ref. 1). (3) Burke, E.A.J., C. Kieft, and M.A. Zakrzewski (1981) The second occurrence of ardaite: Gruvåsen, Bergslagen, Sweden. Can. Mineral., 19, 419–422.

Crystal Data: Cubic. *Point Group:* $4/m\ \overline{3}\ 2/m$. As euhedral crystals with well developed octahedral faces; as patches in other sulfides; massive.

Physical Properties: *Cleavage:* Well developed on {111}. Hardness = n.d. VHN = 132–154 (50 g load). D(meas.) = n.d. D(calc.) = 4.66

Optical Properties: Opaque. *Color:* Bronze-brown; cinnamon-brown in polished section. *Luster:* Metallic.

R: (400) 20.4, (420) 21.0, (440) 21.9, (460) 22.9, (480) 24.2, (500) 25.6, (520) 27.2, (540) 28.8, (560) 30.4, (580) 31.9, (600) 33.3, (620) 34.6, (640) 35.8, (660) 36.8, (680) 37.8, (700) 38.6

Cell Data: *Space Group:* $Fm3m$. $a = 10.521(3)$ Z = 4

X-ray Powder Pattern: Vuonos, Finland.
3.170 (10), 1.858 (10), 2.018 (4), 1.072 (3), 6.06 (2), 5.25 (2), 3.71 (2)

Chemistry:

	(1)	(2)
Ag	13.3	12.1
Fe	34.7	35.6
Ni	21.3	20.0
Cu		0.6
S	31.4	31.5
Total	100.7	99.8

(1) Oktyabr mine, USSR; by electron microprobe. (2) Talnotry mine, Scotland; by electron microprobe.

Occurrence: In pyrite and cubanite-chalcopyrite hydrothermal veins, and in disseminated sulfide deposits in skarn.

Association: Pyrite, pyrrhotite, mackinawite, cubanite, chalcopyrite, stannite, galena, sphalerite, calcite, quartz.

Distribution: In the USSR, in the Oktyabr mine, Talnakh area, Noril'sk region, western Siberia, and the Khovuaksinsk Co-Ni deposit. In the Vuonos, Miihkali, Hietajärvi, and Outokumpu deposits, Finland. In Scotland, at the Talnotry mine, Newton Stewart, Dumfries, and at Galloway. At Koronuda, Macedonia, Greece. From Bird River, and the Agassiz gold deposit, Lynn Lake region, Manitoba, Canada. In the USA, from near Silver City, Ontonogan Co., Michigan. From Windaira, Western Australia.

Name: For the similarity in composition to pentlandite.

Type Material: n.d.

References: (1) Sishkin, M.N., G.A. Mitenkov, V.A. Mikhailova, N.S. Rudashevskii, A.F. Sidarov, A.M. Karpenkov, A.V.Kondrat'ev, and I.A. Bud'ko (1971) Pentlandite variety rich in silver. Zap. Vses. Mineral. Obshch., 100, 184–191 (in Russian). (2) (1973) Mineral. Abs., 24, 71 (abs. ref. 1). (3) Rudashevskii, N.S., G.A. Mintkenov, A.M. Karpenkov, and N.N. Shiskin (1977) Silver-containing pentlandite—the independent mineral species argentopentlandite. Zap. Vses. Mineral. Obshch., 106, 688–691 (in Russian). (4) (1979) Mineral. Abs., 30, 71 (abs. ref. 3). (5) Vuorelainen, Y., T.A. Häkli, and H. Papunen (1972) Argentian pentlandite from some Finnish sulfide deposits. Amer. Mineral., 57, 137–145. (6) Scott, S.D. and E. Gasparini (1973) Argentian pentlandite $(Fe,Ni)_8AgS_8$, from Bird River, Manitoba. Can. Mineral., 12, 165–168. (7) Hall, S.R. and J.M. Stewart (1973) The crystal structure of argentian pentlandite $(Fe,Ni)_8AgS_8$, compared with the refined structure of pentlandite $(Fe,Ni)_9S_8$. Can. Mineral., 12, 169–177.

Crystal Data: Orthorhombic, pseudohexagonal. *Point Group:* $2/m\ 2/m\ 2/m$. As thick tabular pseudohexagonal prisms, sometimes with rough pyramidal terminations {011}. *Twinning:* Pseudohexagonal twins common || [001] by interpenetration of {120}; lamellar twinning is also present.

Physical Properties: *Fracture:* Uneven. *Tenacity:* Brittle. Hardness = 3.5–4 VHN = 225–242 (100 g load). D(meas.) = 4.25 D(calc.) = 4.27

Optical Properties: Opaque. *Color:* Bright gray-white when fresh, commonly tarnished iridescent in greens, blues, yellows, purples, and browns; grayish to yellowish white in reflected light. *Luster:* Metallic. *Anisotropism:* Strong, deep blue to pale blue-gray.
R_1–R_2: (400) 20.8–29.1, (420) 22.1–30.0, (440) 23.4–30.9, (460) 24.5–31.8, (480) 25.3–32.5, (500) 26.0–33.3, (520) 26.3–34.0, (540) 26.8–34.7, (560) 27.3–35.3, (580) 28.0–35.8, (600) 28.6–36.4, (620) 29.3–36.8, (640) 30.0–37.2, (660) 30.8–37.4, (680) 31.4–37.6, (700) 32.0–37.8

Cell Data: *Space Group:* *Pmmn.* $a = 6.639(4)$ $b = 11.463(6)$ $c = 6.452(3)$ Z = 4

X-ray Powder Pattern: Freiberg, Germany.
3.341 (100), 3.318 (100), 1.808 (70), 3.624 (50), 1.931 (50), 3.110 (40), 1.908 (40)

Chemistry:

	(1)	(2)	(3)
Ag	32.89	34.55	34.17
Fe	35.89	35.03	35.37
Cu	0.19	0.17	
As		0.05	
S	30.71	30.15	30.46
Total	99.68	99.95	100.00

(1) St. Andreasberg, Germany. (2) Měděnec, Czechoslovakia; by electron microprobe. (3) $AgFe_2S_3$.

Polymorphism & Series: Dimorphous with sternbergite.

Occurrence: As crusts of euhedral crystals formed in hydrothermal veins.

Association: Arsenic, proustite, pyrargyrite, pyrostilpnite, xanthoconite, sternbergite, stephanite, pyrite, nickel-skutterudite, dolomite, calcite, quartz.

Distribution: From St. Andreasberg, Harz; Marienberg, Schneeberg, Johanngeorgenstadt, and Freiberg, Saxony; and at the Anton mine, Wieden, Black Forest, Germany. In Czechoslovakia, at Jáchymov (Joachimsthal), Příbram, and in the Krušné hory Mountains at Měděnec. At the Tynebottom mine, Garrigill, near Alston, Cumbria, England. From Broken Hill, New South Wales, Australia. In the Omidani mine, Hyogo Prefecture, Japan. At Colquechaca, Potosí, Bolivia.

Name: For its composition and physical similarity to pyrite.

Type Material: Royal Ontario Museum, Toronto, Canada, M13001; National Museum of Natural History, Washington, D.C., USA, R541.

References: (1) Palache, C., H. Berman, and C. Frondel (1944) Dana's system of mineralogy, (7th edition), v. I, 248. (2) Murdoch, J. and L.G. Berry (1954) X-ray measurements on argentopyrite. Amer. Mineral., 39, 475–485. (3) Czamanske, G.K. and R.R. Larsen (1969) The chemical identity and formula of argentopyrite and sternbergite. Amer. Mineral., 54, 1198–1201. (4) Šrein, V., T. Řídkošil, P. Kašpar, and J. Šourek (1986) Argentopyrite and sternbergite from polymetallic veins of the skarn deposit Měděnec, Krušné hory Mts., Czechoslovakia. Neues Jahrb. Mineral., Abh., 154, 207–222.

Crystal Data: Cubic. *Point Group:* $\bar{4}3m$ (by analogy to tetrahedrite–tennantite). As small grains up to 0.1 mm, and surrounding tennantite.

Physical Properties: *Fracture:* Conchoidal. Hardness = 3.4 VHN = 285–320, 305 average. D(meas.) = n.d. D(calc.) = 5.05

Optical Properties: Opaque. *Color:* Gray-black; light gray to greenish gray in polished section. *Streak:* Red-brown to black. *Luster:* Resinous.

R: (400) 30.9, (420) 30.9, (440) 30.8, (460) 30.7, (480) 30.5, (500) 30.4, (520) 30.4, (540) 30.4, (560) 30.4, (580) 30.3, (600) 29.9, (620) 29.4, (640) 28.8, (660) 28.4, (680) 28.2, (700) 27.8.

Cell Data: *Space Group:* $I\bar{4}3m$ (by analogy to tetrahedrite–tennantite). a = 10.584(3). Z = [2]

X-ray Powder Pattern: Kvartsitoviye Gorki deposit, USSR.
3.06 (100), 1.869 (80), 1.595 (40), 2.65 (30), 2.073 (20), 1.230 (20), 1.214 (20)

Chemistry:

	(1)
Ag	33.54
Cu	15.60
Fe	1.13
Zn	5.44
Pb	0.18
Cd	0.10
Sb	12.59
As	8.80
S	22.66
Total	100.04

(1) Kvartsitoviye Gorki deposit, USSR; by electron microprobe, corresponding to $(Ag_{5.67}Cu_{4.48}Zn_{1.52}Fe_{0.37}Pb_{0.01}Cd_{0.01})_{\Sigma=12.06}(As_{2.14}Sb_{1.89})_{\Sigma=4.03}S_{12.90}$.

Polymorphism & Series: Forms series with tennantite and freibergite.

Occurrence: In a polymetallic deposit.

Association: Tennantite–tetrahedrite, freibergite, stibnite, mercurian gold, pyrite, galena, siderite, ankerite, quartz.

Distribution: In the Kvartsitoviye Gorki deposit, northern Kazakhstan, USSR.

Name: For the chemical composition and by analogy to tennantite.

Type Material: A.E. Fersman Mineralogical Museum, Academy of Sciences, Moscow, USSR.

References: (1) Spiridonov, E.M., N.F. Sokolova, A.K. Gapeev, D.M. Dashevskaya, T.L. Evstigneeva, T.N. Chvileva, V.G. Demidov, E.P. Balashov, and V.I. Shul'ga (1986) A new mineral—argentotennantite. Doklady Acad. Nauk SSSR, 290, 206–210 (in Russian). (2) (1988) Amer. Mineral., 73, 439 (abs. ref. 1).

Crystal Data: Orthorhombic, pseudocubic. *Point Group:* $mm2$. Pseudo-octahedra, dodecahedra, cubes, or as combinations of these forms, in crystals as large as 18 cm. Also radiating crystal aggregates, botryoidal crusts, or massive. *Twinning:* Pseudospinel law {111}; repeated interpenetration twins of pseudododecahedra on {111}.

Physical Properties: *Fracture:* Uneven to slightly conchoidal. *Tenacity:* Brittle. Hardness = 2.5–3 VHN = n.d. D(meas.) = 6.29 D(calc.) = 6.32

Optical Properties: Opaque. *Color:* Steel-gray with a red tint, tarnishes black; in polished section, gray-white with a violet tint. *Streak:* Gray-black. *Luster:* Strong metallic. *Pleochroism:* Very weak. *Anisotropism:* Weak.
R: (400) 29.6, (420) 28.7, (440) 27.8, (460) 27.0, (480) 26.4, (500) 25.8, (520) 25.3, (540) 25.0, (560) 24.9, (580) 24.8, (600) 24.8, (620) 24.8, (640) 24.8, (660) 24.9, (680) 25.0, (700) 25.1

Cell Data: *Space Group:* $Pna2_1$ or $Pnam$. $a = 15.149$ $b = 7.476$ $c = 10.589$ Z = 4

X-ray Powder Pattern: Machacamarca, Bolivia.
3.02 (100), 1.863 (50), 2.66 (40), 3.14 (30), 2.44 (30), 2.03 (30), 1.784 (20)

Chemistry:

	(1)	(2)	(3)
Ag	75.78	74.20	76.51
Fe		0.68	
Ge	3.65	4.99	6.44
Sn	3.60	3.36	
Sb	trace	trace	
S	16.92	16.45	17.05
Total	99.95	99.68	100.00

(1) Chocaya, Bolivia. (2) Aullagas, Bolivia. (3) Ag_8GeS_6.

Polymorphism & Series: Forms a series with canfieldite.

Occurrence: In low-temperature polymetallic deposits with silver sulfosalts (Freiberg, Germany); in high-temperature Sn-Ag deposits (Bolivia).

Association: Canfieldite, stephanite, acanthite, pyrargyrite, polybasite, aramayoite, diaphorite, marcasite, pyrite, galena, sphalerite, cassiterite, siderite.

Distribution: From the Himmelsfürst and other mines, Erbisdorf, near Freiberg, Saxony, Germany. In the Fournial mine, Massiac region, Cantal, France. From Silvermines, Co. Tipperary, Ireland. At a number of localities in Bolivia, including Aullagas, Colquechaca, Potosí, Oruro, Porco, Chocaya, and Machacamarca. At the Dolly Varden mine, Alice Arm, British Columbia, Canada. From Rico, Dolores Co., Colorado, USA.

Name: From the Greek for *silver-containing.*

Type Material: n.d.

References: (1) Palache, C., H. Berman, and C. Frondel (1944) Dana's system of mineralogy, (7th edition), v. I, 356–358. (2) Berry, L.G. and R.M. Thompson (1962) X-ray powder data for the ore minerals. Geol. Soc. Amer. Mem. 85, 122. (3) Eulenberger, G. (1977) Die Kristallstruktur der Tieftemperaturmodifikation von Ag_8GeS_6 — synthetischer Argyrodite. Monatshefte Chem., 108, 901–913. (4) Wang, N. (1978) New data for Ag_8SnS_6 (canfieldite) and Ag_8GeS_6 (argyrodite). Neues Jahrb. Mineral., Monatsh., 269–272.

Crystal Data: Hexagonal. *Point Group:* $\overline{3}\ 2/m$. Usually granular, massive, and in concentric layers; sometimes reticulated, reniform, stalactitic; rarely columnar or acicular; also as small rhombohedra. *Twinning:* Rare on twin plane $\{10\overline{1}4\}$; pressure twinning on $\{01\overline{1}2\}$ develops delicate lamellae.

Physical Properties: *Cleavage:* Perfect on {0001}; $\{10\overline{1}4\}$ fair. *Fracture:* Uneven. *Tenacity:* Brittle. Hardness = 3.5 VHN = 72–173 (100 g load). D(meas.) = 5.63–5.78 D(calc.) = 5.778

Optical Properties: Opaque. *Color:* Tin-white, tarnishes to dark gray. *Streak:* Tin-white. *Luster:* Nearly metallic. *Pleochroism:* Feeble. *Anisotropism:* Distinct, yellowish brown and light gray to yellowish gray.

R_1–R_2: (400) 53.9–58.4, (420) 52.2–57.0, (440) 50.5–55.6, (460) 49.6–54.7, (480) 48.9–54.2, (500) 48.3–53.8, (520) 47.8–53.4, (540) 47.4–53.2, (560) 47.2–52.9, (580) 47.0–52.8, (600) 46.9–52.7, (620) 46.6–52.6, (640) 46.4–52.6, (660) 46.3–52.6, (680) 46.4–52.7, (700) 47.0–52.8

Cell Data: *Space Group:* $R\overline{3}m$. $a = 3.768$ $c = 10.574$ Z = 6

X-ray Powder Pattern: Synthetic.
2.771 (100), 3.52 (30), 1.879 (30), 2.05 (20), 1.556 (10), 1.768 (10), 1.757 (7)

Chemistry:

	(1)
As	98.14
Sb	1.65
S	0.16
insol.	0.15
Total	100.10

(1) Mount Royal, near Montreal, Quebec, Canada.

Polymorphism & Series: Dimorphous with arsenolamprite.

Occurrence: In hydrothermal veins and deposits that contain other arsenic minerals; sometimes in Co-Ag veins.

Association: Arsenolite, cinnabar, realgar, orpiment, stibnite, galena, sphalerite, pyrite, barite.

Distribution: Numerous localities are known, but most are of only minor interest. At Sterling Hill, Sussex Co., New Jersey and Washington Camp, Santa Cruz Co., Arizona, USA. In the Akatani mine, Fukui Prefecture, Japan. In the Gabe-Gottes mine, Sainte-Marie-aux-Mines, Haut-Rhin, France. In Germany, from Freiberg, Schneeberg, Johanngeorgenstadt, Marienberg and Annaberg, Saxony; Wolfsberg and St. Andreasberg, Harz; and Wieden, Black Forest. At Jáchymov (Joachimsthal) and Příbram, Czechoslovakia. In Romania, from Săcărâmb (Nagyág), Hunyad, and Cavnic (Kapnik). In the Huallapón mine, Pasta Bueno, Peru. At Bidi, Sarawak Province, Borneo. From Copiapó, Chile.

Name: From the Latin *arsenicum*, earlier Greek *arrenikos*, or *arsenikos*, *masculine*, an allusion to its potent properties.

References: (1) Palache, C., H. Berman, and C. Frondel (1944) Dana's system of mineralogy, (7th edition), v. I, 128–130. (2) (1954) NBS Circ. 539, 3, 6.

Crystal Data: Tetragonal. *Point Group:* $4/m\ 2/m\ 2/m$, 422, $\bar{4}2m$, or $4mm$. Tabular crystals to 20 mm; irregular masses to 10 mm.

Physical Properties: *Fracture:* Conchoidal. Hardness = n.d. VHN = 516–655 (50 g load). D(meas.) = 6.35 D(calc.) = 6.52

Optical Properties: Opaque. *Color:* Bronze. *Luster:* Metallic, brilliant on fresh surfaces. R_1–R_2: (470) 41.6–43.0, (546) 46.2–47.1, (589) 48.2–49.2, (650) 50.8–51.6

Cell Data: *Space Group:* $I4/mmm$, $I422$, $I\bar{4}m2$, $I\bar{4}2m$, or $I4mm$. $a = 10.2711(2)$ $c = 10.8070(4)$ Z = 2

X-ray Powder Pattern: Vermilion mine, Canada.
2.875 (100), 2.393 (66), 2.298 (59), 3.253 (56), 4.343 (53), 3.637 (48), 1.862 (48)

Chemistry:

	(1)	(2)
Ni	44.9	46.52
Fe	1.4	
Co	0.3	
Bi	26.5	27.60
As	4.4	3.30
Sb	0.1	
Te	0.0	
S	22.0	22.58
Total	99.6	100.00

(1) Vermilion mine, Canada; by electron microprobe, corresponding to $(Ni_{8.9}Fe_{0.3}Co_{0.1})_{\Sigma=9.3}Bi_{1.0}(Bi_{0.5}As_{0.7})_{\Sigma=1.2}S_{8.0}$. (2) $Ni_{18}Bi_3AsS_{16}$.

Occurrence: In hydrothermal Ni-Co-Cu veins.

Association: Chalcopyrite, pyrrhotite, gersdorffite, pyrite, gold, nickeline, galena, copper, sperrylite, michenerite, froodite (Vermilion mine, Canada).

Distribution: From the Vermilion mine, Sudbury, Ontario, Canada. At Karagaily, central Kazakhstan, USSR. In the Tsumo mine, near Hiroshima City, Shimane Prefecture, Japan.

Name: Alludes to its chemical relation to the hauchecornite group.

Type Material: Royal Ontario Museum, Toronto, Canada, M29206, M29207, M29208.

References: (1) Gait, R.L., and D.C. Harris (1980) Arsenohauchecornite and tellurohauchecornite: new minerals in the hauchecornite group. Mineral. Mag., 43, 877–878. (2) (1981) Amer. Mineral., 66, 436–437 (abs. ref. 1). (3) Gait, R.L. and D.C. Harris (1972) Hauchecornite—antimonian, arsenian and tellurian varieties. Can. Mineral., 11, 819–825. (4) Grice, J.D. and R.B. Ferguson (1989) The crystal structure of arsenohauchecornite. Can. Mineral., 27, 137–142.

Crystal Data: Orthorhombic. *Point Group:* $2/m\ 2/m\ 2/m$ (synthetic). As needles, to 8 mm, foliated, radial aggregates of plates, and massive.

Physical Properties: *Cleavage:* Perfect in one direction. Hardness = 2 VHN = n.d. D(meas.) = 5.3–5.5 D(calc.) = 5.577

Optical Properties: Opaque. *Color:* Gray-white, altering to a dull black coating. *Streak:* Black. *Luster:* Metallic, brilliant. *Anisotropism:* Weak.
R_1–R_2: (400) —, (420) 42.8–56.8, (440) 43.6–55.4, (460) 44.3–54.0, (480) 44.7–52.8, (500) 44.8–51.6, (520) 44.6–50.5, (540) 44.0–49.6, (560) 43.1–48.7, (580) 42.7–48.0, (600) 42.6–47.4, (620) 42.7–46.8, (640) 43.0–46.3, (660) 43.4–45.8, (680) 43.8–45.5, (700) 44.4–45.0

Cell Data: *Space Group:* *Bmab* (synthetic). $a = 3.63$ $b = 4.45$ $c = 10.96$ Z = 8

X-ray Powder Pattern: Mina Alacrán, Chile
5.76 (10), 2.72 (10), 2.745 (8), 1.875 (7), 1.730 (7), 3.48 (6), 2.230 (5)

Chemistry: Nearly pure As, with up to 3% Bi.

Polymorphism & Series: Dimorphous with arsenic.

Occurrence: As plates and veinlets in carbonate rocks (Černý Důl, Czechoslovakia); in calcite veins (Mackenheim, Germany).

Association: Arsenic, bismuth, silver, sternbergite, emplectite, safflorite, löllingite, pyrite, galena, orpiment, realgar.

Distribution: In Germany, in the Palmbaum mine, Marienberg, Saxony; from Schweisweiler, Palatinate; and at Mackenheim, Odenwald, and Wittichen, Black Forest. From Černý Důl, Krkonoše (Giant Mountains), and Jáchymov (Joachimsthal), Czechoslovakia. At Sainte-Marie-aux-Mines, Haut-Rhin, France. From Mina Alacrán, Pampas Largo, Copiapó, Atacama, Chile.

Name: From its composition and the Greek for *brilliant* in allusion to its reflectance.

Type Material: n.d.

References: (1) Palache, C., H. Berman, and C. Frondel (1944) Dana's system of mineralogy, (7th edition), v. I, 130. (2) Johan, Z. (1959) Arsenolamprit—die rhombische Modifikation des Arsens aus Černý Důl (Schwarzenthal) im Riesengebirge. Chem. Erde, 20, 71–80 (in German). (3) (1960) Amer. Mineral., 45, 479 (abs. ref. 2). (4) Ramdohr, P. (1969) The ore minerals and their intergrowths, (3rd edition), 370. (5) Clark, A.H. (1970) Arsenolamprite confirmed from the Copiapó area, northern Chile. Mineral. Mag., 37, 732–733. (6) Smith, P.M., A.J. Leadbetter, and A.J. Apling (1975) The structures of orthorhombic and vitreous arsenic. Philos. Mag., 31, 57–64. (7) Picot, P. and Z. Johan (1982) Atlas of ore minerals. B.R.G.M., Orleans, France, and Elsevier, Amsterdam, Holland, p. 77.

Crystal Data: Triclinic. *Point Group:* 1. As rounded grains to 1.8 mm.

Physical Properties: *Tenacity:* Malleable. Hardness = n.d. VHN = 379–449 (100 g load). D(meas.) = 10.40 D(calc.) = 11.028

Optical Properties: Opaque. *Color:* White with yellowish creamy tint in polished section. *Luster:* Metallic. *Anisotropism:* Strong, red and golden brown to blue-gray in air, khaki-brown to blue-gray and bright steel-gray in oil.

R_1–R_2: (400) 43.9–44.5, (420) 44.7–45.8, (440) 45.7–46.5, (460) 46.8–47.8, (480) 48.1–49.4, (500) 49.4–50.8, (520) 50.8–52.4, (540) 52.1–53.9, (560) 53.2–55.1, (580) 54.4–56.4, (600) 55.4–57.3, (620) 56.4–58.1, (640) 57.0–58.8, (660) 57.7–59.2, (680) 58.2–59.8, (700) 58.8–60.0

Cell Data: *Space Group:* $P1$. $a = 7.43$ $b = 13.95$ $c = 7.35$ $\alpha = 92°53'$ $\beta = 119°30'$ $\gamma = 87°51'$ Z = 6

X-ray Powder Pattern: Itabira, Brazil.
2.13 (100), 2.34 (60), 1.41 (40), 1.24 (30), 1.21 (30), 2.28 (20), 2.19 (20)

Chemistry:

	(1)
Pb	77.56
Cu	0.02
As	17.08
Sb	5.15
Total	99.81

(1) Itabira, Brazil; by electron microprobe, average of 13 grains.

Occurrence: In gold concentrates.

Association: Hematite, palladian gold, atheneite, stillwaterite, palladseite, isomertieite, quartz.

Distribution: From Itabira, Minas Gerais, Brazil. In the Stillwater Complex, Montana, USA.

Name: For the composition.

Type Material: National Museum of Natural History, Washington, D.C., USA, 142504.

References: (1) Clark, A.M., A.J. Criddle, and E.E. Fejer (1974) Palladium arsenide–antimonides from Itabira, Minas Gerais, Brazil. Mineral. Mag., 39, 528–543. (2) (1974) Amer. Mineral., 59, 1332 (abs. ref. 1). (3) Cabri, L.J., A.M. Clark, and T.T. Chen (1977) Arsenopalladinite from Itabira, Brazil and from the Stillwater Complex, Montana. Can. Mineral., 15, 70–73. (4) (1979) Amer. Mineral., 64, 658 (abs. ref. 3).

Crystal Data: Monoclinic, pseudo-orthorhombic. *Point Group:* $2/m$. Crystals to over 8 cm, flat tabular to blocky to prismatic, striated ∥ [001]. Also compact, granular, columnar. *Twinning:* Common on {100} and {001}; as contact or penetration twins on {101}; on {012} to produce star-shaped trillings or cruciform twins.

Physical Properties: *Cleavage:* Distinct on {101}; {010} in traces. *Fracture:* Uneven. *Tenacity:* Brittle. Hardness = 5.5–6 VHN = 1081 on (001) section (100 g load). D(meas.) = 6.07(15) D(calc.) = 6.18

Optical Properties: Opaque. *Color:* Silver-white to steel-gray; in polished section white with faint yellow tint. *Luster:* Metallic. *Pleochroism:* Weak, in white or bluish tint and faint reddish yellow. *Anisotropism:* Strong, red-violet.

R_1–R_2: (400) 49.9–53.0, (420) 49.2–53.0, (440) 48.5–53.0, (460) 48.3–53.0, (480) 48.5–52.8, (500) 48.9–52.7, (520) 49.4–52.4, (540) 49.7–52.0, (560) 50.2–51.6, (580) 50.5–51.3, (600) 50.8–51.1, (620) 51.0–50.8, (640) 51.1–50.7, (660) 51.1–50.6, (680) 51.1–50.5, (700) 51.0–50.5

Cell Data: *Space Group:* $P2_1/c$. $a = 5.744$ $b = 5.675$ $c = 5.785$ $\beta = 112.17°$ Z = 4

X-ray Powder Pattern: Freiberg, Germany.
2.677 (100), 2.662 (100), 2.418 (95), 2.412 (95), 2.440 (90), 1.814 (90), 1.824 (70)

Chemistry:

	(1)	(2)	(3)
Fe	34.53	32.48	34.30
Co	0.09	1.16	
Bi	0.79		
As	44.34	48.72	46.01
S	20.22	18.80	19.69
Total	99.97	101.16	100.00

(1) O'Brien mine, Cobalt, Canada. (2) Franklin, New Jersey, USA. (3) FeAsS.

Occurrence: Of hydrothermal origin, typically one of the earliest minerals to form. Found in pegmatites, high-temperature gold-quartz and tin veins, and in contact metamorphic sulfide deposits; less commonly of low-temperature hydrothermal origin. Also in gneisses, schists and other metamorphic rocks.

Association: Pyrrhotite, pyrite, chalcopyrite, galena, gold, scheelite, cassiterite, many other species.

Distribution: The most abundant and widespread arsenic mineral; only a few localities for large and fine crystals can be mentioned. In the USA, from Franconia, Grafton Co., New Hampshire and Franklin, Sussex Co., New Jersey. In Canada, in the Cobalt district, Ontario. In Japan, in the Ashio mine, Tochigi Prefecture, and many other localities. From Stratonik, Greece. In Germany, from Altenberg, Ehrenfriedersdorf, and Freiberg, Saxony. At Trepča, Yugoslavia. From Panasqueira, Beira-Baixa Province, Portugal. At Sala, Tunaberg, Stollberg, Boliden, and Nordmark, Sweden. In England, from a number of mines in Cornwall, and in Devonshire, at Tavistock. From Hidalgo del Parral and Santa Eulalia, Chihuahua, Mexico.

Name: A contraction of *arsenical pyrites.*

References: (1) Palache, C., H. Berman, and C. Frondel (1944) Dana's system of mineralogy, (7th edition), v. I, 316–322. (2) Morimoto, N. and L.A. Clark (1961) Arsenopyrite crystal-chemical relations. Amer. Mineral., 46, 1448–1469.

Crystal Data: Cubic. *Point Group:* $\overline{4}3m$. As tiny grains.

Physical Properties: Hardness = 3.5 VHN = n.d. D(meas.) = 4.01–4.2 D(calc.) = 4.39

Optical Properties: Opaque. *Color:* Bronze-yellow. *Luster:* Metallic.
Pleochroism: Distinctive yellowish brown tint.
R: (400) 23.5, (420) 24.9, (440) 26.3, (460) 27.4, (480) 28.6, (500) 29.8, (520) 30.7, (540) 31.7, (560) 32.4, (580) 32.7, (600) 32.4, (620) 31.8, (640) 31.4, (660) 31.1, (680) 30.9, (700) 30.4

Cell Data: *Space Group:* $P\overline{4}3m$. $a = 5.257(3)$ Z = 1

X-ray Powder Pattern: Bor, Yugoslavia.
3.04 (100), 1.867 (100), 1.593 (50), 1.074 (30), 2.77 (10), 1.656 (10), 1.317 (10)

Chemistry:

	(1)	(2)	(3)
Cu	48.84	46.65	52.5
Fe			1.9
As	12.80	11.67	12.5
V	4.16	5.20	
Sb			
S	33.14	31.66	n.d.
insol.	1.01	3.82	
Total	99.95	99.00	n.d.

(1–2) Mongolia. (3) Bor, Yugoslavia; by electron microprobe.

Polymorphism & Series: Forms a series with sulvanite.

Occurrence: In quartz-calcite veins cutting bituminous limestone (Mongolia); in a copper porphyry deposit (Bor, Yugoslavia).

Association: Pyrite, enargite, luzonite, covellite, galena, quartz, calcite.

Distribution: From an undefined locality in Mongolia. In the Tilva Mika deposit, Bor, eastern Serbia, Yugoslavia. From the Osarizawa mine, Akita Prefecture, Honshu, Japan. At Bisbee, Cochise Co., Arizona, USA.

Name: For the chemical relationship with sulvanite.

Type Material: n.d.

References: (1) Betekhtin, A.G. (1941) The new mineral arsenosulvanite. Zap. Vses. Mineral. Obshch., 70, 161–164 (in Russian with English abs.). (2) Mikheev, V.I. (1941) The structure of arsenosulvanite. Zap. Vses. Mineral. Obshch., 70, 165–194 (in Russian with English abs.). (3) (1955) Amer. Mineral., 40, 368–369 (abs. refs. 1 and 2). (4) Sclar, C.B. and M. Drovenik (1960) Lazarevićite [=arsenosulvanite], a new cubic copper-arsenic sulfide from Bor, Jugoslavia. Bull. Geol. Soc. Amer. (abs.), 71, 1971. (5) (1961) Amer. Mineral., 46, 465 (abs. ref. 4).

Crystal Data: Monoclinic. *Point Group:* $2/m$. Thin tabular on {001}; pseudohexagonal. *Twinning:* Twin plane {110}, repeated.

Physical Properties: *Cleavage:* Imperfect on {001}. *Fracture:* Uneven. *Tenacity:* Brittle. Hardness = 2–3 VHN = n.d. D(meas.) = 6.18–6.23 D(calc.) = n.d.

Optical Properties: Opaque, translucent in thin splinters. *Color:* Black; dark red in transmitted light. *Streak:* Black. *Luster:* Metallic.

Optical Class: Biaxial negative (–). *Pleochroism:* In reflected light, weak. *Orientation:* $X = c$; $Y = a$. *Dispersion:* Very strong. $n = > 2.72$ (Li). 2V(meas.) = 22° 2V(calc.) = n.d. *Anisotropism:* Moderate.

R_1–R_2: n.d.

Cell Data: *Space Group:* $C2/m$. $a = 26.08$ $b = 15.04$ $c = 23.84$ $\beta = 90°$ Z = 16

X-ray Powder Pattern: n.d.

Chemistry:

	(1)
Ag	71.20
Cu	3.26
Fe	0.38
As	6.87
Sb	0.80
S	17.37
Total	99.88

(1) Freiberg, Germany.

Occurrence: Formed in silver-bearing hydrothermal veins of low to moderate temperature.

Association: Pyrargyrite, stephanite, tetrahedrite, acanthite, silver, gold, pyrite, quartz, calcite, dolomite.

Distribution: At the Neuer Morgenstern mine, Freiberg, Saxony, Germany. In Peru, at Quespisiza. From Creede, Mineral Co., Colorado, USA.

Name: For the chemical composition and close relationship to polybasite.

Type Material: Harvard University, Cambridge, Massachusetts, USA, 82633, 110527.

References: (1) Frondel, C. (1963) Isodimorphism of the polybasite and pearceite series. Amer. Mineral., 48, 565–572. (2) Harris, D.C., E.W. Nuffield, and M.H. Frohberg (1965) Studies of polybasite, pearceite, antimonpearceite, and arsenpolybasite. Can. Mineral., 8, 172–184. (3) (1965) Amer. Mineral., 50, 1507 (abs. ref. 2).

Crystal Data: Monoclinic (perhaps triclinic). *Point Group:* $2/m$, 2, or m. As complex intergrowths to 0.2 mm.

Physical Properties: Hardness = 2.0–2.5 VHN = 82–103 D(meas.) = n.d. D(calc.) = 7.64

Optical Properties: Transparent. *Color:* Colorless to yellow. *Luster:* Vitreous to adamantine.
R_1–R_2: n.d.

Cell Data: *Space Group:* $P2/m$, $P2$, or Pm. $a = 8.99$ $b = 5.24$ $c = 18.45$ $\beta = 92.28°$ Z = 5

X-ray Powder Pattern: Arzak deposit, USSR. Differs only by intensities from lavrentievite.
3.01 (100b), 3.38 (83), 1.587 (83), 3.96 (67), 2.292 (67), 2.199 (67)

Chemistry:

	(1)	(2)
Hg	75.3	77.02
S	7.98	8.21
Br	11.6	10.23
Cl	3.13	4.54
Total	98.01	100.00

(1) Arzak deposit, USSR; by electron microprobe, corresponding to $Hg_{3.06}S_{2.03}(Br_{1.18}Cl_{0.72})_{\Sigma=1.90}$. (2) $Hg_3S_2(Br, Cl)_2$ with Br:Cl = 1:1.

Polymorphism & Series: Forms a series with lavrentievite.

Occurrence: In the oxidized zone of a hydrothermal deposit.

Association: Lavrentievite, cinnabar, corderoite, quartz, kaolinite.

Distribution: From the Arzak deposit, Tuva ASSR, USSR.

Name: For the occurrence in the Arzak deposit, USSR.

Type Material: Central Siberian Geological Museum, Novosibirsk, USSR.

References: (1) Vasil'ev, V.L., N.A. Pal'chik, and O.K. Grechishchev (1984) Lavrentievite and arzakite, new natural sulfohalogenides of mercury. Geol. i Geofiz., 7, 54–63 (in Russian). (2) (1985) Amer. Mineral., 70, 873–874 (abs. ref. 1). (3) (1984) Chem. Abs., 101, 174794 (abs. ref. 1).

Crystal Data: Monoclinic. *Point Group:* $2/m$, m, or 2. As prismatic, lath-like crystals up to 5 cm, or as thick, slightly bent plates up to 1 cm.

Physical Properties: *Cleavage:* Perfect on {001}. Hardness = n.d. VHN = 150–181 (25 g load). D(meas.) = n.d. D(calc.) = 7.33

Optical Properties: Opaque. *Color:* Lead-gray; creamy white in reflected light. *Luster:* Metallic. *Anisotropism:* Moderate, from gray to red-brown.
R_1–R_2: (470) 45.1–48.1, (546) 43.4–46.3, (589) 42.9–46.3, (650) 42.9–46.3

Cell Data: *Space Group:* $C2/m$, Cm, or $C2$. $a = 13.71$ $b = 4.09$ $c = 31.43$ $\beta = 91.0°$ Z = 4

X-ray Powder Pattern: Ascham Alm, Austria.
3.426 (100), 3.378 (88), 2.941 (54), 2.926 (54), 2.861 (48), 3.525 (42), 2.067 (42)

Chemistry:

	(1)
Pb	62.95
Bi	22.56
S	14.89
Total	100.40

(1) Ascham Alm, Austria; by electron microprobe, corresponding to $Pb_{5.89}Bi_{2.05}S_9$.

Occurrence: In alpine-cleft veins, cutting gneiss.

Association: Cosalite, galena, quartz, albite, orthoclase, calcite, chlorite.

Distribution: From near Ascham Alm, Untersulzbachtal, Austria. At Granite Gap, Hidalgo Co., New Mexico, USA.

Name: For Ascham Alm, Austria.

Type Material: Museum of Natural History, Vienna, Austria; Division of Mineral Chemistry, C.S.I.R.O., Port Melbourne, Victoria, Australia; National Museum of Natural History, Washington, D.C., USA, 160379.

References: (1) Mumme, W.G., G. Niedermayr, P.R. Kelly, and W.H. Paar (1983) Aschamalmite, $Pb_{5.92}Bi_{2.06}S_9$, from Untersulzbach Valley in Salzburg, Austria—"monoclinic heyrovskyite". Neues Jahrb. Mineral., Monatsh., 433–444. (2) (1984) Amer. Mineral., 69, 810 (abs. ref. 1).

Crystal Data: Orthorhombic. *Point Group:* n.d. As crystal laths up to 100 μm long and 40 μm wide; also massive, up to 300 μm in diameter.

Physical Properties: Hardness = n.d. VHN = 78 (15 g load). D(meas.) = n.d. D(calc.) = 6.59

Optical Properties: Opaque. *Color:* Light gray to white. *Luster:* Metallic.
Pleochroism: Distinct, light gray to blue-gray. *Anisotropism:* Strong, creamy white to dark blue.
R_1–R_2: (400) 21.2–25.1, (420) 19.9–25.2, (440) 19.6–25.3, (460) 19.3–25.4, (480) 19.0–25.5, (500) 18.6–25.6, (520) 18.3–25.6, (540) 17.8–25.5, (560) 17.4–25.4, (580) 16.9–25.1, (600) 16.3–25.0, (620) 15.5–24.5, (640) 15.0–24.2, (660) 14.2–23.7, (680) 13.4–23.1, (700) 12.5–22.4

Cell Data: *Space Group:* n.d. $a = 8.227$ $b = 11.982$ $c = 6.441$ Z = 4

X-ray Powder Pattern: Martin Lake, Canada.
3.235 (100), 1.997 (80), 3.015 (60), 1.893 (50), 1.664 (40), 3.44 (30), 1.817 (30)

Chemistry:

	(1)	(2)	(3)
Cu	51.1	52.0	50.15
Se	49.7	44.0	49.85
S		2.7	
Total	100.8	98.7	100.00

(1–2) Martin Lake, Canada; by electron microprobe. (3) Cu_5Se_4.

Occurrence: As inclusions in and replacements of umangite, as stringers and veinlets in carbonate veins cutting basalt.

Association: Umangite, clausthalite, eucairite, berzelianite, sulfurian berzelianite, klockmannite, eskebornite, tyrrellite, copper, silver, uraninite, hematite, pyrite, calcite, barite, quartz, feldspar (Martin Lake, Canada); berzelianite, eucairite, crookesite, tyrrellite, ferroselite, bukovite, krutaite, calcite, dolomite (Petrovice, Czechoslovakia).

Distribution: At the Martin Lake mine, northeast of Martin Lake, Saskatchewan, Canada. In the Petrovice and Předbořice deposits, Czechoslovakia. At Chaméane, Puy-de-Dôme, France. From Skrikerum, near Tryserum, Kalmar, Sweden.

Name: For Lake Athabasca, northern Saskatchewan, Canada.

Type Material: National Museum of Canada, Ottawa; Royal Ontario Museum, Toronto, Canada.

References: (1) Harris, D.C., L.J. Cabri, and S. Kaiman (1970) Athabascaite: a new copper selenide mineral from Martin Lake, Saskatchewan. Can. Mineral., 10, 207–215. (2) (1971) Amer. Mineral., 56, 632 (abs. ref. 1).

Crystal Data: Hexagonal. *Point Group:* $6/m\ 2/m\ 2/m$. As tiny blebs; crystal outlines seldom displayed, grain boundaries being commonly embayed.

Physical Properties: Hardness = n.d. VHN = 419–442, 431 average (100 g load). D(meas.) = 10.2 D(calc.) = 10.16

Optical Properties: Opaque. *Color:* In reflected light, white with bluish tint compared to arsenopalladinite. *Luster:* Metallic. *Pleochroism:* Very weak, in pale yellow-white to bluish gray-white. *Anisotropism:* Distinct.

R_1–R_2: (400) 46.1–48.3, (420) 47.4–49.8, (440) 48.5–50.9, (460) 49.7–52.4, (480) 50.8–53.6, (500) 51.9–54.8, (520) 53.0–55.8, (540) 54.0–56.7, (560) 54.8–57.3, (580) 55.7–57.8, (600) 56.4–58.3, (620) 57.0–58.5, (640) 57.4–58.6, (660) 57.6–58.7, (680) 57.9–59.0, (700) 58.2–59.1

Cell Data: *Space Group:* $P6/mmm$. $a = 6.798$ $c = 3.483$ Z = 2

X-ray Powder Pattern: Itabira, Brazil.
2.423 (vvs), 2.246 (vs), 1.371 (s), 1.302 (s), 1.259 (s), 1.871 (ms), 1.034 (ms)

Chemistry:

	(1)
Pd	66.0
Hg	14.9
Au	0.5
Cu	0.1
As	19.0
Sb	0.1
Total	100.6

(1) Itabira, Brazil; by electron microprobe.

Occurrence: In concentrates from gold washings.

Association: Arsenopalladinite, palladseite, isomertieite, hematite.

Distribution: From Itabira, Minas Gerais, Brazil. From the area around Zlatoust, Ural Mountains, USSR.

Name: After the Greek goddess *Pallas Athene*, in allusion to its palladium content.

Type Material: British Museum (Natural History), London, England; National Museum of Natural History, Washington, D.C., USA, 142504.

References: (1) Clark, A.M., A.J. Criddle, and E.E. Fejer (1974) Palladium arsenide–antimonides from Itabira, Minas Gerais, Brazil. Mineral. Mag., 39, 528–543. (2) (1974) Amer. Mineral., 59, 1330 (abs. ref. 1).

Crystal Data: Cubic. *Point Group:* $4/m\ \overline{3}\ 2/m$. As small grains less than 100 μm in diameter, rarely showing {100}.

Physical Properties: Hardness = n.d. VHN = 357 (25 g load). D(meas.) = n.d. D(calc.) = 14.19

Optical Properties: Opaque. *Color:* In polished section, light cream. *Luster:* Metallic. *Pleochroism:* None to slight. *Anisotropism:* Slight, due to strain.
R: n.d.

Cell Data: *Space Group:* $Fm3m$. $a = 3.991$ Z = 4

X-ray Powder Pattern: Rustenburg and Atok mines, South Africa.
2.295 (100), 1.202 (100), 1.408 (90), 0.9153 (90), 0.8145 (90), 1.992 (80), 0.8922 (80)

Chemistry:

	(1)
Pt	43.74
Pd	38.35
Sn	18.65
Total	100.74

(1) Rustenburg and Atok mines, South Africa; by electron microprobe, corresponding to $(Pd, Pt)_3Sn_{0.81}$.

Occurrence: As very sparse grains in concentrates of ore.

Association: Rustenburgite, unspecified platinum tellurides (Rustenburg and Atok mines, South Africa); keithconnite, palladoarsenide (Stillwater Complex, Montana, USA).

Distribution: In the Rustenburg and Atok platinum mines, on the Merensky Reef, Bushveld Complex, about 48 km southeast of Pietersburg, Transvaal, South Africa. From the Noril'sk region, western Siberia, USSR. In the Stillwater Complex, Montana, USA.

Name: For the Atok mine, South Africa.

Type Material: n.d.

References: (1) Mihálik, P., S.A. Hiemstra, and J.P.R. de Villiers (1975) Rustenburgite and atokite, two new platinum-group minerals from the Merensky Reef, Bushveld Igneous Complex. Can. Mineral., 13, 146–150. (2) (1976) Amer. Mineral., 61, 340 (abs. ref. 1).

Crystal Data: Cubic. *Point Group:* $4/m\ \overline{3}\ 2/m$. Massive, as rims less than 100 μm, irregular patches, and platy aggregations.

Physical Properties: Hardness = ~3.5 VHN = 54 (10 g load). D(meas.) = 11.5 D(calc.) = [13.77]

Optical Properties: Opaque. *Color:* Yellow with a reddish tint; in reflected light, violet-rose. *Luster:* Highly metallic.
R: n.d.

Cell Data: *Space Group:* $Pm3m$. $a = 3.753(5)$ (synthetic). Z = [1]

X-ray Powder Pattern: n.d.

Chemistry:

	(1)	(2)
Au	53.11	50.82
Cu	45.96	49.18
Total	99.07	100.00

(1) Karabash deposit, USSR; by electron microprobe. (2) $AuCu_3$.

Occurrence: In serpentinites as the product of low temperature ordering and unmixing of Au–Cu solid solution alloys.

Association: Gold, copper, other Au–Cu alloys.

Distribution: From Laksia and Pefkos, Cyprus. In the Karabash deposit, Southern Ural Mountains, USSR. From the El Indio mine, east of Coquimbo, Chile.

Name: For the composition.

Type Material: n.d.

References: (1) Ramdohr, P. (1950) Auricuprid. Fortschr. Mineral., 28, 69 (in German). (2) Ramdohr, P. (1967) The wide-spread paragenesis of ore minerals originating during serpentinization (with some data on new and insufficiently described minerals). Geol. Rudn. Mestorozhd., 2, 32–43 (in Russian). (3) (1968) Amer. Mineral., 53, 350 (abs. ref. 2). (4) Novgorodova, M.I., A.I. Tsepin, A.I. Gorshkov, I.M. Kudrevich, and L.N. Vyal'sov (1977) New data on the crystal chemistry and properties of natural intermetallic compounds of the copper-gold system. Zap. Vses. Mineral. Obshch., 106, 540–552 (in Russian). (5) (1977) Chem. Abs., 87, 209755 (abs. ref. 4). (6) Novgorodova, M.I. and A.I. Tsepin (1976) Phase composition of cupriferous gold. Doklady Acad. Nauk SSSR, 227, 184–187 (in Russian).

Crystal Data: Cubic. *Point Group:* $2/m\ \overline{3}$. As sparse elongate grains up to 3.5 cm.

Physical Properties: *Tenacity:* Brittle. Hardness = 3 VHN = 280–292
D(meas.) = 9.98 (synthetic). D(calc.) = 9.91

Optical Properties: Opaque. *Color:* Gray, with a bornite-like tarnish; in polished section, galena-like with pinkish tinge. *Luster:* Metallic.
R: (400) 65.8, (420) 65.0, (440) 64.2, (460) 63.5, (480) 62.8, (500) 62.4, (520) 62.0, (540) 61.8, (560) 61.9, (580) 62.2, (600) 62.6, (620) 63.2, (640) 63.5, (660) 63.6, (680) 63.6, (700) 63.5

Cell Data: *Space Group:* $Pa3$. $a = 6.646$ Z = 4

X-ray Powder Pattern: Giant Yellowknife mine, Canada.
1.999 (100), 3.32 (50), 0.865 (50), 2.97 (40), 0.783 (40), 2.34 (40), 2.71 (30)

Chemistry: Stated to be identical with $AuSb_2$, by X-ray powder diffraction.

Occurrence: In hydrothermal gold-quartz veins, in portions deficient in sulfur and containing other antimony minerals.

Association: Gold, freibergite, stibnite, jamesonite, chalcostibite, bournonite, boulangerite, arsenopyrite, pyrite, chalcopyrite, sphalerite, galena, tetrahedrite.

Distribution: In the Giant Yellowknife mine, Northwest Territories; the Chesterville mine, Larder Lake; and the Hemlo gold deposit, Thunder Bay district, Ontario, Canada. At Krásná Hora, near Milesov, Czechoslovakia. From Costerfield, Victoria, Australia. In the Lone Hand and Jessie gold mines, Gwanda district; and the Indrarama Au-Sb mine, Sebakwe area, Que-Que, Zimbabwe. From the Bestyube goldfield, northern Kazakhstan, USSR.

Name: In reference to its composition.

Type Material: n.d.

References: (1) Graham, A.R. and S. Kaiman (1952) Aurostibite, $AuSb_2$; a new mineral in the pyrite group. Amer. Mineral., 37, 461–469. (2) Naz'mova, G.N., E.M. Spiridonov and Y.S. Shalayev (1975) Aurostibite from the Bestyube deposit, northern Kazakhstan. Doklady Acad. Nauk SSSR, 222, 687–689 (in Russian). (3) (1982) Mineral. Abs., 33, 63 (abs. ref. 2).

Crystal Data: Cubic. *Point Group:* $4/m\ \overline{3}\ 2/m$. Massive; as pebbles, grains and flakes; rare crystals to 4 mm. In meteorites as rims or regular intergrowths with kamacite.

Physical Properties: *Tenacity:* Malleable and flexible. Hardness = 5.5–6 VHN = 320–380 (low Ni); 265–290 (high Ni) (50 g load). D(meas.) = 7.8–8.22 D(calc.) = [6.37–8.53] Strongly magnetic.

Optical Properties: Opaque. *Color:* Silver-white to grayish white. *Luster:* Metallic.
R: (400) 60.4, (420) 61.3, (440) 62.2, (460) 62.3, (480) 63.4, (500) 63.8, (520) 64.0, (540) 64.2, (560) 64.4, (580) 64.8, (600) 65.2, (620) 65.7, (640) 66.2, (660) 66.7, (680) 67.3, (700) 68.0

Cell Data: *Space Group:* $Fm3m$. $a = 3.560$ Z = 4

X-ray Powder Pattern: Oregon, USA. ("josephinite").
2.06 (100), 1.073 (40), 1.783 (30), 1.259 (20), 1.027 (10)

Chemistry:

	(1)	(2)	(3)	(4)
Ni	76.60	67.63	67.77	75.93
Fe	21.45	31.02	32.23	24.07
Co	1.19	0.70		
Cu	0.66			
S	0.06	0.22		
P	0.04			
SiO_2	0.00	0.43		
Total	100.00	100.00	100.00	100.00

(1) South Fork, Smith River, California, USA. (2) Gorge River, New Zealand. (3) Ni_2Fe. (4) Ni_3Fe.

Occurrence: In river placers, from serpentinized peridotites; in meteorites.

Association: Gold, magnetite (placers); copper, heazlewoodite, pentlandite, chromite, millerite (peridotites); kamacite (meteorites).

Distribution: From the Red Hills and the Gorge River draining them, near Awarua Bay, south Westland, New Zealand. At Coolac, New South Wales, and in the Lord Brassey mine, Tasmania, Australia. From the Bobrovka River, Nizhni Tagil, Ural Mountains, USSR. In the Elvo River, Piedmont, Italy. From Poschiavo, Bergell, Switzerland. In the USA, from Josephine and Jackson counties, Oregon; South Fork, Smith River, Del Norte Co., California; and in the Line Pit mine, Cecil Co., Maryland. From the Lillooet district, Fraser River, British Columbia; Holle Canyon, Pelly River, Yukon Territory, Canada. From Oko, Kochi Prefecture, Japan. In the Sakhakot-Qila Complex, Malakand Agency, Pakistan. Additional localities are known.

Name: For the locality near Awarua Bay, New Zealand.

Type Material: n.d.

References: (1) Palache, C., H. Berman, and C. Frondel (1944) Dana's system of mineralogy, (7th edition), v. I, 117–119. (2) Berry, L.G. and R.M. Thompson (1962) X-ray powder data for the ore minerals. Geol. Soc. Amer. Mem. 85, 13. (3) Ramdohr, P. (1969) The ore minerals and their intergrowths, (3rd edition), 356–357. (4) Challis, G.A. (1975) Native nickel from the Jerry River, south Westland, New Zealand; an example of natural refining. Mineral. Mag., 40, 247–251. (5) Ahmed, Z. and J.C.Bevan (1981) Awaruite, iridian awaruite, and a new Ru–OsIr–Ni–Fe alloy from the Sakhakot-Qila Complex, Malakand Agency, Pakistan. Mineral. Mag., 44, 225–230.

Crystal Data: Orthorhombic. *Point Group:* 222, $mm2$, or $2/m\ 2/m\ 2/m$. As rod-like prismatic crystals to 0.2 mm, elongated and striated ∥ [001], and in grains up to 3 mm. *Twinning:* Lamellar in several directions, common.

Physical Properties: Hardness = ~3.5 VHN = 79.4–91.6 (100 g load). D(meas.) = 6.318 (synthetic). D(calc.) = 6.421

Optical Properties: Opaque. *Color:* Steel-gray; white-gray in polished section. *Luster:* Metallic. *Anisotropism:* Strong.
R_1–R_2: (400) — , (420) — , (440) 35.4–30.6, (460) 35.6–31.0, (480) 35.4–31.0, (500) 35.0–30.2, (520) 34.0–38.9, (540) 34.0–28.0, (560) 33.4–27.2, (580) 32.8–26.5, (600) 32.4–16.0, (620) 32.1–25.6, (640) 31.8–25.1, (660) 31.5–24.9, (680) — , (700) —

Cell Data: *Space Group:* $P222$, $Pmm2$, or $Pmmm$. a = 10.62(2) b = 9.42(2) c = 3.92(4) Z = 1

X-ray Powder Pattern: Sedmochislenitsi deposit, Bulgaria.
2.98 (10), 2.55 (10), 1.955 (9), 2.61 (8), 2.45 (8), 1.985 (8), 3.09 (7)

Chemistry:

	(1)	(2)
Cu	37.4	37.1
Ag	33.1	33.3
Hg	14.0	13.8
S	17.2	16.9
Total	101.7	101.1

(1) Sedmochislenitsi deposit, Bulgaria; by electron microprobe, corresponding to $Cu_{8.78}Ag_{4.58}Hg_{1.04}S_{8.00}$. (2) Do.; corresponding to $Cu_{8.76}Ag_{4.69}Hg_{1.04}S_{8.00}$.

Occurrence: In high-grade copper ores in a stratiform Pb-Zn-Cu deposit (Bulgaria).

Association: Bornite, chalcocite, chalcopyrite, djurleite, digenite, tennantite, stromeyerite, mckinstryite, wittichenite, bismuth, rammelsbergite, mercurian silver, cinnabar, pyrite, calcite, barite, aragonite.

Distribution: In the Sedmochislenitsi mine, Vratsa district, western part of the Stara Planina (Balkan Mountains), Bulgaria. From Manhattan, Nye Co., Nevada, USA. From about 20 km southwest of Agua Prieta, Sonora, Mexico.

Name: For the medieval and popular name of the Stara Planina (Balkan Mountains), the main mountain range giving its name to the Balkan Peninsula.

Type Material: Mineralogical Museum of the Higher Institute of Mining and Geology, 619; Museum of the University 'Kl, Ohridski', Sofia, Bulgaria, 1351.

References: (1) Atanassov, V.A. and G.N. Kirov (1973) Balkanite, $Cu_9Ag_5HgS_8$, a new mineral from the Sedmochislenitsi mine, Bulgaria. Amer. Mineral., 58, 11–15.

Crystal Data: Tetragonal, pseudocubic. *Point Group:* $4/m$ or $4/m\ 2/m\ 2/m$. Fine granular aggregates and lenses, intimately intermixed with klockmannite.

Physical Properties: *Fracture:* Fine conchoidal. Hardness = n.d. VHN = n.d. D(meas.) = n.d. D(calc.) = [5.02]

Optical Properties: Opaque. *Color:* In polished section, creamy white. *Anisotropism:* Slight. R_1–R_2: n.d.

Cell Data: *Space Group:* $P4_2/n$ or perhaps $I4_1/amd$. $a = 5.466$ $c = 5.632$ Z = [2]

X-ray Powder Pattern: Moctezuma, Mexico.
3.19 (100), 1.961 (70), 1.653 (50), 1.931 (40), 1.689 (30), 1.270 (20), 1.121 (20)

Chemistry:

	(1)
Cu	25.1
Se	44.5
Te	31.0
Total	100.6

(1) Synthetic material matching the X-ray powder pattern of the natural mineral; by electron microprobe, corresponding to $Cu_{1.00}(Se_{1.428}Te_{0.615})_{\Sigma=2.043}$.

Occurrence: In oxidized Au-Te ore in sub-parallel quartz veins in a hydrothermally altered tuff.

Association: Klockmannite, tellurium, selenium, chalcomenite, tellurite, paratellurite, illite, calcite, quartz.

Distribution: In the Moctezuma mine, near Moctezuma, Sonora, Mexico.

Name: For the Spanish nickname "La Bambolla" of the mine in which it occurs, which roughly translates into "hot air" in allusion to exaggerated tales of rich gold ore.

Type Material: Royal Ontario Museum, Toronto, Canada; National Museum of Natural History, Washington, D.C., USA, 128391.

References: (1) Harris, D.C. and E.W. Nuffield (1972) Bambollaite, a new copper telluro-selenide. Can. Mineral., 11, 738–742. (2) (1973) Amer. Mineral., 58, 805 (abs. ref. 1).

Crystal Data: Tetragonal. *Point Group:* $4/m\ 2/m\ 2/m$. Anhedral masses, intergrown with pyrrhotite; rare single crystals less than 5 μm wide and 5 to 10 times as long.

Physical Properties: *Cleavage:* {112} distinct. *Fracture:* Conchoidal. Hardness = 3.5 VHN = 94–120, 104 average (15 g load). D(meas.) = 3.305 D(calc.) = 3.286 Weakly magnetic.

Optical Properties: Opaque. *Color:* Blackish brown; in polished section, yellowish green-gray to yellowish green-brown. *Streak:* Black. *Luster:* Submetallic. *Pleochroism:* Very weak.
R_1–R_2: (400) 21.7–18.3, (420) 16.7–14.4, (440) 17.7–17.2, (460) 18.4–18.6, (480) 19.4–19.5, (500) 19.7–20.6, (520) 20.3–21.4, (540) 20.9–21.8, (560) 21.9–22.7, (580) 22.7–23.2, (600) 23.4–23.7, (620) 23.9–24.3, (640) 24.7–25.1, (660) 25.0–25.5, (680) 25.2–26.1, (700) 26.4–26.5

Cell Data: *Space Group:* $I4/mmm$. $a = 10.424(1)$ $c = 20.626(1)$ Z = 2

X-ray Powder Pattern: Coyote Peak, California, USA.
2.998 (100), 5.99 (77), 1.833 (40), 9.31 (27), 3.139 (27), 2.379 (25), 1.841 (25)

Chemistry:

	(1)	(2)
K	9.54	10.43
Na	0.05	
Fe	51.2	49.66
Cu	0.62	
Ni	0.19	
Co	0.11	
S	38.4	39.91
Cl	0.02	
Total	100.13	100.00

(1) Coyote Peak, California, USA; by electron microprobe, corresponding to $(K, Na)_{5.70}(Fe, Ni, Co, Cu)_{21.24}S_{26.93}$. (2) $K_3Fe_{10}S_{14}$.

Occurrence: In clots with other sulfides and silicates in an alkalic diatreme which intruded Franciscan rocks.

Association: Rasvumite, djerfisherite, erdite, pyrrhotite, pyrite, sphalerite, löllingite, magnetite, nepheline, phlogopite.

Distribution: From Coyote Peak, Humboldt Co., California, USA.

Name: For Paul B. Barton, Jr., sulfide petrologist with the U.S. Geological Survey.

Type Material: National Museum of Natural History, Washington, D.C., USA, 149498.

References: (1) Czamanske, G.K., R.C. Erd, B.F. Leonard, and J.R. Clark (1981) Bartonite, a new potassium iron sulfide mineral. Amer. Mineral., 66, 369–375. (2) Evans, H.T., Jr. and J.R. Clark (1981) The crystal structure of bartonite, a potassium iron sulfide, and its relationship to pentlandite and djerfisherite. Amer. Mineral., 66, 376–384.

Crystal Data: Triclinic, pseudomonoclinic. *Point Group:* $\overline{1}$, ($2/m$ pseudocell). Crystals short prismatic parallel to [010], tabular on {100}; striated on {100} parallel to [010], as well as on {010} parallel to [001]; crystals sometimes rounded, to 25 mm. Also as aggregates of rounded grains. *Twinning:* Polysynthetic on {100}.

Physical Properties: *Cleavage:* Perfect on {100}. *Fracture:* Conchoidal. Hardness = 3 VHN = 179 D(meas.) = 5.329 D(calc.) = [5.42]

Optical Properties: Opaque. *Color:* Lead-gray to steel-gray, may tarnish to iridescence; in polished section, white with deep red internal reflections. *Streak:* Chocolate-brown. *Luster:* Metallic. *Anisotropism:* Strong, colorful in green, violet, blue-green or yellow.
R_1–R_2: (400) 40.1–43.8, (420) 39.6–43.5, (440) 39.1–43.2, (460) 38.5–42.8, (480) 38.0–42.3, (500) 37.4–41.8, (520) 36.7–41.2, (540) 36.0–40.5, (560) 35.3–39.7, (580) 34.5–38.8, (600) 33.7–37.8, (620) 32.9–36.8, (640) 32.0–35.8, (660) 31.4–35.0, (680) 30.8–34.2, (700) 30.4–33.6

Cell Data: *Space Group:* $P\overline{1}$, ($P2_1/m$ pseudocell). $a = 22.74$ $b = 8.33$ $c = 7.89$ $\alpha = 90°$ $\beta = 97°25'$ $\gamma = 90°$ Z = 4

X-ray Powder Pattern: Binntal, Switzerland.
3.00 (100), 2.94 (100), 2.76 (100), 4.11 (80), 3.43 (80), 3.25 (80), 3.56 (80)

Chemistry:

	(1)	(2)	(3)
Pb	48.86	47.5	51.38
As	26.42	12.0	24.77
Sb		19.5	
S	24.39	21.0	23.85
Total	99.67	100.0	100.00

(1) Binntal, Switzerland. (2) Madoc, Ontario, Canada; by electron microprobe. (3) $Pb_3As_4S_9$.

Occurrence: In sugary dolomite (Binntal, Switzerland).

Association: With other lead sulfarsenides.

Distribution: In the Lengenbach quarry, Binntal, Valais, Switzerland. In Canada, at Madoc, and in the Hemlo gold deposit, Thunder Bay district, Ontario. From the Sterling Hill mine, Ogdensberg, Sussex Co., New Jersey; and at the Zuni mine, Silverton, Colorado, USA. At Novoye, Kaidarkan, Kirgizia, USSR.

Name: For Heinrich Adolph Baumhauer (1848–1926), German mineralogist and Professor, University of Fribourg, Switzerland.

Type Material: British Museum (Natural History), London, England.

References: (1) Palache, C., H. Berman, and C. Frondel (1944) Dana's system of mineralogy, (7th edition), v. I, 460–462. (2) Berry, L.G. and R.M. Thompson (1962) X-ray powder data for the ore minerals. Geol. Soc. Amer. Mem. 85, 154. (3) Berry, L.G. (1953) New data on lead sulpharsenides from Binnental, Switzerland (abs.). Amer. Mineral., 38, 330. (4) Jambor, J.L. (1967) New lead sulfantimonides from Madoc, Ontario; part 2—mineral descriptions. Can. Mineral., 9, 191–213. (5) Engel, P. and W. Nowacki (1969) Die Kristallstruktur von Baumhauerit. Zeits. Krist., 129, 178–202 (in German).

Crystal Data: Tetragonal. *Point Group:* $4/m$. As anhedral grains 0.1 to 0.3 mm in diameter.

Physical Properties: Hardness = < 3 VHN = 33–41 (100 g load). D(meas.) = n.d. D(calc.) = [7.01]

Optical Properties: Opaque. *Color:* Creamy white in reflected light. *Luster:* Metallic. *Pleochroism:* Very weak. *Anisotropism:* Very weak, in gray colors.
R_1–R_2: n.d.

Cell Data: *Space Group:* $P4_2/n$. $a = 11.52$ $c = 11.74$ Z = [32]

X-ray Powder Pattern: Habrí, Czechoslovakia.
2.06 (100), 3.38 (80), 1.763 (70), 2.25 (50), 6.80 (30), 3.50 (10), 2.32 (10)

Chemistry:

	(1)	(2)
Cu	61.5	61.62
Se	38.0	38.38
Total	99.5	100.00

(1) Habrí, Czechoslovakia; by electron microprobe. (2) Cu_2Se.

Polymorphism & Series: Dimorphous with berzelianite.

Occurrence: Formed at moderate to low temperature with other hydrothermal selenides and sulfides.

Association: Berzelianite, umangite, eskebornite, klockmannite, djurleite, digenite, siderite, calcite.

Distribution: From Habrí, near Tisnova, Czechoslovakia.

Name: For Eleodoro Bellido Bravo, Director of Servicio de Geología y Minería, Peru.

Type Material: n.d.

References: (1) De Montreuil, L. A. (1975) Bellidoite, a new copper selenide. Econ. Geol., 70, 384–387. (2) (1975) Amer. Mineral., 60, 736 (abs. ref. 1).

Crystal Data: Monoclinic. *Point Group:* $2/m$. As acicular (200 x 20 μm) crystals or rounded grains (up to 50 μm). *Twinning:* Polysynthetic twinning is common parallel to crystal elongation.

Physical Properties: Hardness = n.d. VHN = 77–116 (15 g load). D(meas.) = n.d. D(calc.) = 5.60

Optical Properties: Opaque. *Color:* Lead-gray; white with a greenish tint under reflected light, with rare dull red internal reflections. *Streak:* Dark gray with a brownish tint. *Luster:* Metallic. *Pleochroism:* Weak. *Anisotropism:* Strong, with tints of brown and blue. R_1–R_2: (400) 42.0–40.7, (420) 42.0–39.6, (440) 42.0–39.3, (460) 42.4–39.5, (480) 43.0–39.9, (500) 42.9–39.7, (520) 42.7–39.5, (540) 42.4–39.0, (560) 42.2–38.9, (580) 41.8–38.7, (600) 41.3–38.3, (620) 40.8–37.9, (640) 40.9–37.5, (660) 39.5–36.8, (680) 38.7–36.1, (700) 37.8–35.6

Cell Data: *Space Group:* $P2_1/a$. $a = 15.74$ $b = 19.14$ $c = 4.06$ $\beta = 91.50$ Z = 2

X-ray Powder Pattern: Uchuc-Chacua, Peru.
3.45 (100), 2.829 (40), 4.10 (30), 2.737 (30), 3.85 (20), 3.169 (20), 3.098 (20)

Chemistry:

	(1)	(2)
Pb	39.9	39.9
Cu		0.1
Zn		0.1
Mn	2.2	1.8
Fe	0.8	1.1
Sb	34.8	33.0
Bi		2.6
S	21.1	21.7
Total	98.8	100.3

(1) Uchuc-Chacua, Peru; by electron microprobe. (2) Sätra mine, Sweden; by electron microprobe, average of five analyses.

Polymorphism & Series: Forms a series with jamesonite.

Occurrence: In a metamorphosed iron sulfide deposit associated with submarine felsic volcanism (Sätra mine, Sweden); in telescoped polymetallic mineralization associated with dacitic intrusions (Uchuc-Chacua, Peru).

Association: Galena, manganoan sphalerite, pyrite, pyrrhotite, alabandite, quartz, bustamite, rhodonite, calcite (Uchuc-Chacua, Peru); galena, freibergite, gudmundite, manganoan sphalerite, bismuth, spessartine (Sätra mine, Sweden); pyrite, calcite (Dachang, China).

Distribution: At the Sätra mine, Dovertorp, Bergslagen, Sweden. From Uchuc-Chacua, Cajatambo Province, Peru. In the Dachang district, northwestern Guangxi Province, China.

Name: To honor the contributions of A. Benavides to the development of mining in Peru.

Type Material: National School of Mines, Paris, France (Uchuc-Chacua, Peru); The Free University, Amsterdam, The Netherlands (Sätra mine, Sweden).

References: (1) Oudin, E., P. Picot, F. Pillard, Y. Moëlo, E.A.J. Burke, and M.A. Zakrzewski (1982) La bénavidésite, $Pb_4(Mn,Fe)Sb_6S_{14}$, un noveau minéral de la série de la jamesonite. Bull. Minéral., 105, 166–169 (in French with English abs.). (2) (1983) Amer. Mineral., 68, 280 (abs. ref. 1). (3) Luke L.Y. Chang, Xilin Li, and Chusheng Zheng (1987) The jamesonite – benavidesite series. Can. Mineral., 25, 667–672.

Crystal Data: Monoclinic. *Point Group:* $2/m$. Massive and as laths to 6 mm which are complex intergrowths of several closely related phases. *Twinning:* Polysynthetic, common.

Physical Properties: *Cleavage:* Fair, parallel to the elongation. Hardness = 3.3–3.5 VHN = 186–232 (50 g load). D(meas.) = n.d. D(calc.) = 6.68

Optical Properties: Opaque. *Color:* Gray on fresh fracture, tarnishing dull or yellow to coppery red; in polished section, creamy white to very pale brownish. *Luster:* Metallic, with greasy appearance. *Anisotropism:* Strong.
R_1–R_2: (400) 44.2–47.2, (420) 44.8–47.8, (440) 45.2–48.6, (460) 45.2–49.0, (480) 45.0–49.2, (500) 44.8–49.2, (520) 44.5–49.0, (540) 44.0–48.6, (560) 43.6–48.0, (580) 43.3–47.6, (600) 43.1–47.2, (620) 43.0–46.8, (640) 42.8–46.3, (660) 42.7–45.9, (680) 42.6–45.6, (700) 42.5–45.2

Cell Data: *Space Group:* $C2/m$. $a = 13.25(2)$ $b = 4.05(1)$ $c = 20.25(3)$ $\beta = 103.14(07)°$ Z = 2

X-ray Powder Pattern: Camsell River, Canada. Easily mistaken for pavonite.
2.851 (100), 3.539 (80), 3.427 (80), 2.813 (60), 2.022 (60), 2.007 (60), 3.302 (50)

Chemistry:

	(1)	(2)
Ag	12.6	14.2
Cu	1.4	0.0
Pb	3.7	0.6
Bi	65.1	64.8
Sb		0.0
S	16.4	18.0
Total	99.2	97.6

(1) Round Mountain, Nevada, USA; by electron microprobe, average of four analyses. (2) AW mine, Australia; by electron microprobe, giving $Ag_{2.82}(Bi_{6.62}Pb_{0.06})_{\Sigma=6.68}S_{12.00}$.

Occurrence: In a quartz vein near the contact of a soda-granite and an intrusive rhyolite (Outlaw mine, Nevada); with arsenides (Camsell River, Canada); in a veinlet in calcite (Cobalt, Canada); in a molybdenite pipe deposit (AW mine, Australia).

Association: Aikinite, chalcopyrite, pyrite, covellite, muscovite, molybdenite, quartz (Outlaw mine, Nevada); safflorite, skutterudite, rammelsbergite, arsenopyrite, nickeline, matildite, bismuthinite, chalcopyrite, pyrite, sphalerite, galena, bismuth, silver (Camsell River, Canada); molybdenite, bismuth, bismuthinite (AW mine, Australia).

Distribution: From the Outlaw mine, about 17 km north of Manhattan, and at Round Mountain, Nye Co., Nevada; in the Alaska mine, Poughkeepsie Gulch, near Ouray, San Juan Co., Colorado, USA. From the Terra Company mine, Camsell River, Northwest Territories; and in the Canadian Keely mine, Cobalt, Ontario, Canada. In the Porvenir mine, Cerro Bente, Potosí, Bolivia. From the AW mine, south of Tenterfield, New South Wales, Australia.

Name: For Dr. Marcus Benjamin (1857–1932), of the U.S. National Museum.

Type Material: National Museum of Natural History, Washington, D.C., 95058; Harvard University, Cambridge, Massachusetts, USA, 85749; Royal Ontario Museum, Toronto, Canada, 13805.

References: (1) Palache, C., H. Berman, and C. Frondel (1944) Dana's system of mineralogy, (7th edition), v. I, 441–442. (2) Nuffield, E.W. (1953) Benjaminite. Amer. Mineral., 38, 550–552. (3) Nuffield, E.W. (1975) Benjaminite — a re-examination of the type material. Can. Mineral., 13, 394–401. (4) Harris, D.C. and T.T. Chen (1975) Benjaminite, reinstated as a valid species. Can. Mineral., 13, 402–407. (5) Herbert, H.K. and W.G. Mumme (1981) Unsubstituted benjaminite from the AW mine, NSW: a discussion of metal substitutions and stability. Neues Jahrb. Mineral., Monatsh., 69–80.

Crystal Data: Tetragonal. *Point Group:* $4/m\ 2/m\ 2/m$, $\bar{4}2m$, $4mm$, 422, $4/m$, $\bar{4}$, or 4. As crusts and fracture fillings, less than 1 mm across, composed of grains and laths to 60 μm. *Twinning:* According to some simple law.

Physical Properties: Hardness = n.d. VHN = 105–125, 117 average (25 g load). D(meas.) = n.d. D(calc.) = 7.76

Optical Properties: Opaque. *Color:* In polished section, very pale blue. *Anisotropism:* Moderate to strong, in tints of brown, blue, and gray.
R_1–R_2: (400) 35.6–36.8, (420) 35.7–36.6, (440) 35.8–36.2, (460) 35.7–35.6, (480) 35.5–34.9, (500) 35.1–34.1, (520) 34.7–33.2, (540) 34.1–32.2, (560) 33.5–31.4, (580) 32.9–30.7, (600) 32.4–30.0, (620) 32.0–29.5, (640) 31.6–29.1, (660) 31.4–28.7, (680) 31.1–28.4, (700) 30.8–28.1

Cell Data: *Space Group:* $P4/mmm$, $P\bar{4}m2$, $P\bar{4}2m$, $P4mm$, $P422$, $P4/m$, $P\bar{4}$, or $P4$. $a = 6.603(5)$ $c = 12.726(6)$ Z = 2

X-ray Powder Pattern: Bambolla mine, Mexico.
2.936 (100), 12.7 (70), 2.608 (35), 2.158 (35), 3.188 (30), 2.863 (25), 2.328 (20)

Chemistry:

	(1)	(2)
Ag	64.5	64.59
Cu	0.1	
Sb	7.3	9.11
As	1.4	
Te	18.7	19.10
S	8.0	7.20
Total	100.0	100.00

(1) Bambolla mine, Mexico; by electron microprobe, average of 14 analyses, corresponding to $Ag_{7.80}Cu_{0.02}(Sb_{0.78}As_{0.24})_{\Sigma=1.02}Te_{1.90}S_{3.25}$. (2) $Ag_8SbTe_2S_3$.

Occurrence: As powdery crusts with other sulfides in fractures in an intensely silicified rhyolitic vitrophyre.

Association: Silver, acanthite, hessite, pyrite, sphalerite, dolomite, quartz.

Distribution: In the Moctezuma mine, Moctezuma, Sonora, Mexico.

Name: For Dr. Benjamin Franklin Leonard (1921–), of the U.S. Geological Survey, Denver, Colorado, USA.

Type Material: British Museum (Natural History), London, England, E.1611, BM 1985,354.

References: (1) Stanley, C.J., A.J. Criddle, and J.E. Chisholm (1986) Benleonardite, a new mineral from the Bambolla mine, Moctezuma, Sonora, Mexico. Mineral. Mag., 50, 681–686.
(2) (1988) Amer. Mineral., 73, 439 (abs. ref. 1).

Crystal Data: Hexagonal. *Point Group:* $\bar{3}\ 2/m$ (C6 polymorph). Tabular hexagonal crystals.

Physical Properties: *Cleavage:* {0001} marked. Hardness = Very soft. VHN = n.d. D(meas.) = 4.5 (synthetic). D(calc.) = [4.55]

Optical Properties: Translucent to transparent. *Color:* Pale yellow; in polished section, gray, with intense brownish to orange-yellow internal reflections. *Luster:* Resinous. *Streak:* Golden yellow. *Pleochroism:* Strong. *Anisotropism:* Dark greenish gray.
R_1–R_2: n.d.

Cell Data: *Space Group:* $P\bar{3}m1$ (C6 polymorph). $a = 3.639$ $c = 5.868$ Z = 1

X-ray Powder Pattern: Synthetic.
5.89 (100), 2.784 (55), 3.162 (30), 1.824 (30), 2.155 (25), 1.743 (20), 1.669 (8)

Chemistry: Sn and S are major constituents.

Polymorphism & Series: Stacking polymorphs C6 and C27 are known.

Occurrence: A secondary mineral in tin sulfide ores (Cerro de Potosí, Bolivia); in a high-temperature deposit, hydrothermally altered (Arandis, Namibia).

Association: Stannite, pyrite, cassiterite, sulfur.

Distribution: From Cerro de Potosí, Bolivia. In the Stiepelmann mine, Arandis, Namibia. In Portugal, from the Lagares-do-Estanho tin pegmatite, and at Panasqueira, Beira-Baixa Province. From Forestville, Northumberland Co., Pennsylvania, USA.

Name: For Dr. Fritz Berndt, German mineralogist, Corporatión Minera de Bolivia, Oruro, Bolivia.

Type Material: n.d.

References: (1) Moh, G.H. and F. Berndt (1964) Two new natural tin sulfides, Sn_2S_3 and SnS_2. Neues Jahrb. Mineral., Monatsh., 94–95. (2) (1965) Amer. Mineral., 50, 2107 (abs. ref. 1). (3) Moh, G.H. (1966) Das binäre System Zinn-Schwefel und seine Minerale. Fortschr. Mineral. (abs.) 42, 211. (4) (1966) Amer. Mineral., 51, 1551 (abs. ref. 3). (5) Clark, A.H. (1972) On the natural occurrence of tin sulfides (berndtite). Naturwissenschaften, 59, 361. (6) (1973) Amer. Mineral., 58, 347 (abs. ref. 5). (7) Hazen, R.M. and L.W. Finger (1978) The crystal structures and compressibilities of layer minerals at high pressure. I. SnS_2, berndtite. Amer. Mineral., 63, 289–292.

Berryite $Pb_3(Ag,Cu)_5Bi_7S_{16}$

Crystal Data: Monoclinic, pseudo-orthorhombic. *Point Group:* $2/m$ or 2. Lath-like aggregates of several not quite parallel individuals, to a maximum length of 1 mm; granular, massive. *Twinning:* Repeated.

Physical Properties: *Cleavage:* Poor. Hardness = n.d. VHN = 131–152 (100 g load), 171 (100 g load). D(meas.) = 6.7, corrected for admixed quartz. D(calc.) = [6.83]

Optical Properties: Opaque. *Color:* White to gray-white in polished section. *Pleochroism:* Weak to distinct. *Anisotropism:* Distinct to strong, from gray-white to red-brown and green.
R_1–R_2: n.d.

Cell Data: *Space Group:* $P\,2_1/m$ or $P2_1$. $a = 12.72$ $b = 4.02$ $c = 58.07$ $\beta = 102.5°$
Z = [4]

X-ray Powder Pattern: Nordmark, Sweden.
3.47 (10), 2.89 (8), 2.80 (7), 2.18 (4), 3.70 (2), 2.86 (2), 3.23 (1)

Chemistry:

	(1)	(2)	(3)
Pb	21.6	20.8	20.9
Ag	4.9	6.8	7.4
Cu	7.1	7.8	5.3
Bi	49.2	47.5	49.5
S	[17.2]	[17.2]	17.0
Total	100.0	100.1	100.1

(1–2) Ivigtut, Greenland; by electron microprobe, sulfur by difference, total calculated to be near 100%. (3) Owen Lake, Canada; by electron microprobe, average of several grains.

Occurrence: In quartz veins with other sulfides and sulfosalts, and in siderite-rich cryolite (Ivigtut, Greenland).

Association: Emplectite, aikinite, cuprobismutite, cupropavonite (Colorado, USA); galena, chalcopyrite, sphalerite, quartz (Nordmark, Sweden); cosalite, galena, ourayite, matildite, aikinite, chalcopyrite, sphalerite, pyrrhotite, marcasite, pyrite, bismuth, gold, fluorite, topaz, weberite, quartz (Ivigtut, Greenland); aikinite, matildite, benjaminite, quartz, barite (Tary Ekan deposit, USSR).

Distribution: From the Missouri mine, Park Co., and the Mike mine, San Juan Co., Colorado, USA. At Nordmark, Värmland, Sweden. From Ivigtut, southern Greenland. In the Adrasman, Kaptar-Hana, and Tary Ekan deposits, eastern Karamazar, Central Asia, and in the Kochbulak deposit, eastern Uzbekistan, USSR. From near Owen Lake, south of Houston, British Columbia, Canada.

Name: For Professor Leonard Gascoigne Berry (1914–1982), mineralogist, Queen's University, Toronto, Canada, who obtained the first X-ray powder pattern of the mineral.

Type Material: National Museum of Natural History, Washington, D.C., USA, 92902.

References: (1) Nuffield, E.W. and D.C. Harris (1966) Studies of mineral sulpho-salts: XX—berryite, a new species. Can. Mineral., 8, 407–413. (2) Karup-Møller, S. (1966) Berryite from Greenland. Can. Mineral., 8, 414–423. (3) (1967) Amer. Mineral., 52, 928 (abs. refs. 1 and 2). (4) Harris, D.C. and D.R. Owen (1973) Berryite, a Canadian occurrence. Can. Mineral., 11, 1016–1018. (5) Vendrell-Saz, M., S. Karup-Møller, and A. Lopez-Soler (1978) Optical and microhardness study of some Ag–Cu–Pb–Bi sulphides. Neues Jahrb. Mineral., Abh., 132, 101–112.

Crystal Data: Orthorhombic. *Point Group:* $2/m\ 2/m\ 2/m$. Prismatic crystals elongated and striated ∥ [001], to 1 cm; usually unterminated; fibrous, plumose or radial, brushes to 6 cm; granular, massive.

Physical Properties: *Cleavage:* Prismatic, indistinct. *Tenacity:* Brittle. Hardness = 2–3 VHN = n.d. D(meas.) = 4.64 D(calc.) = 4.66

Optical Properties: Opaque. *Color:* Dark steel-gray, often tarnished iridescent or pinchbeck-brown; in polished section, gray. *Streak:* Dark brownish gray. *Luster:* Metallic. *Pleochroism:* Marked. *Anisotropism:* Strong.

R_1–R_2: (400) 33.0–43.0, (420) 32.4–42.8, (440) 31.8–42.6, (460) 31.3–42.3, (480) 31.0–42.0, (500) 30.8–41.6, (520) 30.8–41.2, (540) 30.9–40.8, (560) 31.2–40.3, (580) 31.5–39.8, (600) 31.7–39.4, (620) 31.9–39.0, (640) 32.0–38.7, (660) 32.0–38.3, (680) 31.9–38.0, (700) 31.6–37.6

Cell Data: *Space Group:* *Pnam.* $a = 11.44$ $b = 14.12$ $c = 3.76$ Z = 4

X-ray Powder Pattern: Synthetic.
2.601 (100), 2.621 (93), 3.663 (92), 2.862 (78), 3.181 (77), 2.999 (52), 3.628 (51)

Chemistry:

	(1)	(2)
Fe	13.43	13.06
Mn	trace	
Sb	56.06	56.95
S	29.46	29.99
insol.	0.33	
Total	99.28	100.00

(1) Herja, Romania. (2) $FeSb_2S_4$.

Occurrence: In low-temperature hydrothermal antimony veins, in portions deficient in sulfur.

Association: Stibnite, pyrite, barite, quartz.

Distribution: Localities are numerous. In France, from Pontgibaud, Puy-de-Dôme, and Valcros, Provence. At Herja (Kisbánya), and Baia Sprie (Felsőbánya), Romania. From Příbram and Kutná Hora, Czechoslovakia. At Bräunsdorf, near Freiberg, Saxony, Germany. From Ribnova, southern Rhodope Mountains, Bulgaria. In the Niccioleta mine, Tuscany, Italy. In Mexico, from San Martin, Concepción del Oro, Zacatecas; and from mines near Triunfo, Baja California. At Chashan, Guangxi Province, China.

Name: For Pierre Berthier (1782–1861), French chemist.

Type Material: n.d.

References: (1) Palache, C., H. Berman, and C. Frondel (1944) Dana's system of mineralogy, (7th edition), v. I, 481–482. (2) Buerger, M.J. and T. Hahn (1955) The crystal structure of berthierite, $FeSb_2S_4$. Amer. Mineral., 40, 226–238. (3) Barton, P.B., Jr. (1971) The Fe–Sb–S system. Econ. Geol., 66, 121–132.

Crystal Data: Cubic. *Point Group:* n.d. As thin dendritic crusts; fine grained and disseminated as powdery inclusions.

Physical Properties: *Fracture:* Uneven. *Tenacity:* Somewhat malleable. Hardness = 2.7 VHN = 21–24 (100 g load). D(meas.) = 6.71 D(calc.) = [7.28]

Optical Properties: Opaque. *Color:* Silver-white, tarnishing. *Streak:* Shining. *Luster:* Metallic. *Anisotropism:* Isotropic, sometimes slightly anisotropic.
R: (400) 25.8, (420) 27.2, (440) 28.6, (460) 29.4, (480) 29.4, (500) 28.6, (520) 27.8, (540) 25.8, (560) 24.3, (580) 22.7, (600) 21.2, (620) 20.0, (640) 19.0, (660) 18.0, (680) 17.0, (700) 16.2

Cell Data: *Space Group:* n.d. $a = 5.731(8)$ Z = 4

X-ray Powder Pattern: Skrikerum, Sweden.
2.03 (100), 3.33 (90), 1.729 (80), 1.171 (40), 1.434 (30), 1.317 (20), 1.105 (20)

Chemistry:

	(1)	(2)
Cu	57.21	61.62
Ag	3.51	
Se	39.22	38.38
Total	99.94	100.00

(1) Skrikerum, Sweden. (2) Cu_2Se.

Polymorphism & Series: Dimorphous with bellidoite.

Occurrence: With other selenides in hydrothermal veinlets in dolomite (Clausthal, Germany); in iron ores (Lerbach, Germany); with other selenides in calcite veins in serpentine (Skrikerum, Sweden); in a gold-quartz-orthoclase deposit (Redjang-Lebong, Sumatra).

Association: Eucairite, clausthalite, tiemannite, umangite, klockmannite, aguilarite, crookesite, athabascaite, stromeyerite, polybasite, pearceite, gold, uraninite, pyrite, marcasite, calcite.

Distribution: From Lerbach, Tilkerode, Zorge, and Clausthal, Harz, Germany. At Skrikerum, near Tryserum, Kalmar, Sweden. In the Habrí and Bukov mines, near Tisnova, and Petrovice, Czechoslovakia. At the Sierra de Umango, La Rioja Province, Argentina. From Kalgoorlie, Western Australia, and El Sharana, Northern Territory, Australia. In the Pinky Fault uranium deposit, Lake Athabasca, Saskatchewan, Canada. From Aurora, Mineral Co., Nevada, USA. At Redjang-Lebong, Sumatra.

Name: For Jöns Jacob Berzelius (1779–1848), Swedish chemist, who discovered selenium.

Type Material: n.d.

References: (1) Palache, C., H. Berman, and C. Frondel (1944) Dana's system of mineralogy, (7th edition), v. I, 182–183. (2) Earley, J.W. (1950) Description and synthesis of the selenide minerals. Amer. Mineral., 35, 337–364. (3) Sindeeva, N.D. (1964) Mineralogy and types of deposits of selenium and tellurium, 65–67.

Crystal Data: Orthorhombic. *Point Group:* $2/m$ $2/m$ $2/m$. Needles to 2 cm in length, and in irregular masses.

Physical Properties: *Cleavage:* In three directions. Hardness = 3–3.5 VHN = n.d. D(meas.) = 5.96–6.05 D(calc.) = 6.14

Optical Properties: Opaque. *Color:* Black; in polished section, bright cream parallel to elongation, stronger yellowish cream perpendicular, dulls on exposure. *Anisotropism:* Strong, distinctly colored.

R_1–R_2: (400) 34.0–35.4, (420) 33.7–34.6, (440) 33.4–33.8, (460) 33.2–33.3, (480) 32.9–33.0, (500) 32.8–32.6, (520) 32.6–32.3, (540) 32.5–32.1, (560) 32.3–32.0, (580) 32.0–31.8, (600) 31.8–31.8, (620) 31.6–31.9, (640) 31.4–32.2, (660) 31.4–32.4, (680) 31.2–32.4, (700) 31.2–32.4

Cell Data: *Space Group:* *Immm*. $a = 14.67$ $b = 22.80$ $c = 3.85$ Z = 4

X-ray Powder Pattern: Dzhezkazgan, USSR.
1.832 (100), 2.93 (90), 3.08 (80), 1.946 (70), 1.766 (70), 2.35 (60), 2.01 (60)

Chemistry:

	(1)	(2)
Cu	61.39	58.88
Pb	19.20	17.47
Ag		0.79
Fe	1.83	2.81
S	17.25	20.16
Total	99.67	100.11

(1) Mansfeld, Germany. (2) Dzhezkazgan, USSR.

Occurrence: In veins cutting the black cupriferous shale (Mansfeld, Germany).

Association: Bornite, chalcocite, chalcopyrite, galena, silver, celestine, anhydrite, calcite.

Distribution: From the Mansfeld Kupferschiefer, Eisleben, and at Waschenbach, Odenwald, Germany. In the Radka deposit, Pazardzhik, Bulgaria. From near Laird, Sutherlandshire, Scotland. On the Mürtschenalp, Switzerland. In the Bulancak deposit, Giresun, Turkey. At Tsumeb, Namibia. Rich specimens from Kipushi, Shaba Province, Zaire. At the St. Cloud mine, Sierra Co., New Mexico, USA. In the La Leona mine, Santa Cruz Province, Argentina. In the Yoshino mine, Yamagata Prefecture, Japan. At Dzhezkazgan, Kazakhstan, USSR. Probably yet to be recognized in many other mineral deposits.

Name: For Anatolii Georgievich Betekhtin (1897–1962), Russian mineralogist and economic geologist.

Type Material: Mineral Museum, Humboldt University, Berlin, Germany.

References: (1) Schüller, A. and E. Wohlmann (1955) Betechtinit, ein neues Blei-Kupfer-Sulfid aus dem Mansfelder Rücken. Geologie, 4, 535–555 (in German). (2) (1956) Amer. Mineral., 41, 371–372 (abs. ref. 1). (3) Dornberger-Schiff, K. von and E. Höhne (1959) Die Kristallstruktur des Betechtinit $Pb_2(Cu, Fe)_{21}S_{15}$. Acta Cryst., 12, 646–651 (in German). (4) Schüller, A. (1960) Zur Kenntnis des Betechtinit, $(Cu, Fe)_{10}PbS_{>6}$. Neues Jahrb. Mineral., Monatsh., 121–131 (in German).

Crystal Data: Orthorhombic. *Point Group:* n.d. As tabular or irregular grains up to 0.2 x 0.05 mm.

Physical Properties: Hardness = n.d. VHN = 353 (20 g load). D(meas.) = n.d. D(calc.) = 16.3

Optical Properties: Opaque. *Color:* In polished section, bright orange-yellow, resembling gold but having lower reflectivity. *Anisotropism:* Weak, in neutral gray shades.

R: (400) — , (420) 15.1, (440) 15.1, (460) 15.7, (480) 17.5, (500) 21.2, (520) 29.0, (540) 37.5, (560) 45.6, (580) 52.4, (600) 55.7, (620) 57.4, (640) 58.3, (660) 58.7, (680) 58.6, (700) 58.2

Cell Data: *Space Group:* n.d. $a = 4.036$ $b = 4.025$ $c = 4.061$ Z = n.d.

X-ray Powder Pattern: Kamchatka, USSR.
2.33 (10), 1.744 (8-9), 2.61 (8), 3.30 (7), 1.144 (4), 1.405 (4), 1.073 (4)

Chemistry:

	(1)
Au	72.3
Ag	3.77
Cu	6.27
Fe	0.72
Pb	8.95
Te	7.16
Total	98.2

(1) Kamchatka, USSR; by electron microprobe, average of seven analyses of three samples, corresponding to $(Au_{3.59}Ag_{0.34})_{\Sigma=3.93}(Cu_{0.97}Fe_{0.12})_{\Sigma=1.09}(Te_{0.55}Pb_{0.43})_{\Sigma=0.98}$.

Occurrence: Found in the cementation zone of a volcanogenic gold telluride deposit, rarely as rims around grains of gold.

Association: Bilibinskite, gold, tellurides of Cu, Fe and Pb.

Distribution: From an undefined locality in Kamchatka, Far Eastern Region, USSR.

Name: For Dr. M.S. Bezsmertnaya and Dr. V.V. Bezsmertny, Soviet geologists and researchers on ore deposits.

Type Material: Institute of Mineralogy and Geochemistry of Rare Elements; A.E. Fersman Mineralogical Museum, Academy of Sciences, Moscow, USSR.

References: (1) Spiridonov, E.M. and T.N. Chvileva (1979) Bezsmertnovite, $Au_4Cu(Te, Pb)$; a new mineral from the zone of oxidation of deposits of the Far East. Doklady Acad. Nauk SSSR, 249, 185–189 (in Russian). (2) (1981) Amer. Mineral., 66, 878 (abs. ref. 1).

Crystal Data: Cubic (pseudocell). *Point Group:* n.d. Massive.

Physical Properties: Hardness = n.d. VHN = 329–419, 381 average (20 g load). D(meas.) = 12.7 D(calc.) = [14.44]

Optical Properties: Opaque. *Color:* Light brown, rose-brown. *Streak:* Gold-brown to brown. *Luster:* Semimetallic.

R: n.d.

Cell Data: *Space Group:* n.d. $a = 4.095$ Z = [0.5]

X-ray Powder Pattern: USSR.
2.37 (100), 1.232 (80), 2.05 (70), 1.448 (60), 3.06 (40), 1.184 (20), 3.00 (10)

Chemistry:

	(1)	(2)
Au	48.4	50.06
Ag	1.54	
Cu	9.35	10.77
Fe	0.19	
Pb	19.2	17.55
Te	21.6	21.62
Se	0.34	
Total	100.6	100.00

(1) Far Eastern Region, USSR; by electron microprobe, average of eight analyses on two samples, leading to $(Au_{2.90}Ag_{0.17})_{\Sigma=3.07}(Cu_{1.74}Fe_{0.04})_{\Sigma=1.78}Pb_{1.10}(Te_{2.00}Se_{0.05})_{\Sigma=2.05}$. (2) $Au_3Cu_2PbTe_2$.

Occurrence: In the zone of weathering of tellurium deposits.

Association: Gold, sylvanite, krennerite, bezsmertnovite, other tellurides of Au, Cu, Pb, Fe.

Distribution: From undefined localities in Kazakhstan, and Kamchatka, Far Eastern Region, USSR.

Name: For Soviet geologist Yuri A. Bilibin (1901–1952).

Type Material: Institute of Mineralogy and Geochemistry of Rare Elements; A.E. Fersman Mineralogical Museum, Academy of Sciences, Moscow, USSR.

References: (1) Spiridonov, E.M., M.S. Bezsmertnaya, T.N. Chvileva, and V.V. Bezsmertny (1978) Bilibinskite, $Au_3Cu_2PbTe_2$, a new mineral of gold-telluride deposits. Zap. Vses. Mineral. Obshch., 107, 310–315 (in Russian). (2) (1979) Amer. Mineral., 64, 652 (abs. ref. 1). (3) Spiridonov, E.M., M.S. Bezsmertnaya, T.N. Chvileva, and V.V. Bezsmertny (1978) Bilibinskite $Au_3Cu_2PbTe_2$ — a new mineral from gold-telluride deposits. Comments. Zap. Vses. Mineral. Obshch., 107, 501 (in Russian). (4) (1978) Chem. Abs., 89, 200512 (abs. ref. 3).

Crystal Data: Orthorhombic. *Point Group:* n.d. Fine-grained aggregates.

Physical Properties: *Tenacity:* Slightly sectile. Hardness = 2.5 VHN = n.d. D(meas.) = 5.92(2) D(calc.) = 5.90

Optical Properties: Opaque. *Color:* Dark lead-gray. *Luster:* Metallic.
R_1–R_2: n.d.

Cell Data: *Space Group:* n.d. $a = \sim 14.82$ $b = \sim 14.82$ $c = 10.48$ Z = 8

X-ray Powder Pattern: Tintic, Utah, USA.
3.05 (10), 3.19 (7), 3.53 (6), 2.83 (6), 2.49 (6), 3.34 (5), 6.11 (4)

Chemistry:

	(1)
Ag	75.59
Cu	0.02
Fe	0.06
As	5.73
Sb	1.50
S	16.28
insol.	0.61
Total	99.79

(1) Tintic, Utah, USA; corresponding to $Ag_7(As_{0.86}Sb_{0.14})_{\Sigma=1.00}S_6$.

Occurrence: Believed to have occurred in a body of high-grade silver ore.

Association: Acanthite, tennantite, bismuthinite, galena, pyrite.

Distribution: In the North Lily mine, East Tintic district, Juab Co., Utah, USA.

Name: For Paul Billingsley (1887–1962), mining geologist, who discovered the North Lily mine, and collected the type material.

Type Material: Harvard University, Cambridge, Massachusetts; National Museum of Natural History, Washington, D.C., USA, R18987.

References: (1) Frondel, C. and R.M. Honea (1968) Billingsleyite, a new silver sulfosalt. Amer. Mineral., 53, 1791–1798.

Crystal Data: Hexagonal. *Point Group:* $\bar{3}\ 2/m$. Crystals up to 12 cm, but indistinct, often in parallel groupings, or hoppered; reticulated, arborescent, foliated, granular. *Twinning:* Polysynthetic, frequent.

Physical Properties: *Cleavage:* {0001} perfect and easy, $\{10\bar{1}1\}$ good, $\{10\bar{1}4\}$ poor. *Tenacity:* Sectile, brittle. Hardness = 2–2.5 VHN = 16–18 (100 g load). D(meas.) = 9.70–9.83 D(calc.) = 9.753

Optical Properties: Opaque. *Color:* Silver-white, with reddish hue, tarnishes iridescent; in polished section, brilliant creamy white, tarnishing yellow. *Streak:* Silver-white. *Luster:* Metallic. *Pleochroism:* Feeble. *Anisotropism:* Distinct.
R_1–R_2: (400) 61.4–61.6, (420) 60.8–61.6, (440) 60.2–61.0, (460) 60.0–60.8, (480) 60.1–61.0, (500) 60.6–61.7, (520) 61.4–62.7, (540) 62.6–63.8, (560) 63.8–65.0, (580) 64.9–66.2, (600) 65.9–67.2, (620) 66.9–68.8, (640) 67.7–69.0, (660) 68.5–69.7, (680) 69.0–70.4, (700) 69.6–70.8

Cell Data: *Space Group:* $R\bar{3}m$. $a = 4.55$ $c = 11.85$ Z = 6

X-ray Powder Pattern: Synthetic.
3.28 (100), 2.273 (41), 2.37 (40), 1.868 (23), 1.443 (16), 1.491 (11), 1.330 (11)

Chemistry: Nearly pure bismuth, with minor As and Te.

Occurrence: In hydrothermal veins with ores of Co, Ni, Ag, and Sn; in pegmatites and topaz-bearing Sn-W quartz veins.

Association: Chalcopyrite, arsenopyrite, pyrrhotite, pyrite, cobaltite, nickeline, breithauptite, skutterudite, safflorite, löllingite, bismuthinite, silver, cubanite, molybdenite, sphalerite, galena, scheelite, wolframite, calcite, barite, quartz.

Distribution: From numerous localities, but usually as a minor accessory mineral. At Altenberg, Schneeberg, and Annaberg, Saxony, Germany. At Jáchymov (Joachimsthal), Czechoslovakia. From near Villanueva de Córdoba, Córdoba Province, Spain. In the Dolcoath and other mines, Cornwall, England. From Uncia, Chorolque, Llallagua, and Tazna, in Bolivia, economically important. An 11 kg nugget found at Velaque, La Paz, Bolivia. In the Mt. Arthur mine, Queensland, and Kingsgate, New South Wales, Australia. At Natsukidani, Oita Prefecture, Japan, large crystals. From Cobalt, Ontario, Canada.

Name: From the German *weisse masse*, later *wismuth*, *white mass*.

References: (1) Palache, C., H. Berman, and C. Frondel (1944) Dana's system of mineralogy, (7th edition), v. I, 134–135. (2) (1954) NBS Circ. 539, 3, 20.

Crystal Data: Orthorhombic. *Point Group:* $2/m\ 2/m\ 2/m$. Crystals, up to 8 cm, stout prismatic to acicular, elongated and striated ∥ [001]. Usually massive with foliated or fibrous texture.

Physical Properties: *Cleavage:* {010}, perfect and easy, {100} and {110} imperfect. *Tenacity:* Flexible, somewhat sectile. Hardness = 2–2.5 VHN = n.d. D(meas.) = 6.78 D(calc.) = 6.81

Optical Properties: Opaque. *Color:* Lead-gray to tin-white, with a yellowish or iridescent tarnish. *Streak:* Lead-gray. *Luster:* Metallic. *Anisotropism:* Strong, especially in oil.
R_1–R_2: (400) 40.5–49.6, (420) 40.3–49.8, (440) 40.1–50.0, (460) 39.7–50.2, (480) 39.2–50.4, (500) 38.5–50.4, (520) 38.1–50.3, (540) 37.7–50.0, (560) 37.4–49.6, (580) 37.0–49.0, (600) 36.8–48.5, (620) 36.5–48.0, (640) 36.2–47.4, (660) 36.1–46.7, (680) 35.8–45.8, (700) 35.6–45.0

Cell Data: *Space Group:* *Pbnm*. $a = 11.12$ $b = 11.25$ $c = 3.97$ Z = 4

X-ray Powder Pattern: Synthetic.
3.569 (100), 3.118 (80), 3.530 (60), 2.812 (50), 3.967 (40), 2.521 (40), 1.953 (40)

Chemistry:

	(1)	(2)	(3)	(4)
Bi	76.94	79.28	79.47	81.3
Pb		1.68		
Cu		0.48	0.57	
Fe		0.74	0.17	
Te			0.94	
Se	8.80			
S	14.15	18.46	18.42	18.7
insol.			0.50	
Total	99.89	100.64	100.07	100.0

(1) Guanajuato, Mexico. (2) Jonquiére, Quebec, Canada. (3) Riddarhyttan, Västmanland, Sweden. (4) Bi_2S_3.

Occurrence: Typically in low- to high-temperature hydrothermal vein deposits, in tourmaline-bearing copper deposits in granite, in some gold veins formed at high temperatures, and in recent volcanic exhalation deposits.

Association: Bismuth, arsenopyrite, stannite, galena, pyrite, chalcopyrite, tourmaline, wolframite, cassiterite, quartz.

Distribution: Widespread. From the Llallagua, Huanuni, Tazna, and Chorolque districts, Potosí, Bolivia. From Guanajuato, Mexico. At Moravicza and Băiţa (Rézbánya), Romania. At Schneeberg and Altenberg, Saxony, Germany. From a number of mines in the Redruth district, Cornwall, England. In the Mt. Biggenden mine, Queensland, Australia. At Fefena, Madagascar. Exceptional crystals from Spind, Farsum, Norway.

Name: From the composition.

References: (1) Palache, C., H. Berman, and C. Frondel (1944) Dana's system of mineralogy, (7th edition), v. I, 275–278. (2) (1967) NBS Mono. 25, 13. (3) Kupčik, V. and L. Veselá-Nováková (1970) Zur Kristallstruktur des Bismuthinits, Bi_2S_3. Tschermaks Mineral. Petrog. Mitt., 14, 55–59 (in German). (4) Springer, G. (1971) The synthetic solid-solution series Bi_2S_3–$BiCuPbS_3$ (bismuthinite–aikinite). Neues Jahrb. Mineral., Monatsh., 19–24. (5) Mumme, W.G., E. Welin, and B.J. Wuensch (1976) Crystal chemistry and proposed nomenclature for sulfosalts intermediate in the system bismuthinite–aikinite (Bi_2S_3–$CuPbBiS_3$). Amer. Mineral., 61, 15–20.

Crystal Data: Tetragonal. *Point Group:* $4/m\ 2/m\ 2/m$. As elongated tabular crystals; as grains and massive.

Physical Properties: Hardness = n.d. VHN = 360–392 D(meas.) = n.d. D(calc.) = [6.25]

Optical Properties: Opaque. *Color:* Bronze-yellow. *Pleochroism:* Moderate. *Anisotropism:* Weak in reflected light.
R_1–R_2: n.d.

Cell Data: *Space Group:* $P4/mmm$. $a = 7.37$ $c = 5.88$ Z = [1]

X-ray Powder Pattern: 2.80 (10), 1.867 (8), 4.34 (6), 2.40 (6), 2.32 (6), 3.67 (5), 1.808 (4)

Chemistry:

	(1)	(2)	(3)
Ni	43.6	45.85	43.93
Fe	0.2	0.22	
Co	0.8	0.66	
Pb		0.01	
As	0.5	0.72	
Bi	33.9	27.76	34.75
S	21.4	22.46	21.32
Total	100.4	97.68	100.00

(1) Oktyabr mine, USSR; by electron microprobe. (2) Zimmer Lake, Canada: by electron microprobe. (3) $Ni_9Bi_2S_8$.

Occurrence: Of hydrothermal origin; as secondary pore-fillings and replacements of sandstone and conglomerate (Zimmer Lake, Canada).

Association: Chalcocite, galena, pentlandite, altaite (Oktyabr mine, USSR); chalcopyrite, pyrrhotite, millerite (Zimmer Lake, Canada).

Distribution: In the Oktyabr mine, Noril'sk region, western Siberia, USSR. From near Zimmer Lake, northern Saskatchewan, Canada. At the Mihara mine, Okayama Prefecture, and the Tsumo mine, Shimane Prefecture, Japan.

Name: For the relationship to hauchecornite.

Type Material: n.d.

References: (1) Kovalenker, V.A., T.L. Evstigneeva, V.D. Begizov, L.N. Vyal'sov, A.V. Smirnov, Y.K. Krakovetskii, and V.S. Balbin (1978) Hauchecornite from copper-nickel ores of the Oktyabr'skoe deposit. Trudy Min. muzeya Akad. Nauk SSSR, 26, 201–205 (in Russian). (2) (1979) Chem. Abs., 90, 158 (abs. ref. 1). (3) Just, J. (1980) Bismutohauchecornite—new name: hauchecornite redefined. Mineral. Mag., 43, 873–876. (4) (1981) Amer. Mineral., 66, 436 (abs. ref. 2). (5) Watkinson, D.H., J.B. Heslop, and W.D. Ewert (1975) Nickel sulphide-arsenide assemblages associated with uranium mineralization, Zimmer Lake area, northern Saskatchewan. Can. Mineral., 13, 198–204.

Crystal Data: Orthorhombic (probable), pseudocubic. *Point Group:* n.d. Massive.

Physical Properties: Hardness = n.d. –22.5, (540) 5.6–29.5, (560) 6.1–34.5, (580) 8.4–37.6, (600) 13.5–39.2, (620) 21.0–40.0, (640) 29.5–39.8, (660) 36.2–39.0, (680) 41.2–37.7, (700) 44.9–35.7

Cell Data: *Space Group:* n.d. $a = 4.087$ (pseudocell). Z = n.d.

X-ray Powder Pattern: USSR.
2.36 (100), 1.230 (80), 2.045 (60), 1.447 (60), 1.180 (30), 1.092 (25), 0.992 (15)

Chemistry:

	(1)
Au	57.6 – 63.6
Ag	1.67 – 3.39
Cu	3.32 – 15.11
Fe	10.28 – 0.09
Pb	10.7 – 14.4
Te	9.60 – 10.3
Se	0.0 – 0.28
Total	

(1) USSR; by electron microprobe, ranges of 11 analyses. Cu and Fe vary reciprocally from $Cu_{3.36}Fe_{0.32}$ to $Cu_{0.75}Fe_{2.66}$.

Occurrence: In the secondary zone.

Association: Bilibinskite, gold, tellurides of Fe, Cu, Pb.

Distribution: At undefined localities in Kazakhstan and the Far Eastern Region, USSR. From Bisbee, Cochise Co., Arizona, USA.

Name: For Soviet geologist Aleksei A. Bogdanov (1907–1971).

Type Material: A.E. Fersman Mineralogical Museum, Academy of Sciences, Moscow, USSR.

References: (1) Spiridonov, E.M. and T.N. Chvileva (1979) Bogdanovite, $Au_5(Cu,Fe)_3(Te,Pb)_2$, a new mineral of the group of intermetallic compounds of gold. Vestnik Moskva Univ., Ser. Geol., 1, 44–52 (in Russian). (2) (1979) Amer. Mineral., 64, 1329 (abs. ref. 1).

Crystal Data: Hexagonal. *Point Group:* $\overline{3}\ 2/m$. Anhedral grains, to 200 by 600 μm. *Twinning:* Polysynthetic.

Physical Properties: Hardness = 3.2 VHN = 63–96 D(meas.) = n.d. D(calc.) = 7.72

Optical Properties: Opaque. *Color:* Pale creamy yellow to pink, tarnishing more golden. *Anisotropism:* Noticeable, distinctly stronger in oil, in yellowish gray tones.
R_1–R_2: (470) 51.2–52.9, (546) 49.8–51.5, (589) 50.1–51.5, (650) 50.1–51.6

Cell Data: *Space Group:* $P\overline{3}m1$ (most probable). $a = 8.412(6)$ $c = 19.63(3)$ Z = [6]

X-ray Powder Pattern: Kletno, Poland.
2.91 (100), 2.03 (30), 6.54 (20), 3.40 (20), 3.26 (18), 2.09 (18), 1.630 (6)

Chemistry:

	(1)	(2)	(3)	(4)
Ag	22.31	22.6	22.9	22.7
Pb	1.34		0.0	
Cu	0.25		0.2	
Co	0.01			
Ni	0.02			
Bi	44.89	43.2	44.7	44.0
Se	28.46	32.8	30.8	33.3
S	2.47		1.2	
Total	99.75	98.6	99.8	100.0

(1) Kletno, Poland; by electron microprobe, average of three analyses, corresponding to $(Ag_{0.98}Cu_{0.02})_{\Sigma=1.00}(Bi_{0.97}Pb_{0.03})_{\Sigma=1.00}(Se_{1.66}S_{0.34})_{\Sigma=2.00}$. (2) Near Julianehåb, Greenland; by electron microprobe. (3) Kidd Creek mine, Canada; by electron microprobe. (4) $AgBiSe_2$.

Occurrence: In fluorite and quartz in a strongly cracked zone formed in crystalline limestone adjacent to magnetite-bearing skarns (Kletno, Poland).

Association: Clausthalite, tiemannite, umangite, klockmannite, wittichenite, silver, naumannite, bornite, chalcopyrite, chalcocite, uraninite, fluorite, quartz (Kletno, Poland); tennantite, carrollite, cobaltite, bornite, chalcopyrite, chalcocite, naumannite, eucairite, clausthalite (Kidd Creek mine, Canada); hessite, chalcocite, digenite, umangite, naumannite, eucairite, bornite, chalcopyrite, clausthalite, covellite, magnetite, hematite, goethite, malachite, azurite (Julianehåb, Greenland).

Distribution: From Kletno, Sudetes Mountains, Poland. At the Kidd Creek mine, near Timmins, Ontario, Canada. In the Frederik VII's mine, near Julianehåb, southern Greenland. From near Vanos, Chihuahua, Mexico.

Name: For Professor Karol Bohdanowicz (1864–1947), of Cracow, Poland.

Type Material: n.d.

References: (1) Banaś, M., D. Atkin, J.F.W. Bowles, and P.R. Simpson (1979) Definitive data on bohdanowiczite, a new silver bismuth selenide. Mineral. Mag., 43, 131–133. (2) (1979) Amer. Mineral., 64, 1333 (abs. ref. 1). (3) Pringle, G.J. and R.I. Thorpe (1980) Bohdanowiczite, junoite and laitakarite from the Kidd Creek mine, Timmins, Ontario. Can. Mineral., 18, 353–360. (4) Schonwandt, H.K. (1983) Interpretation of ore microstructures from a seleneous Cu-mineralization in South Greenland. Neues Jahrb. Mineral., Abh., 146, 302–332.

Borishanskiite $Pd_{1+x}(As, Pb)_2$ (x = 0 to 0.2)

Crystal Data: Orthorhombic. *Point Group:* $mm2$. Observed usually as 20–30 μm grains, rarely as single grains (6–150 μm), enclosed in sulfides.

Physical Properties: *Cleavage:* Three directions observed. *Tenacity:* Brittle. Hardness = n.d. VHN = 241 (20 g load). D(meas.) = n.d. D(calc.) = 10.2

Optical Properties: Opaque. *Color:* Dark steel-gray; gray-white in polished section. *Luster:* Metallic. *Anisotropism:* Barely noticeable to moderate, in yellowish gray tones.
R_1–R_2: (430) 53.2–51.6, (460) 54.6–54.1, (490) 56.2–55.8, (520) 54.6–56.0, (550) 57.0–56.6, (580) 57.8–57.3, (610) 58.7–58.3, (640) 59.8–59.3, (670) 60.6–60.1, (700) 61.7–61.2

Cell Data: *Space Group:* $Ccm2_1$. a = 7.18(2) b = 8.62(2) c = 10.66(2) Z = 16

X-ray Powder Pattern: Noril'sk region, USSR.
2.65 (100), 2.16 (90), 2.50 (60), 2.25 (60), 1.677 (60), 1.385 (60), 1.169 (60)

Chemistry:

	(1)	(2)
Pd	29.8	31.4
Ag		1.1
Pb	50.4	50.2
As	21.4	19.8
Total	101.6	102.5

(1–2) Talnakh area, USSR; by electron microprobe, corresponding to $Pd_{1.06}(As_{1.08}Pb_{0.92})_{\Sigma=2.00}$ and $(Pd_{1.16}Ag_{0.04})_{\Sigma=1.20}(As_{1.04}Pb_{0.96})_{\Sigma=2.00}$.

Occurrence: In massive and disseminated Cu-Ni sulfide ores.

Association: Nickeline, palladian cuproauride, chalcopyrite, magnetite, pyrrhotite.

Distribution: In the Oktyabr mine, Talnakh area, Noril'sk region, western Siberia, USSR.

Name: For S.S. Borishanski, Soviet mineralogist.

Type Material: A.E. Fersman Mineralogical Museum, Academy of Sciences, Moscow, USSR.

References: (1) Razin, L.V., L.S. Dubakina, V.I. Meschankina, and V.D. Begizov (1975) Borishanskiite—a new plumboarsenide of palladium from the copper-nickel sulfide ores of the Talnakh differentiated intrusive. Zap. Vses. Mineral. Obshch., 104, 57–61. (2) (1976) Amer. Mineral., 61, 502 (abs. ref. 1).

Crystal Data: Cubic. *Point Group:* $4/m\,\overline{3}\,2/m$. Massive.

Physical Properties: Hardness = ~4 VHN = n.d. D(meas.) = n.d. D(calc.) = 6.166

Optical Properties: Opaque. *Color:* Rose-red.
R: (400) 41.3, (420) 42.8, (440) 44.3, (460) 45.4, (480) 46.4, (500) 47.2, (520) 47.8, (540) 42.4, (560) 42.9, (580) 49.2, (600) 49.7, (620) 50.2, (640) 50.8, (660) 57.4, (680) 52.0, (700) 52.7

Cell Data: *Space Group:* $Fd3m$. a = ~10.2 Z = 8

X-ray Powder Pattern: Trogtal quarry, Germany.
2.7 (100), 2.4 (100), 2.3 (100), 2.2 (100), 2.0 (100), 1.96 (100), 1.42 (100)

Chemistry: No analysis appears ever to have been made.

Occurrence: Of hydrothermal origin.

Association: Trogtalite, hastite, clausthalite.

Distribution: At the Trogtal quarry, near Lautenthal, and at Tilkerode, Harz, Germany. From the Pinky Fault uranium deposit, Saskatchewan, Canada. In the Sierra de Cacheuta, Mendoza Province, Argentina.

Name: After Dr. Wilhelm Bornhardt (1864–?), German student of ore deposits.

Type Material: n.d.

References: (1) Ramdohr, P. and M. Schmitt (1955) Vier neue natürliche Kobaltselenide vom Steinbruch Trogtal bei Lautenthal im Harz. Neues Jahrb. Mineral., Monatsh., 133–142 (in German). (2) (1956) Amer. Mineral., 41, 164–165 (abs. ref. 1).

Crystal Data: Orthorhombic, pseudocubic. *Point Group:* $2/m\ 2/m\ 2/m$. Crystals pseudocubic, dodecahedral, octahedral, to 6 cm; granular, compact, or massive. *Twinning:* On {111}; often shows penetration twins.

Physical Properties: *Cleavage:* {111} in traces. *Fracture:* Uneven to subconchoidal. *Tenacity:* Brittle. Hardness = 3–3.25 VHN = n.d. D(meas.) = 5.06–5.08 D(calc.) = 5.074

Optical Properties: Opaque. *Color:* Copper-red to pinchbeck-brown on fresh surfaces, rapidly tarnishes iridescent purplish; in polished section, pinkish brown when fresh. *Streak:* Light grayish black. *Luster:* Metallic. *Pleochroism:* Weak but noticeable. *Anisotropism:* Weak.
R: (400) 19.8, (420) 18.8, (440) 17.8, (460) 17.3, (480) 17.5, (500) 18.6, (520) 19.8, (540) 21.1, (560) 22.7, (580) 24.4, (600) 25.9, (620) 27.3, (640) 28.6, (660) 29.7, (680) 30.3, (700) 30.6

Cell Data: *Space Group:* *Pbca.* $a = 10.950$ $b = 21.862$ $c = 10.950$ Z = 16

X-ray Powder Pattern: Messina, South Africa.
1.937 (100), 3.18 (60), 2.74 (50), 1.258 (50), 1.119 (50), 3.31 (40), 2.50 (40)

Chemistry:

	(1)	(2)
Cu	62.99	63.33
Pb	0.10	
Fe	11.23	11.12
S	25.58	25.55
Total	99.90	100.00

(1) Superior, Arizona, USA. (2) Cu_5FeS_4.

Occurrence: Associated with and disseminated in mafic igneous rocks, in contact metamorphic (skarn) deposits, in pegmatites, in intermediate to high-temperature hydrothermal deposits, and in sedimentary cupriferous shales.

Association: Chalcopyrite, pyrite, other copper and iron sulfides, garnet, calcite, wollastonite, quartz.

Distribution: Important localities for fine crystals include, in the USA, Butte, Silver Bow Co., Montana, and Bristol, Hartford Co., Connecticut. In England, from Redruth, Cornwall. Large crystals from the Mangula mine, Zimbabwe. From the Frossnitz Alpe, eastern Tyrol, Austria. At Dzhezkazgan, Kazakhstan, USSR. In the N'ouva mine, Talate, Morocco. Widespread as an important ore mineral of copper, for example at the Magma mine, Superior, Pinal Co., and Bisbee, calities for fine crystals include, in the USA, Butte, Silver Bow Co., Montana, and Bristol, Hartford Co., Connecticut. In England, from Redruth, Cornwall. Large crystals from the Mangula mine, Zimbabwe. From the Frossnitz Alpe, eastern Tyrol, Austria. At Dzhezkazgan, Kazakhstan, USSR. In the N'ouva mine, Talate, Morocco. Widespread as an important ore mineral of copper, for example at the Magma mine, Superior, Pinal Co., and Bisbee, Cochise Co., Arizona; and Kennecott, Alaska, USA. At Ookiep, Namaqualand, Cape Province; and the Messina mine, Transvaal, South Africa. From Mt. Lyell, Tasmania, Australia.

Name: For Ignatius von Born (1742–1791), distinguished Austrian mineralogist.

References: (1) Palache, C., H. Berman, and C. Frondel (1944) Dana's system of mineralogy, (7th edition), v. I, 195–197. (2) Koto, K. and N. Morimoto (1975) Superstructure investigation of bornite, Cu_5FeS_4, by the modified partial Patterson function. Acta Cryst., B31, 2268–2273. (3) Kanazawa, Y., K. Koto, and N. Morimoto (1978) Bornite Cu_5FeS_4 : stability and crystal structure of the intermediate form. Can. Mineral., 16, 397–404. (4) Pierce, L. and P.R. Buseck (1978) Superstructuring in the bornite–digenite series: a high resolution electron microscope study. Amer. Mineral., 63, 1–16.

Crystal Data: Cubic. *Point Group:* n.d. Irregular grains, to 0.2 mm in size, embedded in sulfides.

Physical Properties: Hardness = n.d. VHN = 88.3 (10 g load). D(meas.) = n.d. D(calc.) = [8.25]

Optical Properties: Opaque. *Color:* Dark gray; in polished section, white. *Luster:* Metallic. *Anisotropism:* Isotropic to slightly anisotropic.
R: n.d.

Cell Data: *Space Group:* n.d. $a = 5.794(8)$ Z = 1

X-ray Powder Pattern: Khautovaarsk deposit, USSR.
2.902 (100), 2.041 (60), 1.550 (50), 1.297 (40), 1.183 (40), 3.34 (30), 1.603 (30)

Chemistry:

	(1)	(2)
Pd	32.39	33.55
Pt	1.23	
Ni	0.25	
Fe	0.04	
Sb	10.98	12.80
Bi	3.34	
Te	51.97	53.65
Total	100.21	100.00

(1) Khautovaarsk deposit, USSR; by electron microprobe, average of four analyses, corresponding to $(Pd_{2.93}Pt_{0.06}Ni_{0.04}Fe_{0.01})_{\Sigma=3.04}(Sb_{0.87}Bi_{0.15})_{\Sigma=1.02}Te_{3.93}$. (2) Pd_3SbTe_4.

Occurrence: In massive pentlandite-chalcopyrite-pyrrhotite ores of hydrothermal origin.

Association: Pyrrhotite, chalcopyrite, altaite, pentlandite.

Distribution: From the Khautovaarsk Cu-Ni deposit, Karelian ASSR, USSR.

Name: For Igor Borisovich Borovskii, Russian pioneer in microprobe analysis.

Type Material: Leningrad Mining Institute, Leningrad, USSR.

References: (1) Yalovoi, A.A., A.F. Siderov, N.S. Rudashevskii, and I.A. Bud'ko (1973) Borovskite, Pd_3SbTe_4, a new mineral. Zap. Vses. Mineral. Obshch., 102, 427–431 (in Russian).
(2) (1974) Amer. Mineral., 59, 873 (abs. ref. 1)

Crystal Data: Monoclinic. *Point Group:* $2/m$. Needle-like crystals, rarely as rings; fibrous, compact masses. Crystals are seldom terminated; striations || [001] strong.

Physical Properties: *Cleavage:* Distinct on {100}. *Tenacity:* Brittle; flexible in thin crystals. Hardness = 2.5–3 VHN = n.d. D(meas.) = ~6.2 D(calc.) = 6.21

Optical Properties: Opaque. *Color:* Dull, lead-gray. *Streak:* Brownish. *Luster:* Metallic, sometimes silky. *Pleochroism:* Weak. *Anisotropism:* Distinct.
R_1–R_2: (400) 41.2–42.8, (420) 40.3–42.4, (440) 39.4–42.0, (460) 39.2–41.9, (480) 38.5–40.4, (500) 37.4–39.3, (520) 37.9–40.1, (540) 38.0–40.3, (560) 37.8–40.2, (580) 37.4–39.9, (600) 37.1–39.6, (620) 36.9–39.5, (640) 36.6–39.1, (660) 36.1–38.6, (680) 35.5–37.9, (700) 35.0–37.1

Cell Data: *Space Group:* $P2_1/a$. $a = 21.56$ $b = 23.51$ $c = 8.09$ $\beta = 100°48'$ Z = 8

X-ray Powder Pattern: Sullivan mine, British Columbia, Canada. (JCPDS 18-688).
3.731 (100), 3.218 (45), 3.025 (40), 2.823 (40), 2.689 (25), 3.669 (20), 3.905 (18)

Chemistry:

	(1)	(2)	(3)
Pb	55.28	51.00	55.42
Fe	0.39		
Sb	25.40	27.82	25.69
S	18.19	18.99	18.89
insol.	0.62	2.54	
Total	99.88	100.35	100.00

(1) Cleveland mine, Stevens Co., Washington, USA. (2) Caspari mine near Arnsberg, North Rhine-Westphalia, Germany. (3) $Pb_5Sb_4S_{11}$.

Occurrence: In hydrothermal veins formed at low to moderate temperatures.

Association: Lead sulfosalts, galena, stibnite, sphalerite, pyrite, arsenopyrite, siderite, quartz.

Distribution: Widespread; only a few localities can be given. In the USA, in the Coeur d'Alene district; and the Wood River district, Blaine Co., Idaho; the Iron Mountain mine, Superior, Mineral Co., Montana; the Echo district, Union Co., Nevada; and at Augusta Mountain, Gunnison Co., Colorado. From Madoc, Ontario, Canada, as rings. In crystals from Sala, Västmanland; Nasafjell, Lappland; and Boliden, Västerbotten, Sweden. At Wolfsberg, Harz Mountains, at Ober-Lahr near Altenkirchen in the Rhineland, and Waldsassen, Bavaria, Germany. From Příbram, Czechoslovakia. In France, at Molières, Gard. In rings from Bottino, Tuscany, Italy. From Trepča, Yugoslavia. In Mexico, at Noche Buena, Zacatecas.

Name: For Charles Louis Boulanger (1810–1849), French mining engineer.

References: (1) Palache, C., H. Berman, and C. Frondel (1944) Dana's system of mineralogy, (7th edition), v. I, 420–422.

Crystal Data: Orthorhombic. *Point Group:* $mm2$. Crystals short prismatic to tabular, commonly striated, as much as 11 cm across; often as subparallel aggregates. Also massive, granular to compact. *Twinning:* On {110} commonly forming cross or cogwheel aggregates.

Physical Properties: *Cleavage:* {010} imperfect, {100} and {001} less perfect. *Fracture:* Subconchoidal to uneven. *Tenacity:* Brittle. Hardness = 2.5–3 VHN = 176–205 (100 g load). D(meas.) = 5.83 D(calc.) = 5.84

Optical Properties: Opaque. *Color:* Steel-gray to iron-black. *Streak:* Steel-gray to iron-black. *Luster:* Brilliant to dull. *Pleochroism:* Very weak. *Anisotropism:* Weak in air. R_1–R_2: (400) 37.3–37.0, (420) 37.4–37.7, (440) 37.5–38.4, (460) 36.9–38.0, (480) 36.2–37.4, (500) 35.6–37.0, (520) 35.0–36.6, (540) 34.5–36.3, (560) 34.0–36.0, (580) 33.6–35.5, (600) 33.0–34.9, (620) 32.6–34.2, (640) 32.2–33.5, (660) 31.8–32.9, (680) 31.4–32.4, (700) 31.2–32.0

Cell Data: *Space Group:* $Pnm2_1$. $a = 8.153(3)$ $b = 8.692(3)$ $c = 7.793(2)$ Z = 4

X-ray Powder Pattern: Neudorf, Germany.
2.74 (100), 3.90 (80), 1.765 (60), 2.59 (50), 4.37 (40), 2.99 (40), 2.69 (40)

Chemistry:

	(1)	(2)	(3)	(4)
Pb	40.21	42.34	39.37	42.40
Cu	15.12	12.80	13.52	13.01
Ag			1.69	
Zn	0.35	0.04	0.09	
Fe	0.35	0.27	0.31	
Sb	18.99	24.44	24.74	24.91
As	2.81			
S	20.04	19.58	19.94	19.68
rem.	1.67	0.37		
Total	99.54	99.84	99.66	100.00

(1) Boggs mine, Yavapai Co., Arizona, USA. (2) Herja (Kisbánya), Romania. (3) Příbram, Czechoslovakia. (4) $PbCuSbS_3$.

Polymorphism & Series: Forms a series with seligmannite.

Occurrence: In hydrothermal veins formed at moderate temperatures.

Association: Galena, tetrahedrite, sphalerite, chalcopyrite, pyrite, stibnite, zinkenite, siderite, quartz, rhodochrosite, dolomite, barite.

Distribution: Numerous localities for fine crystals, of which only a few can be mentioned. From the Georg mine, near Horhausen, Westerwald, and at Clausthal, Neudorf, and Wolfsberg, Harz Mountains, Germany. From Příbram, Czechoslovakia. At Baia Sprie (Felsőbánya), Cavnic (Kapnik), and Săcărâmb (Nagyág), Romania. At Pontgibaud, Puy-de-Dôme; St. Laurent Le Minier, Gard; and Prunières, Isère, France. Exceptional crystals from the Herodsfoot mine, Liskeard, Cornwall, England. In Bolivia, from Chorolque, Colquechaca, Pacuani, Machacamarca, and Oruro. At the Quiruvilca mine, La Libertad, Peru. From Park City, Summit Co., Utah, USA. In Mexico, from Noche Buena, Zacatecas, and Naica, Chihuahua.

Name: After Count Jacques Louis de Bournon (1751–1825), French crystallographer and mineralogist.

References: (1) Palache, C., H. Berman, and C. Frondel (1944) Dana's system of mineralogy, (7th edition), v. I, 406–410. (2) Berry, L.G. and R.M. Thompson (1962) X-ray powder data for the ore minerals. Geol. Soc. Amer. Mem. 85, 133. (3) Edenharter, A., W. Nowacki, and Y. Takéuchi (1970) I. Verfeinerung der Kristallstruktur von Bournonit $[(SbS_3)_2|Cu_2^{IV}Pb^{VII}Pb^{VIII}]$ und von Seligmannit $[(AsS_3)_2|Cu_2^{IV}Pb^{VII}Pb^{VIII}]$. Zeits. Krist., 131, 397–417 (in German).

Crystal Data: Orthorhombic. *Point Group:* $2/m\ 2/m\ 2/m$. As anhedral grains, 0.02 to 0.5 mm across.

Physical Properties: *Tenacity:* Very slightly brittle. Hardness = n.d. VHN = 858–1635, 1288 average (100 g load). D(meas.) = n.d. D(calc.) = 6.91–6.96

Optical Properties: Opaque. *Color:* Pale gray to pale gray-brown in reflected light. *Luster:* Metallic. *Anisotropism:* Weak, dark gray to very dark brown.

R_1–R_2: (400) 43.5–44.5, (420) 44.2–45.2, (440) 44.8–45.8, (460) 45.3–46.5, (480) 45.7–46.9, (500) 45.9–47.3, (520) 46.1–47.6, (540) 46.2–47.8, (560) 46.3–48.0, (580) 46.4–48.2, (600) 46.5–48.3, (620) 46.6–48.4, (640) 46.8–48.5, (660) 46.8–48.6, (680) 46.9–48.6, (700) 47.0–48.7

Cell Data: *Space Group: Pnca.* $a = 8.454(7)$–$8.473(8)$ $b = 5.995(1)$–$6.002(7)$ $c = 6.143(1)$–$6.121(8)$ Z = 4

X-ray Powder Pattern: Goodnews Bay, Alaska, USA.
3.00 (vvs), 0.7877 (vvs), 0.7768 (vvs), 0.7726 (vvs), 1.757 (vs), 2.143 (s), 1.728 (s)

Chemistry:

	(1)	(2)	(3)
Rh	30.8	68.80	29.2
Ir	35.4		44.6
Pt	8.8		
S	25.3	31.05	24.1
Total	100.3	99.85	97.9

(1) Salmon River, Alaska, USA; by electron microprobe. (2) Gaositai, China; by electron microprobe. (3) Gusevogorskii massif, USSR; by electron microprobe.

Occurrence: In platinum alloy nuggets recovered from dredging operations (Salmon River, Alaska, USA); in platinum-bearing ultramafic rocks intruded into a granite gneiss-hornblende gneiss complex (Gaositai, China).

Association: Platinum-iridium, platinum, osmium, laurite, silicate inclusions (Salmon River, Alaska, USA).

Distribution: From the Salmon River, Goodnews Bay, Alaska, USA. At Gaositai, Hebei Province, China. From the Gusevogorskii massif, Ural Mountains, USSR.

Name: For the British scientist, Dr. Stanley Hay Umphray Bowie, of the Institute of Geological Sciences, London, England.

Type Material: British Museum (Natural History), London, England, BM 1983,70; National Museum of Natural History, Washington, D.C., USA, M-Cr-69-1.

References: (1) Desborough, G.A. and A.J. Criddle (1984) Bowieite: a new rhodium–iridium–platinum sulfide in platinum-alloy nuggets, Goodnews Bay, Alaska. Can. Mineral., 22, 543–552.

Crystal Data: Tetragonal. *Point Group:* $4/m$ or 4. As prisms to 2 cm and rounded grains. *Twinning:* Sometimes observed.

Physical Properties: Hardness = n.d. VHN = 946–1064, 997 average (100 g load). D(meas.) = ~10 D(calc.) = 9.383

Optical Properties: Opaque. *Color:* White in reflected light. *Luster:* Metallic. *Anisotropism:* Distinct, in purplish and pinkish shades of gray and brown.
R_1–R_2: (400) 41.3–41.8, (420) 41.8–42.4, (440) 42.1–43.0, (460) 42.4–43.4, (480) 42.5–43.8, (500) 42.7–44.1, (520) 42.7–44.2, (540) 42.6–44.2, (560) 42.5–44.2, (580) 42.4–44.2, (600) 42.3–44.1, (620) 42.2–44.1, (640) 41.9–44.0, (660) 41.9–43.9, (680) 41.9–43.8, (700) 41.5–43.8

Cell Data: *Space Group:* $P4_2/m$. $a = 6.367$ $c = 6.561$ Z = 8

X-ray Powder Pattern: Potgietersrust district, South Africa. Can be confused with vysotskite.
2.86 (100), 2.93 (30), 2.64 (30), 1.852 (30), 1.423 (30), 1.713 (20), 1.595 (2)

Chemistry:

	(1)	(2)
Pt	63.2	62.1
Pd	15.4	19.0
Ni	4.4	2.0
S	17.4	16.9
Total	100.4	100.0

(1) Potgietersrust district, South Africa; by electron microprobe, corresponding to $(Pt_{0.60}Pd_{0.27}Ni_{0.14})_{\Sigma=1.01}S_{1.00}$. (2) Stillwater Complex, Montana, USA; by electron microprobe, corresponding to $(Pt_{0.60}Pd_{0.34}Ni_{0.06})_{\Sigma=1.00}S_{1.00}$.

Polymorphism & Series: Forms a series with vysotskite, dimorphous with cooperite.

Occurrence: In norites occurring in layered mafic intrusives, formed at high magmatic temperatures.

Association: Sperrylite, cooperite, laurite, platinum (Potgietersrust district, South Africa); pentlandite, pyrrhotite, chalcopyrite, cubanite, nickelian mackinawite, gold, cooperite, vysotskite, moncheite, isoferroplatinum, kotulskite, keithconnite, palladian tulameenite (Stillwater Complex, Montana, USA).

Distribution: From the Rustenburg and Potgietersrust districts, Bushveld Complex, Transvaal, South Africa. From the Stillwater Complex, Montana, USA. In the Lac des Iles Complex, Ontario, Canada. From the Noril'sk region, western Siberia, USSR.

Name: To honor Sir William Henry Bragg (1862–1942) and Professor William Lawrence Bragg (1890–1971), pioneers in the X-ray investigation of crystals, as this is the first new mineral to be discovered by X-ray methods alone.

Type Material: British Museum (Natural History), London, England, BM 1932,1304.

References: (1) Palache, C., H. Berman, and C. Frondel (1944) Dana's system of mineralogy, (7th edition), v. I, 259. (2) Berry, L.G. and R.M. Thompson (1962) X-ray powder data for the ore minerals. Geol. Soc. Amer. Mem. 85, 71–72. (3) Childs, J.D. and Hall, S.R. (1973) The crystal structure of braggite, (Pt, Pd, Ni)S. Acta Cryst., B29, 1446–1451. (4) Cabri, L.J., J.H.G. Laflamme, J.M. Stewart, K. Turner, and B.J. Skinner (1978) On cooperite, braggite, and vysotskite. Amer. Mineral., 63, 832–839. (5) Cabri, L.J., Ed. (1981) Platinum group elements: mineralogy, geology, recovery. Can. Inst. Min. & Met., 100. (6) Criddle, A.J. and C.J. Stanley (1985) Characteristic optical data for cooperite, braggite and vysotskite. Can. Mineral., 23, 149–162.

Crystal Data: Cubic. *Point Group:* $2/m\ \overline{3}$. Indistinct crystals with cubic, octahedral, or pyritohedral faces to as large as 1 cm; usually as crusts or nodular masses with a radially fibrous or columnar structure. Often exhibits zonal growth banding or as oscillatory intergrowths with pyrite.

Physical Properties: *Cleavage:* {001}. *Fracture:* Conchoidal to uneven. *Tenacity:* Brittle. Hardness = 5.5–6 VHN = 889–927 (100 g load). D(meas.) = 4.62–4.72 D(calc.) = 4.665

Optical Properties: Opaque. *Color:* Steel-gray, tarnishing brown to lilac. *Luster:* Metallic. R: (400) 36.2, (420) 35.6, (440) 35.0, (460) 34.6, (480) 34.7, (500) 35.0, (520) 35.6, (540) 36.3, (560) 36.8, (580) 37.4, (600) 38.0, (620) 38.7, (640) 39.4, (660) 40.2, (680) 40.8, (700) 41.2

Cell Data: *Space Group:* $Pa3$. $a = 5.57$ Z = 4

X-ray Powder Pattern: Minasragra, Peru.
2.79 (10), 1.687 (60), 2.50 (5), 1.078 (5), 2.28 (4), 1.976 (4), 3.21 (2)

Chemistry:

	(1)	(2)	(3)
Ni	24.73	17.50	24.17
Fe	17.08	21.15	23.00
Co	3.28	6.61	
Cu	0.47		
S	51.15	53.70	52.83
insol.	0.40	1.04	
Total	97.11	100.00	100.00

(1) Mechernich mine, Germany. (2) Müsen, Germany. (3) $(Ni, Fe)S_2$ with Ni:Fe = 1:1.

Occurrence: An early-formed mineral in low-temperature environments; forms in the zone of secondary enrichment, and in deposits of the Missouri lead belt type.

Association: Pyrite, galena, sphalerite, chalcopyrite, marcasite, pyrrhotite, pentlandite, linnaeite, millerite, bismuthinite, parkerite, siegenite, siderite, barite.

Distribution: Prominent localities are: in Germany, from the Mechernich mine, Eifel; the Victoria mine, Müsen, North Rhine-Westphalia; and Manbach, Rhineland. In the Mill Close mine and at Masson Hill, and many other localities in Derbyshire, England. From Spirit Mountain, Lower Copper River Valley, Alaska; Cave-in-Rock, Hardin Co., Illinois; and Fredericktown, Madison Co., Missouri, USA. From the Bushveld Complex, Transvaal, South Africa. In the Langis mine, Cobalt-Gowganda area, Ontario, Canada. From Minasragra, near Cerro de Pasco, Peru. Known from a number of additional occurrences exhibiting only sparse and tiny crystals.

Name: After José J. Bravo (1874–1928), scientist of Lima, Peru.

Type Material: n.d.

References: (1) Palache, C., H. Berman, and C. Frondel (1944) Dana's system of mineralogy, (7th edition), v. I, 290–291. (2) Berry, L.G. and R.M. Thompson (1962) X-ray powder data for the ore minerals. Geol. Soc. Amer. Mem. 85, 88. (3) Petruk, W., D.C. Harris, and J.M. Stewart (1969) Langisite, a new mineral, and the rare minerals cobalt pentlandite, siegenite, parkerite, and bravoite from the Langis mine, Cobalt-Gowganda area, Ontario. Can. Mineral., 9, 597–616.

Crystal Data: Hexagonal. *Point Group:* $6/m\ 2/m\ 2/m$. Crystals rare, thin tabular; arborescent, disseminated, massive. *Twinning:* Twin plane $\{10\bar{1}1\}$.

Physical Properties: *Fracture:* Subconchoidal to uneven. *Tenacity:* Brittle. Hardness = 5.5 VHN = n.d. D(meas.) = 7.591–8.23 D(calc.) = 8.629

Optical Properties: Opaque. *Color:* Light copper-red, often with violet tint. *Streak:* Reddish brown. *Luster:* Metallic. *Pleochroism:* Very distinct. *Anisotropism:* Very noticeable in air and oil.

R_1–R_2: (400) 47.8–51.4, (420) 45.5–49.3, (440) 43.2–47.2, (460) 41.1–45.8, (480) 39.5–45.2, (500) 38.5–45.6, (520) 38.7–46.8, (540) 41.0–48.6, (560) 43.6–51.1, (580) 46.6–53.4, (600) 49.3–55.6, (620) 51.6–57.6, (640) 53.6–59.3, (660) 55.3–60.9, (680) 56.8–62.2, (700) 58.0–63.3

Cell Data: *Space Group:* $P6_3/mmc$. $a = 3.946$ $c = 5.148$ Z = 2

X-ray Powder Pattern: Cobalt, Canada.
2.84 (100), 2.06 (70), 1.965 (70), 1.533 (30), 1.074 (30), 1.610 (20), 1.419 (20)

Chemistry:

	(1)	(2)
Ni	32.09	32.52
Fe	0.04	
Co	0.59	
Sb	66.62	67.48
As	0.58	
S	0.00	
Total	99.92	100.00

(1) Hudson Bay mine, Cobalt, Canada; in part due to nickeline. (2) NiSb.

Occurrence: In hydrothermal calcite veins associated with Co-Ni-Ag ores.

Association: Silver, nickeline, maucherite, cobaltite, ullmannite, tetrahedrite, pyrrhotite, cubanite, chalcopyrite, sphalerite, galena, calcite.

Distribution: From the Cobalt district, near Red Lake; in the Silver Islet mine and the Hemlo gold deposit, Thunder Bay district, Ontario, Canada. In the USA, from Coyote Peak, near Orick, Humboldt Co., California. At St. Andreasberg, Harz Mountains, Germany. From France, at Ar, Basses-Pyrénées, and Pierrefitte, Hautes-Pyrénées. On Monte Narba, Sarrabus, Sardinia, Italy. From Tunaberg, Saxberget, and Långsjön, Sweden. From the Ilímaussaq Intrusion, southern Greenland. At Broken Hill, New South Wales, Australia. From the Noril'sk region, western Siberia, USSR. In the Natsume mine, Hyogo Prefecture, Japan.

Name: After the Saxon mineralogist Johann Friedrich August Breithaupt (1791–1873).

References: (1) Palache, C., H. Berman, and C. Frondel (1944) Dana's system of mineralogy, (7th edition), v. I, 238–239. (2) Berry, L.G. and R.M. Thompson (1962) X-ray powder data for the ore minerals. Geol. Soc. Amer. Mem. 85, 62.

Crystal Data: Monoclinic. *Point Group:* $2/m$ Anhedral grains, from a few μm to 3 mm.

Physical Properties: Hardness = n.d. VHN = n.d. D(meas.) = n.d. D(calc.) = 4.12

Optical Properties: Opaque. *Color:* Brownish gray in polished section; synthetic brezinaite is dull gray.
R_1–R_2: n.d.

Cell Data: *Space Group:* $I2/m$ (synthetic). $a = 5.96(1)$ $b = 3.425(5)$ $c = 11.270(15)$ $\beta = 91.54(3)°$ Z = 2

X-ray Powder Pattern: Tucson iron meteorite.
2.644 (100), 5.67 (70), 2.056 (70), 1.716 (70), 2.978 (65), 2.606 (60), 5.23 (40)

Chemistry:

	(1)	(2)
Cr	48.3	54.88
Fe	3.9	
V	1.61	
Ti	0.96	
Mn	0.86	
Ni	0.08	
S	45.0	45.12
Total	100.71	100.00

(1) Tucson iron meteorite; by electron microprobe, average of 26 grains, corresponding to $(Cr_{2.65}Fe_{0.20}V_{0.09}Ti_{0.06}Mn_{0.04})_{\Sigma=3.04}S_4$. (2) Cr_3S_4.

Occurrence: In the metal matrix and contiguous to silicate inclusions (Tucson iron meteorite).

Association: Forsterite, enstatite, aluminous diopside, anorthite, feldspathic glass, kamacite, taenite, schreibersite (Tucson); troilite, carlsbergite, daubréelite (New Baltimore).

Distribution: In the Tucson and New Baltimore iron meteorites.

Name: For Aristides Brezina (1848–1909), past Director of the Mineralogy-Petrology Section of the Natural History Museum, Vienna, Austria.

Type Material: Meteorite Collection, National Museum of Natural History, Washington, D.C., USA.

References: (1) Bunch, T.E. and L.H. Fuchs (1969) A new mineral: brezinaite, Cr_3S_4, and the Tucson meteorite. Amer. Mineral., 54, 1509–1518. (2) Jellinek, F. (1957) The structure of the chromium sulfides. Acta Cryst., 10. 620. (3) Buchwald, V.F. (1977) The mineralogy of iron meteorites. Phil. Trans. Royal Soc. London, A. 286, 453–491.

Crystal Data: Tetragonal. *Point Group:* $4mm$ or $\overline{4}m2$. Small (0.1–2 mm) grains embedded, interstitial, and in networks in other sulfides. *Twinning:* Commonly shows polysynthetic twinning.

Physical Properties: Hardness = 4.5 VHN = n.d. D(meas.) = n.d. D(calc.) = 4.337

Optical Properties: Opaque. *Color:* Gray to gray-blue in reflected light. *Anisotropism:* Weak in air, distinct in oil.
R: (400) 23.4, (420) 24.2, (440) 25.0, (460) 25.7, (480) 26.4, (500) 27.0, (520) 27.4, (540) 27.6, (560) 27.5, (580) 27.2, (600) 26.8, (620) 26.0, (640) 25.0, (660) 23.8, (680) 22.5, (700) 21.2

Cell Data: *Space Group:* $I4_1md$ or $I\overline{4}d2$. $a = 5.32$ $c = 10.51$ Z = 2

X-ray Powder Pattern: Kipushi, Zaire.
3.06 (100), 1.888 (50), 1.871 (50), 1.608 (50), 1.591 (50), 2.67 (30), 1.533 (30)

Chemistry:

	(1)	(2)
Cu	33.1	32.9
Fe	9.5	2.6
Zn	6.9	12.2
Ge	16.0	13.7
Ga		2.2
Sn	0.5	
S	32.4	36.1
Total	98.4	99.7

(1) Kipushi, Zaire; by electron microprobe, corresponding to $Cu_{2.04}(Fe,Zn)_{1.04}(Ge,Ga)_{0.88}S_{4.04}$.
(2) Tsumeb, Namibia; by electron microprobe, corresponding to $Cu_{1.97}(Fe,Zn)_{0.91}(Ge,Ga)_{0.83}S_{4.29}$.

Occurrence: As inclusions in other Ge-Ga-bearing sulfides.

Association: Chalcopyrite, tennantite, reniérite, germanite, galena, sphalerite.

Distribution: From Kipushi, Shaba Province, Zaire. At Tsumeb, Namibia. In the Radka deposit, Pazardzhik, Bulgaria.

Name: For Gaston Briart, who studied the Kipushi deposit.

Type Material: University of Louvain, Belgium; National School of Mines, Paris, France.

References: (1) Francotte, J., J. Moreau, R. Ottenburgs, and C. Lévy (1965) La briartite, $Cu_2(Fe,Zn)GeS_4$, une nouvelle espèce minérale. Bull. Soc. fr. Minéral., 88, 432–437 (in French). (2) (1966) Amer. Mineral., 51, 1816 (abs. ref. 1). (3) Wintenberger, M. (1979) Etude de la structure cristallographique et magnetique de Cu_2FeGeS_4 et remarque sur la structure magnetique de Cu_2MnSnS_4. Mat. Res. Bull., 14, 1195–1202 (in French with English abs.).

Crystal Data: Tetragonal. *Point Group:* $\bar{4}2m$, $4mm$, or $4/m\ 2/m\ 2/m$. Minute crystals, up to 2 mm, and crystalline, imbedded in other selenides; massive.

Physical Properties: *Cleavage:* {001} good, {100} imperfect. Hardness = 2 VHN = 64 (20 g load). D(meas.) = n.d. D(calc.) = 7.40

Optical Properties: Opaque. *Color:* Gray-brown. *Luster:* Metallic. *Pleochroism:* Weak, creamy gray to gray.
R_1–R_2: (400) 20.0–28.1, (420) 23.0–28.9, (440) 26.0–29.7, (460) 27.3–30.0, (480) 27.9–30.2, (500) 28.2–30.1, (520) 28.0–29.8, (540) 27.7–29.4, (560) 27.2–28.8, (580) 26.5–28.2, (600) 25.7–27.6, (620) 25.0–27.0, (640) 24.2–26.4, (660) 23.3–25.8, (680) 22.5–25.2, (700) 21.9–24.6

Cell Data: *Space Group:* $I\bar{4}2m$, $I4mm$ or $I4/mmm$. $a = 3.976(5)$ $c = 13.70(2)$ Z = 1

X-ray Powder Pattern: Bukov, Czechoslovakia.
2.998 (100), 2.600 (90), 1.771 (80), 2.255 (70), 1.987 (70), 1.656 (60), 3.428 (50)

Chemistry:

	(1)	(2)
Tl	42.3	42.1
Cu	21.8	20.3
Fe	6.0	6.0
Se	30.7	31.0
Total	100.8	99.4

(1) Předbořice, Czechoslovakia; by electron microprobe. (2) Bukov, Czechoslovakia; by electron microprobe.

Occurrence: In calcite veins of hydrothermal origin, in metamorphic rocks.

Association: Clausthalite, eskebornite, eucairite, umangite, berzelianite, klockmannite, sabatierite, crookesite, ferroselite, pyrite, marcasite, chalcopyrite, chalcocite, bornite, uraninite, hematite, goethite, calcite, dolomite, quartz.

Distribution: From Bukov, Petrovice, and Předbořice, Czechoslovakia.

Name: For its occurrence at Bukov, Czechoslovakia.

Type Material: National School of Mines, Paris, France; Charles University, Prague, Czechoslovakia.

References: (1) Johan, Z. and M. Kvaček (1971) La bukovite, $Cu_{3+x}Tl_2FeSe_{4-x}$, une nouvelle espèce minérale. Bull. Soc. fr. Minéral., 94, 529–533 (in French with English abs.). (2) (1972) Amer. Mineral., 57, 1910 (abs. ref. 1). (3) Zemann, J. (1974) Structure type of bukovite. Anz. Oesterr. Akad. Wiss. Math.–Naturwiss. Kl., 110, 126–129 (in German). (4) (1976) Chem. Abs., 84, 20303 (abs. ref. 3). (5) Makovicky, E., Z. Johan, and S. Karup-Møeller (1980) New data on bukovite, thalcusite, chalcothallite and rohaite. Neues Jahrb. Mineral., Abh., 138, 122-146.

Crystal Data: Orthorhombic. *Point Group:* $2/m\ 2/m\ 2/m$. Prismatic, crystals up to 3 mm, and platy elongated grains; partly in radial or polycrystalline aggregates. *Twinning:* Frequent, with twin plane ‖ [001], possibly {110}; lamellar.

Physical Properties: *Cleavage:* Tabular (sic); ‖ [100]. Hardness = Polishing hardness less than chalcopyrite, greater than bismuth. VHN = 126–134 D(meas.) = > 6.2 D(calc.) = n.d.

Optical Properties: Opaque. *Color:* Gray; white in polished section. *Luster:* Metallic. *Pleochroism:* Weak, white with blue tint to brown to light gray. *Anisotropism:* Strong, in gray-bluish and yellowish colors.
R_1–R_2: n.d.

Cell Data: *Space Group:* *Bbmm.* $a = 13.399(20)$ $b = 20.505(10)$ $c = 4.117(5)$ Z = n.d.

X-ray Powder Pattern: Uludağ, Turkey.
2.048 (100), 2.998 (78), 3.670 (67), 3.484 (67), 2.696 (67), 2.794 (44), 3.127 (33)

Chemistry:

	(1)	(2)	(3)
Pb	45.0	45.03	46.57
Ag	1.0	1.41	
Bi	38.5	37.43	37.58
Sb		0.61	
S	14.7	15.53	15.85
Total	99.2	100.01	100.00

(1) Uludağ, Turkey; by electron microprobe, corresponding to $(Pb_{4.72}Ag_{0.20})_{\Sigma=4.92}Bi_{4.00}S_{10.82}$. (2) Shumilovsk deposit, USSR; by electron microprobe, leading to $(Pb_{4.83}Ag_{0.29})_{\Sigma=5.12}(Bi_{3.99}Sb_{0.11})_{\Sigma=4.10}S_{10.78}$. (3) $Pb_5Bi_4S_{11}$.

Occurrence: In a contact metamorphic scheelite deposit (Uludağ, Turkey); in sulfide veinlets around a greisen Sn-W deposit in granite (Shumilovsk deposit, USSR).

Association: Sphalerite, tremolite, pyrite, calcite, garnet, quartz, scheelite, chalcopyrite, bismuth (Uludağ, Turkey); kobellite, tintinaite (Boliden, Sweden); wolframite, cassiterite, bismuthinite, galena, cosalite, heyrovskýite, galenobismutite, cannizzarite, tetradymite, joséite-B, bismuth, quartz (Shumilovsk deposit, USSR).

Distribution: From Uludağ, Bursa Province, Turkey. At Hůrky, Czechoslovakia. From deposits in the Apuseni Mountains, western Romania. At Boliden, Sweden. In the Shumilovsk Sn-W deposit, western Transbaikal, and from the Strezhana pyrite-polymetallic deposit, locality not otherwise defined, USSR. At Organ, Dona Ana County, New Mexico, USA.

Name: For the locality in Bursa Province, Turkey.

Type Material: n.d.

References: (1) Tolun, R. (1954–1955) A study on the concentration tests and beneficiation of the Uludağ tungsten ore. Bull. Mineral Research and Exploration Inst. Turkey, Foreign Ed., 46–47, 106–127 (in English). (2) (1956) Amer. Mineral., 41, 671 (abs. ref. 1). (3) Klominsky, J. M. Rieder, C. Kieft, and L. Mraz (1971) Heyrovskýite, $6(Pb_{0.86}Bi_{0.08}(Ag,Cu)_{0.04})S \bullet Bi_2S_3$, from Hůrky, Czechoslovakia, a new mineral of genetic interest. Mineralium Deposita, 6, 133–147. (4) (1972) Amer. Mineral., 57, 328–329 (abs. ref. 3). (5) Pokrovskaya, I.V., E.M. Muratov, P.S. Bernshtein, and A.P. Slyusarev (1981) Antimonial bursaite and its paragenesis in the Strezhana deposit. Izv. Akad. Nauk Kaz. SSR, Ser. Geol. 68–76 (in Russian). (6) (1982) Chem. Abs., 96, 22497 (abs. ref. 5). (7) Mozgova, N.N., N.I. Organova, Y.S. Borodaev. E.G. Ryabeva, A.V. Sivtsov, T.I. Getmanskaya, and O.V. Kuzmina (1988) New data on cannizzarite and bursaite. Neues Jahrb. Mineral., Abh., 158, 293–309.

Crystal Data: Orthorhombic. *Point Group:* $2/m\ 2/m\ 2/m$. As grains up to 200 μm in size, in massive mooihoekite. *Twinning:* Polysynthetic.

Physical Properties: Hardness = n.d. VHN = 258–282, 272 average (50 g load). D(meas.) = 11.1 D(calc.) = 10.7

Optical Properties: Opaque. *Color:* Pink, with slight lilac tinge. *Luster:* Metallic. *Pleochroism:* Detectable. *Anisotropism:* Strong, in grayish brown to golden colors.
R_1–R_2: (400) — , (420) — , (440) 42.5–44.5, (460) 42.4–45.1, (480) 42.5–46.5, (500) 43.2–48.3, (520) 44.0–50.2, (540) 45.0–52.0, (560) 46.0–54.0, (580) 47.4–56.0, (600) 49.0–58.2, (620) 51.0–60.4, (640) 53.3–63.0, (660) 56.0–65.5, (680) 58.3–68.0, (700) 61.1–70.

Cell Data: *Space Group:* *Pmmm.* $a = 7.88(5)$ $b = 7.88(5)$ $c = 3.94(2)$ Z = 4

X-ray Powder Pattern: Oktyabr mine, USSR.
2.29 (10), 2.17 (9), 1.230 (8), 1.217 (4), 1.840 (3b), 1.434 (3b), 1.182 (3)

Chemistry:

	(1)	(2)	(3)	(4)
Pd	52.1	51.0	49.5	53.87
Pt	2.6	1.0	7.1	
Sn	30.0	30.0	29.2	30.05
Cu	16.2	16.0	15.2	16.08
Ag		2.0		
Sb		1.0		
Total	100.9	101.0	101.0	100.00

(1–3) Oktyabr mine, USSR; by electron microprobe. (4) Pd_2SnCu.

Occurrence: In massive mooihoekite ore, also in galena-chalcopyrite veins.

Association: Mooihoekite, sperrylite, putoranite, paolovite, talnakhite, sobolevskite, polarite, froodite, cassiterite.

Distribution: At the Oktyabr mine, Noril'sk region, western Siberia, USSR.

Name: For Dr. Louis J. Cabri, Canadian Institute of Mining and Metallurgy, describer of a number of platinum group minerals.

Type Material: A.E. Fersman Mineralogical Museum, Academy of Sciences, Moscow, USSR; Canadian Geological Survey, Ottawa, Canada.

References: (1) Evstigneeva, T.L. and A.D. Genkin (1983) Cabriite, Pd_2SnCu, a new mineral species in the mineral group of palladium, tin and copper compounds. Can. Mineral., 21, 481–487. (2) (1984) Amer. Mineral., 69, 1190 (abs. ref. 1).

Crystal Data: Hexagonal. *Point Group:* $6/m\ 2/m\ 2/m$. As smooth flattened grains up to 0.2 mm.

Physical Properties: *Tenacity:* Malleable. Hardness = n.d. VHN = n.d. D(meas.) = 8.65 (synthetic). D(calc.) = [8.65] Diamagnetic.

Optical Properties: Opaque. *Color:* Tin-white with a bluish tint. *Luster:* Metallic. R_1–R_2: n.d.

Cell Data: *Space Group:* $P6_3/mmc$. $a = 2.979$ $c = 5.617$ Z = 2

X-ray Powder Pattern: Synthetic.
2.345 (100), 2.809 (65), 2.580 (32), 1.901 (32), 1.516 (26), 1.490 (19), 1.316 (17)

Chemistry:

	(1)	(2)
Cd	99 – 100	95.70 – 96.73
Total		

(1) Ust'-Khannin intrusive, USSR; by electron microprobe. (2) Verkhoyan'ya, USSR.

Occurrence: Found in the heavy, non-magnetic fraction of a mechanical concentrate from a gabbro intrusive (Ust'-Khannin intrusive, USSR); as a product of post-magmatic activity in mineralized aleurolites, sandstones, dolomites, and mudstones (Verkhoyan'ya, USSR).

Association: Moissanite, iron, copper, lead, tin, zinc, Cu–Zn alloy, Sn–Sb alloy, sulfides, garnet, spinel, kyanite, corundum, rutile (Ust'-Khannin intrusive, USSR); titanite, ilmenite, chalcopyrite, bornite, chalcocite, pyrite, galena, monteponite, otavite (Verkhoyan'ya, USSR).

Distribution: From the Ust'-Khannin intrusive, Vilyui River basin, eastern Siberian platform; and in southern Verkhoyan'ya, USSR.

Name: From the Greek for *calamine*, as the element occurs in slags resulting from smelting smithsonite (formerly calamine) ore.

Type Material: n.d.

References: (1) Oleinikov, B.V., A.V. Okrugin and N.V. Leskova (1979) Native cadmium in traps of the Siberian Platform. Doklady Acad. Nauk SSSR, 248, 1426–1428 (in Russian). (2) (1980) Amer. Mineral., 65, 1065 (abs. ref. 1). (3) Novgorodova, M.I., D.A. Zhivtsov, A.I. Gorshkov, N.V. Trubkin, and A.I. Tsepin (1982) Native cadmium from southern Verkhoyan'ya. Zap. Vses. Mineral. Obshch., 111, 304–315 (in Russian). (4) (1982) Chem. Abs., 97, 112586 (abs. ref. 3). (5) (1954) NBS Circ. 539, 3, 10.

Crystal Data: Hexagonal. *Point Group:* 6*mm*. Fine xenomorphic disseminations cementing sandstone. Crystals, up to 0.1 mm, show the base and horizontally striated hexagonal pyramid.

Physical Properties: *Cleavage:* Perfect, apparently prismatic. *Tenacity:* Brittle. Hardness = 4 VHN = 203–222 D(meas.) = 5.47 D(calc.) = 5.807

Optical Properties: Opaque. *Color:* Black; light gray in reflected light, slightly brown in oil, with brownish internal reflections. *Streak:* Black. *Luster:* Resinous to adamantine. *Anisotropism:* Weak.

R_1–R_2: n.d.

Cell Data: *Space Group:* $P6_3mc$. $a = 4.271$ $c = 6.969$ Z = 2

X-ray Powder Pattern: Tuva ASSR, USSR.
2.13 (10), 1.816 (8), 3.67 (7), 1.96 (7), 1.196 (7), 1.026 (7), 1.433 (6)

Chemistry:

	(1)	(2)
Cd	49.37	58.74
Zn	2.83	
Fe	1.95	
Se	37.50	41.26
S	[8.35]	
Total	100.00	100.00

(1) Tuva ASSR, USSR; Fe attributed to ferroselite, S by difference; total adjusted to 100% after deduction of 4.8% insol. (2) CdSe.

Occurrence: In sedimentary strata, under reducing secondary conditions of medium to high alkalinity.

Association: Ferroselite, clausthalite, cadmian sphalerite, selenium, greenockite, pyrite, laumontite, calcite.

Distribution: From the Givetian sedimentary formations, locality not otherwise defined, Tuva ASSR, USSR.

Name: For the presence of cadmium and selenium.

Type Material: n.d.

References: (1) Bur'yanova, E.Z., G.A. Kovalev, and A.I. Komkov (1957) The new mineral cadmoselite. Zap. Vses. Mineral. Obshch., 86, 626–628 (in Russian). (2) (1958) Amer. Mineral., 43, 623 (abs. ref. 1). (3) (1957) NBS Circ. 539, 7, 12. (4) Vlasov, K.A., Ed. (1966) Mineralogy of rare elements, v. II, 657–659. (5) Sindeeva, N.D. (1964) Mineralogy and types of deposits of selenium and tellurium, 60–61.

Crystal Data: Monoclinic. *Point Group:* $2/m$ or 2. Bladed and short to slender prisms elongated ∥ [010], striated ∥ [010]; also massive, granular. *Twinning:* Commonly on {110}, less frequently on {031} and {111}.

Physical Properties: *Fracture:* Uneven to subconchoidal. *Tenacity:* Brittle.
Hardness = 2.5–3 VHN = 197–213 (100 g load). D(meas.) = 9.10–9.40 D(calc.) = 9.31

Optical Properties: Opaque. *Color:* Grass-yellow to silver-white; white in reflected light. *Streak:* Greenish to yellowish gray. *Luster:* Metallic. *Pleochroism:* Weak. *Anisotropism:* Weak.

R_1–R_2: (400) 49.5–55.6, (420) 51.5–57.5, (440) 53.5–59.4, (460) 55.5–61.0, (480) 57.3–62.5, (500) 58.8–63.8, (520) 60.1–64.9, (540) 61.3–65.9, (560) 62.3–66.8, (580) 63.1–67.6, (600) 63.9–68.3, (620) 64.4–68.9, (640) 65.1–69.5, (660) 65.6–70.0, (680) 66.0–70.4, (700) 66.5–70.9

Cell Data: *Space Group:* $C2/m$ or $C2$. $a = 8.76(1)$ $b = 4.410(5)$ $c = 10.15(1)$ $\beta = 125.2(2)°$ Z = 4

X-ray Powder Pattern: Cripple Creek, Colorado, USA.
3.02 (10), 2.09 (8), 2.20 (4), 2.93 (3), 1.758 (3), 1.689 (3), 1.506 (3)

Chemistry:

	(1)	(2)	(3)
Au	41.66	42.15	43.59
Ag	0.77	0.60	
Te	57.87	57.00	56.41
Total	100.30	99.75	100.00

(1) Cripple Creek, Colorado, USA. (2) Kalgoorlie, Western Australia. (3) $AuTe_2$.

Occurrence: Usually in veins in low-temperature hydrothermal deposits, but also in moderate and high-temperature deposits.

Association: Altaite, coloradoite, krennerite, rickardite, other tellurides, pyrite, arsenopyrite, tetrahedrite, tennantite, sphalerite, stibnite, other sulfides.

Distribution: In the USA, in California, at the Morgan, Melones and Stanislaus mines, Carson Hill district, Calaveras Co.; at the Spotted Horse mine, Marden, Montana; in Colorado, in the Cripple Creek district, Teller Co.; the Central City district, Gilpin Co.; Gold Hill, Boulder Co.; and the Bessie G and Mayday mines, La Plata Co. In several mines in the Kirkland Lake area, and in the Hemlo gold deposit, Thunder Bay district, Ontario; and the Robb-Montbray mine, Quebec, Canada. From the Lake View and North Kalgoorlie mines, Kalgoorlie, Western Australia. From Klyuchi, eastern Siberia, USSR. At Nishizaki, Gifu Prefecture, and in the Date mine, Hokkaido, Japan. In the Emperor mine, Vatukoula, Viti Levu, Fiji Islands.

Name: For its occurrence in Calaveras Co., California, USA.

Type Material: n.d.

References: (1) Palache, C., H. Berman, and C. Frondel (1944) Dana's system of mineralogy, (7th edition), v. I, 335–338. (2) Berry, L.G. and R.M. Thompson (1962) X-ray powder data for the ore minerals. Geol. Soc. Amer. Mem. 85, 112. (3) Sindeeva, N.D. (1964) Mineralogy and types of deposits of selenium and tellurium, 107–110. (4) Pertlik, F. (1984) Kristallchemie natürlicher Telluride. III: Die Kristallstruktur des Minerals Calaverite, $AuTe_2$. Zeits. Krist., 169, 227–236 (in German with English abs.).

Crystal Data: Tetragonal. *Point Group:* $4/m\ 2/m\ 2/m$, $4mm$, or $\bar{4}2m$. As anhedral equidimensional grains up to about 200 μm in diameter.

Physical Properties: *Fracture:* Subconchoidal. *Tenacity:* Brittle. Hardness = 3.5–4 (calculated). VHN = 150–173, 163 average (100 g load). D(meas.) = n.d. D(calc.) = 7.144

Optical Properties: Opaque. *Color:* Gray. *Streak:* Black. *Luster:* Metallic.
Pleochroism: Slight, pale gray to pale brownish gray. *Anisotropism:* Distinct, medium gray to slate-gray to brownish gray.
R_1–R_2: (400) 23.6–28.4, (420) 25.6–30.1, (440) 27.4–31.4, (460) 28.9–32.5, (480) 30.1–33.1, (500) 31.0–33.6, (520) 31.6–33.7, (540) 32.1–33.8, (560) 32.5–33.7, (580) 32.8–33.6, (600) 33.0–33.4, (620) 33.1–33.2, (640) 33.3–33.0, (660) 33.4–32.7, (680) 33.5–32.6, (700) 33.6–32.3

Cell Data: *Space Group:* $P4_2/mmc$, $P4_2mc$, or $P\bar{4}2c$. a = 12.695(2) c = 42.186(6) Z = 16

X-ray Powder Pattern: Good Hope mine, Colorado, USA.
3.45 (100), 2.118 (100), 1.804 (60), 1.377(40), 1.222 (40), 1.815 (30), 1.151 (30)

Chemistry:

	(1)	(2)
Ag	6.29	6.38
Cu	24.45	24.44
Te	68.27	69.94
Total	99.01	100.76

(1–2) Good Hope mine, Colorado, USA; by electron microprobe, average corresponding to $Ag_{1.09}Cu_{7.10}Te_{10}$.

Occurrence: One of a number of metallic tellurium-bearing minerals in a hydrothermal deposit.

Association: Tellurium, rickardite, vulcanite, arsenopyrite, pyrite.

Distribution: From the Good Hope mine, Vulcan, Gunnison Co., Colorado, USA.

Name: For Professor Eugene N. Cameron (1910–), University of Wisconsin, who first recognized the mineral as a new species.

Type Material: National Museum of Canada, Ottawa, Canada, 64956; British Museum (Natural History), London, England, 1984, 356, E.1000, and R933.

References: (1) Roberts, A.C., D.C. Harris, A.J. Criddle, and W.W. Pinch (1986) Cameronite, a new copper-silver telluride from the Good Hope mine, Vulcan, Colorado. Can. Mineral., 24, 379–384. (2) (1987) Amer. Mineral., 72, 1023 (abs. ref. 1).

Crystal Data: Orthorhombic, pseudocubic. *Point Group:* $mm2$. Crystals to 1 cm, with combinations of pseudo-octahedron and dodecahedron; also botryoidal with drusy surface; as small inclusions. *Twinning:* Commonly on {111}, as pseudospinel law twins, or repeated interpenetration twins of pseudododecahedra.

Physical Properties: *Fracture:* Conchoidal to uneven. *Tenacity:* Brittle. Hardness = 2.5 VHN = 90.7–171 (25 g load). D(meas.) = 6.2–6.3 D(calc.) = 6.311

Optical Properties: Opaque. *Color:* Steel-gray with reddish tinge, tarnishes black with blue to purple tint; gray-white with violet tint in reflected light. *Streak:* Gray-black, somewhat shining. *Luster:* Metallic.
R: (400) 29.7, (420) 28.6, (440) 27.5, (460) 26.7, (480) 26.1, (500) 25.6, (520) 25.3, (540) 25.1, (560) 24.9, (580) 24.8, (600) 24.9, (620) 25.0, (640) 25.1, (660) 25.2, (680) 25.2, (700) 25.2

Cell Data: *Space Group:* $Pna2_1$ or $Pnam$. $a = 15.298(2)$ $b = 7.548(1)$ $c = 10.699(1)$ Z = 4

X-ray Powder Pattern: Synthetic.
3.114 (100), 3.085 (90), 1.8960 (90), 2.727 (70), 1.9129 (70), 3.252 (55), 3.232 (55)

Chemistry:

	(1)	(2)	(3)
Ag	74.10	65.12	73.49
Fe	0.21		
Sn	6.94	10.57	10.14
Ge	1.82		
Te		8.69	
S	16.22	13.95	16.37
Total	99.29	98.33	100.00

(1) Aullagas, Bolivia. (2) Revelstoke, Canada; by electron microprobe. (3) Ag_8SnS_6.

Polymorphism & Series: Forms a series with argyrodite; stable below 172°C.

Occurrence: In polymetallic veins, formed very late in the paragenetic sequence.

Association: Argyrodite, pyrargyrite, stephanite, acanthite, polybasite, freibergite, stannite, stannoidite, cassiterite, arsenopyrite, marcasite, pyrrhotite, sphalerite, galena.

Distribution: In the Himmelsfürst mine, Erbisdorf, near Freiberg, Saxony, Germany. From Rejská, near Kutná Hora, and Příbram, Czechoslovakia. Exceptional crystals from the Gallofa vein, Aullagas, near Colquechaca, Potosí; and at Guadaloupe, Chocaya, Bolivia. In Canada, from about 33 km northeast of Revelstoke, British Columbia. From the Campbell mine, Bisbee, Cochise Co., Arizona, and the Leadville district, Lake Co., Colorado, USA. In the Ikuno and Omidani mines, Hyogo Prefecture and the Ashio mine, Honshu, Japan. From the Karamken Au-Ag deposit in the Okhotsk-Chukotka volcanic belt, and the Belukhinsk tungsten deposit, eastern Transbaikalia, USSR. Other minor occurrences are known.

Name: For Frederick Alexander Canfield (1849–1926), mining engineer and mineral collector of Dover, New Jersey, USA.

References: (1) Palache, C., H. Berman, and C. Frondel (1944) Dana's system of mineralogy, (7th edition), v. I, 356–358. (2) Harris, D.C. and D.R. Owens (1971) A tellurium-bearing canfieldite, from Revelstoke, B.C. Can. Mineral., 10, 895–898. (3) Wang, N. (1978) New data for Ag_8SnS_6 (canfieldite) and Ag_8GeS_6 (argyrodite). Neues Jahrb. Mineral., Monatsh., 269–272. (4) Soeda, A., M. Watanabe, K. Hoshino, and K. Nakashima (1984) Mineralogy of tellurium-bearing canfieldite from the Tsumo mine, SW Japan and its implications for ore genesis. Neues Jahrb. Mineral., Abh., 150, 11–23. (5) Sugaki, A., A. Kitakaze, and H. Kitazawa (1985) Synthesized tin and tin-sulfide minerals; Synthetic sulfide minerals (XIII). Sci. Rep., Tohoku Univ., Ser. 3, 16, 199–211 (in English).

Crystal Data: Monoclinic. *Point Group:* $2/m$. As very thin single laths, straight, or often warped, and as felted masses and stellate groups; crystals seldom exceed 2 mm x 0.5 mm in size. *Twinning:* Stellate trillings and as simple V-shapes.

Physical Properties: *Tenacity:* Somewhat malleable. Hardness = ~2 VHN = 132 (20 g load). D(meas.) = 6.7 D(calc.) = 6.95

Optical Properties: Opaque. *Color:* White to silvery gray, tarnishes iridescent. *Luster:* Metallic. *Anisotropism:* Strong, from blue to gray.
R_1–R_2: (400) 49.2–51.3, (420) 49.8–52.4, (440) 50.4–53.5, (460) 50.7–53.7, (480) 50.6–53.3, (500) 50.2–52.8, (520) 49.7–52.2, (540) 49.2–51.6, (560) 48.6–51.0, (580) 48.2–50.4, (600) 47.8–50.0, (620) 47.5–49.5, (640) 47.3–49.2, (660) 47.2–49.0, (680) 47.2–49.0, (700) 47.2–49.2

Cell Data: *Space Group:* $P2_1/m$. $a = 4.13$ $b = 4.09$ $c = 15.48$ $\beta = 98.56°$ Z = 1

X-ray Powder Pattern: Vulcano, Lipari Islands.
3.82 (10), 3.01 (6), 2.68 (6), 2.87 (5), 2.23 (5), 2.03 (5), 1.910 (4)

Chemistry:

	(1)	(2)	(3)	(4)
Pb	36.2	31.8	35.65	33.16
Ag		1.5	0.45	
Bi	44.6	49.0	46.92	50.16
Sb	< 0.1	0.7	0.61	
Se	4.6	0.4	0.00	
Te		0.1		
S	14.1	16.9	16.00	16.68
Total	99.5	100.4	99.63	100.00

(1) Vulcano, Italy; by electron microprobe, average of analyses on eight grains. (2) Vysokogorsk deposit, USSR; by electron microprobe. (3) Shumilovsk deposit, USSR; by electron microprobe, average of three analyses. (4) $Pb_4Bi_6S_{13}$.

Occurrence: Associated with deep fumarolic activity (Vulcano, Lipari Islands); in sulfide veinlets in a greisen Sn-W deposit in granite (Shumilovsk deposit, USSR).

Association: Lillianite, galenobismutite, galena (Vulcano, Lipari Islands); galena, pyrite (Santa Maria, Switzerland); wolframite, cassiterite, bismuthinite, galena, cosalite, heyrovskýite, galenobismutite, bursaite, tetradymite, joséite-B, bismuth, quartz (Shumilovsk deposit, USSR).

Distribution: From Vulcano, in the Lipari Islands, Italy. In the Shumilovsk Sn-W deposit, west Transbaikal, and in the Vysokogorsk deposit, Far Eastern Region, USSR. At Landsman Camp, Graham Co., Arizona, USA. From Santa Maria, Val Medel, Grisons, and at Goppenstein, Lotschental, Valais, Switzerland.

Name: For Stanislao Cannizzaro (1826–1910), celebrated chemist, University of Rome, Italy.

Type Material: n.d.

References: (1) Graham, A.R., R.M. Thompson, and L.G. Berry (1953) Studies of mineral sulfo-salts; XVII—cannizzarite. Amer. Mineral., 38, 536–544. (2) Litavrina, P.F., I.M. Romanenko, and V.M. Chubarov (1978) Cannizzarite ($Pb_4Bi_6S_{13}$) from the Vysokogorsk deposit— first occurrence in USSR. Doklady Acad. Nauk SSSR, 239, 1207–1210 (in Russian). (3) (1979) Amer. Mineral., 64, 244 (abs. ref. 2). (4) Matzat, E. (1979) Cannizzarite. Acta Cryst., B35, 133–136. (5) Mozgova, N.N., O.V. Kuzmina, N.I. Organova, I.P. Laputina, Y.S. Borodaev, and M. Fornaseri (1985) New data on sulphosalt assemblages at Vulcano (Italy). Rend. Soc. Ital. Mineral. Petrol., 40, 277-283. (6) (1987) Amer. Mineral., 72, 229 (abs. ref. 5). (7) Mozgova, N.N., N.I. Organova, Y.S. Borodaev, E.G. Ryabeva, A.V. Sivtsov, T.I. Getmanskaya, and O.V. Kuzmina (1988) New data on cannizzarite and bursaite. Neues Jahrb. Mineral., Abh., 158, 293–309.

Crystal Data: Hexagonal. *Point Group:* $\bar{3}$. Small, 0.005 mm to 0.5 mm, mostly anhedral grains. A few grains show poorly defined rhombohedral and tabular forms.

Physical Properties: *Cleavage:* {0001} perfect and an imperfect prismatic cleavage. *Fracture:* Hackly. Hardness = ~1 VHN = 23.5 (50 g load). D(meas.) = 8.1 D(calc.) = 8.55

Optical Properties: Opaque. *Color:* Dark gray, oxidized surfaces become dull and darker; in reflected light, white with faint bluish cast. *Streak:* Dark gray to black. *Luster:* Bright metallic. *Pleochroism:* Brownish gray to bluish gray. *Anisotropism:* Moderate to distinct.
R_1–R_2: (470) 39.6–41.3, (546) 38.8–40.1, (589) 39.2–40.6, (650) 40.5–42.0

Cell Data: *Space Group:* $R\bar{3}$. $a = 12.12$ $c = 18.175$ Z = 27

X-ray Powder Pattern: Carlin mine, Nevada, USA.
3.030 (100), 2.290 (3), 2.020 (3), 1.749 (3), 2.712 (1), 1.681 (1), 1.513 (1)

Chemistry:

	(1)	(2)
Tl	92.93	92.73
S	7.17	7.27
Total	100.09	100.00

(1) Carlin mine, Nevada, USA; average analysis of three samples. (2) Tl_2S.

Occurrence: As small grains in brecciated fragments of carbonaceous limestone, as a result of epithermal mineralization.

Association: Gold, arsenic, antimony, mercury, organic carbon, quartz.

Distribution: From the east pit of the Carlin mine, Eureka Co., Nevada, USA.

Name: For the Carlin gold deposit, Nevada, USA.

Type Material: Department of Geology, Stanford University, Palo Alto, California, Epithermal Minerals Collection; National Museum of Natural History, Washington, D.C., USA, 132497.

References: (1) Radtke, A.S. and F.W. Dickson (1975) Carlinite, Tl_2S, a new mineral from Nevada. Amer. Mineral., 60, 559–565.

Crystal Data: Cubic. *Point Group:* $4/m\ \overline{3}\ 2/m$. Octahedral crystals to 2 cm; commonly massive, granular to compact. *Twinning:* {111}; may show polysynthetic twin lamellae.

Physical Properties: *Cleavage:* {001} imperfect. *Fracture:* Uneven to subconchoidal. Hardness = 4.5–5.5 VHN = 507–586 (100 g load). D(meas.) = 4.5–4.8 D(calc.) = 4.83

Optical Properties: Opaque. *Color:* Light gray to steel-gray, tarnishing easily to copper-red or violet-gray. *Luster:* Metallic.
R: (400) 40.5, (420) 40.6, (440) 40.7, (460) 40.7, (480) 40.8, (500) 40.9, (520) 41.2, (540) 41.8, (560) 41.8, (580) 41.9, (600) 42.1, (620) 42.7, (640) 43.3, (660) 43.9, (680) 44.6, (700) 45.0

Cell Data: *Space Group:* $Fd3m$. $a = 9.458$ Z = 8

X-ray Powder Pattern: N'Kana, Zambia.
2.86 (100), 1.674 (80), 1.825 (60), 2.37 (50), 0.994 (50), 3.35 (40), 1.234 (30)

Chemistry:

	(1)	(2)	(3)	(4)
Cu	13.90	18.98	9.98	14.2
Co	35.15	35.79	36.08	46.2
Ni	7.01	3.66	7.65	
Fe	2.18	0.93	2.25	
S	40.74	40.64	41.89	39.0
insol.	0.27		0.50	
Total	99.25	100.00	98.35	99.4

(1) Gladhammar, Sweden. (2) Siegen, Germany. (3) Mineral Hill mine, Maryland, USA. (4) Carrizal Alto district, Chile; by electron microprobe.

Occurrence: In hydrothermal vein deposits.

Association: Tetrahedrite, linnaeite, siegenite, polydymite, chalcopyrite, pyrrhotite, pyrite, millerite, gersdorffite, ullmannite, cobaltian calcite.

Distribution: In the USA, in the Patapsco mine, Finksburg, and the Mineral Hill mine, Sykesville, Carroll Co., Maryland. From near Chernomorets, Burgas, Bulgaria. At the Kohlenbach mine, near Eiserfeld, North Rhine-Westphalia, Germany. From Gladhammar, Kalmar, Sweden. At Tsumeb, Namibia. Large crystals from the Kambove and Kolwezi mines, Shaba Province, Zaire. From the Rokana mine, N'Kana, Kitwe, Zambia. In the Sazare mine, Ehime Prefecture, Shikoku Island, and the Shirataki mine, Kochi Prefecture, Japan. From the Kambalda deposit, Kalgoorlie, Western Australia. In the Carrizal Alto copper district, Atacama, Chile. A number of other minor occurrences are known.

Name: For its occurrence in Carroll Co., Maryland, USA.

Type Material: n.d.

References: (1) Palache, C., H. Berman, and C. Frondel (1944) Dana's system of mineralogy, (7th edition), v. I, 262–264. (2) Berry, L.G. and R.M. Thompson (1962) X-ray powder data for the ore minerals. Geol. Soc. Amer. Mem. 85, 77. (3) Clark, A.H. (1974) Hypogene and supergene cobalt-copper sulfides, Carrizal Alto, Atacama, Chile. Amer. Mineral., 59, 302–306.

Crystal Data: Hexagonal. *Point Group:* $\overline{3}\ 2/m$. As anhedral grains up to 1 mm. *Twinning:* The lamellar twinning observed may be the result of pressure-induced deformation.

Physical Properties: Hardness = Very soft. VHN = 17–45 (15 g load) (synthetic). D(meas.) = n.d. D(calc.) = [3.23]

Optical Properties: Opaque. *Color:* Yellow-gray to light gray in reflected light. *Luster:* Metallic. *Pleochroism:* Distinct, pale yellow to gray in air, pale yellow with a greenish tint to gray in oil.
R_1–R_2: (400) 26.7–19.3, (420) 26.2–20.1, (440) 26.3–21.9, (460) 27.4–21.5, (480) 30.0–21.7, (500) 31.6–22.0, (520) 32.3–22.0, (540) 32.8–21.8, (560) 33.0–21.7, (580) 32.5–21.3, (600) 32.2–20.8, (620) 31.4–20.5, (640) 30.7–20.5, (660) 29.4–20.1, (680) 26.7–18.1, (700) 28.8–18.0

Cell Data: *Space Group:* $R\overline{3}m$. $a = 3.55$ $c = 19.5$ Z = [3]

X-ray Powder Pattern: Norton County enstatite achondrite.
2.60 (100), 2.07 (80), 1.910 (80), 1.779 (80), 6.49 (70), 1.465 (60), 1.134 (60)

Chemistry:

	(1)	(2)	(3)
Na	15.7	15.5	16.53
Cr	37.4	37.6	37.38
Fe		0.86	
Zn		0.0	
Ca		0.13	
Mg		0.10	
Mn	0.08	0.10	
Ti	0.18	0.0	
S	46.3	46.0	46.09
Total	99.66	100.29	100.00

(1) Norton County enstatite achondrite; by electron microprobe. (2) Qingzhen enstatite chondrite; by electron microprobe, average of three analyses. (3) $NaCrS_2$.

Occurrence: As inclusions in enstatite crystals and in the brecciated matrix of a meteorite (Norton County); between coarse pyroxene grains in chondrules in a meteorite (Qingzhen).

Association: Daubréelite, titanoan troilite, ferromagnesian alabandite, oldhamite, kamacite, perryite (Norton County); troilite, kamacite, oldhamite (Qingzhen).

Distribution: In the Norton County enstatite achondrite and the Qingzhen enstatite chondrite meteorites.

Name: To honor Dr. Caswell Silver, geologist associated with the University of New Mexico, Albuquerque, New Mexico, USA.

Type Material: n.d.

References: (1) Okada, A. and K. Keil (1982) Caswellsilverite, $NaCrS_2$: a new mineral in the Norton County enstatite achondrite. Amer. Mineral., 67, 132–136. (2) Grossman, J.N., A.E. Rubin, E.R. Rambaldi, R.S. Rajan, and J.T. Wasson (1985) Chondrules in the Qingzhen type-3 enstatite chondrite: Possible precursor components and comparison to ordinary chondrite chondrules. Geochim. Cosmochim. Acta, 49, 1781–1795.

Crystal Data: Cubic. *Point Group:* $2/m\ \overline{3}$. Cubic crystals to 1 cm.; granular intergrowths with other sulfides.

Physical Properties: *Cleavage:* {001} perfect. Hardness = > 4 VHN = 1018–1114 (10 g load). D(meas.) = 4.82 D(calc.) = 4.80

Optical Properties: Opaque. *Color:* Pinkish; in polished section, white. *Luster:* Metallic.
R: (400) 35.8, (420) 35.2, (440) 34.5, (460) 34.3, (480) 34.6, (500) 35.1, (520) 35.8, (540) 36.7, (560) 37.6, (580) 38.4, (600) 39.4, (620) 40.2, (640) 41.1, (660) 41.9, (680) 42.6, (700) 43.3

Cell Data: *Space Group:* *Pa*3. $a = 5.52$ Z = 4

X-ray Powder Pattern: Shinkolobwe mine, Zaire.
2.750 (100), 2.463 (60), 1.663 (55), 1.063 (55), 2.249 (48), 1.950 (34), 1.474 (22)

Chemistry:

	(1)	(2)
Co	42.20	47.90
Ni	3.25	
Fe	2.80	
S	51.75	52.10
Total	100.00	100.00

(1) Shinkolobwe mine, Zaire; recalculated to 100 %. (2) CoS_2.

Polymorphism & Series: Forms series with pyrite and vaesite.

Occurrence: In carbonate rocks (Shinkolobwe mine, Zaire).

Association: Pyrite, chalcopyrite, other linnaeite–polydymite group minerals.

Distribution: In the Shinkolobwe mine, Shaba Province, Zaire. From Gänsberg, near Wiesloch, and Hohensachsen, Black Forest, Germany. At Bald Knob, Sparta, Alleghany Co., North Carolina, USA. From near Filipstad, Wermland, Sweden.

Name: For Felicien Cattier, former Chairman of Union Miniére du Haut Katanga, Belgium.

Type Material: n.d.

References: (1) Kerr, P.F. (1945) Cattierite and vaesite: new Co–Ni minerals from the Belgian Congo. Amer. Mineral., 30, 483–497. (2) Pratt, J.L. and P. Bayliss (1979) Crystal-structure refinement of cattierite. Zeits. Krist., 150, 163–167.

Crystal Data: Tetragonal. *Point Group:* $\overline{4}2m$. Small anhedral grains up to 200 μm, in aggregates with kësterite up to 12 cm.

Physical Properties: Hardness = ~4 VHN = 189 (Tanco mine, Canada); 210 (Hugo mine, USA) (50 g load). D(meas.) = n.d. D(calc.) = 4.776 (Tanco mine, Canada); 4.618 (Hugo mine, USA).

Optical Properties: Opaque. *Color:* Steel-gray. *Streak:* Black. *Luster:* Metallic. *Anisotropism:* Very weak in shades of gray.

R_1–R_2: (470) 24.6 and 25.5, (546) 23.4 and 25.6, (589) 22.3 and 25.0, (650) 22.7 and 24.3. (Tanco mine, Canada and Hugo mine, USA)

Cell Data: *Space Group:* $I\overline{4}2m$. $a = 5.487$ $c = 10.848$ Z = 2

X-ray Powder Pattern: Tanco mine, Canada.
3.167 (100), 1.939 (70), 1.662 (50), 1.954 (40),. 1.639 (40), 1.770 (30), 1.257 (30)

Chemistry:

	(1)	(2)	(3)
Cu	28.1	26.5	26.12
Ag		0.02	
Cd	9.3	18.2	23.11
Fe	3.6	1.1	
Zn	4.9	2.0	
Mn	0.06	n.d.	
Sn	26.3	24.9	24.40
S	28.5	26.9	26.37
Total	100.8	99.8	100.00

(1) Hugo mine, USA; by electron microprobe, corresponding to $Cu_{1.99}(Cd_{0.37}Zn_{0.33}Fe_{0.29}Mn_{0.01})_{\Sigma=1.00}Sn_{1.00}S_{4.00}$. (2) Tanco mine, Canada; by electron microprobe, corresponding to $(Cu_{1.99}Ag_{0.01})_{\Sigma=2.00}(Cd_{0.77}Zn_{0.14}Fe_{0.10})_{\Sigma=1.01}Sn_{1.00}S_{4.00}$. (3) Cu_2CdSnS_4.

Occurrence: A rare constituent of complex zoned pegmatites, as a component of a very minor sulfide mineral suite.

Association: Pyrrhotite, sphalerite, hawleyite, chalcopyrite, stannite, bismuth, kësterite.

Distribution: In the Hugo pegmatite, 1.6 km south of Keystone; and the Peerless mine, Pennington Co., South Dakota, USA. From the Tanco pegmatite at Bernic Lake, southeastern Manitoba, Canada.

Name: For Dr. Petr Černý, mineralogist at the University of Manitoba, Winnipeg, Canada.

Type Material: Royal Ontario Museum, Toronto, Canada; Museum of Geology, South Dakota School of Mines; National Museum of Natural History, Washington, D.C., USA, 136924.

References: (1) Kissin, S.A., D.R. Owens, and W.L. Roberts (1978) Černýite, a copper–cadmium–tin sulfide with the stannite structure. Can. Mineral., 16, 139–146. (2) (1979) Amer. Mineral., 64, 653 (abs. ref. 1). (3) Szymański, J.T. (1978) The crystal structure of černýite, Cu_2CdSnS_4, a cadmium analogue of stannite. Can. Mineral., 16, 147–151.

Cetineite $(K, Na)_{3+x}(Sb_2O_3)_3(SbS_3)(OH)_x \cdot 2.4H_2O$ (x ~0.5)

Crystal Data: Hexagonal. *Point Group:* 6. As tufts of acicular crystals elongated ∥ [0001], up to 0.5 mm in length and 15 μm in diameter. *Cleavage:* On {1000}.

Physical Properties: *Tenacity:* Somewhat sectile. Hardness = n.d. VHN = 127–156 (20 g load). D(meas.) = n.d. D(calc.) = 4.21

Optical Properties: Transparent to translucent. *Color:* Orange-red. *Streak:* Orange. *Luster:* Resinous.

Optical Class: Uniaxial (+). $n = > 1.74$ *Pleochroism:* Weak, from orange to slightly orange-brown. 2V(meas.) = n.d. 2V(calc.) = n.d.

R_1–R_2: n.d.

Cell Data: *Space Group:* $P6_3$. $a = 14.2513(3)$ $c = 5.5900(1)$ Z = 2

X-ray Powder Pattern: Cetine mine, Italy.
2.916 (100), 12.41 (80), 3.000 (74), 2.690 (61), 4.11 (55), 4.67 (54), 3.581 (44)

Chemistry:

	(1)
K_2O	6.66
Na_2O	3.87
Sb_2O_3	81.06
S	7.15
SiO_2	0.67
H_2O	[4.16]
O = S	–3.57
Total	100.00

(1) Cetine mine, Italy; by electron microprobe, average of two analyses, after deduction of SiO_2 giving $(K_{1.78}Na_{1.57})_{\Sigma=3.35}(Sb_2O_3)_{3.03}(SbS_3)_{0.94}(OH)_{0.53} \cdot 2.4H_2O$.

Occurrence: From an antimony deposit in highly silicified evaporites, on ore which has been roasted, then long weathered.

Association: Mopungite, senarmontite.

Distribution: In the Cetine mine, 20 km southwest of Siena, Tuscany, Italy.

Name: For the Cetine mine, Italy.

Type Material: Mineralogical Museum, Florence University, Florence, Italy, 644/RI; National Museum of Natural History, Washington, D.C., USA.

References: (1) Sabelli, C. and G. Vezzalini (1987) Cetineite, a new antimony oxide-sulfide mineral from Cetine mine, Tuscany, Italy. Neues Jahrb. Mineral., Monatsh., 419–425. (2) Sabelli, C., I. Nakai, and S. Katsura (1988) Crystal structures of cetineite and its synthetic Na analogue $Na_{3.6}(Sb_2O_3)_3(SbS_3)(OH)_{0.6} \cdot 2.4H_2O$. Amer. Mineral., 73, 398–404.

Crystal Data: Triclinic. *Point Group:* 1. Rarely in crystals larger than 1 mm; usually in intimate intergrowths with pierrotite, measured in cm.

Physical Properties: *Fracture:* Conchoidal. Hardness = n.d. VHN = 78–124, 95 average (25 g load). D(meas.) = 5.104 D(calc.) = 5.121

Optical Properties: Opaque. *Color:* Black; white in reflected light, red internal reflections noted along cracks. *Streak:* Brown-red. *Luster:* Submetallic to greasy. *Pleochroism:* Weak. *Anisotropism:* Strong, in bluish and greenish colors.

R_1–R_2: (400) 32.4–40.5, (420) 32.1–40.0, (440) 31.8–39.5, (460) 31.5–39.0, (480) 31.3–38.4, (500) 31.1–37.9, (520) 31.0–37.3, (540) 30.8–36.6, (560) 30.6–35.8, (580) 30.2–35.1, (600) 29.6–34.3, (620) 29.0–33.5, (640) 28.2–32.8, (660) 27.5–32.0, (680) 26.8–31.3, (700) 26.4–30.7

Cell Data: *Space Group:* $P1$. $a = 16.346(5)$ $b = 42.602(10)$ $c = 8.534(3)$ $\alpha = 95.86(3)°$ $\beta = 86.91(3)°$ $\gamma = 96.88(3)°$ Z = 1

X-ray Powder Pattern: Jas Roux, France.
3.573 (10), 2.135 (9), 2.808 (8), 3.928 (7), 3.358 (7), 2.853 (7), 2.709 (7)

Chemistry:

	(1)	(2)
Tl	23.87	17.88
Pb	0.00	10.94
Sb	32.92	31.61
As	17.63	14.83
S	26.05	24.82
Total	100.47	100.08

(1) Jas Roux, France; by electron microprobe. (2) Abuta, Japan; by electron microprobe.

Occurrence: In a hydrothermal deposit in dolomitic limestones with other As-Tl minerals (Jas Roux, France).

Association: Pierrotite, parapierrotite, stibnite, pyrite, sphalerite, twinnite, zinkenite, madocite, andorite, smithite, laffittite, routhierite, aktashite, wakabayashilite, realgar, orpiment (Jas Roux, France); getchellite, sphalerite, barite (Abuta, Japan).

Distribution: In France, at the Jas Roux deposit, Hautes-Alpes. In the Toya mine, Abuta, Hokkaido, Japan.

Name: For Chabournéou Glacier, near the Jas Roux deposit, France.

Type Material: National School of Mines, Paris, France.

References: (1) Johan, Z., J. Mantienne, and P. Picot (1981) La chabournéite, un nouveau minéral thallifère. Bull. Minéral., 104, 10–15 (in French with English abs.). (2) (1982) Amer. Mineral., 67, 621 (abs. ref. 1).

Crystal Data: Monoclinic, pseudo-orthorhombic. *Point Group:* $2/m$ or m. Crystals are short prismatic [001], thick to tabular {001}, to 12 cm across, and prismatic [100], to 25 cm long; {001} is striated ∥ [100]. Massive, compact, fine powdery. *Twinning:* Common on {110} yielding pseudohexagonal stellate forms; also on {032}, {112}. Seen as lamellar twinning in polished section.

Physical Properties: *Cleavage:* Indistinct on {110}. *Fracture:* Conchoidal. *Tenacity:* Brittle, somewhat sectile. Hardness = 2.5–3 VHN = 84–87 (100 g load). D(meas.) = 5.5–5.8 D(calc.) = 5.80

Optical Properties: Opaque. *Color:* Blackish lead-gray. *Streak:* Blackish lead-gray. *Luster:* Metallic. *Anisotropism:* Weak.
R: (400) 37.4, (420) 36.7, (440) 36.0, (460) 35.2, (480) 34.4, (500) 33.5, (520) 32.6, (540) 31.8, (560) 31.0, (580) 30.2, (600) 29.6, (620) 29.0, (640) 28.4, (660) 28.0, (680) 27.5, (700) 27.1

Cell Data: *Space Group:* $P2_1/c$ or Pc. $a = 11.82$ $b = 27.05$ $c = 13.43$ $\beta = 90°$ Z = 96

X-ray Powder Pattern: Bristol, Connecticut, USA.
1.8800 (100), 2.4030 (70), 1.9746 (70), 1.8811 (70), 2.4074 (50), 3.276 (35), 2.7256 (35)

Chemistry:

	(1)	(2)	(3)
Cu	79.67	79.50	79.86
Fe	0.14	0.17	
S	20.16	20.05	20.14
SiO_2	0.09	0.17	
Total	100.06	99.89	100.00

(1) Butte, Montana, USA; Fe present as pyrite. (2) New London, Maryland, USA; contains Fe as bornite. (3) Cu_2S.

Occurrence: Uncommon as a primary hydrothermal mineral but important as a secondary mineral. Found in or below the zone of oxidation in hydrothermal veins in and large low-grade porphyry copper orebodies.

Association: Pyrite, chalcopyrite, covellite, bornite, molybdenite, many other sulfides and their alteration products.

Distribution: An important and widely distributed ore mineral of copper. Only a few localities producing exceptional crystals or pure masses can be listed. In the USA, in Arizona, at Bisbee, Cochise Co.; at the Magma mine, Gila Co.; in the United Verde Extension mine, Yavapai Co. From Butte, Silver Bow Co., Montana. At Kennecott, Copper River district, Alaska. Exceptional crystals from Bristol, Hartford Co., Connecticut. In England, fine crystals from Cornwall at St. Just, St. Ives, Camborne, and Redruth. Large crystals from Nababiep West mine, Cape Province, and Messina, Transvaal, South Africa. At M'Passa, Niari Province, Congo Republic. From Turinsk, Bogoslovsk, Ural Mountains, USSR.

Name: For its composition, from the Greek for *chalkos, copper*.

References: (1) Palache, C., H. Berman, and C. Frondel (1944) Dana's system of mineralogy, (7th edition), v. I, 187–190. (2) Evans, H.T., Jr. (1979) The crystal structures of low chalcocite and djurleite. Zeits. Krist., 150, 299–320. (3) Evans, H.T., Jr. (1979) Djurleite ($Cu_{1.94}S$) and low chalcocite (Cu_2S) : new crystal structure studies. Science, 203, 356–358.

Crystal Data: Tetragonal. *Point Group:* $\bar{4}2m$. Equant, tetrahedral-shaped crystals often modified by scalenohedral faces, to as large as 10 cm. Sphenoidal faces {112} commonly large, dull in luster and striated ∥ [1$\bar{1}$0]; {$\bar{1}$12} faces are small and bright. Often massive, compact; sometimes botryoidal. *Twinning:* Twin plane {112}, composition surface usually {112}; twin plane {012}; also by rotation about [001] with composition plane {110}, producing penetration twins.

Physical Properties: *Cleavage:* Poor on {011} and {111}. Hardness = 3.5–4 VHN = 187–203 (basal section); 181–192 (vertical section) (100 g load). D(meas.) = 4.1–4.3 D(calc.) = 4.283

Optical Properties: Opaque. *Color:* Brass-yellow, often tarnished and iridescent. *Streak:* Greenish black. *Luster:* Metallic. *Anisotropism:* Weak.
R_1–R_2: (400) 12.6–14.8, (420) 16.3–18.0, (440) 20.0–21.2, (460) 23.6–25.0, (480) 27.0–28.6, (500) 30.2–31.7, (520) 33.0–34.3, (540) 35.1–36.4, (560) 36.8–38.0, (580) 38.2–39.3, (600) 39.3–40.4, (620) 40.1–41.0, (640) 40.7–41.6, (660) 41.1–41.9, (680) 41.4–42.0, (700) 41.4–41.8

Cell Data: *Space Group:* $I\bar{4}2d$. $a = 5.281$ $c = 10.401$ Z = 4

X-ray Powder Pattern: Merkur mine, Ems, Hesse, Germany.
3.038 (100), 1.8570 (35), 1.5927 (27), 1.8697 (22), 1.5753 (14), 2.644 (5), 1.2025 (5)

Chemistry:

	(1)	(2)
Cu	35.03	34.63
Fe	31.00	30.43
S	34.96	34.94
Total	100.99	100.00

(1) Western mines, Vancouver Island, British Columbia, Canada; by electron microprobe, leading to $Cu_{1.01}Fe_{1.01}S_{2.00}$. (2) $CuFeS_2$.

Polymorphism & Series: Forms a series with eskebornite.

Occurrence: A primary mineral in hydrothermal veins, stockworks, disseminations and massive replacements; as an exsolution product in mafic igneous rocks; of sedimentary origin controlled by redox conditions.

Association: Sphalerite, galena, tetrahedrite, pyrite, many copper sulfides.

Distribution: A very common copper mineral, so only a few outstanding localities can be mentioned. In the USA an important ore mineral at many of the copper mines of Arizona, as at Bisbee, Cochise Co. In crystals from New York, at the Rossie lead mines, St. Lawrence Co.; at French Creek, Chester Co., Pennsylvania; in Missouri at Joplin, Jasper Co. In Canada, in the Rouyn district, Quebec, at the Noranda mine; from Ontario, in the Kidd Creek mine, near Timmins, and at Sudbury. From Czechoslovakia, at Baňská Štiavnica (Schemnitz) and Horní Slavkov (Schlaggenwald). From Freiberg, Saxony; Dillenburg, Hesse; in the Georg mine, near Horhausen, Westerwald; and a number of mines in North Rhine-Westphalia, Germany. At Vinsknoes, Karmoen, Norway. In the Ani and Arakawa mines, Akita Prefecture, Japan. At Huaron, Peru. In the Nababiep mine, Cape Province, South Africa.

Name: From the Greek for *brass* and *pyrite*.

References: (1) Palache, C., H. Berman, and C. Frondel (1944) Dana's system of mineralogy, (7th edition), v. I, 219–224. (2) Hall, S.R. and J.M. Stewart (1973) The crystal structure refinement of chalcopyrite, $CuFeS_2$. Acta Cryst., B29, 579–585. (3) (1985) NBS Mono. 25, 21, 69.

Crystal Data: Orthorhombic. *Point Group:* $2/m\ 2/m\ 2/m$. Bladed crystals to 4 cm flattened ‖ {010}. *Twinning:* Twin and composition planes {104}.

Physical Properties: *Cleavage:* Perfect on {010}; less so on {001} and {100}. *Fracture:* Subconchoidal. *Tenacity:* Brittle. Hardness = 3–4 VHN = 226–279 (010). D(meas.) = 4.95 D(calc.) = 5.011

Optical Properties: Opaque. *Color:* Lead-gray to iron-gray, occasionally with a blue or green tarnish. *Luster:* Metallic. *Pleochroism:* Feeble in air, somewhat stronger in oil. *Anisotropism:* Observed.

R_1–R_2: (400) 37.4–41.8, (420) 38.2–43.8, (440) 39.0–45.8, (460) 39.4–47.0, (480) 39.4–47.2, (500) 39.0–46.7, (520) 38.4–45.2, (540) 37.5–43.7, (560) 36.5–42.3, (580) 35.5–41.0, (600) 34.8–39.8, (620) 34.5–39.0, (640) 34.3–38.4, (660) 34.1–38.3, (680) 33.8–38.2, (700) 33.5–38.2

Cell Data: *Space Group:* *Pnam.* $a = 6.02$ $b = 14.49$ $c = 3.79$ Z = 4

X-ray Powder Pattern: Chocoya la Vieja mine, Potosí, Bolivia.
3.13 (10), 3.00 (9), 1.762 (5), 2.31 (4), 1.831 (4), 2.12 (3), 1.895 (3)

Chemistry:

	(1)	(2)
Cu	24.72	25.48
Sb	48.45	48.81
S	26.20	25.71
Total	99.37	100.00

(1) Pulacayo mine, Bolivia. (2) $CuSbS_2$.

Occurrence: Associated with other sulfosalts and sulfides in hydrothermal veins.

Association: Jamesonite, chalcopyrite, pyrite, tetrahedrite, stibnite, andorite, stannite, dadsonite, siderite, barite, quartz.

Distribution: As large crystals at Rar el Anz, Wadi of Cherrat, east of Casablanca, Morocco. At Wolfsberg, in the Harz Mountains, Germany. From Capileira, Sierra Nevada, Granada Province, Spain. In Bolivia, at the Pulacayo mine, Huanchaca; Tapi near Tupiza; Torapaka, Cacachaca, Challapata, Colquechaca, Uncia, and Oruro. In France, at Saint-Pons, Provence. From Macayan, Philippines. In the Mt. Washington copper mine, Vancouver Island, British Columbia; and the Porter property, Carbon Hill, Wheaton district, Yukon Territory, Canada. From Moctezuma, Sonora, Mexico. Known in small amounts from other localities.

Name: From the Greek *chalkos*, *copper* and *stibium*, *antimony*.

References: (1) Palache, C., H. Berman, and C. Frondel (1944) Dana's system of mineralogy, (7th edition), v. I, 433–435. (2) Berry, L.G. and R.M. Thompson (1962) X-ray powder data for the ore minerals. Geol. Soc. Amer. Mem. 85, 144. (3) Ramdohr, P. (1969) The ore minerals and their intergrowths, (3rd edition), 705–707.

Crystal Data: Tetragonal. *Point Group:* $4/m\ 2/m\ 2/m$. As lamellar aggregates to 2 cm. *Twinning:* A fine grid of lamellae are visible parallel to (110) and $(1\bar{1}0)$.

Physical Properties: *Cleavage:* {001} perfect, {100} and {010} indistinct. *Tenacity:* Plastic, tectonically deformed. Hardness = 2–2.5 VHN = 68–76, average 71, on (001) (50 g load). D(meas.) = 6.6 D(calc.) = [6.74]

Optical Properties: Opaque. *Color:* Lead-gray to iron-black, tarnishes iridescent; creamy white to light gray, with a brownish creamy tint in reflected light. *Streak:* Black. *Luster:* Metallic. *Pleochroism:* Notable in air. *Anisotropism:* Distinct in orange to red-browns.
R_1–R_2: (400) 29.1–31.1, (420) 28.2–31.2, (440) 27.3–31.3, (460) 26.6–31.1, (480) 26.2–30.5, (500) 25.8–30.3, (520) 25.7–29.8, (540) 25.4–29.1, (560) 25.3–28.6, (580) 25.1–28.2, (600) 25.1–27.7, (620) 25.1–27.4, (640) 25.1–27.2, (660) 25.0–27.2, (680) 24.9–27.1, (700) 25.0–27.2

Cell Data: *Space Group:* $I4/mmm$. $a = 3.827(1)$ $c = 34.280(1)$ Z = 2

X-ray Powder Pattern: Ilímaussaq Intrusion, Greenland.
2.447 (10), 1.913 (10), 3.803 (7), 3.015 (7), 3.630(5), 2.704 (5), 2.580 (5)

Chemistry:

	(1)	(2)
Tl	38.07	33.72
K		0.79
Cu	40.58	34.33
Ag	0.19	1.78
Pb	0.13	0.08
Fe	3.79	3.78
Sb	3.93	11.96
Se		0.02
S	12.06	12.52
insol.	1.52	
Total	100.27	98.98

(1) Ilímaussaq Intrusion, Greenland; Au, As, Bi also noted. (2) Do.; by electron microprobe, corresponding to $(Tl_{1.699}K_{0.205}Pb_{0.003})_{\Sigma=1.907}(Cu_{5.517}Fe_{0.712}Ag_{0.125})_{\Sigma=6.354}Sb_{1.021}S_{4.000}$.

Occurrence: In ussingite veins cutting poikilitic sodalite syenite.

Association: Galena, vrbaite, cuprostibite, thalcusite, silver, gudmundite, chalcocite, sphalerite, molybdenite, avicennite, chkalovite, epistolite, niobophyllite, analcime, natrolite, microcline, lithium-mica, tugtupite.

Distribution: From Mt. Nákâlâq, northwest of Taseq Lake, in the Ilímaussaq Intrusion, southern Greenland.

Name: For the composition.

Type Material: n.d.

References: (1) Semenov, E.I., H. Sørensen, M.S. Bezsmertnaya, and L.E. Novorossova (1967) Chalcothallite, a new sulphide of copper and thallium from the Ilímaussaq alkaline intrusion, South Greenland. Medd. Grønland, 181, 13–25. (2) (1968) Amer. Mineral., 53, 1775 (abs. ref. 1). (3) Makovicky, E., Z. Johan, and S. Karup-Møeller (1980) New data on bukovite, thalcusite, chalcothallite and rohaite. Neues Jahrb. Mineral., Abh., 138, 122-146.

Crystal Data: Cubic. *Point Group:* Undetermined, but with a body-centered lattice. Most commonly anhedral, but also intermixed with giraudite in a microscopic myrmekitic texture.

Physical Properties: Hardness = n.d. VHN = 247–292 (25 g load). D(meas.) = n.d. D(calc.) = 6.17

Optical Properties: Opaque. *Color:* Dark gray in reflected light. *Luster:* Metallic.
R: (400) 27.4, (420) 27.1, (440) 26.8, (460) 26.5, (480) 26.1, (500) 26.6, (520) 26.9, (540) 27.1, (560) 27.3, (580) 27.5, (600) 27.7, (620) 27.8, (640) 28.0, (660) 28.2, (680) 28.4, (700) 28.7

Cell Data: *Space Group:* n.d. $a = 11.039$ Z = 8

X-ray Powder Pattern: Chaméane, France.
3.187 (100), 1.951 (90), 1.665 (80), 1.127 (70), 1.266 (60), 1.062 (50), 1.381 (40)

Chemistry:

	(1)	(2)
Cu	33.93	36.38
Fe	5.63	2.58
As	11.05	11.98
Sb	0.48	0.34
Se	47.06	46.99
S	1.59	1.56
Total	99.74	99.83

(1–2) Chaméane, France; by electron microprobe.

Occurrence: As late stage deposits in veins cutting granite.

Association: Giraudite, eskebornite, geffroyite, ankerite.

Distribution: From Chaméane, Puy-de-Dôme, France.

Name: For the occurrence at Chaméane, France.

Type Material: National School of Mines, Paris, France.

References: (1) Johan, Z., P. Picot, and F. Ruhlmann (1982) Evolution paragénétique de la minéralisation uranifère de Chaméane (Puy-de-Dôme) France: chaméanite, geffroyite et giraudite, trois séléniures nouveaux de Cu, Fe, Ag, and As. Tschermaks Mineral. Petrog. Mitt., 29, 151–167 (in French with English abs.). (2) (1982) Amer. Mineral., 67, 1074–1075 (abs. ref. 1).

Crystal Data: Hexagonal. *Point Group:* $6/m\ 2/m\ 2/m$. As thin lamellae, 3–15 μm wide, alternating with graphite.

Physical Properties: Hardness = Slightly harder than graphite (1–2). VHN = n.d. D(meas.) = n.d. D(calc.) = 3.43

Optical Properties: Opaque. *Color:* Black.
R_1–R_2: n.d.

Cell Data: *Space Group:* $P6/mmm$. $a = 8.948$ $c = 14.078$ Z = 168

X-ray Powder Pattern: Ries Crater, Germany.
4.47 (100), 4.26 (100), 4.12 (80), 3.03 (60), 2.55 (60), 2.28 (60), 3.71 (40)

Chemistry:

	(1)
C	99.5
Si	< 0.5
Cl	< 0.5
Total	100.0

(1) Ries Crater, Germany; by electron microprobe, grain containing 35% chaoite.

Polymorphism & Series: Diamond, graphite, and lonsdaleite are polymorphs.

Occurrence: In shock-metamorphosed graphite gneisses and meteorites.

Association: Graphite, zircon, rutile, pseudobrookite, magnetite, nickeliferous pyrrhotite, baddeleyite.

Distribution: From Mottingen, in the Ries Crater, Bavaria, Germany. In the Goalpara and Dyalpur achondrite meteorites.

Name: For Edward Ching-Te Chao (1919–), petrologist with the U.S. Geological Survey.

Type Material: n.d.

References: (1) El Goresy, A. and G. Donnay (1968) A new allotriomorphic form of carbon from the Ries Crater. Science, 161, 363–364. (2) (1969) Amer. Mineral., 54, 326 (abs. ref. 1). (3) El Gorsey, A. (1969) Eine neue Kohlenstoff-Modifikation aus dem Nördlinger Ries. Naturwissenschaften, 56, 493–494 (in German). (4) (1970) Amer. Mineral., 55, 1067 (abs. ref. 3) (5) Smith, P.P.K. and P.R. Buseck (1982) Carbyne forms of carbon: do they exist? Science, 216, 984–986. (6) (1983) Amer. Mineral., 68, 1251 (abs. ref. 5).

Crystal Data: Tetragonal. *Point Group:* $\bar{4}m2$. As rounded grains (30–100 μm) included in tetrahedrite.

Physical Properties: Hardness = n.d. VHN = 258–287 (20 g load). D(meas.) = n.d. D(calc.) = 5.00

Optical Properties: Opaque. *Color:* Pale rose in reflected light. *Anisotropism:* Weak, in shades of brown.
R_1–R_2: (400) — , (420) — , (440) 24.4–25.0, (460) 25.0–25.6, (480) 25.6–26.2, (500) 26.2–26.7, (520) 26.7–27.2, (540) 27.2–27.6, (560) 27.6–28.1, (580) 27.9–28.6, (600) 28.3–29.0, (620) 28.5–29.3, (640) 28.6–29.4, (660) 28.6–29.5, (680) 28.4–29.3, (700) 27.9–28.6

Cell Data: *Space Group:* $P\bar{4}m2$. $a = 7.61(1)$ $c = 5.373(5)$ Z = 1

X-ray Powder Pattern: Uzbekistan, USSR.
1.904 (100), 3.11 (80), 1.625 (40), 1.568 (40), 2.87 (30), 1.058 (30), 2.70 (20)

Chemistry:

	(1)	(2)
Cu	41.17 – 46.87	40.95
Ag	0.00 – 0.68	
Fe	1.44 – 4.31	6.00
Zn	0.82 – 4.89	
Sn	10.75 – 19.11	25.50
Mo	0.41 – 1.08	
Sb	2.97 – 7.25	
As	0.80 – 2.58	
S	27.70 – 28.88	27.55
Total		100.00

(1) Uzbekistan, USSR; by electron microprobe, range of eight grains. (2) $Cu_6FeSn_2S_8$.

Occurrence: As rounded disseminations in tetrahedrite, from a sulfide-bearing quartz vein (Uzbekistan, USSR).

Association: Cassiterite, hemusite, hessite, tetrahedrite (Uzbekistan, USSR); pyrite, sphalerite, marcasite, galena, chalcopyrite, stannite, tetrahedrite–tennantite, canfieldite, arsenopyrite, digenite, covellite, chalcocite, Au–Ag alloy (Cove deposit, USA).

Distribution: From an unspecified locality in the Chatkalo-Kuramin Mountains, eastern Uzbekistan, USSR. In the Cove gold deposit, Lander Co., Nevada, USA.

Name: For the occurrence in the Chatkalo-Kuramin Mountains, USSR.

Type Material: A.E. Fersman Mineralogical Museum, Academy of Sciences, Moscow, USSR.

References: (1) Kovalenker, V.A., T.L. Evstigneeva, V.S. Malov, and L.N. Vyal'sov (1981) Chatkalite, $Cu_6FeSn_2S_8$, a new mineral. Mineralog. Zhurnal., 3, 79–86 (in Russian). (2) (1982) Amer. Mineral., 67, 621–622 (abs. ref. 1).

Crystal Data: Orthorhombic. *Point Group:* $2/m\ 2/m\ 2/m$. As grains up to 20 μm in aggregates up to 100 μm.

Physical Properties: *Cleavage:* One distinct. Hardness = n.d. VHN = 726–754 (50 g load). D(meas.) = n.d. D(calc.) = 9.72

Optical Properties: Opaque. *Color:* Gray-white; in reflected light, white with an orange tint. *Streak:* Grayish black. *Luster:* Metallic. *Pleochroism:* Weak, in yellow to pinkish hue and bluish white. *Anisotropism:* Green to gray-brown.

R_1–R_2: (400) — , (420) 39.2–40.5, (440) 39.6–41.0, (460) 40.3–41.9, (480) 41.0–42.9, (500) 41.9–43.8, (520) 42.7–44.7, (540) 43.6–45.5, (560) 44.1–46.1, (580) 44.5–46.6, (600) 44.9–47.1, (620) 45.1–47.3, (640) 45.2–47.4, (660) 45.3–47.6, (680) 45.5–47.6, (700) 45.4–47.6

Cell Data: *Space Group:* *Pnma*. $a = 5.70(2)$ $b = 3.59(1)$ $c = 6.00(1)$ Z = 4

X-ray Powder Pattern: Koryak-Kamchatka region, USSR.
3.01 (10), 2.10 (6b), 1.813 (2), 1.771 (2b), 1.501 (2b), 1.354 (2b)

Chemistry:

	(1)	(2)
Rh	55.9	57.87
Ru	1.45	
Pt	0.61	
Ni	0.20	
As	41.6	42.13
Total	99.76	100.00

(1) Koryak-Kamchatka region, USSR; by electron microprobe, average of analysis of 16 grains.
(2) RhAs.

Occurrence: In a placer with other platinum group element minerals derived from an ultramafic massif in an ophiolite belt.

Association: Isoferroplatinum, tetraferroplatinum, ferronickelplatinum, rutheniridosmine, laurite, irarsite, cooperite, sperrylite, hollingworthite, chromite, olivine.

Distribution: From an undefined locality in the the Koryak-Kamchatka region, northeastern USSR.

Name: For V.A. Cherepanov (1927–1983), geologist and mineralogist.

Type Material: Mining Museum, Leningrad Mining Institute, Leningrad, USSR.

References: (1) Rudashevskii, N.S., A.G. Mochalov, N.V. Trubkin, N.I. Shumskaya, V.I. Shkurskii, and T.L. Evstigneeva (1985) Cherepanovite RhAs — a new mineral. Zap. Vses. Mineral. Obshch., 114, 464–469 (in Russian). (2) (1986) Amer. Mineral., 71, 1544 (abs. ref. 1). (3) (1986) Mineral. Abs., 37, 529 (abs. ref. 1).

Crystal Data: Monoclinic. *Point Group:* $2/m$. As subhedral grains from 0.5 to 1 mm in length. Some grains are somewhat bladed or flattened, probably on {010}.

Physical Properties: *Cleavage:* Perfect {010}, excellent {110} and {001}, good {$\bar{1}01$}. Hardness = 1–2 VHN = 28–35 (10 g load). D(meas.) = 6.2(2) (synthetic). D(calc.) = 6.37

Optical Properties: Opaque. *Color:* Bright orange to deep red or crimson, darker than realgar. *Streak:* Bright orange. *Luster:* Adamantine.
R_1–R_2: n.d.

Cell Data: *Space Group:* $P2_1/n$. $a = 6.113$ $b = 16.188$ $c = 6.111$ $\beta = 96.71°$ Z = 4

X-ray Powder Pattern: Carlin, Nevada, USA.
2.98 (100), 3.62 (80), 4.03 (60), 3.49 (60), 2.692 (60), 3.36 (50), 2.216 (50)

Chemistry:

	(1)	(2)
Tl	35.2	35.48
Hg	35.1	34.82
As	13.1	13.00
S	16.6	16.70
Total	100.0	100.00

(1) Carlin, Nevada, USA; by electron microprobe, average of four analyses. (2) $TlHgAsS_3$.

Polymorphism & Series: Dimorphous with routhierite.

Occurrence: In hydrothermal barite veins and in mineralized carbonaceous silty dolomite.

Association: Realgar, orpiment, lorandite, barite, getchellite.

Distribution: At the Carlin gold deposit, 50 km northwest of Elko, Eureka Co., Nevada, USA. From Allchar, near Rozdan, Macedonia, Yugoslavia.

Name: For Dr. Charles L. Christ, mineralogist with the U.S. Geological Survey.

Type Material: Geology Department, Stanford University, Palo Alto, California; National Museum of Natural History, Washington, D.C., USA, 144272, 144273.

References: (1) Radtke, A.S., F.W. Dickson, J.F. Slack, and K.L. Brown (1977) Christite, a new thallium mineral from the Carlin gold deposit, Nevada. Amer. Mineral., 62, 421–425.
(2) Brown, K.L. and F.W. Dickson (1976) The crystal structure of synthetic christite, $HgTlAsS_3$. Zeits. Krist., 144, 367–376.

Crystal Data: Cubic. *Point Group:* $4/m\ \overline{3}\ 2/m$. As small grains forming aggregates to several hundred μm.

Physical Properties: Hardness = n.d. VHN = 260 (100 g load). D(meas.) = n.d. D(calc.) = 6.69 Ferromagnetic.

Optical Properties: Opaque. *Color:* Light gray. *Luster:* Metallic.
R: (400) — , (420) — , (440) 50.4, (460) 51.4, (480) 50.9, (500) 52.6, (520) 53.0, (540) 55.3, (560) 56.5, (580) 56.9, (600) 57.9, (620) 58.3, (640) 59.0, (660) 60.0, (680) 60.7, (700) 60.8

Cell Data: *Space Group:* $Pm3m$. $a = 2.859(5)$ Z = [0.5]

X-ray Powder Pattern: Southern Urals, USSR.
2.02 (100), 1.16 (100), 1.43 (80), 1.01 (70), 1.28 (50), 2.87 (20), 1.656 (10)

Chemistry:

	(1)
Fe	88.91
Cr	11.30
Total	100.21

(1) Southern Urals, USSR; by electron microprobe, corresponding to $Fe_{1.5}Cr_{0.5-x}$, with x = 0.3.

Occurrence: In quartz veins within brecciated amphibolites and schist.

Association: Iron, copper, bismuth, gold, ferchromide, graphite, cohenite, halite, sylvite, marialite, quartz.

Distribution: From an unspecified locality in the Southern Ural Mountains, USSR.

Name: For the chemical composition, CHROMium and FERrum, *iron*.

Type Material: A.E. Fersman Mineralogical Museum, Academy of Sciences, Moscow, USSR.

References: (1) Novgorodova, M.I., A.I. Gorshkov, N.V. Trubkin, A.I. Tsepin, and M.T. Dmitrieva (1986) New natural intermetallic compounds of iron and chromium – chromferide and ferchromide. Zap. Vses. Mineral. Obshch., 115, 355–360 (in Russian). (2) (1988) Amer. Mineral., 73, 190 (abs. ref. 1).

Crystal Data: Cubic. *Point Group:* $4/m\ \overline{3}\ 2/m$ (synthetic). Finely granular (20 μm); spherulitic and rounded grains.

Physical Properties: *Tenacity:* Brittle (synthetic). Hardness = n.d. VHN = < 1875–2000 D(meas.) = 7.17 D(calc.) = [7.20]

Optical Properties: Opaque. *Color:* White, with a yellow tint in reflected light. *Luster:* Metallic.
R: (480) 65.3, (546) 67.9, (589) 68.8, (656) 70.0

Cell Data: *Space Group:* $Im3m$ (synthetic). $a = 2.8839$ Z = 2

X-ray Powder Pattern: Synthetic.
2.04 (100), 1.1774 (30), 0.9120 (20), 1.0195 (18), 1.4419 (16), 0.8325 (6)

Chemistry:

	(1)
Cr	98.010
Cu	0.366
Zn	1.400
Fe	0.001
Total	99.777

(1) Sichuan Province, China.

Occurrence: In heavy sands derived from the contact metamorphic zone between a siliceous marble and ultramafic rock (Sichuan Province, China); in ultramafic dike rocks (Gavasai ore field, USSR).

Association: Pyrrhotite, pentlandite, chalcopyrite, platinum group minerals (Sichuan Province, China); cohenite, moissanite, ilmenite, sphene (Gavasai ore field, USSR).

Distribution: From an unspecified locality in Sichuan Province, China. In the Gavasai ore field, location not further given, USSR.

Name: From the Greek *chroma, color*, as all chromium compounds are colored.

Type Material: n.d.

References: (1) Yue Suchin, Wang Wenying, and Sun Sujing (1981) A new mineral—native chromium. Kexue Tongbao, 26, 959–960 (in Chinese). (2) (1982) Amer. Mineral., 67, 854–855 (abs. ref. 1). (3) (1955) NBS Circ. 539, 5, 20. (4) Yusupov, R.G., D.D. Dzhenchuraev, and F.F. Radzhabov (1982) Native accessory chromium and natural iron-chromium-silicon compounds in intrusive rocks from the Gavasai ore field [USSR]. Izv. Akad. Nauk Kirg. SSR, 5, 24–25 (in Russian). (5) (1983) Chem. Abs., 98, 110815 (abs. ref. 4).

Crystal Data: Hexagonal. *Point Group:* $3m$. Tabular, equant, and prismatic grains, up to 0.5 mm, intergrown with other sulfides.

Physical Properties: *Cleavage:* Perfect {0001}. *Tenacity:* Brittle. Hardness = n.d. VHN = 110–153, 135 average (20 g load). D(meas.) = n.d. D(calc.) = 3.94

Optical Properties: Opaque. *Color:* Bronze, tarnishing to a sooty black coating; in reflected light, orange, changing to rosy purple with time. *Luster:* Metallic. *Pleochroism:* Distinct, from light orange to dark gray with a lilac tint. *Anisotropism:* Strong, from black to white.
R_1–R_2: (400) — , (420) 17.9–17.9, (440) 17.7–18.0, (460) 17.5–18.4, (480) 17.4–19.0, (500) 17.3–19.7, (520) 17.4–20.7, (540) 17.5–21.6, (560) 17.7–22.5, (580) 17.8–23.3, (600) 18.0–24.0, (620) 18.2–24.8, (640) 18.5–25.6, (660) 18.8–26.2, (680) 19.1–26.8, (700) 19.4–27.4

Cell Data: *Space Group:* $P3m$. $a = 3.873(1)$ $c = 6.848$ Z = 1

X-ray Powder Pattern: Akatuya deposit, USSR.
3.02 (100), 2.40 (100), 1.945 (100), 3.40 (90), 1.870 (90), 6.85 (60)

Chemistry:

	(1)
Na	10.93
Cu	38.63
Fe	11.64
Zn	6.72
Ca	0.26
Mn	0.06
As	0.55
S	30.83
Total	99.62

(1) Akatuya deposit, USSR; by electron microprobe, corresponding to $Na_{1.01}(Cu_{1.28}Fe_{0.44}Zn_{0.22})_{\Sigma=1.94}S_{2.03}$.

Occurrence: Of hydrothermal origin.

Association: Sphalerite, covellite, chalcocite, galena, pyrite, boulangerite, arsenopyrite, carbonates, quartz.

Distribution: From the Akatuya deposit, Transbaikal, USSR.

Name: For T.N. Chvileva, mineralogist.

Type Material: A.E. Fersman Mineralogical Museum, Academy of Sciences, Moscow, USSR.

References: (1) Kachalovskaya, V.M., B.S. Osipov, N.G. Nazarenko, V.A. Kukoev, A.O. Mazmanyan, I.N. Egorov, and L.N. Kaplunnik (1988) Chvilevaite — a new alkali sulfide with the composition $Na(Cu, Fe, Zn)_2S_2$. Zap. Vses. Mineral. Obshch., 117, 204–207 (in Russian).
(2) (1989) Amer. Mineral., 74, 946 (abs. ref. 1).

Crystal Data: Hexagonal. *Point Group:* 32. Rhombohedral crystals to 10 cm; thick tabular {0001}; stout to slender prismatic ∥ $[10\bar{1}0]$. Also as incrustations, granular and massive. *Twinning:* Twin plane {0001}, twin axis [0001], to form simple contact twins.

Physical Properties: *Cleavage:* $\{10\bar{1}0\}$ perfect. *Fracture:* Subconchoidal, uneven. *Tenacity:* Slightly sectile. Hardness = 2–2.5 VHN = 82–156 (10 g load). D(meas.) = 8.176 D(calc.) = 8.20

Optical Properties: Transparent in thin pieces. *Color:* Cochineal-red, towards brownish red and lead-gray. *Streak:* Scarlet. *Luster:* Adamantine, inclining to metallic when dark; dull in friable material.

Optical Class: Uniaxial (+). $\omega = 2.905$ (598.5) $\epsilon = 3.256$ (598.5) *Anisotropism:* High.

R_1–R_2: (400) 36.3–37.8, (420) 34.4–35.6, (440) 32.5–33.2, (460) 31.2–31.8, (480) 30.3–30.9, (500) 29.6–30.2, (520) 28.9–29.7, (540) 28.4–29.2, (560) 28.0–28.8, (580) 27.7–28.5, (600) 27.5–28.3, (620) 27.2–28.0, (640) 26.9–27.7, (660) 26.6–27.4, (680) 26.3–27.0, (700) 26.2–26.8

Cell Data: *Space Group:* $P3_121$ or $P3_221$. $a = 4.145(2)$ $c = 9.496(2)$ Z = 3

X-ray Powder Pattern: Almadén, Spain.
2.85 (10), 3.34 (9), 1.672 (6), 2.06 (5), 1.969 (5), 1.725 (5), 1.339 (5)

Chemistry: Essentially pure HgS.

Polymorphism & Series: Trimorphous with metacinnabar and hypercinnabar.

Occurrence: Formed from low-temperature hydrothermal solutions in veins, and in sedimentary, igneous, and metamorphic host rocks.

Association: Mercury, realgar, pyrite, marcasite, stibnite, opal, chalcedony, barite, dolomite, calcite.

Distribution: The most common ore of mercury world-wide, so only a few localities for exceptionally abundant or well crystallized material can be mentioned. In the USA, in California, notably at New Almaden, Santa Clara Co. and New Idria, San Benito Co.; in Texas, at Terlingua, Brewster Co.; in Nevada, at Poverty Peak, Humboldt Co.; and near Lovelock, Pershing Co. In Spain, at Almadén, Ciudad Real Province, and Mieres, Asturias. At Hydercahn in the Ferghana Basin, Kazakhstan, USSR. From Tongrin, Wanshanchang, and elsewhere in Guizhou Province, and in exceptional twinned crystals from the Tsar Tien mine, Hunan Province, China. As fine crystals at Mount Avala, near Belgrade; and at Idria, Slovenia, Yugoslavia. At Charcas, San Luis Potosí, Mexico.

Name: From the Medieval Latin *cinnabaris*, traceable to the Persian *zinjifrah*, apparently meaning *dragon's blood*, for the red color.

References: (1) Palache, C., H. Berman, and C. Frondel (1944) Dana's system of mineralogy, (7th edition), v. I, 251–255. (2) Berry, L.G. and R.M. Thompson (1962) X-ray powder data for the ore minerals. Geol. Soc. Amer. Mem. 85, 69. (3) Auvray, P. and F. Genet (1973) Affinement de la structure cristalline du cinabre α–HgS. Bull. Soc. fr. Minéral., 96, 218–219 (in French).

Crystal Data: Cubic. *Point Group:* $4/m\ \overline{3}\ 2/m$. Usually massive, also fine granular or foliated.

Physical Properties: *Cleavage:* {001} good. *Fracture:* Granular. *Tenacity:* Brittle. Hardness = 2.5–3 VHN = 44–49 (100 g load). D(meas.) = 7.8–8.22 D(calc.) = 8.275

Optical Properties: Opaque. *Color:* Lead-gray, bluish; galena-white with a grayish tinge in reflected light. *Streak:* Grayish black. *Luster:* Metallic.
R: (400) 62.0, (420) 60.5, (440) 59.0, (460) 57.3, (480) 55.6, (500) 54.0, (520) 52.6, (540) 51.3, (560) 50.2, (580) 49.2, (600) 48.4, (620) 47.7, (640) 47.3, (660) 47.0, (680) 47.0, (700) 47.2

Cell Data: *Space Group:* $Fm3m$. $a = 6.1243$ Z = 4

X-ray Powder Pattern: Synthetic.
3.062 (100), 2.165 (70), 3.536 (30), 1.369 (25), 1.768 (20), 1.846 (18), 1.250 (16)

Chemistry:

	(1)	(2)
Pb	70.85	72.34
Co	1.10	
Se	27.99	27.66
Total	99.94	100.00

(1) Clausthal, Germany. (2) PbSe.

Polymorphism & Series: Forms a series with galena.

Occurrence: Associated in low sulfur hydrothermal deposits with other selenides, also in mercury deposits. Probably of secondary origin in some deposits.

Association: Tiemannite, klockmannite, berzelianite, umangite, gold, stibiopalladinite, uraninite.

Distribution: Perhaps the most common selenide. In the USA, in the Blue Range Wilderness Area, Greenlee Co., Arizona; and at the Essex mine, Darwin, Inyo Co., California. In the Goldfields district, Saskatchewan; in the Kidd Creek mine, near Timmins, and the Hemlo gold deposit, Thunder Bay district, Ontario, Canada. Found at Clausthal, Tilkerode, Lerbach, and Zorge, in the Harz Mountains, and at Schneeberg, Saxony, Germany. At Skrikerum, Sweden. From the Sierra de Umango and in the Santa Brigida mine, La Rioja Province; and from Sierra de Cacheuta, Mendoza Province, Argentina. At Pacajake, Colquechaca, Bolivia. From El Sharana, Northern Territory, Australia. At Hope's Nose, Torquay, Devon, England. From Bukov and Petrovice, Czechoslovakia. Other minor occurrences are known.

Name: For the locality at Clausthal, Germany.

References: (1) Palache, C., H. Berman, and C. Frondel (1944) Dana's system of mineralogy, (7th edition), v. I, 204–205. (2) Early, J.W. (1950) Description and synthesis of the selenide minerals. Amer. Mineral., 35, 356–358. (3) (1955) NBS Circ. 539, Vol. 5, 38.

Crystal Data: Monoclinic. *Point Group:* $2/m$. Very fine grained; intimately intergrown with other sulfides.

Physical Properties: Hardness = 4.5–5 VHN = 719–875 (100 g load). D(meas.) = n.d. D(calc.) = 7.46

Optical Properties: Opaque. *Color:* Tin-white. *Luster:* Metallic.
R_1–R_2: (400) 53.6–54.3, (420) 54.4–54.8, (440) 55.2–55.3, (460) 55.3–55.5, (480) 54.5–55.4, (500) 53.6–55.2, (520) 52.8–55.0, (540) 52.0–54.7, (560) 51.5–54.5, (580) 51.0–54.2, (600) 50.4–53.8, (620) 50.0–53.5, (640) 49.5–53.2, (660) 49.2–52.8, (680) 49.0–52.4, (700) 48.8–52.0

Cell Data: *Space Group:* $P2_1/n$. $a = 5.040$ $b = 5.862$ $c = 3.139$ $\beta = 90°13'$ Z = 2

X-ray Powder Pattern: Cobalt, Canada.
2.531 (100), 2.427 (80), 2.422 (80), 2.671 (40), 2.657 (40), 1.865 (30), 1.656 (30)

Chemistry:

	(1)
Co	27.4
Ni	0.7
Fe	0.4
Cu	0.0
As	70.1
S	0.6
Total	99.2

(1) Nord mine, Sweden; by electron microprobe, giving $(Co_{0.98}Ni_{0.02}Fe_{0.01})_{\Sigma=1.01}(As_{1.96}S_{0.04})_{\Sigma=2.00}$. (2) Cobalt, Canada; by electron microprobe, analyses of three samples gave: $(Co_{0.76}Fe_{0.14}Ni_{0.10})_{\Sigma=1.00}As_{2.00}$; $(Co_{0.73}Fe_{0.26}Ni_{0.01})_{\Sigma=1.00}As_{2.00}$; and $(Co_{0.70}Fe_{0.21}Ni_{0.09})_{\Sigma=1.00}As_{2.00}$ (actual analyses not given).

Polymorphism & Series: Dimorphous with safflorite.

Occurrence: In hydrothermal Co-Ni ores.

Association: Skutterudite, cattierite, cobaltite, siegenite, bismuth, molybdenite.

Distribution: From Cobalt, Ontario, Canada. In the Aghbar mine, near Bou Azzer, Morocco. From near Filipstad, Wermland, Sweden. At the Nord mine, Nordmark district, Sweden.

Name: In reference to the monoclinic symmetry of this safflorite-like species.

Type Material: Department of Geological Sciences, Queen's University, Kingston, Ontario, Canada.

References: (1) Radcliffe, D. and L.G. Berry (1971) Clinosafflorite: a monoclinic polymorph of safflorite. Can. Mineral., 10, 877–881. (2) (1972) Amer. Mineral., 57, 1552 (abs. ref. 1).
(3) Burke, E.A.J. and M.A. Zakrzewski (1983) A cobalt-bearing sulfide–arsenide assemblage from the Nord mine (Finnshytteberg), Sweden: A new occurrence of clinosafflorite. Can. Mineral., 21, 129–136.

Crystal Data: Orthorhombic, pseudocubic. *Point Group:* $mm2$. Commonly as pseudocubic or pseudopyritohedral crystals, or combinations having striated faces as with pyrite, to as large as 8 cm, also as pseudo-octahedra; granular massive. *Twinning:* {011} and {111} of the pseudocubic habit as twin planes, rare. Twin lamellae common in polished section.

Physical Properties: *Cleavage:* Perfect on {001}. *Fracture:* Uneven. *Tenacity:* Brittle. Hardness = 5.5 VHN = 1095–1346 D(meas.) = 6.33 D(calc.) = 6.335

Optical Properties: Opaque. *Color:* Silver-white. *Streak:* Grayish black. *Pleochroism:* Very weak, on grain boundaries.
R: (400) 48.2, (420) 48.6, (440) 49.0, (460) 48.6, (480) 48.3, (500) 48.7, (520) 49.4, (540) 50.3, (560) 51.2, (580) 52.3, (600) 53.2, (620) 54.1, (640) 54.8, (660) 55.2, (680) 55.4, (700) 55.3

Cell Data: *Space Group:* $Pca2_1$. $a = 5.582$ $b = 5.582$ $c = 5.582$ Z = 4

X-ray Powder Pattern: Cobalt, Canada.
2.49 (100), 1.680 (100), 2.27 (90), 2.77 (80), 1.490 (80), 1.073 (80), 1.973 (60)

Chemistry:

	(1)	(2)
Co	28.64	35.53
Fe	4.11	
Ni	3.06	
As	44.77	45.15
S	19.34	19.32
Total	99.92	100.00

(1) Cobalt, Canada. (2) CoAsS.

Occurrence: In high-temperature hydrothermal deposits, as disseminations, and as veins in contact metamorphosed rocks.

Association: Magnetite, sphalerite, chalcopyrite, skutterudite, allanite, zoisite, scapolite, titanite, calcite (Tunaberg, Sweden); other Co–Ni sulfides and arsenides.

Distribution: In Sweden, fine crystals from Tunaberg, Södermanland; Riddarhyttan and Hånkansboda, Västmanland; and at Vena, Örebro. From Skutterud, near Modum, Buskerud, Norway. At Crown's Engine House, Botallack, St. Just, Cornwall, England. Good crystals from Espanola, and mines in the Cobalt district, Ontario, Canada. In Australia, at Broken Hill and Torrington, New South Wales; Bimbowrie, South Australia; and at Mt. Cobalt and Cloncurry, Queensland. From Bou Azzer, Morocco. Likely from many more localities.

Name: In allusion to the elemental composition.

References: (1) Palache, C., H. Berman, and C. Frondel (1944) Dana's system of mineralogy, (7th edition), v. I, 296–298. (2) Giese, R.F., Jr. and P.F. Kerr (1965) The crystal structures of ordered and disordered cobaltite. Amer. Mineral., 50, 1002–1014. (3) Bayliss, P. (1969) X-ray data, optical anisotropism, and thermal stability of cobaltite, gersdorffite, and ullmannite. Mineral. Mag., 37, 26–33. (4) Bayliss, P. (1982) A further crystal structure refinement of cobaltite. Amer. Mineral., 67, 1048–1057.

Crystal Data: Cubic. *Point Group:* $4/m\ \overline{3}\ 2/m$. As exsolved lamellae and as separate crystals rarely exceeding 4 mm in size.

Physical Properties: *Cleavage:* {100}. Hardness = n.d. VHN = 278–332 (100 g load). D(meas.) = n.d. D(calc.) = 5.22

Optical Properties: Opaque. *Color:* In polished section, slightly lighter bronze-yellow than pentlandite. *Luster:* Metallic.
R: (400) 36.9, (420) 41.9, (440) 45.9, (460) 49.2, (480) 51.7, (500) 53.8, (520) 55.6, (540) 57.0, (560) 58.2, (580) 59.2, (600) 60.0, (620) 60.5, (640) 61.0, (660) 61.2, (680) 61.4, (700) 61.5

Cell Data: *Space Group:* $Fm3m$. $a = 9.973$ Z = 4

X-ray Powder Pattern: Outokumpu mine, Finland.
3.008 (100), 1.763 (100), 1.918 (80), 5.75 (60), 2.878 (60), 1.018 (60), 2.288 (50)

Chemistry:

	(1)	(2)	(3)
Co	54.1	49.33	67.40
Ni	10.4	9.06	
Fe		10.32	
S	34.2	31.29	32.60
Total	98.7	100.00	100.00

(1) Langis mine, Canada; analysis by electron microprobe, giving $Co_{6.9}Ni_{1.3}S_8$. (2) Varislahti, Finland. (3) Co_9S_8.

Polymorphism & Series: Forms a series with pentlandite.

Occurrence: With other sulfides and arsenides in hydrothermal veins.

Association: Pyrite, marcasite, pyrrhotite, chalcopyrite, troilite, cubanite, linnaeite, siegenite, parkerite, bravoite, langisite, magnetite.

Distribution: At the Langis mine, Casey Township, Cobalt-Gowganda area, Ontario, and the Vauze mine, north of Noranda, Quebec, Canada. In the Varislahti, Kuusjärvi, and Savonranta pyrrhotite deposits and at the Outokumpu mine, northern Karelia, Finland. In the Amianthus mine, Kaapsche Hoop, Barberton, Transvaal, South Africa. In Japan, at the Shimokaua mine, Hokkaido. From near Eretria, Othris Mountains, and near Perivoli, Pindos Mountains, Greece. In the Talmessi mine, near Anarak, Iran.

Name: For the cobalt content and the relationship with pentlandite.

Type Material: n.d.

References: (1) Kouva, O., M. Huhma, and Y. Vuorelainen (1959) A natural cobalt analog of pentlandite. Amer. Mineral., 44, 897–900. (2) Gellers, S. (1962) Refinement of the crystal structure of Co_9S_8. Acta Cryst., 15, 1195–1198. (3) Stumpfl, E.F. and E.M. Clark (1964) A natural occurrence of Co_9S_8, identified by x-ray microanalysis. Neues Jahrb. Mineral., Monatsh., 240–245. (4) (1965) Amer. Mineral., 50, 2107–2108 (abs. ref. 3). (5) Petruk, W., D.C. Harris, and J.M. Stewart (1969) Langisite, a new mineral, and the rare minerals cobalt pentlandite, siegenite, parkerite, and bravoite from the Langis mine, Cobalt-Gowganda area, Ontario, Canada. Can. Mineral., 9, 597–605. (6) Rajamini, V. and C.T. Prewitt (1975) Refinement of the structure of Co_9S_8. Can. Mineral., 13, 75–78.

Crystal Data: Cubic. *Point Group:* $\bar{4}3m$. Known only massive, granular.

Physical Properties: *Fracture:* Uneven to subconchoidal. *Tenacity:* Brittle, friable. Hardness = 2.5 VHN = 25–28 (100 g load). D(meas.) = 8.10 D(calc.) = 8.092

Optical Properties: Opaque. *Color:* Iron-black inclining to gray; in polished section, white with slight grayish brown tint, tarnishing to dull purple. *Luster:* Bright metallic. *Anisotropism:* Moderate, dark brownish red to blue-gray.

R: (400) 36.4, (420) 36.0, (440) 35.6, (460) 35.2, (480) 35.0, (500) 34.9, (520) 35.0, (540) 35.4, (560) 36.2, (580) 37.0, (600) 37.3, (620) 37.0, (640) 36.1, (660) 35.4, (680) 34.7, (700) 34.2

Cell Data: *Space Group:* $F\bar{4}3m$. $a = 6.453$ Z = 4

X-ray Powder Pattern: Kalgoorlie, Australia.
3.74 (100), 2.29 (90), 1.949 (70), 1.318 (40), 1.484 (30), 1.244 (30), 1.090 (30)

Chemistry:

	(1)	(2)	(3)
Hg	60.95	58.55	61.12
Pb		1.60	
Te	39.38	39.10	38.88
insol.		0.25	
Total	100.33	99.50	100.00

(1) Kalgoorlie, Australia. (2) Lake Shore mine, Canada. (3) HgTe.

Occurrence: In hydrothermal tellurium-bearing precious metal veins.

Association: Altaite, calaverite, krennerite, petzite, gold, pyrite, chalcopyrite, sphalerite, galena, pyrrhotite, tetrahedrite–tennantite.

Distribution: In the USA, in Colorado, at the Good Hope mine, Vulcan, Gunnison Co.; in the La Plata district, La Plata Co.; and at the Keystone, Mountain Lion, Smuggler, and Ellen mines, Boulder Co. At the Norwegian mine, near Tuttletown, Tuolumne Co., California. In Canada, at a number of mines in the Kirkland Lake district, in the Hemlo gold deposit, Thunder Bay district, and at Hollinger in the Porcupine district, Ontario; also found at the Ardeen mine and the Robb-Montbray mine, Quebec. At Ilova, Czechoslovakia. In the Uzel'ginsk deposit, Southern Ural Mountains, USSR. From Kalgoorlie, Western Australia. In the Emperor mine, Vatukoula, Viti Levu, Fiji Islands.

Name: For the Keystone and Mountain Lion mines, Magnolia, Magnolia district, and Smuggler mine, Ballerat district, Boulder Co., type localities in Colorado, USA.

Type Material: n.d.

References: (1) Palache, C., H. Berman, and C. Frondel (1944) Dana's system of mineralogy, (7th edition), v. I, 218–219. (2) Thompson, R.M. (1949) The telluride minerals and their occurrence in Canada. Amer. Mineral., 34, 342–382. (3) Ramdohr, P. (1969) The ore minerals and their intergrowths, (3rd edition), 520–521.

Crystal Data: Cubic. *Point Group:* $\overline{4}3m$. Granular, massive; also as complex aggregates of modified tetrahedral crystals, to 5 mm. *Twinning:* Rare on {111}.

Physical Properties: *Tenacity:* Brittle. Hardness = 3–4 VHN = 322–379 (100 g load). D(meas.) = 4.2 D(calc.) = [4.78]

Optical Properties: Opaque. *Color:* Bronze, bronze-brown to bronzy gray; cream colored in reflected light. *Streak:* Black. *Luster:* Metallic.
R: (400) 22.0, (420) 23.5, (440) 25.0, (460) 25.9, (480) 26.6, (500) 27.5, (520) 28.8, (540) 30.0, (560) 30.9, (580) 31.6, (600) 31.9, (620) 31.5, (640) 30.8, (660) 29.8, (680) 28.7, (700) 27.8

Cell Data: *Space Group:* $P\overline{4}3n$. a = 10.538(6) Z = [1]

X-ray Powder Pattern: Tramway mine, Butte, Montana, USA.
3.07 (100), 1.881 (60), 1.600 (40), 1.222 (30), 1.085 (30), 2.65 (20), 1.324 (20)

Chemistry:

	(1)	(2)	(3)
Cu	47.99	47.8	50.1
Zn		0.3	
V	2.28	2.9	3.3
Fe	1.09	0.4	
As	9.54	7.0	13.6
Sn	6.71	8.4	0.26
Ge			0.63
Te	1.26		
Sb	0.19	2.8	1.4
S	30.65	30.5	31.2
Total	99.71	100.1	100.49

(1) Butte, Montana, USA. (2) Do.; by electron microprobe. (3) Lorano, Italy; by electron microprobe.

Occurrence: With other sulfides in hydrothermal copper ore veins (Butte, Montana, USA).

Association: Pyrite, tetrahedrite–tennantite, bornite, chalcocite, covellite, enargite, quartz (Butte, Montana, USA); hemusite, enargite, luzonite, stannoidite, reniérite, tennantite, chalcopyrite, pyrite (Chelopech, Bulgaria).

Distribution: In the USA, from the Leonard mine, south of the Colusa claim, and from other mines at Butte, Silver Bow Co., Montana; in Arizona, from the Magma mine, Pioneer district, Gila Co.; and the Campbell mine, Bisbee, Cochise Co. From Red Mountain Pass, San Juan, Ouray Co., and in the Buffalo Boy mine, near Silverton, San Juan Co., Colorado. In the Kidd Creek mine, near Timmins, Ontario, Canada. At Chuquicamata, Antofagasta, Chile. At Cerro de Pasco, Peru. From Lorano, 5 km from Carrara, Tuscany, Italy. At Chizeuil, Saône-et-Loire, France. From Bor, Serbia, Yugoslavia. At Chelopech, Bulgaria. In the Gay Cu-Zn deposit, Southern Ural Mountains, USSR.

Name: For the Colusa claim, Butte, Montana, USA.

Type Material: n.d.

References: (1) Palache, C., H. Berman, and C. Frondel (1944) Dana's system of mineralogy, (7th edition), v. I, 386–387. (2) Berry, L.G. and R.M. Thompson (1962) X-ray powder data for the ore minerals. Geol. Soc. Amer. Mem. 85, 58. (3) Dangel, P.N. and B.J. Wuensch (1970) The crystallography of colusite. Amer. Mineral., 55, 1787–1791. (4) Orlandi, P., S. Merlino, G. Duchi, and G. Vezzalini (1981) Colusite: a new occurrence and crystal chemistry. Can. Mineral., 19, 423–427.

$(Pt, Pd, Ni)S$ **Cooperite**

Crystal Data: Tetragonal. *Point Group:* $4/m\ 2/m\ 2/m$. As distorted crystal fragments elongated ‖ [101] with {110}, {111}, and {001}; as irregular grains to 1.5 mm. *Twinning:* Occasionally.

Physical Properties: *Cleavage:* {011}. *Fracture:* Conchoidal. Hardness = 4–5 VHN = 743–1018 (100 g load). D(meas.) = 9.5 D(calc.) = 10.2

Optical Properties: Opaque. *Color:* Steel-gray; in polished section, brownish. *Luster:* Metallic. *Pleochroism:* From white to creamy white to bluish white. *Anisotropism:* Strong, greenish gray to whitish yellow to brown-gray and brown. R_1–R_2: (400) 40.9–46.6, (420) 42.0–47.3, (440) 42.6–47.9, (460) 42.7–48.3, (480) 42.2–48.4, (500) 41.5–48.2, (520) 40.4–47.8, (540) 39.4–47.2, (560) 38.5–36.6, (580) 37.8–46.1, (600) 37.2–45.5, (620) 36.6–45.1, (640) 36.0–44.7, (660) 35.6–44.2, (680) 35.1–43.7, (700) 34.6–43.3

Cell Data: *Space Group:* $P4_2/mmc$. $a = 3.4700$ $c = 6.1096$ Z = 2

X-ray Powder Pattern: Potgietersrust district, Transvaal, South Africa. 3.013 (10), 1.911 (8), 1.507 (7), 2.450 (6), 1.753 (6), 1.732 (5), 1.231 (5)

Chemistry:

	(1)	(2)	(3)
Pt	85.1	79.7	85.89
Pd	0.6	5.6	
Ni	0.7	0.1	
S	13.9	14.3	14.11
Total	100.3	99.7	100.00

(1) Potgietersrust district, Transvaal, South Africa; by electron microprobe, corresponding to $(Pt_{0.98}Pd_{0.01}Ni_{0.03})_{\Sigma=1.02}S_{0.98}$. (2) Stillwater Complex, Montana, USA; by electron microprobe, corresponding to $(Pt_{0.90}Pd_{0.12}Ni_{0.004})_{\Sigma=1.024}S_{0.98}$. (3) PtS.

Polymorphism & Series: Dimorphous with braggite.

Occurrence: A significant platinum ore mineral, found in concentrates from norites (Bushveld Complex, South Africa); in breccia chalcopyrite-millerite ores (Noril'sk region, USSR).

Association: Braggite, vysotskite, platinum, sperrylite (Bushveld Complex, South Africa); Pt–Fe alloys, chalcopyrite, millerite (Noril'sk region, USSR); pentlandite, pyrrhotite, chalcopyrite, cubanite, nickelian mackinawite, gold, braggite, vysotskite, moncheite, isoferroplatinum, kotulskite, keithconnite, palladian tulameenite (Stillwater Complex, Montana, USA); sperrylite, daomanite (China).

Distribution: At various localities along the Merensky Reef of the Bushveld Complex, Transvaal, South Africa. From the Stillwater Complex, Montana; at the junction of the Trinity and Klamath Rivers, Humboldt Co., and from the Merced River at Snelling and Cressy, Merced Co., California, USA. In Canada, in the Lac des Iles Complex, Ontario, and the Tulameen district, British Columbia. In the Severnyi mine, Noril'sk region, western Siberia, and the Gusevogorskii magnetite deposits, Ural Mountains, USSR. From China, in the "Tao," " Ma," and "Yen" districts — apparently coded names.

Name: For R.A. Cooper, of Johannesburg, South Africa, who first described the mineral.

Type Material: British Museum (Natural History), London, England, BM 1932,1301; Harvard University, Cambridge, Massachusetts, USA, 101935.

References: (1) Palache, C., H. Berman, and C. Frondel (1944) Dana's system of mineralogy, (7th edition), v. I, 258–259. (2) Berry, L.G. and R.M. Thompson (1962) X-ray powder data for the ore minerals. Geol. Soc. Amer. Mem. 85, 71. (3) Cabri, L.J., J.H.G. Laflamme, J.M. Stewart, K. Turner, and B.J. Skinner (1978) On cooperite, braggite, and vysotskite. Amer. Mineral., 63, 832–839. (4) Cabri, L.J., Ed. (1981) Platinum group elements: mineralogy, geology, recovery. Can. Inst. Min. & Met., 100–101. (5) Criddle, A.J. and C.J. Stanley (1985) Characteristic optical data for cooperite, braggite and vysotskite. Can. Mineral., 23, 149–162.

Crystal Data: Cubic. *Point Group:* $4/m\ \overline{3}\ 2/m$. As cubes, dodecahedra, and as tetrahedra; rarely as octahedra and complex combinations. Often flattened {111}, elongated [001]. Also as irregular distortions, in twisted, wirelike shapes; filiform, arborescent, massive; as a coarse powder. Masses weighing hundreds of tons have been found; crystals up to 15 cm. *Twinning:* On {111} to produce simple contact and penetration twins and cyclic groups.

Physical Properties: *Fracture:* Hackly. *Tenacity:* Highly malleable and ductile. Hardness = 2.5–3 VHN = 77–99 (100 g load). D(meas.) = 8.95 D(calc.) = 8.93

Optical Properties: Opaque. *Color:* Light rose on fresh surface, quickly darkens to copper-red, then metallic, shining; in reflected light, rose-white. *Luster:* Metallic.
R: (400) 35.2, (420) 37.7, (440) 40.2, (460) 42.2, (480) 43.8, (500) 45.0, (520) 46.0, (540) 47.6, (560) 52.6, (580) 63.9, (600) 71.0, (620) 75.4, (640) 78.5, (660) 81.0, (680) 83.2, (700) 85.2

Cell Data: *Space Group:* $Fm3m$. $a = 3.615$ Z = 4

X-ray Powder Pattern: Synthetic.
2.088 (100), 1.808 (46), 1.278 (20), 1.0900 (17), 0.8293 (9), 0.8083 (8), 1.0436 (5)

Chemistry: Copper, usually with only small amounts of other metals.

Occurrence: Commonly associated with porous zones in mafic extrusive rocks, less commonly in sandstones and shales, where the copper was probably of hydrothermal origin, precipitated as the result of oxidizing conditions; in the oxidized zone of large, disseminated copper deposits as a result of secondary processes. A rare mineral in some meteorites.

Association: Silver, chalcocite, bornite, cuprite, malachite, azurite, tenorite, iron oxides, many other minerals.

Distribution: Occurs in many districts world-wide. In the USA, as remarkably large masses and excellent, large crystals in deposits of the Keweenaw Peninsula, northern Michigan; in several porphyry copper deposits in Arizona including those at the New Cornelia mine, Ajo, Pima Co.; the Copper Queen and other mines at Bisbee, Cochise Co.; and at Ray, Gila Co.; similarly in the Chino mine at Santa Rita, Grant Co., New Mexico. In Namibia, at the Emke mine, near Onganja, and at Tsumeb. In large crystals from Turinsk, Bogoslovsk, Perm, USSR. In Germany, at Rheinbreitbach, and the Friedrichssegen mine, near Bad Ems, Rheinland-Pfalz. In fine specimens from many mines in Cornwall, England. In Australia, at Broken Hill, New South Wales. In Chile, at Andacolla, near Coquimbo. From Bolivia, at Corocoro.

Name: From the Latin *cuprum*, in turn from the Greek *kyprios*, *Cyprus*, from which island the metal was early produced.

References: (1) Palache, C., H. Berman, and C. Frondel (1944) Dana's system of mineralogy, (7th edition), v. I, 99–102. (2) (1953) NBS Circ. 539, 1, 15.

Crystal Data: Cubic. *Point Group:* $2/m\ \overline{3}$. As rims and replacements of cinnabar, in grains less than 2 μm.

Physical Properties: Hardness = ~3 VHN = 28–61 (25 g load) (synthetic). D(meas.) = n.d. D(calc.) = 6.845

Optical Properties: Transparent. *Color:* Light orange-pink; on exposure to light, darkening to light gray, then black.
Optical Class: Isotropic. $n = > 2.5$
R: 16.7 (470), 15.5 (546), 15.1 (589), 15.1 (650)

Cell Data: *Space Group:* $I2_13$. $a = 8.940(5)$ Z = 4

X-ray Powder Pattern: Synthetic.
3.62 (100), 2.57 (70), 1.749 (60), 2.81 (50), 2.38 (40), 2.10 (40), 6.23 (30)

Chemistry:

	(1)	(2)
Hg	82.6	81.68
S	8.65	8.70
Cl	9.49	9.62
Total	100.74	100.00

(1) Arzak deposit, USSR; by electron microprobe. (2) $Hg_3S_2Cl_2$.

Polymorphism & Series: Dimorphous with lavrentievite.

Occurrence: In lake-bed sediments and underlying silicified rhyolite tuffs, possibly of low-temperature secondary origin (McDermitt mine, Nevada, USA); in an oxidized hydrothermal deposit (Arzak deposit, USSR).

Association: Cinnabar, kleinite, montmorillonite, quartz, cristobalite, orthoclase, plagioclase (McDermitt mine, Nevada, USA); cinnabar, arzakite, lavrentievite, quartz, kaolinite (Arzak deposit, USSR).

Distribution: In the McDermitt mercury mine, Opalite district, Humboldt Co., Nevada, USA. From the Arzak deposit, Tuva ASSR, USSR

Name: For its occurrence in the McDermitt (formerly Cordero) mine, USA.

Type Material: Stanford University, Palo Alto, California; National Museum of Natural History, Washington, D.C., USA, 133354.

References: (1) Foord, E.E., P. Berendsen, and L.O. Storey (1974) Corderoite, first natural occurrence of α–$Hg_3S_2Cl_2$, from the Cordero mercury deposit, Humboldt County, Nevada. Amer. Mineral., 59, 652–655. (2) Carlson, E.H. (1967) The growth of HgS and $Hg_3S_2Cl_2$ single crystals by a vapor phase method. J. Crystal Growth, 1, 271–277. (3) Vasil'ev, V.I. and O.K. Grechishchev (1979) First discovery of corderoite (α–$Hg_3S_2Cl_2$) in mercury ores of the USSR. Doklady Acad. Nauk SSSR, 246, 951–953 (in Russian). (4) (1979) Chem. Abs., 91, 126227 (abs. ref. 3).

Cosalite $Pb_2Bi_2S_5$

Crystal Data: Orthorhombic. *Point Group:* $2/m\ 2/m\ 2/m$. Prismatic crystals elongated ∥ [001]; flexible capillary fibers; usually massive in aggregates of radiating prismatic, fibrous, or feathery forms.

Physical Properties: *Cleavage:* Very rare. *Fracture:* Uneven. Hardness = 2.5–3 VHN = n.d. D(meas.) = 6.86–6.99 D(calc.) = 7.12

Optical Properties: Opaque. *Color:* Lead-gray to steel-gray to silver-white; in polished section, white, with only a trace of cream. *Streak:* Black. *Luster:* Metallic. *Pleochroism:* Weak. *Anisotropism:* Weak, but distinct in oil.
R_1–R_2: (400) 46.9–49.7, (420) 46.2–49.0, (440) 45.5–48.3, (460) 44.7–47.5, (480) 43.9–46.7, (500) 43.0–46.0, (520) 42.2–45.2, (540) 41.5–44.5, (560) 41.0–44.0, (580) 40.5–43.6, (600) 40.2–43.5, (620) 40.1–43.4, (640) 40.0–43.3, (660) 40.0–43.2, (680) 39.9–43.2, (700) 39.9–43.1

Cell Data: *Space Group:* *Pbnm.* $a = 19.09$ $b = 23.87$ $c = 4.055$ Z = 8

X-ray Powder Pattern: Hagidaira mine, Gumma Prefecture, Japan.
3.44 (100), 2.81 (30), 3.37 (25), 2.96 (20), 1.911 (16), 3.72 (14), 4.10 (12)

Chemistry:

	(1)	(2)	(3)
Pb	38.68	39.55	41.75
Cu	2.02	2.71	
Fe		0.25	
Bi	42.38	40.21	42.10
S	16.59	17.20	16.15
rem.		0.78	
Total	99.67	100.70	100.00

(1) McElroy Township, Canada. (2) Vaskö, Hungary (3) $Pb_2Bi_2S_5$.

Occurrence: In hydrothermal deposits formed at moderate temperatures but also found in contact metasomatic replacements, and epithermal replacements as well as in pegmatites.

Association: Sphalerite, chalcopyrite, pyrite, skutterudite, cobaltite, bornite, enargite, stromeyerite, bismuth, bismuthinite, calcite, tremolite, diopside, epidote, quartz.

Distribution: From numerous localities world-wide. In the USA, at Deer Park, Stevens Co., Washington; in Colorado, in the Red Mountain district, San Juan Co.; in Arizona, at the Magma mine, near Superior, and Landsman Camp, Gila Co.; and at Darwin, Inyo Co., California. In Canada, in Ontario, at Boston Creek, McElroy Township; and at Cobalt. From Hůrky, Czechoslovakia. In Sweden, at the Bjelkes mine, Nordmark, and Boliden, Västerbotten. At Carrock Fell mine, Grainsgill, Cumbria, England. From Băiţa (Rézbánya) and Moravicza, Romania. In the New England Range, New South Wales, Australia. From Amparindravato, Madagascar. At the Nuestra Señora mine, Cosalá, Sinaloa, Mexico.

Name: For the locality at Cosalá, Mexico.

Type Material: n.d.

References: (1) Palache, C., H. Berman, and C. Frondel (1944) Dana's system of mineralogy, (7th edition), v. I, 445–447. (2) Hayashi, S. (1961) Cosalite from the Hagidaira mine, Gun'ma Prefecture, Japan. Mineral. J. (Japan), 3, 148–155. (3) Weitz, G. and E. Hellner (1960) Über komplex zusammengesetzte sulfidische Erze VII. Zur Kristallstruktur des Cosalits, $Pb_2Bi_2S_5$. Zeits. Krist., 113, 385–402 (in German with English abs.). (4) Berry, L.G. and R.M. Thompson (1962) X-ray powder data for the ore minerals. Geol. Soc. Amer. Mem. 85, 148. (5) Ramdohr, P. (1969) The ore minerals and their intergrowths, (3rd edition), 767–768. (6) Srikrishnan, T. and W. Nowacki (1974) A redetermination of the crystal structure of cosalite, $Pb_2Bi_2S_5$. Zeits. Krist., 140, 114–136.

Crystal Data: Orthorhombic. *Point Group:* $mm2$. As lamellar inclusions in löllingite, up to 0.2 x 1.4 mm, or intergrown with paracostibite.

Physical Properties: Hardness = n.d. VHN = 781 (15 g load). D(meas.) = n.d. D(calc.) = 6.89

Optical Properties: Opaque. *Color:* In polished section, grayish. *Luster:* Metallic. *Pleochroism:* Weak, gray-white with bluish to brownish tints. *Anisotropism:* Weak, reddish brown or orange, to bluish.
R_1–R_2: n.d.

Cell Data: *Space Group:* $Pmn2_1$. $a = 3.603$ $b = 4.868$ $c = 5.838$ Z = 2

X-ray Powder Pattern: Broken Hill, Australia.
2.596 (100), 2.503 (90), 1.908 (80), 2.902 (60), 4.86 (50), 1.803 (50), 3.08 (40)

Chemistry:

	(1)	(2)	(3)
Co	26.7	25.6	27.70
Fe	0.6	0.8	
Ni	0.2	2.4	
Sb	57.0	56.8	57.23
As	0.3	0.3	
S	15.1	14.7	15.07
Total	99.9	100.6	100.00

(1) Consols mine, Australia; by electron microprobe, average of three grains, giving $(Co, Fe, Ni)_{0.98}(Sb, As)_{1.00}S_{1.00}$. (2) Getön deposit, Sweden; by electron microprobe. (3) CoSbS.

Polymorphism & Series: Dimorphous with paracostibite.

Occurrence: Intimately intergrown with other hydrothermal sulfides (Broken Hill, Australia); in Pb-Zn-Cu-Ag ore deposits remobilized by hydrothermal solutions from later granite emplacement (Bergslagen, Sweden).

Association: Löllingite, willyamite, ullmannite, pyrargyrite (Broken Hill, Australia); nisbite, paracostibite, chalcopyrite, bismuth, pyrrhotite, galena, sphalerite, gersdorffite, ullmannite (Bergslagen, Sweden).

Distribution: In Australia, at the Consols mine, Broken Hill, New South Wales. From the Gruvåsen and Getön deposits, Bergslagen, Sweden.

Name: For the composition.

Type Material: National Museum of Natural History, Washington, D.C., USA, R849.

References: (1) Cabri, L.J., D.C. Harris, and J.M. Stewart (1970) Costibite (CoSbS), a new mineral from Broken Hill, N.S.W., Australia. Amer. Mineral., 55, 10–17. (2) Rowland, J.F., E.J. Gabe, and S.R. Hall (1975) The crystal structures of costibite (CoSbS) and paracostibite (CoSbS). Can. Mineral., 13, 188–196. (3) Zakrzewski, M.A., E.A.J. Burke, and H.W. Nugteren (1980) Cobalt minerals in the Hallëfors area, Bergslagen, Sweden: new occurrences of costibite, paracostibite, nisbite and cobaltian ullmannite. Can. Mineral., 18, 165–171.

Crystal Data: Hexagonal. *Point Group:* $6/m\ 2/m\ 2/m$. Commonly massive and foliated; hexagonal plates as large as 10 cm, flattened on {0001}, which may exhibit hexagonal striae. As rosettes of nearly parallel plates.

Physical Properties: *Cleavage:* Highly perfect on {0001}. *Tenacity:* Flexible in thin leaves. Hardness = 1.5–2 VHN = 128–138 (100 g load). D(meas.) = 4.6–4.76 D(calc.) = 4.602

Optical Properties: Opaque. *Color:* Indigo-blue or darker, commonly highly iridescent, brass-yellow to deep red. *Streak:* Lead-gray, shining. *Luster:* Submetallic, inclining to resinous, somewhat pearly on cleavage; subresinous to dull when massive.
Optical Class: Uniaxial positive (+). *Pleochroism:* Marked, deep blue to blue-white. *Dispersion:* Strong. $\omega = 1.45$ (589). $\epsilon = 2.62$ (589). *Anisotropism:* Strong.
R_1–R_2: (400) 16.6–30.7, (420) 15.7–29.8, (440) 14.8–28.9, (460) 13.7–27.8, (480) 12.2–26.6, (500) 10.6–25.1, (520) 8.8–23.5, (540) 6.8–21.9, (560) 5.2–20.4, (580) 3.8–19.3, (600) 3.0–19.0, (620) 3.2–19.5, (640) 4.6–20.2, (660) 8.7–20.8, (680) 15.2–21.4, (700) 24.2–22.3

Cell Data: *Space Group:* $P6_3mmc$. $a = 3.7938$ $c = 16.341$ Z = 6

X-ray Powder Pattern: Synthetic.
2.813 (100), 1.896 (75), 3.048 (65), 2.724 (55), 1.735 (35), 1.556 (35), 3.220 (30)

Chemistry:

	(1)	(2)	(3)
Cu	66.06	65.49	66.48
Fe	0.14	0.25	
S	33.87	33.45	33.52
insol.	0.11		
Total	100.18	99.19	100.00

(1) Butte, Montana, USA. (2) Bor, Yugoslavia. (3) CuS.

Occurrence: Most commonly of secondary origin in the zone of oxidation in sulfide copper deposits. Rarely of primary hydrothermal origin. Widespread in most copper deposits; common as an iridescent tarnish on other sulfides.

Association: Chalcopyrite, chalcocite, djurleite, bornite, enargite, pyrite, other sulfides.

Distribution: Numerous localities, but with only a few producing rich material. In the USA, at Butte, Silver Bow Co., Montana, as fine crystals in the primary ore; at Kennecott, Alaska; at Summitville, Rio Grande Co., Colorado; in the La Sal district, San Juan Co., Utah; a widespread but minor constituent of the ores in most porphyry copper mines in Arizona. As large crystals from the Calabona mine, Alghero, Sardinia, Italy. In the Sierra de Famatina, La Rioja Province, Argentina. At Bor, Serbia, Yugoslavia. From Leogang, Salzburg, Austria. At Dillenburg, Hesse, and Sangerhausen, Saxony, Germany. From Kedabek, Caucasus Mountains, USSR. In the Bou Skour mine, Morocco.

Name: For Niccolò Covelli (1790–1829), discoverer of Vesuvian covellite.

References: (1) Palache, C., H. Berman, and C. Frondel (1944) Dana's system of mineralogy, (7th edition), v. I, 248–251. (2) (1955) NBS Circ. 539, 4, 13–14. (3) Berry, L.G. (1954) The crystal structure of covellite, CuS, and klockmannite, CuSe. Amer. Mineral., 39, 504–509.
(4) Evans, H.T., Jr. and J.A. Konnert (1976) Crystal structure refinement of covellite. Amer. Mineral., 61, 996–1000. (5) Ohmasa, M., M. Suzuki, and T. Takéuchi (1977) A refinement of the crystal structure of covellite, CuS. Mineral. J. (Japan), 8, 311–319.

Crystal Data: Triclinic. *Point Group:* 1 or $\bar{1}$. As grains up to 0.4 mm showing typically a chevron pattern in polished section, developed by opposing cleavage lamellae.

Physical Properties: *Cleavage:* Perfect on {111}. Hardness = ~1.5 VHN = n.d. D(meas.) = 2.5–2.63 D(calc.) = 2.879 Moderately magnetic.

Optical Properties: Opaque. *Color:* Black; in polished section, pale brownish gray with a pink tint. *Luster:* Metallic. *Pleochroism:* Faint, from gray to pink. *Anisotropism:* Strong, colors from gray to dull golden orange.
R_1–R_2: n.d.

Cell Data: *Space Group:* $P1$ or $P\bar{1}$. $a = 7.409(8)$ $b = 9.881(6)$ $c = 6.441(3)$ $\alpha = 100°25(3)'$ $\beta = 104°37(5)'$ $\gamma = 81°29(5)'$ Z = 2

X-ray Powder Pattern: Near Orick, California, USA.
5.12 (100), 7.13 (90), 3.023 (80), 3.080 (70), 9.6 (60), 5.60 (60), 3.910 (50)

Chemistry:

	(1)	(2)
Na	5.99	5.94
Fe	44.0	43.31
S	41.3	41.44
H_2O	[8.71]	9.31
Total	100.00	100.00

(1) Near Orick, California, USA; by electron microprobe, average of five grains. Water by difference, on independent proof of the presence of oxygen. (2) $NaFe_3S_5 \cdot 2H_2O$.

Occurrence: With rare iron sulfides in small pegmatitic clots thought to have crystallized late in the consolidation of the Coyote Peak intrusive, a mafic alkalic diatreme.

Association: Pyrrhotite, djerfisherite, rasvumite, bartonite, erdite, phlogopite, schorlomite, acmite, sodalite, cancrinite, pectolite, natrolite, magnetite, calcite.

Distribution: From near Orick, Humboldt Co., California, USA.

Name: For Coyote Peak, a local prominence which gives its name to the U.S.G.S. 15 minute quadrangle map of the area.

Type Material: National Museum of Natural History, Washington, D.C., USA, 150335.

References: (1) Erd, R.C. and G.K. Czamanske (1983) Orickite and coyoteite, two new sulfides from Coyote Peak, Humboldt Co., California. Amer. Mineral., 68, 245–254.

Crystal Data: Monoclinic, pseudotetragonal. *Point Group:* $2/m$, m, or 2. *Twinning:* Common, according to some simple law. As lath-like or anhedral grains from 20 to 70 μm.

Physical Properties: Hardness = 3–3.5 VHN = 94–129 (25 g load). D(meas.) = 6.86(7) (synthetic). D(calc.) = 6.57

Optical Properties: Opaque. *Color:* Gray-blue in reflected light. *Streak:* Black. *Luster:* Metallic. *Pleochroism:* Weak, from gray-blue to slightly greenish gray-blue. *Anisotropism:* Distinct to moderate, from buff to slate gray.
R_1–R_2: (400) 40.6–41.8, (420) 40.1–41.0, (440) 39.4–40.1, (460) 38.6–39.4, (480) 38.1–39.2, (500) 37.6–39.1, (520) 37.3–38.9, (540) 36.7–38.4, (560) 36.1–38.0, (580) 35.6–37.6, (600) 35.0–37.2, (620) 34.4–36.6, (640) 33.7–36.0, (660) 33.1–35.3, (680) 32.6–34.6, (700) 32.5–34.1

Cell Data: *Space Group:* $A2/m$, $A2$, or Am. $a = 19.96$ $b = 8.057$ $c = 7.809$ $\beta = 92.08°$ Z = 2

X-ray Powder Pattern: Synthetic.
2.813 (100), 5.63 (90), 2.860 (70), 1.959 (70), 2.018 (60), 3.91 (50), 3.456 (50)

Chemistry:

	(1)	(2)
Tl	7.1 – 7.8	8.02
Ag	7.8 – 8.8	8.46
Au	22.3 – 23.2	23.18
Sb	47.3 – 48.2	47.76
S	12.8 – 13.3	12.58
Total		100.00

(1) Hemlo deposit, Canada; by electron microprobe, range of analyses of five samples, the average leading to $Tl_{0.92}Ag_{1.99}Au_{2.93}Sb_{9.87}S_{10.28}$. (2) $TlAg_2Au_3Sb_{10}S_{10}$.

Occurrence: As a primary hydrothermal mineral in a complex gold deposit.

Association: Aurostibite, chalcostibite, parapierrotite, gold, antimony, stibnite, tetrahedrite, molybdenite, zinkenite, pyrite, quartz.

Distribution: From the Hemlo gold deposit, Thunder Bay district, Ontario, Canada.

Name: For Alan J. Criddle (1944–), mineralogist of the British Museum (Natural History), London, England.

Type Material: British Museum (Natural History), London, England, E.1230, BM 1987,351; Canadian Geological Survey, Ottawa, NMC 65186; Royal Ontario Museum, Toronto, Canada.

References: (1) Harris, D.C., A.C. Roberts, J.H.G. Laflamme, and C.J. Stanley (1988) Criddleite, $TlAg_2Au_3Sb_{10}S_{10}$, a new gold-bearing mineral from Hemlo, Ontario, Canada. Mineral. Mag., 52, 691–697.

Crystal Data: Tetragonal. *Point Group:* $4/m$, 4, or $\bar{4}$. As finely divided, disseminated specks, and as small veinlets.

Physical Properties: *Cleavage:* Two at right angles, fairly well developed. *Tenacity:* Brittle. Hardness = 2.5–3 VHN = 92–123 (100 g load). D(meas.) = 6.90 D(calc.) = 7.443

Optical Properties: Opaque. *Color:* Lead-gray. *Luster:* Metallic. *Anisotropism:* Weak but distinct, in brownish tones.

R_1–R_2: (400) 33.0–37.3, (420) 32.6–37.4, (440) 32.2–37.5, (460) 31.8–37.4, (480) 31.4–37.3, (500) 31.4–37.0, (520) 31.5–36.7, (540) 31.5–36.5, (560) 31.6–36.5, (580) 31.7–36.6, (600) 31.9–36.8, (620) 32.2–37.0, (640) 32.3–37.4, (660) 32.5–37.8, (680) 32.7–38.2, (700) 32.8–38.6

Cell Data: *Space Group:* $I4/m$, $I4$, or $I\bar{4}$. $a = 10.435$ $c = 3.954$ Z = 2

X-ray Powder Pattern: Skrikerum, Sweden.
3.29 (100), 2.59 (100), 3.00 (80), 2.11 (50), 1.833 (40), 1.779 (40), 2.32 (30)

Chemistry:

	(1)	(2)	(3)	(4)
Tl	16.27	18.55	21.03	21.18
Cu	46.55	46.11	46.89	46.09
Ag	5.04	1.44	0.06	
Fe	0.36	0.63		
Se	30.86	33.27	32.43	32.73
Total	99.08	100.00	100.41	100.00

(1–2) Skrikerum, Sweden. (3) Bukov, Czechoslovakia; by electron microprobe, leading to $Tl_{0.98}Cu_{7.07}Ag_{0.01}Se_{3.94}$. (4) $TlCu_7Se_4$.

Occurrence: Of hydrothermal origin, with other selenides.

Association: Umangite, berzelianite, eucairite, klockmannite, clausthalite, sabatierite, selenian linnaeite, calcite, quartz.

Distribution: From Skrikerum, near Tryserum, Kalmar, Sweden. In the Pinky Fault uranium deposit, near Lake Athabasca, Saskatchewan, Canada. From Bukov, near Tisnova, and Petrovice, Czechoslovakia.

Name: After Sir William Crookes (1832–1919), who discovered thallium.

References: (1) Palache, C., H. Berman, and C. Frondel (1944) Dana's system of mineralogy, (7th edition), v. I, 183. (2) Earley, J.W. (1950) Description and synthesis of the selenide minerals. Amer. Mineral., 35, 337–364. (3) Ramdohr, P. (1969) The ore minerals and their intergrowths, (3rd edition), 50–51. (4) Johan, Z. (1987) Crookesite, $TlCu_7Se_4$; new data and isotypism with $NH_4Cu_7S_4$. Comp. Rendu Acad. Sci. Paris, 304 (series II), 1121–1124 (in French). (5) (1988) Amer. Mineral., 73, 933 (abs. ref. 4). (6) Berger, R.A. (1987) Crookesite and sabatierite in a new light — a crystallographer's comment. Zeits. Krist., 181, 241–249.

Crystal Data: n.d. *Point Group:* n.d. Seen only in polished section associated with tetradymite.

Physical Properties: Hardness = n.d. VHN = n.d. D(meas.) = n.d. D(calc.) = n.d.

Optical Properties: Opaque. *Color:* In polished section, gray like tetradymite, but differing from it by lacking its creamy tinge. *Luster:* Metallic. *Anisotropism:* Light to dark bluish gray. R_1–R_2: n.d.

Cell Data: *Space Group:* n.d. Z = n.d.

X-ray Powder Pattern: n.d.

Chemistry:

	(1)	(2)
Bi	67.76	68.56
Fe	trace	
Te	20.41	20.93
Se	1.37	
S	9.97	10.51
rem.	0.40	
Total	99.91	100.00

(1) Ciclova, Romania. (2) Bi_2TeS_2.

Occurrence: With other bismuth tellurides.

Association: Tetradymite, chalcopyrite, calcite.

Distribution: Found on a museum specimen from Ciclova, Romania.

Name: For the occurrence at Ciclova (Csiklova), Romania.

Type Material: n.d.

References: (1) Koch, S. and J. Grasselly (1948) Bismuth minerals in the Carpathian basin. Acta Univ. Szeged., Sec. Sci. Nat., Acta Mineral. Petrog. 2, 1–30 (in English, 17–30). (2) (1950) Amer. Mineral., 35, 333 (abs. ref. 1). (3) Sindeeva, N.D. (1964) Mineralogy and types of deposits of selenium and tellurium, 141–142. (4) Vlasov, K.A., Ed. (1966) Mineralogy of rare elements, v. II, 736–737.

Crystal Data: Orthorhombic. *Point Group:* $2/m\ 2/m\ 2/m$. Crystals elongated ∥ [001]; thick tabular {001} and {110}; {001} striated ∥ [010]; up to 3 cm in maximum dimension; also massive. *Twinning:* Common with twin plane {110}, in pairs, also as fourlings and sixlings; then pseudohexagonal.

Physical Properties: *Cleavage:* None; parting on {110} and $\{1\bar{3}0\}$. *Fracture:* Conchoidal. Hardness = 3.5 VHN = n.d. D(meas.) = 4.03–4.18 D(calc.) = 4.076 Strongly magnetic.

Optical Properties: Opaque. *Color:* Brass- to bronze-yellow. *Anisotropism:* On polished surface, distinctive.

R_1–R_2: (400) 20.6–23.8, (420) 23.6–27.7, (440) 26.6–31.6, (460) 29.2–34.7, (480) 31.4–37.1, (500) 33.4–39.0, (520) 35.0–40.2, (540) 36.3–41.0, (560) 37.3–41.6, (580) 38.3–42.0, (600) 39.0–42.4, (620) 39.8–42.8, (640) 40.7–43.3, (660) 41.6–43.8, (680) 42.4–44.3, (700) 43.3–44.6

Cell Data: *Space Group:* *Pcmn.* $a = 6.467(1)$ $b = 11.117(1)$ $c = 6.231(2)$ Z = 4

X-ray Powder Pattern: Sudbury, Canada.
3.22 (10), 1.867 (8), 1.750 (7), 1.165 (5), 3.49 (4), 3.00 (4), 2.79 (4)

Chemistry:

	(1)	(2)	(3)
Cu	24.32	23.52	23.42
Fe	41.15	41.14	41.15
S	34.37	35.30	35.43
Total	99.84	99.96	100.00

(1) Barracanao, Cuba. (2) Prince William Sound, Alaska, USA. (3) $CuFe_2S_3$.

Polymorphism & Series: Dimorphous with isocubanite.

Occurrence: In hydrothermal deposits formed at relatively high temperature, in pyrrhotite-pentlandite ores in which it commonly occurs as intimate oriented intergrowths with chalcopyrite. Cubanite may exsolve from chalcopyrite below about 200-210°C. A rare constituent of some carbonaceous chondrite meteorites.

Association: Chalcopyrite, pyrite, pyrrhotite, sphalerite.

Distribution: Numerous localities; occasionally an important ore mineral. In the USA, at Fierro, Grant Co., New Mexico; at the Christmas mine, Gila Co., Arizona; at Prince William Sound, Alaska. In the Sudbury district, Ontario; and in exceptional crystals from Chibougamau, Quebec, Canada. From Barracanao, Cuba. In Sweden, at Tunaberg, Södermanland, and at Kaveltorp, near Ljusnarsberg, Örebro. From Virtasalm, Finland. At Traversella, Piedmont, Italy. In Brazil, at the Morro Velho gold mine, Nova Lima, Minas Gerais. At Broken Hill, New South Wales, Australia. From the Noril'sk region, western Siberia, USSR. In the Orgueil and Alais carbonaceous chondrite meteorites.

Name: For its occurrence in Cuba.

References: (1) Palache, C., H. Berman, and C. Frondel (1944) Dana's system of mineralogy, (7th edition), v. I, 243–246. (2) Buerger, M.J. (1947) The crystal structure of cubanite. Amer. Mineral., 32, 415–425. (3) Berry, L.G. and R.M. Thompson (1962) X-ray powder data for the ore minerals. Geol. Soc. Amer. Mem. 85, 65–66. (4) Cabri, L.J., S.R. Hall, J.T. Szymanski, and J.M. Stewart (1973) On the transformation of cubanite. Can. Mineral., 12, 33–38. (5) Szymanski, J.T. (1974) A refinement of the structure of cubanite ($CuFe_2S_3$). Zeits. Krist., 140, 218–239.

Crystal Data: Orthorhombic. *Point Group:* n.d. Myrmekitic and dendritic drop-like grains up to 35 μm within first-generation khatyrkite, and as rounded or irregular grains to 20 μm in cracks and interstices in second-generation khatyrkite.

Physical Properties: Hardness = n.d. VHN = 272–318 (20 and 50 g loads). D(meas.) = n.d. D(calc.) = 5.12

Optical Properties: Opaque. *Color:* Steel-yellow. *Luster:* Metallic. *Anisotropism:* Very weak, from light gray to gray.
R: (400) — , (420) — , (440) 66.8, (460) 66.1, (480) 65.3, (500) 64.5, (520) 63.7, (540) 62.9, (560) 62.1, (580) 61.3, (600) 60.4, (620) 59.7, (640) 58.9, (660) 58.2, (680) 57.7, (700) 57.2

Cell Data: *Space Group:* n.d. $a = 6.95(1)$ $b = 4.16(1)$ $c = 10.04(1)$ Z = 10

X-ray Powder Pattern: Listvenitovii stream, USSR.
5.07 (10), 4.12 (8), 3.59 (2), 2.83 (1), 2.607 (1), 2.316 (1), 2.023 (1)

Chemistry:

	(1)
Cu	59.9 – 61.7
Zn	7.66 – 9.35
Al	29.3 – 30.4
Total	

(1) Listvenitovii stream, USSR; by electron microprobe, ranges of analyses on nine grains, corresponding to (Cu, Zn)Al.

Occurrence: In black slick washed from greenish gray cover weathering from serpentine.

Association: Khatyrkite, two unnamed zinc aluminides.

Distribution: Near the Listvenitovii stream, Khatirskii ultramafic zone of the Koryak–Kamchata fold area, Koryak Mountains, USSR.

Name: For the composition.

Type Material: Mining Museum, Leningrad Mining Institute, Leningrad, USSR.

References: (1) Razin, L.V., N.S. Rudashevskii, and L.N. Vyal'sov (1985) New natural intermetallic compounds of aluminum, copper and zinc — khatyrkite $CuAl_2$, cupalite CuAl and zinc aluminides — from hyperbasites of dunite–harzburgite formation. Zap. Vses. Mineral. Obshch., 114, 90–100 (in Russian). (2) (1986) Amer. Mineral., 71, 1278 (abs. ref. 1).

Crystal Data: Monoclinic. *Point Group:* $2/m$. Massive and as prismatic crystals and thin blades, slightly twisted ∥ [010], the elongation axis.

Physical Properties: Hardness = n.d. VHN = n.d. D(meas.) = 6.36 D(calc.) = [6.24]

Optical Properties: Opaque. *Color:* Gray, commonly tarnished to bluish. *Luster:* Metallic. *Anisotropism:* Distinct.
R_1–R_2: (400) 31.5–38.7, (420) 32.5–39.2, (440) 33.5–39.7, (460) 34.4–40.1, (480) 35.2–40.5, (500) 35.8–40.8, (520) 36.3–41.1, (540) 36.8–41.3, (560) 37.1–41.5, (580) 37.3–41.8, (600) 37.4–42.0, (620) 37.4–42.2, (640) 37.0–42.3, (660) 36.5–42.2, (680) 35.7–42.0, (700) 34.8–41.5

Cell Data: *Space Group:* $C2/m$. a = 17.520(1) b = 3.926(3) c = 15.261(1) β = 100.18(1)° Z = [1]

X-ray Powder Pattern: Hall's Valley, Colorado, USA.
3.10 (10), 2.73 (6), 3.65 (4b), 4.31 (3), 1.961 (3), 1.719 (3), 6.24 (2)

Chemistry:

	(1)	(2)	(3)	(4)
Cu	15.96	12.65	15.1	16.37
Ag	0.89	4.09	1.19	
Pb			0.84	
Zn	0.10	0.07		
Mn			0.14	
Fe	2.13	0.59		
Bi	60.80	63.42	63.8	64.63
Sb			0.08	
Se			0.50	
Te			0.07	
S	[19.94]	[18.83]	18.5	19.00
Total	99.82	99.65	100.2	100.00

(1–2) Hall's Valley, Colorado; analysis by Hillebrand (1884), who assumed the samples to contain: (1) 6.97% chalcopyrite after deducting 4.43% gangue, and (2) 1.91% chalcopyrite, after deducting 59.75% gangue. Sulfur content calculated for the analyses. (3) Ohio mining district, Piute Co., Utah, USA; by electron microprobe, leading to $Cu_{10}Bi_{12}S_{23}$. (4) $Cu_{10}Bi_{12}S_{23}$.

Occurrence: In quartz veins with other sulfides (Colorado, USA).

Association: Emplectite, aikinite, wittichenite, benjaminite, berryite, cupropavonite, wolframite, bismuthinite.

Distribution: In the USA, at the Missouri mine, Hall's Valley, Park Co., and at Silver Cliff, Custer Co., Colorado; in the Fairfax quarry, Centreville, Virginia; and from the Tunnel Extension 2 mine, Marysvale, Ohio district, Piute Co., Utah. From Krupka, Czechoslovakia. At Baicolliou, Switzerland.

Name: For the composition.

Type Material: National Museum of Natural History, Washington, D.C., 92902; Harvard University, Cambridge, Massachusetts, USA, 94989.

References: (1) Palache, C., H. Berman, and C. Frondel (1944) Dana's system of mineralogy, (7th edition), v. I, 437. (2) Nuffield, E.W. (1952) Studies of mineral sulfo-salts: XVI Cuprobismuthite. Amer. Mineral., 37, 447–452. (3) Berry, L.G. and R.M. Thompson (1962) X-ray powder data for the ore minerals. Geol. Soc. Amer. Mem. 85, 143–144. (4) Taylor, C.M., A.S. Radtke, and C.L. Christ (1973) New data on cuprobismutite. J. Res. U.S. Geol. Sur. 1, 99–103. (5) (1973) Amer. Mineral., 58, 967 (abs. ref. 4). (6) Ozawa, T. and W. Nowacki (1975) The crystal structure of, and the bismuth–copper distribution in synthetic cuprobismuthite. Zeits. Krist., 142, 161–176. (7) Mariolacos, K., V. Kupčík, M. Ohmasa, and G. Miehe (1975) The crystal structure of $Cu_4Bi_5S_{10}$ and its relation to the structures of hodrushite and cuprobismutite.

Crystal Data: Cubic. *Point Group:* $4/m\ \overline{3}\ 2/m$. Small (to 300 μm) inclusions in isoferroplatinum.

Physical Properties: *Tenacity:* Very brittle. Hardness = n.d. VHN = 578 (30 g load). D(meas.) = n.d. D(calc.) = 7.24

Optical Properties: Opaque. *Color:* Iron-black; gray in reflected light. *Luster:* Metallic. R_1–R_2: (400) — , (420) — , (440) — , (460) 35.0-37.5, (480) 34.4-37.2, (500) 34.0-37.1, (520) 34.0-37.1, (540) 33.6-37.2, (560) 33.4-37.2, (580) 33.1-37.5, (600) 32.8-37.7, (620) 32.6-37.8, (640) 32.4-37.8, (660) 32.2-37.8, (680) 32.1-38.1, (700) 32.0-38.4

Cell Data: *Space Group:* $Fd3m$. $a = 9.92$ Z = 8

X-ray Powder Pattern: Far Eastern Region, USSR.
3.00 (100), 1.760 (100), 2.489 (90) 1.912 (70), 1.011 (70), 1.290 (60), 1.107 (60)

Chemistry:

	(1)	(2)
Cu	7.41	11.03
Fe	3.17	
Ni	0.27	
Ir	48.9	66.71
Pt	10.5	
Rh	6.05	
S	24.6	22.26
Total	100.9	100.00

(1) Far Eastern Region, USSR; by electron microprobe, corresponding to $(Cu_{0.61}Fe_{0.30}Ni_{0.02})_{\Sigma=0.93}(Ir_{1.33}Pt_{0.28}Rh_{0.31})_{\Sigma=1.92}S_{4.00}$. (2) $CuIr_2S_4$.

Polymorphism & Series: Forms a series with cuprorhodsite and malanite.

Occurrence: In alluvial deposits.

Association: Isoferroplatinum, cuprorhodsite, malanite, osmium, iridosmine, laurite, erlichmanite, cooperite, sperrylite, chalcopyrite, bornite.

Distribution: From undefined localities in the Aldan Shield and Kamchatka, Far Eastern Region, USSR.

Name: For the chemical composition.

Type Material: Mining Museum, Leningrad Mining Institute, Leningrad, USSR.

References: (1) Rudashevskii, N.S., Y.P. Men'shikov, A.G. Mochalov, N.V. Trubkin, N.I. Shumskaya, and V.V. Zhdanov (1985) Cuprorhodsite $CuRh_2S_4$ and cuproiridisite $CuIr_2S_4$ — new natural thiospinels of platinum elements. Zap. Vses. Mineral. Obshch., 114, 187–195 (in Russian). (2) (1986) Amer. Mineral., 71, 1277 (abs. ref. 1).

Crystal Data: Monoclinic. *Point Group:* $2/m$. Fine lamellar intergrowths with pavonite.

Physical Properties: Hardness = n.d. VHN = n.d. D(meas.) = n.d. D(calc.) = [7.04]

Optical Properties: Opaque. *Color:* Lead-gray to tin-white; in polished section, white. *Pleochroism:* Weak in air; weak to distinct in oil. *Anisotropism:* Distinct.

R_1–R_2: (400) 45.1–48.2, (420) 45.0–48.8, (440) 45.0–49.4, (460) 44.6–49.8, (480) 44.4–50.2, (500) 43.9–50.0, (520) 43.4–49.1, (540) 43.0–48.2, (560) 42.5–47.5, (580) 42.2–46.9, (600) 42.1–46.4, (620) 42.0–46.4, (640) 41.9–46.2, (660) 41.7–46.1, (680) 41.6–46.0, (700) 41.2–45.6

Cell Data: *Space Group:* $C2/m$. $a = 13.45$ $b = 4.02$ $c = 33.06$ $\beta = 93.50°$ Z = [4]

X-ray Powder Pattern: Alaska mine, Colorado, USA.
2.892 (vs), 2.257 (vs), 2.193 (vs), 2.118 (vs), 2.019 (vs), 2.007 (vs), 1.8011 (vs)

Chemistry:

	(1)	(2)	(3)	(4)
Ag	5.7	5.9	7.01	5.97
Pb	13.5	13.4	7.65	11.46
Cu	6.2	6.1	7.26	7.03
Bi	56.4	56.8	59.32	57.80
Sb	0.1	0.3		
S	18.0	17.6	18.10	17.74
Total	99.9	100.1	99.34	100.00

(1–2) Alaska mine, Colorado, USA; by electron microprobe, the average corresponding to $AgPb_{1.2}Cu_{1.8}Bi_5S_{10}$. (3) Hall's Valley, Colorado, USA; by electron microprobe. (4) $AgPbCu_2Bi_5S_{10}$.

Occurrence: Found in zoned mesothermal base- and precious-metal veins.

Association: Pavonite, gustavite, cuprobismutite, berryite.

Distribution: In the USA, from the Alaska mine, Poughkeepsie Gulch, near Ouray, San Juan Co., and at the Missouri mine, Hall's Valley, Park Co., Colorado; in the Campbell mine, Bisbee, Cochise Co., Arizona; and from the April Fool mine, Delamar, Lincoln Co., Nevada.

Name: To stress a similarity with pavonite.

Type Material: University of Pennsylvania, Philadelphia, Pennsylvania, USA, 1124, ("alaskaite").

References: (1) Karup-Møller, S. and E. Makovicky (1979) On pavonite, cupropavonite, benjaminite, and "oversubstituted" gustavite. Bull. Minéral., 102, 351–367. (2) (1980) Amer. Mineral., 65, 206 (abs. ref. 1). (3) Nuffield, E.W. (1980) Cupropavonite from Hall's Valley, Park County, Colorado. Can. Mineral., 18, 181–184.

Crystal Data: Cubic. *Point Group:* $4/m\ \overline{3}\ 2/m$. Small inclusions in isoferroplatinum grains, up to 300 μm.

Physical Properties: *Tenacity:* Very brittle. Hardness = n.d. VHN = 498 (50 g load). D(meas.) = n.d. D(calc.) = [5.88]

Optical Properties: Opaque. *Color:* Iron-black; gray in reflected light. *Luster:* Metallic.
R_1–R_2: (400) — , (420) — , (440) — , (460) 35.6–39.5, (480) 35.8–39.3, (500) 36.4–39.3, (520) 36.6–39.2, (540) 36.7–39.2, (560) 36.8–39.1, (580) 36.9–39.1, (600) 37.0–39.1, (620) 37.0–39.0, (640) 37.0–38.9, (660) 36.9–38.8, (680) 36.8–38.8, (700) 36.7–38.7

Cell Data: *Space Group:* $Fd3m$. $a = 9.88$ Z = 8

X-ray Powder Pattern: Far Eastern Region, USSR.
3.00 (100), 1.758 (100), 1.009 (90), 1.904 (80), 2.480 (70), 1.286 (50), 1.102 (50)

Chemistry:

	(1)	(2)
Cu	7.55	15.98
Fe	5.31	
Rh	39.6	51.76
Ir	10.3	
Pt	6.8	
S	29.8	32.26
Total	99.36	100.00

(1) Far Eastern Region, USSR; by electron microprobe, corresponding to $(Cu_{0.51}Fe_{0.41})_{\Sigma=0.92}(Rh_{1.66}Ir_{0.23}Pt_{0.15})_{\Sigma=2.04}S_{4.00}$. (2) $CuRh_2S_4$.

Polymorphism & Series: Forms a series with cuproiridsite.

Occurrence: In alluvial placers.

Association: Isoferroplatinum, cuproiridsite, malanite, osmium, iridosmine, laurite, erlichmanite, cooperite, sperrylite, chalcopyrite, bornite.

Distribution: From undefined localities in the Aldan shield and Kamchatka, Far Eastern Region, USSR.

Name: For the chemical composition.

Type Material: Mining Museum, Leningrad Mining Institute, Leningrad, USSR.

References: (1) Rudashevskii, N.S., Y.P. Men'shikov, A.G. Mochalov, N.V. Trubkin, N.I. Shumskaya, and V.V. Zhdanov (1985) Cuprorhodsite $CuRh_2S_4$ and cuproiridsite $CuIr_2S_4$ — new natural thiospinels of platinum elements. Zap. Vses. Mineral. Obshch., 114, 187–195 (in Russian). (2) (1986) Amer. Mineral., 71, 1277 (abs. ref. 1).

Crystal Data: Tetragonal. *Point Group:* $4/m$ $2/m$ $2/m$. As fine grained aggregates up to 1.5 mm in diameter. *Twinning:* Platy (sic).

Physical Properties: *Cleavage:* One direction. *Fracture:* Uneven. Hardness = n.d. VHN = 216–249 (50 g load). D(meas.) = n.d. D(calc.) = 8.42

Optical Properties: Opaque. *Color:* Steel-gray, violet-red tint on fresh fracture; in reflected light, violet-rose. *Luster:* Metallic. *Pleochroism:* Creamy white to dark rose-violet. *Anisotropism:* Strong.
R_1–R_2: (400) 56.0–60.8, (420) 54.9–59.4, (440) 53.8–58.0, (460) 51.1–55.1, (480) 48.7–52.8, (500) 46.4–50.5, (520) 42.9–48.3, (540) 38.7–45.5, (560) 36.7–43.9, (580) 37.9–44.0, (600) 42.2–45.9, (620) 47.6–48.6, (640) 52.9–51.9, (660) 57.7–54.5, (680) 61.3–56.9, (700) 65.0–57.6

Cell Data: *Space Group:* $P4/nmm$. $a = 3.990$ $c = 6.09$ Z = 2

X-ray Powder Pattern: Mt. Nákâlâq, Greenland.
2.07 (100), 2.56 (50), 2.82 (40), 1.993 (40), 1.167 (40), 1.424 (30), 3.33 (20)

Chemistry:

	(1)	(2)	(3)	(4)
Cu	53.3	50.4	49.24	51.07
Tl	3.5			
Zn			0.29	
Ag	0.1		0.16	
Pb			0.08	
Fe			0.02	
Sb	42.0	48.9	49.36	48.93
As			0.44	
S	1.1		0.16	
Total	100.0	99.3	99.75	100.00

(1) Mt. Nákâlâq, Greenland; by electron microprobe, average of three analyses, yielding $Cu_{2.10}(Sb_{0.86}Tl_{0.04})_{\Sigma=0.90}$. (2) Långban, Sweden; by electron microprobe, average of three analyses. (3) Långsjön, Sweden; by electron microprobe, average of 31 analyses on 16 grains, giving $(Cu_{1.861}Zn_{0.011}Ag_{0.004})_{\Sigma=1.876}(Sb_{0.974}As_{0.014}S_{0.012})_{\Sigma=1.000}$. (4) Cu_2Sb.

Occurrence: As fine-grained aggregates in a vein of ussingite cutting sodalite syenite (Ilímaussaq Intrusion, Greenland); in sulfide veins in metamorphic siliceous dolomite (Långsjön, Sweden).

Association: Löllingite, antimonian silver, chalcopyrite, chalcothallite (Ilímaussaq Intrusion, Greenland); chalcocite, bismuth (Långban, Sweden); sphalerite, galena, dyscrasite, gudmundite, breithauptite, antimony (Långsjön, Sweden).

Distribution: From northwest of Taseq Lake, at Mt. Nákâlâq and other localities in the Ilímaussaq Intrusion, southern Greenland. From Långsjön and Långban, Värmland, Sweden.

Name: For the composition.

Type Material: n.d.

References: (1) Sørensen, H., E.I. Semenov, M.S. Bezsmertnaya, and E.B. Khalezova (1969) Cuprostibite, a new natural compound of copper and antimony. Zap. Vses. Mineral. Obshch., 98, 716–724 (in Russian). (2) (1970) Amer. Mineral., 55, 1810 (abs. ref. 1). (3) Burke, E.A.J. (1980) Cuprostibite, cuprian galena, altaite, cuprian massicot, wittichenite, and bismuthinite from Långban, Sweden. Neues Jahrb. Mineral., Monatsh., 241–246. (4) Hålenius, U. and C. Ålinder (1982) Occurrence and formation of cuprostibite in a Zn–Pb–Ag mineralized siliceous dolomite at Långsjön, central Sweden. Neues Jahrb. Mineral., Monatsh., 201–215.

Crystal Data: Triclinic. *Point Group:* $\bar{1}$. In concentric spherical or tubular shells and aggregates, up to 5 cm across and 2–3 cm in length; also massive.

Physical Properties: *Cleavage:* {100} excellent. *Tenacity:* Slightly malleable. Hardness = 2.5 VHN = 54–93 (100 g load). D(meas.) = 5.42–5.49 D(calc.) = 5.443

Optical Properties: Opaque. *Color:* In reflected light, galena-white. *Streak:* Black. *Luster:* Metallic. *Pleochroism:* Exceptionally weak in air, stronger in oil. *Anisotropism:* Distinct, gray to pale yellowish or brownish gray.
R_1–R_2: (400) 33.8–38.0, (420) 33.8–38.0, (440) 33.8–37.9, (460) 33.7–37.8, (480) 33.4–37.4, (500) 33.0–37.0, (520) 32.6–36.4, (540) 32.0–36.4, (560) 31.5–35.9, (580) 31.0–35.4, (600) 30.6–35.0, (620) 30.3–34.7, (640) 29.9–34.4, (660) 29.6–34.2, (680) 29.3–34.0, (700) 29.0–33.8

Cell Data: *Space Group:* Two subcells are recognized, both $P\bar{1}$: the first (pseudotetragonal) has $a = 11.733(5)$ $b = 5.790(8)$ $c = 5.810(5)$ $\alpha = 90.00(0.20)°$ $\beta = 92.38(0.20)°$ $\gamma = 93.87(0.20)°$ and the second (pseudohexagonal) has $a = 11.709(5)$ $b = 3.670(8)$ $c = 6.320(5)$ $\alpha = 90.00(0.20)°$ $\beta = 92.58(0.20)°$ $\gamma = 90.85(0.20)°$ Z = 2

X-ray Powder Pattern: Poopó, Bolivia.
3.85 (100), 2.885 (100), 3.9 (90), 3.06 (65), 2.849 (65), 2.044 (65), 2.026 (65)

Chemistry:

	(1)	(2)	(3)
Pb	35.41	35.5	33.70
Sn	26.37	26.8	25.74
Fe	3.00	2.7	3.03
Ag	0.62	0.5	
Sb	8.73	12.0	13.20
S	24.50	23.3	24.33
Total	98.63	100.8	100.00

(1) Poopó, Bolivia. (2) Do.; by electron microprobe. (3) $Pb_3Sn_4FeSb_2S_{14}$.

Occurrence: In tin-bearing hydrothermal veins.

Association: Franckeite, stannite, incaite, potosiite, teallite, jamesonite, boulangerite, cassiterite, galena, pyrite, sphalerite.

Distribution: From a number of localities in Bolivia, including Trinacria and other mines at Poopó; at the Porvenir and Maria Francisca mines, Huanuni; from the Nueva Virginia vein, Colquechaca; and from the Purisima vein, all in Oruro Province; also from Llallagua, Potosí. In the Smirnowsk deposit, Transbaikalia, USSR.

Name: In allusion to its typical habit.

Type Material: n.d.

References: (1) Palache, C., H. Berman, and C. Frondel (1944) Dana's system of mineralogy, (7th edition), v. I, 482–483. (2) Makovicky, E. (1974) Mineralogical data on cylindrite and incaite, Neues Jahrb. Mineral., Monatsh., 235–256. (3) (1975) Amer. Mineral., 60, 486–487 (abs. ref. 2). (4) Makovicky, E. (1976) Crystallography of cylindrite. Part I. Crystal lattices of cylindrite and incaite. Neues Jahrb. Mineral., Abh., 126, 304–326. (5) Ramdohr, P. (1969) The ore minerals and their intergrowths, (3rd edition), 739–740. (6) Williams, T.B. and B.G. Hyde (1988) Electron microscopy of cylindrite and franckeite. Phys. Chem. Minerals., 15, 521–544.

Crystal Data: Monoclinic. *Point Group:* $P2$, Pm, or $P2/m$. Minute fibrous needles elongated || [010], resembling steel wool; individual needles are generally multiple crystals with a length of up to 2 mm and a thickness of less than 0.1 mm. Striated || [010].

Physical Properties: Hardness = 2.5 VHN = 226–279 (15 g load). D(meas.) = 5.68 D(calc.) = 5.51

Optical Properties: Opaque. *Color:* Lead-gray; in polished section, white with a greenish tint, with blood-red internal reflections in oil. *Streak:* Black. *Luster:* Metallic. *Anisotropism:* Distinct to strong, in greenish gray.

R_1–R_2: (400) 44.0–46.6, (420) 42.3–45.8, (440) 40.8–45.1, (460) 39.6–44.5, (480) 39.2–44.1, (500) 39.1–44.0, (520) 39.1–44.0, (540) 39.1–43.9, (560) 38.9–43.0, (580) 38.4–42.0, (600) 37.9–41.3, (620) 37.3–40.6, (640) 36.8–39.9, (660) 36.2–39.1, (680) 35.7–38.1, (700) 35.3–37.1

Cell Data: *Space Group:* n.d. $a = 19.041$ $b = 8.226$ $c = 17.327$ $\beta = 96°18'$ Z = 1

X-ray Powder Pattern: Saint-Pons, France.
2.795 (100), 2.069 (92), 3.393 (74), 3.371 (72), 3.713 (51), 2.843 (47), 3.792 (45)

Chemistry:

	(1)	(2)	(3)	(4)
Pb	48.9	49.3	48.8	48.7
Fe	0.04			
Sb	29.1	31.7	30.4	31.3
As	1.0			
S	19.9	20.7	19.5	19.7
Cl	0.19		0.23	0.4
Total	99.13	101.7	98.93	100.1

(1) Madoc, Canada; by electron microprobe. (2) Pershing Co., Nevada, USA; by electron microprobe. (3) Wolfsberg, Germany; by electron microprobe. (4) Saint-Pons, France; by electron microprobe.

Occurrence: In hydrothermal veins with other sulfides and lead sulfosalt minerals.

Association: Jamesonite (Northwest Territories and Ontario, Canada); robinsonite (Pershing Co., Nevada, USA); bournonite, boulangerite, zinkenite, chalcostibite (Saint-Pons, France).

Distribution: In the USA, at the Red Bird mercury mine, Pershing Co., Nevada. In Canada, in the Brock zone of the Giant property, Yellowknife, Northwest Territories; and from Madoc, Ontario. From Wolfsberg, Germany. In France, at Saint-Pons, Provence.

Name: For the Canadian mineralogist, Alexander Stewart Dadson (1906–1958).

Type Material: National Museum of Natural History, Washington, D.C., USA, 123240.

References: (1) Jambor, J.L. (1969) Dadsonite (minerals Q and QM), a new lead sulphantimonide. Mineral. Mag., 37, 437–441. (2) (1970) Amer. Mineral., 55, 1445 (abs. ref. 1). (3) Cervelle, B.D., F.P. Cesbron, and M.-C. Sichére (1979) La chalcostibite et la dadsonite de Saint-Pons, Alpes de Haute Provence, France. Can. Mineral., 17, 601–605 (in French with English abs.). (4) Jambor, J.L., J.H.G. Laflamme, and D.A. Walker (1982) A re-examination of the Madoc sulfosalts. Mineral. Record, 13, 93–100. (5) Makovicky, E. and W.G. Mumme (1984) The crystal structure of izoklakeite, dadsonite and jaskolskiite. Acta Cryst., A40, supplement, C-246.

Crystal Data: Cubic. *Point Group:* n.d. Botryoidal and spherulitic aggregates to 0.2 mm, as rims around chromium.

Physical Properties: Hardness = 4.2 VHN = 234–288 (20 g load). D(meas.) = n.d. D(calc.) = 7.36

Optical Properties: Opaque. *Color:* Silver-white to grayish white; white with a milky blue tint in reflected light, slightly yellowish compared to chromium. *Luster:* Strong metallic.
R: n.d.

Cell Data: *Space Group:* n.d. $a = 7.7615$ Z = 32

X-ray Powder Pattern: Danba, China.
2.0803 (10), 2.3552 (5), 2.1574 (4), 1.3732 (4), 1.5901 (3), 1.2256 (3), 1.2154 (3)

Chemistry:

	(1)	(2)
Cu	33.12	32.52
Zn	66.70	67.47
Total	99.82	99.99

(1) Danba, China; by electron microprobe, corresponding to $Cu_{1.000}Zn_{1.957}$. (2) Do.; corresponding to $Cu_{1.000}Zn_{2.016}$.

Occurrence: In a platinum-bearing Cu-Ni deposit in a highly altered ultramafic intrusion.

Association: Chromium, pyrrhotite, pentlandite, chalcopyrite, violarite, cubanite, bornite, sphalerite, galena, linnaeite, magnetite, testibiopalladite, sudburyite, sperrylite, omeiite, gold.

Distribution: From Danba, Sichuan Province, China.

Name: For the locality at Danba in China.

Type Material: n.d.

References: (1) Yue Shuqin, Wang Wenying, Liu Jinding, Sun Shuqiong, and Che Dianfen (1983) A study on danbaite. Kexue Tongbao, 22, 1383–1386 (in Chinese). (2) (1984) Amer. Mineral., 69, 566 (abs. ref. 1).

Crystal Data: Triclinic (?). *Point Group:* n.d. As ragged polycrystalline masses of grains up to 20 μm in anglesite.

Physical Properties: *Tenacity:* Brittle. Hardness = 2–2.5 VHN = 38 (10 g load). D(meas.) = n.d. D(calc.) = 6.541

Optical Properties: Opaque. *Color:* Gray in polished section *Anisotropism:* Moderate, in shades of gray.
R_1–R_2: (470) 32–34, (546) 30–31, (589) 28–30, (650) 27–29

Cell Data: *Space Group:* n.d. $a = 9.644$ $b = 9.180$ $c = 18.156$ Z = 4

X-ray Powder Pattern: Coppin Pool, Western Australia.
2.622 (10), 1.959 (6), 1.875 (6), 2.392 (5), 2.831 (3), 4.44 (2), 3.648 (2)

Chemistry:

	(1)
Cu	33.9
Ag	36.5
Hg	12.3
S	16.1
Total	98.8

(1) Coppin Pool, Australia; by electron microprobe, average of four analyses, leading to $(Cu_{8.54}Ag_{5.43})_{\Sigma=13.97}Hg_{0.98}S_{8.05}$.

Occurrence: In a gossan pod in a quartz vein, as a weathering product derived from other sulfides.

Association: Anglesite, covellite, stromeyerite, chalcocite, cinnabar, other secondary lead and copper minerals.

Distribution: From near Coppin Pool, about 41 km east-southeast of Mount Tom Price, Western Australia.

Name: For Dr. John L. Daniels, who collected the material in which the species was found.

Type Material: Government Chemical Laboratories, Perth; the Museum of Victoria, Melbourne, Australia.

References: (1) Nickel, E.H. (1987) Danielsite: a new sulfide mineral from Western Australia. Amer. Mineral., 72, 401–403. (2) Kato, A. and E. H. Nickel (1988) Possible unit cell for danielsite. Amer. Mineral., 73, 187–188.

Daomanite $CuPtAsS_2$

Crystal Data: Orthorhombic. *Point Group:* mmm or $mm2$. As tabular crystals, to 0.3 mm across.

Physical Properties: *Cleavage:* Four sets, best to least perfect: {100}, {010}, {001}, {110}. *Fracture:* Step-like or uneven. *Tenacity:* Brittle. Hardness = n.d. VHN = 169–175 (10 to 20 g load). D(meas.) = n.d. D(calc.) = 7.06

Optical Properties: Opaque. *Color:* Steel-gray with yellow tint; in polished section, light greenish yellow. *Luster:* Metallic. *Pleochroism:* Gold-yellow and grayish green. *Anisotropism:* Strong.

R_1–R_2: n.d.

Cell Data: *Space Group:* $Amam$, $Ama2$, or $A2am$. $a = 5.93$ $b = 16.23$ $c = 3.67$ Z = 4

X-ray Powder Pattern: China.
3.02 (10), 3.30 (8), 2.13 (8), 1.82 (8), 8.00 (7)

Chemistry:

	(1)	(2)	(3)	(4)
Cu	15.68	13.36	15.57	15.98
Pt	49.73	51.11	50.64	49.06
As	18.30	18.86	17.74	18.84
S	16.31	16.58	16.43	16.12
Total	100.02	99.91	100.38	100.00

(1–3) China; by electron microprobe, giving $Cu_{0.85-0.98}Pt_{1.02-1.06}As_{0.97-1.01}S_{2.03-2.08}$.
(4) $CuPtAsS_2$.

Occurrence: In pyroxenite-type platinum ores, associated with copper sulfide mineralization.

Association: Bornite, chalcopyrite, magnetite, pyrite, gold, sperrylite, cooperite, moncheite, Pt–Cu and Pt–Fe alloys, olivine, diopside, serpentine, chlorite (in olivine pyroxenite); chalcopyrite, bornite, magnetite, covellite, carrollite, goethite, columbite, cooperite, sperrylite, moncheite, other platinum-group minerals, almandine, hornblende, diopside, augite, plagioclase, titanite, apatite (in garnet-hornblende pyroxenite).

Distribution: In China, in the "Tao" and "Ma" districts — apparently coded names.

Name: Derivation not given.

Type Material: n.d.

References: (1) Yu Tsu-Hsiang, Lin Shu-Jen, Chao Pao, Fang Ching-Sung, and Huang Chi-Shun (1974) A preliminary study of some new minerals of the platinum group and another associated new one in platinum-bearing intrusions in a region of China. Acta Geol. Sinica, 2, 202–218 (in Chinese with English abs.). (2) (1976) Amer. Mineral., 61, 184 (abs. ref. 1). (3) Yu Zuxiang, Ding Kuishou and Zhou Jianxiong (1978) Daomanite, a new platinum mineral. Acta Geol. Sinica, 4, 320–327 (in Chinese with English abs.). [Yu Zuxiang formerly Yu Tsu-Hsiang]. (4) (1980) Amer. Mineral., 65, 408 (abs. ref. 3).

Crystal Data: Cubic. *Point Group:* $4/m\ \overline{3}\ 2/m$. Massive, scaly to platy, as exsolution lamellae parallel to {0001} of troilite, and as discrete grains 10–500 μm across in kamacite.

Physical Properties: *Cleavage:* Distinct. *Fracture:* Uneven. *Tenacity:* Brittle. Hardness = 5 VHN = 260–303 (100 g load). D(meas.) = 3.81 D(calc.) = 3.842

Optical Properties: Opaque. *Color:* Black. *Streak:* Brown. *Luster:* Metallic, brilliant.
R: (400) 33.3, (420) 33.2, (440) 33.1, (460) 33.0, (480) 33.1, (500) 33.1, (520) 33.1, (540) 33.0, (560) 33.0, (580) 32.9, (600) 32.8, (620) 32.7, (640) 32.4, (660) 32.3, (680) 32.1, (700) 31.9

Cell Data: *Space Group:* $Fd3m$. $a = 9.966$ Z = 8

X-ray Powder Pattern: Synthetic. (JCPDS 4-651).
3.01 (100), 1.77 (100), 3.53 (80), 2.50 (80), 1.92 (80), 1.30 (80), 1.25 (80)

Chemistry:

	(1)	(2)
Fe	20.10	19.38
Cr	35.91	36.10
S	42.69	44.52
Total	98.70	100.00

(1) Coahuila meteorite; average of three analyses. (2) $FeCr_2S_4$.

Occurrence: In small amounts in many meteorites.

Association: Troilite, kamacite.

Distribution: In iron meteorites, such as the Cosby's Creek, Toluca, Cranbourne, and Mundrabilla octahedrites, and the Coahuila, North Chile, and Scottsville hexahedrites. In stony meteorites, including the Bustee, Mayo Belwa, Norton County, and Cumberland Falls achondrites, the Hvittis enstatite chondrite, and Odessa olivine-bronzite chondrite. Likely yet to be found in additional meteorites.

Name: For Professor Gabriel Auguste Daubrée (1814–1896), of Paris, France.

Type Material: n.d.

References: (1) Palache, C., H. Berman, and C. Frondel (1944) Dana's system of mineralogy, (7th edition), v. I, 265. (2) Buchwald, V.F. (1977) The mineralogy of iron meteorites. Phil. Trans. Royal Soc. London, A. 286, 453–491.

Crystal Data: Monoclinic. *Point Group:* $2/m$. Minute crystals, to 0.3 mm.

Physical Properties: Hardness = n.d. VHN = 19.5 (100 g load). D(meas.) = n.d. D(calc.) = 5.62

Optical Properties: Opaque. *Color:* White in reflected light. *Anisotropism:* Moderate.
R_1–R_2: (400) 29.7, (420) 25.6, (440) 24.9, (460) 24.2, (480) 23.7, (500) 23.5, (520) 23.1, (540) 22.2, (560) 22.2, (580) 21.9, (600) 21.5, (620) 21.2, (640) 20.9, (660) 20.6, (680) 20.4, (700) 20.4 (averages)

Cell Data: *Space Group:* $P2/a$. $a = 6.833$ $b = 12.932$ $c = 9.638$ $\beta = 99°33'$ Z = [8]

X-ray Powder Pattern: Sainte-Marie-aux-Mines, France.
3.075 (10), 3.019 (8), 2.843 (5), 3.251 (3), 2.659 (3), 3.170 (2), 2.742 (2)

Chemistry:

	(1)	(2)
Ag	61.39	60.81
As	19.03	21.11
S	18.06	18.08
Total	98.48	100.00

(1) Sainte-Marie-aux-Mines, France; by electron microprobe; Pb, Sb and Bi not detected.
(2) Ag_2AsS_2.

Occurrence: In a Co-Ni-Fe-As deposit.

Association: Arsenic, rammelsbergite, safflorite, proustite, calcite, quartz.

Distribution: Found on a museum specimen from the Gabe-Gottes mine, Sainte-Marie-aux-Mines, Haut-Rhin, France.

Name: For Dr. Henri Derville of Strasbourg University, Strasbourg, France.

Type Material: Museum of Natural History, 1531; National School of Mines, Paris, France.

References: (1) Bari, H. (1982) Dervillite Ag_2AsS_2, monoclinic, new definition of the species described by Weil, 1941. Pierres et Terre, 23–24, 62–67 (in French). (2) (1983) Amer. Mineral., 68, 1041 (abs. ref. 1). (3) Bari, H., F. Cesbron, Y. Moëlo, F. Permingeat, P. Picot, R. Perriot, H.-J. Schubnel, and R. Weil (1983) La dervillite, Ag_2AsS_2, nouvelle définition de l'espèce. Bull. Minéral., 106, 519–524 (in French with English abs.).

Crystal Data: Cubic. *Point Group:* $4/m\ \bar{3}\ 2/m$, possibly $\bar{4}3m$. Most commonly octahedral, crystals 6 cm or more, also dodecahedral, tetrahedral, and cubic. Curved and striated faces common; spherical, with internal radial structure. *Twinning:* Contact twins with {111} as twin plane; usually flattened on {111}; as penetration twins, sometimes repeated.

Physical Properties: *Cleavage:* {111} perfect. *Fracture:* Conchoidal. *Tenacity:* Brittle. Hardness = 10 VHN = n.d. D(meas.) = 3.511 D(calc.) = 3.515 Fluorescent and phosphorescent; triboelectric; the highest thermal conductivity of any known substance.

Optical Properties: Transparent to translucent. *Color:* Colorless, pale yellow to deep yellow, brown, white, blue-white; less commonly in oranges, pinks, greens, blues, reds, and black. *Luster:* Adamantine to greasy.
Optical Class: Isotropic. *Dispersion:* Strong. n = 2.4354 (486), 2.4175 (589), 2.4076 (687). *Anisotropism:* Birefringent where strained.
R: n.d.

Cell Data: *Space Group:* $Fd3m$. a = 3.5595 Z = 8

X-ray Powder Pattern: Synthetic.
2.06 (100), 1.261 (25), 1.0754 (16), 0.8182 (16), 0.8916 (8)

Chemistry: Nearly pure carbon.

Polymorphism & Series: Chaoite, graphite, and lonsdaleite are polymorphs.

Occurrence: Primarily formed in pipes, less often in dikes, of deep-seated, igneous origin, composed of kimberlite or lamproite, and in alluvial deposits formed by their weathering. In carbonaceous achondrite and iron meteorites.

Association: Forsterite, phlogopite, pyrope, diopside, ilmenite (kimberlite pipes); ilmenite, garnet, rutile, brookite, anatase, hematite, magnetite, tourmaline, gold, zircon, topaz (placers).

Distribution: Numerous occurrences world-wide, but only a few are of economic importance. Formerly important deposits were in India, in the Golconda region, and near Nágpur and Bundelkhand. From the area around Diamantina, Minas Gerais, and other states in Brazil. In South Africa, formerly obtained from the Orange and Vaal Rivers; still from along the coast north into Namibia. Important current alluvial production from Lunda Norte Province, Angola; Sierra Leone; also from Bakwanga and Tchikapa, Kasai Province, Zaire. From South Africa, in several pipes around Kimberley, Orange Free State; the Premier mine, near Pretoria, Transvaal; and the Finsch mine, near Postmasburg, Cape Province. In the Mwadui pipe, Tanzania. From the Orapa and Jwaneng pipes, Botswana. In the USSR, in Siberia, in the Mir (Peace), Zarnitsa (Thunderflash), Udatchnaya (Success), Aikhal (Glory) and Internatsionalnaya pipes; and at Obnazhennaya, on the River Kuoyaka. In China, from pipes at the Binhai mine, Liaoning Province; and Changma, Shandong Province. The world's richest deposit is the Argyle Pipe, Kimberley, Western Australia.

Name: A corruption of the Greek for *invincible*.

References: (1) Palache, C., H. Berman, and C. Frondel (1944) Dana's system of mineralogy, (7th edition), v. I, 146–151. (2) (1953) NBS Circ. 539, 2, 5.

Crystal Data: Monoclinic. *Point Group:* $P2/m$. Prismatic by elongation ∥ [001]; often striated ∥ [001]; highly complex crystals, to nearly 1 cm. *Twinning:* On {120} and {241}.

Physical Properties: *Fracture:* Subconchoidal to uneven. *Tenacity:* Brittle. Hardness = 2.5–3 VHN = n.d. D(meas.) = 6.04 D(calc.) = 6.019

Optical Properties: Opaque. *Color:* Steel-gray; in polished section, white to grayish white. *Luster:* Metallic. *Anisotropism:* Strong.

R_1–R_2: (400) 41.2–44.4, (420) 40.6–43.6, (440) 40.0–43.0, (460) 39.4–42.5, (480) 38.8–42.0, (500) 38.3–41.6, (520) 37.8–41.1, (540) 37.3–40.7, (560) 36.9–40.3, (580) 36.5–39.8, (600) 36.2–39.5, (620) 35.8–39.1, (640) 35.5–38.6, (660) 35.0–38.2, (680) 34.5–37.6, (700) 33.8–37.0

Cell Data: *Space Group:* $P2_1/a$. $a = 15.849(4)$ $b = 17.914(4)$ $c = 5.901(1)$ $\beta = 116°25.5(2)'$ Z = 4

X-ray Powder Pattern: Machacamarca, Bolivia.
3.29 (10), 2.81 (8), 2.03 (5), 1.704 (4), 2.93 (3), 1.759 (3), 1.653 (2)

Chemistry:

	(1)	(2)
Pb	28.67	30.48
Ag	23.44	23.78
Fe	0.67	
Cu	0.73	
Sb	26.43	26.87
S	20.18	18.87
Total	100.12	100.00

(1) Příbram, Czechoslovakia. (2) $Pb_2Ag_3Sb_3S_8$.

Occurrence: In hydrothermal veins with other sulfides formed at moderate temperatures.

Association: Galena, sphalerite, miargyrite, pyrargyrite, pyrite, siderite, quartz.

Distribution: At Příbram and Kutná Hora, Czechoslovakia. From Baia Sprie (Felsőbánya), Romania. In Germany, at Freiberg, Saxony; in the Alte Hoffnung Gottes mine, Voigtsberg; also the Neues Hoffnung Gottes mine, Bräunsdorf. From Fournial, Cantal, and Pontgibaud, Puy-de-Dôme, France. In the Mangazeika and Bulatsk Pb-Zn deposits, northeastern Yakutia, USSR. In the USA, from the Lake Chelan district, Chelan Co., Washington; the Wood River deposits, Mineral Hill district, Blaine Co., Idaho; and near Morey, Nye Co., Nevada. At Catorze, San Luis Potosí, Mexico. From Zancudo, Colombia. At Machacamarca, Bolivia. Known from a few other localities.

Name: From the Greek for *difference*, as being distinct from freieslebenite.

Type Material: n.d.

References: (1) Palache, C., H. Berman, and C. Frondel (1944) Dana's system of mineralogy, (7th edition), v. I, 414–415. (2) Berry, L.G. and R.M. Thompson (1962) X-ray powder data for the ore minerals. Geol. Soc. Amer. Mem. 85, 136. (3) Hellner, E. (1958) Über komplex zusammengesetzte Spiessglanze III. Zur Struktur des Diaphorits, $Ag_3Pb_2Sb_3S_8$. Zeits. Krist., 110, 169–174 (in German with English abs.).

Crystal Data: Cubic. *Point Group:* n.d. A single cubic crystal, about 0.5 cm on the edge.

Physical Properties: Hardness = n.d. VHN = n.d. D(meas.) = n.d. D(calc.) = n.d.

Optical Properties: Opaque. *Color:* White with a tinge of gray. *Luster:* Bright metallic. R: n.d.

Cell Data: *Space Group:* n.d. a = n.d. Z = n.d.

X-ray Powder Pattern: n.d.

Chemistry:

	(1)	(2)
Ni	67.11	70.16
Co	1.29	
Cu	0.99	
Fe	0.61	
Ag	0.02	
As	30.64	29.84
Total	100.66	100.00

(1) Radstadt, Austria. (2) Ni_3As.

Occurrence: Known only from one loose single crystal.

Association: n.d.

Distribution: From near Radstadt, Salzburg, Austria.

Name: For Professor Karl Diener (1862–1928), the discoverer.

Type Material: Lost.

References: (1) Palache, C., H. Berman, and C. Frondel (1944) Dana's system of mineralogy, (7th edition), v. I, 175. (2) Hackl, O. (1921) Ein neues Nickel–Arsen Mineral. Verh. Geol. Reichs-Anst. Wien, 107–108. (3) (1927) Amer. Mineral., 12, 96 (abs. ref. 2).

Crystal Data: Hexagonal, pseudocubic. *Point Group:* Trigonal most likely. Occasionally in pseudocubic crystals as large as 3 cm; more often as intergrowths with other copper sulfides.

Physical Properties: *Cleavage:* {111} (synthetic). *Fracture:* Conchoidal. *Tenacity:* Brittle. Hardness = 2.5–3 VHN = 86–106 (100 g load). D(meas.) = 5.546 D(calc.) = 5.706

Optical Properties: Opaque. *Color:* Blue to black; distinctly blue in polished section. *Anisotropism:* Mostly isotropic.

R: (400) 28.7, (420) 27.7, (440) 26.7, (460) 25.6, (480) 24.6, (500) 23.6, (520) 22.5, (540) 21.6, (560) 20.8, (580) 20.2, (600) 19.5, (620) 19.0, (640) 18.5, (660) 18.0, (680) 17.6, (700) 17.2

Cell Data: *Space Group:* $Fm3m$ (pseudocubic). $a = 5.57$ Z = 1

X-ray Powder Pattern: Leonard mine, Butte, Montana, USA. Anilite transforms to digenite during grinding.

1.973 (100), 3.21 (40), 2.79 (40), 1.686 (30), 1.139 (20), 3.05 (10), 2.17 (10)

Chemistry:

	(1)	(2)	(3)
Cu	78.11	75.30	78.10
Fe		0.16	
S	21.85	24.54	21.90
Total	99.96	100.00	100.00

(1) Jerome, Arizona, USA. (2) Tsumeb, Namibia. (3) Cu_9S_5.

Occurrence: In hydrothermal copper deposits of primary and secondary origin. Formed under a wide range of conditions; reported from mafic intrusives, as an exhalation product, and in pegmatites.

Association: Chalcocite, djurleite, bornite, chalcopyrite, other copper minerals, pyrite.

Distribution: In the USA, at Butte, Silver Bow Co., Montana; in Arizona, from the United Verde mine, Jerome, Yavapai Co.; the Magma mine, Superior, Gila Co.; and at Bisbee, Cochise Co. Abundant at Kennecott, Alaska. In Namibia, at Tsumeb. From Kiruna, Sweden. At Listulli, Telemark, Norway. In the Botallack mine, Cornwall, and from Seathwaite Tarn, near Coniston, Cumbria, England. At Sangerhausen, Thuringia, Germany. From Cananea, Sonora, Mexico. Probably not yet recognized at many other localities.

Name: From the Greek for *two kinds* or *sexes*, in reference to the presumed presence of both cuprous and cupric ions.

References: (1) Palache, C., H. Berman, and C. Frondel (1944) Dana's system of mineralogy, (7th edition), v. I, 180–182. (2) Buerger, N.W. (1942) X-ray evidence of the existence of the mineral digenite, Cu_9S_5. Amer. Mineral., 27, 712–716. (3) Berry, L.G. and R.M. Thompson (1962) X-ray powder data for the ore minerals. Geol. Soc. Amer. Mem. 85, 40. (4) Morimoto, N. and G. Kullerud (1963) Polymorphism in digenite. Amer. Mineral., 48, 110–123. (5) Morimoto, N. and A. Gyobu (1971) The composition and stability of digenite. Amer. Mineral., 56, 1889–1909.

Crystal Data: Orthorhombic. *Point Group:* $2/m\ 2/m\ 2/m$. Dipyramidal by development of {111}; {101} and {110} are also well developed. Commonly in groups of tiny individuals.

Physical Properties: *Tenacity:* Brittle. Hardness = 1.5 VHN = n.d. D(meas.) = 3.58 D(calc.) = 3.60 Burns without residue.

Optical Properties: Transparent. *Color:* Orange-yellow. *Luster:* Adamantine.
R_1–R_2: n.d.

Cell Data: *Space Group:* *Pnma.* $a = 11.24$ $b = 9.90$ $c = 6.56$ Z = 4

X-ray Powder Pattern: Solfatara, Italy.
4.89 (100), 2.14 (50), 1.620 (50), 3.91 (40), 5.64 (30), 3.07 (25), 2.94 (25)

Chemistry:

	(1)	(2)
As	75.45	75.70
S	24.55	24.30
Total	100.00	100.00

(1) Solfatara, Italy. (2) As_4S_3.

Occurrence: Deposited at a fumerolic vent at temperatures of about 70–80°C (Italy); formed in an ore deposit under secondary conditions (Chile).

Association: Realgar, sal ammoniac, sulfur, various sulfates (Italy); orpiment (Chile).

Distribution: From the Solfatara, Campi Flegrei, near Naples, Italy. In the Alacrán Ag-As-Sb deposit, Pampa Larga district, Copiapó, Atacama, Chile.

Name: From the Greek for *two* and *form,* in reference to the two forms in which the species was thought to occur.

References: (1) Palache, C., H. Berman, and C. Frondel (1944) Dana's system of mineralogy, (7th edition), v. I, 197–198. (2) Frankel, L.S. and T. Zoltai (1973) Crystallography of dimorphites. Zeits. Krist., 138, 161–166. (3) Clark, A.H. (1972) Mineralogy of the Alacrán deposit, Pampa Larga, Chile. Neues Jahrb. Mineral., Monatsh., 423–429.

Crystal Data: Cubic. *Point Group:* $4/m\ \overline{3}\ 2/m$. As rounded grains 0.2–0.4 mm in diameter.

Physical Properties: Hardness = < 3.5 VHN = 172 D(meas.) = n.d. D(calc.) = n.d.

Optical Properties: Opaque. *Color:* Greenish yellow, close to khaki, to olive-drab. *Luster:* Submetallic.

R: (400) 17.5, (420) 18.3, (440) 19.1, (460) 20.0, (480) 20.8, (500) 21.6, (520) 22.3, (540) 22.9, (560) 23.5, (580) 24.0, (600) 24.5, (620) 24.9, (640) 25.2, (660) 25.5, (680) 25.8, (700) 25.9

Cell Data: *Space Group:* $Pm3m$. $a = 10.465(1)$ Z = 2

X-ray Powder Pattern: Kota Kota meteorite.
1.828 (100), 2.985 (70), 2.372 (60), 10.34 (50), 5.97 (50), 3.118 (50), 3.269 (40)

Chemistry:

	(1)	(2)	(3)
K	8.7	8.25	9.00
Na	0.3		0.76
Fe	50.7	43.5	45.4
Cu	4.2	15.5	8.37
Ni	0.8	1.5	1.41
Mg			< 0.05
S	33.8	33.5	33.8
Cl	1.0		1.26
Total	99.5	102.25	100.05

(1) Kota Kota meteorite; by electron microprobe. (2) Talnakh area, USSR; by electron microprobe. (3) Coyote Peak, California, USA; by electron microprobe.

Occurrence: In meteorites, in hydrothermal Cu-Ni ores, in skarns, in pegmatites, in kimberlites, and in a mafic alkalic diatreme.

Association: "Nickel-iron" (kamacite), troilite, schreibersite, clinoenstatite, tridymite, cristobalite, daubréelite, graphite, roedderite, alabandite (Kota Kota meteorite); talnakhite, pentlandite, chalcopyrite, magnetite, valleriite, sphalerite, platinum minerals (Talnakh area, USSR).

Distribution: In the USA, at Coyote Peak, Humboldt Co., California. In the Talnakh area, Noril'sk region, western Siberia; Yakutian kimberlites; the Lovozero massif; the Inagli Complex, Aldan Shield; and in the Khibina massif of the Kola Peninsula, USSR. From the Ilímaussaq Intrusion, southern Greenland. Found in the Kota Kota and St. Marks enstatite chondrites, the Pena Blañca Spring achondrite, also the Toluca and Cape York octahedrite iron meteorites.

Name: For Professor Daniel Jerome Fisher (1896-1988), American mineralogist.

Type Material: n.d.

References: (1) Fuchs, L.H. (1966) Djerfisherite, alkali copper–iron sulfide: a new mineral from enstatite chondrites. Science, 153, 166–167. (2) Genkin, A.D., N.V. Troneva, and M.N. Zhuravlev (1970) The first occurrence in ores of the sulfide of potassium, iron and copper—djerfisherite. Geol. Rudn. Mestorozhd. 11, 57–64 (in Russian). (3) (1970) Amer. Mineral., 55, 1071 (abs. ref. 2). (4) Czamanske, G.K., R.C. Erd, M.N. Sokolova, M.G. Dobrovol'skaya, and M.T. Dmitrieva (1979) New data on rasvumite and djerfisherite. Amer. Mineral., 64, 776–778. (5) Dmitrieva, M.T., V.V. Ilyukhin, and G.B. Bokii (1979) Close packing and cation arrangement in the jerfisherite (sic) structure. Sov. Phys. Crystallog., 24, 683–685 (in English translation).

Crystal Data: Monoclinic. *Point Group:* $2/m$. Crystals are short, prismatic and thick tabular, as large as 1 cm; massive compact. *Twinning:* Common on {110}.

Physical Properties: Hardness = 2.5–3 VHN = 65–85 (100 g load). D(meas.) = n.d. D(calc.) = 5.749

Optical Properties: Opaque. *Color:* Black; gray in reflected light. *Luster:* Metallic. *Anisotropism:* Weak.
R_1–R_2: (400) 34.6–35.8, (420) 34.2–35.2, (440) 33.8–34.6, (460) 33.3–34.0, (480) 32.6–33.2, (500) 31.8–32.3, (520) 30.6–31.2, (540) 29.6–30.2, (560) 28.6–29.3, (580) 27.8–28.6, (600) 27.0–28.1, (620) 26.4–27.6, (640) 25.7–27.1, (660) 25.0–26.7, (680) 24.4–26.4, (700) 23.6–26.1

Cell Data: *Space Group:* $P2_1/n$. $a = 26.897$ $b = 15.745$ $c = 13.565$ $\beta = 90.13°$ Z = 8

X-ray Powder Pattern: Butte, Montana, USA.
1.871 (100), 2.387 (90), 1.964 (90), 1.957 (90), 3.386 (50), 3.282 (30), 3.192 (30)

Chemistry:

	(1)	(2)
Cu	78.6	79.34
Ag	0.1	
Fe	0.1	
S	20.2	20.66
Total	99.0	100.00

(1) Seathwaite Tarn, England; by electron microprobe. (2) $Cu_{31}S_{16}$.

Occurrence: Found with other secondary copper sulfides in enriched zones.

Association: Digenite, bornite, chalcocite, chalcopyrite, anilite, pyrite.

Distribution: In the USA, in many of the large porphyry copper deposits of the Western Cordillera including Butte, Silver Bow Co., Montana; Bisbee, Cochise Co.; Globe-Miami and the Magma mine, Gila Co., Arizona. In isolated crystals from the Gem mine, San Benito Co., California. In Peru, from Morococha. At Tsumeb, Namibia. In the Philippine Islands, at Bagacay, Samar Island. In Mexico, at the Salvadora mine, Milpillas, Chihuahua. In Japan, at the Ani, Osarizawa, and other mines, Akita Prefecture, and elsewhere. In England, from Seathwaite Tarn, near Coniston, Cumbria; Wheal Owles and Dean quarry, Cornwall; Merehead and Cannington Park quarries, Somerset; and Gipsy Lane, Leicester. At Bandaksli, Telemark, Norway. A relatively common mineral, likely still to be identified at other localities.

Name: For S. Djurle, who first synthesized the compound later found in nature.

Type Material: National Museum of Natural History, Washington, D.C., USA, 92349.

References: (1) Roseboom, E.H., Jr. (1962) Djurleite, $Cu_{1.96}S$, a new mineral. Amer. Mineral., 47, 1181–1184. (2) Morimoto, N. (1962) Djurleite, a new copper sulfide mineral. Mineral. J. (Japan), 3, 338–344. (3) (1963) Amer. Mineral., 48, 215 (abs. ref. 2). (4) Evans, H.T., Jr. (1979) Djurleite ($Cu_{1.94}S$) and low chalcocite (Cu_2S) : new crystal structure studies. Science, 203, 356–358.

Crystal Data: Cubic. *Point Group:* $\bar{4}3m$. Reniform, botryoidal; massive.

Physical Properties: *Fracture:* Uneven. Hardness = 3–3.5 VHN = 213–235; 124–140 ("domeykite-beta") (100 g load). D(meas.) = 7.2–7.9 D(calc.) = 7.86

Optical Properties: Opaque. *Color:* Tin-white to steel-gray, tarnishes yellowish, then to brownish, and finally to iridescence. *Luster:* Metallic.

R: (400) 43.6, (420) 45.0, (440) 46.4, (460) 47.8, (480) 49.0, (500) 50.1, (520) 51.0, (540) 51.9, (560) 52.7, (580) 53.2, (600) 53.5, (620) 53.6, (640) 53.7, (660) 53.7, (680) 53.6, (700) 53.6.

R_1–R_2: ("domeykite-beta"): (400) 45.6–47.4, (420) 46.8–48.7, (440) 48.0–50.0, (460) 49.0–51.2, (480) 49.8–52.1, (500) 50.3–52.8, (520) 50.4–53.2, (540) 50.1–53.2, (560) 49.7–52.8, (580) 49.1–52.4, (600) 48.5–51.9, (620) 48.0–51.5, (640) 47.5–51.0, (660) 47.0–50.5, (680) 46.6–50.0, (700) 46.2–49.3

Cell Data: *Space Group:* $I\bar{4}3d$. $a = 9.62$ Z = 16

X-ray Powder Pattern: Keweenaw Co., Michigan, USA.
2.05 (100), 1.888 (70), 1.965 (50), 1.308 (50), 3.95 (40), 3.05 (40), 2.15 (40)

Chemistry:

	(1)	(2)	(3)
Cu	70.56	71.16	71.79
As	29.50	28.27	28.21
Sb		0.12	
S		0.06	
Total	100.06	99.61	100.00

(1) Mohawk mine, Michigan, USA. (2) Wasserfall, France; by electron microprobe. (3) Cu_3As.

Occurrence: Of hydrothermal origin.

Association: Copper (usually arsenian), silver, algodonite.

Distribution: In the USA, at the Mohawk and Ahmeek mines, Keweenaw Co., and in the Sheldon-Columbia mine, Portage Lake, Houghton Co., Michigan; from the Cashin mine, Montrose Co., Colorado. On Michipicoten Island and at the Silver Islet mine in Lake Superior, Ontario, Canada. At San Antonio, near Copiapó, and Chañarcillo, Coquimbo, Chile. From Corocoro, Bolivia. At Zwickau, Saxony, Germany. From Långban, Värmland, Sweden. At Wasserfall, southern Vosges, France. In the Condurrow mine, near Helstone, Cornwall, England. From Cerro de Paracatas, Tlachapa, Guerrero, Mexico. In the Talmessi mine, near Anarak, Iran. Several other occurrences are known.

Name: For the Chilean mineralogist Ignacio Domeyko (1802–1889).

References: (1) Palache, C., H. Berman, and C. Frondel (1944) Dana's system of mineralogy, (7th edition), v. I, 172. (2) Berry, L.G. and R.M. Thompson (1962) X-ray powder data for the ore minerals. Geol. Soc. Amer. Mem. 85, 30. (3) Ramdohr, P. (1969) The ore minerals and their intergrowths, (3rd edition), 393–398. (4) Picot, P. and F. Ruhlmann (1978) Présence d'arséniures de cuivre de haute température dan le granite des Ballons (Vosges méridionales). Bull. Minéral., 101, 563–569 (in French with English abs.).

Crystal Data: Hexagonal. *Point Group:* $6/m\ 2/m\ 2/m$ most probably. Steep pyramidal crystals to 0.5 mm imbedded in other minerals; massive.

Physical Properties: *Cleavage:* {0001} perfect. *Tenacity:* Pliable, difficult to pulverize. Hardness = Very soft. VHN = 46–58 (7.6 g load). D(meas.) = n.d. D(calc.) = 6.248

Optical Properties: Opaque. *Color:* Grayish black with a brownish tint. *Streak:* Brownish black. *Luster:* Metallic. *Pleochroism:* Strong, white to very light gray, to pinkish gray. *Anisotropism:* Very strong, with colors from light bluish gray to purplish brown.

R_1–R_2: (400) 17.0–38.0, (420) 18.2–37.8, (440) 19.4–37.6, (460) 20.4–37.6, (480) 21.4–37.5, (500) 21.8–37.5, (520) 21.8–37.4, (540) 21.6–37.1, (560) 21.4–36.7, (580) 21.1–36.0, (600) 20.8–35.5, (620) 20.6–35.1, (640) 20.5–35.0, (660) 20.4–35.0, (680) 20.4–35.2, (700) 20.4–35.6

Cell Data: *Space Group:* $P6_3mmc$ (synthetic). $a = 3.287(1)$ $c = 12.925(2)$ Z = 2

X-ray Powder Pattern: Kapijimpanga, Zambia.
2.373 (100), 6.46 (75), 2.845 (55), 1.913 (55), 1.643 (40), 1.615 (40), 2.153 (25)

Chemistry:

	(1)	(2)
Mo	35.30	37.79
Se	60.40	62.21
S	3.40	
Total	99.10	100.00

(1) Kapijimpanga, Zambia; by electron microprobe, corresponding to $Mo_{0.85}(Se_{1.76}S_{0.24})_{\Sigma=2.00}$, with spectrographic traces of Si, Mg, Al, Pb, Fe, Bi, Ca, Cu, Ni, and Ti, belonging mostly to impurities. (2) $MoSe_2$.

Occurrence: In the oxidation zone of a uranium deposit in a talc schist.

Association: Uraninite, apatite, masuyite, secondary uranium minerals.

Distribution: From the Kapijimpanga uranium deposit, 16 km southeast of Solwezi, Northwestern Province, Zambia.

Name: For A.R. Drysdall, Director, Geological Survey of Zambia.

Type Material: Charles University, Prague, Czechoslovakia; National Museum of Natural History, Washington, D.C., USA, 145627.

References: (1) Čech, F., M. Rieder, and S. Vrána (1973) Drysdallite, $MoSe_2$, a new mineral. Neues Jahrb. Mineral., Monatsh., 433–442. (2) (1974) Amer. Mineral., 59, 1139 (abs. ref. 1).

Crystal Data: Monoclinic. *Point Group:* $2/m$. Crystals tabular in shape and somewhat elongated [100]; also elongated [010] and striated ∥ [100], to 2.5 cm. *Twinning:* With {001} as twin plane; polysynthetic lamellae are rarely observed in polished section.

Physical Properties: *Cleavage:* {010} perfect. *Fracture:* Conchoidal. *Tenacity:* Brittle. Hardness = 3 VHN = 135–146 (100 g load). D(meas.) = 5.50–5.57 D(calc.) = 5.61

Optical Properties: Sub-translucent. *Color:* Lead-gray to steel-gray; in polished section, white, sometimes showing deep red internal reflections; dark red-brown in transmitted light. *Streak:* Reddish brown to chocolate-brown. *Luster:* Metallic.

Optical Class: n.d. *Pleochroism:* Very weak. $n = > 2.72$ (Li). *Anisotropism:* Strong, in brown-violet and dark green colors.

R_1–R_2: (400) 38.7–42.1, (420) 38.5–41.9, (440) 38.3–41.7, (460) 38.0–41.4, (480) 37.7–41.0, (500) 37.3–40.6, (520) 36.9–40.1, (540) 36.3–39.6, (560) 35.7–38.9, (580) 35.0–38.0, (600) 34.0–37.1, (620) 33.2–36.1, (640) 32.3–35.2, (660) 31.6–34.3, (680) 30.9–33.5, (700) 30.3–32.7

Cell Data: *Space Group:* $P2_1/m$. $a = 7.90$ $b = 25.74$ $c = 8.37$ $\beta = 90°21'$ Z = 2

X-ray Powder Pattern: Binntal, Switzerland.
3.74 (100), 3.00 (90), 2.70 (80), 3.21 (60), 2.36 (60), 2.23 (60), 3.56 (50)

Chemistry:

	(1)	(2)
Pb	57.42	57.20
As	20.89	20.68
S	22.55	22.12
Total	100.86	100.00

(1) Binntal, Switzerland. (2) $Pb_2As_2S_5$.

Occurrence: An uncommon mineral of moderate to low-temperature hydrothermal origin; occurs in crystalline dolomite (Binntal, Switzerland).

Association: Chalcopyrite, sphalerite, realgar, orpiment, tetrahedrite.

Distribution: In Switzerland, at Binntal, Valais, in the Lengenbach quarry, and at Recki Bach. In the USA, from the Silver Star claims, near Fruitland, Stevens Co., Washington. From the Hemlo gold deposit, Thunder Bay district, Ontario, Canada.

Name: For Ours Pierre Armand Petit Dufrénoy (1792–1857), French mineralogist, National School of Mines, Paris, France.

References: (1) Palache, C., H. Berman, and C. Frondel (1944) Dana's system of mineralogy, (7th edition), v. I, 442–445. (2) Berry, L.G. and R.M. Thompson (1962) X-ray powder data for the ore minerals. Geol. Soc. Amer. Mem. 85, 151. (3) Ribár, B., C. Nicca, and W. Nowacki (1969) Dreidimensionale Verfeinerung der Kristallstruktur von Dufrenoysit, $Pb_8As_8S_{20}$. Zeits. Krist., 130, 15–40 (in German with English abs.).

Crystal Data: Orthorhombic. *Point Group:* n.d. As small polycrystalline grains, 0.2 mm maximum; randomly oriented fibers and wiry aggregates.

Physical Properties: Hardness = n.d. VHN = 58 (25 g load). D(meas.) = n.d. D(calc.) = 4.50

Optical Properties: Opaque. *Color:* Red; in polished section, gray-white. *Luster:* Metallic.
R_1–R_2: (400) 39.0–42.0, (420) 36.8–39.4, (440) 34.6–36.8, (460) 33.0–34.8, (480) 32.1–33.6, (500) 31.4–32.7, (520) 30.8–32.2, (540) 30.4–31.8, (560) 30.0–31.5, (580) 29.6–31.0, (600) 29.1–30.6, (620) 28.7–30.3, (640) 28.4–30.0, (660) 28.0–29.7, (680) 27.6–29.3, (700) 27.1–29.0

Cell Data: *Space Group:* n.d. $a = 3.576(2)$ $b = 6.759(2)$ $c = 10.074(5)$ Z = 2

X-ray Powder Pattern: Duranus, France.
2.919 (10), 5.620 (9), 5.037 (9), 1.969 (9), 1.788 (9), 2.682 (8), 3.016 (7)

Chemistry:

	(1)	(2)	(3)
As	90.0	90.8	90.33
S	10.3	10.3	9.67
Total	100.3	101.1	100.00

(1) Duranus, France; by electron microprobe, corresponding to $As_{3.95}S_{1.05}$. (2) Do.; corresponding to $As_{3.96}S_{1.04}$. (3) As_4S.

Occurrence: In calcite veinlets in marls and siliceous limestones.

Association: Arsenic, realgar, orpiment, stibnite, quartz, calcite.

Distribution: From Duranus, Alpes-Maritimes, France. In the Mt. Washington copper mine, Vancouver Island, British Columbia, Canada.

Name: For the locality at Duranus, France.

Type Material: National School of Mines, Paris, France.

References: (1) Johan, Z., C. Laforêt, P. Picot, and J. Ferard (1973) La duranusite, As_4S, un nouveau minéral. Bull. Soc. fr. Minéral., 96, 131–134 (in French). (2) (1975) Amer. Mineral., 60, 945 (abs. ref. 1).

Crystal Data: Orthorhombic. *Point Group:* $mm2$. Pyramidal crystals to 5 cm, also cylindrical, prismatic to platy, striated; granular, foliated or massive. *Twinning:* On {110}, giving pseudohexagonal forms and V-shapes.

Physical Properties: *Cleavage:* Distinct on {001} and {011}; imperfect on {110}. *Fracture:* Uneven. *Tenacity:* Sectile, but brittle. Hardness = 3.5–4 VHN = 153–179 (100 g load). D(meas.) = 9.712 D(calc.) = 9.720

Optical Properties: Opaque. *Color:* Silver-white, tarnishing to lead-gray, yellowish, or black. *Streak:* Silver-white. *Luster:* Metallic. *Pleochroism:* Very weak. *Anisotropism:* Weak.
R_1–R_2: (400) 57.0–57.7, (420) 57.7–59.0, (440) 58.3–60.1, (460) 59.1–61.2, (480) 59.6–61.7, (500) 59.9–62.1, (520) 60.0–62.4, (540) 60.1–62.7, (560) 60.0–62.9, (580) 59.8–63.0, (600) 59.6–63.0, (620) 59.5–63.0, (640) 59.4–63.1, (660) 59.3–63.0, (680) 59.3–63.1, (700) 59.6–63.4

Cell Data: *Space Group:* $Pmm2$. $a = 3.008$ $b = 4.828$ $c = 5.214$ Z = 1

X-ray Powder Pattern: St. Andreasberg, Germany.
2.29 (100), 2.42 (40), 1.370 (40), 2.61 (30), 1.771 (30), 1.506 (30), 1.278 (30)

Chemistry:

	(1)	(2)	(3)
Ag	75.41	74.33	72.66
Hg		0.57	
Sb	24.37	23.86	27.34
Total	99.78	98.76	100.00

(1) St. Andreasberg, Germany; average of seven analyses. (2) Consols mine, Broken Hill, Australia; by electron microprobe. (3) Ag_3Sb.

Occurrence: In hydrothermal veins with other silver minerals as both a primary and secondary mineral.

Association: Silver, pyrargyrite, acanthite, stromeyerite, tetrahedrite, galena, calcite, barite.

Distribution: In Germany, at St. Andreasberg in the Harz Mountains, and at Wolfach and Wittichen, Black Forest. In France, at Wasserfall, southern Vosges; and at Sainte-Marie-aux-Mines, Haut-Rhin. From Příbram, Czechoslovakia. At Långsjön, Sweden. In the Consols mine, Broken Hill, New South Wales, Australia. From Canada, at Cobalt, Ontario, in a number of mines; and in the Tanco pegmatite, Bernic Lake, Manitoba. From the Ilímaussaq Intrusion, southern Greenland. In the Yuzhnyi tin deposit, Tetukhe region, southern Maritime Territory, USSR.

Name: From the Greek for *a bad alloy.*

References: (1) Palache, C., H. Berman, and C. Frondel (1944) Dana's system of mineralogy, (7th edition), v. I, 173–175. (2) Berry, L.G. and R.M. Thompson (1962) X-ray powder data for the ore minerals. Geol. Soc. Amer. Mem. 85, 32. (3) Scott, J.D. (1976) Refinement of the crystal structure of dyscrasite, and its implications for the structure of allargentum. Can. Mineral., 14, 139–142.

Crystal Data: Orthorhombic. *Point Group:* $2/m\ 2/m\ 2/m$. As fan-shaped aggregates of acicular crystals to 1.5 cm long, and granular.

Physical Properties: *Cleavage:* Distinct on {010}, poor on {0*kl*}. Hardness = 2.5 VHN = Markedly anisotropic from 87 (transverse) to 191 (longitudinal) (50 g load). D(meas.) = 6.85 D(calc.) = 6.88

Optical Properties: Opaque. *Color:* Tin-white; in reflected light, whitish gray, bluish gray in oil. *Luster:* Metallic. *Pleochroism:* Moderate to distinct. *Anisotropism:* Distinct in air, strong in oil.

R_1–R_2: (480) 45.2–51.2, (546) 42.9–47.2, (589) 42.0–46.1, (644) 40.2–45.0

Cell Data: *Space Group:* *Pnma*. $a = 54.76(4)$ $b = 4.030(3)$ $c = 22.75(3)$ Z = 4

X-ray Powder Pattern: Bärenbad, Austria.
3.414 (100), 2.014 (80), 2.893 (70), 3.010 (60), 2.141 (50), 2.037 (45), 3.488 (40)

Chemistry:

	(1)	(2)
Pb	34.3	34.99
Cu	0.9	0.60
Fe	0.6	0.52
Ag	0.3	
Bi	45.8	47.05
Sb	1.5	
S	17.3	16.84
Total	100.7	100.00

(1) Bärenbad, Austria; by electron microprobe, average of 16 analyses of two crystals.
(2) $Pb_9Cu_{0.5}Fe_{0.5}Bi_{12}S_{28}$.

Occurrence: In gold-bearing sulfide ores of copper located in quartz veins cutting amphibole-grade meta-mafic rocks.

Association: Pyrite, arsenopyrite, chalcopyrite, pyrrhotite, sphalerite, stannite, bismuth, gold, quartz.

Distribution: From Bärenbad, west of Hollersbachtal, Salzburg, Austria.

Name: To honor Professor E. Clar of Vienna, Austria.

Type Material: The Institute for Geosciences (Mineralogy) of the University of Salzburg, Austria; the Mineralogical–Crystallographical Institute of the University of Göttingen, Germany; British Museum (Natural History), London, England; Royal Ontario Museum, Toronto, Canada; National Museum of Natural History, Washington, D.C., USA, 150482.

References: (1) Paar, W.H., T.T. Chen, V. Kupcik, and K. Hanke (1984) Eclarite, $(Cu,Fe)Pb_9Bi_{12}S_{28}$, ein neues Sulfosalt von Bärenbad, Hollersbachtal, Salzburg, Österreich. Tschermaks Mineral. Petrog. Mitt., 32, 103–110 (in German with English abs.). (2) (1985) Amer. Mineral., 70, 215 (abs. ref. 1). (3) Kupčík, V. (1984) Die Kristallstruktur des Minerals Eclarit $(Cu,Fe)Pb_9Bi_{12}S_{28}$. Tschermaks Mineral. Petrog. Mitt., 32, 259–269 (in German with English abs.).

Crystal Data: Hexagonal. *Point Group:* $3m$ (synthetic). As anhedral to irregular grains, from about 0.5 to 1.3 mm, some having rhombohedral form; also massive.

Physical Properties: *Cleavage:* Excellent to good, rhombohedral. *Fracture:* Hackly. Hardness = n.d. VHN = 36.4–44.0, 39.3 average (50 g load). D(meas.) = 7.10(5) (synthetic). D(calc.) = 7.18

Optical Properties: Opaque. *Color:* Dark gray; light gray with purplish tint in polished section, with deep red to deep orange-red internal reflections. *Streak:* Light brown with a tinge of orange. *Luster:* Metallic. *Pleochroism:* Very weak, in colors from light purplish gray to light pinkish gray. *Anisotropism:* From blue-purple to red-purple to brownish orange.
R_1–R_2: (470) 31.7, (546) 29.2, (589) 28.9, (650) 28.3

Cell Data: *Space Group:* $R3m$ (synthetic). $a = 12.324$ $c = 9.647$ Z = 7

X-ray Powder Pattern: Carlin, Nevada, USA.
2.669 (100), 3.214 (53), 5.333 (37), 2.327 (28), 3.559 (20), 1.780 (15), 2.757 (10)

Chemistry:

	(1)	(2)
Tl	78.2	78.18
As	9.6	9.55
S	12.3	12.27
Total	100.1	100.00

(1) Carlin, Nevada, USA; by electron microprobe, average of five grains. (2) Tl_3AsS_3.

Occurrence: In a hydrothermal gold deposit, in mineralized, argillaceous, carbonaceous dolomite beds.

Association: Gold, pyrite, christite, lorandite, getchellite, realgar, arsenic, carlinite, hydrocarbons.

Distribution: In the Carlin gold deposit, Eureka Co., Nevada, USA.

Name: For Dr. A.J. Ellis, Chemistry Division, D.S.I.R., New Zealand.

Type Material: Department of Geology, Stanford University, Palo Alto, California, USA, Epithermal Minerals Collection.

References: (1) Dickson, F.W., A.S. Radtke, and J.A. Peterson (1979) Ellisite, Tl_3AsS_3, a new mineral from the Carlin gold deposit, Nevada, and associated sulfide and sulfosalt minerals. Amer. Mineral., 64, 701–707. (2) Gostojić, M. (1980) Die Kristallstruktur von synthetishem Ellisit, Tl_3AsS_3. Zeits. Krist., 151, 249–254 (in German with English abs.).

Crystal Data: Orthorhombic. *Point Group:* $2/m\ 2/m\ 2/m$. Crystals prismatic and striated parallel [001], flattened on {010}. *Twinning:* Observed in polished section, of unknown law.

Physical Properties: *Cleavage:* {010} perfect, {001} less so. *Fracture:* Conchoidal to uneven. *Tenacity:* Brittle. Hardness = 2 VHN = n.d. D(meas.) = 6.38 D(calc.) = 6.393

Optical Properties: Opaque. *Color:* Grayish to tin-white; light yellow in polished section. *Luster:* Metallic. *Pleochroism:* Very low.

R_1–R_2: (400) 28.7–31.1, (420) 30.0–32.2, (440) 31.3–33.3, (460) 32.4–34.4, (480) 33.4–35.4, (500) 34.4–36.2, (520) 35.2–37.0, (540) 35.9–37.5, (560) 36.5–38.0, (580) 37.0–38.3, (600) 37.5–38.6, (620) 37.9–38.8, (640) 38.1–38.8, (660) 38.2–38.7, (680) 38.2–38.5, (700) 38.0–38.0

Cell Data: *Space Group:* *Pnma*. $a = 6.1426(3)$ $b = 3.9189(4)$ $c = 14.5282(7)$ Z = 4

X-ray Powder Pattern: Tannenbaum mine, Germany.
3.05 (10), 3.23 (9), 3.13 (7), 7.38 (5), 2.34 (5), 2.17 (4), 1.863 (2)

Chemistry:

	(1)	(2)
Cu	18.80	18.88
Bi	61.95	62.08
S	19.16	19.04
Total	99.91	100.00

(1) Tannenbaum mine, Germany; average of two analyses. (2) $CuBiS_2$.

Occurrence: In hydrothermal veins with other sulfides and sulfosalts formed at moderate temperatures.

Association: Chalcopyrite, pyrite, sphalerite, molybdenite, quartz, fluorite (Horní Slavkov, Czechoslovakia); tetrahedrite–tennantite, luzonite–famatinite, pyrite, mawsonite, nekrasovite, chalcopyrite, laitakarite, bismuth, calcite, quartz, barite (Khayragatsch deposit, USSR).

Distribution: In Czechoslovakia, at Krupka, Horní Slavkov (Schlaggenwald), and Cínvald (Zinnwald). In Germany, in the Tannenbaum mine, near Schwarzenberg; from Schneeberg, Annaberg, and at Sadisdorf; fine crystals from Johanngeorgenstadt, and other localities in Saxony; from Wittichen, Black Forest. In Norway, at the Åmdal copper mines, Telemarken. In Scotland, from Corrie Buie, Meal nan Oighreag, Perthshire. At Cerro Blanco, near Copiapó, Chile. In the Khayragatsch deposit, Kuramin Mountains, eastern Uzbekistan, USSR. From Japan, in the Akenobe mine, Hyogo Prefecture, and the Hade mine, Okayama Prefecture. In Mexico, from the El Cobre mine, Concepción del Oro, Zacatecas. Additional minor occurrences are known.

Name: From the Greek for *interwoven* or *entwined*; in reference to its sometimes intimate association with quartz.

References: (1) Palache, C., H. Berman, and C. Frondel (1944) Dana's system of mineralogy, (7th edition), v. I, 435–437. (2) Berry, L.G. and R.M. Thompson (1962) X-ray powder data for the ore minerals. Geol. Soc. Amer. Mem. 85, 145–146. (3) Portheine, J.C. and W. Nowacki (1975) Refinement of the crystal structure of emplectite, $CuBiS_2$. Zeits. Krist., 141, 387–402.

Empressite $AgTe$

Crystal Data: Orthorhombic. *Point Group:* $2/m$ $2/m$ $2/m$. As compact, granular masses; crystals rare.

Physical Properties: *Fracture:* Uneven to subconchoidal. *Tenacity:* Brittle. Hardness = 3.5 VHN = 108–133 D(meas.) = 7.61(1) D(calc.) = 7.61

Optical Properties: Opaque. *Color:* Pale bronze, tarnishing to darker bronze on exposure. *Luster:* Metallic. *Pleochroism:* Very strong, in gray to creamy white. *Anisotropism:* Strong, gray to yellowish white to grayish blue.
R_1–R_2: n.d.

Cell Data: *Space Group: Pmnm* or *Pmn* (sic – *Pmn*∗ ?). $a = 8.90$ $b = 20.07$ $c = 4.62$ Z = 16

X-ray Powder Pattern: Empress Josephine mine, Colorado, USA.
2.70 (100), 2.23 (80), 3.81 (60), 3.33 (60), 3.18 (50), 2.14 (50), 2.04 (50)

Chemistry:

	(1)	(2)	(3)	(4)
Ag	45.17	44.9	44.9	45.81
Pb		0.05		
Cu		0.35		
Fe	0.22	0.20		
Te	54.75	53.6	55.8	54.19
S		0.1		
rem.	0.39	0.70		
Total	100.53	99.90	100.7	100.00

(1) Empress Josephine mine, Colorado, USA; average of two analyses, remainder insoluble. (2) Do.; remainder includes SiO_2 (0.40%); Al_2O_3 (0.30%). (3) Do.; by electron microprobe. (4) AgTe.

Occurrence: Occurs in low-temperature gold-poor hydrothermal vein deposits.

Association: Tellurium, sylvanite, petzite, hessite, rickardite, altaite, galena (Empress Josephine mine, Colorado, USA); pyrite, rodalquilarite (Tombstone, Arizona, USA); gold, tellurium, hessite, petzite, rickardite (Pitman, Canada).

Distribution: In the USA, from the Empress Josephine mine, Kerber Creek district, and the Klondyke mine, Finley Gulch, both in Saguache Co.; the May Day mine, La Plata Co.; and the Red Cloud mine, Boulder Co., Colorado. From Tombstone, Cochise Co., Arizona. In Canada, from the Grotto group of pits, near Pitman, British Columbia. At Kalgoorlie, Western Australia. From the Emperor mine, Vatukoula, Viti Levu, Fiji Islands. At the Kawazu mine, Shizuoka Prefecture, Japan. In the Kochbulak deposit, eastern Uzbekistan, USSR.

Name: For the Empress Josephine mine, Colorado, USA.

Type Material: University of Colorado, Boulder, Colorado, 1649; National Museum of Natural History, Washington, D.C., R7243; Harvard University, Cambridge, Massachusetts, USA, 106761.

References: (1) Palache, C., H. Berman, and C. Frondel (1944) Dana's system of mineralogy, (7th edition), v. I, 260. (2) Honea, R.M. (1964) Empressite and stuetzite redefined. Amer. Mineral., 49, 325–383. (3) Stumpfl, E.F. and J. Rucklidge (1968) New data on natural phases in the system Ag–Te. Amer. Mineral., 53, 1513–1522. (4) Scott, J.D. (1971) Collecting rare sulphosalts in British Columbia. Mineral. Record, 2, 203–209.

Crystal Data: Orthorhombic. *Point Group:* $mm2$. Tabular on {001}; also prismatic by elongation of [001]; crystals up to 15 cm. Prism zone typically deeply striated ∥ [001]. *Twinning:* Twin plane {320} common, sometimes as interpenetrating pseudohexagonal trillings.

Physical Properties: *Cleavage:* {110} perfect; {100} and {010} distinct; {001} indistinct. *Fracture:* Uneven. *Tenacity:* Brittle. Hardness = 3 VHN = n.d. D(meas.) = 4.45 D(calc.) = 4.40

Optical Properties: Opaque. *Color:* Grayish black to iron-black; in polished section, gray to light pink-brown, with deep red internal reflections sometimes observed. *Streak:* Grayish black. *Luster:* Metallic to dull. *Pleochroism:* Weak. *Anisotropism:* Very strong, dark violet-red or olive-green tones.
R_1–R_2: (400) 27.5–30.0, (420) 27.0–29.3, (440) 26.5–28.6, (460) 26.0–28.0, (480) 25.6–27.4, (500) 25.3–27.0, (520) 25.1–26.6, (540) 25.0–26.4, (560) 25.1–26.4, (580) 25.3–26.6, (600) 25.7–27.0, (620) 26.2–27.6, (640) 26.8–28.3, (660) 27.6–28.9, (680) 28.2–29.6, (700) 28.8–30.3

Cell Data: *Space Group:* $Pmn2_1$. $a = 7.407(1)$ $b = 6.436(1)$ $c = 6.154(1)$ Z = [2]

X-ray Powder Pattern: Ouray, Colorado, USA.
3.22 (100), 1.859 (90), 2.87 (80), 1.731 (60), 1.590 (50), 1.075 (50), 1.049 (50)

Chemistry:

	(1)	(2)	(3)
Cu	47.96	48.67	48.42
Fe	1.22	0.33	
Zn	0.57	0.10	
Sb		1.76	
As	18.16	17.91	19.02
S	32.21	31.44	32.56
rem.		0.11	
Total	100.12	100.32	100.00

(1) Cerro Blanco, Atacama, Chile. (2) Rarus mine, Butte, Montana, USA; remainder 0.11% in solution. (3) Cu_3AsS_4.

Polymorphism & Series: Dimorphous with luzonite.

Occurrence: Found in hydrothermal vein deposits formed at moderate temperatures. Also as a late-stage mineral in low-temperature deposits.

Association: Pyrite, sphalerite, galena, bornite, tetrahedrite–tennantite, chalcocite, covellite, barite, quartz.

Distribution: A widespread mineral, sometimes an important ore of copper, but not often well crystallized. Only a few localities can be mentioned. In the USA, in splendent crystals from Butte, Silver Bow Co., Montana; in Utah, in the Tintic district, Tooele Co.; in Colorado, in fine specimens from a number of mines in the Red Mountain district, Ouray and San Juan Cos. In Austria, at Matzenköpfl, Brixlegg, Tirol. From Bor, Yugoslavia. In Italy, on Sardinia, at Alghero and Calabona. At Tsumeb, Namibia. From Peru, exceptional crystals from the Mina Luz Angelica at Quiruvilca; also from Morococha and Cerro de Pasco. In Argentina, in the Sierra de Famatina, La Rioja Province. In the Philippine Islands, at Mancayan, Luzon. From the Chinkuashi mine, Keelung, Taiwan.

Name: From the Greek for *distinct*, in allusion to its cleavage.

References: (1) Palache, C., H. Berman, and C. Frondel (1944) Dana's system of mineralogy, (7th edition), v. I, 389–391. (2) Berry, L.G. and R.M. Thompson (1962) X-ray powder data for the ore minerals. Geol. Soc. Amer. Mem. 85, 127–128. (3) Springer, G. (1969) Compositional variations in enargite and luzonite. Mineral. Deposita, 4, 72–74. (4) Adiwidjaja, G. and J. Löhn (1970) Strukturverfeinerung von Enargit, Cu_3AsS_4. Acta Cryst., B26, 1878–1879 (in German with English abs.).

Crystal Data: Monoclinic. *Point Group:* $2/m$. In fine granular, impure masses of minute fibers, to several mm; single, fine-bladed crystals up to 1 mm in length.

Physical Properties: *Cleavage:* Good on {110}. Hardness = n.d. VHN = 22–67, 39 average (15 g load). D(meas.) = 2.30 (magnetite-contaminated). D(calc.) = 2.216

Optical Properties: Opaque; transmits light in very fine fibers. *Color:* Copper-red on fresh surface, tarnishes rapidly. *Streak:* Black. *Pleochroism:* Extreme: brilliant reddish orange through greenish grays and pinkish grays to dark gray with a barely perceptible bluish or greenish tint.
R_1–R_2: (400) — , (420) — , (440) 6.6–15.3, (460) 6.8–15.8, (480) 7.0–13.5, (500) 7.4–11.7, (520) 7.6–11.7, (540) 8.8–20.7, (560) 8.5–19.4, (580) 8.8–22.8, (600) 8.7–26.7, (620) 9.6–31.1, (640) 9.9–34.7, (660) 10.0–34.5, (680) 9.9–30.4, (700) 9.9–26.5

Cell Data: *Space Group:* $C2/c$. $a = 10.693$ $b = 9.115$ $c = 5.507$ $\beta = 92°10(2)'$ Z = 4

X-ray Powder Pattern: Coyote Peak, California, USA.
6.935 (100), 5.342 (71), 4.556 (41), 3.467 (28), 2.310 (23), 2.902 (15), 3.317 (12)

Chemistry:

	(1)	(2)	(3)
Na	14.1	11.2	12.84
K		0.12	
Fe	36.0	34.9	31.20
S	40.5	37.6	35.82
O	[9.4]	16.2	17.88
H			2.26
Total	100.0	100.02	100.00

(1) Coyote Peak, California, USA; by electron microprobe, oxygen by difference. (2) Lovozero massif, USSR; by electron microprobe. (3) $NaFeS_2 \cdot 2H_2O$.

Occurrence: Found in abundance, typically associated with other sulfides and fine grained magnetite, in discrete, late segregations within a mafic alkalic diatreme (Coyote Peak, California, USA); in pegmatites in nepheline syenite (Lovozero massif, USSR).

Association: Pyrrhotite, magnetite, rasvumite, djerfisherite, bartonite (Coyote Peak, California, USA); pyrite, murmanite (Lovozero massif, USSR).

Distribution: At Coyote Peak, near Orick, Humboldt Co., California, USA. From the Lovozero massif, Kola Peninsula, USSR.

Name: For Richard C. Erd (1924–), mineralogist with the U.S. Geological Survey.

Type Material: n.d.

References: (1) Czamanske, G.K., B.F. Leonard, and J.R. Clark (1980) Erdite, a new hydrated sodium iron sulfide mineral. Amer. Mineral., 65, 509–515. (2) Konnert, J.A. and H.T. Evans, Jr. (1980) The crystal structure of erdite, $NaFeS_2 \cdot 2H_2O$. Amer. Mineral., 65, 516–521. (3) Khomyakov, A.P., M.F. Korobitsyn, M.G. Dobrovol'skaya, and A.I. Tsepin (1979) Erdite ($NaFeS_2 \cdot 2H_2O$) — first occurrence in the USSR. Doklady Acad. Nauk SSSR, 249, 968–971 (in Russian). (4) (1980) Chem. Abs., 92, 218014 (abs. ref. 3).

Crystal Data: Cubic. *Point Group:* $2/m\ \overline{3}$. Pyritohedral crystals, as large as 1 mm, and as a few minute (ca. 20 μm) grains in Pt–Fe alloy and in a platinum nugget.

Physical Properties: Hardness = n.d. VHN = 1730–1950, 1854 average (100 g load). D(meas.) = 8.28 D(calc.) = 9.59

Optical Properties: Opaque. *Color:* Gray. *Luster:* Metallic.
R: (400) 43.7, (420) 43.7, (440) 43.7, (460) 43.7, (480) 43.6, (500) 43.4, (520) 43.2, (540) 42.9, (560) 42.7, (580) 42.4, (600) 42.1, (620) 41.7, (640) 41.2, (660) 40.8, (680) 40.3, (700) 39.9

Cell Data: *Space Group:* $Pa3$ (synthetic). $a = 5.6196(3)$ Z = 4

X-ray Powder Pattern: Synthetic. (JCPDS 19-882).
3.24 (100), 2.810 (85), 1.694 (85), 0.780 (65), 1.987 (55), 0.787 (55), 1.081 (30)

Chemistry:

	(1)	(2)	(3)
Os	68.0	64.3	74.78
Ir	2.6	3.5	
Rh	3.8	5.5	
Ru	0.4	0.4	
Pd	0.5	0.6	
S	25.2	25.5	25.22
Total	100.5	99.8	100.00

(1) MacIntosh mine, California, USA; by electron microprobe, giving $(Os_{0.89}Rh_{0.09}Ir_{0.04}Pd_{0.01}Ru_{0.01})_{\Sigma=1.04}S_{1.96}$. (2) Western Ethiopia; by electron microprobe, yielding $(Os_{0.84}Ir_{0.05}Rh_{0.13}Ru_{0.01}Pd_{0.01})_{\Sigma=1.04}S_{1.96}$. (3) OsS_2.

Polymorphism & Series: Forms a series with laurite.

Occurrence: Found in noble metal placers.

Association: Pt–Fe alloy, platinum, laurite, hollingworthite, irarsite.

Distribution: In the USA, from a placer at the MacIntosh mine, Willow Creek, Trinity river, Humboldt Co.; and in gravels from the American River, Sacramento Co., California. In a platinum metal nugget from Joubdo stream, Birbir River, Ethiopia. From placers on the Tulameen River, British Columbia, Canada. In the Ural Mountain placers, Siberia, and in the Kondörskii massif, Aldan Shield, USSR. From placers at Guma Water, Sierra Leone. At Tiébaghi, New Caledonia.

Name: For Joseph Erlichman, electron probe analyst, who analyzed a number of new minerals.

Type Material: Stanford University, Palo Alto, California, USA, 51965.

References: (1) Snetsinger, K.G. (1971) Erlichmanite (OsS_2), a new mineral. Amer. Mineral., 56, 1501–1506. (2) Cabri, L.J., Ed. (1981) Platinum group elements: mineralogy, geology, recovery. Can. Inst. Min. & Met., 103–104. (3) Bowles, J.F.W., D. Atkin, J.L.M. Lambert, T. Deans, and R. Phillips (1983) The chemistry, reflectance, and cell size of the erlichmanite (OsS_2) — laurite (RuS_2) series. Mineral. Mag., 47, 465–471.

Eskebornite $CuFeSe_2$

Crystal Data: Cubic, possibly tetragonal, pseudocubic. *Point Group:* $4/m\ \overline{3}\ 2/m$. Crystals thick tabular, up to 1 mm in size, embedded in other selenides.

Physical Properties: *Cleavage:* {001} perfect. Hardness = 3–3.5 VHN = 155–252, 204 average (15 g load). D(meas.) = 5.35 D(calc.) = 5.44 Distinctly magnetic.

Optical Properties: Opaque. *Color:* Brass-yellow, tarnishes dark brown to black; in reflected light, brown-yellow or cream-yellow, sometimes with an orange tint. *Luster:* Metallic. *Pleochroism:* Weak, creamy yellow to yellowish brown. *Anisotropism:* Marked, yellowish to tan. R_1–R_2: (400) 23.5–28.0, (420) 24.5–29.0, (440) 25.5–30.0, (460) 26.4–31.1, (480) 27.2–32.2, (500) 28.0–33.4, (520) 28.7–34.4, (540) 29.4–35.0, (560) 30.0–35.3, (580) 30.6–35.5, (600) 31.0–35.8, (620) 31.5–36.0, (640) 31.9–36.4, (660) 32.2–37.0, (680) 32.6–37.7, (700) 33.0–38.7

Cell Data: *Space Group:* $Pm3m$. $a = 5.53$ Z = 2

X-ray Powder Pattern: Synthetic.
3.17 (100), 1.945 (90), 5.53 (80), 2.459 (80), 1.662 (80), 1.380 (70), 1.266 (70)

Chemistry:

	(1)	(2)
Cu	23.62	23.0
Fe	19.75	19.7
Se	55.96	57.5
Total	99.32	100.2

(1) Martin Lake, Canada; by electron microprobe, average of several analyses, corresponding to $Cu_{1.06}Fe_{1.01}Se_{2.00}$. (2) Eagle Group, Canada; by electron microprobe, average of several analyses, corresponding to $Cu_{0.993}Fe_{0.969}Se_{2.000}$.

Polymorphism & Series: Forms a series with chalcopyrite.

Occurrence: In low-temperature hydrothermal vein deposits.

Association: Chalcopyrite, clausthalite, tiemannite, berzelianite, naumannite, umangite, geffroyite, chaméanite, uraninite, ankerite, dolomite.

Distribution: From the Eskeborn adit, Tilkerode, Harz Mountains, Germany. At Martin Lake and the Eagle Group, Lake Athabasca area, Saskatchewan, Canada. From Chaméane, Puy-de-Dôme, France. At Sierra de Cacheuta, Mendoza Province, and Sierra de Umango, La Rioja Province, Argentina. In Czechoslovakia, from Bukov, Petrovice, Předbořice, and Slavkovice.

Name: For the Eskeborn adit, Tilkerode, Germany, where first discovered.

Type Material: Harvard University, Cambridge, Massachusetts, USA, 98902.

References: (1) Ramdohr, P. (1949) Neue Erzmineralien. Fortschr. Mineral., 28, 69–70 (in German). (2) Harris, D.C. and E.A.J. Burke (1971) Eskebornite, two Canadian occurrences. Can. Mineral., 10, 786–796. (3) (1972) Amer. Mineral., 57, 1560 (abs. ref. 2). (4) Kvaček, M. (1973) Selenides from the uranium deposits of western Moravia, Czechoslovakia. Part 1. Berzelianite, umangite, eskebornite. Acta. Univ. Carol., Geol., 1-2, 23–36. (5) Sindeeva, N.D. (1964) Mineralogy and types of deposits of selenium and tellurium, 61–62.

Crystal Data: Monoclinic. *Point Group:* $2/m$ or m. As lamellar grains, the average size of which is 0.5 mm. *Twinning:* On {010}, twin lamellae || [001] often present.

Physical Properties: Hardness = n.d. VHN = 162–223, 191 average (50 g load). D(meas.) = n.d. D(calc.) = 7.12

Optical Properties: Opaque. *Color:* In polished section, galena-white. *Pleochroism:* Absent in air, absent to weak in oil. *Anisotropism:* Distinct to strong, light gray to steel bluish black. R_1–R_2: n.d.

Cell Data: *Space Group:* $B2/m$ or Bm. $a = 13.459$ $b = 30.194$ $c = 4.100$ $\beta = 93.35°$ Z = 1

X-ray Powder Pattern: Ivigtut, Greenland.
3.36 (100), 2.87 (60), 2.96 (50), 2.05 (50), 1.754 (50), 2.08 (40), 1.667 (40)

Chemistry:

	(1)	(2)	(3)
Ag	9.64	9.68	10.61
Pb	30.42	28.80	29.12
Cu	0.21	0.30	
Bi	44.79	46.70	44.05
S	16.40	16.17	16.22
Total	101.46	101.65	100.00

(1–2) Ivigtut, Greenland; by electron microprobe, average of analysis on 9 and 15 grains respectively. (2) $Ag_7Pb_{10}Bi_{15}S_{36}$.

Occurrence: In a cryolite deposit (Ivigtut, Greenland).

Association: Berryite, aikinite, galena (Ivigtut, Greenland); pyrite, enargite (Flathead mine, Montana, USA).

Distribution: From Ivigtut, southern Greenland. In the USA, at the Flathead mine, Niarada, Montana, and Manhattan, Nye Co., Nevada. In the Kochbulak deposit, eastern Uzbekistan, USSR.

Name: For the Eskimos, early settlers of Greenland.

Type Material: n.d.

References: (1) Karup-Møller, S. (1977) Mineralogy of some Ag–(Cu–Pb–Bi) sulfide associations. Bull. Geol. Soc. Denmark, 26, 41–68. (2) Makovicky, E. and S. Karup-Møller (1977) Chemistry and crystallography of the lillianite homologous series II: Definition of new minerals: eskimoite, vikingite, ourayite, and treasurite. Redefinition of schirmerite and new data on the lillianite–gustavite solid solution series. Neues Jahrb. Mineral., Abh., 131, 56–82. (3) (1979) Amer. Mineral., 64, 243–244 (abs. refs. 1 and 2).

Crystal Data: Orthorhombic, pseudotetragonal. *Point Group:* $4/m\ 2/m\ 2/m$ (pseudotetragonal). In granular masses and as disseminated blebs.

Physical Properties: *Fracture:* Uneven to subconchoidal. *Tenacity:* Brittle, somewhat sectile. Hardness = 2.5 VHN = 27–31 (25 g load). D(meas.) = 7.6–7.8 D(calc.) = 7.91

Optical Properties: Opaque. *Color:* Brilliant creamy white, tarnishing to bright orange; in polished section, tin-white with a faint creamy tinge. *Streak:* Shining. *Luster:* Metallic. *Pleochroism:* Weak. *Anisotropism:* Strong.

R_1–R_2: (400) 34.7–35.0, (420) 34.8–35.1, (440) 34.9–35.2, (460) 35.3–35.4, (480) 36.2–35.8, (500) 37.0–36.4, (520) 37.2–36.9, (540) 37.1–37.2, (560) 36.8–37.3, (580) 36.3–37.0, (600) 35.6–36.5, (620) 35.0–35.7, (640) 34.4–35.0, (660) 33.8–34.4, (680) 33.3–34.0, (700) 32.9–33.8

Cell Data: *Space Group:* $P4/nmm$ (pseudotetragonal). $a = 4.105$ $b = 20.35$ $c = 6.31$ Z = 8

X-ray Powder Pattern: Skrikerum, Sweden.
2.12 (100), 2.61 (70), 2.88 (50), 2.48 (40), 2.02 (20), 3.14 (10), 1.861 (10)

Chemistry:

	(1)	(2)
Ag	42.20	43.04
Cu	25.41	25.36
Se	32.43	31.60
Total	100.04	100.00

(1) Argentina. (2) CuAgSe.

Occurrence: Widely distributed in selenium deposits of hydrothermal origin; locally abundant with other selenides.

Association: Berzelianite, weissite, crookesite, clausthalite, umangite, klockmannite, tiemannite, chalcomenite, malachite, calcite.

Distribution: In Canada, in the Martin Lake mine near Lake Athabasca, Saskatchewan, and in the Kidd Creek mine, near Timmins, Ontario. From the Cougar mine, Slick Rock Canyon, San Miguel Co., Colorado, USA. In Germany, at Lerbach, Harz. From Kletno, Poland. At Bukov, Petrovice, Lasovice and Jáchymov (Joachimsthal), Czechoslovakia. From Hope's Nose, Torquay, Devon, England. In Sweden, at the Skrikerum copper mine near Tryserum, Kalmar. From the Frederik VII's mine, near Julianehåb, southern Greenland. At Sierra de Umango and in the Santa Brigida mine, La Rioja Province, Argentina. From Aguas Blancas, Copiapó, Chile. At Kalgoorlie, Western Australia.

Name: From the Greek for *opportunity*, because it was discovered shortly after discovery of the element selenium.

References: (1) Palache, C., H. Berman, and C. Frondel (1944) Dana's system of mineralogy, (7th edition), v. I, 183–184. (2) Earley, J.W. (1950) Description and synthesis of the selenide minerals. Amer. Mineral., 35, 337–364. (3) Sindeeva, N.D. (1964) Mineralogy and types of deposits of selenium and tellurium, 48–49. (4) Frueh, A.J., Jr., G.K. Czamanske, and C. Knight (1957) The crystallography of eucairite, CuAgSe. Zeits. Krist., 108, 389–391.

Crystal Data: Tetragonal. *Point Group:* $\bar{4}2m$. Massive, granular; reniform; as crusts of tiny crystals. *Twinning:* Polysynthetic, very commonly seen in polished section.

Physical Properties: *Cleavage:* {101} good; {100} distinct. *Fracture:* Uneven, conchoidal. *Tenacity:* Brittle. Hardness = 3.5 VHN = n.d. D(meas.) = 4.635 D(calc.) = 4.66

Optical Properties: Opaque. *Color:* Deep pinkish brown; in polished section, pale brownish pink. *Streak:* Black. *Luster:* Dull metallic. *Pleochroism:* Weak, light orange, pink-violet. *Anisotropism:* Strong, greenish yellow and purplish red.

R_1–R_2: (400) 24.4–30.0, (420) 24.2–25.5, (440) 24.0–25.0, (460) 23.8–24.1, (480) 23.1–23.4, (500) 22.7–23.1, (520) 22.9–23.6, (540) 23.8–25.0, (560) 24.3–26.0, (580) 24.5–26.5, (600) 24.8–27.0, (620) 25.4–27.5, (640) 26.0–28.1, (660) 26.8–28.6, (680) 27.3–29.0, (700) 27.5–29.4

Cell Data: *Space Group:* $I\bar{4}2m$. $a = 5.38$ $c = 10.76$ Z = 2

X-ray Powder Pattern: Synthetic.
3.07 (100), 1.895 (80), 1.614 (70), 1.232 (60), 1.099 (60), 1.342 (50), 1.037 (50)

Chemistry:

	(1)	(2)	(3)
Cu	43.94	42.98	43.27
Fe	0.48	0.26	
Sb	13.19	24.36	27.63
As	9.08	3.31	
Bi	1.79		
S	30.86	28.97	29.10
rem.	0.17		
Total	99.51	99.88	100.00

(1) Goldfield, Nevada, USA. (2) Sierra Famatina, Argentina. (3) Cu_3SbS_4.

Occurrence: Found in copper deposits formed at low to moderate temperatures.

Association: Pyrite, enargite, tetrahedrite–tennantite, chalcopyrite, covellite, sphalerite, bismuthinite, silver, gold, marcasite, quartz, barite.

Distribution: From the Sierra de Famatina, La Rioja Province, Argentina. In the USA, at Goldfield, Esmeralda Co., Nevada; and in Arizona, from Tombstone and the Campbell mine, Bisbee, Cochise Co.; from the Magma mine, Superior, Pinal Co. Also at Butte, Silver Bow Co., Montana; Red Mountain, Ouray Co., Colorado; and Darwin, Inyo Co., California. At Hokuetsu, Japan. Well crystallized in the Chinkuashi mine, Keelung, Taiwan. From Cerro de Pasco and Morococha, Peru. In Mexico, at Cananea, Sonora. In minor amounts from other localities.

Name: For the locality, Sierra de Famatina, Argentina.

References: (1) Palache, C., H. Berman, and C. Frondel (1944) Dana's system of mineralogy, (7th edition), v. I, 387–388. (2) Gaines, R. (1957) Luzonite, famatinite, and some related minerals. Amer. Mineral., 42, 766–777. (3) Ramdohr, P. (1969) The ore minerals and their intergrowths, (3rd edition), 578–582.

Crystal Data: Cubic. *Point Group:* $4/m\ \overline{3}\ 2/m$. As small grains forming aggregates to several hundred μm.

Physical Properties: Hardness = n.d. VHN = 900 (100 g load). D(meas.) = n.d. D(calc.) = 6.18 Ferromagnetic.

Optical Properties: Opaque. *Color:* Light gray. *Luster:* Metallic.
R: (400) — , (420) — , (440) 55.2, (460) 55.4, (480) 56.2, (500) 56.9, (520) 58.0, (540) 58.8, (560) 59.5, (580) 60.4, (600) 61.0, (620) 61.0, (640) 61.8, (660) — , (680) 62.8, (700) 63.2

Cell Data: *Space Group:* *Pm3m*. a = 2.882(5) Z = [0.5]

X-ray Powder Pattern: Southern Ural Mountains, USSR.
2.04 (100), 1.17 (90), 0.77 (80), 1.02 (70), 1.44 (60), 1.66 (50), 1.29 (50)

Chemistry:

	(1)
Fe	12.60
Cr	87.58
Total	100.18

(1) Southern Ural Mountains, USSR; by electron microprobe, corresponding to $Cr_{1.5}Fe_{0.5-x}$, with x = 0.3.

Occurrence: In quartz veins within brecciated amphibolites and schist.

Association: Iron, copper, bismuth, gold, chromferide, graphite, cohenite, halite, sylvite, marialite, quartz.

Distribution: From an unspecified locality in the Southern Ural Mountains, USSR.

Name: For the chemical composition, FERrum, *iron*, and CHROMium.

Type Material: A.E. Fersman Mineralogical Museum, Academy of Sciences, Moscow, USSR.

References: (1) Novgorodova, M.I., A.I. Gorshkov, N.V. Trubkin, A.I. Tsepin, and M.T. Dmitrieva (1986) New natural intermetallic compounds of iron and chromium – chromferide and ferchromide. Zap. Vses. Mineral. Obshch., 115, 355–360 (in Russian). (2) (1988) Amer. Mineral., 73, 191 (abs. ref. 1).

Crystal Data: Tetragonal. *Point Group:* n.d. Intergrown with other platinum group minerals, as unrounded or slightly rounded, nodule-like grains up to 4.5 mm, but monomineralic areas do not exceed 0.15 mm.

Physical Properties: *Tenacity:* Ductile. Hardness = n.d. VHN = 481 (50 g load). D(meas.) = n.d. D(calc.) = n.d.

Optical Properties: Opaque. *Color:* Silvery white; in polished section, rosy cream. *Luster:* Metallic. *Anisotropism:* Weak.

R_1–R_2: (400) — , (420) 56.5, (440) 56.4, (460) 56.8, (480) 57.4, (500) 58.2, (520) 58.7, (540) 59.0, (560) 59.3, (580) 59.7, (600) 60.2, (620) 60.5, (640) 60.9, (660) 61.4, (680) 62.0, (700) 62.8

Cell Data: *Space Group:* n.d. $a = 3.871(4)$ $c = 3.635(5)$ Z = n.d.

X-ray Powder Pattern: Koryak-Kamchatka region, USSR.
2.192 (10), 1.935 (5), 1.324 (4), 1.699 (3), 1.157 (3b)

Chemistry:

	(1)	(2)
Pt	75.7 – 77.6	77.30
Ir	0.27 – 0.69	
Fe	10.4 – 11.0	11.07
Ni	10.2 – 11.7	11.63
Cu	0.33 – 0.66	
Total		100.00

(1) Koryak-Kamchatka region, USSR; by electron microprobe, range of analyses of six grains, corresponding to $(Pt_{2.016}Ir_{0.012})_{\Sigma=2.028}Fe_{0.983}(Ni_{0.962}Cu_{0.027})_{\Sigma=0.989}$. (2) Pt_2FeNi.

Polymorphism & Series: Forms a series with tulameenite.

Occurrence: In the heavy fraction of Quaternary alluvial deposits associated with ultramafics from an ophiolite belt (Koryak-Kamchatka region, USSR).

Association: Isoferroplatinum, tetraferroplatinum, rutheniridosmine, laurite, irarsite, cooperite, sperrylite, hollingworthite, cherepanovite, chromite, olivine (Koryak-Kamchatka region, USSR).

Distribution: At an undefined locality in the Koryak-Kamchatka folded belt, northeastern USSR. From Goodnews Bay, Alaska, USA.

Name: For the composition.

Type Material: Mining Museum, Leningrad Mining Institute, Leningrad, USSR.

References: (1) Rudashevskii, N.S., A.G. Mochalov, Y.P. Men'shikov, and N.I. Shumskaya (1983) Ferronickelplatinum, Pt_2FeNi, a new mineral species. Zap. Vses. Mineral. Obshch., 112, 487–494 (in Russian). (2) (1984) Amer. Mineral., 69, 1190–1191 (abs. ref. 1).

Crystal Data: Orthorhombic. *Point Group:* $2/m$ $2/m$ $2/m$. As slender prismatic crystals to 1 mm long and up to 0.1 mm across; striated on {110} parallel to [001]. Cross sections are rhombic in shape. *Twinning:* As stellate and cruciform twins; twin lamellae noted in polished section.

Physical Properties: *Tenacity:* Very brittle. Hardness = 6–6.5 VHN = 858–933, 897 average (25 g load). D(meas.) = 7.20 D(calc.) = 7.139 Becomes magnetic on heating.

Optical Properties: Opaque. *Color:* Steel-gray to tin-white with a rose tint, also brass-yellow, tarnishes to a brassy tone; in polished section, white with a gray-brown hue. *Streak:* Black. *Luster:* Metallic. *Pleochroism:* Distinct. *Anisotropism:* Strong, from greenish gray to lilac-gray, much like marcasite.
R_1–R_2: (400) 43.8–46.6, (420) 43.8–46.8, (440) 43.8–47.0, (460) 43.9–47.6, (480) 44.0–48.3, (500) 44.4–49.2, (520) 44.9–50.4, (540) 45.6–51.4, (560) 46.4–52.5, (580) 47.3–53.8, (600) 48.0–54.9, (620) 48.5–55.8, (640) 48.9–56.6, (660) 49.2–57.3, (680) 49.6–57.7, (700) 49.9–57.8

Cell Data: *Space Group:* *Pnnm.* $a = 4.8007$ $b = 5.776$ $c = 3.5850$ Z = 2

X-ray Powder Pattern: Powder River Basin, Wyoming, USA. Cannot be distinguished from rammelsbergite.
2.568 (100), 2.474 (100), 1.885 (70), 2.871 (50), 1.695 (40), 2.400 (35), 1.541 (35)

Chemistry:

	(1)	(2)	(3)	(4)
Fe	27.87	25.2	23.8	26.13
Co		0.0	1.1	
Se	72.13	73.8	73.2	73.87
Total	100.00	99.0	98.1	100.00

(1) Tuva ASSR, USSR; recalculated to 100% after deduction of 4% insoluble residue, leading to $FeSe_{1.83}$. (2) Virgin no. 3 mine, Montrose Co., Colorado, USA; corresponds to $FeSe_{2.07}$. (3) A.E.C. no. 8 mine, Utah, USA; giving $(Fe, Co)Se_{2.08}$ (4) $FeSe_2$.

Occurrence: In red-bed deposits of the Colorado Plateau type, peripheral to pyrite and uranium concentrations (Utah, USA); in U-V ores in sandstone, with coalified wood (Colorado, USA); with other sulfides and selenides cementing sandstones and pelites (Tuva ASSR, USSR).

Association: Uraninite, pyrite, marcasite, chalcopyrite, sphalerite, selenium, clausthalite, cadmoselite, bornite, cobaltomenite, laumontite, barite.

Distribution: In the USA, at several localities in Utah, on the Colorado Plateau, including the A.E.C. no. 8 mine, Temple Mountain, Emery Co.; in Wyoming, from sandstones of the Powder River Basin; from the Kermac Sec. 10 mine, McKinley Co., New Mexico; and from a number of mines in the Urvan district, Montrose Co., Colorado. At an undefined locality in the Tuva ASSR, USSR. From Finland, at Kuusamo. In the Petrovice and Předbořice deposits, Czechoslovakia.

Name: For the composition.

Type Material: n.d.

References: (1) Bur'yanova, E.Z. and A.I. Komkov (1955) A new mineral—ferroselite. Doklady Acad. Nauk SSSR, 105, 812–813 (in Russian). (2) (1956) Amer. Mineral., 41, 671 (abs. ref. 1). (3) Kullerud, G. and G. Donnay (1958) Natural and synthetic ferroselite. Geochim. Cosmochim. Acta, 15, 73–79. (4) Coleman, R.G. (1959) New occurrences of ferroselite (FeS_2). Geochim. Cosmochim. Acta, 16, 296–301. (5) Granger, H.C. (1966) Ferroselite in a roll-type uranium deposit, Powder River Basin, Wyoming. U.S. Geol. Sur. Prof. Paper 550-C, C133–C137. (6) Santos, E.S. (1968) Reflectivity and microindentation hardness of ferroselite from Colorado and New Mexico. Amer. Mineral., 53, 2075–2077. (7) Vlasov, K.A., Ed. (1966) Mineralogy of rare elements, v. II, 667–669.

Crystal Data: Cubic. *Point Group:* 432. Massive; in xenomorphic grains.

Physical Properties: Hardness = n.d. VHN = 32.7 (50 g load). D(meas.) = n.d. D(calc.) = 9.05

Optical Properties: Opaque. *Color:* Light pink in reflected light. *Luster:* Metallic.
R: (400) 33.4, (420) 33.4, (440) 33.4, (460) 33.0, (480) 32.3, (500) 31.2, (520) 30.1, (540) 29.8, (560) 30.2, (580) 31.0, (600) 32.0, (620) 33.1, (640) 34.0, (660) 34.8, (680) 35.4, (700) 36.0

Cell Data: *Space Group:* $I4_132$. $a = 9.967$ Z = 8

X-ray Powder Pattern: Předbořice, Czechoslovakia.
2.662 (100), 2.229 (80), 2.035 (80), 1.820 (80), 1.266 (70), 7.08 (60), 1.954 (60)

Chemistry:

	(1)	(2)	(3)	(4)
Ag	47.5	48.6	52.3	47.70
Cu		0.3	0.9	
Au	27.4	27.2	23.4	29.03
Se	24.4	22.8	22.8	23.27
Total	99.3	98.9	99.4	100.00

(1) Předbořice, Czechoslovakia; by electron microprobe, corresponding to $Ag_{2.97}Au_{0.94}Se_{2.09}$. (2) Do.; corresponding to $(Ag_{3.07}Cu_{0.03})_{\Sigma=3.10}Au_{0.94}Se_{1.97}$. (3) Hope's Nose, England; by electron microprobe. (4) Ag_3AuSe_2.

Occurrence: In carbonate veins in epithermal precious metal deposits (Předbořice, Czechoslovakia).

Association: Naumannite, clausthalite, permingeatite, gold, calcite, quartz (Předbořice, Czechoslovakia).

Distribution: From Předbořice, Czechoslovakia. At Hope's Nose, Torquay, Devon, England. From the Kidd Creek mine, near Timmins, Ontario, Canada. In the De Lamar mine, Silver City district, Owyhee Co., Idaho, USA. At Flamenco, Atacama, Chile.

Name: For Raymond Fischesser, Director of the National School of Mines, Paris, France.

Type Material: National School of Mines, Paris, France.

References: (1) Johan, Z., P. Picot, R. Pierrot, and M. Kvaček (1971) La fischesserite, Ag_3AuSe_2, premier séléniurs d'or, isotype de la petzite. Bull. Soc. fr. Minéral., 94, 381–384 (in French with English abs.). (2) (1972) Amer. Mineral., 57, 1554 (abs. ref. 1).

Fizélyite $Pb_{14}Ag_5Sb_{21}S_{48}$(?)

Crystal Data: Monoclinic. *Point Group:* $2/m$. Crystals small, deeply striated and lacking terminal faces. *Twinning:* On {010}, with two sets of polysynthetic twins, || to [001] and to {001}, giving a chess-board pattern.

Physical Properties: *Cleavage:* {010}. *Tenacity:* Very brittle. Hardness = 2 VHN = n.d. D(meas.) = 5.56 D(calc.) = 5.224

Optical Properties: Opaque. *Color:* Dark lead-gray to steel-gray. *Streak:* Dark gray. *Luster:* Metallic. *Anisotropism:* Shown, but without distinct color effects.
R_1–R_2: (400) 41.6–48.3, (420) 40.6–47.4, (440) 39.6–46.5, (460) 39.3–46.1, (480) 39.0–45.8, (500) 38.8–45.6, (520) 38.4–45.2, (540) 38.0–44.9, (560) 37.8–44.5, (580) 37.4–43.8, (600) 36.9–43.2, (620) 36.3–42.4, (640) 35.6–41.4, (660) 34.8–40.3, (680) 34.0–39.0, (700) 33.0–37.8

Cell Data: *Space Group:* $P2_1/n$. $a = 13.21$ $b = 19.27$ $c = 8.68$ $\beta = 90.4°$ Z = 2

X-ray Powder Pattern: Herja, Romania.
3.34 (10), 3.49 (5), 2.96 (5), 2.89 (5), 3.80 (4), 2.80 (3), 2.77 (3)

Chemistry:

	(1)	(2)
Pb	37.48	38.49
Ag	7.70	7.16
Fe	0.62	
Sb	34.02	33.93
As	0.32	
S	20.10	20.42
insol.	0.30	
Total	100.54	100.00

(1) Herja, Romania. (2) $Pb_{14}Ag_5Sb_{21}S_{48}$.

Occurrence: In hydrothermal veins.

Association: Semseyite, galena, sphalerite, pyrite, pyrrhotite, quartz, dolomite (Herja, Romania).

Distribution: From Herja (Kisbánya) and Băiţa (Rézbánya), Romania. In the Gabe-Gottes mine, at Třebsko, near Příbram, Czechoslovakia. In the USA, from Morey, Nye Co., Nevada. From the Les Farges mine, near Ussel, Corrèze; and Bournac, Montagne Noire, France. At the Inakuraishi mine, Hokkaido, Japan.

Name: For Sandor Fizély, mining engineer, who discovered the mineral.

Type Material: n.d.

References: (1) Palache, C., H. Berman, and C. Frondel (1944) Dana's system of mineralogy, (7th edition), v. I, 450. (2) Berry, L.G. and R.M. Thompson (1962) X-ray powder data for the ore minerals. Geol. Soc. Amer. Mem. 85, 156–157. (3) Moëlo, Y., E. Makovicky, and S. Karup-Møller (1984) New data on the minerals of the andorite series. Neues Jahrb. Mineral., Monatsh., 175–182. (4) (1985) Amer. Mineral., 70, 219–220 (abs. ref. 3).

Crystal Data: Cubic. *Point Group:* $4/m\ \overline{3}\ 2/m$. As minute (1 to 200 μm) crystals.

Physical Properties: Hardness = n.d. VHN = 446–464 (25 g load). D(meas.) = n.d. D(calc.) = 4.76

Optical Properties: Opaque. *Color:* Steel-gray; in polished section, creamy white. *Luster:* Metallic.
R: (420) 36.4, (460) 40.9, (500) 42.5, (546) 43.4, (589) 45.3, (640) 43.8

Cell Data: *Space Group:* $Fd3m$ (assumed). $a = 9.520$ Z = 8

X-ray Powder Pattern: Fletcher mine, Missouri, USA.
1.68 (vs), 1.83 (s), 2.87 (m), 2.39 (m), 1.37 (m), 1.24 (m), 1.19 (m)

Chemistry:

	(1)	(2)	(3)	(4)
Cu	19.5	14.8	21.2	20.53
Ni	25.9	34.9	19.7	18.97
Co	13.6	9.5	17.0	19.05
Fe	0.9	0.2	1.2	
S	41.6	38.5	40.4	41.45
Total	101.5	97.9	99.5	100.00

(1–3) Fletcher mine, Missouri, USA; by electron microprobe. (4) $Cu(Ni,Co)_2S_4$ with Ni:Co = 1:1.

Occurrence: Found disseminated in copper sulfides, in copper-rich pods replacing dolostone.

Association: Vaesite, pyrite, covellite, chalcopyrite, bornite, digenite.

Distribution: At the Fletcher mine, Reynolds Co., Missouri, USA.

Name: For the Fletcher mine in Missouri, USA.

Type Material: National Museum of Natural History, Washington, D.C., USA, 137072; British Museum (Natural History), London, England; Royal Ontario Museum, Toronto, Canada.

References: (1) Craig, J.R. and A.B. Carpenter (1977) Fletcherite, $Cu(Ni,Co)_2S_4$, a new thiospinel from the Viburnum Trend (New Lead Belt), Missouri. Econ. Geol., 72, 480–486.
(2) (1977) Amer. Mineral., 62, 596 (abs. ref. 1).

Crystal Data: Triclinic. *Point Group:* $\bar{1}$. Thin crystals, tabular on {010}; elongated ∥ [100], as much as 6 cm long, striated on {010}; usually massive, radiated or foliated; often in spherical, rosette- or cauliflower-like aggregates of thin plates, then up to 1 cm in diameter. Crystals often warped or bent. *Twinning:* Complex twinning has been observed.

Physical Properties: *Cleavage:* {010} perfect. *Tenacity:* Flexible, inelastic; slightly malleable. Hardness = 2.5–3. VHN = n.d. D(meas.) = 5.88–5.92 D(calc.) = 5.88

Optical Properties: Opaque. *Color:* Grayish black. *Streak:* Grayish black. *Luster:* Metallic. *Anisotropism:* Weak.
R_1–R_2: (400) 40.9–42.0, (420) 40.2–41.4, (440) 39.5–40.8, (460) 38.8–40.2, (480) 38.2–39.7, (500) 37.6–39.1, (520) 37.0–38.6, (540) 36.4–38.2, (560) 35.8–37.7, (580) 35.3–37.3, (600) 34.8–36.8, (620) 34.4–36.5, (640) 33.9–36.0, (660) 33.5–35.7, (680) 33.1–35.4, (700) 32.8–35.1

Cell Data: *Space Group:* $P\bar{1}$. $a = 46.9$ $b = 5.82$ $c = 17.3$ $\alpha = 90°$ $\beta = 94°40'$ $\gamma = 90°$ Z = 8

X-ray Powder Pattern: Bolivia.
3.44 (100), 2.91 (100), 2.86 (100), 2.82 (100), 2.05 (75), 4.30 (50), 3.11 (50)

Chemistry:

	(1)	(2)	(3)	(4)
Pb	50.57	46.23	46.11	49.71
Fe	2.48	2.69	2.55	
Zn	1.22	0.57	0.79	
Ag		0.97	0.88	
Sn	12.34	17.05	16.08	17.09
Sb	10.51	11.56	10.98	11.69
S	21.04	21.12	21.14	21.51
rem.	0.71		0.72	
Total	98.87	100.19	99.25	100.00

(1) Chocaya, Bolivia. (2) Poopó, Bolivia. (3) Porvenir mine, Huanuni, Bolivia. (4) $Pb_5Sn_3Sb_2S_{14}$.

Occurrence: In hydrothermal Ag-Sn deposits (Bolivia) and in a limestone contact metamorphic deposit (Kalkar quarry, California, USA).

Association: Cylindrite, teallite, plagionite, zinkenite, cassiterite, wurtzite, pyrrhotite, marcasite, arsenopyrite, galena, pyrite, sphalerite, siderite (Bolivia); cassiterite, galena, stannite, teallite, cylindrite (USSR).

Distribution: From Poopó, Oruro, Llallagua, Huanuni, Chocaya, and Colquechaca, Bolivia. In the Thompson mine, Darwin district, Inyo Co., and the Kalkar quarry, Santa Cruz Co., California, USA. From near the headwaters of the east branch of the Coal River, Yukon Territory, Canada. At Vens Haut, Cantal, France. In the Sinantscha zinc deposit, Sichota-Alin, and at Smirnowsk, Transbaikalia, USSR. At Dachang, Guangxi Province, China. At the Hoei mine, Oita Prefecture, Japan. From the Renison Bell mine, Tasmania, and the Wallah Wallah mine, Rye Park, New South Wales, Australia.

Name: For the mining engineers Carl and Ernest Francke.

References: (1) Palache, C., H. Berman, and C. Frondel (1944) Dana's system of mineralogy, (7th edition), v. I, 448–450. (2) Coulon, M., F. Heitz, and M.T. Le Bihan (1961) Contribution à l'étude structurale d'un sulfure de plomb, d'antimoine et d'étain: la frankeite (sic). Bull. Soc. fr. Minéral., 84, 350–354 (in French). (3) Moh, G. (1987) Mutual Pb^{2+}/Sn^{2+} substitution in sulfosalts. Mineral. Petrol., 36, 191–204. (4) Williams, T.B. and B.G. Hyde (1988) Electron microscopy of cylindrite and franckeite. Phys. Chem. Minerals., 15, 521–544.

Crystal Data: Hexagonal. *Point Group:* $6/m\ 2/m\ 2/m$. Massive; as imbedded grains.

Physical Properties: Hardness = < 6, much softer than trogtalite and hastite. VHN = n.d. D(meas.) = n.d. D(calc.) = 7.70

Optical Properties: Opaque. *Color:* Similar to nickeline, but more intense violet in oil. *Anisotropism:* Less anisotropic than nickeline.
R_1–R_2: n.d.

Cell Data: *Space Group:* $P6/mmc$. $a = 3.61$ $c = 5.28$ Z = 2

X-ray Powder Pattern: Steinbruch, Germany.
2.05 (100), 1.066 (100), 0.963 (100), 2.4 (50), 1.71 (50), 1.08 (50), 0.992 (50)

Chemistry: Composition established by analogy to synthetic material.

Occurrence: In dolomite veinlets.

Association: Hematite, clausthalite, nickeline, guanajuatite, hastite, trogtalite, bornhardtite, chalcopyrite, millerite, sphalerite.

Distribution: From the Trogtal quarry, Steinbruch, and at Rote Berg, St. Andreasberg, Harz; also from Hartenstein, Saxony, Germany. In the Pinky Fault uranium deposit, Saskatchewan, Canada. At Temple Mountain, Emery Co., Utah, USA.

Name: For Professor Georg Frebold of Hannover, Germany.

Type Material: n.d.

References: (1) Ramdohr, P. and Schmitt, M. (1955) Vier neue natüraliche kobaltselenide von Steinbruch Trogtal bei Lautenthal im Harz. Neues Jahrb. Mineral., Monatsh., 133–142 (in German). (2) (1956) Amer. Mineral., 41, 164–165 (abs. ref. 1). (3) Strunz, H. (1957) Mineralogische Tabellen (3rd edition), 98 (in German). (4) (1959) Amer. Mineral., 44, 907 (abs. ref. 3).

Freibergite $(Ag, Cu, Fe)_{12}(Sb, As)_4S_{13}$

Crystal Data: Cubic. *Point Group:* $\overline{4}3m$. Tetrahedral crystals; massive.

Physical Properties: Hardness = n.d. VHN = 263–340 (100 g load). D(meas.) = n.d. D(calc.) = [5.41]

Optical Properties: Opaque. *Color:* Gray to black.
R: (400) 35.2, (420) 35.1, (440) 35.0, (460) 34.7, (480) 34.3, (500) 33.7, (520) 33.1, (540) 32.6, (560) 32.2, (580) 31.8, (600) 31.4, (620) 30.9, (640) 30.5, (660) 30.0, (680) 29.5, (700) 29.0

Cell Data: *Space Group:* $I\overline{4}3m$. $a = 10.560$ Z = 2

X-ray Powder Pattern: n.d.

Chemistry:

	(1)	(2)	(3)
Ag	35.6	36.0	48.9
Cu	13.4	12.4	3.2
Fe	6.5	4.1	0.2
Zn	0.1	0.6	6.1
Hg			0.1
Sb	23.1	25.1	21.0
As	0.2	0.5	0.6
Bi			0.1
S	20.1	21.8	19.9
Total	99.0	100.5	100.1

(1) Hi-Ho mine, Canada; by electron microprobe. (2) Mt. Isa, Australia; by electron microprobe, corresponding to $(Ag_{6.37}Cu_{3.72}Fe_{1.41}Zn_{0.18})_{\Sigma=11.68}(Sb_{3.95}As_{0.13})_{\Sigma=4.08}S_{13.00}$. (3) Knappenstube mine, Austria; by electron microprobe, average of analyses of 14 samples.

Polymorphism & Series: Forms series with tetrahedrite and argentotennantite.

Occurrence: In hydrothermal deposits.

Association: A wide variety of sulfides and sulfosalts, as for tetrahedrite.

Distribution: In the Hi-Ho mine, Cobalt-Gowganda region, Ontario; and the Keno Hill-Galena Hill area, Yukon Territory, Canada. In Japan, in the Inakuraishi, Koryu, and Sanru mines, Hokkaido. At Mt. Isa, Queensland; and in the Meerschaum mine, north of Omeo, Victoria, Australia. In the Himmelsfürst mine, Erbisdorf, near Freiberg, Saxony, Germany. From Kutná Hora, Czechoslovakia. In Austria, from the Knappenstube mine, Hochtor, Salzburg. From Yukhondzha, Yakutia, USSR. At Slädekärr and Vena, Sweden. In Scotland, from Tyndrum, Perthshire. Increasingly recognized at additional localities.

Name: For the long-known occurrence of silver-rich tetrahedrite at Freiberg, Germany.

Type Material: n.d.

References: (1) Palache, C., H. Berman, and C. Frondel (1944) Dana's system of mineralogy, (7th edition), v. I, 374–384 (tetrahedrite–tennantite). (2) Petruk, W. and staff (1971) The silver–arsenic deposits of the Cobalt–Gowganda region, Ontario: characteristics of the sulfides. Can. Mineral., 11, 196–231. (3) Riley, J.F (1974) The tetrahedrite–freibergite series with reference to the Mount Isa Pb–Zn–Ag orebody. Mineral. Deposita, 9, 117–124. (4) Peterson, R.C. and I. Miller (1986) Crystal structure and cation distribution in freibergite and tetrahedrite. Mineral. Mag., 50, 717–721. (5) Paar, W.H., T.T. Chen, and W. Guenther. (1978) Extremely silver-rich freibergite in lead-zinc-copper ores of the "Knappenstube" mine, Hochtor, Salzburg. Carinthia 2, 88, 35–42. (6) (1980) Chem. Abs., 93, 189277 (abs. ref. 5).

Crystal Data: Monoclinic. *Point Group:* $2/m$. Crystals prismatic and striated parallel to [001], up to 2 cm. *Twinning:* Twin plane {100}; lamellae seen in polished section.

Physical Properties: *Cleavage:* Imperfect on {110}. *Fracture:* Subconchoidal to uneven. Hardness = 2–2.5 VHN = n.d. D(meas.) = 6.20–6.23 D(calc.) = 6.22

Optical Properties: Opaque. *Color:* Light steel-gray to silver-white or lead-gray. *Streak:* Light steel-gray to silver-white or lead-gray. *Luster:* Metallic. *Pleochroism:* Very weak. *Anisotropism:* Just noticeable.

R_1–R_2: (400) 39.5–46.7, (420) 39.4–46.0, (440) 39.3–45.3, (460) 39.1–44.6, (480) 38.7–44.0, (500) 38.2–43.4, (520) 37.6–42.8, (540) 37.0–42.3, (560) 36.6–41.8, (580) 36.2–41.4, (600) 35.8–41.0, (620) 35.4–40.5, (640) 35.0–40.0, (660) 34.5–39.5, (680) 34.1–38.9, (700) 33.6–38.3

Cell Data: *Space Group:* $P2_1/a$. $a = 7.518$ $b = 12.809$ $c = 5.940$ $\beta = 92.25°$ Z = 4

X-ray Powder Pattern: Hiendelaencina, Spain.
2.83 (100), 3.48 (80), 2.98 (70), 1.784 (50), 2.08 (40), 3.25 (30), 2.01 (30)

Chemistry:

	(1)	(2)	(3)
Ag	22.45	23.08	20.24
Pb	31.90	30.77	38.87
Sb	26.83	27.11	22.84
S	17.60	18.41	18.05
rem.		0.63	
Total	98.78	100.00	100.00

(1) Hiendelaencina, Spain. (2) Příbram, Czechoslovakia; remainder is Fe. (3) $AgPbSbS_3$.

Occurrence: Of hydrothermal origin.

Association: Acanthite, pyrargyrite, silver, andorite, galena, siderite.

Distribution: At the Himmelsfürst mine, Erbisdorf, and Braünsdorf, near Freiberg; and Marienberg, Saxony, Germany. From Příbram, Czechoslovakia. As fine crystals from Hiendelaencina, Guadalajara Province, Spain. In the Les Farges mine, near Ussel, Corrèze; Pontgibaud, Puy-de-Dôme; and Vialas, Lozère, France. From Dossena, Bergamo, Italy. At Hällefors, Bergslagen, Sweden. In the USA, from the Garfield mine, Cascade Mountains, Gunnison Co., Colorado, and the Castle Dome Mountains, Yuma Co., Arizona. In Canada, at Cobalt, Ontario, and at the Weber prospect of Mount Nansen Mines, Yukon Territory.

Name: For Johann Karl Freiesleben (1774–1846), Mining Commissioner of Saxony, Germany.

References: (1) Palache, C., H. Berman, and C. Frondel (1944) Dana's system of mineralogy, (7th edition), v. I, 416–418. (2) Berry, L.G. and R.M. Thompson (1962) X-ray powder data for the ore minerals. Geol. Soc. Amer. Mem. 85, 137. (3) Hellner, E. (1957) Über komplex zusammengesetzte sulfidische Erze. II. Struktur des Freieslebenits, $PbAgSbS_3$. Zeits. Krist., 109, 4 (in German with English abs.). (4) Ito, T. and W. Nowacki (1974) The crystal structure of freieslebenite, $PbAgSbS_3$. Zeits. Krist., 139, 85–102.

Crystal Data: Orthorhombic. *Point Group:* $mm2$. As isolated columnar crystals or as granular aggregates.

Physical Properties: Hardness = n.d. VHN = 224 (50 g load). D(meas.) = 6.98 D(calc.) = 7.06

Optical Properties: Opaque. *Color:* In polished section, creamy yellowish white in air, more pinkish in oil. *Luster:* Metallic. *Anisotropism:* Strong in oil, distinct in air.
R_1–R_2: n.d.

Cell Data: *Space Group:* $Pb2_1m$. $a = 33.84$ $b = 11.65$ $c = 4.010$ Z = 2

X-ray Powder Pattern: Salzburg, Austria.
3.644 (100), 3.584 (100), 3.161 (100), 2.850 (80), 4.05 (40), 2.577 (40), 1.979 (40)

Chemistry:

	(1)	(2)	(3)
Pb	29.7	28.0	30.53
Cu	9.1	10.2	9.36
Bi	44.2	44.7	43.10
S	17.2	17.2	17.01
Total	100.2	100.1	100.00

(1) Salzburg, Austria; by electron microprobe, average of six analyses, corresponding to $Cu_{9.61}Pb_{9.62}Bi_{14.19}S_{36.00}$. (2) Do.; by electron microprobe, average of 12 analyses. (3) $Pb_5Cu_5Bi_7S_{18}$.

Occurrence: As aggregates of granular crystals, in boulders of vein quartz which occur in the scree of a landslip (Salzburg, Austria).

Association: Chalcopyrite, covellite, cerussite, chlorite, mica, quartz (Salzburg, Austria).

Distribution: Near the emerald beryl deposit in the "Sedl" region, east of the Habachtal, Salzburg, Austria. In the USA, from the Fremont mine, Apache Hills, east of Hachita, Grant Co., New Mexico, and near Panguitch, Garfield Co., Utah.

Name: For Professor Dr.-Ing O.M. Friedrich, of the Mining University, Leoben, Styria, Austria.

Type Material: Institute of Mineralogy, University of Salzburg, Austria; Museum of Landeskunde, Joanneum, Graz, Styria, Austria; Royal Ontario Museum, Toronto, Canada; National Museum of Natural History, Washington, D.C., 144072, 144185; Harvard University, Cambridge, Massachusetts, USA, 117007.

References: (1) Chen, T.T., E. Kirchner, and W. Paar (1978) Friedrichite, $Cu_5Pb_5Bi_7S_{18}$, a new member of the aikinite–bismuthinite series. Can. Mineral., 16, 127–130. (2) (1979) Amer. Mineral., 64, 654 (abs. ref. 1). (3) Pring, A. (1989) Structural disorder in aikinite and krupkaite. Amer. Mineral., 74, 250–255.

Crystal Data: Orthorhombic. *Point Group:* $2/m\ 2/m\ 2/m$. As tiny equant grains forming reaction rims around chalcopyrite, and as inclusions.

Physical Properties: Hardness = 3–4 VHN = 250–297 D(meas.) = n.d. D(calc.) = 8.057

Optical Properties: Opaque. *Color:* In polished section, white with bluish or pinkish tint. *Luster:* Metallic. *Pleochroism:* Extremely weak. *Anisotropism:* Strong, orange-red to inky blue.
R_1–R_2: (400) 56.7–57.1, (420) 54.8–56.2, (440) 52.9–55.3, (460) 51.9–54.5, (480) 51.1–53.8, (500) 50.6–53.1, (520) 50.1–52.6, (540) 50.0–52.4, (560) 50.2–52.5, (580) 50.7–53.0, (600) 51.4–53.8, (620) 52.2–54.5, (640) 53.0–55.4, (660) 53.9–56.3, (680) 55.0–57.4, (700) 56.2–58.5

Cell Data: *Space Group:* *Pnnm.* $a = 5.29$ $b = 6.27$ $c = 3.86$ Z = 2

X-ray Powder Pattern: Robb-Montbray mine, Canada.
2.81 (100), 2.71 (70), 2.07 (50), 1.846 (40), 3.29 (30), 1.940 (30), 1.577 (30)

Chemistry:

	(1)	(2)	(3)	(4)
Fe	18.1	17.1	18.3	17.94
Te	82.7	80.4	82.5	82.06
Total	100.8	97.5	100.8	100.00

(1) Robb-Montbray mine, Canada; by electron microprobe. (2) Lindquist Lake, Canada; by electron microprobe. (3) Noranda, Canada; by electron microprobe. (4) $FeTe_2$.

Polymorphism & Series: Forms a series with mattagamite.

Occurrence: In hydrothermal ore deposits, rimming chalcopyrite grains, and as inclusions in gold, petzite, or chalcopyrite.

Association: Chalcopyrite, altaite, melonite, montbrayite, petzite, sylvanite, tellurobismuthite, gold, pyrite, sphalerite, marcasite, chalcocite, covellite.

Distribution: From the Robb-Montbray and Noranda mines, Quebec; and at Lindquist Lake, British Columbia, Canada. From Gold Hill, north of San Simon, Cochise Co., Arizona, USA. From Faţa Băii (Facebánya) and Săcărâmb (Nagyág), Romania. At Bodennec, Finistère, France. From Jabal Sayid, Saudi Arabia.

Name: For Dr. Max Hans Frohberg, mining geologist, Toronto, Canada.

Type Material: n.d.

References: (1) Thompson, R.M. (1947) Frohbergite, $FeTe_2$: a new member of the marcasite group. Univ. Toronto Studies, Geol. Ser., 51, 35–40. (2) Thompson, R.M. (1947) Frohbergite, $FeTe_2$, a new member of the marcasite group (abs.). Amer. Mineral., 32, 210. (3) Berry, L.G. and R.M. Thompson (1962) X-ray powder data for the ore minerals. Geol. Soc. Amer. Mem. 85, 103–104. (4) Rucklidge, J. (1969) Frohbergite, montbrayite and a new Pb–Bi telluride. Can. Mineral., 9, 709–715. (5) Ramdohr, P. (1969) The ore minerals and their intergrowths, (3rd edition), 845–847.

Crystal Data: Monoclinic. *Point Group:* $2/m$. As flat cleavage fragments and rounded grains. *Twinning:* Sometimes present.

Physical Properties: *Cleavage:* {100} perfect; {001} less perfect. *Fracture:* Uneven. *Tenacity:* Brittle. Hardness = 2.5 VHN = 84 (25 g load). D(meas.) = 12.5–12.6, 11.5 (synthetic). D(calc.) = [11.42]

Optical Properties: Opaque. *Color:* Gray; creamy white in reflected light. *Streak:* Black. *Luster:* Metallic, splendent, tarnishes quickly. *Anisotropism:* Evident, in light and dark grays.
R_1–R_2: (400) — , (420) 50.7–52.9, (440) 51.9–54.3, (460) 53.1–55.5, (480) 54.3–56.4, (500) 55.2–57.2, (520) 55.7–57.8, (540) 56.1–58.4, (560) 56.5–59.0, (580) 56.9–59.6, (600) 57.4–60.2, (620) 57.7–60.7, (640) 58.0–61.1, (660) 58.5–61.4, (680) 58.8–62.0, (700) 59.3–63.0

Cell Data: *Space Group:* $C2/m$. $a = 12.74$ $b = 4.29$ $c = 5.71$ $\beta = 102°27'$ Z = [4]

X-ray Powder Pattern: Frood mine, Canada.
2.77 (100), 1.556 (80), 2.97 (70), 2.48 (70), 2.21 (70), 1.637 (60), 1.419 (60)

Chemistry:

	(1)	(2)	(3)
Pd	20.3	19.3	20.29
Bi	79.7	81.5	79.71
Te		0.39	
Total	100.0	101.19	100.00

(1) Frood mine, Canada; by electron microprobe, corresponding to $Pd_{1.0}Bi_{2.0}$. (2) Oktyabr mine, USSR; by electron microprobe, corresponding to $Pd_{0.95}(Bi_{2.03}Te_{0.02})_{\Sigma=2.05}$. (3) $PdBi_2$.

Occurrence: In mill concentrates of As- and Pb-Cu-rich ores (Frood mine, Canada).

Association: Chalcopyrite, cubanite, pyrite, pentlandite, altaite, hessite, galena, pyrrhotite, Bi-tellurides, parkerite, sudburyite, michenerite, sperrylite, insizwaite, niggliite, cabriite, mooihoekite.

Distribution: From the Frood, Creighton and Coleman mines, Sudbury, Ontario; and the Pipe mine, Manitoba, Canada. From Fox Gulch, Goodnews Bay, Alaska, USA. At the Oktyabr mine, Talnakh area, Noril'sk region, western Siberia; and the Karik'yavr Cu-Ni deposit, Kola Peninsula, USSR.

Name: For the Frood mine in Canada, in which it occurs.

Type Material: Harvard University, Cambridge, Massachusetts, USA, 108371.

References: (1) Hawley, J.E. and L.G. Berry (1958) Michenerite and froodite, palladium bismuth minerals. Can. Mineral., 6, 200–209. (2) (1959) Amer. Mineral., 44, 207 (abs. ref. 1). (3) Cabri, L.J., D.C. Harris, and R.I. Gait (1973) Michenerite (PdBiTe) redefined and froodite ($PdBi_2$) confirmed from the Sudbury area. Can. Mineral., 11, 903–912. (4) Cabri, L.J., Ed. (1981) Platinum group elements: mineralogy, geology, recovery. Can. Inst. Min. & Met., 104–105, 155.

Crystal Data: Cubic. *Point Group:* n.d. In a uniform, nearly eutectic-like intergrowth with pyrite and covellite.

Physical Properties: Hardness = ~4. VHN = n.d. D(meas.) = 4.86 D(calc.) = 4.90

Optical Properties: Opaque. *Color:* Dark brownish gray; in polished section, pinkish brown, very similar to bornite. *Luster:* Submetallic.
R: n.d.

Cell Data: *Space Group:* n.d. $a = 5.58$ Z = 1

X-ray Powder Pattern: Hanawa mine, Japan.
2.789 (vs), 3.21 (s), 1.685 (s), 2.281 (m), 2.497 (w), 1.971 (w), 1.545 (w)

Chemistry:

	(1)	(2)
Cu	37.9 – 40.6	37.90
Fe	10.5 – 12.9	11.10
S	49.2 – 53.3	51.00
Total		100.00

(1) Hanawa mine, Japan; by X-ray fluorescence, range of analyses. (2) Cu_3FeS_8.

Occurrence: Found in an ore body of gypsum-anhydrite, occurring in interstices of small masses consisting of barite, covellite, and pyrite.

Association: Pyrite, covellite, barite, gypsum.

Distribution: From the Hanawa mine, Akita Prefecture, Honshu, Japan.

Name: For Nobuyo Fukuchi (1877–1934), Japanese mineralogist and geologist.

Type Material: National Museum of Natural History, Washington, D.C., USA, 135971.

References: (1) Kajiwara, Y. (1969) Fukuchilite, Cu_3FeS_8, a new mineral from the Hanawa mine, Akita Prefecture, Japan. Mineral. J. (Japan), 5, 399–416. (2) (1970) Amer. Mineral., 55, 1811 (abs. ref. 1). (3) Shimazaki, H. and L.A. Clark (1970) Synthetic FeS_2—CuS_2 solid solution and fukuchilite-like minerals. Can. Mineral., 10, 648–664.

Crystal Data: Monoclinic. *Point Group:* $2/m$. Crystals short prismatic [201] (up to 2 mm), and pyramidal. Striated on {100} ∥ [$0\bar{1}0$] and on {$\bar{1}12$} ∥ [110]. Curved crystals common.

Physical Properties: *Fracture:* Uneven. *Tenacity:* Brittle. Hardness = 2.5 VHN = 140–157 (100 g load). D(meas.) = 5.22 D(calc.) = 5.19

Optical Properties: Opaque. *Color:* Lead-gray; may tarnish steel-blue or bronzy white in polished section. *Streak:* Reddish gray. *Luster:* Metallic. *Anisotropism:* Moderate in blue-green to red-brown.

R_1–R_2: (400) 33.7–42.6, (420) 33.6–42.2, (440) 33.5–41.8, (460) 33.1–41.4, (480) 32.5–40.8, (500) 32.2–40.4, (520) 32.0–40.2, (540) 31.9–40.1, (560) 31.7–40.0, (580) 31.2–39.6, (600) 30.5–38.8, (620) 29.5–37.7, (640) 28.4–36.4, (660) 27.6–35.1, (680) 26.9–34.3, (700) 26.5–33.8

Cell Data: *Space Group:* $C2/c$. a = 13.441 b = 11.726 c = 16.930 β = 94.71(8)° Z = 4

X-ray Powder Pattern: Baia Mare, Romania.
3.89 (100), 3.20 (90), 2.919 (80), 2.822 (80), 3.23 (70), 2.749 (70), 6.25 (50)

Chemistry:

	(1)	(2)
Pb	28.29	29.93
Sb	47.50	46.91
S	24.10	23.16
rem.	0.19	
Total	100.08	100.00

(1) Baia Mare, Romania; remainder is SiO_2. (2) $Pb_3Sb_8S_{15}$.

Occurrence: Of hydrothermal origin.

Association: Zinkenite, semseyite, fizélyite, andorite, freieslebenite, geocronite, boulangerite, jamesonite, sphalerite, quartz, dolomite.

Distribution: From Delaul de Crucii, Baia Mare (Kereszthegy mine, Nagybánya), Romania. At Wet Swine Gill, Caldbeck Fells, Cumbria, England. From Auliac, Cantal; Bodennec, Finistère; and Bournac, Montagne Noire, France.

Name: For Dr. Béla Fülöpp, mineral collector of Hungary.

Type Material: n.d.

References: (1) Palache, C., H. Berman, and C. Frondel (1944) Dana's system of mineralogy, (7th edition), v. I, 463–464. (2) Jambor, J.L. (1969) Sulfosalts of the plagionite group. Mineral. Mag., 37, 442–446. (3) Nuffield, E.W. (1975) The crystal structure of fülöppite, $Pb_3Sb_8S_{15}$. Acta Cryst., B31, 151–157. (4) Swinnea, J.S., A.J. Tenorio, and H. Steinfink (1985) $Sb_{10}S_{15}$, a Pb-free analog of fülöppite, $Pb_3Sb_8S_{15}$. Amer. Mineral., 70, 1056–1058.

Crystal Data: Monoclinic. *Point Group:* m, 2, or $2/m$. Granular (10–300 μm).

Physical Properties: Hardness = n.d. VHN = 100–108 (25 g load). D(meas.) = n.d. D(calc.) = 6.74

Optical Properties: Opaque. *Color:* Gray, with a creamy yellow tint in reflected light. *Luster:* Metallic. *Pleochroism:* Weak. *Anisotropism:* Moderate, from light yellow to dark brown.
R_1–R_2: (400) 35.7–34.2, (420) 35.2–34.3, (440) 35.0–34.5, (460) 34.7–34.9, (480) 34.5–35.3, (500) 34.4–35.6, (520) 34.4–35.9, (540) 34.6–36.2, (560) 34.7–36.3, (580) 34.8–36.4, (600) 34.9–36.4, (620) 34.9–36.3, (640) 34.9–36.3, (660) 34.8–36.2, (680) 34.5–36.1, (700) 34.3–36.0

Cell Data: *Space Group:* Cm, $C2$, or $C2/m$. $a = 20.025(13)$ $b = 3.963(2)$ $c = 9.705(4)$ $\beta = 101.57(4)°$ Z = 4

X-ray Powder Pattern: Furutobe mine, Japan.
2.50 (100), 2.95 (90), 2.55 (70), 3.43 (50), 2.61 (50), 2.14 (40), 2.09 (40)

Chemistry:

	(1)	(2)	(3)
Cu	40.4	42.4	41.75
Ag	15.7	13.9	14.17
Pb	26.6	27.6	27.23
S	16.8	16.0	16.85
Total	99.5	99.9	100.00

(1) Furutobe mine, Japan; by electron microprobe, average of seven analyses. (2) Tsumeb, Namibia; by electron microprobe. (3) Cu_5AgPbS_4.

Occurrence: In veinlets of stromeyerite that cut bornite in the Kuroko zone of a stratabound, Kuroko-type, massive sulfide deposit (Furutobe mine, Japan).

Association: Stromeyerite, bornite, galena, sphalerite, tennantite, digenite, argentian gold.

Distribution: In the Daikokuzawa-Higashi deposit of the Furutobe mine, and the No. 11 deposit of the Shakanai mine, near Ohdate City, Akita Prefecture, Japan. At Tsumeb, Namibia. In the Erasmus mine, Leogang, Austria.

Name: For the type locality at the Furutobe mine, Japan.

Type Material: Institute of Mineralogy, Petrology and Economic Geology, Faculty of Science, Tohoku University, Sendai, Japan.

References: (1) Sugaki, A., A. Kitakaze, and Y. Odashima (1981) Furutobeite, a new copper–silver–lead sulfide mineral. Bull. Minéral., 104, 737–741. (2) (1982) Amer. Mineral., 67, 1075 (abs. ref. 1).

Crystal Data: Cubic. *Point Group:* $4/m\ \bar{3}\ 2/m$. Most commonly cubic, crystals to a meter on an edge are known; less often cubo-octahedral or octahedral. Rarely tabular on {001}; also forms reticulated masses and skeletal crystals. As cleavable masses, coarse to very fine granular; fibrous, plumose. *Twinning:* Twin plane {111}, as both contact and penetration twins; twin plane {114}, lamellar.

Physical Properties: *Cleavage:* Perfect, {001}; parting or cleavage on {111}. *Fracture:* Subconchoidal. *Tenacity:* Brittle. Hardness = 2.5–2.75 VHN = 79–104 (100 g load). D(meas.) = 7.58 D(calc.) = 7.57

Optical Properties: Opaque. *Color:* Lead-gray; in polished section, white. *Streak:* Lead-gray. *Luster:* Metallic.
R: (400) 52.8, (420) 50.5, (440) 48.2, (460) 46.4, (480) 45.0, (500) 43.9, (520) 43.0, (540) 42.4, (560) 41.9, (580) 41.6, (600) 41.5, (620) 41.6, (640) 41.8, (660) 41.9, (680) 42.1, (700) 42.2

Cell Data: *Space Group:* $Fm3m$. $a = 5.936$ Z = 4

X-ray Powder Pattern: Synthetic.
2.969 (100), 3.429 (84), 2.099 (57), 1.790 (35), 1.327 (17), 1.714 (16), 1.484 (10)

Chemistry:

	(1)	(2)	(3)
Pb	86.50	86.22	86.60
Ag		0.12	
Cu	0.07	trace	
Fe		0.01	
Bi		0.06	
S	13.31	13.96	13.40
Total	99.88	100.37	100.00

(1) Shaba Province, Zaire; Cu assumed present in covellite and S equivalent deducted. (2) Sadon Formation, Caucasus Mountains, USSR; Ag presumed in acanthite. (3) PbS.

Polymorphism & Series: Forms a series with clausthalite.

Occurrence: Occurs in many different types of environments. In hydrothermal veins, formed under a wide range of temperatures; in contact metamorphic deposits, in pegmatites, rarely; limestones and dolomites are common host rocks.

Association: Sphalerite, marcasite, pyrite, chalcopyrite, tetrahedrite, silver minerals, siderite, calcite, dolomite, barite, quartz, many other hydrothermal minerals.

Distribution: The most important ore mineral of lead; only a few deposits of note can be mentioned. In the USA, in Idaho, in the Coeur d'Alene region; in Colorado, at Leadville, Lake Co., and from Breckenridge, Summit Co. In the Mississippi Valley region, in the Tri-State district, at Joplin, Jasper Co., Missouri; Galena, Cherokee Co., Kansas; and Picher, Ottawa Co., Oklahoma. At the Miliken (Sweetwater) mine, Reynolds Co., Missouri, and other mines north along the Viburnum Trend district, brilliant crystals. From Příbram, Czechoslovakia. At Freiberg, Saxony; from near Bad Ems, Hesse; and at St. Andreasberg, Neudorf, and Clausthal, in the Harz, Germany. In France, from Pontgibaud, Puy-de-Dôme. In England, at Alston Moor, Cumbria. In Scotland, at Wanlockhead, Dumfries. In Ireland, from the Mogul mine, Co. Tipperary. At Naica and Santa Eulalia, Chihuahua, Mexico.

Name: A Latin word *galena*, given to lead ore or the dross from melted lead.

References: (1) Palache, C., H. Berman, and C. Frondel (1944) Dana's system of mineralogy, (7th edition), v. I, 200–204. (2) (1953) NBS Circ. 539, 2, 18. (3) Ramdohr, P. (1969) The ore minerals and their intergrowths, (3rd edition), 635–649. (4) Deer, W.A., R.A. Howie and J. Zussman (1963) Rock forming minerals. v. 5, non-silicates, 180–185.

Crystal Data: Orthorhombic. *Point Group:* $2/m\ 2/m\ 2/m$. Crystals lath-like, elongated [001], flattened {100}, sometimes needle-like [001], and as very thin plates, up to 1 cm. Usually crystalline, massive, indistinctly columnar to fibrous.

Physical Properties: *Cleavage:* Good on {110}. *Tenacity:* Flexible. Hardness = 2.5–3.5 VHN = 142–194 (100 g load). D(meas.) = 7.04 D(calc.) = 7.195

Optical Properties: Opaque. *Color:* Light gray to tin-white or lead-gray, may tarnish yellow or iridescent. *Streak:* Black. *Luster:* Metallic. *Anisotropism:* Strong.
R_1–R_2: (400) 48.0–51.1, (420) 47.9–51.1, (440) 47.8–51.1, (460) 47.7–51.0, (480) 47.5–50.8, (500) 47.1–50.5, (520) 46.6–49.8, (540) 45.9–48.8, (560) 45.3–48.2, (580) 44.8–47.7, (600) 44.6–47.2, (620) 44.4–46.8, (640) 44.4–46.7, (660) 44.4–46.7, (680) 44.5–46.8, (700) 44.7–47.1

Cell Data: *Space Group:* *Pnam.* $a = 11.669$ $b = 14.533$ $c = 4.090$ Z = 4

X-ray Powder Pattern: Cariboo mine, Canada.
3.45 (10), 1.961 (5), 3.03 (4), 2.46 (4), 2.05 (4), 3.65 (3), 2.76 (3)

Chemistry:

	(1)	(2)
Pb	27.65	27.50
Fe	trace	
Bi	54.69	55.48
S	17.35	17.02
Total	99.69	100.00

(1) Nordmark, Sweden. (2) $PbBi_2S_4$.

Occurrence: Of hydrothermal origin.

Association: Galena, pyrite, cosalite, gold, quartz.

Distribution: In the USA, in the Belzazzar mine, Quartzburg district, Boise Co., Idaho; and in the Hatfield mine, Okanogan Co., and the Germania mine, Fruitland, Stevens Co., Washington. From the Cariboo mine, Barkerville, Cariboo district, British Columbia; in the Siscoe mine, Gowganda, Ontario; and at Dublin Gulch, Yukon Territory, Canada. In Sweden, at Gladhammar, Kopparberg, and Falun. From Oberpinzgau, Salzburg, Austria. At Pechtelsgrön, Saxony, Germany. In the Gupworthy mine, Brendon Hills, Somerset, England. At Corrie Buie, Meal nan Oighreag, Perthshire, Scotland. From Kingsgate, New South Wales; and at Mt. Farrell, Tasmania, Australia. Known from a number of other minor occurrences.

Name: In allusion to the composition.

Type Material: Natural History Museum, Stockholm, Sweden, 3987.

References: (1) Palache, C., H. Berman, and C. Frondel (1944) Dana's system of mineralogy, (7th edition), v. I, 471–473. (2) Berry, L.G. (1940) Studies of mineral sulfo-salts: IV — galenobismutite and "lillianite". Amer. Mineral., 25, 726–734. (3) Berry, L.G. and R.M. Thompson (1962) X-ray powder data for the ore minerals. Geol. Soc. Amer. Mem. 85, 161–162. (4) Iitaka, Y. and W. Nowacki (1962) A redetermination of the crystal structure of galenobismutite, $PbBi_2S_4$. Acta Cryst., 15, 691–698. (5) Takéuchi, Y. and J. Takagi (1974) Structure of galenobismutite. Proc. Jap. Acad., 50, 222-225 (in English). (6) (1974) Chem. Abs., 81, 66346 (abs. ref. 5).

Crystal Data: Cubic. *Point Group:* $\bar{4}3m$. As crystals up to 1 cm, and granular aggregates.

Physical Properties: *Fracture:* Uneven to fine conchoidal. *Tenacity:* Brittle. Hardness = 3 VHN = 186–202 (100 g load). D(meas.) = 5.4 D(calc.) = 5.44

Optical Properties: Opaque. *Color:* Dark orange-red. *Streak:* Orange-yellow. *Luster:* Vitreous to adamantine.

R: (400) 30.5, (420) 29.7, (440) 29.3, (460) 28.8, (480) 28.0, (500) 26.8, (520) 26.6, (540) 24.6, (560) 23.8, (580) 23.1, (600) 22.6, (620) 22.2, (640) 21.9, (660) 21.6, (680) 21.4, (700) 21.2

Cell Data: *Space Group:* $I\bar{4}3m$. $a = 10.365(3)$ Z = 12

X-ray Powder Pattern: Gal-Khaya, USSR.
3.01 (100), 2.78 (80), 4.27 (70), 7.40 (50), 1.841 (50), 2.604 (30), 1.569 (30)

Chemistry:

	(1)	(2)	(3)	(4)
Cs			3.3	5.1
Tl	0.46	2.90	0.8	2.4
Hg	47.60	49.02	51.7	50.7
Cu	3.49	2.85	3.4	3.2
Zn	3.00	0.60	1.6	1.8
Fe	0.31			
As	23.60	19.49	14.5	15.2
Sb	0.59	5.51	3.0	0.3
Se	0.0003	0.015		
S	21.00	19.31	22.3	22.0
Total	100.05	99.70	100.6	100.7

(1) Gal-Khaya, USSR; by electron microprobe. (2–3) Khaidarkan, USSR; by electron microprobe. (4) Getchell mine, Nevada, USA; by electron microprobe.

Occurrence: Found in hydrothermal Hg–Au deposits.

Association: Pyrite, stibnite, cinnabar, metacinnabar, aktashite, wakabayashilite, orpiment, realgar, getchellite, calcite, fluorite, quartz (USSR); pyrite, realgar, stibnite, orpiment, getchellite, fluorite (Getchell mine, Nevada, USA).

Distribution: In the mercury deposits of Gal-Khaya, Yakutia, and Khaidarkan, Kirgizia, USSR. At the Getchell mine, Humboldt Co., and the Carlin mine, Elko Co., Nevada, USA. From the Hemlo gold deposit, Thunder Bay district, Ontario, Canada.

Name: For the Gal-Khaya deposit, USSR.

Type Material: A.E. Fersman Mineralogical Museum, Academy of Sciences, Moscow, USSR.

References: (1) Gruzdev, V.S., V.I. Stepanov, N.G. Shumkova, N.M. Chernitsova, R.N. Yudin, and I.A. Bryzgalov (1972) Galkhaite, $HgAsS_2$, a new mineral from arsenic–antimony–mercury deposits of the U.S.S.R. Doklady Acad. Nauk SSSR, 205, 1194–1197 (in Russian). (2) (1974) Amer. Mineral., 59, 208–209 (abs. ref. 1). (3) Chen, T.T. and J.T. Szymański (1981) The structure and chemistry of galkhaite, a mercury sulfosalt containing Cs and Tl. Can. Mineral., 19, 571–581. (4) (1983) Amer. Mineral., 68, 474 (abs. ref. 3). (5) Chen, T.T. and J.T. Szymański (1982) A comparison of galkhaite from Nevada and from the type locality, Khaydarkan, Kirgizia, U.S.S.R. Can. Mineral., 20, 575–577.

Crystal Data: Tetragonal. *Point Group:* $\bar{4}2m$. Massive, as minute embedded grains and as exsolution lamellae, up to 1–2 mm. *Twinning:* Parallel to {112} and {111}.

Physical Properties: Hardness = 3.0–3.5 VHN = n.d. D(meas.) = 4.2 D(calc.) = 4.35

Optical Properties: Metallic. *Color:* Gray. *Streak:* Gray-black. *Anisotropism:* Low.
R: (400) 22.0, (420) 21.8, (440) 21.6, (460) 21.3, (480) 20.8, (500) 20.4, (520) 19.7, (540) 19.2, (560) 18.8, (580) 18.6, (600) 18.7, (620) 18.8, (640) 18.9, (660) 19.1, (680) 19.4, (700) 19.5

Cell Data: *Space Group:* $I\bar{4}2d$. $a = 5.360$ $c = 10.49$ Z = 4

X-ray Powder Pattern: Synthetic.
3.07 (100), 1.876 (90), 1.610 (90), 1.089 (90), 1.894 (60), 1.215 (60), 1.076 (60)

Chemistry:

	(1)	(2)	(3)
Cu	30.8	31.2	32.19
Pb	2.3		
Zn	4.4	1.7	
Fe	4.9	2.3	
Ga	29.2	32.4	35.32
Ge		0.3	
S	28.4	32.8	32.49
Total	100.0	100.7	100.00

(1) Tsumeb, Namibia; by electron microprobe, recalculated after deduction of impurity phases. (2) Do.; by electron microprobe. (3) $CuGaS_2$.

Occurrence: In base-metal vein deposits with relatively high gallium content.

Association: Reniérite, germanite, bornite, chalcocite, digenite, pyrite, sphalerite, galena, tetrahedrite.

Distribution: At Tsumeb, Namibia. In Zaire, at Kipushi, Shaba Province. From the Radka deposit, Pazardzhik; and the Chelopech deposit, Sofia, Bulgaria. In Cuba, in northwestern Pinar del Rio Province.

Name: For the mineral's gallium content.

Type Material: n.d.

References: (1) Strunz, H., B.H. Geier, and E. Seeliger (1958) Gallit, $CuGaS_2$, das erste selbständige Galliummineral, und seine Verbreitung in den Erzen der Tsumeb- und Kipushi-Mine. Neues Jahrb. Mineral., Monatsh., 241–264 (in German). (2) (1959) Amer. Mineral., 44, 906 (abs. ref. 1). (3) Springer, G. (1969) Microanalytical investigations into germanite, renierite, briartite, and gallite. Neues Jahrb. Mineral., Monatsh., 435–441.

Crystal Data: Orthorhombic. *Point Group:* n.d. Small aggregates of anhedral crystals, up to 200 μm across.

Physical Properties: Hardness = n.d. VHN = 212–222 (50 g load). D(meas.) = n.d. D(calc.) = 5.64

Optical Properties: Opaque. *Color:* In reflected light, gray with olive-brown tint. *Luster:* Metallic. *Pleochroism:* Distinct. *Anisotropism:* Strong, in yellowish green to bluish gray.
R_1–R_2: n.d.

Cell Data: *Space Group:* n.d. $a = 11.439$ $b = 14.093$ $c = 3.754$ Z = 4

X-ray Powder Pattern: Valle del Frigido, Italy.
3.62 (vs), 3.20 (vs), 2.63 (vs), 2.51 (vs), 2.98 (s), 2.89 (s), 14.0 (m)

Chemistry:

	(1)	(2)
Fe	9.41	10.85
Cu	0.30	
Sb	29.90	23.65
Bi	33.28	40.59
As	0.16	
S	27.43	24.91
Total	100.48	100.00

(1) Valle del Frigido, Italy. (2) $FeSbBiS_4$.

Occurrence: In a hydrothermal copper deposit with disseminated chalcopyrite in siderite gangue.

Association: Chalcopyrite, tetrahedrite, pyrrhotite, pyrite, marcasite, galena, sphalerite, meneghinite, ullmannite, pentlandite, vaesite, bismuthinite, chalcanthite, siderite, quartz.

Distribution: At Valle del Frigido, one km east of Massa, northern Tuscany, Italy.

Name: For Professor C.L. Garavelli, Italian mineralogist.

Type Material: Universities of Florence and Bari, Italy; National Museum of Natural History, Washington, D.C., USA, 145868.

References: (1) Gregorio, F., P. Lattanzi, G. Tanelli, and F. Vurro (1979) Garavellite, $FeSbBiS_4$, a new mineral from the Cu–Fe deposit of Valle del Frigido in the Apuane Alps, northern Tuscany, Italy. Mineral. Mag., 43, 99–102. (2) (1979) Amer. Mineral., 64, 1329–1330 (abs. ref. 1).

Crystal Data: Hexagonal, pseudocubic. *Point Group:* $\overline{3}\,2/m$, $3m$ or 32. As coatings and thin (15 μm) platelets.

Physical Properties: Hardness = Softer than sphalerite, < 3.5–4. VHN = 75–96 (25 g load). D(meas.) = n.d. D(calc.) = 5.61

Optical Properties: Opaque. *Color:* Bluish white in reflected light. *Luster:* Metallic. *Pleochroism:* Weak. *Anisotropism:* Moderate, in yellows.
R_1–R_2: n.d.

Cell Data: *Space Group:* $R\overline{3}m$, $R3m$, or $R32$. $a = 3.83$ $c = 46.84$ Z = 1

X-ray Powder Pattern: DeKalb Township, New York, USA.
3.128 (100), 1.918 (50), 1.637 (30), 1.109 (20), 2.712 (10), 1.870 (10), 1.683 (10)

Chemistry:

	(1)	(2)	(3)	(4)
Cu	61.44	76.94	78.1	76.02
Fe			0.4	
Zn			1.0	
S	21.47	26.04	21.1	23.98
Total	82.91	102.98	100.6	100.00

(1) DeKalb Township, New York, USA; by electron microprobe, corresponding to $Cu_{1.44}S_{1.00}$. (2) Do.; corresponding to $Cu_{1.49}S_{1.00}$. (3) Eretria, Greece; by electron microprobe. (4) Cu_8S_5.

Occurrence: Replacing sphalerite along cleavage traces (DeKalb Township, New York, USA); in a serpentine-hosted magnetite-chromite deposit (Eretria, Greece).

Association: Spionkopite, sphalerite, tetrahedrite, chalcopyrite, malachite, azurite, brochantite, chrysocolla, cervantite, stibiconite, hemimorphite, calcite (DeKalb Township, New York, USA); spionkopite, chalcopyrite, cobalt pentlandite, magnetite, chromite, andradite, chlorite, diopside (Eretria, Greece).

Distribution: In DeKalb Township, St. Lawrence Co., New York, and from Cuchillo, Winston district, Sierra Co., New Mexico, USA. From near Eretria, Othris Mountains, Greece.

Name: To honor the original collector, Adam Geer, of Utica, New York, USA.

Type Material: New York State Museum, Albany, New York; National Museum of Natural History, Washington, D.C., USA, 144186; Queen's University, Kingston, Ontario, Canada.

References: (1) Goble, R.J. and G. Robinson (1980) Geerite, $Cu_{1.60}S$, a new copper sulfide from Dekalb Township, New York. Can. Mineral., 18, 519–523. (2) (1981) Amer. Mineral., 66, 1274 (abs. ref. 1). (3) Economou, M.I. (1981) A second occurrence of the copper sulfides geerite and spionkopite in Eretria area, central Greece. Neues Jahrb. Mineral., Monatsh., 489–494.
(4) Goble, R.J. (1985) The relationship between crystal structure, bonding and cell dimensions in the copper sulfides. Can. Mineral., 23, 61–76.

Crystal Data: Cubic. *Point Group:* $4/m\ \overline{3}\ 2/m$. Commonly in a micro-myrmekitic intergrowth (0.2–0.7 mm) with clausthalite and eskebornite.

Physical Properties: Hardness = n.d. VHN = 67.5–72.4 (15 g load). D(meas.) = n.d. D(calc.) = 5.39

Optical Properties: Opaque. *Color:* Brown with a creamy tint in reflected light. *Luster:* Metallic.
R: (400) — , (420) 19.0, (440) 21.4, (460) 23.5, (480) 25.5, (500) 27.5, (520) 29.1, (540) 30.1, (560) 31.7, (580) 32.7, (600) 33.6, (620) 34.4, (640) 35.1, (660) 35.8, (680) 36.3, (700) 36.9

Cell Data: *Space Group:* $Fm3m$. $a = 10.889$ Z = 4

X-ray Powder Pattern: Chaméane, France.
1.925 (100), 3.282 (90), 3.145 (90), 2.094 (60), 1.112 (60), 1.660 (50), 2.436 (40)

Chemistry:

	(1)	(2)
Ag	7.05	5.16
Cu	28.57	28.81
Fe	19.01	19.38
Se	34.42	35.76
S	10.06	9.87
Total	99.11	98.98

(1–2) Chaméane, France; by electron microprobe.

Occurrence: As late stage deposits in veins cutting granite (Chaméane, France).

Association: Eskebornite, clausthalite, chaméanite, ankerite (Chaméane, France).

Distribution: From Chaméane, Puy-de-Dôme, France. In the San Miguel mine, Moctezuma, Sonora, Mexico.

Name: To honor Jacques Geffroy, metallurgist for the French Atomic Energy Commission.

Type Material: National School of Mines, Paris, France.

References: (1) Johan, Z., P. Picot, and F. Ruhlmann (1982) Evolution paragénétique de la minéralisation uranifère de Chaméane (Puy-de-Dôme) France: chaméanite, geffroyite et giraudite, trois séléniures nouveaux de Cu, Fe, Ag, and As. Tschermaks Mineral. Petrog. Mitt., 29, 151–167 (in French with English abs.). (2) (1982) Amer. Mineral., 67, 1074–1075 (abs. ref. 1).

Crystal Data: Tetragonal. *Point Group:* 422. As irregular grains, up to about 165 μm in size.

Physical Properties: Hardness = n.d. VHN = 603–677 (25 g load). D(meas.) = n.d. D(calc.) = [8.83]

Optical Properties: Opaque. *Color:* Pale brown or tan with a yellowish tinge. *Pleochroism:* Weak in oil for some grains. *Anisotropism:* Moderate to strong; from gray to extinction.
R_1–R_2: (400) 41.1–42.1, (420) 43.0–44.0, (440) 44.9–45.9, (460) 46.6–47.3, (480) 48.0–48.5, (500) 49.0–49.5, (520) 50.0–50.3, (540) 50.8–51.1, (560) 51.6–52.0, (580) 52.5–53.0, (600) 53.6–54.0, (620) 54.4–55.1, (640) 55.1–56.1, (660) 55.7–56.9, (680) 56.2–57.4, (700) 56.6–57.3

Cell Data: *Space Group:* n.d. $a = 7.736$ $c = 24.161$ Z = 8

X-ray Powder Pattern: Onverwacht mine, South Africa.
2.265 (100), 3.020 (90), 1.934 (60), 1.910 (50), 0.9043 (50b), 0.9025 (50b), 2.146 (40)

Chemistry:

	(1)	(2)
Pt	44.4	41.86
Pd	9.0	7.01
Rh	6.6	8.23
Ni	2.0	3.41
Cu	0.25	
Sb	35.8	38.96
Bi	1.7	
As	0.89	
Total	100.64	99.47

(1) Onverwacht mine, South Africa; by electron microprobe, giving $(Pt_{2.17}Pd_{0.81}Rh_{0.61}Ni_{0.32}Cu_{0.04})_{\Sigma=3.95}(Sb_{2.81}As_{0.11}Bi_{0.8})_{\Sigma=3.00}$. (2) Shetland Islands, Scotland; by electron microprobe, leading to $(Pt_{2.04}Rh_{0.76}Pd_{0.62}Ni_{0.55})_{\Sigma=3.97}Sb_{3.03}$.

Occurrence: In ultramafics or ophiolites mineralized with Pt-Fe-Cu-Ni, and placers derived from them.

Association: Sperrylite, platarsite, ruthenarsenite, stibiopalladinite, mertieite-II, Pt–Fe alloy, chromite (Onverwacht mine, South Africa); osmium, Pt–Pd–Cu alloy, hollingworthite, irarsite, laurite, ruthenian pentlandite, chromite (Shetland Islands, Scotland).

Distribution: In the Onverwacht and Driekop mines, Transvaal, South Africa. From the Birbir River, Ethiopia. On Unst and Fetlar, Shetland Islands, Scotland. From Fox Gulch, Goodnews Bay, Alaska, USA.

Name: For Dr. A.D. Genkin, Soviet mineralogist.

Type Material: National Museum of Canada, Ottawa; Royal Ontario Museum, Toronto, Canada; National Museum of Natural History, Washington, D.C., USA, 136485; A.E. Fersman Mineralogical Museum, Academy of Sciences, Moscow, USSR, N79000.

References: (1) Cabri, L.J., J.M. Stewart, J.H.G. Laflamme, and J.T. Szymański (1977) Platinum group minerals from Onverwacht. III. Genkinite, $(Pt, Pd)_4Sb_3$, a new mineral. Can. Mineral., 15, 389–392. (2) (1979) Amer. Mineral., 64, 654 (abs. ref. 1). (3) Cabri, L.J., Ed. (1981) Platinum group elements: mineralogy, geology, recovery. Can. Inst. Min. & Met., 105–107. (4) Prichard, H.M. and M. Tarkian (1988) Platinum and palladium minerals from two PGE-rich localities in the Shetland ophiolite complex. Can. Mineral., 26, 979–990.

Crystal Data: Monoclinic, pseudo-orthorhombic. *Point Group:* $2/m$. Crystals rare, but occasionally large, to 8 cm from several localities; tabular on {001}. Massive, as rounded and irregular grains; earthy. *Twinning:* Common on {001}; lamellar.

Physical Properties: *Cleavage:* Distinct on {011}. *Fracture:* Uneven. Hardness = 2.5 VHN = 144–160 (100 g load). D(meas.) = 6.46 D(calc.) = 6.44

Optical Properties: Opaque. *Color:* Light lead-gray to grayish; in polished section, pure white. *Streak:* Light lead-gray to grayish blue. *Luster:* Metallic. *Pleochroism:* Weak. *Anisotropism:* Moderate.

R_1–R_2: (400) 44.7–44.8, (420) 43.5–44.0, (440) 42.3–43.2, (460) 41.3–42.3, (480) 40.5–41.6, (500) 39.9–41.0, (520) 39.3–40.5, (540) 38.8–40.1, (560) 38.4–39.7, (580) 38.1–39.3, (600) 37.8–39.0, (620) 37.4–38.5, (640) 37.0–38.1, (660) 36.5–37.6, (680) 36.0–37.0, (700) 35.4–36.3

Cell Data: *Space Group:* $P2_1/m$. $a = 8.963$ $b = 31.93$ $c = 8.500$ $\beta = 118.02(1)°$ Z = [2]

X-ray Powder Pattern: Park City, Utah, USA.
3.54 (100), 3.06 (90), 2.89 (90), 3.39 (80), 3.18 (86), 2.98 (70), 3.71 (60)

Chemistry:

	(1)	(2)	(3)	(4)
Pb	67.52	68.90	68.49	68.60
Sb	11.48	9.27	9.13	8.64
As	3.65	4.54	4.59	5.32
S	17.45	17.13	17.20	17.44
Total	100.10	99.84	99.41	100.00

(1) Silver King mine, Park City, Utah, USA. (2) Sala, Sweden. (3) Kilbricken, Ireland. (4) $Pb_{14}(Sb, As)_6S_{23}$ with Sb:As = 1:1.

Polymorphism & Series: Forms a series with jordanite.

Occurrence: Found in hydrothermal veins with sulfides and other sulfosalt minerals.

Association: Galena, pyrite, tetrahedrite, barite, fluorite, quartz.

Distribution: Apparently occurs at a number of localities; only some of the best authenticated are mentioned here. In the USA, in Utah, in the Silver King mine, Park City, Summit Co.; and in the Tintic district, Juab Co.; in California, at Darwin, Inyo Co.; at the Inexco #1 mine, Jamestown, Boulder Co., Colorado. From the Hemlo gold deposit, Thunder Bay district, Ontario, Canada. In Sweden, at Sala, Västmanland, and Falun. In Ireland, in the Kilbricken mine, Co. Clare. In Italy, at Pietrasanta, Val de Castello, Tuscany. From the Xanda mine, Virgem da Lapa, Minas Gerais, Brazil. In the Treore mine, St. Treath, Cornwall, England. From Noche Buena, Zacatecas, Mexico.

Name: From the Greek for *Earth* and *Saturn*, the alchemistic name for lead.

References: (1) Palache, C., H. Berman, and C. Frondel (1944) Dana's system of mineralogy, (7th edition), v. I, 395–396. (2) Douglass, R.M., M.J. Murphy, and A. Pabst (1954) Geocronite. Amer. Mineral., 39, 908–928. (3) Birnie, R. and C.W. Burnham (1976) The crystal structure and extent of solid solution of geocronite. Amer. Mineral., 61, 963–970.

Crystal Data: Cubic. *Point Group:* $\overline{4}3m$. Minute cubic crystals; usually massive, commonly intergrown with reniérite.

Physical Properties: *Tenacity:* Brittle. Hardness = 4 VHN = n.d. D(meas.) = 4.46–4.59 D(calc.) = 4.30

Optical Properties: Opaque. *Color:* Reddish gray, tarnishes to a dull brown; in polished section, pinkish gray. *Streak:* Dark gray to black. *Luster:* Metallic, dull.
R: (400) 21.5, (420) 21.0, (440) 20.5, (460) 20.2, (480) 20.0, (500) 19.9, (520) 20.1, (540) 20.6, (560) 21.2, (580) 21.8, (600) 22.6, (620) 23.5, (640) 24.0, (660) 24.6, (680) 25.0, (700) 25.4

Cell Data: *Space Group:* $F\overline{4}3n$. a = 10.585 Z = [1]

X-ray Powder Pattern: Tsumeb, Namibia.
3.054 (100), 1.87 (80), 1.596 (70), 2.64 (40), 1.080 (40), 1.214 (30), 0.8956 (30)

Chemistry:

	(1)	(2)	(3)
Cu	45.1	48.9	51.76
Fe	7.4	1.7	7.00
Zn	1.3	0.1	
Pb			
Mo		1.8	
V		2.2	
Ge	9.7	5.1	9.10
Ga		0.8	
As	2.6	7.6	
S	33.4	32.1	32.14
Total	99.5	100.3	100.00

(1) Tsumeb, Namibia: by electron microprobe, degree of contamination in question. (2) Do.; by electron microprobe. (3) $Cu_{26}Fe_4Ge_4S_{32}$.

Occurrence: In primary Cu-Pb-Zn ores (Tsumeb, Namibia).

Association: Reniérite, pyrite, tennantite, enargite, galena, sphalerite, digenite, bornite, chalcopyrite.

Distribution: From Tsumeb, Namibia. At the Inexco # 1 mine, Jamestown, Boulder Co., Colorado; and in the Ruby Creek deposit, Brooks Range, near Bornite, Alaska, USA. In Bulgaria, from the Radka deposit, Pazardzhik. From the Bancairoun mine, Var, Alpes-Maritimes, France. At M'Passa, 150 km west of Brazzaville, Congo Republic. In Zaire, at Kipushi, Shaba Province. In the USSR, at Vaygach, Arkhangel'sk, and in the Noril'sk region, western Siberia. At the Shakanai mine, Akita Prefecture, Japan. In Cuba, from Pinar del Rio Province.

Name: For the germanium content of the mineral.

Type Material: n.d.

References: (1) Palache, C., H. Berman, and C. Frondel (1944) Dana's system of mineralogy, (7th edition), v. I, 385–386. (2) Murdoch, J. (1953) X-ray investigation of colusite, germanite, and reniérite. Amer. Mineral., 38, 794–801. (3) Sclar, C.B. and B.H. Geier (1954) The paragenetic relationships of germanite and reniérite from Tsumeb, South West Africa. Econ. Geol., 52, 612–631. (4) Springer, G. (1969) Microanalytical investigations into germanite, renierite, briartite, and gallite. Neues Jahrb. Mineral., Monatsh., 435–441. (5) Tettenhorst, R.T. and C.E. Corbató (1984) Crystal structure of germanite, $Cu_{26}Ge_4Fe_4S_{32}$, determined by x-ray powder diffraction. Amer. Mineral., 69, 943–947.

Gersdorffite $NiAsS$

Crystal Data: Cubic. *Point Group:* $2/m\ \overline{3}$. Crystals are octahedral, sometimes modified by the cube, up to 3–4 cm; pyritohedral and occasionally striated as with pyrite. Often internally zoned.

Physical Properties: *Cleavage:* Perfect on {100}. *Fracture:* Uneven. *Tenacity:* Brittle. Hardness = 5.5 VHN = 657–767 (100 g load). D(meas.) = 5.9 D(calc.) = 5.966

Optical Properties: Opaque. *Color:* Silver-white to steel-gray; may tarnish gray or grayish black; in polished section, white. *Streak:* Grayish black. *Luster:* Metallic.
R: (400) 47.4, (420) 46.6, (440) 45.8, (460) 45.4, (480) 45.1, (500) 45.0, (520) 45.0, (540) 45.0, (560) 45.1, (580) 45.3, (600) 45.5, (620) 45.9, (640) 46.2, (660) 46.5, (680) 46.9, (700) 47.3

Cell Data: *Space Group:* $Pa3$. $a = 5.60$–5.72, 5.594 (synthetic). Z = 4

X-ray Powder Pattern: Synthetic.
2.545 (100), 2.325 (90), 1.716 (80), 2.848 (60), 2.013 (35), 1.521 (35), 1.096 (20)

Chemistry:

	(1)	(2)	(3)
Ni	28.92	35.7	35.42
Fe	8.89	0.2	
Co	1.72	0.1	
Cu	1.10		
As	43.54	44.3	45.23
Sb	1.63	1.6	
S	13.84	19.0	19.35
Total	99.64	97.7	100.00

(1) Coburg mine, Dobšiná, Czechoslovakia; Cu present as chalcopyrite. (2) Cochabamba, Bolivia; by electron microprobe. (3) NiAsS.

Polymorphism & Series: Modifications –Pa3, –$P2_13$, and –$Pca2_1$ are known.

Occurrence: In hydrothermal vein deposits formed at moderate temperatures.

Association: Nickeline, nickel-skutterudite, ullmannite, other sulfides, siderite.

Distribution: A number of new localities have been recognized in recent years, only a few of which can be listed. In the USA, large crystals at the Snowbird mine, Mineral Co., Montana. In Canada, from several mines in Sudbury and Cobalt, Ontario. In Germany, at Müsen, Wissen, and Ramsbeck, in North Rhine-Westphalia; at Lobenstein, in Thuringia; at Bad Ems and Dillenburg, in Hesse, and from Rammelsberg and Wolfsberg, in the Harz Mountains. At Dobšiná (Dobschau), Czechoslovakia. In the Aït Ahmane mine, 10 km east of Bou Azzer, Morocco, as fine crystals. From the Craignure mine, Inverary, Strathclyde region, Scotland. At Olsa, Freisach, Carinthia, Austria. From Cochabamba, Bolivia. At the Mt. Ogilvie and Nichol Nob mines, Flinders Ranges, South Australia.

Name: For the von Gersdorffs, owners of the nickel mine at Schladming, Austria (ca. 1842).

References: (1) Palache, C., H. Berman, and C. Frondel (1944) Dana's system of mineralogy, (7th edition), v. I, 298–300. (2) (1961) NBS Monograph, 25, 35. (3) Yund, R.A. (1962) The system Ni–As–S: phase relations and mineralogical significance. Amer. J. Sci., 260, 761–782. (4) Bayliss, P. (1982) A further crystal structure refinement of gersdorffite. Amer. Mineral., 67, 1058–1064. (5) Bayliss, P. (1986) Subdivision of the pyrite group, and a chemical and X-ray diffraction investigation of ullmannite. Can. Mineral., 24, 27–33. (6) Criddle, A.J. and C.J. Stanley, Eds. (1986) The quantitative data file for ore minerals. British Museum (Natural History), London, England, 132.

Crystal Data: Monoclinic. *Point Group:* m. As spherules with a fibrous to platy structure, up to 2.5 cm in diameter; also as fine granular aggregates and groups of small thick plates.

Physical Properties: *Cleavage:* Perfect on {010} and {100}; poor on {001}. Hardness = 2.5 VHN = n.d. D(meas.) = 3.62 D(calc.) = 3.723

Optical Properties: Transparent. *Color:* Cinnabar-red. *Streak:* Bright cinnabar-red; powder darkens on exposure. *Luster:* Weakly adamantine.
Optical Class: n.d. *Pleochroism:* Weak, X = salmon-red, Y and Z = deep blood-red.
$n = > 2.01$
R_1–R_2: n.d.

Cell Data: *Space Group:* Cm. $a = 9.911(8)$ $b = 23.05(2)$ $c = 7.097(8)$ $\beta = 127.85(7)°$ Z = 2

X-ray Powder Pattern: Baker mine, California, USA.
11.85 (100), 3.05 (90), 5.64 (70), 4.03 (70), 2.81 (50), 2.739 (40), 1.934 (40)

Chemistry:

	(1)	(2)	(3)
Na	4.28	4.65	3.29
Li	0.135	0.15	
Sb	47.80	51.91	55.74
As	7.38	8.02	8.57
S	27.01	29.33	29.82
H_2O	5.47	5.94	2.58
gangue	n.d.		
Total	92.07	100.00	100.00

(1) Baker mine, California, USA. (2) Analysis (1) recalculated to 100%.
(3) $Na_2(Sb, As)_8S_{13} \cdot 2H_2O$ with Sb:As = 4:1.

Occurrence: A low-temperature mineral found embedded in massive borates and clay.

Association: Borax, probertite, tincalconite, realgar, stibnite.

Distribution: From the Baker mine, Boron, Kern Co., California, USA.

Name: For James Mack Gerstley (1907–), President of the Pacific Coast Borax Company.

Type Material: National Museum of Natural History, Washington, D.C., 106916; Harvard University, Cambridge, Massachusetts, USA, 111307.

References: (1) Frondel, C. and V. Morgan (1956) Inderite and gerstleyite from the Kramer borate district, Kern County, California. Amer. Mineral., 41, 839–843. (2) Nakai, I. and D.E. Appleman (1981) The crystal structure of gerstleyite $Na_2(Sb, As)_8S_{13} \cdot 2H_2O$: the first sulfosalt mineral of sodium. Chem. Lett., 10, 1327-1330. (3) (1981) Chem. Abs., 95, 195617 (abs. ref. 2).

Crystal Data: Monoclinic. *Point Group:* $2/m$. Massive and as subhedral crystals up to 2 cm, commonly bent.

Physical Properties: *Cleavage:* Perfect micaceous on {001}, yielding flexible, inelastic lamellae. *Fracture:* Splintery. Hardness = 1.5–2 VHN = n.d. D(meas.) = 3.92 D(calc.) = 3.98 (synthetic).

Optical Properties: Transparent. *Color:* Dark blood-red, tarnishes to green to purple iridescence. *Streak:* Orange-red. *Luster:* Pearly to vitreous on cleavage surfaces, otherwise resinous.

Optical Class: Biaxial positive (+). *Orientation:* $Z = b$; $Y \wedge a = 15(5)°$, $Y \wedge c = 101(5)°$. *Dispersion:* $r > v$, strong, crossed. $\alpha = > 2.72$ (Li). 2V(meas.) = < 46° 2V(calc.) = n.d. *Anisotropism:* Weak.

R_1–R_2: (400) 34.8–37.9, (420) 34.2–37.4, (440) 33.6–36.9, (460) 32.7–36.0, (480) 31.8–34.8, (500) 30.7–33.5, (520) 29.5–31.9, (540) 28.1–30.6, (560) 27.0–29.4, (580) 26.2–28.6, (600) 25.7–28.0, (620) 25.2–27.4, (640) 24.9–27.0, (660) 24.5–26.8, (680) 24.2–26.5, (700) 23.8–26.1

Cell Data: *Space Group:* $P2_1/a$. $a = 11.8568$ $b = 9.0152$ $c = 10.1938$ $\beta = 116.365(4)°$ Z = 8

X-ray Powder Pattern: Zarehshuran, Iran.
3.66 (100), 2.915 (100), 2.880 (100), 4.96 (80), 4.46 (80), 2.815 (80), 4.08 (60)

Chemistry:

	(1)	(2)	(3)
As	25.09	26.50	25.59
Sb	42.04	41.80	41.57
S	32.82	34.30	32.84
Total	99.95	102.60	100.00

(1) Getchell mine, Nevada, USA; average of several analyses. (2) Zarehshuran, Iran. (3) $AsSbS_3$.

Occurrence: In an epithermal gold deposit formed in a narrow, steeply-dipping fault zone cutting interbedded Paleozoic (?) shales, argillites, and limestones, near a granodiorite intrusive (Getchell mine, Nevada, USA).

Association: Orpiment, realgar, stibnite, cinnabar, galkhaite, laffittite, chabournéite, marcasite, quartz, barite, fluorite, calcite.

Distribution: At the Getchell mine, about 32 km northeast of Golconda, Humboldt Co., Nevada, USA. From Iran, at Zarehshuran, near Takap, Azerbaijan. In the Gal-Khaya deposit, Yakutia, and at Khaidarkan, Kirgizia, USSR. From the Toya mine, Abuta, Hokkaido, Japan.

Name: For the Getchell mine, Nevada, USA, in which it was discovered.

Type Material: National Museum of Natural History, Washington, D.C., USA, 118159, 118160.

References: (1) Weissberg, B.G. (1965) Getchellite, $AsSbS_3$, a new mineral from Humboldt County, Nevada. Amer. Mineral., 50, 1817–1826. (2) Bariand, P., F. Cesbron, H. Agrinier, J. Geffroy, and V. Issakhanian (1968) La getchellite $AsSbS_3$ de Zarehshuran, Afshar, Iran. Bull. Soc. fr. Minéral., 91, 403–406 (in French). (3) Guillermo, T.R. and B.J. Wuensch (1973) The crystal structure of getchellite. Acta Cryst., B29, 2536–2541.

Crystal Data: Cubic. *Point Group:* $2/m\ \overline{3}$. As tiny grains.

Physical Properties: Hardness = 4.5–5 VHN = 726–766 (50 g load). D(meas.) = n.d. D(calc.) = 10.97

Optical Properties: Opaque. *Color:* In polished section, light gray. *Luster:* Metallic.
R: (400) 58.9, (420) 59.1, (440) 59.3, (460) 59.4, (480) 59.4, (500) 59.4, (520) 59.2, (540) 58.9, (560) 58.4, (580) 57.8, (600) 57.4, (620) 57.3, (640) 57.2, (660) 57.2, (680) 57.4, (700) 57.8

Cell Data: *Space Group:* $Pa3$. $a = 6.428$ Z = 4

X-ray Powder Pattern: Synthetic. (JCPDS 14-141).
1.240 (100), 1.940 (83), 1.140 (83), 1.194 (58), 1.720 (50), 1.310 (50), 1.174 (50)

Chemistry:

	(1)	(2)
Pt	45.0	44.48
Sb	51.5	55.52
Total	96.5	100.00

(1) Driekop mine, South Africa; by electron microprobe. (2) $PtSb_2$.

Occurrence: In concentrates of platinum minerals derived from Pt-Fe-Ni-Cu deposits in ultramafic rocks (Driekop mine, South Africa).

Association: Platinum, gold, hollingworthite, sperrylite, stibiopalladinite, chalcopyrite, pyrrhotite (Driekop mine, South Africa); vozhminite, heazlewoodite, tučekite, magnetite, copper (Vozhmin massif, USSR).

Distribution: In the Driekop mine, eastern Transvaal, South Africa. From the Morozova mine, Noril'sk region, western Siberia; and the Vozhmin massif, northeastern Karelia, USSR. From Fox Gulch, Goodnews Bay, Alaska, USA.

Name: For Professor T.W. Gevers, eminent South African geologist.

Type Material: n.d.

References: (1) Thomassen, L. (1929) Über Kristallstrukturen einiger binärer Verbindungen der Platinmetalle. Zeitschrift Physikal Chem., 2, 349–379 (in German). (2) Stumpfl, E.F. (1961) Some new platinoid-rich minerals, identified with the electron microanalyser. Mineral. Mag., 32, 833–846. (3) (1961) Amer. Mineral., 46, 1518 (abs. ref. 2). (4) Cabri, L.J., Ed. (1981) Platinum group elements: mineralogy, geology, recovery. Can. Inst. Min. & Met., 107–108.

Crystal Data: Monoclinic, pseudo-orthorhombic. *Point Group:* $2/m$. As fine needles to 1.5 mm. *Twinning:* Intimately on (100).

Physical Properties: Hardness = ~2.5 VHN = 65 D(meas.) = n.d. D(calc.) = 7.45

Optical Properties: Opaque. *Color:* Grayish black. *Streak:* Grayish black. *Luster:* Metallic. R_1–R_2: n.d.

Cell Data: *Space Group:* $P2_1/n$. $a = 34.51(3)$ $b = 38.18(5)$ $c = 4.080(8)$ $\beta = 90.33(5)^\circ$ Z = 2

X-ray Powder Pattern: Bjørkåsen, Norway.
2.0271 (100), 3.436 (90), 3.404 (90), 2.1514 (90), 2.9061 (70), 2.8867 (70), 2.8413 (70)

Chemistry:

	(1)	(2)	(3)	(4)
Pb	44.4	47.5	48.3	48.9
Cu	1.3	1.2	0.7	0.88
Ag				0.14
Fe				0.09
Bi	28.5	29.8	31.7	31.2
Sb	4	4.2	3.5	3.1
S	22.2	16.5	16.0	16.4
Total	100.4	99.2	100.2	100.8

(1–2) Binntal, Switzerland; by electron microprobe. (3–4) Bjørkåsen, Norway; by electron microprobe, averages of three and eight analyses respectively.

Occurrence: Of hydrothermal origin, with other sulfides.

Association: Galena, pyrite, pyrrhotite, sphalerite, tennantite, seligmannite, geocronite, quartz, dolomite.

Distribution: From Turtschi, between Giessen and Binn, about 2 km from the Lengenbach quarry, Binntal, Valais; and at Lake Zervreila, Vals, Graubünden, Switzerland. From the Bjørkåsen sulfide deposit, Otoften, Norway. From Vena, Sweden. In the Otome mine, Yamanashi Prefecture, Japan.

Name: For Giessen, a village nearby the Binntal, Switzerland.

Type Material: n.d.

References: (1) Graeser, S. (1963) Giessenit — ein neues Pb–Bi–Sulfosalz aus dem Dolomit des Binnatals. Schweiz. Mineral. Petrog. Mitt., 43, 471–478. (2) (1965) Amer. Mineral., 50, 264 (abs. ref. 1). (3) Graeser, S. and D.C. Harris (1986) Giessenite from Giessen near Binn, Switzerland: new data. Can. Mineral., 24, 19–20. (4) (1987) Amer. Mineral., 72, 229 (abs. ref. 3). (5) Makovicky, E. and S. Karup-Møller (1986) New data on giessenite from the Bjørkåsen sulfide deposit at Otoften, northern Norway. Can. Mineral., 24, 21–25. (6) (1987) Amer. Mineral., 72, 229 (abs. ref. 5).

Crystal Data: Cubic. *Point Group:* $\bar{4}3m$. As anhedral individuals up to 400 μm; more rarely as micro-myrmekitic intergrowths with chaméanite.

Physical Properties: Hardness = n.d. VHN = 233–333 (25 g load). D(meas.) = n.d. D(calc.) = 5.75

Optical Properties: Opaque. *Color:* Light gray, with a creamy tint in reflected light. *Luster:* Metallic.

R: (400) — , (420) 32.2, (440) 32.1, (460) 31.8, (480) 31.7, (500) 31.6, (520) 31.5, (540) 31.7, (560) 31.8, (580) 31.7, (600) 31.7, (620) 31.8, (640) 31.6, (660) 31.5, (680) 31.3, (700) 30.8

Cell Data: *Space Group:* $I\bar{4}3m$. $a = 10.578$ Z = 2

X-ray Powder Pattern: Chaméane, France.
3.050 (100), 1.868 (90), 1.593 (70), 1.932 (60), 2.497 (50), 2.076 (40), 1.714 (40)

Chemistry:

	(1)	(2)
Cu	30.06	32.78
Zn	3.37	3.19
Ag	3.89	1.73
Hg	0.57	0.30
Fe	0.19	0.03
As	7.61	9.56
Sb	10.66	7.50
Se	41.09	40.62
S	2.73	4.12
Total	100.17	99.83

(1–2) Chaméane, France; by electron microprobe.

Occurrence: As late stage deposits in veins cutting granite.

Association: Geffroyite, chaméanite, ankerite.

Distribution: From Chaméane, Puy-de-Dôme, France.

Name: To honor Roger Giraud, of the electron microprobe laboratory of the B.R.G.M.–C.N.R.S. in Orléans, France.

Type Material: National School of Mines, Paris, France.

References: (1) Johan, Z., P. Picot, and F. Ruhlmann (1982) Evolution paragénétique de la minéralisation uranifère de Chaméane (Puy-de-Dôme) France: chaméanite, geffroyite et giraudite, trois séléniures nouveaux de Cu, Fe, Ag, and As. Tschermaks Mineral. Petrog. Mitt., 29, 151–167 (in French with English abs.). (2) (1982) Amer. Mineral., 67, 1074–1075 (abs. ref. 1).

Crystal Data: Orthorhombic. *Point Group:* $2/m$ $2/m$ $2/m$. Prismatic crystals up to 2 cm in length and 2–6 mm thick.

Physical Properties: *Cleavage:* Good on {010}; fair on {100}. Hardness = 2–3 VHN = n.d. D(meas.) = 6.96 D(calc.) = 6.9

Optical Properties: Opaque. *Color:* Lead-gray. *Streak:* Black. *Luster:* Metallic. R_1–R_2: n.d.

Cell Data: *Space Group:* *Pbnm.* $a = 33.531$ $b = 11.486$ $c = 4.003$ Z = 4

X-ray Powder Pattern: Gladhammar, Sweden. (JCPDS 25-1422).
3.56 (100), 1.308 (90b), 1.101 (90), 1.082 (90), 2.81 (70), 1.948 (70), 1.916 (70)

Chemistry:

	(1)	(2)
Pb	12.40	12.92
Cu	3.98	3.96
Fe	0.19	
Bi	64.96	65.14
S	18.04	17.98
insol.	0.12	
Total	99.69	100.00

(1) Gladhammar, Sweden. (2) $PbCuBi_5S_9$.

Occurrence: Of hydrothermal origin.

Association: Galenobismutite, quartz.

Distribution: At Gladhammar, Kalmar, Sweden. From Bleka, Telemark, Norway. In Canada, from the Tanco deposit, Bernic Lake, Manitoba. From Krupka, Czechoslovakia. In the Comstock mine, Dos Cabezas, Cochise Co., Arizona, USA.

Name: For the locality, Gladhammar, Sweden.

Type Material: n.d.

References: (1) Palache, C., H. Berman, and C. Frondel (1944) Dana's system of mineralogy, (7th edition), v. I, 483. (2) Welin, E. (1968) Notes on the mineralogy of Sweden. 5. Bismuth-bearing sulphosalts from Gladhammar, a revision. Arkiv. Min. Geol., 4, 377–386. (3) (1968) Amer. Mineral., 53, 351 (abs. ref. 2). (4) Kohatsu, I. and B.J. Wuensch (1976) The crystal structure of gladite, $PbCuBi_5S_9$, a superstructure intermediate in the series Bi_2S_3–$PbCuBiS_3$ (bismuthinite–aikinite). Acta Cryst., B32, 2401–2409.

Crystal Data: Orthorhombic. *Point Group:* $2/m\ 2/m\ 2/m$. Crystals prismatic by elongation ‖ [001]; less commonly ‖ [010], to 2 cm. Striated ‖ [001]. Massive. *Twinning:* On {101}; on {012}, forming cruciform penetration twins; also as trillings.

Physical Properties: *Cleavage:* Perfect on {010}, less so on {101}. *Fracture:* Uneven. *Tenacity:* Brittle. Hardness = 5 VHN = n.d. D(meas.) = 6.055 D(calc.) = 6.155

Optical Properties: Opaque. *Color:* Grayish tin-white to reddish silver-white. *Streak:* Black. *Luster:* Metallic. *Pleochroism:* Less distinct than for arsenopyrite. *Anisotropism:* Distinct, less strong than arsenopyrite.

R_1–R_2: (400) 47.9–52.0, (420) 48.3–51.8, (440) 48.7–51.6, (460) 49.1–51.5, (480) 49.4–51.4, (500) 49.6–51.3, (520) 49.9–51.2, (540) 50.2–51.1, (560) 50.5–51.1, (580) 50.6–51.0, (600) 50.7–50.9, (620) 51.0–50.7, (640) 51.1–50.6, (660) 51.1–50.5, (680) 51.1–50.5, (700) 51.6–50.5

Cell Data: *Space Group:* *Cmmm.* $a = 6.64$ $b = 28.39$ $c = 5.64$ Z = 24

X-ray Powder Pattern: Håkansbö, Sweden.
2.72 (100), 1.827 (90), 2.45 (80), 2.42 (70), 1.125 (60), 1.006 (60), 0.981 (50)

Chemistry:

	(1)	(2)
Co	16.68	11.99
Fe	19.60	22.72
Ni	trace	
As	44.01	45.72
S	20.18	19.57
insol.	0.20	
Total	100.67	100.00

(1) Håkansbö, Sweden. (2) (Fe, Co)AsS with Fe:Co = 2:1.

Polymorphism & Series: Dimorphous with alloclasite.

Occurrence: Found in deep-seated deposits of high-temperature hydrothermal origin.

Association: Cobaltite, pyrite, chalcopyrite.

Distribution: In Sweden, at Håkansbö, Västmanland, as large twinned and untwinned crystals; also at Tunaberg, Södermanland. From Skutterud, Norway. At Oraviţa (Oravicza), Banat, Romania. From Gosenbach, North Rhine-Westphalia, Germany. In the Tynebottom mine, Garrigill, near Alston, Cumbria, England. From Tsumeb, Namibia. At Huasco, Atacama, Chile. In the USA, at the Standard Consolidated mine, Sumpter, Grant Co., Oregon, and at Franconia, Grafton Co., New Hampshire. From the Cobalt-Gowganda district, Ontario, Canada.

Name: From the Greek for *blue*, in reference to its use in the dark blue glass called smalt.

References: (1) Palache, C., H. Berman, and C. Frondel (1944) Dana's system of mineralogy, (7th edition), v. I, 322–325. (2) Ferguson, R.B. (1947) The unit cell of glaucodot. University of Toronto Studies, Geol. Series, 51, 41–47.

Godlevskite $(Ni, Fe)_9S_8$

Crystal Data: Orthorhombic, pseudotetragonal. *Point Group:* 222. As aggregates and single grains typically under 0.4 mm in size; as fine disseminations. *Twinning:* Common and complex, ⊥ (101), sometimes resulting in geniculated forms.

Physical Properties: Hardness = n.d. VHN = 383–415 (40–50 g load). D(meas.) = n.d. D(calc.) = 5.273

Optical Properties: Opaque. *Color:* In polished section, pale yellow. *Luster:* Metallic. *Pleochroism:* Weak, pale cream to pinkish cream. *Anisotropism:* Strong, from bluish to reddish. R_1–R_2: (400) 27.2–29.4, (420) 32.6–36.0, (440) 38.0–42.6, (460) 42.2–46.3, (480) 45.2–48.7, (500) 47.4–50.3, (520) 49.1–51.4, (540) 50.5–52.2, (560) 51.5–52.7, (580) 52.4–53.0, (600) 53.3–53.3, (620) 54.2–53.7, (640) 55.0–54.1, (660) 55.8–54.7, (680) 56.5–55.2, (700) 57.2–55.8

Cell Data: *Space Group:* *C*222. $a = 9.3359(7)$ $b = 11.2185(10)$ $c = 9.4300(6)$ Z = 4

X-ray Powder Pattern: Zapolyarnyi mine, USSR.
2.85 (100), 1.803 (90), 1.795 (80), 1.654 (80), 3.28 (50), 2.10 (50), 2.33 (40)

Chemistry:

	(1)	(2)	(3)	(4)
Ni	61.5	64.45	68.2	67.32
Fe	3.0	2.08		
Co	0.6	0.19		
S	35.0	32.85	32.7	32.68
Total	100.1	99.57	100.9	100.00

(1–2) Zapolyarnyi mine, USSR; by electron microprobe. (3) Barberton district, South Africa; by electron microprobe. (4) Ni_9S_8.

Occurrence: In hydrothermal veins, or in peridotite with other nickel sulfides.

Association: Millerite, heazlewoodite, pentlandite, pyrrhotite, pyrite, bornite, chalcopyrite, magnetite.

Distribution: In the USSR, at the Zapolyarnyi mine, Talnakh area, Noril'sk region, western Siberia. At the Amianthus mine, Kaapsche Hoop, Barberton, Transvaal, South Africa. From Bou Azzer, Morocco. In Canada, at the Texmont mine, about 50 km south of Timmins, Ontario. From near Moapa, Clark Co., Nevada, USA. At Mount Clifford, 54 km north-northwest of Leonora, Western Australia.

Name: For M.N. Godlevskii, Russian economic geologist.

Type Material: Institute of Geology; A.E. Fersman Mineralogical Museum, Academy of Sciences, Moscow, USSR.

References: (1) Kulagov, E.A., T.L. Evstigneeva, and O.E. Yushko-Zakharova (1969) The new nickel sulfide godlevskite. Geol. Rudn. Mestorozhd. 11, 115–121 (in Russian). (2) (1970) Amer. Mineral., 55, 317 (abs. ref. 1). (3) Fleet, M.E. (1987) Structure of godlevskite, Ni_9S_8. Acta Cryst., C43, 2255–2257. (4) Fleet, M.E. (1988) Stoichiometry, structure and twinning of godlevskite and synthetic low-temperature Ni-excess nickel sulfide. Can. Mineral., 26, 283–291.

Crystal Data: Cubic. *Point Group:* $4/m\,\bar{3}\,2/m$. As octahedra, dodecahedra, and cubes, usually crude or rounded, up to about 2 cm; also elongated ∥ [111]. In twinned and parallel crystal groups; reticulated, dendritic, arborescent, platy, filiform, spongy; massive, and in rounded nuggets; scales and flakes. *Twinning:* Common on {111}; repeated to form reticulated and dendritic aggregates.

Physical Properties: *Fracture:* Hackly. *Tenacity:* Very malleable and ductile. Hardness = 2.5–3 VHN = 30–34, 44-58 (argentian) (10 g load). D(meas.) = 19.3 D(calc.) = 19.302

Optical Properties: Opaque in all but thinnest foils. *Color:* Gold-yellow when pure, silver-white to copper-red when impure; blue and green in transmitted light. *Luster:* Metallic.
R: (400) 24.9, (420) 26.5, (440) 28.1, (460) 31.6, (480) 39.0, (500) 49.5, (520) 57.8, (540) 63.4, (560) 67.8, (580) 71.0, (600) 73.8, (620) 76.1, (640) 78.2, (660) 80.1, (680) 81.9, (700) 83.6

Cell Data: *Space Group:* $Fm3m$. $a = 4.0786$ Z = 4

X-ray Powder Pattern: Synthetic.
2.355 (100), 2.039 (52), 1.230 (36), 1.442 (32), 0.9357 (23), 0.8325 (23), 0.9120 (22)

Chemistry: Gold, with some Ag; also Cu, Fe; rarely Bi, Sn, Pb, Zn, Pt, Pd, Ir, Rh.

Polymorphism & Series: Forms a series with silver.

Occurrence: Widespread in very small quantities in rocks of many kinds throughout the world, and in sea water. In veins of epithermal origin, usually in quartz with pyrite and other sulfides, and with tellurides; in pegmatites; in contact metamorphic deposits. Common in placers.

Association: Pyrite, chalcopyrite, arsenopyrite, pyrrhotite, sylvanite, krennerite, calaverite, altaite, scheelite, ankerite, tourmaline, quartz.

Distribution: Many localities for fine specimens. In the USSR, in Siberia, eastern Ural Mountains; important localities near Sverdlovsk (Ekaterinburg); in the Miass district; large crystal groups from the Lena River, Yakutia. Sharply crystallized from Romania, at Roşia Montană (Verespatak) and Săcărâmb (Nagyág). In Australia, many occurrences, as at Bendigo and Ballarat, in Victoria; the Palmer River and Gympie, in Queensland; and Kalgoorlie, Western Australia, with gold telluride ores, also very large alluvial nuggets. The world's most important gold district is the Witwatersrand, Transvaal, South Africa, which, however, only rarely produces crystalline material. In Canada, especially in Ontario, in the Porcupine and Hemlo districts. In the USA, in California, in the Mother Lode belt of the Sierra Nevada, with fine examples from both lode and placer deposits. In South Dakota, from the Homestake mine at Lead, Lawrence Co.; in Alaska, in lode mines in the Juneau district and placers in the Yukon. In Colorado, wire and leaf gold from Breckenridge, Summit Co.; in Gilpin Co., at Leadville. Near Santa Elena, in the Grand Savannah River region, Venezuela, a placer producing exceptional skeletal crystals.

Name: An Old English word for the metal; perhaps related to the Sanskrit *jval*; chemical symbol from the Latin *aurum*, *shining dawn*.

References: (1) Palache, C., H. Berman, and C. Frondel (1944) Dana's system of mineralogy, (7th edition), v. I, 90–95. (2) (1953) NBS Circ. 539, 1, 33.

Crystal Data: Cubic. *Point Group:* $\overline{4}3m$. Massive; as crusts.

Physical Properties: *Tenacity:* Brittle. Hardness = 3–3.5 VHN = 291–342 (100 g load). D(meas.) = n.d. D(calc.) = 4.95

Optical Properties: Opaque. *Color:* Dark lead-gray. *Luster:* Metallic.
R: (400) 31.1, (420) 30.8, (440) 30.5, (460) 30.5, (480) 30.2, (500) 30.1, (520) 30.1, (540) 30.1, (560) 30.1, (580) 30.2, (600) 30.3, (620) 30.5, (640) 30.8, (660) 30.9, (680) 31.0, (700) 31.1

Cell Data: *Space Group:* $I\overline{4}3m$. $a = 10.304$ Z = 2

X-ray Powder Pattern: Goldfield, Nevada, USA.
2.97 (100), 1.819 (60), 3.64 (30), 1.552 (30), 2.58 (20), 2.43 (20), 5.15 (10)

Chemistry:

	(1)	(2)	(3)
Cu	44.3	33.49	44.61 – 45.73
Ag	0.06	0.18	
Fe	0.1		
Zn	0.2		3.04 – 3.35
Te	14.5	17.00	9.98 – 10.38
Sb	7.1	19.26	1.15 – 2.46
As	5.3	0.68	12.57 – 13.12
Bi		6.91	
S	27.8	21.54	26.41 – 26.61
other		2.51	
Total	99.36	101.57	

(1–2) Goldfield, Nevada, USA; by electron microprobe. (3) Bittibulakhsk deposit, USSR; ranges of analyses.

Occurrence: Occurs in epithermal precious metal veins.

Association: Marcasite, famatinite, tennantite, hessite, petzite, sylvanite, altaite, chalcostibite, emplectite, gold, kuramite, bismuthinite, pyrite, sphalerite.

Distribution: In the USA, from the Mohawk and Claremont mines, Goldfield, Esmeralda Co., Nevada; the Tramway mine, Butte, Silver Bow Co., Montana; and in the Lucky Cuss mine, Tombstone, Cochise Co., Arizona. In the USSR, in the Kuramin Mountains, eastern Uzbekistan, and in the Bittibulakhsk Cu-As deposit, Lesser Caucasus Mountains. In the Kawazu mine, Shizuoka Prefecture, Japan.

Name: After the type locality, Goldfield, Nevada, USA.

Type Material: n.d.

References: (1) Palache, C., H. Berman, and C. Frondel (1944) Dana's system of mineralogy, (7th edition), v. I, 384. (2) Levy, C. (1968) Contribution à la minéralogie des sulfures de cuivre du type Cu_3XS_4. Mém. Bur. Rech. Géol. Minières, 54, 1–178 (in French). (3) (1968) Amer. Mineral., 53, 2105–2106 (abs. ref. 2). (4) Loginov, V.P., A.A. Magribi, V.L. Rusinov, S.E. Borisovskii, and L.P. Nosik (1984) First occurrence of goldfieldite in a pyritic deposit. Doklady Acad. Nauk SSSR, 273, 437–440 (in Russian). (5) (1984) Chem. Abs., 100, 88925 (abs. ref. 4).

Crystal Data: Orthorhombic. *Point Group:* n.d. As anhedral grains (200 x 50 μm). *Twinning:* Fine lamellar twinning in some grains.

Physical Properties: Hardness = n.d. VHN = 186–230 (10 g load). D(meas.) = n.d. D(calc.) = [6.80]

Optical Properties: Opaque. *Color:* Blackish lead-gray, resembling chalcocite. *Luster:* Metallic. *Anisotropism:* Strong, with colors from grayish white with a bluish tint to blue.
R_1–R_2: (400) 28.0–31.4, (420) 28.1–31.4, (440) 28.0–31.3, (460) 27.8–31.2, (480) 27.6–31.1, (500) 27.4–31.1, (520) 27.0–30.8, (540) 26.7–30.5, (560) 26.2–30.0, (580) 25.9–29.5, (600) 25.6–29.0, (620) 25.4–28.6, (640) 25.3–28.4, (660) 25.3–28.2, (680) 25.3–28.1, (700) 25.2–28.0

Cell Data: *Space Group:* n.d. $a = 14.96$ $b = 7.90$ $c = 24.1$ Z = [12]

X-ray Powder Pattern: Gortdrum deposit, Ireland.
4.58 (100), 3.38 (70), 2.88 (50), 2.78 (50), 6.03 (40), 3.08 (30), 3.02 (30)

Chemistry:

	(1)	(2)
Cu	38.33	39.59
Fe	2.40	1.41
Hg	41.37	41.35
S	16.87	16.61
Total	98.97	98.96

(1–2) Gortdrum deposit, Ireland; by electron microprobe.

Occurrence: Admixed with other sulfide minerals in a vein cutting dolomitized limestone; mineral associations suggest formation at less than 200°C.

Association: Chalcopyrite, bornite, chalcocite, cinnabar, ferroan dolomite, barite.

Distribution: From the Gortdrum deposit, Co. Tipperary, Ireland.

Name: For the type locality, the Gortdrum deposit, in Ireland.

Type Material: British Museum (Natural History), London, England.

References: (1) Steed, G.M. (1983) Gortdrumite, a new sulphide mineral containing copper and mercury, from Ireland. Mineral. Mag., 47, 35–36. (2) (1984) Amer. Mineral., 69, 407 (abs. ref. 1).

Crystal Data: Hexagonal. *Point Group:* $6/m\ 2/m\ 2/m$ (2H polymorph); $\bar{3}\ 2/m$ (3R polymorph). As hexagonal platy {0001} crystals having triangular striations on the base, to 20 cm. Massive, foliated; also scaly, columnar, granular, compact, earthy; in globular aggregates having radial structure. *Twinning:* Twin plane probably $\{11\bar{2}1\}$; twinning due to pressure gliding, produces trigonal or hexagonal striae on {0001}; also by 30° (90°) rotation about [0001].

Physical Properties: *Cleavage:* Perfect and easy on {0001}. *Tenacity:* Flexible but not elastic; sectile. Hardness = 1–2 VHN = 7–11 (10 g load). D(meas.) = 2.09–2.23 D(calc.) = 2.26 Greasy feel.

Optical Properties: Opaque; transparent only in extremely thin flakes. *Color:* Iron-black to steel-gray; deep blue in transmitted light. *Streak:* Black to steel-gray, shining. *Luster:* Metallic; may be dull, earthy.

Optical Class: Uniaxial negative (–). *Pleochroism:* Strong. ω = 1.93–2.07 (red). ϵ = n.d. (extreme birefringence). *Anisotropism:* Extreme.

R_1–R_2: (400) 7.9–18.7, (420) 7.8–18.8, (440) 7.7–18.9, (460) 7.6–19.2, (480) 7.5–19.5, (500) 7.5–19.8, (520) 7.5–20.1, (540) 7.4–20.5, (560) 7.4–20.8, (580) 7.4–21.3, (600) 7.5–21.7, (620) 7.5–22.2, (640) 7.6–22.8, (660) 7.6–23.4, (680) 7.6–24.0, (700) 7.6–24.6

Cell Data: *Space Group:* $6_3/mmc$ (2H polymorph), with a = 2.463 c = 6.714 Z = 4, and $R\bar{3}m$ (synthetic 3R polymorph), with a = 2.456 c = 10.044 Z = 6

X-ray Powder Pattern: Sri Lanka.
3.36 (10), 1.678 (8), 2.03 (5), 1.158 (5), 0.994 (4), 0.829 (4), 1.232 (3)

Chemistry: Carbon.

Polymorphism & Series: Diamond, lonsdaleite, and chaoite are polymorphs.

Occurrence: Formed by metamorphism of sedimentary carbonaceous material, by the reduction of carbon compounds; as a primary constituent in igneous rocks.

Association: A wide variety of minerals stable in the metamorphic conditions under which graphite forms. In meteorites, in nodules with troilite, silicates.

Distribution: Numerous localities, but only a few afford well-crystallized examples. In the USA, at Monroe and Ticonderoga, Essex Co., New York; at Franklin and Sterling Hill, Sussex Co., New Jersey. In Canada, commercially significant occurrences in Quebec, at Buckingham and Grenville, and in adjacent parts of Ontario. In Siberia, USSR, from Nizhni Tunguski, east of Turukhansk, near the Yenisei River, and at Shunga, Karelia. Around Ratnapura, Matara, and Kurunegale, Sri Lanka, large deposits of pure material. At Passau, Bavaria, Germany. From Pargas, Finland. In England, at Barrowdale, near Keswick, Cumbria. In Mexico, at Santa Maria, Sonora, formed by metamorphism of coal beds.

Name: From the Greek *to write*, in allusion to its use as a crayon.

References: (1) Palache, C., H. Berman, and C. Frondel (1944) Dana's system of mineralogy, (7th edition), v. I, 152–154. Berry, L.G. and R.M. Thompson (1962) X-ray powder data for the ore minerals. Geol. Soc. Amer. Mem. 85, 23.

Crystal Data: Hexagonal. *Point Group:* $3m$. Crystals to 5 mm, prismatic by elongation $\parallel$ [0001] with $\{11\bar{2}0\}$ prominent; $\{02\bar{2}1\}$ is well developed and frequently the only terminal form. Massive.

Physical Properties: *Tenacity:* Brittle. Hardness = 2.5 VHN = 130–146 (100 g load). D(meas.) = 6.22(2) D(calc.) = 6.17

Optical Properties: Opaque. *Color:* Dark lead-gray; in polished section, white, distinctly yellow against galena. *Streak:* Black. *Luster:* Metallic. *Pleochroism:* Very weak.
R_1–R_2: (400) 36.9–37.8, (420) 36.3–37.2, (440) 35.7–36.6, (460) 35.3–36.1, (480) 35.1–35.7, (500) 34.8–35.4, (520) 34.5–35.0, (540) 34.2–34.8, (560) 34.1–34.6, (580) 34.0–34.5, (600) 33.9–34.5, (620) 33.8–34.4, (640) 33.7–34.2, (660) 33.5–33.8, (680) 33.0–33.4, (700) 32.4–32.7

Cell Data: *Space Group:* $R3m$. a = 17.758(14) c = 7.807(6) Z = 3

X-ray Powder Pattern: Cerro de Pasco, Peru.
3.43 (100), 3.74 (80), 2.86 (70), 2.71 (70), 2.20 (60), 2.05 (60), 1.746 (40)

Chemistry:

	(1)	(2)
Pb	71.12	70.49
Fe	0.39	
As	10.82	11.33
Sb	0.21	
S	17.38	18.18
Total	99.92	100.00

(1) Cerro de Pasco, Peru. (2) $Pb_9As_4S_{15}$.

Occurrence: In vugs in hydrothermal copper ores.

Association: Realgar, pyrite, sphalerite, enargite, tetrahedrite, jordanite, hutchinsonite.

Distribution: In the Excelsior mine, Cerro de Pasco, Peru, in large crystals. In abundance, in large crystals in the Blei-Scharley mine, Upper Silesia, Poland. From Wiesloch, Black Forest, Germany. At Rio Tinto, Huelva Province, Spain. From Tsumeb, Namibia.

Name: For Louis Caryl Graton (1880–1970), Professor of Economic Geology, Harvard University, Cambridge, Massachusetts, USA.

Type Material: Harvard University, Cambridge, Massachusetts, USA, 94611.

References: (1) Palache, C., H. Berman, and C. Frondel (1944) Dana's system of mineralogy, (7th edition), v. I, 397–398. (2) Ribar, B. and W. Nowacki (1969) Neubestimmung der kristallstruktur von gratonit, $Pb_9As_4S_{15}$. Zeits. Krist., 128, 321–338 (in German with English abs.).

Crystal Data: Hexagonal. *Point Group:* $6mm$. Crystals hemimorphic pyramidal, micro to over 1 cm; most commonly as earthy coatings. *Twinning:* Twin plane $\{11\bar{2}2\}$ rare, forming trillings.

Physical Properties: *Cleavage:* Distinct on $\{11\bar{2}2\}$; $\{0001\}$ imperfect. *Fracture:* Conchoidal. *Tenacity:* Brittle. Hardness = 3–3.5 VHN = n.d. D(meas.) = 4.820 D(calc.) = 4.824

Optical Properties: Nearly opaque to translucent. *Color:* Shades of yellow and orange, rarely deep red; in polished section, very light lemon-yellow to red-brown and blood-red. *Streak:* Orange-yellow to brick-red. *Luster:* Adamantine to resinous.
Optical Class: Uniaxial positive (+) for red to blue-green, uniaxial negative (–) for blue-green to blue; isotropic at 523 nm. *Pleochroism:* Weak. $\omega = 2.506$ (Na); 2.431 (Li). $\epsilon = 2.529$ (Na); 2.456 (Li).
R: (400) 20.1, (420) 21.0, (440) 21.9, (460) 22.2, (480) 21.8, (500) 20.7, (520) 20.0, (540) 19.6, (560) 19.1, (580) 18.7, (600) 18.5, (620) 18.2, (640) 18.0, (660) 17.8, (680) 17.7, (700) 17.5

Cell Data: *Space Group:* $P6_3mc$. $a = 4.136$ $c = 6.713$ Z = 2

X-ray Powder Pattern: Synthetic.
3.160 (100), 3.583 (75), 3.367 (60), 2.068 (55), 1.761 (45), 1.898 (40), 2.450 (25)

Chemistry: Essentially pure CdS.

Polymorphism & Series: Dimorphous with hawleyite.

Occurrence: As earthy coatings, especially on sphalerite, in which it also occurs as a substituent; also rarely as crystals in cavities in mafic igneous rocks.

Association: Sphalerite, smithsonite, prehnite, other zeolites.

Distribution: In the USA, in New Jersey, at Franklin and Sterling Hill, Sussex Co., as coatings, and from Paterson, Passaic Co., crystals with zeolites; in the Arlington quarry, Leesburg, Loudoun Co., Virginia, as crystals; at Friedensville, Lehigh Co., Pennsylvania; and in the Joplin district, Jasper Co., Missouri, on sphalerite. As bright yellow coatings on smithsonite in Marion Co., Arkansas. From Bishopton, Renfrewshire, crystals on prehnite; as coatings on sphalerite in the Leadhills–Wanlockhead district, Dumfries, Scotland. At Příbram, Czechoslovakia. From Pierrefitte, Hautes-Pyrénées, France. In Greece, at Laurium. Pigmenting smithsonite stalactites from Masua, Sardinia, Italy. In the USSR, from the Kti-Teberda deposit, northern Caucasus Mountains. Complex twins from the Asunta Ag-Sn mine, 4 km northeast of San Vicente, Potosí, and at Llallagua, Bolivia. Numerous other minor occurrences are known.

Name: For Charles Murray Cathcart (1783–1859), Lord Greenock.

References: (1) Palache, C., H. Berman, and C. Frondel (1944) Dana's system of mineralogy, (7th edition), v. I, 228–230.

Crystal Data: Cubic. *Point Group:* $4/m\ \overline{3}\ 2/m$. As balls of intergrown octahedra with curved faces averaging 0.3 mm on an edge; infrequently as cubes; also as minute grains.

Physical Properties: Hardness = 4–4.5 VHN = 401–423 (50 g load). D(meas.) = 4.049 D(calc.) = 4.079 Strongly magnetic.

Optical Properties: Opaque. *Color:* Pinkish, tarnishing to a metallic blue; sooty black when very fine grained; in polished section, pale creamy white. *Luster:* Metallic.
R: (400) 28.8, (420) 28.9, (440) 29.0, (460) 29.2, (480) 29.4, (500) 29.8, (520) 30.3, (540) 30.7, (560) 31.3, (580) 31.9, (600) 32.4, (620) 33.2, (640) 33.9, (660) 34.7, (680) 35.5, (700) 36.3

Cell Data: *Space Group:* $Fd3m$. $a = 9.876$ Z = 8

X-ray Powder Pattern: Kramer area, California, USA.
2.980 (100), 1.746 (77), 2.470 (55), 3.498 (32), 1.008 (31), 1.901 (29), 1.105 (16)

Chemistry:

	(1)	(2)	(3)
Fe	56.5	55.9	56.64
Cu	0.08	0.2	
Ni	0.10		
Zn	0.007		
Mn		0.1	
Cr	0.14		
Sb		1.3	
As	0.38		
S	42.2	42.2	43.36
Total	99.407	99.7	100.00

(1) Zacatecas, Mexico; leading to Fe:S = 3:3.903. (2) Cornwall, England; by electron microprobe. (3) $Fe^{+2}Fe_2^{+3}S_4$.

Occurrence: Formed in lacustrine beds consisting of interbedded calcareous clays, silts, and fine- to medium-grained arkosic sands; in varve-like laminae of grayish black sulfide-bearing clays and fine silts; possibly as the result of seasonal variation in the composition and physical condition of lake waters. Also in hydrothermal vein deposits.

Association: Montmorillonite, chlorite, calcite, colemanite, veatchite (Kramer deposit, California, USA); sphalerite, pyrite, marcasite, galena, calcite, dolomite (Zacatecas, Mexico).

Distribution: In the USA, in California, from the Kramer borate deposit, Kern Co., and at Coyote Peak, near Orick, Humboldt Co. In Mexico, near Zacatecas. From the bed of the Black Sea, and in the Gal-Khaya deposit, northeast Yakutia, USSR. At Montemesola, Taranto, Italy. In the Lojane chromium deposit, 40 km northeast of Skopje, Macedonia, Yugoslavia. From the Treore mine, St. Teath, Cornwall, England. In the Hanoaka mine, Odate, Akita Prefecture, and the Akagane mine, Iwate Prefecture, Japan.

Name: For Dr. Joseph Wilson Greig (1895–1977), mineralogist and physical chemist, Pennsylvania State University, State College, Pennsylvania, USA.

Type Material: National Museum of Natural History, Washington, D.C., USA, 117502, 136416.

References: (1) Skinner, B.J., R.C. Erd, and F.S. Grimaldi (1964) Greigite, the thio-spinel of iron; a new mineral. Amer. Mineral., 49, 543–555. (2) Criddle, A.J. and C.J. Stanley, Eds. (1986) The quantitative data file for ore minerals. British Museum (Natural History), London, England, 148.

Crystal Data: Hexagonal. *Point Group:* n.d. Massive, presumably.

Physical Properties: Hardness = n.d. VHN = 295(5) (30 g load). D(meas.) = n.d. D(calc.) = 5.88

Optical Properties: Opaque. *Color:* Gray-black; white in reflected light. *Luster:* Metallic. *Anisotropism:* Weak.
R: (460) 33.1, (540) 32.8, (580) 32.7, (660) 31.9

Cell Data: *Space Group:* n.d. $a = 13.90(2)$ $c = 9.432$ Z = 3

X-ray Powder Pattern: Chauvai deposit, USSR.
3.16 (100), 1.929 (90), 1.645 (80), 1.113 (60), 1.251 (50)

Chemistry:

	(1)	(2)
Cu	19.99	20.55
Fe	0.29	
Hg	32.73	32.45
Sb	26.21	26.26
As	0.37	
S	20.44	20.74
Total	100.03	100.00

(1) Chauvai deposit, USSR; by electron microprobe, average of six samples, corresponding to $(Cu_{5.87}Fe_{0.10})_{\Sigma=5.97}Hg_{3.04}(Sb_{4.01}As_{0.09})_{\Sigma=4.10}S_{11.89}$. (2) $Cu_6Hg_3Sb_4S_{12}$.

Polymorphism & Series: Forms a series with aktashite.

Occurrence: In veinlets of probable low-temperature hydrothermal origin.

Association: Aktashite, cinnabar, metacinnabar, wurtzite, fluorite, calcite, barite, stilbite.

Distribution: At the Chauvai Sb-Hg deposit, southern Kirgizia, USSR. From the San Miguel mine, Moctezuma, Sonora, Mexico.

Name: To honor the Russian mineralogist, V.S. Gruzdev (1938–1977).

Type Material: A.E. Fersman Mineralogical Museum, Academy of Sciences, Moscow, USSR.

References: (1) Spiridonov, E.P., L.Y. Krapiva, A.K. Gapeev, V.I. Stepanov, E.Y. Prushinskaya, and V.Y. Volgin (1981) Gruzdevite, $Cu_6Hg_3Sb_4S_{12}$, a new mineral from the Chauvai antimony–mercury deposit, Central Asia. Doklady Acad. Nauk SSSR, 261, 971–976 (in Russian). (2) (1982) Amer. Mineral., 67, 855 (abs. ref. 1).

Crystal Data: Orthorhombic. *Point Group:* $2/m\ 2/m\ 2/m$. Acicular crystals to 2 cm, striated lengthwise, forming semicompact masses. Also massive, having granular, foliated, or fibrous texture.

Physical Properties: *Cleavage:* Distinct on {010}, indistinct on {001}. *Tenacity:* Somewhat sectile. Hardness = 2.5–3 VHN = 53–82 (10 g load). D(meas.) = 6.25–6.98 D(calc.) = 7.54

Optical Properties: Opaque. *Color:* Bluish gray; in polished section, white with a yellowish hue. *Streak:* Shiny gray. *Luster:* Metallic. *Pleochroism:* Distinct. *Anisotropism:* Strong.
R_1–R_2: (400) 41.2–51.8, (420) 41.8–51.8, (440) 42.3–51.9, (460) 42.8–51.9, (480) 43.3–52.0, (500) 43.3–52.0, (520) 43.3–52.0, (540) 43.1–52.0, (560) 42.9–52.0, (580) 42.7–52.2, (600) 42.4–52.2, (620) 42.2–52.1, (640) 41.9–51.9, (660) 41.5–51.6, (680) 41.2–51.2, (700) 40.8–50.7

Cell Data: *Space Group:* *Pnma.* $a = 11.37$ $b = 11.55$ $c = 4.054$ Z = 4

X-ray Powder Pattern: Guanajuato, Mexico.
3.19 (100), 3.65 (90), 1.989 (70), 2.88 (60), 2.58 (50), 2.31 (50), 5.16 (40)

Chemistry:

	(1)	(2)	(3)
Bi	67.38	68.0	63.8
Se	24.13	24.5	36.2
S	6.60	6.6	
Total	98.11	99.1	100.0

(1) Guanajuato, Mexico. (2) Do.; by electron microprobe. (3) Bi_2Se_3.

Polymorphism & Series: Dimorphous with paraguanajuatite.

Occurrence: In hydrothermal deposits of low to moderate temperatures.

Association: Bismuthinite, bismuth, clausthalite, nevskite, galena, pyrite, calcite.

Distribution: In the Santa Catarina and La Industrial mines, Sierra de Santa Rosa, near Guanajuato, Mexico. In the USA, near Salmon, Lemhi Co., Idaho; and at the Thomas and Essex mines, Darwin, Inyo Co., California. From St. Andreasberg, Harz Mountains, Germany.

Name: For the locality in Guanajuato, Mexico.

References: (1) Palache, C., H. Berman, and C. Frondel (1944) Dana's system of mineralogy, (7th edition), v. I, 278–279. (2) Berry, L.G. and R.M. Thompson (1962) X-ray powder data for the ore minerals. Geol. Soc. Amer. Mem. 85, 85. (3) Criddle, A.J. and C.J. Stanley, Eds. (1986) The quantitative data file for ore minerals. British Museum (Natural History), London, England, 150.

Gudmundite $FeSbS$

Crystal Data: Monoclinic, pseudo-orthorhombic. *Point Group:* $2/m$. Prismatic ∥ [001]; prism zone typically tapered. *Twinning:* On {101} forming penetration and contact twins, with cruciform and butterfly shapes.

Physical Properties: *Tenacity:* Brittle. Hardness = ~5–6 VHN = 572–673 (100 g load). D(meas.) = 6.72 D(calc.) = 6.95

Optical Properties: Opaque. *Color:* Silver-white to steel-gray in polished section. *Luster:* Metallic. *Pleochroism:* Distinct, in pure white and pinkish white. *Anisotropism:* Strong, in yellow and red tints.

R_1–R_2: (400) 45.1–52.6, (420) 46.0–53.4, (440) 46.9–54.2, (460) 47.7–54.8, (480) 48.3–55.3, (500) 48.7–55.5, (520) 49.3–55.5, (540) 49.9–55.3, (560) 50.4–55.1, (580) 50.8–55.0, (600) 51.1–54.7, (620) 51.3–54.4, (640) 51.5–54.0, (660) 51.9–53.6, (680) 52.4–53.1, (700) 53.1–52.7

Cell Data: *Space Group:* $P2_1/c$. $a = 10.00$ $b = 5.95$ $c = 6.73$ $\beta = 90°00'$ Z = 8

X-ray Powder Pattern: Gudmundstorp, Sweden.
2.55 (100), 1.917 (70), 4.09 (50), 1.416 (50), 2.81 (40), 1.098 (40), 1.024 (40)

Chemistry:

	(1)	(2)
Fe	26.79	26.63
Ni	trace	
Sb	57.31	58.08
S	15.47	15.29
Total	99.57	100.00

(1) Gudmundstorp, Sweden. (2) FeSbS.

Occurrence: A late stage hydrothermal mineral formed in sulfide deposits.

Association: Pyrite, pyrrhotite, chalcopyrite, lead sulfantimonides, antimony, bismuth, kermesite, arsenopyrite, argentian gold.

Distribution: In Sweden, at Gudmundstorp, 3 km north of Sala, Västmanland; also from Boliden, Holmtjärn, Malånäs, and at Vena. In Finland, at Kalliolampi, near Nurma. At Waldsassen, Bavaria, Germany. From the Lac Nicolet antimony mine, South Ham, Wolfe Co., Quebec; in the Hemlo gold deposit, Thunder Bay district, and near Red Lake, Ontario, Canada. At Broken Hill, New South Wales, Australia. In Czechoslovakia, at Vlastějovice, Sáxova, and Pezinok. In Ireland, at the Shallee and Gortnadyne mines, Silvermines, Co. Tipperary. From the Ilímaussaq Intrusion, southern Greenland. At Askot, Pithoragarh, India. A number of other minor occurrences are known.

Name: For the occurrence at Gudmundstorp, Sweden.

Type Material: n.d.

References: (1) Palache, C., H. Berman, and C. Frondel (1944) Dana's system of mineralogy, (7th edition), v. I, 325–326.

Crystal Data: Monoclinic. *Point Group:* $2/m$. As minute acicular crystals, the faces of which are irregularly streaked parallel to the axis of elongation; also as small anhedral grains. *Twinning:* Polysynthetic twinning on {100}, common in polished section.

Physical Properties: *Cleavage:* Perfect on {001}. *Fracture:* Conchoidal. *Tenacity:* Very brittle. Hardness = n.d. VHN = 180–197 (50 g load). D(meas.) = n.d. D(calc.) = [5.29]

Optical Properties: Opaque. *Color:* Grayish black; in polished section, white with reddish internal reflections. *Luster:* Metallic. *Pleochroism:* Relatively strong.
R_1–R_2: (470) 37.6–42.6, (546) 36.1–41.2, (589) 34.8–39.3, (650) 32.8–36.7

Cell Data: *Space Group:* $P2_1/a$. $a = 20.17$ $b = 7.94$ $c = 8.72$ $\beta = 101°35(30)'$ Z = [8]

X-ray Powder Pattern: Madoc, Canada.
3.52 (100), 2.795 (90), 4.19 (50), 3.90 (50), 2.670 (50), 2.653 (50), 2.335 (40)

Chemistry:

	(1)	(2)	(3)	(4)
Pb	44.5	43.5	38.50	38.94
Cu			0.49	
Sb	22.0	22.0	23.57	22.88
As	12.0	12.0	13.61	14.08
S	21.5	20.5	23.46	24.10
Total	100.0	98.0	99.63	100.00

(1–2) Madoc, Canada; by electron microprobe. (3) Pitone quarry, Italy; by electron microprobe. (4) $Pb(Sb, As)_2S_4$ with Sb:As = 1:1.

Polymorphism & Series: Dimorphous with twinnite.

Occurrence: Of low-temperature hydrothermal origin, in marbles.

Association: Pyrite, sphalerite, wurtzite, galena, stibnite, orpiment, realgar, enargite, tetrahedrite, zinkenite, jordanite, bournonite, sterryite, boulangerite, jamesonite, sartorite (Madoc, Canada); zinkenite, boulangerite, semseyite, jordanite, enargite (Silverton, Colorado, USA).

Distribution: At Madoc, Ontario, Canada. In the Pitone marble quarry, near Seravezza, Tuscany, Italy. At the Jas Roux deposit, Hautes-Alpes, France. From Novoye, Khaidarkan, Kirgizia, USSR. From Silverton, and at the Brobdingnag mine, San Juan Co., Colorado, USA.

Name: For Jean Etienne Guettard (1715–1786), French geologist.

Type Material: n.d.

References: (1) Jambor, J.L. (1967) New lead sulfantimonides from Madoc, Ontario. Part 2 — Mineral descriptions. Can. Mineral., 9, 191–213. (2) (1968) Amer. Mineral., 53, 1425 (abs. ref. 1). (3) Bracci, G., D. Dalena, P. Orlandi, G. Duchi, and G. Vezzalini (1980) Guettardite from Tuscany, Italy: a second occurrence. Can. Mineral., 18, 13–15. (4) Mozgova, N.N., N.S. Bortnikov, Y.S. Borodaev, and A.I. Tzépine (1982) Sur la non-stoechiométrie des sulfosels antimonieux arséniques de plomb. Bull. Minéral., 105, 3–10 (in French with English abs.).

Crystal Data: Orthorhombic. *Point Group:* $2/m\ 2/m\ 2/m$. As tabular and sometimes slightly bent crystals up to 2 x 2 x 0.5 mm; as oriented intergrowths with benjaminite.

Physical Properties: *Cleavage:* Rare, parallel to the tabular faces; even more rarely perpendicular to the tabular faces. Hardness = n.d. VHN = 175–218 (100 g load). D(meas.) = n.d. D(calc.) = 7.01

Optical Properties: Opaque. *Color:* In polished section, white to grayish white. *Pleochroism:* White to gray. *Anisotropism:* Weak in air, distinct in oil, bluish black to grayish white.
R_1–R_2: n.d.

Cell Data: *Space Group:* *Bbmm* or $Bb2_1m$. $a = 13.510(8)$ $b = 20.169(15)$ $c = 4.092(8)$ Z = 4

X-ray Powder Pattern: Ivigtut, Greenland.
3.363 (100), 2.996 (100), 2.895 (100), 3.640 (80), 3.401 (80), 3.376 (80), 3.977 (50)

Chemistry:

	(1)	(2)	(3)	(4)
Pb	22.82	18.9	19.9	18.26
Ag	7.39	8.9	9.4	9.51
Bi	51.15	52.0	53.2	55.27
Sb		3.3		
S	17.13	17.7	17.4	16.96
Total	98.49	100.8	99.9	100.00

(1) Ivigtut, Greenland; by electron microprobe. (2) Bernic Lake, Canada; by electron microprobe, corresponding to $Ag_{0.85}Pb_{1.03}Bi_{2.69}Sb_{0.31}S_{6.04}$. (3) Camsell River, Canada; by electron microprobe. (4) $PbAgBi_3S_6$.

Polymorphism & Series: Forms a series with lillianite.

Occurrence: A rare mineral of hydrothermal origin; also found in pegmatites.

Association: Aikinite, bismuthinite, cosalite, pavonite, benjaminite, bismuth, chalcopyrite, tetrahedrite, pyrrhotite, stannite, arsenopyrite.

Distribution: In the USA, from the Silver Bell mine, Red Mountain, Ouray Co., and the Old Lout mine, Poughkeepsie Gulch, near Ouray, San Juan Co., Colorado; from South Mountain, Owyhee Co., Idaho; and at Darwin, Inyo Co., and Randsburg, San Bernardino Co., California. At Ivigtut, southern Greenland. In Canada, from the deposit of the Terra Mining and Exploration Co., Camsell River, Northwest Territories, and from the Tanco pegmatite, Bernic Lake, Manitoba. In the Monteneme Sn-W deposit, Galicia, Spain. From the Kti-Teberda deposit, northern Caucasus Mountains, and several other poorly specified localities in the Far Eastern Region, USSR.

Name: For Gustav Adolf Hageman, chemical engineer for the Cryolite Firm, Ivigtut, Greenland.

Type Material: National Museum of Natural History, Washington, D.C., USA, 136172.

References: (1) Karup-Møller, S. (1970) Gustavite, a new sulfosalt mineral from Greenland. Can. Mineral., 10, 173–190. (2) (1971) Amer. Mineral., 56, 633-634 (abs. ref. 1). (3) Harris, D.C. and T.T. Chen (1975) Gustavite: two Canadian occurrences. Can. Mineral., 13, 411–414.

Crystal Data: Hexagonal. *Point Group:* n.d. As thin scales.

Physical Properties: Hardness = Very soft, less than that of graphite. VHN = 9–11 (3 g load). D(meas.) = n.d. D(calc.) = 3.57

Optical Properties: Opaque. *Color:* Bronze-red; in polished section, brown. *Luster:* Metallic, strong. *Pleochroism:* Distinct, light brown to grayish brown. *Anisotropism:* Very strong, bronze-red to grayish white.
R_1–R_2: n.d.

Cell Data: *Space Group:* n.d. $a = 3.64(2)$ $c = 34.02(2)$ Z = [1.5]

X-ray Powder Pattern: Near Outokumpu, Finland.
11.34 (100), 5.67 (90), 3.18 (80b), 1.841 (70), 1.821 (70), 2.835 (40), 2.59 (40)

Chemistry:

	(1)
Fe	24.13
Ni	14.85
Cu	0.12
Co	0.01
MgO	18.77
FeO	6.39
CaO	0.00
Al_2O_3	0.15
S	22.04
H_2O	[13.82]
Total	100.28

(1) Near Outokumpu, Finland; by electron microprobe, average of three analyses, corresponding to $2(Fe_{0.63}Ni_{0.37})S \cdot 1.610(Mg_{0.84}Fe^{+2}_{0.16})(OH)_2$.

Occurrence: As thin scales in the Kokka serpentine and in carbonate; rarely in chrysotile.

Association: Pentlandite, maucherite, chalcopyrite, magnetite, chromite, lizardite, calcite.

Distribution: In the Kokka serpentinite body, about 33 km north-northwest from the Outokumpu mine, Tapiola, Finland.

Name: For Paavo Haapala, Chief Geologist of the Outokumpu Company, Finland.

Type Material: Preserved by the Outokumpu Company, Finland.

References: (1) Huhma, M., Y. Vuorelainen, T.A. Hakli, and H. Papunen (1973) Haapalaite, a new nickel–iron sulfide of the valleriite type from East Finland. Bull. Geol. Soc. Finland, 45, 103–106. (2) (1973) Amer. Mineral., 58, 1111 (abs. ref. 1).

Crystal Data: Cubic. *Point Group:* $\bar{4}3m$. As anhedral grains up to 0.3 mm in size.

Physical Properties: Hardness = n.d. VHN = 306 (40 g load). D(meas.) = n.d. D(calc.) = 6.3

Optical Properties: Opaque. *Color:* Gray-brown; in polished section, creamy white to clear brown. *Luster:* Metallic.

R: (400) 31.8, (420) 32.5, (440) 33.2, (460) 33.6, (480) 33.7, (500) 33.7, (520) 33.5, (540) 33.4, (560) 33.4, (580) 33.5, (600) 33.6, (620) 33.6, (640) 33.6, (660) 33.5, (680) 33.5, (700) 33.4

Cell Data: *Space Group:* $I\bar{4}3m$ (probable). $a = 10.83$ Z = 8

X-ray Powder Pattern: Předbořice, Czechoslovakia.
3.140 (100), 1.925 (90), 1.639 (80), 2.910 (70), 1.985 (70), 2.568 (60), 1.764 (60)

Chemistry:

	(1)	(2)
Cu	26.6	26.6
Hg	15.3	14.3
Sb	15.5	19.1
As	3.2	0.7
Se	34.0	38.5
S	3.5	
Total	98.1	99.2

(1) Předbořice, Czechoslovakia; by electron microprobe, corresponding to $(Cu_{2.52}Hg_{0.46})_{\Sigma=2.98}(Sb_{0.77}As_{0.26})_{\Sigma=1.03}(Se_{2.60}S_{0.66})_{\Sigma=3.26}$. (2) Do.; corresponding to $(Cu_{2.65}Hg_{0.45})_{\Sigma=3.10}(Sb_{0.95}As_{0.06})_{\Sigma=1.01}Se_{3.09}$.

Occurrence: Found in epithermal calcite veins.

Association: Berzelianite, clausthalite, umangite, chalcopyrite, pyrite, uraninite, hematite, goethite, gold.

Distribution: At Bukov and Předbořice, Czechoslovakia. From Moctezuma, Sonora, Mexico.

Name: For Jaroslav Hak, mineralogist, Institute of Ore Research, Kutná Hora, Czechoslovakia.

Type Material: Charles University, Prague, Czechoslovakia; National School of Mines, Paris, France.

References: (1) Johan, Z. and M. Kvaček (1971) La hakite, un nouveau minéral du groupe de la tétraédrite. Bull. Soc. fr. Minéral., 94, 45–48 (in French with English abs.). (2) (1972) Amer. Mineral., 57, 1553 (abs. ref. 1).

Crystal Data: Orthorhombic. *Point Group:* $2/m$ $2/m$ $2/m$ or $mm2$. As short prisms or needles.

Physical Properties: *Cleavage:* Good on {010}. *Fracture:* Flat, conchoidal. Hardness = 3–4 VHN = n.d. D(meas.) = 6.734 D(calc.) = 7.05

Optical Properties: Opaque. *Color:* Steel-gray with red tint. *Streak:* Black. *Luster:* Metallic.
R_1–R_2: (400) 40.1–47.0, (420) 39.9–47.0, (440) 39.7–47.0, (460) 39.6–47.1, (480) 39.6–47.2, (500) 39.5–47.4, (520) 39.4–47.5, (540) 39.3–47.5, (560) 39.2–47.3, (580) 39.0–47.0, (600) 38.8–46.5, (620) 38.5–46.0, (640) 38.2–45.5, (660) 37.9–45.0, (680) 37.6–44.5, (700) 37.5–44.0

Cell Data: *Space Group:* *Pbnm* or $Pbn2_1$. a = 33.772 b = 11.5857 c = 4.01 Z = 4

X-ray Powder Pattern: Dzhida deposit, USSR. (JCPDS 22-240).
3.63 (100b), 3.14 (100b), 4.01 (90b), 3.56 (80), 2.83 (80b), 2.00 (80), 1.973 (80)

Chemistry:

	(1)	(2)	(3)
Pb	27.40	23.13	24.87
Cu	7.60	7.69	7.63
Bi	47.59	49.00	50.18
S	17.01	17.57	17.32
rem.	0.04	2.71	
Total	99.64	100.10	100.00

(1) Gladhammar, Sweden. (2) Dzhida deposit, USSR. (3) $Pb_2Cu_2Bi_4S_9$.

Occurrence: Of hydrothermal origin.

Association: Paděraite, pekoite, bismuthinite, chalcopyrite, grossular, andradite (Băiţa, Romania).

Distribution: From Gladhammar, Kalmar, Sweden. In the Dzhida Mo-W deposit, western Transbaikal, USSR. At Băiţa (Rézbánya), Romania. In the Victoria district, Dona Ana Co., New Mexico, USA.

Name: For the locality at Gladhammar, Sweden.

Type Material: n.d.

References: (1) Johansson, K. (1924) Bidrag till Gladhammar - gruvornas mineralogi (Contributions to the mineralogy of the Gladhammar mine). Arkiv. Kemi, Mineral., Geol., 9, 11. (2) (1925) Amer. Mineral., 10, 157 (abs. ref. 1). (3) Welin, E. (1966) Notes on the mineralogy of Sweden. 5. Bismuth-bearing sulphosalts from Gladhammar, a revision. Arkiv Min. Geol., 4, 377–386. (4) (1968) Amer. Mineral., 53, 351 (abs. ref. 3). (5) Povilatis, M.M., N.N. Nozgova, Y.S. Borodaev, V.M. Senderova, and G.N. Ronami (1969) The first occurrence of hammarite in the USSR. Doklady Acad. Nauk SSSR, 187, 886–889 (in Russian). (6) (1970) Mineral. Abs., 21, 254 (abs. ref. 5). (7) Mumme, W.G., E. Welin, and B.J. Wuensch (1976) Crystal chemistry and proposed nomenclature for sulfosalts intermediate in the system bismuthinite–aikinite (Bi_2S_3 – $CuPbBiS_3$). Amer. Mineral., 61, 15–20. (8) Horiuchi, H. and B.J. Wuensch (1976) The ordering scheme for metal atoms in the crystal structure of hammarite, $Cu_2Pb_2Bi_4S_9$. Can. Mineral., 14, 536–539.

Crystal Data: Orthorhombic. *Point Group:* $2/m\ 2/m\ 2/m$. As small idiomorphic grains; also radiating aggregates. *Twinning:* Frequent.

Physical Properties: Hardness = ~6 VHN = n.d. D(meas.) = n.d. D(calc.) = 7.22

Optical Properties: Opaque. *Color:* Light brownish red to dark reddish violet; yellowish red in radiating aggregates. *Pleochroism:* Strong, especially in oil. *Anisotropism:* Strong.
R_1–R_2: (400) 46.6–47.2, (420) 46.4–47.7, (440) 46.2–48.2, (460) 46.1–48.8, (480) 46.0–49.6, (500) 46.2–50.4, (520) 46.6–51.2, (540) 46.9–52.2, (560) 47.4–53.3, (580) 47.9–54.2, (600) 48.5–55.4, (620) 49.1–56.4, (640) 49.7–57.2, (660) 50.4–58.0, (680) 51.0–58.6, (700) 51.4–59.0

Cell Data: *Space Group:* *Pmnn.* $a = 3.60$ $b = 4.84$ $c = 5.72$ Z = 2

X-ray Powder Pattern: Trogtal quarry, Germany.
2.6 (100), 2.5 (100), 1.9 (100), 1.02 (100), 2.9 (30), 1.61 (30), 1.44 (30)

Chemistry: Composition inferred by similarity of X-ray pattern with marcasite.

Polymorphism & Series: Dimorphous with trogtalite.

Occurrence: In hydrothermal veins.

Association: Trogtalite, bornhardtite, gold, clausthalite, selenium.

Distribution: In Germany, from the Trogtal quarry, near Lautenthal, in the upper Harz. In the USA, at the La Sal mine, Beaver Mesa, Montrose Co., Colorado. From the San Francisco mine, La Rioja Province, Argentina.

Name: For Dr. P.F. Hast, mining engineer.

Type Material: National Museum of Natural History, Washington, D.C., USA, 112811.

References: (1) Ramdohr, P. and M. Schmitt (1955) Vier neue natürliche Kobaltselenide vom Steinbruch Trogtal bei Lautenthal im Harz. Neues Jahrb. Mineral., Monatsh., 133–142 (in German). (2) (1956) Amer. Mineral., 41, 164 (abs. ref. 1).

Crystal Data: Triclinic. *Point Group:* $\bar{1}$ or 1. As small (to 0.7 mm) euhedral crystals.

Physical Properties: Hardness = n.d. VHN = n.d. D(meas.) = n.d. D(calc.) = 5.8

Optical Properties: Opaque to sub-translucent. *Color:* Lead-gray; dark red in transmitted light. *Streak:* Chocolate-brown.
R_1–R_2: n.d.

Cell Data: *Space Group:* $P1$ or $P\bar{1}$. $a = 9.37(1)$ $b = 7.84(1)$ $c = 8.06(1)$
$\alpha = 66°25(15)'$ $\beta = 63°20(15)'$ $\gamma = 84°58(15)'$ Z = 2

X-ray Powder Pattern: Binntal, Switzerland.
3.314 (100), 2.863 (90), 3.401 (80), 3.569 (40), 4.54 (30), 4.28 (30), 3.750 (20)

Chemistry:

	(1)	(2)	(3)
Pb	25.5	24.6	24.97
Tl	25.5	24.5	24.64
Ag	8.8	10.9	13.00
Cu	3.0	0.95	
As	17.0	20.7	18.06
Sb	1.5		
S	18.6	19.0	19.33
Total	99.9	100.65	100.00

(1–2) Binntal, Switzerland; by electron microprobe. (3) $(Pb, Tl)_2AgAs_2S_5$ with Pb:Tl = 1:1.

Occurrence: Of hydrothermal origin, in crystalline dolomite.

Association: Rathite, probably.

Distribution: From the Lengenbach quarry, Binntal, Valais, Switzerland.

Name: For Dr. Frederick Henry Hatch (1864–1932), American geologist and mining engineer.

Type Material: n.d.

References: (1) Palache, C., H. Berman, and C. Frondel (1944) Dana's system of mineralogy, (7th edition), v. I, 487. (2) Nowacki, W., G. Burri, P. Engel, and F. Marumo (1967) Über einige Mineralstufen aus dem Lengenbach (Binnatal) II. Neues Jahrb. Mineral., Monatsh., 43–48 (in German). (3) (1971) Amer. Mineral., 56, 361–362 (abs. ref. 2). (4) Marumo, F., W. Nowacki, C. Bahezre, and G. Burri (1967) The crystal structure of hatchite. Zeits. Krist., 125, 249–265. (5) Berry, L.G. (1969) The crystallography of hatchite. Indian Mineralogist, 10, 165–173.

Hauchecornite $Ni_9Bi(Sb, Bi)S_8$

Crystal Data: Tetragonal. *Point Group:* $4/m$ $2/m$ $2/m$. Tabular on {001}, also short prismatic and dipyramidal, to 4 mm.

Physical Properties: *Fracture:* Flat, conchoidal. Hardness = 5 VHN = 447–655 (50 g load). D(meas.) = 6.35–6.47 D(calc.) = 6.58

Optical Properties: Opaque. *Color:* Light bronze-yellow, tarnishing darker. *Streak:* Gray-black. *Luster:* Lively metallic on fresh break. *Pleochroism:* Very weak in air; more distinct in oil. *Anisotropism:* Distinct.

R_1–R_2: (400) 35.8–37.9, (420) 37.2–39.3, (440) 38.6–40.7, (460) 40.0–42.2, (480) 41.4–43.8, (500) 42.7–45.4, (520) 44.0–46.9, (540) 45.2–48.2, (560) 46.3–49.3, (580) 47.3–50.2, (600) 48.2–51.0, (620) 49.0–51.7, (640) 49.7–52.3, (660) 50.4–52.9, (680) 51.0–53.4, (700) 51.5–53.9

Cell Data: *Space Group:* $P4/mmm$. $a = 7.300(3)$ $c = 5.402(2)$ Z = 1

X-ray Powder Pattern: Friedrich mine, Germany.
2.79 (100), 2.39 (60), 2.30 (60), 4.34 (50), 1.861 (50), 3.65 (40), 3.25 (40)

Chemistry:

	(1)	(2)
Ni	45.883	45.58
Co	0.82	
Fe	0.17	
Bi	23.72	27.04
Sb	6.226	5.25
As	0.45	
S	22.625	22.13
Total	99.894	100.00

(1) Friedrich mine, Germany. (2) $Ni_9Bi(Sb, Bi)S_8$ with Sb:Bi = 0.5:1.5.

Occurrence: Of hydrothermal origin.

Association: Millerite, bismuthian-arsenian ullmannite, antimonian gersdorffite, siegenite, bismuthinite, gold, galena, sphalerite, quartz.

Distribution: From the Friedrich mine, near Wissen, North Rhine-Westphalia, Germany.

Name: For William Hauchecorn (1828–1900), Director of the Geological Survey and the Mining Academy, Berlin, Germany.

Type Material: Harvard University, Cambridge, Massachusetts, USA, 89710.

References: (1) Peacock, M.A. (1950) Hauchecornite. Amer. Mineral., 35, 440–446. (2) Gait, R.I. and D.C. Harris (1972) Hauchecornite—antimonian arsenian, and tellurian varieties. Can. Mineral., 11, 819–825. (3) Kocman, V. and E.W. Nuffield (1974) The crystal structure of antimonian hauchecornite from Westphalia. Can. Mineral., 12, 269–274. (4) Just, J. (1980) Bismutohauchecornite—new name: hauchecornite redefined. Mineral. Mag., 43, 873–876. (5) (1981) Amer. Mineral., 66, 436 (abs. ref. 4).

Crystal Data: Cubic. *Point Group:* $2/m\ \overline{3}$. Octahedral crystals common, also cubo-octahedral, up to 5 cm; and as globular aggregates.

Physical Properties: *Cleavage:* Perfect on {001}. *Fracture:* Uneven to subconchoidal. *Tenacity:* Brittle. Hardness = 4 VHN = n.d. D(meas.) = 3.463 D(calc.) = 3.444

Optical Properties: Opaque to sub-translucent. *Color:* Reddish brown to brownish black; in polished section grayish white with very light brownish hue, with red internal reflections; deep red in transmitted light. *Streak:* Brownish red. *Luster:* Metallic-adamantine.

Optical Class: Isotropic. $n = 2.69$ (Li)

R: (400) 26.2, (420) 26.2, (440) 26.2, (460) 26.0, (480) 25.9, (500) 25.6, (520) 25.2, (540) 24.8, (560) 24.3, (580) 23.8, (600) 23.2, (620) 22.8, (640) 22.5, (660) 22.2, (680) 22.0, (700) 21.9

Cell Data: *Space Group:* $Pa3$. $a = 6.107$ Z = 4

X-ray Powder Pattern: Radussa, Sicily, Italy.
3.07 (100), 1.175 (80), 1.843 (70), 0.990 (70), 2.75 (50), 2.51 (50), 2.04 (50)

Chemistry:

	(1)	(2)	(3)
Mn	46.47	46.28	46.14
Fe	0.03		
S	53.27	53.51	53.86
SiO_2	0.16		
Total	99.93	99.79	100.00

(1–2) Raddusa, Sicily, Italy. (3) MnS_2.

Occurrence: A low-temperature mineral commonly associated with solfataric waters, in clay deposits rich in sulfur, and from decomposed extrusive rocks.

Association: Sulfur, realgar, gypsum, calcite.

Distribution: In the USA, in Texas, at the Gulf and Big Hill salt domes, Matagorda Co.; the High Island salt dome, Galveston Co.; the Boling salt dome, Wharton Co.; and the Fannett salt dome, Jefferson Co. In good crystals at the Destricella mine, Raddusa, Sicily, Italy. From Czechoslovakia, at Kalinka, near Baňská Bystrica (Neusohl), and at Baňská Štiavnica (Schemnitz). From Jesiórko and Grzybow, also the Tarnobrzeg region, Poland. In the Yazovsk and Podgornensk deposits, Aktyubinsk, Ural Mountains, USSR. From the Lake Wakatipu district, Collingwood, New Zealand.

Name: After Joseph Ritter von Hauer (1778–1863), and his son Franz von Hauer (1822–1899), Austrian geologists.

References: (1) Palache, C., H. Berman, and C. Frondel (1944) Dana's system of mineralogy, (7th edition), v. I, 293–294. (2) Berry, L.G. and R.M. Thompson (1962) X-ray powder data for the ore minerals. Geol. Soc. Amer. Mem. 85, 89–90.

Crystal Data: Cubic. *Point Group:* $\bar{4}3m$. Very fine-grained powdery coatings.

Physical Properties: Hardness = n.d. VHN = n.d. D(meas.) = n.d. D(calc.) = 4.87

Optical Properties: Opaque to sub-translucent. *Color:* Bright yellow.
R: n.d.

Cell Data: *Space Group:* $F\bar{4}3m$. $a = 5.818$ Z = 4

X-ray Powder Pattern: Hector-Calumet mine, Galena Hill, Canada.
3.36 (100), 2.058 (80), 1.753 (60), 2.90 (40), 1.337 (30), 1.186 (30), 1.120 (30)

Chemistry: Composition established by semiquantitative spectroscopic analysis.

Polymorphism & Series: Dimorphous with greenockite.

Occurrence: As coatings on fine-grained sphalerite and siderite in vugs; probably of secondary origin, deposited from meteoric waters in vugs and along late fractures.

Association: Sphalerite, siderite, greenockite.

Distribution: In Canada, in the Hector-Calumet mine, Keno-Galena Hill area, Yukon Territory. In the USA, near Hanover, Grant Co., New Mexico; at the Crestmore quarry, Riverside Co., California; and from near Bethel Church, Pike Co., Indiana. From near Komna, Czechoslovakia. In the Los Blancos mine, Sierra de Cartagena, Murcia Province, Spain. From Ragada, Greece. At Tynagh, near Killimor, Co. Galway, Ireland. From the Noril'sk region, western Siberia, USSR. In the Coquimbana mine, Cerro Blanco district, Atacama, Chile. From the Tui mine, Mt. Te Aroha, New Zealand. Likely yet to be discovered at additional localities, as easily mistaken for greenockite.

Name: For Professor James Edwin Hawley (1897–1965), Canadian mineralogist, Queen's University, Kingston, Ontario, Canada.

Type Material: n.d.

References: (1) Traill, R.J. and R.W. Boyle (1955) Hawleyite, isometric cadmium sulfide, a new mineral. Amer. Mineral., 40, 555–559.

Crystal Data: Orthorhombic, pseudotetragonal. *Point Group:* 222. Observed in polished section as tiny grains up to about 500 μm. *Twinning:* On {103}, polysynthetic, sometimes visible in polished section.

Physical Properties: Hardness = n.d. VHN = 206–231 (100 g load). D(meas.) = n.d. D(calc.) = 4.33

Optical Properties: Opaque. *Color:* Brass-yellow, much like chalcopyrite; does not tarnish easily. *Anisotropism:* Weak to moderate.

R_1–R_2: (400) 13.9–13.5, (420) 15.5–15.9, (440) 18.3–19.4, (460) 21.7–23.2, (480) 25.1–26.8, (500) 28.2–29.9, (520) 31.0–32.5, (540) 33.3–34.5, (560) 35.1–36.1, (580) 36.6–37.3, (600) 37.8–38.3, (620) 38.6–38.9, (640) 39.2–39.4, (660) 39.6–39.6, (680) 39.8–39.7, (700) 40.0–39.8

Cell Data: *Space Group:* $P222$, $P222_1$, $P2_12_12$, or $P2_12_12_1$. a = 10.705(5) b = 10.734(5) c = 31.630(15) Z = 12

X-ray Powder Pattern: Mooihoek Farm, South Africa.
3.07 (100), 1.876 (80), 1.889 (60), 1.612 (60), 1.089 (60), 0.937 (60), 1.214 (50)

Chemistry:

	(1)	(2)	(3)
Cu	32.16	31.26	32.18
Fe	35.03	36.12	35.35
Ni	0.40	0.23	
S	32.41	32.86	32.47
Total	100.14	100.46	100.00

(1) Mooihoek Farm, South Africa; by electron microprobe, average of fifteen analyses, leading to $Cu_{1.00}Fe_{1.24}Ni_{0.01}S_{2.00}$. (2) Minnesota, USA; by electron microprobe, average of four analyses. (3) $Cu_4Fe_5S_8$.

Occurrence: In a hortonolite dunite (replacement) pegmatite in the Bushveld Complex (Mooihoek Farm, South Africa).

Association: Mooihoekite, copper, troilite, pentlandite, cubanite, magnetite (Minnesota, USA).

Distribution: In the USA, in the basal Duluth Gabbro Complex, Minnesota. In the Bushveld Complex, at Mooihoek Farm, Lydenburg district, Transvaal, South Africa. From the Talnakh area, Noril'sk region, western Siberia, USSR. At Krzemianka, Poland. From near Madera, Sonora, Mexico.

Name: For M.H. Haycock, former Head, Mineralogy Section, Mineral Sciences Division, Department of Energy, Mines, and Resources, Mines Branch, Ottawa, Canada.

Type Material: National Museum of Canada, Ottawa; Royal Ontario Museum, Toronto, Canada; Princeton University, Princeton, New Jersey; National Museum of Natural History, Washington, D.C., USA, 124965; Heidelberg University, Heidelberg, Germany.

References: (1) Cabri, L.J. and S.R. Hall (1972) Mooihoekite and haycockite, two new copper–iron sulfides and their relationship to chalcopyrite and talnakhite. Amer. Mineral., 57, 689–708. (2) Rowland, J.F. and S.R. Hall (1975) Haycockite, $Cu_4Fe_5S_8$: a superstructure in the chalcopyrite series. Acta Cryst., B31, 2105–2112.

Crystal Data: Hexagonal. *Point Group:* 32. Massive, fine-grained, compact; rarely in minute crystals. *Twinning:* Possibly the cause of mosaic structure seen in polished section.

Physical Properties: Hardness = 4 VHN = 230–254 (100 g load). D(meas.) = 5.82 D(calc.) = 5.87

Optical Properties: Opaque. *Color:* Pale bronze; in polished section, yellowish cream colored. *Luster:* Metallic. *Anisotropism:* Strong, brown to bluish gray.

R: (400) 34.2, (420) 40.0, (440) 45.8, (460) 49.9, (480) 52.9, (500) 55.0, (520) 56.2, (540) 57.0, (560) 57.4, (580) 57.6, (600) 58.0, (620) 58.4, (640) 59.0, (660) 59.5, (680) 60.1, (700) 60.8

Cell Data: *Space Group:* *R*32. $a = 5.741$ $c = 7.139$ Z = 3

X-ray Powder Pattern: Heazlewood, Tasmania, Australia.
1.828 (10), 1.817 (10), 2.88 (9), 1.661 (8), 4.11 (5), 2.04 (5), 2.39 (4)

Chemistry:

	(1)	(2)
Ni	72.13	73.31
Fe	0.55	
S	25.96	26.69
insol.	0.59	
Total	99.23	100.00

(1) Heazlewood, Tasmania, Australia. (2) Ni_3S_2.

Occurrence: In serpentine, probably of hydrothermal origin.

Association: Pentlandite, zaratite, shandite, awaruite, magnetite.

Distribution: In the USA, from Josephine Co., Oregon; from the Cedar Hill quarry, Lancaster Co., Pennsylvania; and in California, in crystals from the Dorleska mine, Trinity Co., California. In Canada, at Miles Ridge, Yukon Territory; and at the Marbridge mine, Malartic, La Motte Township, and the Jeffrey mine, Asbestos, Quebec. From Heazlewood and Trial Harbour, Tasmania, Australia. In Iran, at Tschagal, near Rabad Sefid. At the Kop Krom mine, near Askale, Turkey. In Switzerland, at Poschiavo, in Graubünden. At Hirt, near Friesach, Carinthia, Austria. In the Amianthus mine, Kaapsche Hoop, Barberton, Transvaal, South Africa. In small amounts from a number of other localities.

Name: For the locality at Heazlewood, Tasmania, Australia.

Type Material: National Museum of Natural History, Washington, D.C., USA, R641.

References: (1) Peacock, M.A. (1947) On heazlewoodite and the artificial compound Ni_3S_2. University of Toronto Studies, Geol. Ser., 51, 59–69. (2) (1947) Amer. Mineral., 32, 484 (abs. ref. 1). (3) Berry, L.G. and R.M. Thompson (1962) X-ray powder data for the ore minerals. Geol. Soc. Amer. Mem. 85, 45. (4) Fleet, M.E. (1977) The crystal structure of heazlewoodite, and the metallic bonds in sulfide minerals. Amer. Mineral., 62, 341–345. (5) Parise, J.B. (1980) Structure of heazlewoodite (Ni_3S_2). Acta Cryst., B36, 1179–1180.

Crystal Data: Hexagonal. *Point Group:* $\overline{3}\ 2/m$. As platy masses up to 6 mm wide and 1 mm thick.

Physical Properties: *Cleavage:* Perfect on {0001}, warped. *Tenacity:* Flexible lamellae. Hardness = 2 VHN = 31–42 (25 g load). D(meas.) = 8.91 D(calc.) = 8.93

Optical Properties: Opaque. *Color:* Tin-white, tarnishing iron-black; in polished section, white, between that of galena and silver. *Luster:* Metallic. *Anisotropism:* Slight, light gray to dark gray.
R_1–R_2: (400) 68.5–69.6, (420) 67.5–68.5, (440) 66.5–67.4, (460) 65.9–66.8, (480) 65.7–66.6, (500) 65.8–66.5, (520) 66.0–66.6, (540) 66.2–66.7, (560) 66.4–67.0, (580) 66.6–67.2, (600) 66.9–67.6, (620) 67.2–67.9, (640) 67.3–68.2, (660) 68.0–68.6, (680) 68.4–69.0, (700) 68.8–69.5

Cell Data: *Space Group:* $R\overline{3}m$. $a = 4.47$ $c = 119.0$ Z = 6

X-ray Powder Pattern: Good Hope claim, Hedley, Canada.
3.26 (10), 2.37 (5), 2.24 (4), 1.626 (4), 1.484 (4), 1.989 (3), 1.845 (3)

Chemistry:

	(1)	(2)	(3)
Bi	80.60	81.6	79.26
Te	18.52	17.5	20.74
Se		0.3	
S	0.12	0.1	
Total	99.24	99.5	100.00

(1) Good Hope claim, Hedley, Canada. (2) Millapaya, Bolivia; by electron microprobe. (3) Bi_7Te_3.

Occurrence: In irregular quartz veins and stringers cutting garnet-epidote-pyroxene skarn.

Association: Bismuth, gold, joséite, pyrrhotite, arsenopyrite, calcite (Hedley, Canada); pyrite, chalcopyrite, galena, tellurides, gold, bismuth, bismuthinite, scheelite (Kumbel' deposit, USSR).

Distribution: In Canada, from the Good Hope mineral claim, about 6 km southeast of Hedley, Osoyoos mining division; also the Oregon mine, near Hedley, British Columbia; and in the Upper Burwash Creek placer, Kluane Lake district, Yukon Territory. From Millapaya, Sorata, La Paz, Bolivia. In the Kumbel' skarn deposit, northern Kirgizia; Ugat, western Uzbekistan; and Vostok-2, Maritime Territory, USSR.

Name: For the locality at the Good Hope claim, near Hedley, Canada.

Type Material: n.d.

References: (1) Warren, H.V. and M.A. Peacock (1945) Hedleyite, a new bismuth telluride from British Columbia, with notes on wehrlite and some bismuth–tellurium alloys. University of Toronto Studies, Geol. Ser., 49, 55–69. (2) (1945) Amer. Mineral., 30, 644 (abs. ref. 1).
(3) Criddle, A.J. and C.J. Stanley, Eds. (1986) The quantitative data file for ore minerals. British Museum (Natural History), London, England, 156.

Heideite $(Fe, Cr)_{1+x}(Ti, Fe)_2S_4$ (x = 0.15)

Crystal Data: Monoclinic, probable. *Point Group:* $2/m$. As minute anhedral grains up to 100 μm.

Physical Properties: Hardness = 3.5–4.5 VHN = n.d. D(meas.) = 3.942 (synthetic) D(calc.) = 4.10 (synthetic)

Optical Properties: Opaque. *Color:* In polished section, creamy white.
Pleochroism: Moderately strong, from purple-gray to cream-gray.
R_1–R_2: n.d.

Cell Data: *Space Group:* $I2/m$. $a = 5.97$ $b = 3.42$ $c = 11.4$ $\beta = 90.2°$ Z = 2

X-ray Powder Pattern: Synthetic heideite.
2.068 (100), 2.644 (90), 1.719 (50), 2.975 (15), 1.445 (10), 1.051 (5), 1.010 (5)

Chemistry:

	(1)	(2)
Ti	28.5	29.5
Fe	25.1	25.1
Cr	2.9	
S	44.9	45.2
Total	101.4	99.8

(1) Bustee meteorite; by electron microprobe, average of analyses of five grains, corresponding to $(Fe^{2+}_{0.99}Cr^{2+}_{0.16})_{\Sigma=1.15}(Ti^{3+}_{1.70}Fe^{2+}_{0.30})_{\Sigma=2.00}S_{4.00}$. (2) Synthetic heideite.

Occurrence: As minute anhedral grains in the Bustee enstatite achondrite.

Association: Titanian troilite, ferroan alabandite, daubréelite.

Distribution: In the Bustee enstatite achondrite meteorite.

Name: For Professor Fritz Heide (1891–1973), meteoriticist of Jena, Germany.

Type Material: n.d.

References: (1) Keil, K. and R. Brett (1974) Heideite, $(Fe, Cr)_{1+x}(Ti, Fe)_2S_4$, a new mineral in the Bustee enstatite achondrite. Amer. Mineral., 59, 465–470.

Crystal Data: Cubic. *Point Group:* $4/m\ \bar{3}\ 2/m$, 432, or $\bar{4}3m$. As rounded grains and aggregates of irregular shape about 0.05 mm in diameter.

Physical Properties: Hardness = ~4 VHN = 210–215 D(meas.) = n.d. D(calc.) = 4.469

Optical Properties: Opaque. *Color:* Gray; in polished section, violet-gray to ash-gray.
R: (469) 23.4, (518) 24.9, (589) 24.2, (668) 23.4

Cell Data: *Space Group:* $Fm3m$, $F432$, or $F\bar{3}m$. $a = 10.82$ Z = 4

X-ray Powder Pattern: Chelopech, Bulgaria.
3.11 (100), 1.919 (50), 1.858 (30), 1.632 (30) 3.61 (20), 2.87 (20), 2.55 (20)

Chemistry:

	(1)	(2)	(3)
Cu	43.6	43.70 – 45.39	44.73
Sn	12.8	12.26 – 13.44	13.92
Fe	2.6	0.26 – 0.50	
Mo	11.7	11.06 – 11.48	11.26
As	0.1	0.11 – 0.11	
Se		0.41 – 0.53	
Te		0.15 – 0.27	
S	28.0	29.26 – 30.07	30.09
Total	98.8		100.00

(1) Chelopech, Bulgaria; by electron microprobe. (2) Kochbulak deposit, USSR; by electron microprobe, ranges of analyses. (3) Cu_6SnMoS_8.

Occurrence: Of hydrothermal origin, early formed in the mineral association (Chelopech, Bulgaria); in a polymetallic deposit in carboniferous porphyritic andesites (Kochbulak deposit, USSR).

Association: Enargite, luzonite, colusite, stannoidite, reniérite, tennantite, chalcopyrite, pyrite (Chelopech, Bulgaria).

Distribution: At Chelopech, Bulgaria. From the Kochbulak deposit, eastern Uzbekistan, USSR.

Name: After an ancient name for the Balkan Mountains, on the southern slope of which the Chelopech deposit occurs; probably of Thracian origin.

Type Material: University of Sofia; Geological Institute, Bulgarian Academy of Sciences, Sofia, Bulgaria.

References: (1) Terziev, G.I. (1971) Hemusite—a complex copper–tin–molybdenum sulfide from the Chelopech ore deposit, Bulgaria. Amer. Mineral., 56, 1847–1854. (2) Kovalenker, V.A., N.V. Troneva, T.L. Evstigneeva, and L.N. Vyal'sov (1980) First occurrence of hemusite in the USSR. Doklady Acad. Nauk SSSR, 252, 699–703 (in Russian). (3) (1980) Chem. Abs., 93, 153290 (abs. ref. 1).

Crystal Data: Cubic. *Point Group:* n.d. As anhedral grains ranging in size from 0.1 to 0.8 mm.

Physical Properties: Hardness = n.d. VHN = 109–115 (100 g load). D(meas.) = n.d. D(calc.) = 7.86

Optical Properties: Opaque. *Color:* In reflected light, pale blue.
R: (400) 41.8, (420) 40.2, (440) 38.7, (460) 37.4, (480) 36.2, (500) 35.1, (520) 34.0, (540) 33.1, (560) 32.3, (580) 31.4, (600) 30.4, (620) 29.4, (640) 28.4, (660) 27.4, (680) 26.5, (700) 25.7

Cell Data: *Space Group:* n.d. $a = 12.20(2)$ Z = 8

X-ray Powder Pattern: Bisbee, Arizona, USA.
2.157 (10), 3.050 (8), 7.04 (6), 2.348 (6), 4.31 (5), 3.522 (5), 2.728 (5)

Chemistry:

	(1)	(2)
Cu	22.3	23.36
Ag	30.2	29.74
Te	47.5	46.90
Total	100.0	100.00

(1) Bisbee, Arizona, USA; by electron microprobe, average of five analyses, corresponding to $Cu_{3.77}Ag_{3.01}Te_{4.00}$. (2) $Cu_4Ag_3Te_4$.

Occurrence: Of hydrothermal origin, with other Cu-Ag tellurides.

Association: Hessite, petzite, sylvanite, altaite, rickardite, pyrite.

Distribution: From the Campbell mine, Bisbee, Cochise Co., Arizona, USA.

Name: For Professor Norman Fordyce McKerron Henry (1909–1983), Cambridge University, Cambridge, England.

Type Material: British Museum (Natural History), London, England.

References: (1) Criddle, A.J., C.J. Stanley, J.E. Chisholm, and E.E. Fejer (1983) Henryite, a new copper–silver telluride from Bisbee, Arizona. Bull. Minéral., 106, 511–517. (2) (1985) Amer. Mineral., 70, 216 (abs. ref. 1).

Crystal Data: Orthorhombic. *Point Group:* $2/m\ 2/m\ 2/m$. Fine-grained, massive.

Physical Properties: *Cleavage:* {001}. Hardness = ~2 VHN = n.d. D(meas.) = 5.22 D(calc.) = 5.197

Optical Properties: Opaque. *Color:* Black; in polished section, white, with internal reflections in dark red-brown. *Streak:* Black. *Luster:* Metallic. *Pleochroism:* Very low, light to darker blue-white. *Anisotropism:* Intense in oil, with yellow to red colors.
R_1–R_2: (400) 45.9–52.9, (420) 45.6–53.6, (440) 45.2–53.8, (460) 44.6–53.2, (480) 44.0–52.2, (500) 43.7–51.2, (520) 43.6–50.6, (540) 43.2–49.9, (560) 42.3–48.8, (580) 41.4–47.6, (600) 40.7–46.6, (620) 40.3–45.8, (640) 40.0–45.0, (660) 39.9–44.2, (680) 39.6–43.7, (700) 39.3–43.1

Cell Data: *Space Group:* *Pbnm.* $a = 4.328$ $b = 11.190$ $c = 3.978$ Z = 4

X-ray Powder Pattern: Synthetic.
2.793 (100), 1.399 (70), 2.831 (25), 3.243 (15), 2.305 (15), 0.9326 (15), 3.421 (10)

Chemistry:

	(1)	(2)
Sn	80.0	78.73
S	20.2	21.27
Total	100.2	100.00

(1) Maria-Teresa mine, Bolivia; by electron microprobe. (2) SnS.

Occurrence: Of hydrothermal origin, later than the deposition of cassiterite.

Association: Cassiterite, stannite, romarchite, ottemannite, berndtite, pyrite, sphalerite, arsenopyrite, quartz.

Distribution: In the Maria-Teresa mine, near Huari, between Oruro and Uyuni, Bolivia. At the Stiepelmann mine, near Arandis, Namibia. From Shinkiura and in the Hoei tin mine, Oita Prefecture, Japan. From the Golub deposit, Primor'ye, USSR. At Järkvissle, Västernorrland, Sweden.

Name: For Robert Herzenberg (1885–?), German chemist of Oruro, Bolivia.

Type Material: Harvard University, Cambridge, Massachusetts, USA, 92713.

References: (1) Palache, C., H. Berman, and C. Frondel (1944) Dana's system of mineralogy, (7th edition), v. I, 259–260. (2) Mosberg, S., D.R. Ross, P.M. Bethke, and P. Toulmin (1961) X-ray powder data for herzenbergite, teallite, and tin trisulfide. U.S. Geol. Sur. Prof. Paper 424C, C347–C348. (3) Hofmann, W. (1935) Ergebnisse der Strukturbestimmung komplexer Sulfide. Zeits. Krist., 92, 161–185 (in German). (4) Ramdohr, P. (1969) The ore minerals and their intergrowths, (3rd edition), 659–660.

Crystal Data: Monoclinic. *Point Group:* $2/m$. Crystals pseudocubic, highly modified and generally irregularly developed and distorted, to 1.7 cm. Also massive, compact, or fine-grained. *Twinning:* Twin lamellae visible in polished section.

Physical Properties: *Cleavage:* {100} indistinct. *Fracture:* Even, smooth. *Tenacity:* Sectile. Hardness = 2–3 VHN = 22–24 (100 g load). D(meas.) = 8.24–8.45 D(calc.) = 8.4

Optical Properties: Opaque. *Color:* Lead-gray to steel-gray, tarnishing to black. *Luster:* Metallic. *Anisotropism:* Distinct, from dark orange to dark blue.
R_1–R_2: (400) 42.1–43.5, (420) 41.0–42.6, (440) 39.9–41.7, (460) 39.3–40.8, (480) 38.9–40.3, (500) 38.7–40.2, (520) 38.5–40.5, (540) 38.4–41.1, (560) 38.4–41.6, (580) 38.4–42.0, (600) 38.4–42.4, (620) 38.4–42.7, (640) 38.4–43.0, (660) 38.5–43.1, (680) 38.4–43.2, (700) 38.3–43.2

Cell Data: *Space Group:* $P2_1/c$. $a = 8.13$ $b = 4.48$ $c = 8.09$ $\beta = 112°55'$ Z = 4

X-ray Powder Pattern: Botés, Romania.
2.31 (100), 2.87 (80), 2.25 (70), 3.01 (60), 2.14 (60), 1.389 (40), 3.19 (20)

Chemistry:

	(1)	(2)	(3)	(4)
Ag	62.42	61.16	61.52	62.86
Au			1.01	
Pb		1.90		
Te	36.96	36.11	37.77	37.14
rem.	0.24	0.83		
Total	99.62	100.00	100.30	100.00

(1) Savodinskii mine, USSR. (2) San Sebastián, Mexico. (3) Botés, Romania. (4) Ag_2Te.

Polymorphism & Series: Cubic above 155°C.

Occurrence: Found in hydrothermal moderate and low-temperature veins; also in small quantities in some massive pyrite deposits.

Association: Calaverite, sylvanite, altaite, petzite, empressite, rickardite, gold, tellurium, pyrite, galena, tetrahedrite, chalcopyrite.

Distribution: In the USSR, from the Savodinskii mine, near Ziryanovskii, Altai Mountains, Siberia. From Săcărâmb (Nagyág), Băiţa (Rézbánya), and in exceptional crystals and masses from Botés, near Zlatna, Romania. At Kalgoorlie, Western Australia. In the USA, from Gold Hill, Boulder Co., Colorado. In California, from the Stanislaus mine, Calaveras Co.; the McAlpine mine, Sonora, Tuolumne Co.; and Nevada City, Nevada Co. From Arizona, in the Campbell mine, Bisbee, and Tombstone, Cochise Co. In small amounts at numerous localities in Ontario, British Columbia, and Quebec, Canada, and other localities world-wide.

Name: For Germain Henri Hess (1802–1850), Swiss-Russian chemist, of Leningrad (St. Petersburg), Russia.

References: (1) Palache, C., H. Berman, and C. Frondel (1944) Dana's system of mineralogy, (7th edition), v. I, 184–186. (2) Thompson, R.M. (1949) The telluride minerals and their occurrence in Canada. Amer. Mineral., 34, 342–382. (3) Rowland, J.F. and L.G. Berry (1951) The structural lattice of hessite. Amer. Mineral., 36, 471–479.

Crystal Data: Monoclinic. *Point Group:* $2/m$. Pyramidal, tending to be elongated parallel to pyramid edges; also rounded, striated forms; twisted, distorted, and composed of subparallel individuals. Massive.

Physical Properties: *Cleavage:* Good on {112}. *Fracture:* Uneven. *Tenacity:* Brittle. Hardness = 2.5–3 VHN = n.d. D(meas.) = 5.73 D(calc.) = 5.86

Optical Properties: Opaque. *Color:* Iron-black. *Streak:* Black. *Luster:* Metallic. R_1–R_2: n.d.

Cell Data: *Space Group:* $C2/c$. $a = 13.628(5)$ $b = 11.943(4)$ $c = 21.285(8)$ $\beta = 90°55(7)'$ Z = 4

X-ray Powder Pattern: Wolfsberg, Germany.
3.30 (100), 3.40 (80), 3.25 (80), 3.75 (70), 3.10 (70), 2.970 (70b), 2.884 (70)

Chemistry:

	(1)	(2)
Pb	47.86	47.81
Zn	0.60	
Sb	31.20	32.11
S	19.90	20.08
Total	99.56	100.00

(1) Arnsberg, Germany. (2) $Pb_7Sb_8S_{19}$.

Occurrence: Of hydrothermal origin.

Association: Sphalerite, plagionite, semseyite.

Distribution: From Arnsberg, North Rhine-Westphalia, and Wolfsberg, in the Harz Mountains, Germany. At Příbram, Czechoslovakia. In the Kara Kamar deposit, Tadzhikistan, USSR. From Yecora, Sonora, Mexico.

Name: From the Greek for *different* and *form*, in allusion to the difference between this mineral and a proposed dimorphous species.

References: (1) Palache, C., H. Berman, and C. Frondel (1944) Dana's system of mineralogy, (7th edition), v. I, 465–466. (2) Jambor, J.L. (1969) Sulphosalts of the plagionite group. Mineral. Mag., 37, 442–446. (3) Edenharter, A. (1980) Die kristallstruktur von heteromorphit, $Pb_7Sb_8S_{19}$. Zeits. Krist., 151, 193–202 (in German with English abs.).

Crystal Data: Hexagonal. *Point Group:* n.d. Massive.

Physical Properties: Hardness = 2.0–2.2 VHN = 75–108 D(meas.) = n.d. D(calc.) = 8.904

Optical Properties: Opaque. *Color:* Pale yellow to yellowish white. R_1–R_2: n.d.

Cell Data: *Space Group:* n.d. $a = 3.98$ $c = 5.35$ Z = [1]

X-ray Powder Pattern: Locality "Y", China.
2.890 (100), 2.109 (80), 1.990 (70), 1.108 (60), 1.580 (50), 1.635 (40),1.452 (30)

Chemistry:

	(1)
Ni	20.
Pd	16.
Sb	31.
Bi	0.1
Te	33.
Total	100.1

(1) Locality "Y", China; by electron microprobe, corresponding to $(Ni_{1.32}Pd_{0.58})_{\Sigma=1.90}(Sb_{0.99}Bi_{0.03})_{\Sigma=1.02}Te_{1.00}$.

Occurrence: In ore concentrates from Cu-Ni-sulfide deposits.

Association: Testibiopalladite.

Distribution: In southwestern China, at locality "Y" – a code name.

Name: Presumably named for the crystal system and composition.

Type Material: n.d.

References: (1) Platinum Metal Mineral Research Group, Microprobe Analysis Laboratory, X-Ray Powder Photograph Laboratory, and Mineral Dressing Laboratory, Kweiyang Institute of Geochemistry, Academia Sinica (1974) Tellurostibnide of palladium and nickel and other new minerals and varieties of platinum metals. Geochimica, 3, 169–181 (in Chinese with English abs.). (2) (1976) Amer. Mineral., 61, 182 (abs. ref. 1).

Crystal Data: Orthorhombic. *Point Group:* $2/m\ 2/m\ 2/m$ or $mm2$. As prismatic to acicular crystals, elongate [001] and flattened (010), striated on {010} parallel to [001]; up to 20 mm long and from 0.1 to 0.5 mm thick.

Physical Properties: *Cleavage:* Poor ∥ [001]. *Fracture:* Conchoidal. *Tenacity:* Brittle. Hardness = n.d. VHN = 166–234 (50 g load), strongly anisotropic. D(meas.) = 7.17 D(calc.) = 7.18

Optical Properties: Opaque. *Color:* Tin-white. *Streak:* Grayish black. *Luster:* Metallic, tarnishes black. *Pleochroism:* Weak to distinct, white to light gray. *Anisotropism:* Strong, dark gray to white.
R_1–R_2: (470) 43.3–47.3, (546) 40.6–44.6, (589) 39.8–43.9, (650) 39.6–43.7

Cell Data: *Space Group:* *Pbmm* or *Pbm*2. $a = 13.705$ $b = 31.194$ $c = 4.121$ Z = 4

X-ray Powder Pattern: Hůrky, Czechoslovakia.
3.435 (100), 2.962 (80), 2.098 (35), 1.790 (22), 3.343 (20), 3.006 (15), 2.067 (15)

Chemistry:

	(1)	(2)
Pb	53.65	54.50
Ag	2.5	2.84
Cu	0.05	
Bi	28.35	27.48
S	14.4	15.18
Total	98.95	100.00

(1) Hůrky, Czechoslovakia; by electron microprobe, corresponding to $Pb_{4.98}(Ag, Cu)_{0.42}Bi_{2.60}S_{9.00}$.
(2) $Pb_{10}AgBi_5S_{18}$.

Occurrence: In high-temperature quartz veins.

Association: Pyrite, sphalerite, galena, molybdenite, cosalite, bismuth, galenobismutite, chalcopyrite, covellite, quartz, albite, microcline.

Distribution: In Czechoslovakia, near Hůrky, about 65 km west of Prague. From the Furka Pass, Canton Uri, and at Goppenstein, Lotschentnal, Canton Valais, Switzerland. At the Clara mine, Black Forest, Germany. In the Balikesir Balya deposit, Turkey. From the Spokoinoe deposit, eastern Transbaikal, and the Shumilovsk Sn-W deposit, western Transbaikal, USSR. From 27 km west of Castlegar, British Columbia, Canada. In the USA, from near Austin, Lander Co., Nevada, and in the Idarado mine, Ouray Co., Colorado.

Name: For Jaroslav Heyrovský, Czech Nobel laureate in chemistry.

Type Material: Department of Mineralogy, Charles University, Prague, Czechoslovakia, 14265.

References: (1) Klomínský, J., M. Rieder, C. Kieft, and L. Mráz (1971) Heyrovskýite, $6(Pb_{0.86}Bi_{0.08}(Ag, Cu)_{0.04})S \bullet Bi_2S_3$ from Hůrky, Czechoslovakia, a new mineral of genetic interest. Mineral. Deposita, 6, 133–147. (2) (1972) Amer. Mineral., 57, 325 (abs. ref. 1). (3) Mozgova, N.N., Y.S. Borodaev, L.E. Syritso, and D.P. Romanov (1976) New data on goongarrite (warthaite) and about the identity of heyrovskýite with goongarrite. Neues Jahrb. Mineral., Abh., 127, 62–83.

Crystal Data: Tetragonal. *Point Group:* $\bar{4}2m$. Grains less than 1 mm in diameter, and intergrowths. *Twinning:* Polysynthetic twinning common.

Physical Properties: Hardness = ~4 VHN = 209–234 (25 g load). D(meas.) = n.d. D(calc.) = 4.77

Optical Properties: Opaque. *Color:* In polished section, brownish gray. *Pleochroism:* Weak, gray-brown to violet-gray. *Anisotropism:* Distinct, orange to greenish.
R_1–R_2: (400) 27.0–27.7, (420) 26.5–27.2, (440) 26.0–26.7, (460) 25.5–26.3, (480) 25.2–26.0, (500) 24.8–25.7, (520) 24.7–25.5, (540) 24.5–25.4, (560) 24.4–25.3, (580) 24.5–25.4, (600) 24.6–25.5, (620) 24.8–25.6, (640) 25.2–25.2, (660) 25.4–24.8, (680) 24.8–24.4, (700) 23.7–24.5

Cell Data: *Space Group:* $I\bar{4}2m$. $a = 5.74$ $c = 10.96$ Z = 2

X-ray Powder Pattern: Tacama, Bolivia.
3.26 (100), 1.98 (80), 1.72 (70), 2.03 (50), 2.87 (40), 2.74 (30), 1.666 (30)

Chemistry:

	(1)	(2)	(3)
Ag	36.0	39.6	41.61
Fe	7.6	4.8	10.77
Zn	4.2	6.7	
Cu	1.8	1.5	
Sn	25.0	23.0	22.89
S	26.0	24.0	24.73
Total	100.6	99.6	100.00

(1) Tacama, Bolivia; by electron microprobe. (2) Pirquitas deposit, Argentina; by electron microprobe. (3) Ag_2FeSnS_4.

Polymorphism & Series: Forms a series with pirquitasite.

Occurrence: In tin veins.

Association: Sphalerite, wurtzite, stannite, siderite, fluorite.

Distribution: In Bolivia, from Tacama, Hocaya, Colquechaca, Potosí, and Chocaya. From the Pirquitas deposit, Jujuy Province, Argentina. In the USA, from Bisbee, Cochise Co., Arizona, and in the Dean mine, southeast of Battle Mountain, Lander Co., Nevada. At Fournial, Cantal, France. In the Chat-Karagai tin deposit, Tallas Alatan, USSR.

Name: For Raymond Hocart, Professor of Mineralogy, University of Paris, France.

Type Material: National School of Mines, Paris, France.

References: (1) Caye, R., Y. Laurent, P. Picot, R. Pierrot, and C. Lévy (1968) La hocartite, Ag_2SnFeS_4, une nouvelle espèce minérale. Bull. Soc. fr. Minéral., 91, 383–387 (in French).
(2) (1969) Amer. Mineral., 54, 573 (abs. ref. 1).

Crystal Data: Monoclinic. *Point Group:* $2/m$. Rarely as needle-like crystals less than 1 mm long, and as columnar and platy crystals, vertically striated, to 5 mm. Usually as irregular grains and fine-grained aggregates. *Twinning:* Polysynthetic.

Physical Properties: *Tenacity:* Very brittle. Hardness = 3–4 VHN = 187–213, 200 average. D(meas.) = 6.35 D(calc.) = 6.451

Optical Properties: Opaque. *Color:* Steel-gray with slightly yellowish tint on fresh surface, tarnishes quickly to brownish orange; in polished section, creamy with very slight pinkish tint. *Luster:* Metallic. *Pleochroism:* Very weak. *Anisotropism:* Very feeble, in grays.

R_1–R_2: (400) 36.2–39.8, (420) 36.2–40.4, (440) 36.2–41.0, (460) 36.3–41.6, (480) 36.4–42.0, (500) 36.6–42.4, (520) 36.9–42.6, (540) 37.4–42.8, (560) 37.6–43.0, (580) 37.8–43.2, (600) 37.9–43.3, (620) 38.0–43.5, (640) 38.0–43.7, (660) 38.0–43.7, (680) 37.9–43.6, (700) 37.8–43.4

Cell Data: *Space Group:* $A2/m$. $a = 27.205$ $b = 3.927$ $c = 17.575$ $\beta = 92°9(10)'$ Z = 2

X-ray Powder Pattern: Banská Hodruša, Czechoslovakia.
3.102 (vs), 3.62 (s), 2.715 (s), 3.22 (m/s), 3.48 (m), 1.722 (m), 1.450 (w/m)

Chemistry:

	(1)	(2)	(3)
Cu	12.83	13.88	13.66
Fe		0.44	
Bi	54.49	64.92	67.39
S	18.04	18.98	18.95
Fe_2O_3	15.09		
Total	100.45	98.22	100.00

(1) Banská Hodruša, Czechoslovakia. (2) Do.; by electron microprobe, corresponding to $Cu_{8.12}Fe_{0.29}Bi_{11.4}S_{22}$. (3) $Cu_8Bi_{12}S_{22}$.

Occurrence: In polymetallic ore deposits of subvolcanic type, in a quartz vein system developed in propylitized pyroxenite andesite (Banská Hodruša, Czechoslovakia).

Association: Quartz, hematite, chalcopyrite, wittichenite (?), pavonite (Banská Hodruša, Czechoslovakia).

Distribution: In the Rosalia mine, Banská Hodruša, near Banská Štiavnica (Schemnitz), Czechoslovakia. From Pioche, Lincoln Co., Nevada, USA.

Name: For the mining community of Banská Hodruša, Czechoslovakia.

Type Material: n.d.

References: (1) Koděra, M., V. Kupčík, and E. Makovický (1970) Hodrushite—a new sulfosalt. Mineral. Mag., 37, 641–648. (2) (1971) Amer. Mineral., 56, 633 (abs. ref. 1). (3) Makovicky, E. and W.H. McLean (1972) Electron microprobe analysis of hodrushite. Can. Mineral., 11, 504–513. (4) Kupčik, V. and E. Makovický (1968) Die Kristallstruktur des Minerales $(Pb,Ag,Bi)Cu_4Bi_5S_{11}$. Neues Jahrb. Mineral., Monatsh., 236–237 (in German).

Crystal Data: Cubic. *Point Group:* $2/m\ \overline{3}$. Euhedral to anhedral grains, up to 40 μm.

Physical Properties: Hardness = > 6, greater than that of sperrylite. VHN = ~650–700 (100 g load). D(meas.) = n.d. D(calc.) = [7.86]

Optical Properties: Opaque. *Color:* In polished section, medium gray, slightly bluish in oil. R: (400) — , (420) 50.0, (440) 51.3, (460) 51.9, (480) 51.9, (500) 51.6, (520) 51.3, (540) 50.8, (560) 50.2, (580) 49.7, (600) 49.2, (620) 48.5, (640) 47.8, (660) 47.2, (680) 46.4, (700) 45.6

Cell Data: *Space Group:* *Pa*3. *a* = 5.769(5)–5.797(5) Z = [4]

X-ray Powder Pattern: n.d.

Chemistry:

	(1)	(2)	(3)	(4)
Rh	30.8	25.0	35.2	47.62
Pt	10.3	20.0	10.6	
Pd	8.7		7.1	
Ir	3.1	5.0	0.1	
Ru		4.0		
Fe				0.56
As	32.6	35.0	35.0	37.02
S	13.9	11.0	12.7	14.99
Total	99.4	100.0	100.7	100.19

(1) Driekop mine, South Africa; by electron microprobe, corresponding to $(Rh_{0.68}Pd_{0.18}Pt_{0.12}Ir_{0.04})_{\Sigma=1.02}As_{0.99}S_{0.99}$. (2) Noril'sk, USSR; by electron microprobe, corresponding to $(Rh_{0.60}Ru_{0.10}Pt_{0.25}Ir_{0.06})_{\Sigma=0.98}As_{1.15}S_{0.84}$. (3) Stillwater Complex, Montana, USA; by electron microprobe, corresponding to $(Rh_{0.77}Pd_{0.15}Pt_{0.12}Ir_{0.01})_{\Sigma=1.05}As_{1.05}S_{0.90}$. (4) Shetland Islands; by electron microprobe, giving $(Rh_{0.97}Fe_{0.02})_{\Sigma=0.99}As_{1.03}S_{0.98}$.

Polymorphism & Series: Forms a series with irarsite.

Occurrence: In a dunite pipe deposit (Driekop mine, South Africa) and platinum deposits in layered ultramafic intrusives; in Cu-Ni sulfide ores.

Association: Sperrylite, geversite, erlichmanite, irarsite, Pt–Fe alloys (South Africa); chalcopyrite, cobaltite, gersdorffite, pyrrhotite (Canada); Au–Ag alloy, braggite, sperrylite, cobaltite, pyrrhotite, pentlandite, pyrite, chalcopyrite (Stillwater Complex, Montana, USA).

Distribution: In the Driekop and Onverwacht mines, Transvaal, South Africa. From the Hitura Ni-Cu deposit, Finland. In the Upper and Banded zones of the Stillwater Complex, Montana; from the Yuba River, Nevada Co., California; and at Goodnews Bay, Alaska, USA. In Canada, at Sudbury and Werner Lake, Ontario. From the Noril'sk region, western Siberia, USSR. At the Cliff and Harold's Grave quarries, Shetland Islands.

Name: For Professor Sidney Ewart Hollingworth (1899–1966), eminent British geologist, University College, London, England.

Type Material: n.d.

References: (1) Stumpfl, E.F. and A.M. Clark (1965) Hollingworthite, a new rhodium mineral, identified by electron probe microanalysis. Amer. Mineral., 50, 1068–1074. (2) Cabri, L.J., Ed. (1981) Platinum group elements: mineralogy, geology, recovery. Can. Inst. Min. & Met., 108–109, 154. (3) Volborth, A., M. Tarkian, E.F. Stumpfl, and R.M. Housley (1986) A survey of the Pd–Pt mineralization along the 35-km strike of the J-M reef, Stillwater Complex, Montana. Can. Mineral., 24, 329–346. (4) Tarkian, M. and H.M. Prichard (1987) Irarsite-hollingworthite solid-solution series and other associated Ru-, Os-, Ir-, and Rh-bearing PGM's from the Shetland ophiolite complex. Mineral. Deposita, 22, 178–184.

Crystal Data: Hexagonal. *Point Group:* 32, $3m$ or $\overline{3}\ 2/m$. Rarely platy; massive, irregular in grains typically 0.1 to 0.5 mm.

Physical Properties: *Tenacity:* Brittle, somewhat ductile and malleable. Hardness = 4 VHN = 204.3 ∥ [10$\overline{1}$0], 276.8 ∥ [0001] (50 g load). D(meas.) = n.d. D(calc.) = 15.63

Optical Properties: Opaque. *Color:* Lead-gray, sometimes tarnished; in reflected light, bright white with a slight brownish tint. *Streak:* Black. *Luster:* Metallic. *Anisotropism:* Weak to moderate, dull gray to dull reddish brown.

R_1–R_2: (480) 58.9–56.7, (546) 62.8–60.5, (589) 65.8–63.8

Cell Data: *Space Group:* $R32$, $R3m$, or $R\overline{3}m$. a = 10.713 c = 13.192 Z = n.d.

X-ray Powder Pattern: "Yen" district, China.
2.199 (10), 1.895 (8), 1.148 (8), 1.350 (5), 1.325 (5), 0.856 (5)

Chemistry:

	(1)	(2)
Pt	74.93	75.43
Cu	24.52	24.57
Total	99.45	100.00

(1) "Yen" district, China; by electron microprobe, average of analyses on four grains. (2) PtCu.

Occurrence: In an actinolized diopside-type of platinum deposit ("Yen" district, China).

Association: Cooperite, sperrylite, vysotskite, isomertieite, magnetite, bornite, polydymite, diopside, actinolite, epidote ("Yen" district, China).

Distribution: From the "Yen" district (apparently a code name), in an unspecified region of China. [Previously given as the "Hong", "Huang", or "Hung" district]. From Fox Gulch, Goodnews Bay, Alaska, USA.

Name: For the region in China in which it occurs.

Type Material: Institute of Geology, Chinese Academy of Geological Sciences, Beijing, China.

References: (1) Yu Tsu-Hsiang, Lin Shu-Jen, Chao Pao, Fang Ching-Sung, and Huang Chi-Shun (1974) A preliminary study of some new minerals of the platinum group and another associated new one in platinum-bearing intrusions in a region in China. Acta Geol. Sinica, 2, 202–218 (in Chinese with English abs.). (2) (1976) Amer. Mineral., 61, 185 (abs. ref. 1). (3) Peng Zhiizhong, Chang Chiehung, and Ximen Lovlov (1978) Discussion on published articles in the research of new minerals of the platinum-group discovered in China in recent years. Acta Geol. Sinica, 4, 326–336 (in Chinese with English abs.). [Peng Zhiizhong formerly Pen Chih-Zhong]. (4) (1980) Amer. Mineral., 65, 408 (abs. ref. 3). (5) Ding, K. (1980) Further studies of the minerals "isoplatincopper" and "hongshiite". Scientia Geol. Sinica, 2, 168–171 (in Chinese with English abs.). (6) Yu Zuxiang (1982) New data for hongshiite. Bull. Institute of Geology, Chinese Academy of Geological Sciences, 4, 78–81 (in Chinese with English abs.). (7) (1984) Amer. Mineral., 69, 411–412 (abs. ref. 6). (8) Cabri, L.J., Ed. (1981) Platinum group elements: mineralogy, geology, recovery. Can. Inst. Min. & Met., 109.

Crystal Data: n.d. *Point Group:* n.d. Massive.

Physical Properties: *Fracture:* Uneven. *Tenacity:* Brittle. Hardness = 4–5 VHN = n.d. D(meas.) = 8.81 D(calc.) = n.d.

Optical Properties: Opaque. *Color:* Silver-white, easily tarnished; in polished section, slightly cream-colored to silver-blue and grayish white. *Luster:* Metallic, brilliant.
R_1–R_2: n.d.

Cell Data: *Space Group:* n.d. Z = n.d.

X-ray Powder Pattern: n.d.

Chemistry:

	(1)	(2)
Cu	73.37	72.29
Sb	26.86	27.71
Total	100.23	100.00

(1) Near Mytilene. (2) Cu_5Sb.

Occurrence: n.d.

Association: n.d.

Distribution: Said to occur in some abundance on the Turkish mainland, near the Greek island of Mytilene.

Name: For Eben Norton Horsford (1818–1893), Professor of Chemistry, Harvard University, Cambridge, Massachusetts, USA.

Type Material: n.d.

References: (1) Dana, E.S. (1892) Dana's system of mineralogy, (6th edition), 44.
(2) Palache, C., H. Berman, and C. Frondel (1944) Dana's system of mineralogy, (7th edition), v. I, 173.

Crystal Data: Orthorhombic. *Point Group:* $2/m\ 2/m\ 2/m$. Prismatic to acicular ∥ [001], crystals to 1 cm; also as radiating tufts.

Physical Properties: *Cleavage:* Good on {010}. *Fracture:* Conchoidal. *Tenacity:* Brittle. Hardness = 1.5–2 VHN = 170 D(meas.) = 4.6–4.7 D(calc.) = 4.58

Optical Properties: Opaque; translucent in thin splinters. *Color:* Scarlet-vermilion to deep cherry-red, with strong red internal reflections; transmits red light in thin splinters. *Luster:* Adamantine to submetallic.

Optical Class: Biaxial negative (–). *Orientation:* $X = a$; $Y = b$; $Z = c$. *Dispersion:* $r < v$, extreme. $\alpha = 3.078(18)$ $\beta = 3.176(3)$ $\gamma = 3.188(3)$ 2V(meas.) = 37° (589). 2V(calc.) = n.d. *Anisotropism:* Strong.

R_1–R_2: n.d.

Cell Data: *Space Group:* *Pbca.* $a = 10.80$ $b = 35.35$ $c = 8.16$ Z = 4

X-ray Powder Pattern: Binntal, Switzerland.
2.74 (100), 3.79 (70), 3.05 (60), 4.44 (50), 2.39 (30), 2.22 (30), 1.907 (30)

Chemistry:

	(1)	(2)	(3)
Pb	12.5	19.3	19.28
Tl	25.0	17.3	19.02
Ag	9.0	0.0	
As	30.5	31.1	34.85
Sb		1.9	
S	26.0	29.3	26.85
Total	103.0	98.9	100.00

(1) Binntal, Switzerland. (2) Quiruvilca, Peru; by electron microprobe. (3) $PbTlAs_5S_9$.

Occurrence: Of hydrothermal origin.

Association: Orpiment, realgar, pyrite, sphalerite, other As and Tl sulfide minerals.

Distribution: In Switzerland, in the Lengenbach quarry, Binntal, Valais. At the Segen Gottes mine, near Wiesloch, Black Forest, Germany. Very well crystallized from the La Libertad mine, Quiruvilca, Peru.

Name: For Arthur Hutchinson (1866–1937), Professor of Mineralogy, Cambridge University, England.

Type Material: n.d.

References: (1) Palache, C., H. Berman, and C. Frondel (1944) Dana's system of mineralogy, (7th edition), v. I, 468–469. (2) Berry, L.G. and R.M. Thompson (1962) X-ray powder data for the ore minerals. Geol. Soc. Amer. Mem. 85, 160–161. (3) Takéuchi, Y., S. Ghose, and W. Nowacki (1965) The crystal structure of hutchinsonite, $(Tl, Pb)_2As_5S_9$. Zeits. Krist., 121, 321–348. (4) White, J.S. and J.A. Nelen (1985) Hutchinsonite from Quiruvilca, Peru. Mineral. Record, 16, 459–460.

Crystal Data: Hexagonal. *Point Group:* n.d. Intimately associated with metacinnabar.

Physical Properties: *Fracture:* Subconchoidal to uneven. Hardness = 3 VHN = 51.5(8.0) (25 g load). D(meas.) = 7.43 D(calc.) = 7.54

Optical Properties: Translucent. *Color:* Black with purple cast; in polished section, grayish white. *Streak:* Dark black-purple. *Luster:* Adamantine. *Anisotropism:* Distinct in oil.
R_1–R_2: n.d.

Cell Data: *Space Group:* n.d. $a = 7.01(3)$ $c = 14.13(7)$ Z = 12

X-ray Powder Pattern: Synthetic. (JCPDS 19-798).
3.08 (100), 1.980 (100), 1.892 (100), 2.80 (90), 3.43 (80), 3.70 (70), 2.97 (60)

Chemistry: Composition established by similarity of X-ray pattern with synthetic material.

Polymorphism & Series: Trimorphous with cinnabar and metacinnabar.

Occurrence: Intimately associated with metacinnabar, which it generally resembles.

Association: Metacinnabar.

Distribution: In the USA, at the Mt. Diablo mine, Contra Costa Co., California; at Manhattan, Nye Co., Nevada.

Name: In allusion to the fact that the stability field of the species extends to higher temperatures than that of cinnabar or metacinnabar.

Type Material: n.d.

References: (1) Potter, R.W. and H.L. Barnes (1978) Phase relations in the binary Hg–S. Amer. Mineral., 63, 1143–1152.

Crystal Data: Hexagonal. *Point Group:* n.d. Massive, as rare discrete anhedral grains, as rims on chalcopyrite and as fracture fillings in chalcopyrite; also as exsolution lamellae in bornite.

Physical Properties: Hardness = 2.5–3.5 VHN = n.d. D(meas.) = n.d. D(calc.) = 4.2

Optical Properties: Opaque. *Color:* Coppery red to pinchbeck-brown, bornite-like but not tarnished. *Luster:* Metallic. *Anisotropism:* Strong, reddish orange to reddish brown.
R_1–R_2: (400) 13.9–17.7, (420) 14.8–18.5, (440) 15.7–19.3, (460) 16.9–21.0, (480) 18.4–23.2, (500) 20.5–25.5, (520) 23.2–27.8, (540) 26.1–29.4, (560) 28.9–30.6, (580) 30.6–31.4, (600) 31.9–31.6, (620) 32.8–31.9, (640) 33.5–32.2, (660) 34.1–32.8, (680) 34.7–33.2, (700) 35.2–33.7

Cell Data: *Space Group:* n.d. $a = 3.90$ $c = 16.95$ Z = 1

X-ray Powder Pattern: Ida mine, Khan, Namibia.
3.14 (100), 2.82 (100), 1.89 (100), 1.85 (100), 1.564 (100), 2.70 (80), 1.317 (80)

Chemistry:

	(1)	(2)
Cu	51.34	50.87
Fe	14.5	14.90
S	33.66	34.23
Total	99.50	100.00

(1) Skouriotissa, Cyprus; by electron microprobe, average of five analyses. (2) Cu_3FeS_4.

Occurrence: A lamellar decomposition product of bornite, commonly associated with fine spindles of chalcopyrite; apparently of secondary origin, a first product of enrichment.

Association: Chalcopyrite, bornite, pyrite, sphalerite, chalcocite, pyrrhotite, mackinawite (Skouriotissa, Cyprus).

Distribution: Recognized at numerous localities world-wide. In the USA, at upper White Canyon, San Juan Co., Utah. From Canada, at Algoma, Jarvis Township, Ontario. In Namibia, at the Ida mine, Khan, and at Tsumeb. Near Bou Azzer, Morocco. From Skouriotissa, Cyprus. In Switzerland, near Grimentz. From the Radka deposit, Pazardzhik, Bulgaria. In Brazil, at Caraiba, Bom Fin, Bahia. From several mines in the Copiapó district, northern Chile. In the Kaikita mine, Aomori Prefecture, the Ojamine mine, Yamagata Prefecture, and other localities in Japan.

Name: For the Ida mine in Namibia.

Type Material: n.d.

References: (1) Frenzel, G. (1958) Ein neues Mineral: Idait. Neues Jahrb. Mineral., Monatsh., 142 (in German). (2) (1958) Amer. Mineral., 43, 1219 (abs. ref. 1). (3) Frenzel, G. (1959) Ein neues Mineral: Idait, natürliches Cu_5FeS_6. Neues Jahrb. Mineral., Abh., 93, 87–114 (in German). (4) (1959) Amer. Mineral., 44, 1327 (abs. ref. 3). (5) Yund, R.A. (1963) Crystal data for synthetic $Cu_{5.5x}Fe_xS_{6.5x}$ (idaite). Amer. Mineral., 48, 472–476. (6) Sillitoe, R.H. and A.H. Clark (1969) Copper and copper-iron sulfides as the initial products of supergene oxidation, Copiapó mining district, northern Chile. Amer. Mineral., 54, 1684–1710. (7) Constantinou, G. (1975) Idaite from the Skouriotissa massive sulfide orebody, Cyprus: Its composition and conditions of formation. Amer. Mineral., 60, 1013–1018. (8) Rice, C.M., D. Atkin, J.F.W. Bowles, and A.J. Criddle (1979) Nukundamite, a new mineral, and idaite. Mineral. Mag., 43, 194–200.

Crystal Data: Hexagonal. *Point Group:* $\overline{3}\ 2/m$. Plates to 3 cm, foliated massive; as minute grains.

Physical Properties: *Cleavage:* {0001} perfect. *Tenacity:* Very flexible. Hardness = 2 VHN = n.d. D(meas.) = 7.8 D(calc.) = 7.97

Optical Properties: Opaque. *Color:* Lead-gray; in polished section, white with faintly creamy tint. *Streak:* Dark gray. *Luster:* Metallic. *Anisotropism:* Nearly isotropic in basal section, moderately anisotropic, pale gray to gray, in diagonal section.

R_1–R_2: (400) 46.6–48.8, (420) 46.3–49.0, (440) 46.0–49.2, (460) 45.7–49.0, (480) 45.5–48.7, (500) 45.3–48.6, (520) 45.2–48.6, (540) 45.0–48.8, (560) 45.0–48.9, (580) 44.9–49.0, (600) 44.9–49.0, (620) 44.8–48.9, (640) 44.6–48.7, (660) 44.2–48.5, (680) 43.8–48.1, (700) 43.4–47.6

Cell Data: *Space Group:* $R\overline{3}m$. $a = 4.15$ $c = 39.19$ Z = 3

X-ray Powder Pattern: Ikuno mine, Japan.
3.022 (100), 4.34 (50), 2.205 (30), 6.56 (20), 3.536 (20), 2.076 (20), 1.865 (20)

Chemistry:

	(1)
Bi	79.69
Se	1.98
S	8.89
rem.	[9.44]
Total	100.00

(1) Ikuno mine, Japan; remainder is ferberite and quartz.

Occurrence: In a quartz vein (Ikuno mine, Japan).

Association: Ferberite, bismuth, bismuthinite, joséite, molybdenite, arsenopyrite, cassiterite, chalcopyrite, quartz.

Distribution: At the Ikuno and Akenobe mines, Hyogo Prefecture, and the Ashio mine, Tochigi Prefecture, Japan. At Kingsgate, New South Wales, Australia. In Sweden, from the Falu mine, Falun. In the Vysokogorsk tin deposit, Primor'ye, and in the Kara-Oba tungsten deposit, central Kazakhstan, USSR.

Name: For the Ikuno mine, Japan, in which it was first found.

Type Material: n.d.

References: (1) Kato, A. (1959) Ikunolite, a new bismuth mineral from the Ikuno mine, Japan. Mineral. J. (Japan), 2, 397–407 (in English). (2) (1960) Amer. Mineral., 45, 477–478. (abs. ref. 1).

Crystal Data: Monoclinic. *Point Group:* $2/m$. As thin plates, sometimes aggregated into radial or spherical forms, less than 1 mm in size.

Physical Properties: Hardness = ~2 VHN = 38 D(meas.) = n.d. D(calc.) = 4.39

Optical Properties: Translucent. *Color:* Copper-red; in polished section, pure white, inclining to cream colored compared to galena, with strong bright red internal reflections. *Anisotropism:* Very strong.
R_1–R_2: n.d.

Cell Data: *Space Group:* $P2_1/n$. $a = 8.755(5)$ $b = 24.425(15)$ $c = 5.739(3)$ $\beta = 108.28(5)°$ Z = 2

X-ray Powder Pattern: Binntal, Switzerland.
2.675 (100), 2.88 (90), 3.96 (70), 3.67 (65), 4.10 (40), 3.58 (40), 3.18 (40)

Chemistry:

	(1)	(2)
Tl	33.6	32.52
Cu	1.67	1.68
As	30.2	31.79
S	33.7	34.01
Total	99.17	100.00

(1) Binntal, Switzerland; by electron microprobe. (2) $Tl_6CuAs_{16}S_{40}$.

Occurrence: Of hydrothermal origin.

Association: Realgar, lead sulfantimonides.

Distribution: From the Lengenbach quarry, Binntal, Valais, Switzerland.

Name: For Josef Imhof (1902–1969), professional mineral collector of Binn, Switzerland.

Type Material: n.d.

References: (1) Burri, G., F. Graeser, F. Marumo, and W. Nowacki (1965) Imhofit, ein neues Thallium–Arsenosulfosalz aus dem Lengenbach (Binnatal, Kanton Wallis): Chimia (Switzerland), 19, 499–500. (2) (1966) Amer. Mineral., 51, 531–532 (abs. ref. 1). (3) Nowacki, W. (1967) Über neue Mineralien aus dem Lengenbach. Jahrb. Naturhist. Mus. Bern, 1963–1965, 293–299 (in German). (4) (1969) Amer. Mineral., 54, 1498 (abs. ref. 3). (5) Divjaković, V. and W. Nowacki (1976) Die Kristallstruktur von Imhofit, $Tl_{5.6}As_{15}S_{25.3}$. Zeits. Krist., 144, 323–333 (in German with English abs.).

Crystal Data: Monoclinic. *Point Group:* $2/m$. As crystals to 3 mm, and anhedral grains up to 1 mm.

Physical Properties: Hardness = n.d. VHN = 86 (100 g load). D(meas.) = n.d. D(calc.) = 7.846(4)

Optical Properties: Opaque. *Color:* White-gray. *Pleochroism:* Bluish to pinkish. *Anisotropism:* Strong, from blue to reddish brown.
R_1–R_2: (400) — , (420) 29.8–33.8, (440) 33.4–38.3, (460) 32.1–36.4, (480) 31.3–34.8, (500) 30.6–33.7, (520) 29.9–32.8, (540) 29.5–32.1, (560) 29.2–31.5, (580) 28.9–31.1, (600) 28.8–30.8, (620) 28.6–30.5, (640) 28.5–30.3, (660) 28.2–29.9, (680) 27.9–29.7, (700) 27.5–29.4

Cell Data: *Space Group:* $P2_1/c$. a = 4.0394(8) b = 8.0050(6) c = 6.5812(8) β = 107.12(2)° Z = [2]

X-ray Powder Pattern: Imiter mine, Morocco.
2.768 (100), 2.746 (100), 2.461 (80), 3.466 (50), 1.467 (35), 4.88 (30), 3.138 (30)

Chemistry:

	(1)	(2)
Ag	42.89	44.90
Hg	42.64	41.75
S	13.34	13.35
Total	98.87	100.00

(1) Imiter mine, Morocco; by electron microprobe, average of 20 analyses, corresponding to $Ag_{1.91}Hg_{1.02}S_{2.00}$. (2) Ag_2HgS_2.

Occurrence: In a pyritic deposit.

Association: Chalcopyrite, sphalerite, acanthite, polybasite, galena, arsenopyrite.

Distribution: In the Imiter mine, Anti-Atlas Mountains, Morocco.

Name: For the Imiter mine locality in Morocco.

Type Material: B.R.G.M., Orléans; National School of Mines, Paris, France.

References: (1) Guillou, J.-J., J. Monthel, P. Picot, F. Pillard, J. Protas, and J.-C. Samama (1985) L'imitérite, Ag_2HgS_2, nouvelle espèce minérale; propriététes structure cristalline. Bull. Minéral., 108, 457–464 (in French with English abs.). (2) (1986) Amer. Mineral., 71, 1277–1278 (abs. ref. 1).

Crystal Data: Hexagonal. *Point Group:* 6, $6/m\ 2/m\ 2/m$, 622, or $\bar{6}m2$. Irregular grains up to 150 μm, as inclusions in isoferroplatinum, intergrown with erlichmanite and malanite, and rimming iridosmine and laurite.

Physical Properties: *Cleavage:* In two directions. *Tenacity:* Brittle. Hardness = n.d. VHN = 347–726, 575 average. D(meas.) = n.d. D(calc.) = n.d.

Optical Properties: Opaque. *Color:* Steel-gray. *Luster:* Metallic. *Anisotropism:* Weak. R_1–R_2: n.d.

Cell Data: *Space Group:* $P6_3$, $P6$, $P6/mmm$, $P622$, or $P\bar{6}2m$. $a = 7.03(1)$ $c = 16.44(1)$ Z = n.d.

X-ray Powder Pattern: Inagli massif, USSR.
2.98 (10), 5.7 (9), 2.438 (8), 1.753 (7)

Chemistry:

	(1)
Ir	38.7 – 50.5
Pt	11.5 – 21.1
Rh	0.0 – 2.54
Pb	8.39 – 9.15
Cu	6.69 – 7.36
Ni	0.0 – 0.08
S	19.7 – 21.4
Total	

(1) Inagli massif, USSR; by electron microprobe, ranges of seven analyses.

Occurrence: With other platinum group minerals in ultramafic intrusives.

Association: Isoferroplatinum, erlichmanite, malanite, iridosmine, laurite.

Distribution: From an undefined locality in the Inagli massif, Yakutia; and the Nizhni Tagil massif, Ural Mountains, USSR.

Name: For the occurrence in the Inagli massif, USSR.

Type Material: Mining Museum, Leningrad Mining Institute, Leningrad, USSR.

References: (1) Rudashevskii, N.S., A.G. Mochalov, V.D. Begizov, Y.P. Men'shikov, and N.I. Shumskaya (1984) Inaglyite, $PbCu_3(Ir,Pt)_8S_{16}$, a new mineral. Zap. Vses. Mineral. Obshch., 113, 712–717 (in Russian). (2) (1986) Amer. Mineral., 71, 228 (abs. ref. 1).

Crystal Data: Monoclinic, perhaps triclinic. *Point Group:* 2, m, or $2/m$, perhaps 1 or $\bar{1}$. As very fine lamellae.

Physical Properties: *Cleavage:* Excellent on {100}. Hardness = Very soft. VHN = n.d. D(meas.) = n.d. D(calc.) = n.d.

Optical Properties: Opaque. *Color:* In polished section, grayish white. *Anisotropism:* Distinct in greenish brown and brownish gray.
R_1–R_2: (480) 29.8–33.7, (540) 29.0–33.2, (600) 28.6–32.9, (640) 28.1–32.5

Cell Data: *Space Group:* Two subcells are recognized, both $A2$, Am, or $A2/m$; the first (pseudotetragonal) has: $a = 17.29(2)$ $b = 5.79(1)$ $c = 5.83(2)$ $\beta = 94.14(0.30)°$ and the second (pseudohexagonal) has: $a = 17.25(2)$ $b = 3.66(1)$ $c = 6.35(2)$ $\beta = 91.13(0.20)°$ Z = n.d.

X-ray Powder Pattern: Poopó, Bolivia.
2.862 (100), 2.029 (80), 3.43 (70), 4.31 (60), 3.83 (40), 3.23 (40), 3.12 (40)

Chemistry:

	(1)
Pb	36.53
Ag	2.15
Sn	23.09
Fe	2.35
Sb	12.70
S	23.20
Total	100.02

(1) Poopó, Bolivia; by electron microprobe, average of five analyses, corresponding to $(Pb_{3.67}Ag_{0.37})_{\Sigma=4.04}Sn_{4.07}Fe_{0.88}Sb_{2.23}S_{15.00}$.

Occurrence: In hydrothermal Ag-Sn ores (Poopó, Bolivia).

Association: Stannite, miargyrite, cylindrite (Poopó, Bolivia).

Distribution: At Poopó, Bolivia. From near Mejillones, north of Antofagasta, Chile.

Name: For the Incas, first recorded miners of the Ag-Sn ores in Bolivia.

Type Material: Redpath Museum, McGill University, Montreal, Canada, 898.

References: (1) Makovicky, E. (1974) Mineralogical data on cylindrite and incaite. Neues Jahrb. Mineral., Monatsh., 235–256. (2) (1975) Amer. Mineral., 60, 486–487 (abs. ref. 1). (3) Makovicky, E. (1976) Crystallography of cylindrite. Part I. Crystal lattices of cylindrite and incaite. Neues Jahrb. Mineral., Abh., 126, 304–326.

$Fe^{+2}In_2S_4$ **Indite**

Crystal Data: Cubic. *Point Group:* $4/m\ \overline{3}\ 2/m$. Massive, as grains typically less than 0.2 mm, rarely to 0.5 mm.

Physical Properties: Hardness = n.d. VHN = 309 D(meas.) = n.d. D(calc.) = 4.588

Optical Properties: Opaque. *Color:* Iron-black; in polished section, white. *Luster:* Metallic.
R: (400) 26.3, (420) 26.9, (440) 27.5, (460) 27.9, (480) 28.0, (500) 27.8, (520) 27.4, (540) 27.0, (560) 26.6, (580) 26.2, (600) 25.9, (620) 25.6, (640) 25.4, (660) 25.1, (680) 24.8, (700) 24.5

Cell Data: *Space Group:* $Fd3m$. $a = 10.618(3)$ Z = [8]

X-ray Powder Pattern: Dzhalindin deposit, USSR.
3.20 (100), 1.877 (90), 1.085 (80), 2.05 (70), 1.028 (70), 3.76 (50), 1.384 (50)

Chemistry:

	(1)	(2)
Fe	8.84	13.50
In	59.3	55.50
S	31.85	31.00
Total	99.99	100.00

(1) Dzhalindin deposit, USSR; by microspectrographic analysis, corresponding to $Fe_{0.6}In_2S_4$.
(2) $FeIn_2S_4$.

Occurrence: Of primary hydrothermal origin, replacing botryoidal cassiterite.

Association: Cassiterite, dzhalindite.

Distribution: In the Dzhalindin deposit, Little Khingan Ridge, Far Eastern Region, USSR.

Name: For the indium in its composition.

Type Material: Leningrad Mining Institute, Leningrad; A.E. Fersman Mineralogical Museum, Academy of Sciences, Moscow, USSR.

References: (1) Genkin, A.D. and I.V. Murav'eva (1963) Indite and dzhalindite, new indium minerals. Zap. Vses. Mineral. Obshch., 92, 445–457 (in Russian). (2) (1964) Amer. Mineral., 49, 439 (abs. ref. 1). (3) Hill, R.J., J.R. Craig, and G.V. Gibbs (1978) Cation ordering in the tetrahedral sites of the thiospinel $FeIn_2S_4$ (indite). J. Phys. Chem. Solids, 39, 1105–1111.

Crystal Data: Tetragonal. *Point Group:* $4/m\ 2/m\ 2/m$. Massive, as grains up to 1 mm.

Physical Properties: Hardness = n.d. VHN = 130–159, 142 average. D(meas.) = n.d. D(calc.) = 7.29

Optical Properties: Opaque. *Color:* Gray with yellowish tint; in polished section, light rose-white. *Luster:* Metallic. *Anisotropism:* Weak.
R_1–R_2: n.d.

Cell Data: *Space Group:* $I4/mmm$. $a = 3.2517$ $c = 4.9459$ Z = 2

X-ray Powder Pattern: Synthetic.
2.715 (100), 2.298 (36), 1.683 (24), 1.395 (23), 2.471 (21), 1.470 (16), 1.0904 (12)

Chemistry: Composition established by microspectrographic analysis.

Occurrence: In greisenized and albitized granite.

Association: Silver.

Distribution: From undefined localities in the eastern Transbaikal and the Ukrainian Shield, USSR.

Name: From its indigo blue emission spectrum.

Type Material: n.d.

References: (1) Ivanov, V.V. (1964) Native indium. In: Geochemistry, mineralogy, and genetic types of deposits of rare elements, 2, 568–569. Izdatelstvo "Nauka" Moscow (1964) (in Russian). (2) (1967) Amer. Mineral., 52, 299 (abs. ref. 1). (3) Vlasov, K.A., Ed. (1966) Mineralogy of rare elements, v. II, 584–585. (4) (1954) NBS Circ. 539, 3, 12.

Crystal Data: Hexagonal. *Point Group:* $\overline{3}\ 2/m$, $3m$, or 32. Massive, presumably.

Physical Properties: *Cleavage:* Perfect. *Tenacity:* Brittle. Hardness = Soft. VHN = 60.9–64.6 D(meas.) = n.d. D(calc.) = 7.88

Optical Properties: Opaque. *Color:* Silver-white. *Luster:* Metallic. *Pleochroism:* Noticeable in air, from creamy to grayish white. *Anisotropism:* Moderate.
R_1–R_2: (460) 55.5–51.4, (540) 57.7–52.5, (580) 57.8–52.4, (660) 56.9–51.6

Cell Data: *Space Group:* $P\overline{3}m$, $P3m1$ or $P321$. $a = 4.248$ $c = 23.22$ Z = 9

X-ray Powder Pattern: Brandy Gill, England; or Ingoda deposit, USSR.
3.11 (100), 2.28 (60), 2.13 (50), 1.942 (40)

Chemistry:

	(1)	(2)	(3)
Bi	68.8 – 73.3	51.7 – 73.3	72.36
Pb	0.0 – 1.2	0.6 – 10.6	
Sb	0.0 – 0.15		
Te	19.3 – 25.3	19.3 – 32.1	22.09
Se	0.0 – 0.4	0.0 – 0.4	
S	5.1 – 6.3	5.2 – 6.3	5.55
Total			100.00

(1) Brandy Gill, England and Ingoda, USSR; by electron microprobe, ranges of sample values of five analyses. (2) Range of 12 analyses, by electron microprobe, on specimens from several localities. (3) Bi_2TeS.

Occurrence: n.d.

Association: Bismuthinite, joséite.

Distribution: From the Ingoda deposit, near the source of the Ingoda River, central Transbaikal; and otherwise unspecified localities in Kamchatka and the Southern Ural Mountains, USSR. At Brandy Gill, Cumbria, England. From the Bluebird mine, Little Dragoon Mountains, Cochise Co., Arizona, USA.

Name: For the Ingoda deposit, USSR.

Type Material: n.d.

References: (1) Zav'yalov, E.N. and V.D. Begizov (1981) The new bismuth mineral ingodite, Bi_2TeS. Zap. Vses. Mineral. Obshch., 110, 594–600 (in Russian). (2) Zav'yalov, E.N. and V.D. Begizov (1981) Once again on the problem of grünlingite. Zap. Vses. Mineral. Obshch., 110, 633–635. (3) (1982) Amer. Mineral., 67, 855 (abs. refs. 1 and 2). (4) Zav'yalov, E.N., V.D. Begizov, and V.Y. Tedchuk (1984) Additional data on the chemical composition of ingodite. Zap. Vses. Mineral. Obshch., 113, 31–35 (in Russian). (5) (1985) Amer. Mineral., 70, 220 (abs. ref. 4).

Crystal Data: Cubic. *Point Group:* $2/m\ \overline{3}$. As small rounded grains up to about 120 μm.

Physical Properties: Hardness = n.d. VHN = 488–540, 519 average (25 g load). D(meas.) = n.d. D(calc.) = 12.8

Optical Properties: Opaque. *Color:* In polished section, white. *Luster:* Metallic.
R: (400) — , (420) 49.3, (440) 50.5, (460) 52.5, (480) 52.8, (500) 53.6, (520) 53.8, (540) 54.0, (560) 54.4, (580) 54.4, (600) 54.6, (620) 55.0, (640) 55.2, (660) 55.7, (680) 56.0, (700) 56.5

Cell Data: *Space Group:* *Pa*3. *a* = 6.691 Z = 4

X-ray Powder Pattern: Synthetic.
2.996 (100), 2.017 (90), 2.732 (80), 1.788 (60), 3.343 (50), 1.459 (50), 0.772 (50)

Chemistry:

	(1)	(2)
Pt	36.1	31.4
Pd		1.7
Ni		0.1
Sn		0.8
Bi	52.0	56.7
Sb	12.9	2.1
Te		7.7
Total	101.0	100.5

(1) Insizwa deposit, South Africa; by electron microprobe, average of five analyses. (2) Sudbury, Canada; by electron microprobe.

Occurrence: Of hydrothermal origin, in a vein cutting massive pyrrhotite ore (Insizwa deposit, South Africa) and in Cu-Ni ore (Sudbury, Canada).

Association: Pentlandite, cubanite, hessite, altaite, argentopentlandite, chalcopyrite, mackinawite, niggliite, froodite, parkerite, galena, sphalerite, magnetite, pyrrhotite.

Distribution: From the Insizwa deposit, Waterfall Gorge, Pondoland and East Griqualand areas, Transkei, South Africa. In the Coleman mine, Sudbury, Ontario, Canada. At Fox Gulch, Goodnews Bay, Alaska, USA. From the Noril'sk region, western Siberia, USSR.

Name: For the Insizwa mineral deposit, South Africa, where it was discovered.

Type Material: National Museum of Canada, Ottawa, Canada.

References: (1) Cabri, L.J. and D.C. Harris (1972) The new mineral insizwaite ($PtBi_2$) and new data on niggliite (PtSn). Mineral. Mag., 38, 794–800. (2) (1973) Amer. Mineral., 58, 805 (abs. ref. 1). (3) Cabri, L.J., Ed. (1981) Platinum group elements: mineralogy, geology, recovery. Can. Inst. Min. & Met., 109–110.

Crystal Data: Cubic. *Point Group:* $2/m\ \overline{3}$. Grains up to 1 mm, and intergrown with hollingworthite or laurite.

Physical Properties: *Tenacity:* Brittle. Hardness = Not scratched with a steel needle. VHN = 976 D(meas.) = n.d. D(calc.) = 11.92

Optical Properties: Opaque. *Color:* Iron-black; in polished section, grayish white with a bluish tint. *Luster:* Metallic.
R: (400) — , (420) 44.0, (440) 43.7, (460) 43.3, (480) 42.6, (500) 42.1, (520) 41.2, (540) 40.6, (560) 39.7, (580) 39.1, (600) 38.6, (620) 38.1, (640) 37.8, (660) 37.2, (680) 36.8, (700) 36.6

Cell Data: *Space Group:* *Pa*3. $a = 5.777$ Z = 1

X-ray Powder Pattern: Onverwacht deposit, Transvaal, South Africa.
3.32 (100), 2.87 (100), 1.74 (100), 2.04 (90), 1.112 (90), 2.57 (80), 1.021 (80)

Chemistry:

	(1)	(2)	(3)
Ir	23.0	63.24	62.35
Ru	9.4	1.05	
Rh	7.2		
Pt	12.6		
Os		0.85	
Fe		0.45	
As	34.5	21.67	26.60
S	11.6	11.67	11.47
Total	98.3	98.93	100.42

(1) Driekop mine, Transvaal, South Africa; by electron microprobe, corresponding to $(Ir_{1.45}Ru_{1.13}Rh_{0.84}Pt_{0.78})_{\Sigma=4.20}As_{5.60}S_{4.40}$. (2) Pletene deposit, Bulgaria; by electron microprobe. (3) Shetland Islands; by electron microprobe, giving $Ir_{0.94}As_{1.03}S_{1.03}$.

Polymorphism & Series: Forms a series with hollingworthite.

Occurrence: In chromite in hortonolite dunite (South Africa); as inclusions in chrompicotites (Bulgaria); and as rims around Pt-group alloys.

Association: Platinum, ruthenian hollingworthite, sperrylite, laurite, iridarsenite, ruthenarsenite, rutheniridosmine, chalcopyrite, chalcocite, pyrrhotite, cobaltite, gersdorffite, pentlandite, nickeline, magnetite, chromite, olivine.

Distribution: In the Onverwacht and Driekop mines, Transvaal, South Africa. From the Hitura Ni-Cu deposit, western Finland. In the USA, from the American River, Sacramento Co., and from the Stanislaus River near Knight's Ferry, Stanislaus Co., California; from Fox Gulch, Goodnews Bay, Alaska. In Canada, at Werner Lake and Sudbury, Ontario, and the Pipe mine, in Manitoba. From Vourinos, Greece. In the Pletene chromite deposit, Bulgaria. At the Cliff and Harold's Grave quarries, Shetland Islands.

Name: For the IRidium and ARSenic in the composition.

Type Material: n.d.

References: (1) Genkin, A.D., N.N. Zhuravlev, N.V. Troneva, and I.V. Murav'eva (1966) Irarsite, a new sulfarsenide of iridium, rhodium, ruthenium, and platinum. Zap. Vses. Mineral. Obshch., 95, 700–712 (in Russian). (2) (1967) Amer. Mineral., 52, 1580 (abs. ref. 1). (3) Cabri, L.J., Ed. (1981) Platinum group elements: mineralogy, geology, recovery. Can. Inst. Min. & Met., 110–111. (4) Naidenova, E., M. Zhelyazkova-Panaiotova, G.P. Kudryavtseva, and I.P. Laputina (1984) First discovery of laurite and irarsite in Bulgaria. Dokl. Bolg. Akad. Nauk, 37, 183–187 (in Russian). (5) (1984) Chem. Abs., 101, 41185 (abs. ref. 4). (6) Tarkian, M. and H.M. Prichard (1987) Irarsite-hollingworthite solid-solution series and other associated Ru-, Os-, Ir-, and Rh-bearing PGM's from the Shetland ophiolite complex. Mineral. Deposita, 22, 178–184.

Crystal Data: Monoclinic. *Point Group:* $2/m$. As irregular inclusions (up to 60 μm) in a matrix of rutheniridosmine.

Physical Properties: Hardness = n.d. VHN = 488 and 686 (100 g load) on two grains. D(meas.) = n.d. D(calc.) = 10.9

Optical Properties: Opaque. *Color:* In polished section, medium gray with brownish tint. *Luster:* Metallic. *Pleochroism:* Weak to none. *Anisotropism:* Weak but distinct, from medium gray to pale orange-brown.
R_1–R_2: (470) 47.2–46.9, (546) 45.4–46.1, (589) 44.9–46.6, (650) 41.4–44.0

Cell Data: *Space Group:* $P2_1/c$ (synthetic). $a = 6.05$ $b = 6.06$ $c = 6.18$ $\beta = 113°17'$ Z = 4

X-ray Powder Pattern: Papua New Guinea.
3.90 (100), 2.840 (70), 2.069 (60), 2.610 (50), 1.910 (50), 1.943 (40), 1.875 (40)

Chemistry:

	(1)
Ir	52.2
Ru	1.7
Os	0.4
Pt	1.1
Rh	0.2
Pd	0.1
As	44.0
S	0.2
Total	99.9

(1) Papua New Guinea; by electron microprobe, average of four grains, leading to $(Ir_{0.92}Ru_{0.06}Os_{0.01}Pt_{0.02}Rh_{0.01})_{\Sigma=1.02}(As_{1.97}S_{0.03})_{\Sigma=2.00}$.

Occurrence: In nuggets or fragments of natural Os–Ir–Ru alloys.

Association: Irarsite, ruthenarsenite, rutheniridosmine.

Distribution: From Papua New Guinea. In the Witwatersrand, South Africa. From near Zlatoust, Ural Mountains, USSR. In placers in Tibet, China. From Fox Gulch, Goodnews Bay, Alaska, USA.

Name: For the IRIDium and ARSENic in the composition.

Type Material: National Museum of Canada, Ottawa, Canada.

References: (1) Harris, D.C. (1974) Ruthenarsenite and iridarsenite, two new minerals from the Territory of Papua and New Guinea and associated irarsite, laurite, and cubic iron-bearing platinum. Can. Mineral., 12, 280–284. (2) (1976) Amer. Mineral., 61, 177 (abs. ref. 1). (3) Quensel, J.C. and R.D. Heyding (1962) Transition metal arsenides V. A note on the rhodium/arsenic system and the monoclinic diarsenides of the cobalt family. Canadian Jour. Chem., 40, 814–818.

Crystal Data: Cubic. *Point Group:* $4/m\ \bar{3}\ 2/m$. Generally cubic, commonly in rounded or angular grains. *Twinning:* On {111} in polysynthetic groups.

Physical Properties: *Cleavage:* Indistinct on {001}. *Fracture:* Hackly. *Tenacity:* Somewhat malleable. Hardness = 6–7 VHN = n.d. D(meas.) = 22.65–22.84 D(calc.) = 22.66

Optical Properties: Opaque. *Color:* Silver-white with yellow tinge, gray on fracture. *Luster:* Very high metallic.

R: (400) 66.7, (420) 67.6, (440) 68.5, (460) 69.2, (480) 69.9, (500) 70.4, (520) 70.8, (540) 71.2, (560) 71.6, (580) 72.0, (600) 72.3, (620) 72.7, (640) 73.2, (660) 73.7, (680) 74.2, (700) 74.8

Cell Data: *Space Group:* $Fm3m$. $a = 3.8394$ Z = 4

X-ray Powder Pattern: Synthetic.
2.2170 (100), 1.9197 (50), 1.1574 (45), 0.7838 (45), 1.3575 (40), 0.8808 (40), 0.8586 (40)

Chemistry:

	(1)	(2)
Ir	76.85	49.6
Pt	19.64	21.8
Pd	0.89	
Ru		23.9
Os		2.1
Cu	1.78	0.5
Fe		0.2
Total	99.16	98.1

(1) Nizhni Tagil, USSR. (2) Sorashigawa placer, Japan; by electron microprobe.

Occurrence: As exsolution particles in Pt–Fe alloys.

Association: Platinum, Pt–Fe alloys.

Distribution: From Nizhni Tagil, Ural Mountains, USSR. In Japan, at the Sorashigawa placer. In Canada, from Bear Creek, in the Tulameen River district, British Columbia. From Goodnews Bay, Alaska, USA. In the Witwatersrand, South Africa.

Name: From the Latin *iris, rainbow*, as the element's compounds are often highly colored.

Type Material: n.d.

References: (1) Palache, C., H. Berman, and C. Frondel (1944) Dana's system of mineralogy, (7th edition), v. I, 110–111. (2) (1955) NBS Circ. 539, 4, 9–10. (3) Harris, D.C. and L.J. Cabri (1973) The nomenclature of the natural alloys of osmium, iridium and ruthenium based on new compositional data of alloys from world-wide occurrences. Can. Mineral., 12, 104–112. (4) (1975) Amer. Mineral., 60, 946 (abs. ref. 2). (5) Cabri, L.J., Ed. (1981) Platinum group elements: mineralogy, geology, recovery. Can. Inst. Min. & Met., 111–112.

Crystal Data: Hexagonal. *Point Group:* $6/m$ $2/m$ $2/m$. Rounded to sub-rounded grains, rarely hexagonal in cross-section; as laths in iron-bearing platinum.

Physical Properties: *Cleavage:* Perfect but difficult on {0001}. *Tenacity:* Slightly malleable to brittle. Hardness = 6–7 VHN = 974–1064 (100 g load). D(meas.) = n.d. D(calc.) = 22.15

Optical Properties: Opaque. *Color:* White, with a bluish gray tinge in reflected light. *Streak:* Gray. *Luster:* Metallic. *Anisotropism:* Weak to moderate.

R_1–R_2: (400) 62.6–63.6, (420) 63.2–64.2, (440) 63.8–64.8, (460) 64.3–65.3, (480) 64.6–65.6, (500) 64.8–66.0, (520) 64.6–66.2, (540) 64.3–66.2, (560) 64.0–66.0, (580) 63.5–65.7, (600) 63.1–65.4, (620) 62.8–65.4, (640) 62.6–65.4, (660) 62.4–65.5, (680) 62.2–65.6, (700) 62.0–65.8

Cell Data: *Space Group:* $P6_3/mmc$. a = 2.724(3) c = 4.333(7) Z = [2]

X-ray Powder Pattern: Spruce Creek, Canada.
2.168 (100), 2.071 (80), 1.231 (80), 0.8478 (70), 0.8136 (70), 0.9135 (60), 0.8736 (60)

Chemistry:

	(1)	(2)
Os	55.5	55.7
Ir	42.0	43.5
Ru	1.3	1.5
Pt	0.3	
Pd		
Rh	0.2	
Fe	0.3	0.2
Cu		0.4
Total	99.6	101.3

(1) Ruby Creek, Canada; by electron microprobe. (2) Sorashigawa placer, Japan; by electron microprobe.

Occurrence: In magmatic segregation deposits, and concentrated into placers on their reduction by weathering.

Association: Osmiridium, irarsite, laurite, isoferroplatinum, platinum, rutheniridosmine.

Distribution: In Canada, in British Columbia, from Ruby Creek, Atlin district; Spruce Creek; and Granite Creek, in the Tulameen River area. From the Sorashigawa placer and other localities in Japan. From the Heazlewood Complex and the Adamsfield district, Tasmania, Australia. In the Guma Water placers, Sierra Leone. From near Vershilo, Bourgas district, Bulgaria. From Vourinos, Greece. At Anduo, Tibet, China. In the Inagli and Konder massifs, Aldan Shield, Yakutia; and a number of other less well defined localities in the USSR.

Name: For the composition, intermediate between iridium and osmium.

Type Material: n.d.

References: (1) Palache, C., H. Berman, and C. Frondel (1944) Dana's system of mineralogy, (7th edition), v. I, 111–113. (2) Harris, D.C. and L.J. Cabri (1973) The nomenclature of the natural alloys of osmium, iridium and ruthenium based on new compositional data of alloys from world-wide occurrences. Can. Mineral., 12, 104–112. (3) (1975) Amer. Mineral., 60, 946 (abs. ref. 2). (4) Cabri, L.J., Ed. (1981) Platinum group elements: mineralogy, geology, recovery. Can. Inst. Min. & Met., 112–113.

Crystal Data: Cubic. *Point Group:* $4/m\ \overline{3}\ 2/m$. In small blebs but also in masses up to 20 tons; crystals rare. *Twinning:* On {111}; also on {112} when in lamellar masses.

Physical Properties: *Cleavage:* {001}; parting on {112}. *Fracture:* Hackly. *Tenacity:* Malleable. Hardness = 4 VHN = 160 (100 g load). D(meas.) = 7.3–7.87 D(calc.) = 7.874 Magnetic.

Optical Properties: Opaque. *Color:* Steel-gray to iron-black; in polished section, white. *Luster:* Metallic.

R: (400) 51.3, (420) 52.2, (440) 53.1, (460) 54.0, (480) 54.8, (500) 55.6, (520) 56.2, (540) 56.8, (560) 57.3, (580) 57.8, (600) 58.4, (620) 59.0, (640) 59.6, (660) 60.2, (680) 61.0, (700) 61.8

Cell Data: *Space Group:* $Im3m$. $a = 2.8664$ Z = 2

X-ray Powder Pattern: Synthetic.
2.0268 (100), 1.1702 (30), 1.4332 (20), 0.9064 (12), 1.0134 (10), 0.8275 (6)

Chemistry:

	(1)	(2)
Fe	93.16	99.16
Ni	2.01	
Co	0.80	
Cu	0.12	
C	2.34	0.065
P	0.32	0.207
S	0.41	
Cl	0.02	
SiO_2		0.37
Total	99.18	99.802

(1) Blaafjeld, Greenland. (2) Cameron, Missouri, USA.

Occurrence: In igneous rocks, especially basalts; in carbonaceous sediments; and in petrified wood, mixed with limonite and organic matter.

Association: Pyrite, magnetite, troilite.

Distribution: In Greenland, at Fortune Bay, Mellemfjord, Asuk, and elsewhere on the west coast; on Disko Island, near Uivfaq and Kitdlît. From Ben Bhreck, Scotland. At Bühl, near Weimar, Hesse, Germany. In Poland, near Rouno, Wolyn district. In the USSR, at Grushersk, in the Don district, and from the Hatanga region, Siberia. In the USA, at Cameron, Clinton Co., Missouri; and near New Brunswick, Somerset Co., New Jersey. In Ontario, Canada, in Cameron Township, Nipissing district; and on St. Joseph Island, Lake Huron. Noted in small amounts at a number of additional localities.

Name: An Old English word for the metal; the chemical symbol from the Latin *ferrum*.

References: (1) Palache, C., H. Berman, and C. Frondel (1944) Dana's system of mineralogy, (7th edition), v. I, 114–116. (2) (1955) NBS Circ. 539, 4, 3. (3) Ulff-Møller, F. (1985) Solidification history of the Kitdlît lens: immiscible metal and sulphide liquids from a basaltic dyke on Disko, central west Greenland. J. Petrol., 26, 64–91.

Crystal Data: Cubic. *Point Group:* $4/m\ \overline{3}\ 2/m$. Euhedral grains and cubo-octahedra from a few μms to 400 μm, intergrown with chalcopyrite; rimming cubanite.

Physical Properties: Hardness = n.d. VHN = 175 (100 g load). D(meas.) = n.d. D(calc.) = n.d.

Optical Properties: Opaque. *Color:* Bronze; in reflected light, pinkish brown. *Luster:* Metallic.
R: (400) — , (420) 22.7, (440) 24.6, (460) 26.9, (480) 29.2, (500) 31.3, (520) 33.2, (540) 34.8, (560) 36.2, (580) 37.4, (600) 38.2, (620) 39.1, (640) 39.7, (660) 40.3, (680) 40.9, (700) 41.3

Cell Data: *Space Group:* $Fm3m$. $a = 5.303(3)$ Z = n.d.

X-ray Powder Pattern: East Pacific Rise (21°N).
3.059 (100), 1.876 (70), 1.602 (50), 2.647 (20), 1.327 (20)

Chemistry:

	(1)	(2)
Cu	21.23	23.42
Fe	41.64	41.15
Zn	0.96	
S	35.57	35.43
Total	99.40	100.00

(1) East Pacific Rise (21°N); by electron microprobe. (2) $CuFe_2S_3$.

Polymorphism & Series: Dimorphous with cubanite, from which it can be formed by heating to between 200°C and 270°C.

Occurrence: In mixtures of sulfides formed around modern undersea "black smoker" chimneys (East Pacific Rise and Red Sea); in hydrothermal copper sulfide deposits (Talnakh area, USSR).

Association: Chalcopyrite, pyrrhotite, pyrite, sphalerite, wurtzite, anhydrite.

Distribution: From along the East Pacific Rise (21°N) and in the Atlantis II Deep of the Red Sea. From the Talnakh area, Noril'sk region, western Siberia, USSR.

Name: For the cubic structure and relation to cubanite.

Type Material: National School of Mines, Paris, France.

References: (1) Caye, R., B. Cervelle, F. Cesbron, E. Oudin, P. Picot, and F. Pillard (1988) Isocubanite, a new definition of the cubic polymorph of cubanite $CuFe_2S_3$. Mineral. Mag., 52, 509–514.

Crystal Data: Cubic. *Point Group:* n.d. Crystals up to 3 mm, also skeletal and cubic grains, sometimes exhibiting stepped facets, and as worm-like intergrowths and veinlets.

Physical Properties: Hardness = n.d. VHN = 503–572 (100 g load). D(meas.) = n.d. D(calc.) = 18.41

Optical Properties: Opaque. *Color:* Bright white in reflected light. *Luster:* Metallic. R: n.d.

Cell Data: *Space Group:* n.d. $a = 3.866$ Z = 1

X-ray Powder Pattern: Gusevogorskii pluton, USSR.
1.164 (10), 2.22 (7), 1.364 (6), 1.926 (4), 1.114 (4)

Chemistry:

	(1)	(2)	(3)	(4)
Pt	80.0	87.8	88.1	91.29
Pd			0.8	
Ir	2.0	0.18	1.0	
Os	< 0.1	0.19	0.0	
Ru	0.6	1.1		
Rh	3.8	0.77		
Fe	8.5	8.7	7.5	8.71
Cu	0.6	0.28	1.0	
Ni	0.3	0.06		
Total	95.8	99.08	98.4	100.00

(1) Anglo America Corporation gold mines, Witwatersrand, South Africa; by electron microprobe. (2) Western Holding Mines, Witwatersrand, South Africa; by electron microprobe, analysis of heavy concentrates. (3) Gusevogorskii pluton, USSR; by electron microprobe, average of ten analyses. (4) Pt_3Fe.

Occurrence: In Pt-Fe and Cu-Ni deposits in ultramafic rocks, and in placers derived from them.

Association: Platinum, gold, pyrite, iridium, osmium.

Distribution: From the Tulameen River, British Columbia, Canada. In the Transvaal and Orange Free State, South Africa. From the Stillwater Complex, Montana, and Fox Gulch, Goodnews Bay, Alaska, USA. In the Inagli and Konder massifs, Aldan Shield, Yakutia; the Noril'sk region, western Siberia; along the Vilyui River and on the Kamchatka Peninsula; in the Gusevogorskii pluton, Ural Mountains; and at a number of less well defined localities in the USSR. From the Birbir River, Ethiopia. From the Department of Chocó, Cauca, Colombia. On the Riam Kanan River, Borneo.

Name: For the structure and composition.

Type Material: n.d.

References: (1) Cabri, L.J. and C.E. Feather (1975) Platinum–iron alloys: a nomenclature based on a study of natural and synthetic alloys. Can. Mineral., 13, 117–126. (2) (1976) Amer. Mineral., 61, 338 (abs. ref. 1). (3) Begizov, V.D., L.F. Borisenko, and Y.D. Uskov (1975) Sulfides and natural solid solutions of platinum metals from ultramafic rocks of the Gusevogorskii pluton, Urals. Doklady Acad. Nauk SSSR, 225, 1408–1411 (in Russian). (4) Cabri, L.J., Ed. (1981) Platinum group elements: mineralogy, geology, recovery. Can. Inst. Min. & Met., 113–114.

Crystal Data: Cubic. *Point Group:* $4/m\ \overline{3}\ 2/m$. Grains 0.4 to 0.8 mm in size, rarely showing crystal outlines.

Physical Properties: Hardness = n.d. VHN = 587–597, 592 average (100 g load). D(meas.) = n.d. D(calc.) = 10.33

Optical Properties: Opaque. *Color:* In polished section, pale yellow-white. *Anisotropism:* Weak, in shades of dull brown.
R: (400) 43.2, (420) 44.1, (440) 45.5, (460) 47.2, (480) 49.2, (500) 51.2, (520) 53.1, (540) 54.8, (560) 56.2, (580) 57.4, (600) 58.6, (620) 59.4, (640) 60.1, (660) 60.5, (680) 60.8, (700) 61.1

Cell Data: *Space Group:* $Fd3m$. $a = 12.283$ Z = 16

X-ray Powder Pattern: Itabira, Brazil.
2.167 (100), 2.356 (90), 1.533 (70), 1.253 (70), 1.188 (70), 0.996 (70), 1.446 (60)

Chemistry:

	(1)	(2)	(3)	(4)
Pd	72.4	72.53	74.1	74.85
Au		0.37	1.6	
Cu	1.1	1.08	0.1	
Sb	15.6	14.61	15.6	15.57
As	10.9	10.42	9.5	9.58
Total	100.0	99.01	100.9	100.00

(1) Itabira, Brazil; by electron microprobe, average of four grains, corresponding to $(Pd_{4.88}Cu_{0.12})_{\Sigma=5.00}As_{1.04}Sb_{0.92}$. (2) Do. (3) Hope's Nose, England; by electron microprobe. (4) $Pd_{11}Sb_2As_2$.

Polymorphism & Series: Dimorphous with mertieite-I.

Occurrence: In heavy-metal concentrates from Precambrian iron-precious metal deposits (Itabira, Brazil).

Association: Arsenopalladinite, atheneite, palladseite, hematite, gold (Itabira, Brazil); chalcopyrite, millerite, kotulskite, arsenopalladinite, hematite (Lac des Iles Complex, Canada); hongshiite, cooperite, sperrylite, vysotskite, magnetite, bornite, polydymite, diopside, actinolite, epidote (Yen district, China)

Distribution: From Itabira, Minas Gerais, Brazil. At Hope's Nose, Torquay, Devon, England. From the Lac des Iles Complex, Ontario, Canada. In the Konttijärvi Intrusion, northern Finland. In China, from the Yen district – a code name. From around Zlatoust, Ural Mountains, USSR.

Name: For its structural and compositional relationship to mertieite-I.

Type Material: British Museum (Natural History), London, England, BM 1934,72; National Museum of Natural History, Washington, D.C., USA, 142504.

References: (1) Clark, A.M., A.J. Criddle, and E.E. Fejer (1974) Palladium arsenide–antimonides from Itabira, Minas Gerais, Brazil. Mineral. Mag., 39, 528–543. (2) (1974) Amer. Mineral., 59, 1330 (abs. ref. 1). (3) Shi, Ni-Cheng, Zhe-Sheng Ma, Nai-Xian Zhang, and Xui-Shou Ding (1978) Crystal structure of isomertieite (fengluanite). K'o Hsueh T'ung Pao, 23, 499–501 (in Chinese). (4) (1978) Chem. Abs., 89, 187292 (abs. ref. 3). (5) Clark, A.M. and A.J. Criddle (1982) Palladium minerals from Hope's Nose, Torquay, Devon. Mineral. Mag., 46, 371–377. (6) (1983) Amer. Mineral., 68, 851 (abs. ref. 5).

Crystal Data: Orthorhombic. *Point Group:* $2/m$ $2/m$ $2/m$. Acicular aggregates up to 1 mm.

Physical Properties: *Cleavage:* {001} good. *Fracture:* Conchoidal, distinctive. *Tenacity:* Brittle. Hardness = 3.7–4.2 VHN = 150–212 (50 gm load). D(meas.) = 6.47 D(calc.) = 6.505

Optical Properties: Opaque. *Color:* Lead-gray. *Streak:* Gray-black. *Luster:* Metallic. *Pleochroism:* From pale greenish white to darker greenish white or gray. *Anisotropism:* Distinct, from greenish gray to dark gray to brownish gray.
R_1–R_2: (400) 44.6–48.5, (420) 44.1–48.4, (440) 43.5–48.0, (460) 42.5–47.4, (480) 42.2–46.9, (500) 41.6–46.2, (520) 41.0–45.7, (540) 40.6–45.2, (560) 40.1–44.8, (580) 39.8–44.4, (600) 39.6–44.1, (620) 39.3–43.8, (640) 39.1–43.5, (660) 38.8–43.2, (680) 38.5–42.9, (700) 38.2–42.5

Cell Data: *Space Group:* *Pnnm.* $a = 33.88(2)$ $b = 38.02(2)$ $c = 4.070(2)$ Z = 1

X-ray Powder Pattern: Izok Lake, Canada.
3.398 (100), 2.149 (60), 3.305 (40), 2.038 (40b), 2.878 (40), 1.745 (40), 3.159 (30)

Chemistry:

	(1)	(2)
Pb	46.6	50.01
Cu	1.0	0.88
Fe	0.2	0.19
Ag	2.0	0.59
Sb	13.3	11.67
Bi	20.5	19.20
S	17.0	16.91
Total	100.6	99.45

(1) Izok Lake, Canada; by electron microprobe, average of three analyses. (2) Vena, Sweden; by electron microprobe, average of 19 analyses.

Occurrence: In a massive Zn-Cu-Pb sulfide deposit (Izok Lake, Canada); from a Cu-Co deposit in skarn (Vena, Sweden).

Association: Galena, pyrrhotite, pyrite, jaskólskiite (Izok Lake, Canada); galena, jaskólskiite, pyrrhotite, bismuth, antimony, antimonian bismuthinite, kobellite (Vena, Sweden).

Distribution: From Izok Lake, Northwest Territories, Canada. In the Vena Co-Cu mine, Bergslagen, central Sweden. From near Lake Zervreila, Graubünden, Switzerland.

Name: For the locality at Izok Lake, Canada.

Type Material: British Museum (Natural History), London, England; Canadian Geological Survey, Ottawa; Royal Ontario Museum, Toronto, Canada; National Museum of Natural History, Washington, D.C., USA, 165272.

References: (1) Harris, D.C., A.C. Roberts, and A.J. Criddle (1986) Izoklakeite, a new mineral species from Izok Lake, Northwest Territories. Can. Mineral., 24, 1–5. (2) (1987) Amer. Mineral., 72, 222 (abs. ref. 1). (3) Zakrzewski, M.A. and E. Makovicky (1986) Izoklakeite from Vena, Sweden, and the kobellite homologous series. Can. Mineral., 24, 7–18. (4) (1987) Amer. Mineral., 72, 229 (abs. ref. 3). (5) Makovicky, E. and W.G. Mumme (1986) The crystal structure of izoklakeite, $Pb_{51.3}Sb_{20.4}Bi_{19.5}Ag_{1.2}Cu_{2.9}Fe_{0.7}S_{114}$. The kobellite homologous series and its derivatives. Neues Jahrb. Mineral., Abh., 153, 121-145. (6) Armbruster, T. and W. Hummel (1987) (Sb,Bi,Pb) ordering in sulfosalts: crystal-structure refinement of a Bi-rich izoklakeite. Amer. Mineral., 72, 821–831.

Crystal Data: Tetragonal. *Point Group:* $4/m\ 2/m\ 2/m$. As coarse, irregular and foliated masses and as grains and veinlets in other sulfides.

Physical Properties: *Cleavage:* Prismatic, good. *Fracture:* Subconchoidal. *Tenacity:* Sectile, malleable. Hardness = 2–2.5 VHN = 29 (100 g load). D(meas.) = 6.820 D(calc.) = 6.827

Optical Properties: Opaque. *Color:* Light metallic gray; tarnished lustrous dark gray, locally to a chalcopyrite-like iridescence. *Luster:* Metallic. *Pleochroism:* Distinct in oil, brownish gray to pure gray. *Anisotropism:* Distinct, in greenish blues and light green.

R_1–R_2: (400) 34.4–35.6, (420) 34.1–36.0, (440) 33.8–36.4, (460) 33.2–36.4, (480) 32.4–35.6, (500) 31.8–34.6, (520) 31.2–33.9, (540) 30.7–33.3, (560) 30.3–32.8, (580) 29.8–32.3, (600) 29.5–31.9, (620) 29.3–31.5, (640) 29.0–31.1, (660) 28.8–30.7, (680) 28.6–30.2, (700) 28.2–29.8

Cell Data: *Space Group:* $I4_1/amd$. $a = 8.663$ $c = 11.743$ Z = 8

X-ray Powder Pattern: Boulder Co., Colorado, USA.
2.345 (100), 2.794 (90), 2.740 (80), 2.422 (80), 2.114 (70), 4.298 (60), 2.006 (60)

Chemistry:

	(1)	(2)	(3)
Ag	71.51	71.7	71.71
Cu	13.12	14.05	14.08
Fe	0.79		
S	14.36	[14.2]	14.21
Total	99.78	100.0	100.00

(1) Jalpa, Mexico. (2) Silver Plume, Colorado, USA; average of four analyses, S by difference. (3) Ag_3CuS_2.

Occurrence: Formed under low-temperature (below 117°C) hydrothermal conditions.

Association: Galena, sphalerite, pyrite, chalcopyrite, stromeyerite, polybasite, tetrahedrite–tennantite, silver, acanthite.

Distribution: In the USA, in Colorado, at the Payrock mine in Silver Plume, Clear Creek Co.; the Bulldog Mountain mine, Creede district, Mineral Co.; the Hock Hocking mine, near Alma, Park Co.; also from Boulder Co., at the E.C. Hite mine, and the Caribou mine, near Nederland. From Mogollon, Catron Co., New Mexico. In Mexico, at the La Leonora mine, Jalpa, Zacatecas, and at La Mesa, Chihuahua. From Kongsberg, Norway. In France, at Barreins, Aude. At Bohutin, near Příbram, Czechoslovakia. From Keel, Aedagh, Co. Longford, Ireland. In the Savodinskii mine, Altai Mountains, USSR. From the Sado mine, Niigata Prefecture, Japan. In Australia, from Broken Hill, New South Wales.

Name: For the Mexican locality at Jalpa.

Type Material: Mining Institute, Freiberg, Saxony, Germany.

References: (1) Johan, Z. (1967) Etude de la jalpaite. Acta Univ. Carolinae, Geol., 2, 113–122. (2) (1968) Amer. Mineral., 53, 1778 (abs. ref. 1). (3) Grybeck, D. and J.J. Finney (1968) New occurrences of and data for jalpaite. Amer. Mineral., 53, 1530–1542.

Crystal Data: Monoclinic. *Point Group:* $2/m$. Crystals elongated ∥ [001], acicular and fibrous, to 10 cm; striated ∥ [001]. As felt-like masses; massive, columnar, also radial and plumose. *Twinning:* On {100}; lamellae very commonly observed in polished section.

Physical Properties: *Cleavage:* Good on {001}; reported on {010} and {120}. *Tenacity:* Brittle. Hardness = 2.5 VHN = 66–86 (100 g load). D(meas.) = 5.63 D(calc.) = 5.76

Optical Properties: Opaque. *Color:* Gray-black, sometimes tarnishes iridescent; in polished section, gray-black. *Luster:* Metallic. *Pleochroism:* Distinct. *Anisotropism:* Strong. R_1–R_2: (400) 37.6–42.5, (420) 36.8–42.2, (440) 36.0–41.9, (460) 35.6–41.5, (480) 35.4–41.2, (500) 35.0–40.7, (520) 34.6–40.3, (540) 34.0–39.8, (560) 33.6–39.3, (580) 33.1–38.8, (600) 32.7–38.2, (620) 32.2–37.7, (640) 31.8–37.0, (660) 31.5–36.4, (680) 31.2–35.7, (700) 31.0–35.0

Cell Data: *Space Group:* $P2_1/a$. $a = 15.57$ $b = 18.98$ $c = 4.03$ $\beta = 91°48'$ Z = 2

X-ray Powder Pattern: Itos mine, Oruro, Bolivia.
3.44 (100), 2.84 (90), 2.75 (80), 3.18 (50), 3.09 (50), 2.06 (50), 3.87 (40)

Chemistry:

	(1)	(2)	(3)	(4)
Pb	40.32	40.14	42.79	40.15
Fe	3.68	2.64	2.83	2.71
Cu		0.18	1.01	
Zn			1.84	
Sb	32.92	34.25	31.94	35.39
S	21.40	22.34	20.86	21.75
rem.	1.24			
Total	99.56	99.55	101.27	100.00

(1) Slate Creek, Idaho, USA. (2) Cornwall, England; average of three analyses. (3) Selke Valley, Germany. (4) $Pb_4FeSb_6S_{14}$.

Polymorphism & Series: Dimorphous with parajamesonite; forms a series with benavidesite.

Occurrence: Usually a late-stage hydrothermal mineral in Pb-Ag-Zn veins formed at low to moderate temperatures.

Association: Other lead sulfosalts, pyrite, sphalerite, galena, tetrahedrite, stibnite, quartz, siderite, calcite, dolomite, rhodochrosite.

Distribution: From numerous localities, but only a few have provided rich material. In the USA, in Idaho, at Slate Creek, Shoshone Co. From Cornwall, England, at Endillion. In Germany, at Freiberg, Saxony; and near Magdesprung, Selke Valley, and Clausthal, Harz Mountains. From Příbram, Czechoslovakia. At Aranyidka and Baia Sprie (Felsőbánya), Romania. From Trepča, Yugoslavia. At Sala, Västmanland, Sweden. From Machacamarca, Poopó, and Huanuni, Oruro, Bolivia. In Mexico, from Noche Buena, near Mazapil; the Santa Rita mine, Nieves; and in relatively thick crystals from the Noria mine, near Sombrerete, Zacatecas. From Dachang, Guangxi Autonomous Region, China.

Name: For Robert Jameson (1774–1854), mineralogist of Edinburgh, Scotland.

References: (1) Palache, C., H. Berman, and C. Frondel (1944) Dana's system of mineralogy, (7th edition), v. I, 451–455. (2) Niizeki, N. and M.J. Buerger (1957) The crystal structure of jamesonite, $FePb_4Sb_6S_{14}$. Zeits. Krist., 109, 161–183. (3) Berry, L.G. and R.M. Thompson (1962) X-ray powder data for the ore minerals. Geol. Soc. Amer. Mem. 85, 150. (4) Luke L.Y. Chang, Xilin Li, and Chusheng Zheng (1987) The jamesonite – benavidesite series. Can. Mineral., 25, 667–672.

Crystal Data: Orthorhombic. *Point Group:* $2/m\ 2/m\ 2/m$. As irregular small aggregates up to a few mm in diameter.

Physical Properties: Hardness = 4 VHN = 165–179 (100 g load). D(meas.) = n.d. D(calc.) = 6.50

Optical Properties: Opaque. *Color:* Lead-gray; gray in polished section. *Streak:* Dark gray. *Luster:* Metallic. *Pleochroism:* Moderate, in yellow tints. *Anisotropism:* Strong, without distinct colors.

R_1–R_2: (400) 38.4–44.9, (420) 38.4–44.6, (440) 38.1–44.1, (460) 37.8–43.6, (480) 37.4–43.1, (500) 37.0–42.6, (520) 36.7–42.3, (540) 36.5–42.0, (560) 36.3–41.7, (580) 36.1–41.5, (600) 36.0–41.3, (620) 35.8–41.0, (640) 35.5–40.7, (660) 35.2–40.2, (680) 34.8–39.8, (700) 34.4–39.3

Cell Data: *Space Group:* *Pbnm.* $a = 11.331(1)$ $b = 19.871(2)$ $c = 4.100(1)$ Z = 4

X-ray Powder Pattern: Vena, Sweden.
3.710 (100), 2.970 (80), 3.333 (60), 2.761 (60), 3.595 (50), 2.751 (50), 2.050 (50)

Chemistry:

	(1)
Pb	50.74
Cu	1.31
Sb	15.74
Bi	14.35
S	17.52
Total	99.65

(1) Vena, Sweden; by electron microprobe, average of 16 analyses, corresponding to $Pb_{2.22}Cu_{0.19}(Sb_{1.17}Bi_{0.62})_{\Sigma=1.79}S_{5.00}$.

Occurrence: In aggregates of sulfosalts in Cu-Co ores of the fahlband type.

Association: Galena, pyrrhotite, chalcopyrite, arsenopyrite, cobaltite, pyrite, sphalerite, bismuth, antimony, cubanite, freibergite, izoklakeite, gudmundite.

Distribution: In the Vena deposit, Närke, Bergslagen metallic province, Sweden. From Izok Lake, Northwest Territories, Canada.

Name: For Professor Stanislaw Jaskólski (1896–1981), of the Akademia Górniczo-Hutnicza, Kraków, Poland.

Type Material: Institute of Earth Sciences, Free University, Amsterdam, The Netherlands, 150-J-2.

References: (1) Zakrzewski, M.A. (1984) Jaskólskiite, a new Pb–Cu–Sb sulfosalt from the Vena deposit, Sweden. Can. Mineral., 22, 481–485. (2) Makovicky, E. and W.G. Mumme (1984) The crystal structure of izoklakeite, dadsonite and jaskolskiite. Acta Cryst., A40, supplement, C-246. (3) (1985) Amer. Mineral., 70, 872 (abs. refs. 1 and 2). (4) Harris, D.C., A.C. Roberts, and A.J. Criddle (1984) Jaskólskiite from Izok Lake, Northwest Territories. Can. Mineral., 22, 486–491. (5) Makovicky, E. and R. Nørrestam (1985) The crystal structure of jaskolskiite, $Cu_xPb_{2+x}(Sb,Bi)_{2-x}S_5$ (x=0.2), a member of the meneghinite homologous series. Zeits. Krist., 171, 179–194.

Crystal Data: Amorphous to X-rays. *Point Group:* n.d. Globular coating on rock fragments.

Physical Properties: *Fracture:* Conchoidal. Hardness = n.d. VHN = n.d. D(meas.) = n.d. D(calc.) = n.d.

Optical Properties: Opaque; translucent in thin edges. *Color:* Black; cherry-red in transmitted light.
R: n.d.

Cell Data: *Space Group:* n.d. Z = n.d.

X-ray Powder Pattern: n.d.

Chemistry:

	(1)
As	46.8
Sb	trace
Se	7.5
Te	trace
S	40.8
insol.	4.9
Total	100.0

(1) Jerome, Arizona, USA.

Occurrence: Coating rock fragments beneath iron hoods placed over vents from which gasses rich in SO_2 issued, a product of burning sulfide ores.

Association: n.d.

Distribution: From the United Verde mine, Jerome, Yavapai Co., Arizona, USA.

Name: For Jerome, Arizona, USA.

Type Material: n.d.

References: (1) Lausen, C. (1928) Hydrous sulphates formed under fumarolic conditions at the United Verde mine. Amer. Mineral., 13, 227–229.

Crystal Data: Monoclinic. *Point Group:* $2/m$. Crystals tabular with pseudohexagonal aspect, to 4 cm. Rarely reniform. *Twinning:* On {001} common, often lamellar on {$\bar{2}$01}, rare on {$\bar{1}$01} and {101}.

Physical Properties: *Cleavage:* {010} perfect, {001} parting. *Fracture:* Conchoidal. *Tenacity:* Brittle. Hardness = 3 VHN = 106–141 (50 g load). D(meas.) = 6.44 D(calc.) = 6.38

Optical Properties: Opaque. *Color:* Lead-gray, commonly tarnished, iridescent. *Streak:* Black. *Luster:* Metallic.
R_1–R_2: (400) 43.2–47.4, (420) 42.0–46.2, (440) 40.8–45.0, (460) 40.1–44.1, (480) 39.6–43.5, (500) 39.2–43.0, (520) 39.0–42.6, (540) 38.6–42.2, (560) 38.2–41.7, (580) 38.0–41.1, (600) 37.6–40.6, (620) 37.2–40.2, (640) 36.7–39.6, (660) 36.1–39.1, (680) 35.3–38.5, (700) 34.4–37.9

Cell Data: *Space Group:* $P2_1/m$. $a = 8.92$ $b = 31.88$ $c = 8.457$ $\beta = 117°43'$ Z = 2

X-ray Powder Pattern: Binntal, Switzerland.
3.51 (10), 3.16 (9), 3.04 (8), 2.87 (8), 3.68 (7), 3.35 (7), 2.96 (5)

Chemistry:

	(1)	(2)	(3)
Pb	69.22	72.2	70.96
Tl		1.0	
As	12.42	7.6	11.00
Sb		2.3	
S	18.36	16.4	18.04
Total	100.00	99.5	100.00

(1) Binntal, Switzerland; recalculated from 99.12% total. (2) Do.; by electron microprobe. (3) $Pb_{14}As_6S_{23}$.

Polymorphism & Series: Forms a series with geocronite.

Occurrence: In metamorphosed Pb-As occurrences in dolomite (Binntal, Switzerland); low-temperature epithermal veins (Aomori Prefecture, Japan); epithermal gold-quartz veins (Săcărâmb, Romania).

Association: Tennantite, sphalerite, galena, dolomite, barite (Binntal, Switzerland); boulangerite, semseyite, guettardite, zinkenite, enargite (Silverton, Colorado, USA).

Distribution: In the Lengenbach quarry, Binntal, Valais, Switzerland. From Beuthen, Upper Silesia, Poland. In Germany, from Wiesloch, near Heidelberg. At Săcărâmb (Nagyág), Romania. From Horni Benesov, Czechoslovakia. In the Pitone marble quarry, near Seravezza, Tuscany, Italy. In Japan, from the Yunosawa and Okoppe mines, Aomori Prefecture. In the USA, at the Zuni mine and Brobdingnag prospect, near Silverton, San Juan Co., Colorado.; in the Keystone mine, Birmingham, Blair Co., Pennsylvania; and in the Balmat-Edwards mine, Balmat, St. Lawrence Co., New York. From the Penberthy Croft mine, St. Hilary, Cornwall, England.

Name: For Dr. H. Jordan (1808–1887), of Saarbrücken, Germany, who provided the original specimens for study.

References: (1) Palache, C., H. Berman, and C. Frondel (1944) Dana's system of mineralogy, (7th edition), v. I, 398–401. (2) Fisher, D.J. (1940) Discussion of the formula of jordanite. Amer. Mineral., 25, 297–298. (3) Berry, L.G. and R.M. Thompson (1962) X-ray powder data for the ore minerals. Geol. Soc. Amer. Mem. 85, 131–132. (4) Jambor, J.L. (1968) New lead sulfantimonides from Madoc, Ontario. Part 3—syntheses, paragenesis, origin. Can. Mineral., 9, 505–521. (5) Ito, T. and W. Nowacki (1974) The crystal structure of jordanite, $Pb_{28}As_{12}S_{46}$. Zeits. Krist., 139, 161–185. (6) Criddle, A.J. and C.J. Stanley, Eds. (1986) The quantitative data file for ore minerals. British Museum (Natural History), London, England, 181.

Crystal Data: Amorphous to X-rays. *Point Group:* n.d. Massive, as associations of colloidal-sized particles and disseminations.

Physical Properties: *Tenacity:* Sectile. Hardness = Soft. VHN = n.d. D(meas.) = n.d. D(calc.) = n.d.

Optical Properties: Opaque. *Color:* Gray-black to lead-gray. *Luster:* Submetallic, dull. R: n.d.

Cell Data: *Space Group:* n.d. Z = n.d.

X-ray Powder Pattern: n.d.

Chemistry: Established as MoS_2 on very impure materials.

Polymorphism & Series: Trimorphous with molybdenite and molybdenite-3R.

Occurrence: As veinlets and coatings of probable moderate to low-temperature hydrothermal origin.

Association: Cinnabar, ilsemannite, molybdenite, stilbite, calcite, quartz.

Distribution: In the USA, in the Sun Valley mine, east of Jacob Lake, Coconino Co., Arizona; in Oregon, at the Kiggins mercury mine on the Oak fork of the Clackamas River, about 80 km southeast of Portland, Clackamas Co.; and at Ambrosia Lake, McKinley Co., New Mexico. In Germany, at the Himmelsfürst mine, Erbisdorf, near Freiberg, Saxony. In Austria, from Bleiberg, Carinthia. In Chile, from Carrizal Alto, Atacama. From Campo do Agostinho, Brazil.

Name: For Eduard Friedrich Alexander Jordis (1868–1917), colloidal chemist.

Type Material: n.d.

References: (1) Palache, C., H. Berman, and C. Frondel (1944) Dana's system of mineralogy, (7th edition), v. I, 331. (2) Staples, L.W. (1951) Ilsemannite and jordisite. Amer. Mineral., 36, 609–614. (3) Clark, A.H. (1971) Molybdenite-2H, molybdenite-3R and jordisite from Carrizal Alto, Atacama, Chile. Amer. Mineral., 56, 1832–1835.

Crystal Data: Hexagonal. *Point Group:* $\overline{3}$ $2/m$. Sheets and plates, up to 5 cm long, with occasional straight edges.

Physical Properties: *Cleavage:* {0001}, perfect. *Tenacity:* Flexible. Hardness = 2 VHN = 29–43 (25 g load). D(meas.) = 8.18 D(calc.) = [8.67]

Optical Properties: Opaque. *Color:* Silver-white; white in polished section. *Luster:* High metallic, tarnishing lead-gray, iridescent, steel-blue to iron-black. *Anisotropism:* Nearly isotropic in basal sections, transverse sections moderately anisotropic in greenish gray to dark greenish gray. R_1–R_2: (400) 57.2–58.6, (420) 56.6–58.4, (440) 56.0–58.2, (460) 55.7–58.3, (480) 55.6–58.5, (500) 55.7–59.0, (520) 56.0–59.4, (540) 56.2–60.0, (560) 56.3–60.6, (580) 56.3–61.1, (600) 56.3–61.4, (620) 56.3–61.5, (640) 56.2–61.6, (660) 56.2–61.5, (680) 56.0–61.4, (700) 55.9–61.1

Cell Data: *Space Group:* $R\overline{3}m$. $a = 4.24$ $c = 39.69$ Z = [3]

X-ray Powder Pattern: Glacier Gulch, Canada.
3.07 (10), 2.24 (5), 2.11 (5), 1.744(3), 1.537 (3), 4.38 (2), 3.61 (2)

Chemistry:

	(1)	(2)	(3)	(4)
Bi	79.15	81.23	82.7	81.34
Te	15.93	14.67	12.0	12.42
Se	1.48	2.84		
S	3.15	1.46	6.0	6.24
Total	99.71	100.20	100.7	100.00

(1-2) San José, Brazil. (3) Glacier Gulch, Canada. (4) Bi_4TeS_2.

Occurrence: In granular limestone (San José, Brazil).

Association: Gold, bismuth, bismuthinite, hessite, tetradymite, joséite-B (Glacier Gulch, Canada).

Distribution: From San José, near Marianna, Minas Gerais, Brazil. In England, at Carrock Fell and at Coniston, English Lake district, Cumbria. From Kingsgate, New South Wales, and Maldon, Victoria, Australia. In Spain, at Serrania de Ronda, Málaga Province. In Canada, at Glacier Gulch, Hudson Bay Mountain, near Smithers; and the Windpass mine, near Chu Chua, British Columbia; and in Yukon Territory, from placers in Clear Creek, McQuesten district, and in Highet Creek, Mayo district. In the Sosukchan deposit, northeastern Yakutia, USSR.

Name: For the locality at San José in Brazil.

Type Material: National Museum of Natural History, Washington, D.C., USA, R400.

References: (1) Palache, C., H. Berman, and C. Frondel (1944) Dana's system of mineralogy, (7th edition), v. I, 166–167. (2) R.M. Thompson (1949) The telluride minerals and their occurrence in Canada. Amer. Mineral., 34, 342–382.

Crystal Data: Hexagonal. *Point Group:* $\overline{3}\ 2/m$. Coarse plates.

Physical Properties: *Cleavage:* {0001}, perfect. *Tenacity:* Ductile. Hardness = n.d. VHN = n.d. D(meas.) = 8.3 D(calc.) = 8.39

Optical Properties: Opaque. *Color:* Steel-gray; creamy white in reflected light. *Anisotropism:* Distinct, gray to dark brown.
R_1–R_2: n.d.

Cell Data: *Space Group:* $R\overline{3}m$. $a = 4.33$ $c = 40.75$ Z = 3

X-ray Powder Pattern: Glacier Gulch, Canada.
3.16 (10), 2.16 (5), 2.29 (4), 1.779 (3), 4.52 (2), 1.943 (2), 1.566 (2)

Chemistry:

	(1)	(2)
Bi	75.14	74.42
Pb	0.68	
Fe	0.52	
Te	19.25	22.72
S	3.64	2.86
insol.	0.30	
Total	99.53	100.00

(1) Glacier Gulch, Canada. (2) Bi_4Te_2S.

Occurrence: In quartz and skarn (Hedley, Canada).

Association: Bismuth, gold, joséite, hedleyite, pyrrhotite, arsenopyrite, molybdenite.

Distribution: From Glacier Gulch, Hudson Bay Mountain, near Smithers; and the Good Hope claim, near Hedley, Osoyoos mining division, British Columbia, Canada. At the Stepnyak gold deposit, Kazakhstan; in the Sosukchan deposit, northeastern Yakutia; in the Shumilovsk Sn-W deposit, and the gold placer Krasnyi Klyuch, Baikal region, USSR. From Natsukidani, Oita Prefecture, and in the Tsumo mine, about 50 km northwest of Hiroshima City, Akita Prefecture, Japan. From Smolotely, near Příbram, and in the Kasejovice-Belčice district, Czechoslovakia.

Name: For the relationship to joséite; "B" to distinguish it from joséite, formerly joséite-A.

Type Material: n.d.

References: (1) Peacock, M.A. (1941) On joseite, grünlingite, oruetite. Univ. Toronto Studies, Geol. Ser. 46, 83–105. (2) R.M. Thompson (1949) The telluride minerals and their occurrence in Canada. Amer. Mineral., 34, 342–382.

Crystal Data: Monoclinic. *Point Group:* $2/m$. As tabular crystals to 150 μm, and irregular grains to 0.5 mm.

Physical Properties: *Cleavage:* One good cleavage. Hardness = n.d. VHN = 114–213 D(meas.) = n.d. D(calc.) = 6.77

Optical Properties: Opaque. *Color:* In polished section, creamy white. *Pleochroism:* Strong, from cream to gray with green and mauve tints. *Anisotropism:* Strong.
R_1–R_2: n.d.

Cell Data: *Space Group:* $C2/m$. $a = 26.66$ $b = 4.06$ $c = 17.03$ $\beta = 127.20(2)°$
Z = n.d.

X-ray Powder Pattern: Juno mine, Australia.
3.55 (100), 3.90 (80), 2.919 (70), 3.23 (50), 2.970 (50), 3.39 (40), 2.213 (40)

Chemistry:

	(1)	(2)	(3)
Pb	20.6	20.9	18.8
Cu	3.9	3.9	3.9
Ag			0.4
Bi	57.2	59.0	52.3
Se	7.6	5.6	14.7
S	12.8	12.6	10.1
Total	102.1	102.0	100.2

(1–2) Juno mine, Australia; by electron microprobe, leading to an average composition of $Pb_{0.73}Cu_{0.45}Bi_{2.00}Se_{0.60}S_{2.90}$. (3) Kidd Creek mine, Canada; by electron microprobe, corresponding to $Pb_{2.89}Cu_{1.86}Ag_{0.12}Bi_{7.99}Se_{5.94}S_{10.06}$.

Occurrence: Of hydrothermal origin.

Association: Gold, selenian heyrovskýite, krupkaite, proudite, chalcopyrite, magnetite (Juno mine, Australia); chalcopyrite, sphalerite, cobaltite, kësterite, mawsonite (Kidd Creek, Canada).

Distribution: In the Juno mine, Tennant Creek, Northern Territory, Australia. At the Kidd Creek mine, near Timmins, Ontario, Canada. From near Potts, Lander Co., Nevada, USA. In the Kochbulak deposit, eastern Uzbekistan, USSR.

Name: For the Juno mine, Australia, in which it was first found.

Type Material: n.d.

References: (1) Mumme, W.G. (1975) Junoite, $Cu_2Pb_3Bi_8(S,Se)_{16}$, a new sulfosalt from Tennant Creek, Australia: its crystal structure and relationship with other bismuth sulfosalts. Amer. Mineral., 60, 548–558. (2) Large, R.R. and W.G. Mumme (1975) Junoite, "wittite" and related seleniferous bismuth sulfosalts from Juno mine, Northern Territory, Australia. Econ. Geol., 70, 369–383. (3) (1975) Amer. Mineral., 60, 737 (abs. ref. 2). (4) Pringle, G.J. and R.I. Thorpe (1980) Bohdanowiczite, junoite and laitakarite from the Kidd Creek mine, Timmins, Ontario. Can. Mineral., 18, 353–360.

Crystal Data: Cubic. *Point Group:* $4/m\ \overline{3}\ 2/m$. Irregular slab-like aggregates up to 0.5 mm.

Physical Properties: Hardness = n.d. VHN = 468 (50 g load). D(meas.) = n.d. D(calc.) = 4.045 Strongly magnetic.

Optical Properties: Opaque. *Color:* Black, sometimes tarnishes to colors; in reflected light, light creamy. *Luster:* Adamantine.
R: (400) — , (420) — , (440) 34.4, (460) 35.2, (480) 35.3, (500) 35.4, (520) 35.4, (540) 35.2, (560) 35.3, (580) 35.2, (600) 34.8, (620) 34.1, (640) 33.8, (660) 32.7, (680) 32.5, (700) —

Cell Data: *Space Group:* $Fd3m$. $a = 9.997$ Z = 8

X-ray Powder Pattern: Southern Baikal region, USSR.
1.757 (10), 1.019 (10), 2.98 (9), 1.910 (8), 2.47 (7), 3.48 (5), 1.153 (5)

Chemistry:

	(1)	(2)
Zn	18.89	21.96
Cu	2.73	
Cr	34.10	34.94
V	0.61	
Sb	0.73	
S	42.22	43.10
Total	99.28	100.00

(1) Southern Baikal region, USSR; by electron microprobe, average of nine analyses on three grains, corresponding to $Zn_{0.870}Cu_{0.130}Cr_{1.977}V_{0.036}Sb_{0.019}S_{3.968}$. (2) $ZnCr_2S_4$.

Occurrence: In a garnet-pyroxene mixture in diopside-quartz-calcite rocks.

Association: Karelianite, eskolaite, diopside, quartz, calcite, barite, zircon, chromian-vanadian tremolite, goldmanite–uvarovite, pyrite.

Distribution: From an undefined locality in the southern Baikal region, USSR.

Name: For P.I. Kalinin, Russian mineralogist and petrologist, investigator of the southern Baikal region.

Type Material: A.E. Fersman Mineralogical Museum, Academy of Sciences, Moscow, USSR.

References: (1) Reznickii, L.Z., E.V. Skl'arov, and Z.F. Ustschapovskaya (1985) Kalininite, $ZnCr_2S_4$ — a new natural sulphospinel. Zap. Vses. Mineral. Obshch., 114, 622–627 (in Russian). (2) (1987) Amer. Mineral., 72, 223 (abs. ref. 1).

Kamacite (Fe, Ni)

Crystal Data: Cubic. *Point Group:* $4/m\ \overline{3}\ 2/m$. As plates and lamellar masses and in regular intergrowth with taenite. May occur in crystals from mm size to 30 cm; in extended plates and ribbons in Widmanstätten bands.

Physical Properties: Hardness = n.d. VHN = 145–165 (100 g load). D(meas.) = n.d. D(calc.) = [7.90] Magnetic.

Optical Properties: Opaque. *Color:* Steel-gray to iron-black. *Luster:* Metallic.
R: n.d.

Cell Data: *Space Group:* $Fm3m$ (disordered phase). $a = \sim 8.60$ Z = 54

X-ray Powder Pattern: Linville nickel-rich ataxite.
2.031 (100), 1.170 (70), 1.967 (60), 1.435 (30), 3.032 (10), 2.953 (10), 1.481 (10)

Chemistry:

	(1)	(2)
Fe	93.75	93.09
Ni	5.43	6.69
Co	0.58	0.25
C		0.02
P	0.19	
S	0.08	
Total	100.03	100.05

(1) North Chile hexahedrite. (2) Welland octahedrite.

Occurrence: A major constituent of iron meteorites (siderites) and present in varying amounts in most other meteorites except certain of the stony meteorites (aerolites).

Association: Taenite, graphite, cohenite, moissanite, schreibersite, troilite, daubréelite, oldhamite, other meteorite minerals.

Distribution: Terrestrial occurrences at Blaafjeld, near Ovifak, Disko Island, Greenland. In Germany, from Bühl, near Weimar, Hesse. Otherwise from meteorites.

Name: From the Greek for *shaft* or *lath.*

Type Material: n.d.

References: (1) Palache, C., H. Berman, and C. Frondel (1944) Dana's system of mineralogy, (7th edition), v. I, 114–116. (2) Ramsden, A.R. and E.N. Cameron (1966) Kamacite and taenite superstructures and a metastable tetragonal phase in iron meteorites. Amer. Mineral., 51, 37–55.

Crystal Data: Orthorhombic. *Point Group:* $2/m$ $2/m$ $2/m$. As subhedral granular crystals (0.05–0.2 mm); on the rims of platinum or iridosmine grains.

Physical Properties: *Fracture:* Flinty. *Tenacity:* Brittle, weakly elastic. Hardness = n.d. VHN = 1529(477) (50 g load). D(meas.) = n.d. D(calc.) = 9.10

Optical Properties: Opaque. *Color:* Grayish black; light gray in reflected light. *Luster:* Metallic. *Pleochroism:* Weak. *Anisotropism:* Moderate to strong, with reddish violet tints.
R_1–R_2: (440) 43.7–45.9, (460) 45.1–45.5, (480) 46.1–45.5, (500) 46.8–45.6, (520) 47.2–45.7, (540) 47.5–45.8, (560) 47.6–46.0, (580) 47.6–46.2, (600) 47.4–46.3, (620) 47.1–46.5, (640) 46.8–46.6, (660) 46.6–46.8, (680) 46.3–46.9, (700) 46.0–47.0

Cell Data: *Space Group:* *Pbcn*. $a = 8.464$ $b = 6.001$ $c = 6.146$ Z = 4

X-ray Powder Pattern: Nizhni Tagil, USSR.
2.99 (100), 1.736 (60b), 2.14 (40), 1.758 (40), 1.136 (40b), 1.028 (40), 0.982 (40)

Chemistry:

	(1)	(2)
Ir	57.9	52.6 – 75.8
Rh	15.7	1.7 – 22.9
Pt	0.2	
Fe	0.1	
Cu	0.7	0.0 – 0.8
S	23.0	22.7 – 23.9
Total	97.6	

(1) A USSR placer deposit; by electron microprobe, average of analyses on four grains. (2) Nizhni Tagil, USSR; by electron microprobe, range of 18 analyses.

Occurrence: In placer deposits derived from ultramafic rocks.

Association: Iridosmine, platinum, osmiridium, laurite, erlichmanite, chromite, Fe–Pt alloy, sulfides of Fe, Cu, Ir, Rh.

Distribution: From placers around Nizhni Tagil, Ural Mountains, and at an unidentified ultramafic massif in the Far Eastern Region, USSR. From Goodnews Bay, Alaska, USA.

Name: To honor S.A. Kashin, a Russian investigator of ore deposits in the Ural Mountains.

Type Material: Institute of Mineralogy and Geochemistry of Rare Elements; A.E. Fersman Mineralogical Museum, Academy of Sciences, Moscow, USSR.

References: (1) Begizov, V.D., E.N. Zabyalov, N.S. Rudashevskii, and L.N. Vyal'sov (1985) Kashinite $(Ir, Rh)_2S_3$ —A new iridium and rhodium sulphide. Zap. Vses. Mineral. Obshch., 114, 617–622 (in Russian). (2) (1987) Amer. Mineral., 72, 223 (abs. ref. 1). (3) Anon. (1985) A study of kashinite $(Ir, Rh)_2S_3$. Acta Mineralogica Sinica, 5, 6–9 (in Chinese with English abs.). (4) (1986) Mineral. Abs., 37, 236 (abs. ref. 3).

Crystal Data: Hexagonal. *Point Group:* $\overline{3}\ 2/m$. As very thin foils, up to 4 mm across, with maximum thickness of 50 μm.

Physical Properties: *Cleavage:* Perfect on {0001}. *Tenacity:* Flexible. Hardness = 1.5 VHN = 100 (50 g load). D(meas.) = > 7.5 D(calc.) = 8.08

Optical Properties: Opaque. *Color:* Silver-white to tin-white. *Streak:* Light steel-gray. *Luster:* Metallic. *Pleochroism:* Pinkish creamy white to white with creamy tint. *Anisotropism:* Strong, bluish gray to brownish gray to gray.
R_1–R_2: n.d.

Cell Data: *Space Group:* $R\overline{3}m$, $R3m$, $R32$. $a = 4.240$ $c = 29.66$ Z = 3

X-ray Powder Pattern: Kawazu mine, Japan.
3.12 (100), 2.31 (50), 2.12 (50), 4.92 (40), 3.64 (30), 2.61 (20), 1.757 (20)

Chemistry:

	(1)	(2)	(3)
Bi	55.4	56.3	55.57
Fe		0.37	
Te	31.9	32.4	33.93
Se	9.9	9.92	10.50
S	0.1		
rem.		0.34	
Total	97.3	99.33	100.00

(1) Kawazu mine, Japan; by electron microprobe, corresponding to $Bi_{2.07}Te_{1.95}Se_{0.97}S_{0.03}$.
(2) Mazenod Lake, Canada; by electron microprobe, remainder is Si and U, leading to $Bi_{2.13}Fe_{0.05}Te_{2.01}Se_{0.79}$. (3) Bi_2Te_2Se.

Occurrence: Of hydrothermal origin, in a quartz vein (Kawazu mine, Japan); in a breccia pipe cutting dacitic ignimbrites (Mazenod Lake, Canada).

Association: Selenian tellurium (Kawazu mine, Japan); tellurobismuthite, uraninite, hematite, yarrowite (Mazenod Lake, Canada).

Distribution: In the Kawazu mine, Shizuoka Prefecture, Japan. From the Dianne Cu-U claims, Mazenod Lake, Northwest Territories, Canada. In the USA, from the Lone Pine mine, near Silver City, Grant Co., New Mexico, and in the Ward mine, south of Ely, White Pine Co., Nevada.

Name: For the Kawazu mine, Japan.

Type Material: Sakurai Museum, Tokyo, Japan; National Museum of Natural History, Washington, D.C., USA, 121926, 160136.

References: (1) Kato, A. (1970) In: Introduction to Japanese minerals. Geol. Surv. Japan, 1970, 87–88. (2) (1972) Amer. Mineral., 57, 1312 (abs. ref. 1). (3) Miller, R. (1981) Kawazulite Bi_2Te_2Se, related bismuth minerals and selenian covellite from the Northwest Territories. Can. Mineral., 19, 341–348.

Crystal Data: Hexagonal. *Point Group:* $\overline{3}$. As grains up to to 145 x 220 μm.

Physical Properties: Hardness = n.d. VHN = 394–424, 410 average (15 g load). D(meas.) = n.d. D(calc.) = [10.16]

Optical Properties: Opaque. *Color:* In polished section, cream colored, where isolated; dull cream with brownish tint with kotulskite; gray next to telluropalladinite. *Pleochroism:* Not detectable in air; very slight in oil. *Anisotropism:* Moderate to strong.
R_1–R_2: (400) 40.8–40.6, (420) 41.6–41.1, (440) 41.8–41.5, (460) 42.0–42.0, (480) 42.3–42.6, (500) 42.9–43.4, (520) 43.6–44.1, (540) 44.4–44.9, (560) 45.3–45.8, (580) 46.0–46.6, (600) 46.8–47.5, (620) 48.0–48.8, (640) 49.3–49.9, (660) 50.6–50.9, (680) 52.0–52.1, (700) 53.5–53.4

Cell Data: *Space Group:* $R\overline{3}$. $a = 11.45$ $c = 11.40$ Z = 18(?)

X-ray Powder Pattern: Stillwater Complex, Montana, USA.
2.26 (100), 2.16 (90), 0.791 (40), 1.32 (30), 0.885 (30), 2.22 (20), 1.40 (20)

Chemistry:

	(1)
Pd	68.7
Pb	1.6
Bi	1.0
Te	29.1
Total	100.4

(1) Stillwater Complex, Montana, USA; by electron microprobe.

Occurrence: With other platinum-group minerals.

Association: Merenskyite, kotulskite, telluropalladinite, moncheite, vysotskite, gold, magnetite.

Distribution: In the Stillwater Complex, Montana, USA.

Name: For H. Keith Conn, the geologist largely responsible for the discovery of Pt-Pd mineralization of parts of the Stillwater Complex, Montana.

Type Material: National Museum of Natural History, Washington, D.C., USA, 114975; Canadian Geological Survey, Ottawa, 12188; Royal Ontario Museum, Toronto, Canada, M35867.

References: (1) Cabri, L.J., J.F. Rowland, J.H.G. Laflamme, and J.M. Stewart (1979) Keithconnite, telluropalladinite, and other Pd–Pt tellurides from the Stillwater Complex, Montana. Can. Mineral., 17, 589–594. (2) (1981) Amer. Mineral., 66, 1275 (abs. ref. 1).

Crystal Data: Triclinic, pseudomonoclinic. *Point Group:* $\bar{1}$. Crystals elongated ∥ [010], lath-shaped on {001}, in radiating aggegates to 5 cm, sometimes tufted and hair-like.

Physical Properties: *Cleavage:* Perfect on {001}; parting on {010}. *Tenacity:* Sectile, thin splinters flexible. Hardness = 1–1.5 VHN = 30–90 (5 g load). D(meas.) = 4.68 D(calc.) = 4.69

Optical Properties: Translucent. *Color:* Cherry-red; red in thin splinters. *Streak:* Brownish red. *Luster:* Adamantine to semimetallic.

Optical Class: Biaxial (+). *Orientation:* $Z = b$ = elongation. $\alpha = > 2.72$ $\beta = > 2.72$ $\gamma = > 2.72$ *Anisotropism:* Very strong to extreme.

R_1–R_2: (400) 31.1–37.5, (420) 30.0–36.4, (440) 29.0–35.2, (460) 28.0–34.0, (480) 27.2–33.0, (500) 26.4–32.2, (520) 25.7–31.3, (540) 25.2–30.6, (560) 24.8–30.0, (580) 24.4–29.4, (600) 24.0–29.0, (620) 23.8–28.6, (640) 23.5–28.2, (660) 23.4–28.0, (680) 23.2–27.8, (700) 23.1–27.7

Cell Data: *Space Group:* $C2/m$ (apparent). $a = 10.784$ $b = 4.065$ $c = 10.206$ $\beta = 101°31'$ Z = 4

X-ray Powder Pattern: Zimbabwe. (JCPDS 11-91).
2.92 (100), 3.13 (90), 2.69 (70), 4.06 (60), 1.782 (60), 5.29 (50), 2.49 (50)

Chemistry:

	(1)	(2)
Sb	75.2	75.24
S	19.9	19.82
O	4.8	4.94
Total	99.9	100.00

(1) Globe and Phoenix mine, Zimbabwe; by electron microprobe. (2) Sb_2S_2O.

Occurrence: A secondary mineral, as an alteration of stibnite, in antimony deposits.

Association: Stibnite, antimony, senarmontite, valentinite, cervantite, stibiconite.

Distribution: In small amounts in many deposits. At Braunsdorf, near Freiberg, Saxony, Germany. From Pernek, Pezinok, and Příbram, Czechoslovakia. In the Chalanches mine, near Allemont, Iseré, France. From the Cetine mine, near Siena, Tuscany, Italy. At Kadamdja, Kirgizia, USSR. In the Santa Cruz and San Francisco mines, Poopó, Oruro, Bolivia. From Broken Hill, New South Wales, Australia. Exceptional radiating groups from the Globe and Phoenix mine, Que Que, Zimbabwe. At Sombrerete, Zacatecas, Mexico. From Canada, in the Lac Nicolet mine, South Ham Township, Wolfe Co., Quebec, and at other localities.

Name: From *kermes* (after Persian *qurmizq, crimson*) for red amorphous antimony trisulfide.

References: (1) Palache, C., H. Berman, and C. Frondel (1944) Dana's system of mineralogy, (7th edition), v. I, 279–280. (2) Kupčik, V. (1967) Die Kristallstruktur des Kermesits, Sb_2S_2O. Naturwiss., 54, 114 (in German). (3) Cervelle, B. (1971) Détermination par microréflectométrie de propriétés optiques d'un cristal monoclinique absorbant (kermésite Sb_2S_2O). Bull. Soc. fr. Minéral., 94, 486–491 (in French). (4) Criddle, A.J. and C.J. Stanley, Eds. (1986) The quantitative data file for ore minerals. British Museum (Natural History), London, England, 184.

Crystal Data: Tetragonal, pseudocubic. *Point Group:* $\overline{4}$. Massive.

Physical Properties: Hardness = 4.5 VHN = 328–348 (100 g load). D(meas.) = 4.54–4.59 D(calc.) = 4.524

Optical Properties: Opaque. *Color:* Greenish black. *Streak:* Black. *Luster:* Metallic.
R_1–R_2: (400) 23.6–24.2, (420) 24.0–24.8, (440) 24.4–25.4, (460) 24.5–25.6, (480) 24.4–25.2, (500) 24.1–24.8, (520) 23.8–24.5, (540) 23.4–24.3, (560) 23.2–24.2, (580) 23.2–24.1, (600) 23.2–24.2, (620) 23.3–24.2, (640) 23.4–24.4, (660) 23.6–24.6, (680) 23.8–24.8, (700) 24.2–25.2

Cell Data: *Space Group:* $I\overline{4}$. $a = 5.427$ $c = 10.871$ Z = 2

X-ray Powder Pattern: Snowflake mine, Canada.
3.15 (100), 1.929 (70), 1.645 (50), 1.113 (40), 2.73 (30), 1.248 (30), 0.860 (30)

Chemistry:

	(1)	(2)
Cu	26.69	29.40
Ag	0.49	
Zn	10.32	10.18
Fe	2.62	3.12
Sn	31.80	28.07
In	0.02	
S	27.58	29.19
insol.	0.36	
Total	99.88	99.96

(1) Këster deposit, USSR; corresponding to $Cu_{1.98}(Zn, Fe)_{0.95}Sn_{1.25}S_{4.00}$. (2) St. Michael's Mount, Cornwall, England; by electron microprobe.

Occurrence: In quartz-sulfide hydrothermal veinlets in tin deposits.

Association: Arsenopyrite, stannoidite.

Distribution: In the Këster deposit, Yano-Adychansk region, Yakutia, USSR. At Cínvald (Zinnwald), Czechoslovakia. From St. Michael's Mount, and in the Cligga mine, Perranzabuloe, Cornwall, England. From Chizeuil, Saône-et-Loire, France. At Kirki, Thrace, Greece. In Canada, from the Snowflake mine, Revelstoke mining division, British Columbia; the Brunswick tin mine, 55 km southwest of Fredericton, New Brunswick; and in the Kidd Creek mine, near Timmins, Ontario. In the USA, at the Hugo mine, Custer Co., South Dakota. From the Pirquitas deposit, Jujuy Province, Argentina. At Oruro, Bolivia. A few other minor occurrences are known.

Name: For the locality in the USSR at the Këster deposit.

Type Material: n.d.

References: (1) Ivanov, V.V. and Y.A. Pyatenko (1958) On the so-called kësterite. Zap. Vses. Mineral. Obshch., 88, 165–168 (in Russian). (2) (1959) Amer. Mineral., 44, 1329 (abs. ref. 1). (3) Hall, S.R., J.T. Szymański, and J.M. Stewart (1978) Kesterite, $Cu_2(Zn, Fe)SnS_4$, and stannite, $Cu_2(Fe, Zn)SnS_4$, structurally similar but distinct minerals. Can. Mineral., 16, 131–137. (4) Kissin, S.A. and D.R. Owens (1979) New data on stannite and related tin sulfide minerals. Can. Mineral., 17, 125–135. (5) Moore, F. and R.A. Howie (1984) Tin-bearing sulphides from St Michael's Mount and Cligga Head, Cornwall. Mineral. Mag., 48, 389–396.

Kharaelakhite $(Pt, Cu, Pb, Fe, Ni)_9S_8$

Crystal Data: Orthorhombic (?). *Point Group:* 222, $mm2$, or $2/m\ 2/m\ 2/m$. As thin rims on braggite-cooperite.

Physical Properties: Hardness = n.d. VHN = n.d. D(meas.) = n.d. D(calc.) = 7.78

Optical Properties: Opaque. *Color:* In reflected light, a brownish lilac tint. *Luster:* n.d. *Anisotropism:* Distinct, from pink-lilac to bluish.
R_1–R_2: n.d.

Cell Data: *Space Group:* $P222$, $Pmm2$, or $Pmmm$. $a = 9.713$ $b = 8.333$ $c = 14.500$ Z = 4

X-ray Powder Pattern: Talnakh deposit, USSR.
2.80 (10), 1.813 (5), 3.04 (4), 6.30 (2), 1.518 (2)

Chemistry:

	(1)
Pt	32.70 – 34.70
Pd	0.12 – 0.24
Cu	12.20 – 12.90
Pb	23.40 – 25.00
Fe	5.35 – 5.80
Ni	3.90 – 4.40
S	18.70 – 19.70
Total	

(1) Talnakh deposit, USSR; by electron microprobe, ranges of values.

Occurrence: In hydrothermal chalcopyrite ores.

Association: Braggite–cooperite, chalcopyrite, bornite, millerite.

Distribution: From the Talnakh deposit, Noril'sk region, western Siberia, USSR.

Name: For the Kharaelakh Plateau, Noril'sk region, USSR.

Type Material: n.d.

References: (1) Genkin, A.D., T.L. Evstigneeva, L.N. Vyal'sov, and I.P. Laputina (1985) Kharaelakhite $(Pt, Cu, Pb, Fe, Ni)_9S_8$, a new sulfide of platinum, copper and lead. Mineral. Zhurnal, 7, 78–83 (in Russian with English abs.). (2) (1986) Mineral. Abs., 37, 530 (abs. ref. 1). (3) (1985) Chem. Abs., 103, 56903 (abs. ref. 1).

Crystal Data: Tetragonal. *Point Group:* $4/m\ 2/m\ 2/m$. Intimately intergrown with cupalite in small grains, up to 1.5 mm, and as prismatic crystals up to 400 μm in maximum dimension.

Physical Properties: *Cleavage:* Parallel to {100}. *Tenacity:* Malleable. Hardness = n.d. VHN = 433–474 (100 g load). D(meas.) = n.d. D(calc.) = 4.42

Optical Properties: Opaque. *Color:* Steel-gray yellow; in reflected light, isotropic sections are bluish, anisotropic sections are blue to creamy pink. *Luster:* Metallic. *Anisotropism:* Distinct, grayish brown to red.
R_1–R_2: (400) — , (420) — , (440) 75.8–70.1, (460) 75.6–70.3, (480) 75.9–71.5, (500) 76.5–73.0, (520) 76.4–74.5, (540) 76.1–76.1, (560) 75.3–77.4, (580) 74.3–78.4, (600) 73.2–79.2, (620) 72.0–79.7, (640) 70.7–79.8, (660) 69.5–79.9, (680) 68.5–79.8, (700) 67.5–79.5

Cell Data: *Space Group:* $I4/mcm$. $a = 6.07(1)$ $c = 4.89(1)$ Z = 4

X-ray Powder Pattern: Listvenitovii stream, USSR.
4.27 (10), 2.119 (8), 2.372 (7), 1.920 (7), 3.04 (5), 1.894 (4), 2.156 (2)

Chemistry:

	(1)
Cu	53.4 – 54.9
Zn	0.87 – 1.91
Al	44.4 – 45.9
Total	

(1) Listvenitovii stream, USSR; by electron microprobe, ranges of analyses on nine grains, corresponding to $(Cu_{1.015}Zn_{0.024})_{\Sigma=1.039}Al_{2.000}$.

Occurrence: In black slick washed from greenish gray cover weathering from serpentine.

Association: Cupalite, two unnamed zinc aluminides.

Distribution: Near the Listvenitovii stream, Khatirskii ultramafic zone of the Koryak–Kamchata fold area, Koryak Mountains, USSR.

Name: For the occurrence in the Khatirskii ultramafic zone, USSR.

Type Material: Mining Museum, Leningrad Mining Institute, Leningrad, USSR.

References: (1) Razin, L.V., N.S. Rudashevskii, and L.N. Vyal'sov (1985) New natural intermetallic compounds of aluminum, copper and zinc—khatyrkite $CuAl_2$, cupalite CuAl and zinc aluminides—from hyperbasites of dunite–harzburgite formation. Zap. Vses. Mineral. Obshch., 114, 90–100 (in Russian). (2) (1986) Amer. Mineral., 71, 1278 (abs. ref. 1). (3) (1985) Chem. Abs., 102, 223538 (abs. ref. 1).

Crystal Data: Cubic. *Point Group:* n.d. As irregular grains, up to 100 μm across, as rims on scheelite, and included in colusite.

Physical Properties: Hardness = n.d. VHN = 183 (100 g load). D(meas.) = n.d. D(calc.) = 4.88

Optical Properties: Opaque. *Color:* Pale gray-brown; pale gray in reflected light. *Luster:* Metallic.
R: (400) 22.7, (420) 23.2, (440) 23.6, (460) 23.7, (480) 23.3, (500) 22.6, (520) 22.1, (540) 22.5, (560) 23.3, (580) 23.9, (600) 24.1, (620) 24.0, (640) 23.7, (660) 23.2, (680) 22.6, (700) 22.2

Cell Data: *Space Group:* n.d. $a = 10.856(2)$ Z = 4

X-ray Powder Pattern: Kidd Creek mine, Canada.
6.29 (100), 1.919 (60), 3.138 (50), 5.41 (30), 3.270 (30), 1.839 (30), 2.712 (20)

Chemistry:

	(1)	(2)	(3)
Cu	39.6	40.6	40.55
Fe		0.7	
Sn	12.3	11.9	12.62
V		0.1	
W	20.4	19.0	19.55
Sb		0.5	
Se	2.1		
S	24.7	26.8	27.28
Total	99.1	99.6	100.00

(1) Kidd Creek mine, Canada; by electron microprobe. (2) Bisbee, Arizona, USA; by electron microprobe. (3) Cu_6SnWS_8.

Occurrence: In massive copper sulfide ores.

Association: Scheelite, carrollite, clausthalite, tennantite, tungstenite, sphalerite (Kidd Creek mine, Canada); pyrite, colusite, stützite, altaite (Bisbee, Arizona, USA).

Distribution: From the Kidd Creek mine, near Timmins, Ontario, Canada. In the Campbell mine, Bisbee, Cochise Co., Arizona, USA.

Name: For the Kidd Creek mine, Canada.

Type Material: British Museum (Natural History), London, England, 1982,2 and 1982,3; Canadian Geological Survey, Ottawa, 64076-78; Royal Ontario Museum, Toronto, Canada, M39791.

References: (1) Harris, D.C., A.C. Roberts, R.I. Thorp, A.J. Criddle, and C.S. Stanley (1984) Kiddcreekite, a new mineral species from the Kidd Creek mine, Timmins, Ontario and from the Campbell orebody, Bisbee, Arizona. Can. Mineral., 22, 227–232. (2) (1985) Amer. Mineral., 70, 437 (abs. ref. 1).

Crystal Data: Hexagonal, probably orthorhombic or monoclinic, pseudohexagonal. *Point Group:* 622 ($6/m$ $2/m$ $2/m$ apparent). Subhedral grains to 200 μm. *Twinning:* Polysynthetically twinned.

Physical Properties: Hardness = n.d. VHN = 150 (100 g load). D(meas.) = n.d. D(calc.) = 6.82

Optical Properties: Opaque. *Color:* White in reflected light. *Anisotropism:* Weak. R_1–R_2: n.d.

Cell Data: *Space Group:* $P6_322$ $a = 8.69(5)$ $c = 26.06(10)$ Z = 2

X-ray Powder Pattern: Kirki, Greece.
3.260 (100), 3.070 (70), 3.475 (60), 2.854 (60), 2.190 (50), 3.65 (50), 1.815 (50)

Chemistry:

	(1)	(2)
Pb	59.4	58.65
Bi	15.2	17.75
As	6.2	6.36
Sb	0.5	
S	17.4	17.24
Total	98.7	100.00

(1) Kirki, Greece; by electron microprobe, average of six analyses, leading to the formula $Pb_{10.08}Bi_{2.55}As_{2.91}Sb_{0.13}S_{19.00}$. (2) $Pb_{10}Bi_3As_3S_{19}$.

Occurrence: In a hydrothermal Pb-Zn deposit.

Association: Cosalite, bismuthinite, bismuthian jordanite, seligmannite, sphalerite, pyrite, galena.

Distribution: In the Aghios Philippos Pb-Zn deposit, near Kirki, Thrace, Greece.

Name: For Kirki, Greece.

Type Material: n.d.

References: (1) Moëlo, Y., E. Oudin, E. Makovicky, S. Karup-Møller, F. Pillard, M. Bornuat, and E. Evanghelou (1985) La kirkiite, $Pb_{10}Bi_3As_3S_{19}$, une nouvelle espèce minérale homologue de la jordanite. Bull. Minéral., 108, 667–677 (in French with English abs.). (2) (1986) Amer. Mineral., 71, 1278–1279 (abs. ref. 1).

Crystal Data: Hexagonal. *Point Group:* $\overline{3}\ 2/m$. As crystals ranging up to 5 mm; massive.

Physical Properties: *Cleavage:* Good on {001}. Hardness = n.d. VHN = 110
D(meas.) = 7.19 D(calc.) = 7.22

Optical Properties: Opaque. *Color:* Pale yellow. *Luster:* Metallic. *Anisotropism:* Distinct, from pale gray to pink or purple.

R_1–R_2: (400) 45.8–47.0, (420) 46.6–48.5, (440) 47.4–50.0, (460) 48.2–51.7, (480) 49.1–53.5, (500) 50.2–55.0, (520) 51.2–56.5, (540) 52.3–57.9, (560) 53.5–59.1, (580) 54.7–60.3, (600) 55.8–61.4, (620) 56.9–62.3, (640) 51.8–63.2, (660) 58.7–63.9, (680) 59.6–64.4, (700) 60.4–64.8

Cell Data: *Space Group:* $P\overline{3}m1$. $a = 3.716$ $c = 5.126$ Z = 1

X-ray Powder Pattern: Kuusamo, Finland.
2.729 (100), 2.007 (45), 1.510 (35), 2.567 (25), 1.860 (25), 1.535 (20), 5.13 (15)

Chemistry:

	(1)	(2)	(3)
Ni	22.42	23.33	22.13
Co	0.40	0.17	
Cu	0.07	0.02	
Ag	0.005	0.005	
Bi		0.02	
Te	47.46	48.65	48.10
Se	30.22	27.58	29.77
S		0.43	
Total	100.57	100.20	100.00

(1) Kuusamo, Finland; corresponding to $(Ni_{1.01}Co_{0.02})_{\Sigma=1.03}Te_{0.99}Se_{1.01}$. (2) Do.; corresponding to $(Ni_{1.06}Co_{0.01})_{\Sigma=1.07}Te_{1.01}(Se_{0.93}S_{0.04})_{\Sigma=0.97}$. (3) NiTeSe.

Occurrence: In narrow carbonate-bearing fissure veinlets in albite.

Association: Clausthalite, hematite, selenian polydymite, selenian linnaeite, pyrite.

Distribution: From Kuusamo, northeastern Finland.

Name: For the river Kitka, in Finland, in the valley of which the mineral was found.

Type Material: n.d.

References: (1) Häkli, T.A., Y. Vuorelainen, and T.G. Sahama (1965) Kitkaite (NiTeSe), a new mineral from Kuusamo, northeast Finland. Amer. Mineral., 50, 581–586.

Crystal Data: Hexagonal. *Point Group:* $6/m\ 2/m\ 2/m$. As granular aggregates, some with thick and thin tabular grains.

Physical Properties: *Cleavage:* Perfect on {0001}. Hardness = 2–3 VHN = 82–104 (100 g load). D(meas.) = 5.99 (synthetic). D(calc.) = 6.12

Optical Properties: Opaque. *Color:* Slate-gray tarnishing to blue-black. *Luster:* Metallic when freshly broken, soon becoming dull. *Anisotropism:* Strong; brownish gray to gray-white.
R_1–R_2: (400) 17.2–43.7, (420) 17.2–42.8, (440) 17.2–41.9, (460) 17.0–41.0, (480) 16.9–40.0, (500) 16.6–39.2, (520) 16.2–38.3, (540) 15.8–37.4, (560) 15.1–36.6, (580) 14.4–35.8, (600) 13.8–35.0, (620) 13.6–34.2, (640) 13.8–33.2, (660) 14.7–32.4, (680) 16.2–31.6, (700) 18.8–30.8

Cell Data: *Space Group:* $P6_3/mmc$. $a = 3.938$ $c = 17.25$ Z = 6

X-ray Powder Pattern: Synthetic.
2.88 (100), 3.18 (90), 1.969 (80), 3.34 (60), 1.821 (60), 1.623 (50), 2.00 (40)

Chemistry:

	(1)	(2)
Cu	44.7	44.59
Ag	0.3	
Se	54.1	55.41
Total	99.1	100.00

(1) Locality uncertain; by electron microprobe. (2) CuSe.

Occurrence: Of hydrothermal origin.

Association: Clausthalite, umangite, eucairite, berzelianite, crookesite, chalcomenite.

Distribution: From the Sierra de Umango, and in the Santa Brigida mine, La Rioja Province; and the Sierra de Cacheuta, Mendoza Province, Argentina. At Lerbach and Tilkerode, Harz Mountains, Germany. From Skrikerum, Kalmar, Sweden. In the Pinky Fault uranium deposit, Lake Athabasca, Saskatchewan, Canada. From Mexico, at Moctezuma, Sonora. At Bukov and Petrovice, Czechoslovakia.

Name: For Professor Friedrich Klockmann (1858–1937), German mineralogist of Aachen.

Type Material: n.d.

References: (1) Palache, C., H. Berman, and C. Frondel (1944) Dana's system of mineralogy, (7th edition), v. I, 251. (2) Earley, J.W. (1949) Studies of natural and artificial selenides: I. Klockmannite, CuSe. Amer. Mineral., 34, 435–440. (3) Berry, L.G. and R.M. Thompson (1962) X-ray powder data for the ore minerals. Geol. Soc. Amer. Mem. 85, 68. (4) Berry, L.G. (1954) The crystal structure of covellite, CuS and klockmannite, CuSe. Amer. Mineral., 39, 504–509. (5) Effenberger, H. and F. Pertlik (1981) Ein beitrag zur Kristallstruktur von α–CuSe (Klockmannit). Neues Jahrb. Mineral., Monatsh., 197–205 (in German with English abs.).

Crystal Data: Orthorhombic. *Point Group:* $2/m\ 2/m\ 2/m$. Bladed, fibrous, massive, granular.

Physical Properties: *Cleavage:* Good on {010}. Hardness = 2.5–3 VHN = n.d. D(meas.) = 6.48 D(calc.) = 6.51

Optical Properties: Opaque. *Color:* Blackish lead-gray to steel-gray; in polished section, white. *Streak:* Black. *Luster:* Metallic. *Pleochroism:* Weak. *Anisotropism:* Distinct.
R_1–R_2: (400) 43.4–50.6, (420) 44.3–50.3, (440) 45.2–50.0, (460) 45.1–49.5, (480) 44.4–48.8, (500) 43.6–48.0, (520) 42.7–47.2, (540) 41.9–46.4, (560) 41.2–45.7, (580) 40.6–45.0, (600) 40.0–44.4, (620) 39.6–44.0, (640) 39.4–43.5, (660) 39.2–43.1, (680) 39.0–43.0, (700) 39.0–42.8

Cell Data: *Space Group:* *Pnnm* or *Pnmm*. $a = 22.62$ $b = 34.08$ $c = 4.02$ Z = 4

X-ray Powder Pattern: Vena, Sweden.
3.54 (100), 3.41 (90), 2.72 (50), 3.98 (40), 3.27 (40), 2.85 (20), 2.150 (20)

Chemistry:

	(1)	(2)
Pb	33.2	38.0
Cu	1.0	
Ag	0.5	
Fe	0.6	
Bi	37.6	28.5
Sb	9.6	15.0
S	18.6	18.0
Total	101.1	99.5

(1) Vena, Sweden; by electron microprobe. (2) Raleigh, North Carolina, USA; by electron microprobe.

Polymorphism & Series: Forms a series with tintinaite.

Occurrence: A high-temperature hydrothermal mineral.

Association: Cobaltite, arsenopyrite, chalcopyrite (Vena, Sweden); bismuthinite, jamesonite, tetrahedrite (Raleigh, North Carolina, USA).

Distribution: In the USA, from the Superior stone quarry, Raleigh, Wake Co., North Carolina. In Colorado, at the Silver Bell mine, Red Mountain, San Juan Co., and near Leadville, Lake Co. In Canada, at the Deer Park mine, in the Rossland area, and the Dodger tungsten mine, Salmo, British Columbia; and from the Tintina silver mines, Yukon Territory. In Sweden, at the Vena mines, near Askersund, Örebro. From Smolotely, near Příbram, Czechoslovakia. At Ciclova, Romania. In the Ustarasai deposit, Uzbekistan, USSR. From Zeehan, Tasmania, Australia. Becoming known in small amounts from an increasing number of localities.

Name: For Wolfgang Franz von Kobell (1803–1882), German mineralogist.

Type Material: n.d.

References: (1) Palache, C., H. Berman, and C. Frondel (1944) Dana's system of mineralogy, (7th edition), v. I, 447–448. (2) Harris, D.C., J.L. Jambor, G.R. Lachance, and R.I. Thorpe (1968) Tintinaite, the antimony analogue of kobellite. Can. Mineral., 9, 371–382. (3) Miehe, G. (1971) The crystal structure of kobellite. Nature, Phys. Sci., 231, 133–134. (4) Moëlo, Y., J.L. Jambor, and D.C. Harris (1984) Tintinaïte et sulfosels associés de Tintina (Yukon): la cristallochimie de la série de la kobellite. Can. Mineral., 22, 219–226 (in French with English abs.). (5) (1985) Amer. Mineral., 70, 441 (abs. ref. 4). (6) Zakrzewski, M.A. and E. Makovicky (1986) Izoklakeite from Vena, Sweden, and the kobellite homologous series. Can. Mineral., 24, 7–18.

Crystal Data: Orthorhombic. *Point Group:* mmm, $mm2$, or 222. Flame-like aggregates and veinlets up to 20 μm across, in rims and intergrowths with other tellurides and sulfides.

Physical Properties: Hardness = n.d. VHN = n.d. D(meas.) = n.d. D(calc.) = 9.14

Optical Properties: Opaque. *Color:* Gray in reflected light.
R_1–R_2: 25.0–26.1 (400), 25.3–26.2 (420), 25.6–26.5 (440), 25.9–27.0 (460), 26.5–27.5 (480), 27.0–27.9 (500), 27.4–28.3 (520), 27.7–28.6 (540), 28.1–29.0 (560), 28.3–29.2 (580), 28.5–29.4 (600), 28.6–29.6 (620), 28.6–29.6 (640), 28.6–29.5 (660), 28.6–29.5 (680), 28.6–29.5 (700)

Cell Data: *Space Group:* $Pmmm$, $Pmm2$, or $P222$. $a = 5.93(5)$ $b = 3.25(5)$ $c = 3.89(5)$ Z = 1

X-ray Powder Pattern: Champion Reef mine, India. Electron diffraction.
3.27 (100), 2.35 (50), 3.91 (40), 2.00 (40), 1.86 (30), 1.79 (30), 1.50 (30)

Chemistry:

	(1)	(2)
Pb	50.05	51.07
Te	32.87	31.45
S	0.00	
Cl	16.88	17.48
Total	99.60	100.00

(1) Champion Reef mine, India; by electron microprobe, leading to $Pb_{1.00}Te_{1.06}Cl_{1.94}$.
(2) $PbTeCl_2$.

Occurrence: Of primary origin, in a hydrothermal gold-quartz vein containing sulfides and selenides.

Association: Galena, altaite, cotunnite, radhakrishnaite, volynskite, hessite, pyrrhotite, chalcopyrite, ullmannite, hawleyite, cadmian tetrahedrite, cadmian sphalerite.

Distribution: From the Champion Reef mine, Kolar Gold Fields, Karnataka, India.

Name: For its occurrence in the Kolar deposit, India.

Type Material: Institute of Geology of Ore Deposits, Petrography, Mineralogy and Geochemistry; A.E. Fersman Mineralogical Museum, Academy of Sciences, Moscow, USSR.

References: (1) Genkin, A.D., Y.G. Safonov, V.N. Vasudev, B. Krishna Rao, V.A. Boronikhin, L.N. Vyalsov, A.I. Gorshkov, and A.V. Mokhov (1985) Kolarite $PbTeCl_2$ and radhakrishnaite $PbTe_3(Cl,S)_2$, new mineral species from the Kolar gold deposit, India. Can. Mineral., 23, 501–506. (2) (1986) Amer. Mineral., 71, 1545–1546 (abs. ref. 1).

Crystal Data: Cubic. *Point Group:* $4/m\ \overline{3}\ 2/m$, $\overline{4}3m$, or 432. Intergrown with copper aggregates, with individual crystals of cubo-octahedral form, less than 5 μm in size.

Physical Properties: *Tenacity:* Brittle. Hardness = n.d. VHN = 220–267, 247 average (20 g load). D(meas.) = 13.0 D(calc.) = 13.1

Optical Properties: Opaque. *Color:* Tin-white on fresh fracture, quickly altering in moist air to brownish black. *Luster:* Metallic.

R: (400) — , (420) 59.0, (440) 61.4, (460) 63.8, (480) 66.8, (500) 69.4, (520) 71.2, (540) 72.1, (560) 72.8, (580) 73.6, (600) 74.0, (620) 74.6, (640) 75.3, (660) 76.0, (680) 76.9, (700) 77.8

Cell Data: *Space Group:* $Im3m$, $I432$, or $I\overline{4}3m$. Synthetic material is $I\overline{4}3m$. $a = 9.418(4)$ Z = 4

X-ray Powder Pattern: Magadan region, USSR.
2.22 (100), 2.52 (42), 2.98 (25), 2.09 (25), 2.01 (25), 1.279 (25b), 1.524 (18b)

Chemistry:

	(1)	(2)
Cu	26.6	26.98
Hg	72.6	73.02
Total	99.2	100.00

(1) Magadan region, USSR; by electron microprobe, average of nine analyses, corresponding to $Cu_{6.97}Hg_{6.03}$. (2) Cu_7Hg_6.

Occurrence: Found in the heavy mineral fraction of concentrates (Magadan region, USSR).

Association: Copper, stibnite, berthierite, pyrite, arsenopyrite, quartz (Magadan region, USSR).

Distribution: At the Krokhalin antimony deposit, Magadan region, basin of the Kolyma River, Yakutia, USSR. At a prospect near Marcelita, about 70 km southeast of Copiapó, Chile.

Name: For the locality near the Kolyma River, USSR.

Type Material: Institute of Mineralogy and Geochemistry of Rare Elements; A.E. Fersman Mineralogical Museum, Academy of Sciences, Moscow, USSR.

References: (1) Markova, E.A., N.M. Chernitsova, Y.S. Borodaev, L.S. Dubakina, and O.E. Yushko-Zakharova (1980) The new mineral kolymite, Cu_7Hg_6. Zap. Vses. Mineral. Obshch., 109, 206–211 (in Russian). (2) (1981) Amer. Mineral., 66, 218 (abs. ref. 1).

Crystal Data: Hexagonal. *Point Group:* $6/m$, 6, $6/m\ 2/m\ 2/m$, 622, or $\bar{6}m2$. As inclusions in Pt–Fe alloy.

Physical Properties: *Cleavage:* In two directions, average. *Tenacity:* Brittle. Hardness = n.d. VHN = 372–793, 592 average. D(meas.) = n.d. D(calc.) = n.d.

Optical Properties: Opaque. *Color:* Steel-gray. *Luster:* Metallic.
R_1–R_2: n.d.

Cell Data: *Space Group:* $P6/m$, $P6$, $P6/mmm$, $P622$, or $P\bar{6}m2$. $a = 7.024(20)$ $c = 16.48(2)$ Z = n.d.

X-ray Powder Pattern: Konder massif, USSR.
2.98 (10), 1.763 (10), 2.459 (9), 2.85 (5), 1.715 (5), 1.291 (3), 5.10 (2)

Chemistry:

	(1)
Rh	14.2
Pt	25.2
Ir	19.2
Pb	9.53
Cu	8.25
Fe	0.28
Ni	0.38
S	23.7
Total	100.74

(1) Konder massif, USSR; by electron microprobe, average of 10 analyses, corresponding to $Pb_{1.00}(Cu_{2.81}Ni_{0.14}Fe_{0.11})_{\Sigma=3.06}(Rh_{2.99}Pt_{2.80}Ir_{2.16})_{\Sigma=7.95}S_{16.00}$.

Occurrence: As inclusions in a Pt–Fe alloy from an alkali-ultramafic massif.

Association: Pt–Fe alloy, erlichmanite.

Distribution: From the Konder massif, Aldan Shield, Yakutia, USSR.

Name: For the occurrence in the Konder massif, USSR.

Type Material: Mining Museum, Leningrad Mining Institute, Leningrad, USSR.

References: (1) Rudashevskii, N.S., A.G. Mochalov, N.V. Trubkin, A.I. Gorshkov, Y.P. Men'shikov, and N.I. Shumskaya (1984) Konderite, $PbCu_3(Rh, Pt, Ir)_8S_{16}$, a new mineral. Zap. Vses. Mineral. Obshch., 113, 703–712 (in Russian). (2) (1986) Amer. Mineral., 71, 229 (abs. ref. 1).

Crystal Data: Orthorhombic. *Point Group:* n.d. As small (20–50 μm) equant to slightly elongated grains. *Twinning:* Fine lamellar, 0.1 to 20 μm thick.

Physical Properties: *Cleavage:* Distinct in one direction. *Tenacity:* Brittle. Hardness = 2–2.5 VHN = 180–186 (25 g load). D(meas.) = n.d. D(calc.) = [7.94]

Optical Properties: Opaque. *Color:* Grayish white; in polished section, creamy white. *Luster:* Metallic. *Anisotropism:* Distinct.

R_1–R_2: (400) 38.7–54.0, (420) 41.5–55.9, (440) 44.0–57.7, (460) 46.0–59.2, (480) 47.5–60.3, (500) 48.8–61.1, (520) 49.7–61.6, (540) 50.5–61.7, (560) 51.2–61.6, (580) 51.7–61.4, (600) 52.1–61.2, (620) 52.3–61.0, (640) 52.5–60.8, (660) 52.7–60.7, (680) 52.9–60.7, (700) 53.3–60.7

Cell Data: *Space Group:* n.d. $a = 16.50$ $b = 8.84$ $c = 4.42$ Z = [4]

X-ray Powder Pattern: Chelopech, Bulgaria.
3.03 (100), 2.10 (90), 2.93 (60), 5.03 (50), 3.36 (50), 3.24 (50), 2.23 (40)

Chemistry:

	(1)	(2)	(3)	(4)
Au	25.2	27.2	27.37	25.55
Ag	0.4	0.5	4.03	
Cu	7.7	7.3	4.85	8.24
Fe		0.1		
Sb			< 0.34	
Te	67.6	64.4	63.97	66.21
S		0.1		
Total	100.9	99.6	100.56	100.00

(1) Chelopech, Bulgaria; by electron microprobe, corresponding to $Au_{0.97}Ag_{0.03}Cu_{0.92}Te_{4.00}$. (2) Bisbee, Arizona, USA; by electron microprobe. (3) Kochbulak deposit, USSR. (4) $CuAuTe_4$.

Occurrence: From gold- and tellurium-bearing replacement copper deposits (Chelopech, Bulgaria).

Association: Këesterite, tellurium, chalcopyrite, tennantite, barite (Chelopech, Bulgaria); tellurium, hessite, sylvanite (Buckeye Gulch, Colorado, USA).

Distribution: From Chelopech, Bulgaria. In the Campbell mine, Bisbee, Cochise Co., Arizona, and at Buckeye Gulch, near Leadville, Lake Co., Colorado, USA. From the Kochbulak deposit, eastern Uzbekistan, USSR.

Name: For Professor Ivan Kostov (1913–), Bulgarian mineralogist.

Type Material: Geological Institute, Bulgarian Academy of Sciences; University of Sophia, Bulgaria; Institute of Geology of Ore Deposits, Petrography, Mineralogy and Geochemistry, Moscow, USSR.

References: (1) Terziev, G. (1966) Kostovite, a gold–copper telluride from Bulgaria. Amer. Mineral., 51, 29–36. (2) Kovalenker, V.A., N.V. Troneva, O.V. Kuz'mina, L.N. Vyal'sov, and P.M. Goloshchukov (1979) First occurrence of kostovite in the USSR. Doklady Acad. Nauk SSSR, 247, 1249–1252 (in Russian). (3) (1980) Chem. Abs., 92, 44792 (abs. ref. 2).

Crystal Data: Hexagonal. *Point Group:* n.d. As minute grains in other minerals.

Physical Properties: Hardness = n.d. VHN = 277–322, 291 average (15 g load). D(meas.) = n.d. D(calc.) = n.d.

Optical Properties: Opaque. *Color:* Steel-gray; in polished section, cream or pale yellow. *Luster:* Metallic. *Pleochroism:* Distinct, from light cream to darker grayish cream. *Anisotropism:* Strong, from gray to dark bluish gray.
R_1–R_2: (400) 36.4–33.6, (420) 41.5–41.5, (440) 46.6–49.4, (460) 50.2–54.8, (480) 53.4–59.5, (500) 56.0–63.1, (520) 58.2–65.6, (540) 59.6–67.4, (560) 60.7–68.5, (580) 61.2–69.0, (600) 61.5–69.3, (620) 61.9–69.7, (640) 62.3–70.2, (660) 62.8–70.8, (680) 63.3–71.4, (700) 63.9–72.0

Cell Data: *Space Group:* n.d. $a = 4.145$ $c = 5.67$ Z = n.d.

X-ray Powder Pattern: Rustenburg mine, South Africa.
3.03 (100), 2.22 (90), 2.08 (70), 1.52 (30), 1.72 (20), 1.67 (20), 1.32 (10)

Chemistry:

	(1)	(2)	(3)
Pd	31.1	45.9	44.3
Pt			2.31
Bi	24.9	17.2	1.80
Te	44.0	38.0	53.5
Total	100.0	101.1	101.91

(1) Monchegorsk deposit, USSR; by electron microprobe, corresponding to $Pd_{1.0}(Te_{1.2}Bi_{0.4})_{\Sigma=1.6}$. (2) Rustenburg mine, South Africa; by electron microprobe. (3) Thierry mine, Canada; by electron microprobe, corresponding to $(Pd_{0.41}Pt_{0.012})_{\Sigma=0.422}(Te_{0.42}Bi_{0.009})_{\Sigma=0.429}$.

Occurrence: As grains in chalcopyrite veins cutting magnetite (Monchegorsk deposit, USSR).

Association: Moncheite, michenerite, hessite, merenskyite, chalcopyrite, magnetite (Monchegorsk deposit, USSR); leadamalgam, chromite, ilmenite, magnetite, gersdorffite, pyrite, chalcopyrite, violarite, millerite, galena, stibnite, argentian gold, niggliite, sperrylite, iridosmine, platinum, merenskyite (Shiaonanshan, China); merenskyite, moncheite, stützite, chalcopyrite, pyrrhotite, pentlandite (Thierry mine, Canada).

Distribution: From the Monchegorsk deposit, Kola Peninsula, and the Noril'sk region, western Siberia, USSR. In South Africa, at the Rustenburg platinum mine, on the Merensky Reef of the Bushveld Complex; and in the Artonvilla mine, Messina, Transvaal. In Canada, in the Levak West and Creighton mines, Sudbury; the Lac des Iles Complex; and from the Thierry mine, near Pickle Lake, Ontario. In the Stillwater Complex, Montana; at the New Rambler mine, Medicine Bow Mountains, Albany Co., Wyoming; in the Key West mine, east of Moapa, Clark Co., Nevada; and on the Yuba River, Nevada Co., California, USA. In China, at Shiaonanshan, Inner Mongolia Autonomous Region, and at Danba, Sichuan Province.

Name: For Vladimir Klement'evich Kotul'skii, economic geologist of the USSR, an authority on Cu-Ni sulfide deposits.

Type Material: n.d.

References: (1) Genkin, A.D., N.N. Zhuravlev, and E.M. Smirnova (1963) Moncheite and kotulskite – new minerals – and the composition of michenerite. Zap. Vses. Mineral. Obshch., 92, 33–50 (in Russian). (2) (1963) Amer. Mineral., 48, 1181 (abs. ref. 1). (3) Kingston, G.A. (1966) The occurrence of platinoid bismuthotellurides in the Merensky Reef at Rustenburg platinum mine in the western Bushveld. Mineral. Mag., 35, 815–834. (4) (1967) Amer. Mineral., 52, 928 (abs. ref. 3). (5) Patterson, G.C. and D.H. Watkinson (1984) Metamorphism and supergene alteration of Cu–Ni sulfides, Thierry mine, northwestern Ontario. Can. Mineral., 22, 13–21.

Crystal Data: Hexagonal. *Point Group:* n.d. In fine grains as an intergrowth.

Physical Properties: Hardness = ~3.5 VHN = 136–157 (25 g load). D(meas.) = 8.48 D(calc.) = 8.437

Optical Properties: Opaque. *Color:* In polished section, bluish gray. *Streak:* Black. *Luster:* Metallic. *Anisotropism:* Strong.

R_1–R_2: (400) 52.6–50.8, (420) 51.0–50.3, (440) 49.4–49.8, (460) 47.5–48.7, (480) 45.3–47.5, (500) 43.3–46.1, (520) 41.4–44.5, (540) 39.7–43.1, (560) 38.3–41.8, (580) 37.0–40.8, (600) 36.0–40.4, (620) 35.4–40.2, (640) 34.8–40.2, (660) 34.4–40.3, (680) 34.1–40.4, (700) 34.0–40.7

Cell Data: *Space Group:* n.d. $a = 11.54$ $c = 14.36$ Z = 18

X-ray Powder Pattern: Daluis, France.
2.090 (100), 2.020 (100), 1.998 (100), 2.45 (80), 2.16 (50), 1.791 (40), 1.750 (40)

Chemistry:

	(1)	(2)	(3)
Cu	68.2	67.98	67.95
Fe		0.10	
As	32.5	31.69	32.05
Total	100.7	99.77	100.00

(1) Daluis, France; by electron microprobe. (2) Wasserfall, France; by electron microprobe. (3) Cu_5As_2.

Occurrence: In arsenical copper deposits.

Association: Arsenic, silver, skutterudite, löllingite, chalcocite, algodonite, domeykite, allargentum, kutinaite, paxite, calcite.

Distribution: From Černý Důl, Krkonoše (Giant Mountains), Czechoslovakia. In France, from Daluis, Vallée du Var; Lautaret, Hautes-Alpes; and Wasserfall, southern Vosges. At Långban, and from the Harstigen mine, Pajsberg, Värmland, Sweden. From the Talmessi and Meskani deposits, Anarak district, Iran. At Mohawk, Keeweenaw Co., Michigan, USA.

Name: For J. Koutek, Professor of Economic Geology, Charles University, Prague, Czechoslovakia.

Type Material: n.d.

References: (1) Johan, Z. (1958) Koutekite; a new mineral. Nature, 181, 1553–1554. (2) (1958) Amer. Mineral., 43, 794 (abs. ref. 1). (3) Johan, Z. (1960) Koutekite—Cu_2As, ein neues Mineral. Chem. Erde, 20, 217–226. (4) (1961) Amer. Mineral., 46, 467 (abs. ref. 3). (5) Picot, P. and J. Vernet (1967) Un nouveau gisement de koutekite: Le dôme du Barrot (Alpes Maritimes). Bull. Soc. fr. Minéral., 60, 82–89. (6) Picot, P. and F. Ruhlmann (1978) Présence d'arséniures de cuivre de haute température dan le granite des Ballons (Vosges méridionales). Bull. Minéral., 101, 563–569 (in French with English abs.). (7) Makovicky, M. and Z. Johan (1978) Reflectivity and microhardness of synthetic and natural koutekite, kutinaite and beta-domeykite. Neues Jahrb. Mineral., Monatsh., 421–432.

Crystal Data: Orthorhombic. *Point Group: mm*2. Crystals typically short prismatic and striated || [001].

Physical Properties: *Cleavage:* Perfect on {001}. *Fracture:* Subconchoidal to uneven. *Tenacity:* Brittle. Hardness = 2–3 VHN = 166–186 (100 g load). D(meas.) = 8.62 D(calc.) = 8.86

Optical Properties: Opaque. *Color:* Silver-white to light brass-yellow (tarnish ?); in polished section, creamy white. *Luster:* Very high metallic. *Pleochroism:* Weak. *Anisotropism:* Strong, to dark brown.

R_1–R_2: (400) 47.4–52.3, (420) 50.2–54.7, (440) 53.0–57.1, (460) 55.5–59.0, (480) 57.5–60.4, (500) 58.8–61.3, (520) 59.8–62.0, (540) 60.4–62.4, (560) 60.9–62.8, (580) 61.1–63.0, (600) 61.5–63.4, (620) 62.0–63.9, (640) 62.3–64.4, (660) 62.6–64.8, (680) 62.8–65.2, (700) 62.8–65.5

Cell Data: *Space Group: Pma*2. $a = 16.58(1)$ $b = 8.849(5)$ $c = 4.464(3)$ Z = 8

X-ray Powder Pattern: Cripple Creek, Colorado, USA.
3.03 (100), 2.11 (70), 2.94 (60), 2.23 (50), 2.07 (40), 1.78 (40), 1.69 (40)

Chemistry:

	(1)	(2)	(3)
Au	36.19	34.77	43.59
Ag	4.87	5.87	
Fe	0.05		
Te	58.50	58.60	56.41
insol.	0.09	1.58	
Total	99.70	100.82	100.00

(1) Moose mine, Cripple Creek, Colorado, USA; Fe is from pyrite. (2) Săcărâmb, Romania. (3) $AuTe_2$.

Occurrence: Found in hydrothermal veins with other tellurides.

Association: Calaverite, coloradoite, sylvanite, krennerite, petzite, hessite, tellurium, gold, pyrite, quartz.

Distribution: In the USA, in Colorado, at several mines in Cripple Creek, Teller Co.; the Smoky Hill mine, Boulder Co.; and the Central City district, Gilpin Co. In the Campbell mine, Bisbee; and at Tombstone, Cochise Co., Arizona. In Canada, at the Bencourt mine, Louvricourt, Quebec. At Kalgoorlie and Mulgabbie, Western Australia. At Săcărâmb (Nagyág) and Facebánya, Romania. In the Konstantin-Ovkoye gold deposit, Krasnoyarskii region, USSR. From the Emperor mine, Vatukoula, Viti Levu, Fiji Islands.

Name: For Joseph A. Krenner (1839–1920), mineralogist of Hungary.

References: (1) Palache, C., H. Berman, and C. Frondel (1944) Dana's system of mineralogy, (7th edition), v. I, 333–335. (2) Ramdohr, P. (1969) The ore minerals and their intergrowths, (3rd edition), 427–428. (3) Pertlik, F. (1984) Crystal chemistry of natural tellurides. II: Redetermination of the crystal structure of krennerite, $(Au_{1-x}Ag_x)Te_2$ with x ~0.2. Tschermaks Mineral. Petrog. Mitt., 33, 253–262.

Crystal Data: Orthorhombic. *Point Group:* *mm*2. Fibrous aggregates of microscopic parallel intergrowths with bismuthinite.

Physical Properties: Hardness = n.d. VHN = 187 (100 g load). D(meas.) = n.d. D(calc.) = 6.98

Optical Properties: Opaque. *Color:* Steel-gray; in polished section, grayish white. *Luster:* Metallic.

R_1–R_2: n.d.

Cell Data: *Space Group:* $Pmc2_1$. $a = 4.003(3)$ $b = 11.200(9)$ $c = 11.560(9)$ Z = 2

X-ray Powder Pattern: Juno mine, Australia.
3.74 (100), 2.84 (37), 3.65 (30), 3.16 (27), 2.66 (26), 1.970 (26), 2.55 (24)

Chemistry:

	(1)	(2)	(3)
Pb	19.3	19.0	19.01
Cu	5.95	5.8	5.83
Bi	59.8	57.7	57.51
Se	0.95		
S	17.25	17.7	17.65
Total	103.25	100.2	100.00

(1) Krupka, Czechoslovakia; by electron microprobe. (2) Juno mine, Australia; by electron microprobe. (3) $PbCuBi_3S_6$.

Occurrence: In a molybdenite-feldspar deposit, in a vein between gneiss and an orthoclase pegmatite (Krupka, Czechoslovakia).

Association: Bismuthinite, bismuth, gladite, cassiterite, quartz, lithium mica (Krupka, Czechoslovakia).

Distribution: From the Barbora gallery, Krupka, Kruňé Hory Mountains, and at Dobšiná (Dobschau), Czechoslovakia. In Australia, in the Juno mine, Tennant Creek, Northern Territory. From Temiskaming, Quebec, Canada. In the USA, from Cucomungo Canyon, Esmeralda Co., Nevada, and in Ball's mine, Little Cottonwood district, Salt Lake Co., Utah.

Name: For the locality in Czechoslovakia at Krupka.

Type Material: Charles University, Prague, Czechoslovakia.

References: (1) Žák, L., V. Syneček, and J. Hybler (1974) Krupkaite, $CuPbBi_3S_6$, a new mineral of the bismuthinite–aikinite group. Neues Jahrb. Mineral., Monatsh., 533–541. (2) Syneček, V. and J. Hybler (1974) The crystal structure of krupkaite, $CuPbBi_3S_6$, and of gladite, $CuPbBi_5S_9$, and the classification of superstructures in the bismuthinite–aikinite group. Neues Jahrb. Mineral., Monatsh., 541–560. (3) (1975) Amer. Mineral., 60, 737 (abs. refs. 1 and 2). (4) Mumme, W.G. (1975) The crystal structure of krupkaite, $CuPbBi_3S_6$, from the Juno mine at Tennant Creek, Northern Territory, Australia. Amer. Mineral., 60, 300–308.

Crystal Data: Cubic. *Point Group:* $2/m\ \bar{3}$. As zoned hypidiomorphic crystals up to 10 mm included in clausthalite.

Physical Properties: Hardness = n.d. VHN = 248 (25 g load). D(meas.) = n.d. D(calc.) = 6.53

Optical Properties: Opaque. *Color:* Gray. *Luster:* Metallic. *Anisotropism:* Zones rich in Co and Ni are anisotropic in yellow-brown tints.

R: (400) 36.0, (420) 36.2, (440) 36.4, (460) 36.3, (480) 36.1, (500) 35.8, (520) 35.4, (540) 34.9, (560) 34.3, (580) 33.6, (600) 33.0, (620) 32.4, (640) 32.0, (660) 31.7, (680) 31.5, (700) 31.4

Cell Data: *Space Group:* $Pa3$. $a = 6.056$ Z = 4

X-ray Powder Pattern: Petrovice, Czechoslovakia.
2.712 (100), 3.023 (90), 2.477 (80), 1.827 (60), 1.678 (40), 1.166 (40), 1.618 (30)

Chemistry:

	(1)	(2)	(3)
Cu	20.6	21.5	28.69
Co	4.8	5.5	
Ni	1.5	4.5	
Fe	0.6	0.5	
Hg	0.2	0.2	
Se	69.5	67.1	71.31
Total	97.2	99.3	100.00

(1) Petrovice, Czechoslovakia; by electron microprobe, corresponding to $(Cu_{0.74}Co_{0.18}Ni_{0.06}Fe_{0.03})_{\Sigma=1.01}Se_{1.99}$. (2) Do.; corresponding to $(Cu_{0.74}Co_{0.20}Ni_{0.17}Fe_{0.02})_{\Sigma=1.13}Se_{1.87}$. (3) $CuSe_2$.

Occurrence: Of hydrothermal origin.

Association: Clausthalite, eskebornite, uraninite, hematite, ferroselite, bukovite, umangite, berzelianite, chalcopyrite, goethite.

Distribution: From Petrovice, Czechoslovakia.

Name: For Dr. Tomáš Kruťa, Director of the Mineralogy Laboratory, Moravian Museum, Brno, Czechoslovakia.

Type Material: National School of Mines, Paris, France.

References: (1) Johan, Z., P. Picot, R. Pierrot, and M. Kvaček (1972) La krutaïte, $CuSe_2$, un nouveau minéral du groupe de la pyrite. Bull. Soc. fr. Minéral., 95, 475–481 (in French with English abs.). (2) (1974) Amer. Mineral., 59, 210 (abs. ref. 1).

Crystal Data: Cubic. *Point Group:* 23. As grains up to 0.1 mm in close intergrowth with other sulfides.

Physical Properties: Hardness = 5.5 VHN = n.d. D(meas.) = n.d. D(calc.) = [7.08]

Optical Properties: Opaque. *Color:* Grayish white, paler than associated nickel-skutterudite; in polished section, bright white with a rose tint. *Luster:* Metallic.
R: n.d.

Cell Data: *Space Group:* $P2_13$. $a = 5.794$ Z = 4

X-ray Powder Pattern: Jáchymov, Czechoslovakia.
2.593 (100), 2.365 (80), 1.746 (80), 2.897 (60), 1.548 (60), 1.0242 (60), 2.051 (50)

Chemistry:

	(1)	(2)	(3)	(4)
Ni	25.0	26.7	25.04	28.15
Co	0.22	0.30	1.65	
Fe	0.15	0.21	0.26	
Cu	0.86	0.85		
As	73.4	71.9	72.37	71.85
Sb			0.09	
S	0.19	0.15		
Total	99.82	100.11	99.41	100.00

(1–2) Jáchymov, Czechoslovakia; by electron microprobe, averages of 63 total analyses of 2 samples. Ratios (As + S)/total metals, 2.21 and 2.02, yielding $Ni_{1-x}As_2$ (x = 0–0.1). (3) Khovuaksinsk deposit, USSR. (4) $NiAs_2$.

Polymorphism & Series: Trimorphous with rammelsbergite and pararammelsbergite.

Occurrence: In Cu-Ni-As veins of hydrothermal origin.

Association: Nickel-skutterudite, tennantite (Jáchymov, Czechoslovakia); nickeline, breithauptite, löllingite, rammelsbergite, silver (Khovuaksinsk deposit, USSR).

Distribution: In the Geshiber vein, Jáchymov (Joachimsthal), Czechoslovakia. From the Khovuaksinsk As-Ni-Co deposit, Tuva ASSR, USSR.

Name: For Georgi Alekseevich Krutov, Professor of Mineralogy, Moscow Gosdarst University, Moscow, USSR.

Type Material: The Mining Institute, Leningrad; A.E. Fersman Mineralogical Museum, Academy of Sciences, Moscow, USSR; National Museum, Prague, Czechoslovakia.

References: (1) Vinogradova, R.A., N.S. Rudashevskii, I.A. Bud'ko, L.I. Bochek, P. Kaspar, and K. Padera (1976) Krutovite, a new cubic nickel diarsenide. Zap. Vses. Mineral. Obshch., 105, 59–71 (in Russian). (2) (1977) Amer. Mineral., 62, 173–174 (abs. ref. 1). (3) Vinogradova, R.A., N.S. Rudashevskii, L.I. Bochek, and I.A. Bud'ko (1976) First discovery of krutovite in the USSR. Doklady Acad. Nauk SSSR, 230, 938–941 (in Russian). (4) (1977) Chem. Abs., 86, 75940 (abs. ref. 3).

Crystal Data: Orthorhombic. *Point Group:* $2/m\ 2/m\ 2/m$. Fine-grained, massive.

Physical Properties: Hardness = Soft. VHN = n.d. D(meas.) = n.d. D(calc.) = 6.72

Optical Properties: Opaque. *Color:* Lead-gray; in polished section, creamy grayish white, paler than penroseite. *Pleochroism:* Distinct in oil, gray to pale gray. *Anisotropism:* Very strong, yellowish gray to gray to almost black.
R_1–R_2: n.d.

Cell Data: *Space Group:* $Pnnm$, probably. $a = 4.89$ $b = 5.96$ $c = 3.67$ Z = 2

X-ray Powder Pattern: Kuusamo, Finland.
2.64 (100), 2.545 (100), 2.935 (80), 1.925 (80), 1.84 (80), 1.648 (60), 2.095 (40)

Chemistry:

	(1)	(2)
Ni	23.1	27.10
Co	1.4	
Fe	1.91	
Cu	0.5	
Se	73.1	72.90
Total	100.01	100.00

(1) Kuusamo, Finland; by X-ray fluorescence. (2) $NiSe_2$.

Occurrence: In calcite veins in sills of albite diabase in schist, associated with low-grade uranium mineralization, almost exclusively as an alteration product of wilkmanite.

Association: Wilkmanite, sederholmite, penroseite, selenium, ferroselite.

Distribution: From Kuusamo, northeastern Finland.

Name: For Gunnar Kullerud, Geophysical Laboratory, Washington, D.C., USA.

Type Material: n.d.

References: (1) Vuorelainen, Y., A. Huhma, and A. Häkli (1964) Sederholmite, wilkmanite, kullerudite, mäkinenite, and trüstedtite, five new nickel selenide minerals. Compt. Rendus Soc. Géol. Finlande, 36, 113–125. (2) (1965) Amer. Mineral., 50, 519–520 (abs. ref. 1).

Crystal Data: Tetragonal. *Point Group:* $\bar{4}2m$. As minute inclusions, 5–80 μm in diameter, aggregated with goldfieldite in round to elongated grains.

Physical Properties: Hardness = n.d. VHN = 390 (20 g load). D(meas.) = n.d. D(calc.) = 4.56

Optical Properties: Opaque. *Color:* In polished section, neutral gray. *Streak:* Metallic. *Anisotropism:* Distinct, in shades of brown.

R: (400) — , (420) — , (440) 25.2, (460) 25.3, (480) 25.5, (500) 26.0, (520) 26.3, (540) 26.5, (560) 26.6, (580) 26.7, (600) 26.8, (620) 26.7, (640) 26.5, (660) 26.4, (680) 26.4, (700) 26.3

Cell Data: *Space Group:* $I\bar{4}2m$. $a = 5.445(5)$ $c = 10.75(2)$ Z = 2

X-ray Powder Pattern: Kuramin Mountains, USSR.
3.13 (10), 1.914 (8), 1.640 (6), 1.108 (4), 1.244 (3), 2.70 (2), 1.044 (2)

Chemistry:

	(1)	(2)
Cu	37.27	43.56
Fe	2.36	
Zn	1.88	
Sn	30.12	27.13
In	0.32	
S	27.86	29.31
Total	99.81	100.00

(1) Kuramin Mountains, USSR; by electron microprobe, average of five analyses. (2) Cu_3SnS_4.

Occurrence: In gold-sulfide-quartz veins, as inclusions in goldfieldite (Kuramin Mountains, USSR).

Association: Goldfieldite, hessite, petzite, sylvanite, altaite, gold, chalcostibite, emplectite (Kuramin Mountains, USSR).

Distribution: From an undefined locality in the Kuramin Mountains, eastern Uzbekistan, USSR. At Bisbee, Cochise Co., Arizona, USA.

Name: For the occurrence in the Kuramin Mountains, USSR.

Type Material: Institute of Geology of Ore Deposits, Petrography, Mineralogy and Geochemistry; A.E. Fersman Mineralogical Museum, Academy of Sciences, Moscow, USSR.

References: (1) Kovalenker, V.A., T.L. Evstigneeva, N.V. Traneva, and L.N. Vyal'sov (1979) Kuramite, Cu_3SnS_4, a new mineral of the stannite group. Zap. Vses. Mineral. Obshch., 108, 564–569 (in Russian). (2) (1980) Amer. Mineral., 65, 1067 (abs. ref. 1).

Crystal Data: Cubic. *Point Group:* n.d. Fine-grained, as intergrowths with novákite.

Physical Properties: *Tenacity:* Malleable. Hardness = ~4.5 VHN = 237–270 (25 g load). D(meas.) = 8.38(7) (synthetic). D(calc.) = 8.37

Optical Properties: Opaque. *Color:* Silvery gray; in polished section, grayish white to light blue-gray. *Luster:* Metallic.

R: (400) 49.6, (420) 48.8, (440) 48.0, (460) 46.8, (480) 45.6, (500) 44.5, (520) 43.5, (540) 42.6, (560) 42.0, (580) 41.4, (600) 41.1, (620) 40.8, (640) 40.5, (660) 40.4, (680) 40.5, (700) 40.9

Cell Data: *Space Group:* n.d. $a = 11.76$ Z = 4

X-ray Powder Pattern: Černý Důl, Czechoslovakia.
2.259 (100), 2.078 (100), 2.702 (90), 2.398 (80), 1.991 (70), 1.959 (70), 1.776 (70)

Chemistry:

	(1)	(2)	(3)
Cu	43.68	43.89	43.16
Ag	31.63	31.63	31.40
As	25.16	24.26	25.44
Total	100.47	99.78	100.00

(1) Černý Důl, Czechoslovakia; by electron microprobe. (2) Wasserfall, France; by electron microprobe. (3) $Cu_{14}Ag_6As_7$.

Occurrence: In carbonate-rich hydrothermal veins (Černý Důl, Czechoslovakia).

Association: Novákite, koutekite, paxite, arsenolamprite, löllingite (Černý Důl, Czechoslovakia); allargentum, koutekite, domeykite (Wasserfall, France); koutekite, domeykite (Meskani, Iran); paxite, lautite (Niederbeerbach, Germany).

Distribution: At Černý Důl, Krkonoše (Giant Mountains), Czechoslovakia. From Wasserfall, southern Vosges, France. In the Meskani mine, Anarak district, Iran. From Niederbeerbach, Odenwald, Germany.

Name: For Dr. J. Kutina, Lecturer at the Charles University, Prague, Czechoslovakia.

Type Material: Charles University, Prague, Czechoslovakia; Yale University, New Haven, Connecticut, USA.

References: (1) Hak, J., Z. Johan, and B.J. Skinner (1970) Kutinaite, a new copper–silver arsenide mineral from Černý Důl, Czechoslovakia. Amer. Mineral., 55, 1083–1087. (2) Picot, P. and F. Ruhlmann (1978) Présence d'arséniures de cuivre de haute température dan le granite des Ballons (Vosges méridionales). Bull. Minéral., 101, 563–569 (in French with English abs.). (3) Makovicky, M. and Z. Johan (1978) Reflectivity and microhardness of synthetic and natural koutekite, kutinaite and beta-domeykite. Neues Jahrb. Mineral., Monatsh., 421–432. (4) Johan, Z. (1985) The Černy Důl deposit (Czechoslovakia): an example of Ni-, Fe-, Ag-, Cu-arsenide mineralization with extremely high activity of arsenic; new data on paxite, novakite and kutinaite. Tschermaks Mineral. Petrog. Mitt., 34, 167–182.

Crystal Data: Monoclinic, pseudocubic. *Point Group:* $2/m$. As tiny grains up to 1 mm.

Physical Properties: *Cleavage:* Imperfect in one direction. Hardness = n.d. VHN = 92–138 (25 g load). D(meas.) = 6.11 (synthetic). D(calc.) = 6.195

Optical Properties: Opaque. *Color:* Dark red; in polished section, bluish white with numerous purple-red internal reflections. *Streak:* Brownish red. *Luster:* Metallic, adamantine. *Pleochroism:* Strong. *Anisotropism:* Strong, from light gray-blue to dark gray.

R_1–R_2: (400) 34.5–38.4, (420) 34.6–37.7, (440) 34.7–37.0, (460) 34.4–36.1, (480) 33.8–35.2, (500) 32.9–34.3, (520) 31.8–33.4, (540) 30.7–32.5, (560) 29.5–31.7, (580) 28.6–30.8, (600) 28.0–30.1, (620) 27.4–29.7, (640) 27.1–29.4, (660) 26.8–29.4, (680) 26.6–29.2, (700) 26.3–28.9

Cell Data: *Space Group:* $A2/a$. $a = 7.732(3)$ $b = 11.285(7)$ $c = 6.643(4)$ $\beta = 115.16(3)°$ Z = 4

X-ray Powder Pattern: Jas Roux mine, France.
2.682 (100), 3.198 (80), 3.005 (80), 3.505 (70), 2.133 (60), 5.31 (50), 1.905 (50)

Chemistry:

	(1)
Ag	21.9
Hg	41.0
As	15.6
S	20.3
Total	98.8

(1) Jas Roux mine, France; by electron microprobe, corresponding to $Ag_{0.97}Hg_{0.98}As_{1.01}S_{3.04}$.

Occurrence: With other arsenic sulfides, in hydrothermal deposits.

Association: Smithite, stibnite, pierrotite, realgar, sphalerite, pyrite, routhiérite (Jas Roux mine, France); orpiment, realgar, getchellite (Getchell mine, Nevada, USA).

Distribution: In the Jas Roux mine, Hautes-Alpes, France. In the USA, at the Getchell mine, Humboldt Co., Nevada.

Name: For Pierre Laffitte, Director, National School of Mines, Paris, France.

Type Material: National School of Mines, Paris, France.

References: (1) Johan, Z., J. Mantienne, and P. Picot (1974) La routhiérite, $TlHgAsS_3$, et la laffittite, $AgHgAsS_3$, deux nouvelles espèces minérales. Bull. Soc. fr. Minéral., 97, 48–53 (in French with English abs.). (2) (1975) Amer. Mineral., 60, 945–946 (abs. ref. 1). (3) Nakai, I. and D.E. Appleman (1983) Laffittite, $AgHgAsS_3$; crystal structure and second occurrence from the Getchell mine, Nevada. Amer. Mineral., 68, 235–244.

Crystal Data: Hexagonal. *Point Group:* $\bar{3}2/m$. As foliated plates and sheets; grains 0.5–2 mm across.

Physical Properties: *Cleavage:* {001} excellent. Hardness = n.d. VHN = 54–78 (100 g load). D(meas.) = 8.12 D(calc.) = n.d.

Optical Properties: Opaque. *Color:* In polished section, galena-white. *Luster:* Metallic. *Anisotropism:* Moderate, dark brownish gray.
R_1–R_2: (400) 46.7–53.6, (420) 46.6–53.8, (440) 46.5–54.0, (460) 46.2–54.1, (480) 46.0–53.9, (500) 45.7–53.4, (520) 45.4–53.1, (540) 45.0–53.0, (560) 44.7–53.0, (580) 44.3–52.8, (600) 44.0–52.5, (620) 43.7–52.0, (640) 43.3–51.7, (660) 43.0–51.1, (680) 42.5–50.6, (700) 42.1–50.0

Cell Data: *Space Group:* $R\bar{3}m$. $a = 4.225$ $c = 39.93$ Z = 3

X-ray Powder Pattern: Orijärvi mine, Finland.
3.072 (100), 2.246 (80), 2.112 (80), 1.741 (70), 4.425 (60), 3.586 (60), 1.538 (60)

Chemistry:

	(1)	(2)
Bi	78.28	77.9
Pb	0.78	3.4
Ag	0.71	
Cu	0.26	
Zn	0.14	
Se	15.50	15.4
S	3.28	3.3
insol.	0.93	
Total	99.88	100.0

(1) Orijärvi mine, Finland; after deducting 4% of galena, sphalerite, and chalcopyrite this yields $Bi_{3.90}(Se_{2.04}S_{0.92})_{\Sigma=2.96}$. (2) Falun, Sweden; by electron microprobe.

Occurrence: In veinlets in quartz-anthophyllite-cordierite-biotite rocks (Orijärvi mine, Finland).

Association: Chalcopyrite, bismuth, sphalerite, molybdenite, silver, pyrite, galena (Orijärvi mine, Finland); tetrahedrite–tennantite, luzonite–famatinite, pyrite, mawsonite, nekrasovite, chalcopyrite, emplectite, bismuth, calcite, quartz, barite (Kuramin Mountains, USSR); nevskite, wolframite, natanite, cassiterite, guanajuatite (Nevskii tin deposit, USSR).

Distribution: In southwestern Finland, at the Orijärvi mine. From Falun, Sweden. At the Khayragatsch deposit, Kuramin Mountains, eastern Uzbekistan; and at the Nevskii tin deposit, northeastern USSR. From the Akenobe mine, Hyogo Prefecture, Japan. In the Bonser vein, Coniston, Cumbria, England. From the Kidd Creek mine, near Timmins, Ontario, Canada.

Name: For Aarne Laitakari, Director, Geological Survey of Finland, who collected the original material.

Type Material: n.d.

References: (1) Vorma, A. (1959) Laitakarite, a new Bi–Se mineral in Orijärvi. Geologi, 3, 11 (in Finnish). (2) (1959) Amer. Mineral., 44, 908 (abs. ref. 1). (3) Vorma, A. (1960) Laitakarite, a new Bi–Se mineral. Bull. Comm. Geol. Finlande, 188, 1–10 (in English). (4) (1962) Amer. Mineral., 47, 806 (abs. ref. 3).

Crystal Data: Hexagonal. *Point Group:* $6/m\ 2/m\ 2/m$. As irregular grains and lamellae in safflorite.

Physical Properties: Hardness = n.d. VHN = 780–857 (50 g load). D(meas.) = n.d. D(calc.) = 8.174

Optical Properties: Opaque. *Color:* In polished section, pinkish buff. *Luster:* Metallic. *Pleochroism:* Weak. *Anisotropism:* Moderate, in bluish gray and light brown.
R_1–R_2: n.d.

Cell Data: *Space Group:* $P6_3/mmc$. $a = 3.538$ $c = 5.127$ Z = 2

X-ray Powder Pattern: Langis mine, Canada.
2.631 (10), 1.966 (9), 1.770 (8), 1.493 (4), 1.470 (3), 1.315 (3), 1.141 (3)

Chemistry:

	(1)
Co	35.5
Ni	7.0
As	56.0
Total	98.5

(1) Langis mine, Canada; by electron microprobe, corresponding to $(Co_{0.84}Ni_{0.16})_{\Sigma=1.00}As_{1.04}$.

Occurrence: In pockets of ore minerals, mineralized fault gouge, and breccia cemented with calcite and quartz.

Association: Cobalt pentlandite, siegenite, parkerite, bravoite, safflorite, maucherite, pyrite, marcasite.

Distribution: From the Langis mine, Casey Township, Cobalt-Gowganda area, Ontario, Canada.

Name: For the Langis mine in Canada, where it was discovered.

Type Material: n.d.

References: (1) Petruk, W., D.C. Harris, and J.M. Stewart (1969) Langisite, a new mineral, and the rare minerals cobalt pentlandite, siegenite, parkerite and bravoite from the Langis mine, Cobalt–Gowganda area, Ontario. Can. Mineral., 9, 597–616. (2) (1972) Amer. Mineral., 57, 1910–1911 (abs. ref. 1).

Crystal Data: Monoclinic. *Point Group:* $2/m$. As prismatic crystals, < 5 mm, elongated [100] and tabular (010). Crystals are commonly resorbed to give incomplete individuals.

Physical Properties: *Cleavage:* Perfect on {010}. *Tenacity:* Flexible but not elastic, extremely malleable. Hardness = Soft. VHN = n.d. D(meas.) = 4.5(1) D(calc.) = 4.60

Optical Properties: Nearly opaque. *Color:* Dark red; gray-white in reflected light, strong fiery red internal reflections, golden yellow internal reflections occur along scratches. *Streak:* Reddish orange. *Luster:* Resinous. *Pleochroism:* Moderate, from white to gray. *Anisotropism:* Moderate, in tints of gray.

R_1–R_2: (400) 34.4–42.1, (420) 33.9–41.0, (440) 33.8–39.9, (460) 33.8–38.8, (480) 33.2–37.4, (500) 32.3–36.3, (520) 30.9–35.1, (540) 29.5–34.0, (560) 28.4–33.1, (580) 27.6–32.3, (600) 26.9–31.6, (620) 26.4–31.1, (640) 26.4–30.9, (660) 26.3–30.6, (680) 26.1–30.4, (700) 25.9–30.1

Cell Data: *Space Group:* $P2_1/n$. $a = 11.86(1)$ $b = 9.756(9)$ $c = 4.265(9)$ $\beta = 90.17(30)^\circ$ Z = 4

X-ray Powder Pattern: Burnside, Pennsylvania, USA.
2.833(100), 2.773 (80), 4.87 (70), 2.905 (60), 1.777(50), 1.709 (50), 3.72 (40)

Chemistry:

	(1)
As	47.0
Se	43.7
S	8.7
Total	99.4

(1) Burnside, Pennsylvania, USA; by electron microprobe.

Occurrence: As a secondary encrustation, probably by sublimation, on clinker adjacent to a surface vent on a burning pile of waste material from an anthracite coal mine.

Association: Arsenolite, orpiment.

Distribution: At Burnside, Northumberland Co., Pennsylvania, USA.

Name: To honor Dr. Davis M. Lapham (1931–1974), former Chief Mineralogist of the Pennsylvania Geological Survey.

Type Material: National Museum of Natural History, Washington, D.C., USA, 163039; British Museum (Natural History), London, England, E.1036, 1984,843.

References: (1) Dunn, P.J., D.R. Peacor, A.J. Criddle, and R.B. Finkelman (1986) Laphamite, an arsenic selenide analogue of orpiment, from burning anthracite deposits in Pennsylvania. Mineral. Mag., 50, 279–282. (2) (1987) Amer. Mineral., 72, 1024–1025 (abs. ref. 1).

Crystal Data: Orthorhombic. *Point Group:* 222. As subhedral grains up to 150 μm in length.

Physical Properties: *Cleavage:* Fair, ∥ [100]. Hardness = 4.5–5 VHN = n.d. D(meas.) = n.d. D(calc.) = 4.966

Optical Properties: Opaque. *Color:* Gray, presumably. *Luster:* Metallic. *Pleochroism:* In shades of greenish gray to gray. *Anisotropism:* Strong, with very pale blue to intense yellowish pink tints.
R_1–R_2: (400) 33.4–32.1, (420) 34.0–32.2, (440) 34.6–32.5, (460) 34.9–33.3, (480) 34.8–34.5, (500) 34.5–35.8, (520) 34.1–36.5, (540) 33.9–36.8, (560) 34.1–36.7, (580) 34.3–36.7, (600) 34.3–36.8, (620) 34.4–37.0, (640) 34.8–37.3, (660) 35.6–37.8, (680) 36.6–38.6, (700) 37.8–39.5

Cell Data: *Space Group:* $P2_12_12_1$. $a = 7.422(2)$ $b = 12.508(3)$ $c = 4.900(1)$ Z = 4

X-ray Powder Pattern: Lapie River, Canada.
2.959 (100), 3.178 (90), 1.837 (90), 1.855 (60), 1.601 (30), 2.769 (20), 2.637 (20)

Chemistry:

	(1)	(2)	(3)
Cu	18.6	18.5	18.68
Ni	17.0	16.9	17.26
Fe		0.2	
Sb	35.2	35.3	35.79
As		0.2	
S	27.9	27.6	28.27
Total	98.7	98.7	100.00

(1–2) Lapie River, Canada; by electron microprobe, average corresponding to $Cu(Ni_{1.00}Fe_{0.01})_{\Sigma=1.01}(Sb_{0.98}As_{0.01})_{\Sigma=0.99}S_{2.97}$. (3) $CuNiSbS_3$.

Occurrence: In a highly altered and mineralized glacial erratic.

Association: Nickelian pyrite, gersdorffite, polydymite, millerite, marcasite, tetrahedrite, chalcopyrite, spinel, magnetite, mica, quartz.

Distribution: Near the confluence of Glacier Creek and the Lapie River, St. Cyr Ranges, Yukon Territory, Canada.

Name: For the Lapie River, Canada.

Type Material: Canadian Geological Survey, Ottawa, Canada, 63844, 63845, and 63846.

References: (1) Harris, D.C., A.C. Roberts, R.I. Thorp, I.R. Jonasson, and A.J. Criddle (1984) Lapieite $CuNiSbS_3$, a new mineral species from the Yukon Territory. Can. Mineral., 22, 561–564. (2) (1985) Amer. Mineral., 70, 1329–1330 (abs. ref. 1).

Crystal Data: Orthorhombic. *Point Group:* n.d. As small acicular to flame-like crystals, the largest being about 25 μm wide and 300 μm long.

Physical Properties: Hardness = n.d. VHN = 87–124 (25 g load). D(meas.) = n.d. D(calc.) = 6.19

Optical Properties: Opaque. *Color:* In polished section, whitish buff. *Luster:* Metallic. *Pleochroism:* Weak. *Anisotropism:* Moderate, in gray and greenish buff.
R_1–R_2: n.d.

Cell Data: *Space Group:* n.d. $a = 22.15$ $b = 24.03$ $c = 11.67$ Z = 10

X-ray Powder Pattern: Foster mine, Canada.
1.982 (100), 2.917 (90), 2.846 (60), 2.471 (60), 3.206 (50), 2.555 (40), 2.162 (40)

Chemistry:

	(1)	(2)
Cu	49.7	48.3
Ag	15.6	15.6
Pb	9.0	8.5
Bi	8.9	9.6
S	18.5	19.8
Total	101.7	101.8

(1–2) Foster mine, Canada; by electron microprobe.

Occurrence: As small crystals in an assemblage of chalcocite and stromeyerite in the central portion of a layered hydrothermal vein.

Association: Chalcocite, stromeyerite, arsenopyrite, galena, tetrahedrite, polybasite.

Distribution: In the Foster mine, Cobalt, Ontario, Canada.

Name: For Fred LaRose, one of the discoverers of silver ore at Cobalt.

Type Material: Canadian Geological Survey, Ottawa, Canada.

References: (1) Petruk, W. (1972) Larosite, a new copper–lead–bismuth sulfide. Can. Mineral., 11, 886–891. (2) (1974) Amer. Mineral., 59, 382 (abs. ref. 1).

Crystal Data: Monoclinic. *Point Group:* 2, m or $2/m$. Material dug from polished surfaces tends to be somewhat fibrous by elongation || [010], resulting from the two perfect cleavages.

Physical Properties: *Cleavage:* Perfect on {100} and {001}. Hardness = n.d. VHN = 179 D(meas.) = n.d. D(calc.) = 5.83

Optical Properties: Opaque. *Color:* Lead-gray. *Streak:* Black. *Luster:* Metallic. *Pleochroism:* Fairly strong, from white to gray. *Anisotropism:* Strong.
R_1–R_2: (470) 42.6–38.6, (546) 43.8–36.9, (589) 42.7–36.2, (650) 40.9–35.5

Cell Data: *Space Group:* $C2$, Cm, or $C2/m$ (pseudocell). $a = 42.6$ $b = 8.04$ $c = 32.3$ $\beta = 102°5(45)'$ Z = 2

X-ray Powder Pattern: Madoc, Canada.
3.45 (100), 4.17 (80), 2.92 (80), 2.010 (70), 3.40 (60), 2.836 (50), 3.97 (30)

Chemistry:

	(1)	(2)
Pb	48.5	47.09
Sb	29.5	32.70
As	1.5	
S	21.25	20.21
Total	100.75	100.00

(1) Madoc, Canada; by electron microprobe, average of two analyses. Agrees well with $Pb_9Sb_{10}S_{24}$ but $Pb_{22}Sb_{26}S_{61}$ is considered more satisfactory. (2) $Pb_{22}Sb_{26}S_{61}$.

Occurrence: Of hydrothermal origin, in marbles.

Association: Veenite, boulangerite.

Distribution: From Madoc, Ontario, Canada.

Name: For Louis de Launay (1860–1938), French student of mineral deposits.

Type Material: National Museum of Canada, Ottawa, Canada.

References: (1) Jambor, J.L. (1967) New lead sulfantimonides from Madoc, Ontario. Part 2—mineral descriptions. Can. Mineral., 9, 191–194. (2) (1968) Amer. Mineral., 53, 1423 (abs. ref. 1).

Crystal Data: Cubic. *Point Group:* $2/m\ \bar{3}$. As octahedral, cubic, and pyritohedral crystals to 1 mm, and as rounded grains and inclusions in other minerals.

Physical Properties: *Cleavage:* Perfect on {111}. *Fracture:* Subconchoidal. *Tenacity:* Very brittle. Hardness = 7.5 VHN = 2760–2898, 2870 average (25 g load). D(meas.) = 6.43 D(calc.) = 6.39

Optical Properties: Opaque. *Color:* Iron-black, in polished section, white to gray, bluish. *Streak:* Dark gray. *Luster:* Metallic.

R: (400) 46.0, (420) 45.0, (440) 44.0, (460) 43.1, (480) 42.2, (500) 41.5, (520) 40.8, (540) 40.2, (560) 39.6, (580) 39.0, (600) 38.4, (620) 38.0, (640) 37.5, (660) 37.0, (680) 36.6, (700) 36.2

Cell Data: *Space Group:* *Pa*3. *a* = 5.6135 Z = 4

X-ray Powder Pattern: Borneo.
2.81 (vs), 3.29 (vs), 3.24 (s), 1.982 (s), 1.082 (s), 1.287 (ms), 1.256 (ms)

Chemistry:

	(1)	(2)	(3)	(4)
Ru	65.18	61.0	48.9	61.18
Os	n.d.		8.9	
Rh			4.7	
Ir		1.0	0.1	
Fe			0.8	
S	31.79	38.0	33.0	38.82
Total	96.67	100.0	96.4	100.00

(1) Borneo; placer material. (2) Goodnews Bay, Alaska, USA; by electron microprobe. (3) Rustenburg mine, South Africa; by electron microprobe. (4) RuS_2.

Polymorphism & Series: Forms a series with erlichmanite.

Occurrence: In ultramafic complexes and in placer deposits.

Association: Cooperite, braggite, sperrylite, other minerals of the platinum group elements.

Distribution: In South Africa, from the Merensky Reef of the Bushveld Complex, at Potgietersrust; in the Rustenburg and Union mines; also the Driekop and Onverwacht mines. In the USA, in the Stillwater Complex, Montana; on the Yuba River, Nevada Co., and from the Trinity River near Douglas City, Trinity Co., California; at Goodnews Bay, Alaska. In Canada, in the Tulameen area, British Columbia. In placers in Borneo, as at Kalimantan (formerly Pontijn, Tanah Laut). In Sierra Leone, from placers at Guma Water, and the Senduma chrome mines. From the Konder massif, Aldan Shield, Yakutia, USSR. Known in small amounts from other localities.

Name: For Laura R. Joy, wife of Charles A. Joy, American chemist, Columbia University, New York City, USA.

Type Material: n.d.

References: (1) Palache, C., H. Berman, and C. Frondel (1944) Dana's system of mineralogy, (7th edition), v. I, 291–292. (2) Leonard, B.F. and G.A. Desborough (1969) Ore microscopy and chemical composition of some laurites. Amer. Mineral., 54, 1330–1346. (3) Bowles, J.F.W., D. Atkin, J.L.M. Lambert, T. Deans, and R. Phillips (1983) The chemistry, reflectance, and cell size of the erlichmanite (OsS_2) — laurite (RuS_2) series. Mineral. Mag., 47, 465–471.

Crystal Data: Orthorhombic. *Point Group:* $mm2$. As crystals, short prismatic [100] and tabular [001]; striated on {001} parallel to [100]. May also be massive, fine granular or radiating. *Twinning:* On {110}.

Physical Properties: *Cleavage:* {001}. *Tenacity:* Brittle. Hardness = 3–3.5 VHN = 239–259 (100 g load). D(meas.) = 4.9 D(calc.) = 4.878

Optical Properties: Opaque. *Color:* Black, steel-gray with reddish tone. *Streak:* Black. *Luster:* Metallic to semimetallic. *Pleochroism:* Very weak. *Anisotropism:* Weak.
R_1–R_2: (400) 36.0–37.6, (420) 34.7–36.4, (440) 33.4–35.2, (460) 32.3–34.0, (480) 31.4–33.1, (500) 30.7–32.3, (520) 30.4–31.7, (540) 30.2–31.4, (560) 30.1–31.2, (580) 30.2–31.1, (600) 30.3–31.2, (620) 30.5–31.3, (640) 30.8–31.5, (660) 31.1–31.8, (680) 31.5–32.0, (700) 31.9–32.3

Cell Data: *Space Group:* $Pna2_1$. $a = 11.350$ $b = 5.456$ $c = 3.749$ Z = 4

X-ray Powder Pattern: Lauta, Germany.
3.10 (100), 1.903 (80), 1.610 (60), 1.232 (50), 1.095 (50), 1.030 (50), 1.797 (40)

Chemistry:

	(1)	(2)	(3)
Cu	36.10	37.07	37.28
As	45.66	44.53	43.92
S	17.88	18.30	18.80
Total	99.64	99.90	100.00

(1) Lauta, Germany. (2) Rauenthal, France. (3) CuAsS.

Occurrence: In hydrothermal veins formed at moderate temperatures.

Association: Arsenic, tennantite, proustite, chalcopyrite, galena, barite (Lauta, Germany); kutinaite, paxite (Niederbeerbach, Germany); arsenic, bismuth, tennantite, löllingite, rammelsbergite, proustite, quartz (Gabe Gottes mine, France).

Distribution: In Germany, at Lauta, near Marienberg, Saxony; and at Niederbeerbach, Odenwald. From the Gabe Gottes mine, Rauenthal, near Sainte-Marie-aux-Mines, Haut-Rhin, France.

Name: For the Lauta, Germany occurrence.

References: (1) Palache, C., H. Berman, and C. Frondel (1944) Dana's system of mineralogy, (7th edition), v. I, 327–328. (2) Berry, L.G. and R.M. Thompson (1962) X-ray powder data for the ore minerals. Geol. Soc. Amer. Mem. 85, 108–109. (3) Craig, D.C. and N.C. Stevenson (1965) The crystal structure of lautite, CuAsS. Acta Cryst., 19, 543–547.

Crystal Data: Monoclinic (perhaps triclinic). *Point Group:* $2/m$, 2, or m. As complex intergrowths to 0.2 mm.

Physical Properties: Hardness = 2.0–2.5 VHN = 82–103 D(meas.) = n.d. D(calc.) = 7.26–7.5

Optical Properties: Transparent. *Color:* Colorless to yellow. *Luster:* Vitreous to adamantine.
R_1–R_2: n.d.

Cell Data: *Space Group:* $P2/m$, $P2$, or Pm. $a = 8.94$ $b = 5.194$ $c = 18.33$ $\beta = 92.44°$ Z = 5

X-ray Powder Pattern: Arzak deposit, USSR. Differs only by intensities from arzakite.
3.38 (100), 3.01 (100), 3.96 (80), 2.292 (80), 1.587 (80), 2.199 (60)

Chemistry:

	(1)	(2)	(3)
Hg	81.4	78.48	77.02
S	8.51	8.35	8.21
Br	1.29	6.94	10.23
Cl	9.13	6.04	4.54
Total	100.33	99.81	100.00

(1) Arzak deposit, USSR; by electron microprobe, corresponding to $Hg_{3.01}S_{1.96}(Cl_{1.91}Br_{0.12})_{\Sigma=2.03}$. (2) Do.; corresponding to $Hg_{3.01}S_{2.00}(Cl_{1.31}Br_{0.67})_{\Sigma=1.98}$. (3) $Hg_3S_2(Cl, Br)_2$ with Cl:Br = 1:1.

Polymorphism & Series: Forms a series with arzakite; dimorphous with corderoite.

Occurrence: In the oxidized zone of a hydrothermal deposit (Arzak deposit, USSR).

Association: Arzakite, cinnabar, corderoite, quartz, kaolinite (Arzak deposit, USSR); calomel, eglestonite, mercury (Kadyrelskii deposit, USSR)

Distribution: From the Arzak deposit, Tuva ASSR, also the Kadyrelskii deposit, USSR.

Name: For M.A. Lavrentiev, founder of the Siberian Academy of Science.

Type Material: Central Siberian Geological Museum, Novosibirsk, USSR.

References: (1) Vasil'ev, V.L., N.A. Pal'chik, and O.K. Grechishchev (1984) Lavrentievite and arzakite, new natural sulfohalogenides of mercury. Geol. i Geofiz., 7, 54–63 (in Russian). (2) (1985) Amer. Mineral., 70, 873–874 (abs. ref. 1). (3) (1984) Chem. Abs., 101, 174794 (abs. ref. 1).

Crystal Data: Cubic. *Point Group:* $4/m\ \overline{3}\ 2/m$. Crystals rare; octahedral, cubic, and dodecahedral; commonly as rounded masses and plates, up to 60 kg in weight. Assumes wirelike and dendritic forms. *Twinning:* On {111}.

Physical Properties: *Tenacity:* Very malleable; moderately sectile. Hardness = 1.5 VHN = 5 (100 g load). D(meas.) = 11.37 D(calc.) = 11.341

Optical Properties: Opaque. *Color:* Gray-white, tarnishing to dull lead-gray; in polished section, gray-white, quickly tarnishing. *Streak:* Lead-gray. *Luster:* Metallic.
R: (400) 45.6, (420) 45.9, (440) 46.2, (460) 46.5, (480) 46.9, (500) 47.4, (520) 48.1, (540) 48.9, (560) 49.7, (580) 50.4, (600) 51.0, (620) 51.7, (640) 52.5, (660) 53.6, (680) 55.2, (700) 57.9

Cell Data: *Space Group:* $Fm3m$. $a = 4.9506$ Z = 4

X-ray Powder Pattern: Synthetic.
2.855 (100), 2.475 (50), 1.493 (32), 1.750 (31), 1.1359 (10), 1.429 (9), 0.8369 (9)

Chemistry: Nearly pure lead, sometimes with a little Ag, Sn or Sb.

Occurrence: A rare mineral of hydrothermal origin, and found in placers; possibly also formed by authigenic processes and known to replace tree roots.

Association: Hydrocerussite, caryopilite, sarkinite, brandtite (Harstigen mine, Sweden); galena, minium, cerussite (Jay Gould mine, Idaho, USA; Red Cap mine, Australia); willemite, andradite, axinite (Franklin, New Jersey, USA).

Distribution: In the USA, in the Parker shaft at Franklin, Sussex Co., New Jersey; in Idaho, at the Jay Gould mine, Wood River district, Blaine Co., and near Hailey, Mineral Hill district, Alturas Co.; in Arizona, from near Tubac, Santa Cruz Co., where it replaced tree roots; and in the Shafter district, Presidio Co., Texas. At Keno Hill, Yukon Territory, Canada. From the Ilímaussaq Intrusion, southern Greenland. In Sweden, in fine crystals from Långban, and from the Harstigen mine, Pajsberg, Värmland. From the Red Cap mine, near Chilliago, Queensland; and as minute balls in the Hawkesbury Sandstone, New South Wales, Australia. Reported from numerous other localities, but confirmation by modern methods is desirable.

Name: An Old English word for the metal; the chemical symbol from the Latin *plumbum*.

References: (1) Palache, C., H. Berman, and C. Frondel (1944) Dana's system of mineralogy, (7th edition), v. I, 102–103. (2) (1953) NBS Circ. 539, 1, 141–143.

Crystal Data: Tetragonal. *Point Group:* $4/m\ 2/m\ 2/m$. Known only as a single grain of maximum dimension 50 μm.

Physical Properties: Hardness = 1.6 VHN = 12 (100 g load). D(meas.) = n.d. D(calc.) = 11.96

Optical Properties: Opaque. *Color:* Silver-white; yellowish white in reflected light. *Luster:* Strong metallic. *Pleochroism:* Weak, bluish gray to pale bluish gray.
R_1–R_2: n.d.

Cell Data: *Space Group:* $I4/mmm$. $a = 3.545(16)$ $c = 4.525(20)$ Z = n.d.

X-ray Powder Pattern: Shiaonanshan, China.
1.49 (vs), 1.40 (vs), 2.49 (m), 2.25 (m), 1.78 (m), 1.68 (m), 2.78 (w)

Chemistry:

	(1)	(2)	(3)
Hg	33.03	31.48	32.62
Pb	66.96	68.42	67.38
Total	99.99	99.90	100.00

(1–2) Shiaonanshan, China; by electron microprobe. (3) $HgPb_2$.

Occurrence: In heavy concentrates of crushed ores from a platinum-bearing Cu-Ni-sulfide deposit.

Association: Gersdorffite, pyrite, chalcopyrite, violarite, millerite, galena, stibnite, argentian gold, niggliite, sperrylite, iridosmine, platinum, merenskyite, kotulskite, chromite, ilmenite, magnetite.

Distribution: From Shiaonanshan, Inner Mongolia Autonomous Region, China.

Name: For the composition.

Type Material: n.d.

References: (1) Chen Keqiao, Yang Huifang, Ma Letian, and Peng Zhizhong (1981) The discovery of two new minerals— γ–goldamalgam and leadamalgam. Dizhi Pinglun, 27, 107–115 (in Chinese with English abs.). (2) (1985) Amer. Mineral., 70, 215–216 (abs. ref. 1).

Crystal Data: Triclinic. *Point Group:* 1 or $\bar{1}$. As thin, bladed crystals which are commonly bent or curled, to 4 cm; the large flat face {100} is striated parallel to its elongation direction. *Twinning:* Twin lamellae seen in polished section.

Physical Properties: *Cleavage:* Perfect on {100}; two others across the flat {100} face. *Tenacity:* Malleable, flexible, but not elastic. Hardness = 1.5–2 VHN = 35
D(meas.) = 5.80–5.85 D(calc.) = n.d.

Optical Properties: Opaque. *Color:* Steel-gray, may tarnish to iridescence. *Streak:* Brown-black. *Luster:* Metallic. *Anisotropism:* Weak.
R_1–R_2: (400) 38.9–40.0, (420) 38.5–39.8, (440) 38.2–39.4, (460) 37.6–39.0, (480) 37.3–38.7, (500) 36.8–38.3, (520) 36.3–37.9, (540) 35.9–37.4, (560) 35.5–36.9, (580) 35.3–36.5, (600) 35.1–36.1, (620) 34.7–35.7, (640) 34.2–35.2, (660) 33.5–34.4, (680) 32.8–33.7, (700) 32.0–32.9

Cell Data: *Space Group:* $P1$ or $P\bar{1}$. $a = 18.45$ $b = 11.68$ $c = 70.16$ $\alpha = 90.0°$ $\beta = 91.01°$ $\gamma = 90.0°$ Z = 30

X-ray Powder Pattern: Binntal, Switzerland.
3.06 (100), 2.84 (90), 4.60 (30), 2.93 (20), 2.04 (20), 9.31 (10), 7.69 (10)

Chemistry:

	(1)	(2)
Pb	57.89	54.5
Ag	5.64	9.4
Cu	2.36	2.5
Fe	0.17	
Sb	0.77	
As	13.46	13.5
S	19.33	20.8
Total	99.62	100.7

(1) Binntal, Switzerland. (2) Do.; by electron microprobe.

Occurrence: Of hydrothermal origin.

Association: Pyrite, jordanite.

Distribution: In the Lengenbach quarry, Binntal, Valais, Switzerland.

Name: For the locality at the Lengenbach quarry, Switzerland.

Type Material: n.d.

References: (1) Palache, C., H. Berman, and C. Frondel (1944) Dana's system of mineralogy, (7th edition), v. I, 398. (2) Berry, L.G. and R.M. Thompson (1962) X-ray powder data for the ore minerals. Geol. Soc. Amer. Mem. 85, 130. (3) Nowacki, W. (1968) Über Hatchit, Lengenbachit und Vrbait. Neues Jahrb. Mineral., Monatsh., 69–75 (in German). (4) Williams, T.B. and A. Pring (1988) Structure of lengenbachite: a high-resolution transmission electron microscope study. Amer. Mineral., 73, 1426–1433.

Crystal Data: Orthorhombic. *Point Group:* $2/m\ 2/m\ 2/m$. Elongated platy crystals; also fibrous.

Physical Properties: *Fracture:* Conchoidal. Hardness = 2–3 VHN = n.d. D(meas.) = 7.02–7.16 D(calc.) = 7.09

Optical Properties: Opaque. *Color:* Steel-gray; in polished section, white. *Streak:* Black. *Luster:* Metallic. *Pleochroism:* Very weak. *Anisotropism:* Distinct.
R_1–R_2: n.d.

Cell Data: *Space Group:* *Bbmm* (synthetic). $a = 13.522$ $b = 20.608$ $c = 4.112$ Z = 4

X-ray Powder Pattern: Synthetic.
3.52 (100), 2.913 (80), 3.42 (70), 1.777 (70), 3.01 (60), 2.778 (60), 2.070 (60)

Chemistry:

	(1)	(2)
Pb	49.61	50.46
Ag	2.17	
Cu	0.14	
Bi	31.31	33.93
Sb	0.57	
S	15.35	15.61
Total	99.15	100.00

(1) Spokoinoe deposit, USSR; corresponds to $(Pb_{2.94}Ag_{0.25}Cu_{0.03})_{\Sigma=3.22}(Bi_{1.84}Sb_{0.06})_{\Sigma=1.90}S_{5.89}$. (2) $Pb_3Bi_2S_6$.

Polymorphism & Series: Forms a series with gustavite.

Occurrence: In hydrothermal deposits of probable high-temperature origin.

Association: At a number of early reported occurrences the "lillianite" was shown to be mixtures, especially of galena, galenobismutite.

Distribution: The occurrence at the "type locality" - the Lillian mine, Leadville, Colorado, USA, has been discredited. At Twin Lakes, Fresno Co., California, and from a prospect north of Minersville, Beaver Co., Utah, USA. In the USSR, at the Spokoinoe wolframite deposit, eastern Transbaikal; the Kochbulak deposit, eastern Uzbekistan; and elsewhere. In Czechoslovakia, from the Bukvinsk deposit. In Finland, at Jilijärvi, near Orijärvi. From the Oberpinzgau, Salzburg, Austria. Near Narechen, southern Rhodope Mountains, Bulgaria. From Vulcano, in the Lipari Islands, Italy. At Yecora, 5 km west of Iglesia, Sonora, Mexico.

Name: For the Lillian mine, Leadville, Colorado, USA, the discredited type locality.

Type Material: n.d.

References: (1) Palache, C., H. Berman, and C. Frondel (1944) Dana's system of mineralogy, (7th edition), v. I, 404–406. (2) Syritso, L.F. and V.M. Senderova (1964) The problem of the existence of lillianite. Zap. Vses. Mineral. Obshch., 93, 468–471 (in Russian). (3) (1965) Amer. Mineral., 50, 811 (abs. ref. 2). (4) Klyakhin, V.A. and M.T. Dmitrieva (1969) New data on synthetic and natural lillianite. Doklady Acad. Nauk SSSR, 178, 173–175 (in Russian). (5) Otto, H.H. and H. Strunz (1968) Zur Kristallchemie synthetischer Blei–Wismut-Speissglanze. Neues Jahrb. Mineral., Abh., 108, 1–9 (in German). (6) (1969) Amer. Mineral., 54, 579 (abs. refs. 4 and 5). (7) Takagi, J. and Y. Takéuchi (1972) The crystal structure of lillianite. Acta Cryst., B28, 649–651.

Crystal Data: Orthorhombic. *Point Group:* $2/m\ 2/m\ 2/m$. Crystals prismatic, elongated [001], to 1 cm long and several mm thick, striated parallel to [001].

Physical Properties: *Cleavage:* Good on {100} and {010}; also on {230}. *Fracture:* Conchoidal to uneven. Hardness = 3–3.5 VHN = n.d. D(meas.) = 7.01 D(calc.) = 7.03

Optical Properties: Opaque. *Color:* Lead-gray. *Streak:* Black. *Luster:* Metallic.
R_1–R_2: (400) 42.3–46.0, (420) 42.2–45.8, (440) 42.1–45.6, (460) 42.0–45.4, (480) 41.8–45.3, (500) 41.6–45.2, (520) 41.4–45.3, (540) 41.1–45.4, (560) 41.0–45.5, (580) 40.7–45.3, (600) 40.3–45.1, (620) 40.1–45.0, (640) 39.8–44.7, (660) 39.6–44.4, (680) 39.3–44.0, (700) 39.0–43.6

Cell Data: *Space Group:* *Pbnm*. $a = 56.115$ $b = 11.5695$ $c = 4.001$ Z = 4

X-ray Powder Pattern: n.d.

Chemistry:

	(1)	(2)	(3)
Pb	18.95	22.1	22.57
Cu	5.84	6.5	6.92
Fe	trace	trace	
Bi	57.13	53.0	53.04
S	[17.88]	[18.4]	17.47
insol.	0.20		
Total	100.00	100.0	100.00

(1–2) Gladhammar, Sweden. (3) $Pb_3Cu_3Bi_7S_{15}$.

Occurrence: Of hydrothermal origin.

Association: Bismuth, quartz (Gladhammar, Sweden).

Distribution: At Gladhammar, Kalmar, Sweden. In Canada, in the Silver Miller mine, Cobalt, Ontario. From the Beresovsk district, Ural Mountains, USSR. At Yecora, 5 km west of Iglesia, Sonora, Mexico. In the Beaver Mountains, near Milford, Beaver Co., Utah, and at Manhattan, Nye Co., Nevada, USA.

Name: For Gustav Lindström (1838–1916), Swedish mineral chemist of the Swedish Museum of Natural History, Stockholm, Sweden.

Type Material: n.d.

References: (1) Palache, C., H. Berman, and C. Frondel (1944) Dana's system of mineralogy, (7th edition), v. I, 459–460. (2) Mumme, W.G., E. Welin, and B.J. Wuensch (1976) Crystal chemistry and proposed nomenclature for sulfosalts intermediate in the system bismuthinite–aikinite (Bi_2S_3–$CuPbBiS_3$). Amer. Mineral., 61, 15–20. (3) Welin, E. (1968) Notes on the mineralogy of Sweden. 5. Bismuth-bearing sulphosalts from Gladhammar, a revision. Amer. Mineral., 53, 351. (4) Horiuchi, H. and B.J. Wuensch (1977) Lindströmite, $Cu_3Pb_3Bi_7S_{15}$: its space group and ordering scheme for metal atoms in the crystal structure. Can. Mineral., 15, 527–535.

Crystal Data: Cubic. *Point Group:* $4/m\ \overline{3}\ 2/m$. Dominantly octahedral, crystals to 5 mm; usually massive, granular and compact. *Twinning:* On {111}.

Physical Properties: *Cleavage:* Imperfect on {001}. *Fracture:* Uneven to subconchoidal. Hardness = 4.5–5.5 VHN = n.d. D(meas.) = 4.5–4.8 D(calc.) = 4.85

Optical Properties: Opaque. *Color:* Light gray to steel-gray. *Luster:* Metallic, easily tarnished.
R: (400) 44.1, (420) 43.6, (440) 43.1, (460) 42.0, (480) 42.5, (500) 42.5, (520) 43.1, (540) 43.6, (560) 43.9, (580) 44.7, (600) 45.2, (620) 45.8, (640) 46.4, (660) 46.9, (680) 47.5, (700) 47.9

Cell Data: *Space Group:* $Fd3m$. $a = 9.43$ Z = 8

X-ray Powder Pattern: Musen, Germany.
2.83 (100), 1.670 (80), 2.36 (70), 1.815 (60), 0.988 (50), 3.34 (40), 1.090 (40)

Chemistry:

	(1)	(2)	(3)
Co	40.71	48.70	57.96
Ni	7.35	4.75	
Cu	8.79	2.40	
Fe	1.30	2.36	
S	41.43	41.70	42.04
insol.	0.14	0.40	
Total	99.72	100.31	100.00

(1) Gladhammar, Sweden. (2) Carroll Co., Maryland, USA. (3) $Co^{+2}Co_2^{+3}S_4$.

Polymorphism & Series: Forms a series with polydymite.

Occurrence: In hydrothermal veins with other cobalt and nickel sulfides.

Association: Chalcopyrite, pyrrhotite, millerite, bismuthinite, gersdorffite, ullmannite, galena, sphalerite.

Distribution: In the USA, at the Springfield and Mineral Hill mines, Carroll Co., Maryland; and at the Mine La Motte, Madison Co., Missouri. In Germany, at Müsen and Altenberg, North Rhine-Westphalia; at Grube Georg, near Horhausen, Westerwald. At Kladno, Czechoslovakia. In Sweden, at Gladhammar, Kalmar, and at the Bastnäs mine, near Riddarhyttan, Västmanland. In Zaire, at the Musonoi mine, Shaba Province. At the N'Kana mine, Kitwe, Zambia. From Tsumeb, Namibia. Known from a number of other minor occurrences.

Name: For Carolus Linnaeus (1707–1778), Swedish taxonomist.

References: (1) Palache, C., H. Berman, and C. Frondel (1944) Dana's system of mineralogy, (7th edition), v. I, 262–265. (2) Berry, L.G. and R.M. Thompson (1962) X-ray powder data for the ore minerals. Geol. Soc. Amer. Mem. 85, 103.

Crystal Data: Monoclinic. *Point Group:* 2. As tiny crystals.

Physical Properties: Hardness = 3 VHN = 169–180 (100 g load). D(meas.) = 5.3 D(calc.) = [5.43]

Optical Properties: Opaque. *Color:* Lead-gray, with deep red internal reflections. *Streak:* Chocolate-brown. *Luster:* Metallic. *Anisotropism:* Distinct.

R_1–R_2: (400) 40.0–43.0, (420) 39.3–42.6, (440) 38.6–42.2, (460) 38.0–41.8, (480) 37.4–41.5, (500) 36.9–41.2, (520) 36.3–40.7, (540) 35.6–40.1, (560) 34.9–39.3, (580) 34.2–38.5, (600) 33.6–37.6, (620) 32.8–36.7, (640) 32.1–35.8, (660) 31.3–34.8, (680) 30.6–34.0, (700) 30.0–33.4

Cell Data: *Space Group:* $P2_1$. $a = 8.44$ $b = 69.119$ $c = 7.929$ $\beta = 90°$ Z = 4

X-ray Powder Pattern: Binntal, Switzerland. (JCPDS 31-678).
2.990 (100), 4.11 (90), 3.423 (90), 2.350 (90), 2.093 (90), 1.931 (80), 1.917 (80)

Chemistry:

	(1)	(2)	(3)
Pb	47.58	49.1	49.90
As	26.93	26.5	26.07
S	24.91	24.0	24.03
Total	99.42	99.6	100.00

(1) Binntal, Switzerland. (2) Do.; by electron microprobe. (3) $Pb_9As_{13}S_{28}$.

Occurrence: Of hydrothermal origin.

Association: Pyrite, sphalerite.

Distribution: From the Lengenbach quarry, Binntal, Valais, Switzerland.

Name: For George D. Liveing (1827–1924), Professor of Chemistry, Cambridge University, Cambridge, England.

Type Material: n.d.

References: (1) Palache, C., H. Berman, and C. Frondel (1944) Dana's system of mineralogy, (7th edition), v. I, 462. (2) Le Bihan, M.-T. (1962) Etude structurale de quelques sulfures de plomb et d'arsenic naturels du gisement de Binn. Bull. Soc. fr. Minéral., 85, 15–47 (in French). (3) Nowacki, W. (1967) Über die mögliche Identität von "liveingit" mit Rathite-II. Neues Jahrb. Mineral., Monatsh., 353–354 (in German). (4) (1969) Amer. Mineral., 54, 1498 (abs. refs. 2 and 3). (5) Engel, P. and W. Nowacki (1970) Die Kristallstruktur von Rathite II [$As_{25}S_{56}Pb_{6.5}Pb_{12}$]. Zeits. Krist., 131, 356–375 (in German).

Crystal Data: Monoclinic. *Point Group:* $2/m$. As needles elongated [010], up to 12 cm; also fibrous, massive, columnar, and in globular masses and interlaced needles. *Twinning:* Twin gliding postulated. Polysynthetic-twinning-like features noted in polished section.

Physical Properties: *Cleavage:* Perfect on {001}; poor on {010} and {100}. *Tenacity:* Flexible. Hardness = 2 VHN = 96–125 (100 g load). D(meas.) = 4.8(2); 4.88(2) (synthetic). D(calc.) = 4.98

Optical Properties: Opaque, translucent in thin splinters. *Color:* Blackish gray; in polished section, white; red in transmitted light, with deep red internal reflections. *Streak:* Red. *Luster:* Adamantine to metallic.

Optical Class: Biaxial negative (–). *Pleochroism:* Weak. *Orientation:* $Z = b$. $n = > 2.72$ *Anisotropism:* Strong.

R_1–R_2: (400) 33.2–42.2, (420) 33.0–42.6, (440) 32.8–43.0, (460) 32.4–43.1, (480) 31.9–43.0, (500) 31.4–42.4, (520) 30.8–41.5, (540) 30.4–40.5, (560) 30.1–39.7, (580) 29.7–39.0, (600) 29.5–38.4, (620) 29.2–37.7, (640) 28.8–37.0, (660) 28.4–36.1, (680) 27.8–35.1, (700) 27.2–34.2

Cell Data: *Space Group:* $A2/a$. $a = 30.567(6)$ $b = 4.015(1)$ $c = 21.465(3)$ $\beta = 103.39(1)°$ Z = 8

X-ray Powder Pattern: Guerrero, Mexico.
3.00 (100), 3.74 (70), 3.48 (70), 2.28 (60), 1.734 (50), 5.16 (40), 2.67 (30)

Chemistry:

	(1)	(2)	(3)
Hg	22.52	21.0	21.25
Sb	53.75	51.6	51.58
S	23.73	28.6	27.17
Total	100.00	101.2	100.00

(1) Guerrero, Mexico; recalculated after deduction of 13-16% impurities. (2) Do.; by electron microprobe. (3) $HgSb_4S_8$.

Occurrence: In low-temperature hydrothermal veins.

Association: Cinnabar, stibnite, sulfur, gypsum.

Distribution: In Mexico, from the La Cruz mine, Huitzuco, Guerrero; and at the Trinidad mine, Guadalcázar, San Luis Potosí. From the Matsuo mine, Iwate Prefecture, Honshu, Japan. In the USSR, at Khaidarkan, Kirgizia. At Pedrosa del Rey, León, Spain.

Name: After David Livingstone (1813–1873), Scottish explorer and missionary in Africa.

References: (1) Palache, C., H. Berman, and C. Frondel (1944) Dana's system of mineralogy, (7th edition), v. I, 485–486. (2) Richmond, W.E. (1936) Crystallography of livingstonite. Amer. Mineral., 21, 719–720. (3) Gorman, D.H. (1951) An x-ray study of the mineral livingstonite. Amer. Mineral., 36, 480–483. (4) Berry, L.G. and R.M. Thompson (1962) X-ray powder data for the ore minerals. Geol. Soc. Amer. Mem. 85, 170. (5) Craig, J.R. (1970) Livingstonite, $HgSb_4S_8$: synthesis and stability. Amer. Mineral., 55, 919–924. (6) Srikrishnan, T. and W. Nowacki (1975) A redetermination of the crystal structure of livingstonite, $HgSb_4S_8$. Zeits. Krist., 141, 174–192.

Crystal Data: Orthorhombic. *Point Group:* $2/m$ $2/m$ $2/m$. Prismatic ∥ [001]; also pyramidal and doubly terminated, crystals to as much as 4 cm; massive. *Twinning:* On {011}, may produce trillings; polysynthetic on {101}.

Physical Properties: *Cleavage:* Sometimes distinct on {010}, {101}. Hardness = 5–5.5 VHN = 859–920 (100 g load). D(meas.) = 7.43 (for $FeAs_{1.83}$). D(calc.) = 7.472

Optical Properties: Opaque. *Color:* Steel-gray to silver-white. *Streak:* Grayish black. *Luster:* Metallic. *Anisotropism:* Strong.

R_1–R_2: (400) 49.2–55.5, (420) 50.3–56.2, (440) 51.4–56.9, (460) 52.2–56.6, (480) 52.6–55.8, (500) 52.9–54.8, (520) 53.2–54.0, (540) 53.6–53.1, (560) 54.0–52.4, (580) 54.2–51.7, (600) 54.3–51.1, (620) 54.2–50.5, (640) 53.9–50.0, (660) 53.5–49.6, (680) 53.0–49.2, (700) 52.6–48.9

Cell Data: *Space Group:* *Pnnm*. $a = 5.16$ $b = 5.93$ $c = 3.05$ Z = 2

X-ray Powder Pattern: Synthetic.
2.605 (100), 2.331 (65), 2.599 (60), 2.532 (60), 1.634 (35), 2.421 (30), 1.854 (30)

Chemistry:

	(1)	(2)	(3)	(4)
Fe	27.93	29.40	27.83	27.15
As	70.83	69.80	70.24	72.85
Bi			0.05	
S	0.77	0.21	0.57	
Total	99.53	99.41	98.69	100.00

(1) Drum's Farm, Alexander Co., North Carolina, USA. (2) Franklin, New Jersey, USA. (3) Västersele, Ångermanland, Sweden. (4) $FeAs_2$.

Occurrence: In mesothermal deposits associated with other sulfides and with calcite gangue; also found in pegmatites.

Association: Skutterudite, bismuth, nickeline, nickel-skutterudite, siderite, calcite.

Distribution: In the USA, from Mt. Mica, South Paris, Oxford Co., Maine; from Franklin and the Sterling Hill mine, Ogdensburg, Sussex Co., New Jersey; in Colorado, from several mines in Gunnison Co. In Canada, at a number of mines at Cobalt, Ontario. At Lölling, near Hüttenberg, Carinthia; and at Schladming, Styria, Austria. From Reichenstein, Silesia, Poland. At St. Andreasberg, in the Harz Mountains, Germany. From Guadalcanal, in the Sierra Morena, Seville Province, Spain. In Norway, from the Langesund district. From Broken Hill, New South Wales, Australia. Also from numerous minor localities.

Name: For the Austrian locality at Lölling.

References: (1) Palache, C., H. Berman, and C. Frondel (1944) Dana's system of mineralogy, (7th edition), v. I, 303–307. (2) (1960) NBS Circ. 539, 10, 34. (3) Radcliffe, D. and L.G. Berry (1968) The safflorite–loellingite solid solution series. Amer. Mineral., 53, 1856–1881.

Crystal Data: Hexagonal. *Point Group:* $6/m\ 2/m\ 2/m$. Very fine-grained aggregates, forming cubes and cubo-octahedra up to 0.7 mm; in polycrystalline aggregates, mixed with diamond.

Physical Properties: Hardness = 3 VHN = n.d. D(meas.) = > 3.20 D(calc.) = 3.51

Optical Properties: Transparent. *Color:* Grayish in crystals; also pale yellowish or brown in broken fragments. *Luster:* Adamantine.
Optical Class: n.d. n = Slightly above 2.404. *Anisotropism:* Birefringence slight, probably due to strain.
R_1–R_2: n.d.

Cell Data: *Space Group:* $P6_3/mmc$. $a = 2.51$ $c = 4.12$ Z = 4

X-ray Powder Pattern: Cañon Diablo meteorite.
2.061 (100), 1.257 (60), 2.18 (40), 1.075 (30), 1.933 (20), 1.50 (10), 1.17 (10)

Chemistry:

	(1)
C	100.0
N	0.0
Total	100.0

(1) Cañon Diablo meteorite; by electron microprobe.

Polymorphism & Series: Diamond, graphite, and chaoite are polymorphs.

Occurrence: Discovered in the residue (ca. 200 mg) from the dissolution of 5 kg of Cañon Diablo meteorite. The mineral may be formed by impact shock, or be a product of direct crystallization in the parent body.

Association: Schreibersite, cohenite, taenite, graphite, chromite, kosmochlor, sphalerite, black diamond (Cañon Diablo); troilite, graphite, diamond, schreibersite, cohenite (Allen Hills 77283).

Distribution: In the Cañon Diablo, Goalpara, and Allen Hills 77283 meteorites. From placers in northern Yakutia; found in soil at the Tunguska explosion site, Federated SSR, USSR.

Name: For Professor Kathleen Y. Lonsdale (1903–1971), distinguished British crystallographer.

Type Material: Harvard University, Cambridge, Massachusetts, USA, 130245.

References: (1) Frondel, C. and U.B. Marvin (1967) Lonsdaleite, a hexagonal polymorph of diamond. Nature, 214, 587–589. (2) (1967) Amer. Mineral., 52, 1579 (abs. ref. 1). (3) Hanneman, R.E., H.M. Strong, and F.P. Bundy (1967) Hexagonal diamonds in meteorites: Implications. Science, 155, 995–997. (4) Kaminskii, F.V., G.K. Blinova, E.M. Galimov, G.A. Gurkina, Y.A. Klyuev, L.A. Kodina, V.I. Koptil, V.F. Krivonos, L.N. Frolova, and A.Y. Khrenov (1985) Polycrystalline aggregates of diamond with lonsdaleite from Yakutian placers. Mineral. Zhurnal., 7, 27–36 (in Russian). (5) (1985) Chem. Abs., 103, 9146 (abs. ref. 4).

Crystal Data: Monoclinic. *Point Group:* $2/m$. Short prisms or tabular on {201}, pyramidal, to 2 mm; striated parallel to [001]; in grains to 4 mm.

Physical Properties: *Cleavage:* {100} excellent; {201} very good; {001} good. *Tenacity:* Flexible, forming cleavage lamellae and fibers. Hardness = 2–2.5 VHN = n.d. D(meas.) = 5.53 D(calc.) = 5.53

Optical Properties: Translucent to transparent. *Color:* Cochineal-red to carmine-red surface, often dark lead-gray, sometimes coated by an ocher-yellow powder. *Streak:* Cherry-red. *Luster:* Metallic, adamantine.

Optical Class: Biaxial positive (+) (?). *Pleochroism:* Weak, Y = purple-red, Z = orange-red. *Orientation:* X, $Y \sim\perp$ {100}; $Z = b$. *Dispersion:* $r > v$, strong. $n = > 2.72$ (Li) 2V(meas.) = Large. 2V(calc.) = n.d. *Anisotropism:* Strong.

R_1–R_2: (400) 28.8–30.6, (420) 28.7–30.7, (440) 28.6–30.8, (460) 28.4–30.9, (480) 28.2–31.2, (500) 27.8–31.1, (520) 27.5–30.6, (540) 27.0–29.8, (560) 26.4–28.8, (580) 25.8–27.7, (600) 25.0–26.7, (620) 24.4–26.0, (640) 23.9–25.5, (660) 23.5–25.0, (680) 23.2–24.7, (700) 22.9–24.5

Cell Data: *Space Group:* $P2_1/a$. a = 12.28(1) b = 11.30(1) c = 6.101(6) β = 104°5(2)′ Z = [8]

X-ray Powder Pattern: Allchar, Yugoslavia.
3.59 (10), 2.88 (8), 2.97 (7), 1.867 (5), 1.484(5), 5.28 (4), 2.64 (4)

Chemistry:

	(1)	(2)	(3)	(4)
Tl	59.76	58.75	59.5	59.46
As	22.30	21.65	21.6	21.87
S	18.99	19.26	18.8	18.67
rem.		0.08		
Total	101.05	99.74	99.9	100.00

(1–2) Allchar, Yugoslavia. (3) Carlin mine, Nevada, USA; by electron microprobe, average of five analyses. (4) $TlAsS_2$.

Occurrence: A hydrothermal mineral, probably formed at relatively low temperatures.

Association: Stibnite, realgar, orpiment, greigite, marcasite, pyrite, barite.

Distribution: At Allchar, Macedonia, Yugoslavia. In the Dzhizhikrut Sb-Hg deposit, Tadzhikistan, USSR. From the Lengenbach quarry, Binntal, Valais, Switzerland. In the USA, at the Rambler mine near Encampment, Albany Co., Wyoming, and in the Carlin gold deposit, Eureka Co., Nevada.

Name: For Eötvös Lorand (1848–1919), physicist of Budapest, Hungary.

References: (1) Palache, C., H. Berman, and C. Frondel (1944) Dana's system of mineralogy, (7th edition), v. I, 437–439. (2) Vlasov, K.A., Ed. (1966) Mineralogy of rare elements, v. II, 592–597. (3) Berry, L.G. and R.M. Thompson (1962) X-ray powder data for the ore minerals. Geol. Soc. Amer. Mem. 85, 146–147. (4) Fleet, M.E. (1973) The crystal structure and bonding of lorandite, $Tl_2As_2S_4$. Zeits. Krist., 138, 147–160. (5) Radtke, A.S., C.M. Taylor, R.C. Erd, and F.W. Dickson (1974) Occurrence of lorandite, $TlAsS_2$, at the Carlin gold deposit, Nevada. Econ. Geol., 69, 121–174.

Crystal Data: Hexagonal. *Point Group:* n.d. Granular to tabular grains, to 10 μm, in irregular spherical aggregates, 0.1 to 0.6 mm; intergrown with mercurian silver and silicates.

Physical Properties: *Tenacity:* Malleable. Hardness = ~2.5 VHN = 44–75
D(meas.) = 12.5 D(calc.) = [12.6]

Optical Properties: Opaque. *Color:* Milky white in reflected light; tarnishes black. *Streak:* Black. *Luster:* Metallic. *Pleochroism:* Pinkish to milky white. *Anisotropism:* Very weak.
R_1–R_2: (546) 64.2–70.5, (589) 64.9–74.0

Cell Data: *Space Group:* n.d. $a = 6.61$ $c = 10.98$ Z = 6

X-ray Powder Pattern: Hebei Province, China.
1.495 (100), 1.204 (90), 2.830 (70), 1.134 (70), 2.000 (60), 1.105 (60), 1.010 (6)

Chemistry:

	(1)	(2)
Ag	62.4	61.73
Fe	0.0	
Co	0.05	
Ni	0.0	
Cu	0.01	
Hg	37.9	38.27
Te	0.10	
Total	100.46	100.00

(1) Hebei Province, China; by electron microprobe, average of five analyses, ranging from $Ag_{3.08}Hg_{0.92}$ to $Ag_{2.96}Hg_{1.04}$. (2) Ag_3Hg.

Occurrence: In a placer containing gold and other elements.

Association: Gold, lead, zinc, mercurian silver, silicates.

Distribution: From a poorly specified locality in Hebei Province, China.

Name: For the Luan He river, Hebei Province, China, in which it is found.

Type Material: The National Geological Museum, Chinese Academy of Geological Sciences, Beijing, China.

References: (1) Shao Dianxin, Zhou Jianxiong, Zhang Jianhong, and Bao Daxi (1984) Luanheite—a new mineral. Acta Mineral. Sinica, 4, 97–101 (in Chinese with English abs.). (2) (1988) Amer. Mineral., 73, 192–193 (abs. ref. 1).

Crystal Data: Tetragonal. *Point Group:* $\bar{4}2m$. Usually fine-grained, massive, rarely coarse-grained; crystals rare, less than 2 mm across. *Twinning:* Common on {112}, also polysynthetic in polished section.

Physical Properties: *Cleavage:* {101} good; {100} distinct. Hardness = 3.5 VHN = n.d. D(meas.) = 4.38 D(calc.) = 4.53

Optical Properties: Opaque. *Color:* Deep pinkish brown, similar to bornite, although darker; in polished section, pale brownish pink. *Streak:* Black. *Luster:* Metallic, dull. *Pleochroism:* Weak. *Anisotropism:* Strong, in greenish yellow and purplish.
R_1–R_2: (400) 26.5–29.2, (420) 25.5–27.7, (440) 24.5–26.2, (460) 23.8–25.0, (480) 23.6–24.6, (500) 23.6–25.0, (520) 23.9–26.0, (540) 24.4–27.0, (560) 25.0–27.9, (580) 25.6–28.7, (600) 26.2–29.3, (620) 27.0–29.9, (640) 27.7–30.3, (660) 28.4–30.5, (680) 28.9–30.5, (700) 29.0–30.3

Cell Data: *Space Group:* $I\bar{4}2m$. $a = 5.332(5)$ $c = 10.57(1)$ Z = 2

X-ray Powder Pattern: Mancayan, Philippines.
3.046 (100), 1.855 (90), 1.592 (70), 1.578 (60), 1.204 (60), 1.078 (60), 1.321 (50)

Chemistry:

	(1)	(2)
Cu	48.32	48.42
As	17.35	19.02
Sb	1.48	
S	32.85	32.56
Total	100.00	100.00

(1) Mancayan, Philippines; analysis recalculated after deduction of 0.97% insol. and 0.56% pyrite. (2) Cu_3AsS_4.

Polymorphism & Series: Dimorphous with enargite.

Occurrence: Found in hydrothermal deposits formed at low to moderate temperatures.

Association: Enargite, colusite, stannoidite, tetrahedrite–tennantite, pyrite, chalcopyrite, covellite, sphalerite, bismuthinite, silver sulfosalts, silver, gold, marcasite, alunite, barite, quartz.

Distribution: In the USA, at Goldfield, Esmeralda Co., Nevada; Butte, Silver Bow Co., Montana; and at Summitville, Rio Grande Co., Colorado. In Japan, at the Teine, Morino, and Date mines, Hokkaido; the Kasuga and Akeshi mines, Kagoshima Prefecture; and other localities. From the Lepanto mine, Mancayan, Luzon, Philippines. In the Chinkuashi mine, Keelung, Taiwan. From the Sierra Famatina, La Rioja Province, Argentina. In the Julcani district, at Cerro de Pasco, and at Huaron, Peru. At Chelopech, Bulgaria. From Calabona, Sardinia, Italy.

Name: For Luzon Island in the Philippines.

References: (1) Gaines, R.V. (1957) Luzonite, famatinite and some related minerals. Amer. Mineral., 42, 766–779. (2) Marumo, F. and W. Nowacki (1967) A refinement of the crystal structure of luzonite, Cu_3AsS_4. Zeits. Krist., 124, 1–8. (3) Springer, G. (1969) Compositional variations in enargite and luzonite. Mineral. Deposita, 4, 72–74.

Crystal Data: Tetragonal. *Point Group:* $4/m\ 2/m\ 2/m$. As well-formed thin tabular crystals; massive, fine-feathery.

Physical Properties: *Cleavage:* Very good on {001}. Hardness = Soft. VHN = 63–89 (15 g load). D(meas.) = n.d. D(calc.) = 4.30

Optical Properties: Opaque. *Color:* Bronzy. *Streak:* Black. *Luster:* Metallic. *Pleochroism:* Strong. *Anisotropism:* Extreme.

R_1–R_2: (400) 21.0–35.5, (420) 21.0–37.2, (440) 21.0–38.9, (460) 21.0–40.4, (480) 21.1–41.8, (500) 21.2–43.1, (520) 21.3–44.3, (540) 21.6–45.3, (560) 21.9–46.1, (580) 22.2–46.9, (600) 22.6–47.4, (620) 22.9–47.8, (640) 23.3–48.2, (660) 23.8–48.5, (680) 24.3–48.8, (700) 24.9–49.1

Cell Data: *Space Group:* $P4/nmm$. $a = 3.673$ $c = 5.035$ Z = 2

X-ray Powder Pattern: Mackinaw mine, Washington, USA.
5.03 (100), 2.31 (90), 1.809 (80), 2.96 (70), 1.838 (50), 1.729 (50), 1.055 (40)

Chemistry:

	(1)	(2)	(3)
Fe	63.0	58.8	54.9
Cu		2.0	0.1
Co			0.4
Ni	3.1	3.8	8.8
S	34.0	35.9	35.8
Total	100.1	100.5	100.0

(1) Mackinaw mine, Washington, USA; by electron microprobe, corresponding to $Fe_{1.06}Ni_{0.05}S$ or $(Fe, Ni)_9S_8$. (2) Talnakh area, USSR; by electron microprobe. (3) Scotia deposit, Australia; by electron microprobe, corresponding to $(Fe_{0.86}Ni_{0.13}(Co + Cu)_{0.01})_{\Sigma=1.00}S_{1.00}$.

Occurrence: Formed by hydrothermal activity in mineral deposits, during serpentinization of peridotites, and in the reducing environment of river bottom muds. Rarely in iron meteorites.

Association: Chalcopyrite, pentlandite, maucherite (Mackinaw mine, Washington, USA); troilite, chalcopyrite, pentlandite, alabandite.

Distribution: Now known from a number of localities in addition to those listed here. In the USA, at the Mackinaw mine, Snohomish Co., Washington; in muds from the bottom of the Mystic River, Boston, Massachusetts; from the Howard Montgomery quarry, Howard Co., Maryland; in California, from the Kramer borate deposit, Kern Co. In Canada, in the Muskox Intrusion, Northwest Territories. From the Outokumpu mine and Hitura, Finland. At the Vena mine, Sweden. In the Fiat mine, Piedmont, Italy. From England, at The Rill, Kynance Cliff, Lizard, Cornwall. In the Talnakh area, Noril'sk region, western Siberia, USSR. From Broken Hill, New South Wales, and the Scotia nickel deposit, Western Australia. On Cyprus, at Skouriotissa.

Name: For the Mackinaw mine in Washington, USA.

Type Material: n.d.

References: (1) Evans, H.T., Jr., C. Milton, E.C.T. Chao, I. Adler, C. Mead, B. Ingram, and R.A. Berner (1964) Valleriite and the new iron sulfide, mackinawite. U.S. Geol. Sur. Prof. Paper 475-D, D64–D69. (2) Kouvo, O. and Y. Vuorelainen (1963) A tetragonal iron sulfide. Amer. Mineral., 48, 511–524. (3) Ramdohr, P. (1969) The ore minerals and their intergrowths, (3rd edition), 673–679. (4) Buchwald, V.F. (1977) The mineralogy of iron meteorites. Phil. Trans. Royal Soc. London, A. 286, 453–491. (5) Ostwald, J. (1978) A note on the occurrences of nickeliferous and cupriferous mackinawite. Mineral. Mag., 42, 516–517.

Crystal Data: Orthorhombic. *Point Group:* $2/m\ 2/m\ 2/m$ or $mm2$. Elongated and striated along [001].

Physical Properties: *Cleavage:* Perfect on {010}. *Fracture:* Conchoidal. Hardness = n.d. VHN = 155 (50 g load). D(meas.) = n.d. D(calc.) = 5.98

Optical Properties: Opaque. *Color:* Gray-black. *Streak:* Gray-black, shining. *Luster:* Metallic. *Pleochroism:* Strong, from white to gray. *Anisotropism:* Strong. R_1–R_2: n.d.

Cell Data: *Space Group:* *Pbam* or *Pba*2. $a = 27.2$ $b = 34.1$ $c = 8.12$ Z = 4

X-ray Powder Pattern: Madoc, Canada.
3.396 (100), 3.355 (90), 2.720 (80), 3.67 (70), 2.925 (60), 3.87 (50), 3.110 (40)

Chemistry:

	(1)
Pb	55.0
Sb	22.8
As	3.1
S	19.9
Total	100.8

(1) Madoc, Canada; by electron microprobe, average of three analyses.

Occurrence: In the marbles of the Precambrian Grenville Limestone, at the margin of an intrusive.

Association: Jamesonite, boulangerite (Madoc, Canada); chabournéite, pierrotite, parapierrotite, stibnite, pyrite, sphalerite, twinnite, zinkenite, andorite, smithite, laffittite, routhierite, aktashite, wakabayashilite, realgar, orpiment (Jas Roux, France).

Distribution: From near Madoc, Huntington Township, Ontario, Canada. At Jas Roux, Hautes-Alpes, France. From Novoye, Khaidarkan, Kirgizia, USSR. At Boliden, Västerbotten, Sweden.

Name: For the locality at Madoc, Canada.

Type Material: National Museum of Natural History, Washington, D.C., USA, 160247.

References: (1) Jambor, J.L. (1967) New lead sulfantimonides from Madoc, Ontario—Part 1. Can. Mineral., 9, 7–24. (2) (1968) Amer. Mineral., 53, 1421 (abs. ref. 1).

Crystal Data: Hexagonal. *Point Group:* n.d. As rounded or oval inclusions to 100 μm in polarite and as intergrowths with stannopalladinite.

Physical Properties: Hardness = ~5 VHN = 520 (50 g load). D(meas.) = 9.33 (synthetic). D(calc.) = 10.5

Optical Properties: Opaque. *Color:* Grayish white in reflected light. *Anisotropism:* Weak.
R_1–R_2: (400) — , (420) — , (440) 47.4, (460) 49.0, (480) 50.5, (500) 51.2, (520) 52.3, (540) 52.8, (560) 54.4, (580) 55.2, (600) 55.8, (620) 56.1, (640) 56.7, (660) 58.0, (680) 57.9, (700) 58.5

Cell Data: *Space Group:* n.d. $a = 6.066$ $c = 7.20$ Z = 6

X-ray Powder Pattern: Majak mine, USSR.
2.65 (100), 1.988 (100), 2.19 (70), 2.30 (50), 3.04 (40), 2.40 (40), 1.800 (40b)

Chemistry:

	(1)	(2)	(3)
Pd	41.3	41.2	44.33
Ni	27.0	27.4	24.46
As	31.0	30.9	31.21
Total	99.3	99.5	100.00

(1–2) Majak mine, USSR; by electron microprobe, corresponding to $Pd_{0.93}Ni_{1.09}As_{0.98}$ and $Pd_{0.92}Ni_{1.10}As_{0.98}$ respectively. (3) PdNiAs.

Occurrence: As intergrowths with other platinum-group minerals in chalcopyrite and talnakhite ores.

Association: Chalcopyrite, talnakhite, polarite, stanopalladinite, silver, other platinum-group minerals.

Distribution: In the Majak mine, Talnakh area, Noril'sk region, western Siberia, and near Nizhni Tagil, Ural Mountains, USSR. In the Konttijärvi Intrusion, northern Finland.

Name: For the Majak mine, USSR.

Type Material: Institute of Geology of Ore Deposits, Petrography, Mineralogy and Geochemistry; A.E. Fersman Mineralogical Museum, Academy of Sciences, Moscow, USSR.

References: (1) Genkin, A.D., T.L. Evstigneeva, N.V. Troneva, and L.N. Vyal'sov (1976) Majakite, PdNiAs, a new mineral from copper–nickel sulfide ores. Zap. Vses. Mineral. Obshch., 105, 698–703. (2) (1977) Amer. Mineral., 62, 1260 (abs. ref. 1).

Crystal Data: Hexagonal. *Point Group:* $3m$. Massive.

Physical Properties: Hardness = 2.5–3 VHN = n.d. D(meas.) = n.d. D(calc.) = 7.22

Optical Properties: Opaque. *Color:* In polished section, orange-yellow, pure yellow in oil. *Pleochroism:* Strong, yellow to greenish yellow. *Anisotropism:* Strong, in oil pale green to pale orange-yellow, in air glowing cinder-red to blue-green or green.

R_1–R_2: (400) 24.0–25.0, (420) 27.5–30.0, (440) 31.0–35.0, (460) 35.3–38.7, (480) 39.4–41.6, (500) 43.0–43.8, (520) 46.0–45.8, (540) 48.8–47.4, (560) 51.1–48.6, (580) 53.2–49.6, (600) 54.9–50.5, (620) 56.3–51.2, (640) 57.4–51.9, (660) 58.5–52.4, (680) 59.4–52.9, (700) 60.2–53.4

Cell Data: *Space Group:* n.d. $a = 10.01$ $c = 3.28$ Z = 9

X-ray Powder Pattern: Kuusamo, Finland.
2.88 (100), 2.63 (100), 2.35 (100), 1.95 (100), 4.99 (60), 1.81 (40), 1.71 (40)

Chemistry:

	(1)	(2)
Ni	41.1	42.65
Co	1.0	
Cu	trace	
Se	57.9	57.35
Total	100.0	100.00

(1) Kuusamo, Finland; by X-ray fluorescence analysis. (2) NiSe.

Occurrence: In sills of albite diabase in schist, associated with low-grade uranium mineralization.

Association: Clausthalite, selenian melonite.

Distribution: From Kuusamo and Hitura, Finland.

Name: For Eero Mäkinen, Finnish geologist, and former President of the Outokumpu Company.

Type Material: n.d.

References: (1) Vuorelainen, Y., A. Huhma, and A. Häkli (1964) Sederholmite, wilkmanite, kullerudite, mäkinenite, and trüstedtite, five new nickel selenide minerals. Compt. Rendus Soc. Géol. Finlande, 36, 113–125. (2) (1965) Amer. Mineral., 50, 519–520 (abs. ref. 1).

Crystal Data: Cubic. *Point Group:* n.d. Granular.

Physical Properties: Hardness = n.d. VHN = n.d. D(meas.) = n.d. D(calc.) = [7.40]

Optical Properties: Opaque. *Color:* In polished section, bright white with a faint brown tint. R: n.d.

Cell Data: *Space Group:* n.d. $a = 9.910$ Z = [8]

X-ray Powder Pattern: "Tao" district, China.
2.50 (10), 1.75 (10), 1.90 (8), 0.783 (8), 1.105 (7), 5.35 (6), 2.95 (5)

Chemistry:

	(1)
Cu	9.95
Ni	0.33
Co	2.60
Fe	1.00
Pt	36.77
Ir	25.23
Pd	0.45
S	23.47
Total	99.80

(1) "Tao" district, China; giving $(Pt_{1.02}Ir_{0.71}Co_{0.24}Pd_{0.02})_{\Sigma=1.99}(Cu_{0.85}Fe_{0.10}Ni_{0.03})_{\Sigma=0.98}S_{3.98}$.

Polymorphism & Series: Forms a series with cuproiridisite.

Occurrence: In peridotite-type platinum ores with Cu-Ni sulfides ("Tao" district, China); in dunites, with sulfides (Konder massif, USSR).

Association: Pentlandite, pyrrhotite, bornite, magnetite, cooperite, sperrylite, platinum, olivine, pyroxenes, serpentine, chlorite ("Tao" district, China); iridosmine, laurite, isoferroplatinum (Konder massif, USSR).

Distribution: In China, from the "Tao" district – a code name. From the Konder massif, Aldan Shield, Yakutia, USSR.

Name: Presumably for a vaguely specified type locality.

Type Material: Geological Institute, Academy of Geological Sciences of China, (Beijing ?), China.

References: (1) Yu Tsu-Hsiang, Lin Shu-Jen, Chao Pao, Fang Ching-Sung, and Huang Chi-Shun (1974) A preliminary study of some new minerals of the platinum group and another associated new one in platinum-bearing intrusions in a region in China. Acta Geol. Sinica, 2, 202–218 (in Chinese with English abs.). (2) (1976) Amer. Mineral., 61, 185 (abs. ref. 1). (3) Peng Zhiizhong, Chang Chiehung, and Ximen Lovlov (1978) Discussion on published articles in the research of new minerals of the platinum-group discovered in China in recent years. Acta Geol. Sinica, 4, 326–336 (in Chinese with English abs.) [Peng Zhiizhong formerly Pen Chih-Zhong]. (4) (1980) Amer. Mineral., 65, 408 (abs. ref. 3). (5) Yu Zuxiang (1981) A restudy of malanite and cobalt-malanite (dayingite). Geol. Rev., 27, 55–71 (in Chinese with English abs.). [Yu Zuxiang formerly Yu Tsu-hsiang]. (6) (1982) Amer. Mineral., 67, 1081–1082 (abs. ref. 5). (7) Rudashevskii, N.S., A.G. Mochalov, V.V. Shkurskii, N.I. Shumskaya, and Y.P. Men'shikov (1984) First discovery of malanite ($Cu(Pt,Ir,Rh)_2S_4$) in the USSR. Mineral. Zhurnal., 6, 93–97 (in Russian). (8) (1984) Chem. Abs., 101, 10130 (abs. ref. 7).

Crystal Data: Cubic. *Point Group:* $4/m\ \overline{3}\ 2/m$. Octahedral crystals; massive granular; also in thin coatings; typically seen in polished section as "decomposed" into predominately coarse myrmekitic intergrowth of Au and Bi in which are wedges of maldonite.

Physical Properties: *Cleavage:* Distinct on {001} and {110}. *Fracture:* Conchoidal. *Tenacity:* Malleable and sectile. Hardness = 1.5–2 VHN = 147–264 (100 g load). D(meas.) = 15.46 D(calc.) = 15.70

Optical Properties: Opaque. *Color:* Silver-white with pinkish tinge on fresh surface, tarnishes copper-red to black. *Luster:* Metallic.
R: (400) 47.0, (420) 48.6, (440) 50.2, (460) 51.6, (480) 52.8, (500) 54.0, (520) 55.0, (540) 55.8, (560) 56.5, (580) 56.9, (600) 57.3, (620) 57.4, (640) 57.5, (660) 57.7, (680) 58.1, (700) 58.6

Cell Data: *Space Group:* $Fd3m$. $a = 7.971$ Z = 8

X-ray Powder Pattern: Nuggety Reef, Australia.
2.41 (100), 1.537 (60), 2.30 (50), 1.412 (50), 2.82 (40), 1.038 (40), 1.629 (30)

Chemistry:

	(1)	(2)	(3)	(4)
Au	65.12	63.80	65.92	65.36
Bi	34.88	36.80	34.07	34.64
Total	100.00	100.60	99.99	100.00

(1) Nuggety Reef, Australia. (2) Salsigne deposit, France; by electron microprobe. (3) Tyrnyauz district, USSR; by electron microprobe, trace elements Pb, Cu, Mo, Fe, Ti, Ca, Mg, Al, and Si. (4) Au_2Bi.

Occurrence: Formed within what is probably a restricted high-temperature range under hydrothermal conditions in gold-quartz veins.

Association: Gold, scheelite, apatite (Nuggety Reef, Australia); arsenopyrite, pyrite, pyrrhotite, chalcopyrite, stannite, hübnerite–ferberite, bismuthinite, bismuth, gold, siderite, quartz (Salsigne deposit, France).

Distribution: In Australia, in Victoria, near Maldon, at Nuggety Reef, and in the Eagle Hawk mine, Union Reef. In the Salsigne gold deposit, north of Carcassonne, Aude, and at Scoufour, Cantal, France. From the Tyrnyauz district, northern Caucasus Mountains, USSR.

Name: For the locality at Maldon, Australia.

References: (1) Palache, C., H. Berman, and C. Frondel (1944) Dana's system of mineralogy, (7th edition), v. I, 95–96. (2) Berry, L.G. and R.M. Thompson (1962) X-ray powder data for the ore minerals. Geol. Soc. Amer. Mem. 85, 15. (3) Boyer, F. and P. Picot (1963) Sur la présence de maldonite (Au_2Bi) à Salsigne (Aude). Bull. Soc. fr. Minéral., 86, 429 (in French). (4) Prokuronov, P.V., Y.I. Dryzhak, and V.I. Shkurskii (1976) First find of maldonite in the USSR. Zap. Vses. Mineral. Obshch., 105, 453–456 (in Russian). (5) (1976) Chem. Abs., 85, 180204 (abs. ref. 4).

Crystal Data: Cubic. *Point Group:* $4/m\ \overline{3}\ 2/m$ (probable). As irregular grains, hundredths of a mm to 0.4 mm in size, and as veinlets less than 0.1 mm thick, mostly in cubanite. *Twinning:* Polysynthetic twinning seen in polished section.

Physical Properties: Hardness = n.d. VHN = 195 (20 g load). D(meas.) = n.d. D(calc.) = 4.44

Optical Properties: Opaque. *Color:* In polished section, grayish yellow, darker than cubanite. *Luster:* Metallic. *Pleochroism:* Weak. *Anisotropism:* Weak.

R: (400) — , (420) — , (440) 21.2, (460) 22.7, (480) 24.7, (500) 26.5, (520) 27.9, (540) 29.0, (560) 29.9, (580) 30.8, (600) 31.2, (620) 32.3, (640) 32.9, (660) 33.8, (680) 34.5, (700) 34.6

Cell Data: *Space Group:* $Fm3m$ (probable). $a = 10.73$ Z = 4

X-ray Powder Pattern: Oktyabr mine, USSR.
3.23 (100), 1.894 (90), 1.097 (40), 3.08 (30), 2.07 (30), 3.78 (20), 2.46 (20)

Chemistry:

	(1)	(2)
Mn	3.2	4.6
Pb	8.2	4.8
Cd	1.3	1.0
Cu	29.8	31.19
Fe	26.4	27.21
S	29.4	31.46
Total	98.3	100.26

(1) Mayak mine, USSR; by electron microprobe. (2) Oktyabr mine, USSR; by electron microprobe.

Occurrence: In Cu-Ni ores (Noril'sk region, USSR).

Association: Cubanite.

Distribution: In the Oktyabr and Mayak mines, Talnakh area, Noril'sk region, western Siberia, USSR.

Name: For Tatyana Shadlun, Russian ore mineralogist.

Type Material: A.E. Fersman Mineralogical Museum, Academy of Sciences, Moscow, USSR.

References: (1) Evstigneeva, T.L., A.D. Genkin, N.V. Troneva, A.A. Filimonova, and A.I. Tsepin (1973) Shadlunite, a new sulfide of copper, iron, lead, manganese, and cadmium from copper–nickel ores. Zap. Vses. Mineral. Obshch., 102, 63–74 (in Russian). (2) (1973) Amer. Mineral., 58, 1114 (abs. ref. 1).

Crystal Data: Orthorhombic. *Point Group:* $2/m\ 2/m\ 2/m$. Crystals typically tabular on {010}, also pyramidal, prismatic, and, rarely, capillary; curved faces common. Stalactitic, reniform, fine-granular massive; cockscomb and spear shapes due to twinning on {101}. *Twinning:* Common and repeated on {101}; less common on {011}. Intense twin lamellae development observed in polished section.

Physical Properties: *Cleavage:* {101} rather distinct, {110} in traces. *Tenacity:* Brittle. Hardness = 6–6.5 VHN = 915–1099 (200 g load). D(meas.) = 4.887 D(calc.) = 4.875

Optical Properties: Opaque. *Color:* Tin-white on fresh surface, pale bronze-yellow, darkening on exposure, iridescent tarnish. *Streak:* Grayish or brownish black. *Luster:* Metallic. *Pleochroism:* [100] creamy white; [010] light yellowish white; [001] white with rose-brown tint. *Anisotropism:* Very strong, yellow through light green to dark green.
R_1–R_2: (400) 40.6–45.1, (420) 42.2–46.8, (440) 43.8–48.5, (460) 44.8–50.3, (480) 45.5–52.1, (500) 46.2–54.1, (520) 47.4–55.4, (540) 48.6–55.5, (560) 49.2–54.9, (580) 49.1–54.1, (600) 48.9–53.5, (620) 48.6–52.8, (640) 47.8–52.2, (660) 47.2–51.6, (680) 46.6–51.1, (700) 46.2–50.6

Cell Data: *Space Group:* *Pnnm.* $a = 4.436$ $b = 5.414$ $c = 3.381$ Z = 2

X-ray Powder Pattern: Webb City, Missouri, USA.
2.70 (100), 1.755 (90), 3.43 (60), 2.41 (60), 2.32 (60), 1.911 (50), 1.094 (50)

Chemistry:

	(1)	(2)	(3)
Fe	46.53	46.55	46.55
Cu	trace		
As		trace	
S	53.30	53.05	53.45
rem.	0.20		
Total	100.03	99.60	100.00

(1) Joplin, Missouri, USA; remainder is SiO_2. (2) Jasper, Wyoming, USA. (3) FeS_2.

Polymorphism & Series: Dimorphous with pyrite.

Occurrence: Typically formed under low-temperature highly acidic conditions, both in sedimentary environments (shales, limestones, and low rank coals) and in hydrothermal veins formed by ascending solutions. Stalactitic development and fossil pseudomorphism may be spectacular.

Association: Pyrite, pyrrhotite, galena, sphalerite, fluorite, dolomite, calcite.

Distribution: From numerous localities world-wide. In the USA, at Joplin, Jasper Co., Missouri; Picher and Cardin, Ottawa Co., Oklahoma, and Baxter Springs, Cherokee Co., Kansas. At Galena, Jo Davies Co., Illinois; and in Wisconsin, at Mineral Point, Iowa Co., and Racine, Racine Co. In Czechoslovakia, at Vinitrov, Litmice, Most, Osek, and other places. In Germany, at Clausthal, in the Harz Mountains, and Freiberg and Annaberg, Saxony. In France, at Cap Blanc-Nez, Pas-de-Calais. From England, in the chalk at Kent, and between Folkestone and Dover; at Tavistock, Devonshire. In Mexico, in the Santa Eulalia district, Chihuahua.

Name: A word of Arabic or Moorish origin, early applied to pyrite and other substances.

References: (1) Palache, C., H. Berman, and C. Frondel (1944) Dana's system of mineralogy, (7th edition), v. I, 311–315. (2) Berry, L.G. and R.M. Thompson (1962) X-ray powder data for the ore minerals. Geol. Soc. Amer. Mem. 85, 102. (3) Brostigen, G., A. Kjekshus, and C. Rømming (1973) Compounds with the marcasite type crystal structure VIII. Redetermination of the prototype. Acta Chem. Scand., 27, 2791–2796.

Crystal Data: Monoclinic. *Point Group:* $2/m$. Crystals equant to tabular on {010}; striated parallel to [001]; nearly globular crystals up to 5 mm. *Twinning:* Partly bent twin lamellae seen in polished section, in two sets crossing at oblique angles.

Physical Properties: *Fracture:* Conchoidal. *Tenacity:* Brittle. Hardness = 3 VHN = 168 D(meas.) = n.d. D(calc.) = 5.822

Optical Properties: Opaque. *Color:* Lead-gray to steel-gray, often tarnishes to iridescence; white in polished section, with red internal reflections. *Luster:* Metallic. *Anisotropism:* Distinct, strong in oil.
R_1–R_2: n.d.

Cell Data: *Space Group:* $P2_1a$. $a = 7.291$ $b = 12.68$ $c = 5.998$ $\beta = 91°13'$ Z = 4

X-ray Powder Pattern: Binntal, Switzerland.
3.45 (100), 2.75 (100), 3.00 (70), 2.05 (50), 2.91 (40), 2.01 (40), 0.996 (40)

Chemistry:

	(1)	(2)
Pb	41.0	42.62
Ag	23.7	22.19
As	17.9	15.41
S	18.8	19.78
Total	101.4	100.00

(1) Binntal, Switzerland; by electron microprobe. (2) $PbAgAsS_3$.

Occurrence: Of hydrothermal origin, in dolomite.

Association: Lengenbachite, rathite.

Distribution: At the Lengenbach quarry, Binntal, Valais, Switzerland.

Name: For Dr. John Edward Marr (1857–1933), geologist of Cambridge, England.

Type Material: n.d.

References: (1) Palache, C., H. Berman, and C. Frondel (1944) Dana's system of mineralogy, (7th edition), v. I, 487–488. (2) Wuensch, B.J. and W. Nowacki (1963) Zur Kristallchemie des Sulfosalzes Marrite. Chimia (Switzerland), 17, 381–382. (3) (1965) Amer. Mineral., 50, 812 (abs. ref. 2). (4) Wuensch, B.J. and W. Nowacki (1967) The crystal structure of marrite, $PbAgAsS_3$. Zeits. Krist., 125, 459–488.

Maslovite $(Pt, Pd)BiTe$

Crystal Data: Cubic. *Point Group:* 23. As elongated, sometimes rounded grains up to 200 μm in diameter.

Physical Properties: Hardness = n.d. VHN = 262–388 (20 g load), inversely proportional to Pd content. D(meas.) = n.d. D(calc.) = 11.51–11.74

Optical Properties: Opaque. *Color:* In polished section, light gray with a lilac hue. *Luster:* Metallic.

R: (400) — , (420) — , (440) 55.9, (460) 55.2, (480) 55.0, (500) 55.5, (520) 56.0, (540) 56.1, (560) 55.9, (580) 55.9, (600) 56.1, (620) 56.3, (640) 56.8, (660) 57.1, (680) 57.5, (700) 58.4

Cell Data: *Space Group:* $P2_13$. $a = 6.689(7)$ Z = 4

X-ray Powder Pattern: Oktyabr mine, USSR.
3.01 (10), 2.71 (8), 2.02 (6), 1.788 (3), 2.36 (2), 1.852 (2), 1.286 (2)

Chemistry:

	(1)	(2)	(3)
Pt	26.6	21.0	36.69
Pd	5.6	10.6	
Pb	0.0	0.0	
Bi	53.0	49.0	39.31
Sb	1.6	1.2	
Te	15.7	19.05	24.00
Total	102.5	101.9	100.00

(1) Oktyabr mine, USSR; by electron microprobe, corresponding to $(Pt_{0.71}Pd_{0.27})_{\Sigma=0.98}Bi_{1.31}Te_{0.64}Sb_{0.07}$. (2) Do.; corresponding to $(Pt_{0.53}Pd_{0.49})_{\Sigma=1.02}Bi_{1.16}Te_{0.78}Sb_{0.05}$. (3) PtBiTe.

Occurrence: In galena-rich portions of massive cubanite-chalcopyrite and mooihoekite ores.

Association: Altaite, sobolevskite, moncheite, michenerite, hessite, froodite, sperrylite, galena, chalcopyrite, cubanite, mooihoekite.

Distribution: From the Oktyabr mine, Noril'sk region, western Siberia, USSR. In the Rustenburg mine, on the Merensky Reef, Bushveld Complex, Transvaal, South Africa.

Name: For G.D. Maslov (1915–1968).

Type Material: Institute of Geology of Ore Deposits, Petrography, Mineralogy and Geochemistry; A.E. Fersman Mineralogical Museum, Academy of Sciences, Moscow, USSR.

References: (1) Kovalenker, V.A., V.D. Begizov, T.L. Evstigneeva, N.V. Troneva, and V.A. Ryabikin (1979) Maslovite, PtBiTe : a new mineral from the Oktyabr copper-nickel deposit. Geolog. rud. mestor., 21, 94–104 (in Russian). (2) (1980) Amer. Mineral., 65, 406–407 (abs. ref. 1).

Crystal Data: Hexagonal. *Point Group:* n.d. Massive, granular; subhedral prismatic crystals rare. Characteristically in intimate intergrowth with galena, sometimes forming Widmanstätten-like textures.

Physical Properties: *Fracture:* Uneven. *Tenacity:* Brittle. Hardness = 2.5 VHN = 72–85 (100 g load). D(meas.) = 6.9 D(calc.) = 6.99

Optical Properties: Opaque. *Color:* Iron-black to gray; white in polished section. *Streak:* Light gray. *Luster:* Metallic. *Pleochroism:* Weak. *Anisotropism:* Weak.
R_1–R_2: (400) 41-4–47.2, (420) 41.9–47.8, (440) 42.2–48.3, (460) 43.1–48.8, (480) 43.4–49.3, (500) 43.4–49.3, (520) 43.0–49.0, (540) 42.8–48.4, (560) 42.5–47.8, (580) 42.2–47.3, (600) 41.7–46.9, (620) 41.3–46.7, (640) 40.8–46.7, (660) 40.3–46.7, (680) 39.9–46.3, (700) 39.3–46.1

Cell Data: *Space Group:* n.d. $a = 8.12$ $c = 19.02$ Z = 12

X-ray Powder Pattern: Camsell River area, Canada.
2.827 (100), 3.302 (80), 1.966 (60), 2.029 (50), 6.311 (30), 1.709 (30), 3.453 (20)

Chemistry:

	(1)	(2)	(3)
Ag	28.76	27.3	28.33
Bi	54.50	56.0	54.84
S	17.24	16.9	16.83
Total	100.50	100.2	100.00

(1) Morococha, Peru; average of three analyses, after deduction of galena. (2) Camsell River area, Canada; by electron microprobe. (3) $AgBiS_2$.

Occurrence: Formed in hydrothermal deposits at moderate to high temperatures and in pegmatites.

Association: Galena, pyrite, chalcopyrite, sphalerite, arsenopyrite, tetrahedrite.

Distribution: In the USA, from the Mayflower mine, Boise Basin area, Ada Co., Idaho, and in the Darwin mine, Inyo Co., California. In Colorado, near Lake City, Hinsdale Co.; in the Revell 1 mine, north of Silver Cliff, Custer Co.; and at Leadville, Lake Co. In Canada, at the O'Brien and other mines, Cobalt, Ontario; at Glacier Gulch, British Columbia; and on the Camsell River, about six km south of Great Bear Lake, Northwest Territories. In Germany, from Schapbach, Black Forest. From Bustarviejo, near Madrid, Spain. At Panasqueira, Beira Baixa Province, Portugal. From southern Greenland, at Ivigtut. In the Matilda mine, near Morococha, Peru. Also known from a few other occurrences with less well-defined locality information.

Name: For the Matilda mine, Peru.

References: (1) Palache, C., H. Berman, and C. Frondel (1944) Dana's system of mineralogy, (7th edition), v. I, 429–430. (2) Harris, D.C. and R.I. Thorpe (1969) New observations on matildite. Can. Mineral., 9, 655–662.

Crystal Data: Hexagonal. *Point Group:* 3*m*. In pyramidal crystal aggregates consisting of oriented sceptre-shaped overgrowths of matraite and sphalerite.

Physical Properties: Hardness = n.d. VHN = n.d. D(meas.) = n.d. D(calc.) = 4.13

Optical Properties: Transparent. *Color:* Brownish yellow. *Luster:* Vitreous.
Anisotropism: Pronounced in some crystals.
R_1–R_2: n.d.

Cell Data: *Space Group:* *R*3*m*. $a = 3.8$ $c = 9.4$ Z = 3

X-ray Powder Pattern: n.d.

Chemistry:

	(1)	(2)
Zn	61.70	67.10
Fe	5.10	
S	33.22	32.90
Total	100.02	100.00

(1) Matra Mountains, Hungary. (2) ZnS.

Polymorphism & Series: Trimorphous with sphalerite and wurtzite.

Occurrence: Of hydrothermal origin.

Association: Wurtzite, sphalerite, galena, chalcopyrite, pyrite.

Distribution: From an undefined locality in the the Matra Mountains, Hungary. At Telluride, San Miguel Co., Colorado, USA.

Name: For the Matra Mountains locality in Hungary.

Type Material: n.d.

References: (1) Koch, S. (1958) The associated occurrence of three ZnS modifications in Gyöngyösoroszi. Acta mineralog. petrog. Univ. Szegediensis, 11, 11–12. (2) Sasvari, K. (1958) ZnS mineral with ZnS – 3R crystal structure. Acta mineralog. petrog. Univ. Szegediensis, 11, 23–27. (3) (1960) Amer. Mineral., 45, 1131 (abs. refs. 1 and 2). (4) Buck, D.C. and L.W. Strock (1955) Trimorphism in zinc sulfide. Amer. Mineral., 40, 192–200.

Crystal Data: Orthorhombic. *Point Group:* $2/m\ 2/m\ 2/m$. As equidimensional and rarer blade-like grains isolated in altaite, and as irregular rims a few μm thick on pyrrhotite and chalcopyrite in contact with altaite. *Twinning:* In polished section twinning commonly observed perpendicular to elongation axis of the laths.

Physical Properties: Hardness = n.d. VHN = 383, 404 (25 g load). D(meas.) = n.d. D(calc.) = 8.00

Optical Properties: Opaque. *Color:* In polished section, violet. *Pleochroism:* Weak, colors from pink to cream. *Anisotropism:* Weak, pinkish violet to grayish violet.

R_1–R_2: (400) 50.2–62.2, (420) 51.0–60.3, (440) 51.8–58.4, (460) 51.0–56.6, (480) 49.9–55.6, (500) 48.7–54.7, (520) 48.2–54.2, (540) 47.9–53.9, (560) 47.8–53.9, (580) 47.7–54.0, (600) 47.8–54.3, (620) 48.0–54.7, (640) 48.3–55.2, (660) 48.6–55.6, (680) 48.9–56.3, (700) 49.1–56.9

Cell Data: *Space Group:* *Pnnm.* $a = 5.31$ $b = 6.31$ $c = 3.89$ Z = 2

X-ray Powder Pattern: Mattagami Lake mine, Canada.
2.805 (10), 2.703 (8), 2.066 (6), 1.843 (4), 3.31 (3), 1.583 (3), 1.514 (2)

Chemistry:

	(1)	(2)
Co	10.3	18.76
Fe	6.7	
Te	82.4	81.24
Total	99.4	100.00

(1) Mattagami Lake mine, Canada; by electron microprobe, analyses of three grain sizes, corresponding to $Co_{0.54}Fe_{0.37}Te_{2.00}$. (2) $CoTe_2$.

Polymorphism & Series: Forms a series with frohbergite.

Occurrence: In a small telluride zone in a massive zinc-rich stratiform deposit in Archaen volcanics.

Association: Frohbergite, altaite, pyrrhotite, chalcopyrite, magnetite, talc, chlorite.

Distribution: From the Mattagami Lake mine, Galinee Township, near Matagami, Quebec, Canada.

Name: For its occurrence at Mattagami Lake (mine?), Canada.

Type Material: Canadian Geological Survey, Ottawa; Royal Ontario Museum, Toronto, Canada.

References: (1) Thorpe, R.I. and D.C. Harris (1973) Mattagamite and tellurantimony, two new telluride minerals from Mattagami Lake mine, Matagami area, Quebec. Can. Mineral., 12, 55–60. (2) (1974) Amer. Mineral., 59, 382 (abs. ref. 1).

Crystal Data: Tetragonal. *Point Group:* 422. Frequently tabular {001}; also pyramidal with faces striated parallel to their intersections with {001}. Massive, granular, and radiating fibrous. *Twinning:* On {203} and {106}.

Physical Properties: *Fracture:* Uneven. *Tenacity:* Brittle. Hardness = 5 VHN = 623–723 (100 g load). D(meas.) = 8.00 D(calc.) = 8.02

Optical Properties: Opaque. *Color:* Platinum-gray with reddish tint, tarnishes coppery red; in polished section, pinkish gray. *Streak:* Blackish gray. *Luster:* Metallic. *Anisotropism:* Weak.
R_1–R_2: (400) 45.6–48.0, (420) 45.8–48.0, (440) 46.0–48.0, (460) 46.4–48.2, (480) 46.8–48.6, (500) 47.2–49.2, (520) 47.8–49.9, (540) 48.4–50.8, (560) 49.3–51.5, (580) 50.3–52.5, (600) 51.4–53.5, (620) 52.7–54.8, (640) 53.8–56.0, (660) 55.0–57.2, (680) 56.0–58.3, (700) 57.0–59.3

Cell Data: *Space Group:* $P4_12_12$. $a = 6.872(4)$ $c = 21.821(1)$ Z = 4

X-ray Powder Pattern: Eisleben, Germany.
2.01 (100), 1.713 (100), 2.69 (90), 1.212 (60), 1.449 (50), 1.108 (50), 1.083 (50)

Chemistry:

	(1)	(2)	(3)	(4)
Ni	49.96	50.03	51.7	51.85
Co	0.20	0.84	0.3	
Fe	0.84	trace		
Cu	0.69	0.13		
As	45.88	45.90	48.5	48.15
Sb			0.1	
S	0.97	0.18		
rem.	0.68	1.66		
Total	99.22	98.74	100.6	100.00

(1) Sudbury, Canada; remainder is H_2O 0.36%, gangue 0.32%. (2) Eisleben, Germany; remainder is gangue 1.66%. (3) Elk Lake, Canada. (4) $Ni_{11}As_8$.

Occurrence: In hydrothermal veins with other nickel arsenides and sulfides.

Association: Nickeline, nickel-skutterudite, chalcopyrite (Eisleben, Germany); millerite, uvarovite, pyroxene, calcite (Orford, Canada).

Distribution: In Canada, at several mines in the Cobalt district, and at the Moose Horn mine, Elk Lake, Timiskaming district, Ontario; and in Quebec, at Orford, and at the Jeffrey mine, Asbestos. In the USA, at the Gem mine, northwest of Silver Cliff, Fremont Co., Colorado; in the Mohawk mine, Keweenaw Co., Michigan; and in the Mackinaw mine, Monte Cristo, Snohomish Co., Washington. In Germany, at Eisleben, Saxony, and Mansfeld, Thuringia. From near Schladming, Styria, Austria. From the Littleham Cove area, Budleigh Salterton, Devon, England. In Spain, from Los Jarales, Málaga Province; and Vimbodi, Tarragona Province. From Bou Azzer, Morocco. In the Talmessi mine, near Anarak, Iran. Also known from a few other localities.

Name: For William Maucher (1879–1930), mineral dealer of Munich, Germany.

References: (1) Palache, C., H. Berman, and C. Frondel (1944) Dana's system of mineralogy, (7th edition), v. I, 192–194. (2) Berry, L.G. and R.M. Thompson (1962) X-ray powder data for the ore minerals. Geol. Soc. Amer. Mem. 85, 42–43. (3) Fleet, M.E. (1973) The crystal structure of maucherite ($Ni_{11}As_8$). Amer. Mineral., 58, 203–210. (4) Ramdohr, P. (1969) The ore minerals and their intergrowths, (3rd edition), 400–402.

Crystal Data: Tetragonal. *Point Group:* $\overline{4}2m$. As minute rounded to irregular inclusions (0.05 to 1.3 mm) in bornite.

Physical Properties: *Cleavage:* Two imperfect at right angles suggested by fractures during hardness tests. Hardness = 3.5–4 VHN = n.d. D(meas.) = n.d. D(calc.) = 4.65 Magnetic.

Optical Properties: Opaque. *Color:* In polished section, brownish orange.
Pleochroism: Strong, orange to brown with slight orange tint. *Anisotropism:* Very strong, from bright straw-yellow to bright royal blue to dark blue.
R_1–R_2: (400) 9.5–12.2, (420) 10.5–13.5, (440) 11.5–14.8, (460) 13.0–16.3, (480) 14.6–18.0, (500) 16.5–19.8, (520) 18.5–21.6, (540) 20.6–23.0, (560) 23.0–24.5, (580) 25.6–25.8, (600) 28.2–27.0, (620) 30.0–28.0, (640) 31.5–29.0, (660) 32.9–29.9, (680) 34.2–30.7, (700) 35.2–31.4

Cell Data: *Space Group:* $P\overline{4}m2$. $a = 7.603(2)$ $c = 5.358(1)$ Z = 1

X-ray Powder Pattern: Mt. Lyell, Tasmania, Australia.
3.09 (100), 1.895 (80), 1.618 (60), 2.680 (50), 1.063 (50), 1.232 (30), 5.37 (20)

Chemistry:

	(1)	(2)	(3)	(4)
Cu	44.3	45.0	44.0	43.91
Fe	12.5	11.8	12.9	12.87
Zn		< 0.1		
Sn	10.4	11.8	13.7	13.67
Se			1.1	
S	33.0	30.3	28.8	29.55
Total	100.2	98.9	100.5	100.00

(1) North Lyell mine, Australia; by electron microprobe. (2) Tingha, Australia. (3) Kidd Creek mine, Canada. (4) $Cu_6Fe_2SnS_8$.

Occurrence: In massive to disseminated hydrothermal copper ores within highly altered volcanic rocks of Cambrian age (Mt. Lyell, Australia).

Association: Pyrite, chalcopyrite, bornite, chalcocite, digenite, tetrahedrite–tennantite, enargite, sphalerite, galena (Mt. Lyell, Australia); tetrahedrite–tennantite, luzonite–famatinite, pyrite, mawsonite, nekrasovite, chalcopyrite, emplectite, laitakarite, bismuth, calcite, quartz, barite (Kuramin Mountains, USSR).

Distribution: In Australia, at the North Lyell and Crown Lyell mines, Mt. Lyell district, Queenstown, Tasmania; and from New South Wales, at the Royal George mine, near Tingha, New England district. From Bolivia, at Vila Apacheta. In Peru, at Cholquijirca, Junin. From the New Brunswick Tin Mines deposit, New Brunswick; the Maggie copper deposit, British Columbia; and at the Kidd Creek mine, near Timmins, Ontario, Canada. From Chizeuil, Saône-et-Loire, France. At Tsumeb, Namibia. In the Khayragatsch deposit, Kuramin Mountains, eastern Uzbekistan, USSR. At the Akenobe, Tada, and Ikuno mines, Hyogo Prefecture, the Ashio mine, Tochigi Prefecture, the Fukoku mine, Kyoto Prefecture, and the Konjo mine, Okayama Prefecture, Japan. In the Ulsan mine, Kyongsangnam Province, Korea. Also known from a few other localities.

Name: For Sir Douglas Mawson (1882–1958), noted Australian geologist and Antarctic explorer.

Type Material: British Museum (Natural History), London, England; Harvard University, Cambridge, Massachusetts, USA.

References: (1) Markham, N.L. and L.J. Lawrence (1965) Mawsonite, a new copper–iron–tin sulfide from Mt. Lyell, Tasmania and Tingha, New South Wales. Amer. Mineral., 50, 900–908. (2) Szymanski, J.T. (1976) The crystal structure of mawsonite, $Cu_6Fe_2SnS_8$. Can. Mineral., 14, 529–535.

Crystal Data: Orthorhombic. *Point Group:* $2/m\ 2/m\ 2/m$ or $mm2$. As granular aggregates of interlocked 0.2–3 mm crystals.

Physical Properties: *Cleavage:* One, poorly developed. *Fracture:* Subconchoidal. Hardness = n.d. VHN = 43–45 (25 g load). D(meas.) = 6.61 D(calc.) = 6.57

Optical Properties: Opaque. *Color:* Steel-gray on fresh surface, becoming dark gray to black on long exposure; in polished section, light grayish white. *Streak:* Dark steel-gray. *Anisotropism:* Strong, in gray, pale grayish blue, and light tan.

R_1–R_2: (400) 24.9–31.4, (420) 25.3–31.1, (440) 25.7–30.8, (460) 25.7–30.4, (480) 25.4–29.8, (500) 25.0–29.1, (520) 24.3–28.4, (540) 24.0–27.9, (560) 23.5–27.4, (580) 23.3–27.1, (600) 23.2–26.9, (620) 23.0–26.6, (640) 23.0–26.4, (660) 23.0–26.1, (680) 23.0–25.9, (700) 22.9–25.8

Cell Data: *Space Group:* $Pnam$ or $Pna2_1$. $a = 14.043$ $b = 15.677$ $c = 7.803$ Z = 32

X-ray Powder Pattern: Foster mine, Cobalt, Canada.
2.606 (100), 2.070 (70), 3.51 (60), 3.06 (60), 2.862 (60), 1.948 (50), 2.567 (40)

Chemistry:

	(1)	(2)
Ag	60.0	60.7
Cu	24.9	24.8
S	15.1	14.8
Total	100.0	100.3

(1) Foster mine, Cobalt, Canada; by electron microprobe, corresponding to $(Ag_{1.18}Cu_{0.82})_{\Sigma=2.00}S$.
(2) Godejord, Norway.

Occurrence: Of hydrothermal origin; probably formed below 94.4°C, the upper stability limit for mckinstryite.

Association: Silver, arsenopyrite, stromeyerite, actinolite, calcite (Foster mine, Canada); bornite, chalcocite, chalcopyrite, djurleite, digenite, tennantite, stromeyerite, wittichenite, bismuth, rammelsbergite, balkanite, mercurian silver, cinnabar, pyrite, calcite, barite, aragonite (Sedmochislenitsi mine, Bulgaria).

Distribution: In Canada, from the Foster mine, Coleman Township, about two km southeast of Cobalt, Ontario, and from the Echo Bay mine, Great Bear Lake, Northwest Territories. In the USA, at the Colorado Central mine, near Georgetown, Clear Creek Co., and the Bulldog Mountain mine, Creede district, Mineral Co., Colorado; and at Mogollon, Catron Co., New Mexico. From Jalpa, Zacatecas, Mexico. At Godejord, Grong area, Norway. From Příbram, Czechoslovakia. At the Sedmochislenitsi mine, Vratsa district, western part of the Stara Planina (Balkan Mountains), Bulgaria. In the Tort Kudak Au-Ag deposit, Kazakhstan, USSR. From Broken Hill, New South Wales, Australia. In the Sado mine, Niigata Prefecture, Japan.

Name: For Hugh Exton McKinstry (1896–1961), Professor of Geology at Harvard University, Cambridge, Massachusetts, USA.

Type Material: National Museum of Natural History, Washington, D.C., 120056; Harvard University, Cambridge, Massachusetts, USA, 108804.

References: (1) Skinner, B.J., J.L. Jambor, and M. Ross (1966) Mckinstryite, a new copper–silver sulfide. Econ. Geol., 61, 1383–1389. (2) (1967) Amer. Mineral., 52, 1253 (abs. ref. 1).

Crystal Data: Hexagonal. *Point Group:* $\overline{3}\ 2/m$. As hexagonal plates to as much as 1 cm, and small laminated particles.

Physical Properties: *Cleavage:* Perfect {0001}. Hardness = 1–1.5 VHN = 46–59 (50 g load). D(meas.) = 7.72 D(calc.) = 7.73

Optical Properties: Opaque. *Color:* Reddish white, tarnished yellowish brown; in polished section white to faint pink. *Streak:* Dark gray. *Luster:* Metallic. *Pleochroism:* Very weak, cream-white to slightly more pinkish. *Anisotropism:* Moderate to strong.
R_1–R_2: (400) 52.9–69.9, (420) 53.6–64.2, (440) 54.3–58.5, (460) 55.1–56.0, (480) 56.1–55.4, (500) 57.3–55.7, (520) 58.5–56.5, (540) 59.8–57.7, (560) 61.0–58.9, (580) 62.0–60.1, (600) 63.0–61.2, (620) 63.8–62.3, (640) 64.6–63.2, (660) 65.4–64.0, (680) 66.0–65.0, (700) 66.6–66.0

Cell Data: *Space Group:* $P\overline{3}m1$. $a = 3.843$ $c = 5.265$ Z = 1

X-ray Powder Pattern: Robb-Montbray mine, Canada.
2.82 (100), 1.549 (60), 2.06 (50), 1.918 (50), 2.64 (30), 1.590 (20), 1.227 (20)

Chemistry:

	(1)	(2)	(3)
Ni	18.31	11.3	18.70
Pd		7.4	
Pt		3.7	
Ag	0.86		
Bi		14.4	
Te	80.75	65.2	81.30
Total	99.92	102.0	100.00

(1) Stanislaus mine, California, USA. (2) Strathcona mine, Canada; by electron microprobe, corresponding to $(Ni_{0.66}Pd_{0.24}Pt_{0.07})_{\Sigma=0.97}(Te_{1.52}Bi_{0.48})_{\Sigma=2.00}$. (3) $NiTe_2$.

Occurrence: Found with other tellurium minerals in the late stages of hydrothermal veins formed at moderate to low temperatures. Also in high temperature magmatic sulfide deposits.

Association: Altaite, petzite, hessite, calaverite, coloradoite, krennerite, tellurobismuthite, montbrayite, gold, pyrite, chalcopyrite.

Distribution: In the USA, at the Melones and Stanislaus mines, Calaveras Co., California. In Colorado, at the Cresson mine, Cripple Creek, Teller Co., and from a number of mines in the Magnolia district, Boulder Co. In Ontario, Canada, from the Strathcona mine at Sudbury; in the Hemlo gold deposit, Thunder Bay district; and at the Wright-Hargreaves mine, Kirkland Lake; in Quebec, from the Robb-Montbray mine, Montbray Township. At Kalgoorlie and Kambalda, Western Australia, and the Worturpa mine, Flinders Ranges, South Australia. From the Emperor mine, Vatukoula, Viti Levu, Fiji Islands. In South Africa, at Ookiep, Namaqualand. From the USSR, in the Monchegorsk deposit, Kola Peninsula. In the Yokozuru mine, north Kyushu, Japan. Additional minor occurrences are known.

Name: For the Melones mine in California, USA, where it was discovered.

References: (1) Palache, C., H. Berman, and C. Frondel (1944) Dana's system of mineralogy, (7th edition), v. I, 341. (2) Peacock, M.A. and R.M. Thompson (1945) Melonite from Quebec and the crystal structure of $NiTe_2$. University of Toronto Studies, Geol. Series, 50, 63–73. (3) Ramdohr, P. (1969) The ore minerals and their intergrowths, (3rd edition), 419–420. (4) Berry, L.G. and R.M. Thompson (1962) X-ray powder data for the ore minerals. Geol. Soc. Amer. Mem. 85, 145. (5) Rucklidge, J. (1968) Electron microprobe investigations of platinum metal minerals from Ontario. Can. Mineral., 9, 617–628.

Crystal Data: Orthorhombic. *Point Group:* $mm2$ ($2/m$ $2/m$ $2/m$ pseudocell). Crystals are slender prismatic and striated || [001]. Also massive, fibrous to compact.

Physical Properties: *Cleavage:* Perfect but interrupted on {010}; difficult on {001}. *Fracture:* Conchoidal. *Tenacity:* Brittle. Hardness = 2.5 VHN = n.d. D(meas.) = 6.36 D(calc.) = 6.391

Optical Properties: Opaque. *Color:* Blackish lead-gray; white in polished section. *Streak:* Shining black. *Luster:* Bright metallic. *Pleochroism:* Weak. *Anisotropism:* Strong. R_1–R_2: (400) 38.5–44.5, (420) 39.2–45.2, (440) 39.9–45.9, (460) 40.2–46.4, (480) 40.4–46.6, (500) 40.4–46.5, (520) 40.3–46.3, (540) 40.0–46.0, (560) 39.8–45.6, (580) 39.5–45.1, (600) 39.2–44.5, (620) 38.8–43.8, (640) 38.5–43.0, (660) 38.2–42.1, (680) 37.8–41.2, (700) 37.4–40.2

Cell Data: *Space Group:* $Pnm2_1$ ($Pbnm$ pseudocell). a = 11.36 b = 24.06 c = 99.07 Z = [24]

X-ray Powder Pattern: Bottino, Italy.
3.30 (100), 3.71 (90), 2.92 (80), 2.08 (50), 1.199 (50), 2.75 (40), 1.973 (40)

Chemistry:

	(1)	(2)	(3)
Pb	61.52	62.44	61.51
Cu	1.20	1.52	1.45
Sb	19.53	19.47	19.46
S	17.59	17.49	17.58
Total	99.84	100.92	100.00

(1) Bottino, Italy; by electron microprobe. (2) Marmora, Canada; by electron microprobe. (3) $Pb_{13}CuSb_7S_{24}$.

Occurrence: In hydrothermal veins and contact metasomatic deposits.

Association: Galena, chalcopyrite, sphalerite, bournonite, cubanite, pyrite, pyrrhotite, boulangerite, jamesonite, franckeite, tetrahedrite, gudmundite.

Distribution: In Canada, in Ontario, at Marble Lake, Barrie Township, Frontenac Co.; in British Columbia, at the Bluebird-Mayflower mine, Trail Creek; and other localities. In the USA, at the Kalkar Quarry, Santa Cruz Co., California. In Italy, at Bottino, near Seravezza; and in Val di Castello, near Pietrasanta, Tuscany. In England, from Shallowford Bridge, near South Molton, Devon, and in the Pengenna mine, St. Kew, Cornwall. From the Ochsenkopf, near Schwarzenberg, Saxony; and from Goldkronach, in the Fichtelgebirge, Bavaria, Germany. From Hällefors, Bergslagen, Sweden. At Pundung, Korea. From Broken Hill, New South Wales, Australia. Additional minor occurrences are known.

Name: For Professor Giuseppi Meneghini (1811–1889) of Pisa, Italy, who first observed the species.

References: (1) Palache, C., H. Berman, and C. Frondel (1944) Dana's system of mineralogy, (7th edition), v. I, 402–404. (2) Berry, L.G. and R.M. Thompson (1962) X-ray powder data for the ore minerals. Geol. Soc. Amer. Mem. 85, 170–171. (3) Jambor, J.L. (1975) Synthetic copper-free meneghinite. Geol. Survey of Canada Paper 75-1B, 71–72. (4) Hicks, W.D. and E.W. Nuffield (1978) Natural and synthetic meneghinite. Can. Mineral., 16, 393–395.

Crystal Data: Hexagonal. *Point Group:* $\overline{3}\ 2/m$. Commonly as liquid spheres or globules; crystallizes in rhombohedra.

Physical Properties: Hardness = n.d. VHN = n.d. D(meas.) = 13.596 (liquid). D(calc.) = n.d. Liquid above −39°C; volatile; vapor highly toxic.

Optical Properties: Opaque. *Color:* Tin-white. *Luster:* Metallic, brilliant.
R_1–R_2: n.d.

Cell Data: *Space Group:* $R\overline{3}m$. $a = 3.463$ $c = 6.706$ Z = 3

X-ray Powder Pattern: n.d.

Chemistry: Composition essentially mercury, occasionally with a little Ag or Au.

Occurrence: In hydrothermal deposits formed at low temperature and associated with hot springs.

Association: Cinnabar, metacinnabar, calomel, terlinguaite, eglestonite, mercurian silver.

Distribution: From a number of localities, but rarely in significant amounts. In the USA, in California, especially at New Almaden, Santa Clara Co., and Mt. Diablo, Contra Costa Co.; in Texas, at Terlingua, Brewster Co. A noted locality at Almadén, Ciudad Real Province, Spain. In Yugoslavia, from Idrija (Idria), Slovenia, and Mount Avala, near Belgrade. In Germany, at Landsberg, near Obermoschel, Rhineland-Palatinate. At Brezina, Czechoslovakia. In the USSR, at Nagolnii Krjasch, Donetz Basin. From Sala, Västmanland, Sweden.

Name: From the Latin *Mercurius*, the mythological messenger of the gods, in allusion to its mobility in liquid form; the chemical symbol from the Latin *hydrargyrum*, *liquid silver*.

References: (1) Palache, C., H. Berman, and C. Frondel (1944) Dana's system of mineralogy, (7th edition), v. I, 103.

Merenskyite $Pd(Te, Bi)_2$

Crystal Data: Hexagonal. *Point Group:* $\bar{3}2/m$. As minute grains, intimately intergrown with other Pt minerals.

Physical Properties: Hardness = ~3.5 VHN = 82–128 (10 g load). D(meas.) = n.d. D(calc.) = 8.547

Optical Properties: Opaque. *Color:* In polished section, white. *Pleochroism:* Weak, white to grayish white in air, more distinct in oil, from white to slightly creamy to light grayish white. *Anisotropism:* Distinct to strong, from dark brown to light greenish gray.
R_1–R_2: (400) 58.2–59.1, (420) 59.4–60.4, (440) 60.6–61.7, (460) 61.8–63.0, (480) 63.3–64.4, (500) 64.5–65.8, (520) 65.6–67.2, (540) 66.6–68.2, (560) 67.4–68.9, (580) 68.1–69.4, (600) 68.5–69.7, (620) 69.0–70.3, (640) 69.4–70.8, (660) 69.7–71.2, (680) 69.6–71.4, (700) 69.3–71.2

Cell Data: *Space Group:* $P\bar{3}m1$. $a = 3.978(1)$ $c = 5.125(2)$ Z = 1

X-ray Powder Pattern: Merensky Reef, South Africa.
2.92 (100), 2.10 (60), 3.07 (30), 2.02 (30), 1.67 (30), 1.54 (30), 2.51 (20)

Chemistry:

	(1)	(2)	(3)
Pd	33.2	29.2	29.42
Bi	15.1	2.6	
Sb		0.3	
Te	56.3	67.8	70.58
Total	104.6	99.9	100.00

(1) Rustenburg mine, South Africa; by electron microprobe. (2) Temagami, Canada; by electron microprobe. (3) $PdTe_2$.

Polymorphism & Series: Forms a series with moncheite.

Occurrence: With a variety of minerals containing platinum group elements, in disseminated interstitial segregations of Cu-Fe-Ni sulfides (Merensky Reef, South Africa). Most likely formed as a consequence of *in situ* fractional crystallization and sulfide immiscibility in an oxide-silicate melt.

Association: Moncheite, kotulskite, temagamite, michenerite, niggliite, sperrylite, iridosmine, platinum, argentian gold, leadamalgam, hessite, chalcopyrite, pentlandite, violarite, millerite, galena, stibnite, pyrrhotite, pyrite, gersdorffite, chromite, ilmenite, magnetite.

Distribution: In the Rustenburg mine, western Bushveld Complex; and at Messina, Transvaal, South Africa. In Canada, in Ontario, at Werner Lake, Rathbun Lake, Shebandowan, Temagami Island in Temagami Lake, the Lac des Iles Complex, in the Thierry mine, near Pickle Lake, and in the Sudbury district. From the Stillwater Complex, Montana; the New Rambler mine, Medicine Bow Mountains, Albany Co., Wyoming; and near Moapa, Clark Co., Nevada, USA. At Shiaonanshan, Inner Mongolia Autonomous Region, China. From the Noril'sk region, western Siberia, USSR.

Name: For Dr. Hans Merensky (1871–1952), who was instrumental in the discovery of the "Reef" also named for him.

Type Material: n.d.

References: (1) Kingston, G.A. (1966) The occurrence of platinoid bismuthotellurides in the Merensky Reef at Rustenburg platinum mine in the western Bushveld. Mineral. Mag., 35, 815–834. (2) (1967) Amer. Mineral., 52, 926 (abs. ref. 1). (3) Cabri, L.J., Ed. (1981) Platinum group elements: mineralogy, geology, recovery. Can. Inst. Min. & Met., 118–119, 156.

Crystal Data: Hexagonal, possibly monoclinic, pseudohexagonal. *Point Group:* n.d. As small grains, to 0.5 mm.

Physical Properties: Hardness = n.d. VHN = 561–593, 578 average (50 g load). D(meas.) = n.d. D(calc.) = n.d.

Optical Properties: Opaque. *Color:* In polished section, brassy yellow. *Anisotropism:* Distinct.
R_1–R_2: n.d.

Cell Data: *Space Group:* n.d. $a = 15.04$ $c = 22.41$ Z = 36

X-ray Powder Pattern: Goodnews Bay, Alaska, USA.
2.278 (vs), 2.171 (vs), 2.232 (m), 2.017 (m), 1.918 (m), 1.861 (m), 1.572 (m)

Chemistry:

	(1)
Pd	72.9
Cu	< 1.2
Sb	15.3
As	9.2
Total	< 98.6

(1) Goodnews Bay, Alaska, USA; by electron microprobe, average of four grains, corresponding to $(Pd_{5.04}Cu_{0.14})_{\Sigma=5.18}(Sb_{0.92}As_{0.90})_{\Sigma=1.82}$.

Polymorphism & Series: Dimorphous with isomertieite.

Occurrence: As fine grains in precious metal placer concentrates, apparently derived from ultramafic source rock.

Association: Gold, chromite, laurite, mertieite-II, Pt–Ir–Os alloys.

Distribution: In the USA, from the placer dredgings at Goodnews Bay, Alaska.

Name: For John B. Mertie, Jr. (1888–1980), geologist, U.S. Geological Survey, who provided the original material; "I" to distinguish its unique composition and crystallography from that of mertieite-II and isomertieite.

Type Material: National Museum of Natural History, Washington, D.C., USA, 132499.

References: (1) Desborough, G.A., J.J. Finney, and B.F. Leonard (1973) Mertieite, a new palladium mineral from Goodnews Bay, Alaska. Amer. Mineral., 58, 1–10. (2) Cabri, L.J., J.H.G. Laflamme, J.M. Stewart, J.F. Rowland, and T.Z. Chen (1975) New data on some platinum arsenides and antimonides. Can. Mineral., 13, 321–335.

Crystal Data: Hexagonal. *Point Group:* n.d. Anhedral grains, up to about 0.5 mm.

Physical Properties: Hardness = n.d. VHN = 561–593, 578 average (50 g load). D(meas.) = n.d. D(calc.) = 11.2 for Sb:As = 5:1.

Optical Properties: Opaque. *Color:* In polished section, brassy yellow. *Luster:* Metallic. *Anisotropism:* Distinct, dark blue-gray to dark brown.
R: (400) — , (420) 45.0, (440) 44.4, (460) 44.7, (480) 45.6, (500) 47.0, (520) 48.7, (540) 50.1, (560) 51.3, (580) 52.3, (600) 53.2, (620) 54.0, (640) 54.9, (660) 55.7, (680) 56.5, (700) 57.6

Cell Data: *Space Group:* n.d. $a = 7.546(2)$ $c = 43.18(1)$ Z = 12

X-ray Powder Pattern: Goodnews Bay, Alaska, USA.
2.286 (100), 2.177 (90), 1.226 (70), 1.200 (70), 0.9355 (60), 2.022 (50), 1.577 (40)

Chemistry:

	(1)	(2)	(3)
Pd	72.3	68.9	69.3
Cu	< 0.1	1.8	2.3
Sn		0.3	
Sb	24.7	27.4	28.0
As	3.3	1.9	2.2
Total	100.4	100.3	101.8

(1) Goodnews Bay, Alaska, USA; by electron microprobe, average of four grains and five areas, corresponding to $(Pd_{5.13}Cu_{<0.02})_{\Sigma=5.14}(Sb_{1.53}As_{0.33})_{\Sigma=1.86}$. (2) Farm Tweefontein, South Africa; by electron microprobe, corresponding to $(Pd_{7.67}Cu_{0.34})_{\Sigma=8.01}(Sb_{2.66}As_{0.30}Sn_{0.03})_{\Sigma=2.99}$. (3) Oktyabr mine, USSR; by electron microprobe, corresponding to $(Pd_{7.57}Cu_{0.42})_{\Sigma=7.99}(Sb_{2.67}As_{0.34})_{\Sigma=3.01}$.

Occurrence: As fine grains in precious metal placer concentrates, apparently derived from ultramafic source rock (Goodnews Bay, Alaska, USA); above massive cubanite-mooihoekite ore (Oktyabr mine, USSR).

Association: Gold, sperrylite, laurite, platarsite, ruthenarsenite, mertieite-I, genkinite, Pt–Ir–Os alloys, Pt–Fe alloy, stibiopalladinite, chalcocite, bornite, heazlewoodite, galena, chalcopyrite, pentlandite, valleriite, hauchecornite, parkerite, chromite.

Distribution: In the USA, from the placers at Goodnews Bay, Alaska; in the Stillwater Complex, Montana. At Farm Tweefontein, 26 km from Potgietersrust, and in the Onverwacht mine, Transvaal, South Africa. From the Oktyabr mine, Noril'sk area, western Siberia, and around Zlatoust, Ural Mountains, USSR. At Hope's Nose, Torquay, Devon, England. From the Konttijärvi Intrusion, northern Finland.

Name: For John B. Mertie, Jr. (1888–1980), geologist, U.S. Geological Survey, who provided the original material; "II" to distinguish its unique composition and crystallography from that of mertieite-I and isomertieite.

Type Material: n.d.

References: (1) Desborough, G.A., J.J. Finney, and B.F. Leonard (1973) Mertieite, a new palladium mineral from Goodnews Bay, Alaska. Amer. Mineral., 58, 1–10. (2) Cabri, L.J., J.H.G. Laflamme, J.M. Stewart, J.F. Rowland, and T.Z. Chen (1975) New data on some platinum arsenides and antimonides. Can. Mineral., 13, 321–335. (3) Cabri, L.J. and T.Z. Chen (1976) Stibiopalladinite from the type locality. Amer. Mineral., 61, 1249–1254. (4) Cabri, L.J., Ed. (1981) Platinum group elements: mineralogy, geology, recovery. Can. Inst. Min. & Met., 119–120, 159. (5) Clark, A.M. and A.J. Criddle (1982) Palladium minerals from Hope's Nose, Torquay, Devon. Mineral. Mag., 46, 371–377. (6) (1983) Amer. Mineral., 68, 851 (abs. ref. 5).

Crystal Data: Cubic. *Point Group:* $\bar{4}3m$. Usually massive; rarely as small tetrahedral crystals having rough faces. *Twinning:* Common on {111}, forming lamellae in polished section.

Physical Properties: *Fracture:* Subconchoidal. *Tenacity:* Brittle. Hardness = 3 VHN = n.d. D(meas.) = 7.65 D(calc.) = 7.63

Optical Properties: Opaque. *Color:* Grayish black; in polished section, grayish white. *Streak:* Black. *Luster:* Metallic. *Pleochroism:* Weak, occasionally. *Anisotropism:* Very weak, occasionally.

R: (400) 28.8, (420) 28.2, (440) 27.6, (460) 27.1, (480) 26.7, (500) 26.3, (520) 25.9, (540) 25.6, (560) 25.3, (580) 25.2, (600) 25.0, (620) 25.0, (640) 24.9, (660) 25.0, (680) 25.1, (700) 25.2

Cell Data: *Space Group:* $F\bar{4}3m$. $a = 5.8717(5)$ Z = 4

X-ray Powder Pattern: Synthetic.
3.378 (100), 2.068 (55), 1.7644 (45), 2.926 (35), 1.3424 (12), 1.6891 (10), 1.3085 (10)

Chemistry:

	(1)	(2)	(3)	(4)
Hg	79.73	67.45	81.33	86.22
Zn	4.23	3.10		
Cd		11.72		
Fe	trace	0.2		
Se	1.08		6.49	
S	14.58	15.63	10.30	13.78
Total	99.62	98.10	98.12	100.00

(1) Guadalcázar, Mexico. (2) Uland area, USSR. (3) San Onofre, Mexico. (4) HgS.

Polymorphism & Series: Trimorphous with cinnabar and hypercinnabar.

Occurrence: Found with cinnabar in mercury deposits formed near-surface, under low-temperature conditions.

Association: Cinnabar, mercury, wurtzite, stibnite, marcasite, realgar, calcite, barite, chalcedony, hydrocarbons.

Distribution: Not uncommon in mercury deposits; occasionally a principal ore mineral. In the USA, in California, in the Redington mine and other properties, near Knoxville, Napa Co.; Ryne mine, Mt. Diablo, Contra Costa Co.; at New Almaden, Santa Clara Co.; at Skaggs Springs, Sonoma Co.; and at New Idria, San Benito Co. At Marysvale, Piute Co., Utah; and in the Reward mines, Franklin, King Co., Washington. In Canada, from Read Island, Nanaimo mining division, British Columbia. In Yugoslavia, at Idrija (Idria), Slovenia. In Italy, at Levigliani, Tuscany. In Romania, at Baia Sprie (Felsőbánya). In Mexico, from San Onofre, near Plateros, Zacatecas; also from Guadalcázar, San Luis Potosí. In the Uland area, Kurai ore zone, Altai Mountains, USSR.

Name: From the Greek for *with*, plus *cinnabar*, in allusion to its common association.

References: (1) Palache, C., H. Berman, and C. Frondel (1944) Dana's system of mineralogy, (7th edition), v. I, 215–217. (2) (1955) NBS Circ. 539, 4, 21–23. (3) Aurivillius, K. (1964) An x-ray and neutron diffraction study of metacinnabarite. Acta Chem. Scand., 18, 1552–1553.

Crystal Data: Amorphous to X-rays. *Point Group:* n.d. Primarily as stains, mammilary and powdery coatings, and impregnations of siliceous sinter.

Physical Properties: Hardness = 2–3 VHN = n.d. D(meas.) = n.d. D(calc.) = n.d.

Optical Properties: Translucent; transparent in thin splinters. *Color:* Red, reddish brown, purplish; in polished section, bluish white, with intense red internal reflections in oil. *Streak:* Red. *Luster:* Submetallic. *Anisotropism:* Moderate in air.

R: n.d.

Cell Data: *Space Group:* n.d. Z = n.d.

X-ray Powder Pattern: n.d.

Chemistry:

	(1)	(2)	(3)
Sb	71.50	71.58	71.69
S	26.56	27.57	28.31
Total	98.06	99.15	100.00

(1) Copiapó, Chile; by electron microprobe. (2) Rujevac deposit, Yugoslavia; by electron microprobe. (3) Sb_2S_3.

Polymorphism & Series: Dimorphous with stibnite.

Occurrence: Appears to be of primary hydrothermal origin around fumarolic centers (The Geysers, California; Steamboat Springs, Nevada, USA); and the result of oxidation of stibnite (Copiapó, Chile).

Association: Cinnabar (The Geysers, California, USA); stibnite (Copiapó, Chile; Rujevac deposit, Yugoslavia).

Distribution: In the USA, at The Geysers, Sonoma Co., California; and at Steamboat Springs, Washoe Co., Nevada. In northern Chile, at the Alacrán mine, Pampa Larga district, Copiapó. In Bolivia, at the San José mine, Oruro. From the Rujevac Sb-Zn-Pb deposit, western Serbia, Yugoslavia

Name: From the Greek for *with*, plus *stibnite*, in allusion to its composition.

Type Material: National Museum of Natural History, Washington, D.C., USA, 79185.

References: (1) Palache, C., H. Berman, and C. Frondel (1944) Dana's system of mineralogy, (7th edition), v. I, 275. (2) Brookins, D.G. (1970) Metastibnite from The Geysers, Sonoma County, California. Amer. Mineral., 55, 2103–2104. (3) Clark, A.H. (1970) Supergene metastibnite from Mina Alacrán, Pampa Larga, Copiapó, Chile. Amer. Mineral., 55, 2104–2106. (4) Mozgova, N.N., Y.S. Borodaev, A.I. Tsepin, and S. Jankovič (1977) New data on metastibnite. Doklady Acad. Nauk SSSR, 237, 937–940 (in Russian). (5) (1978) Chem. Abs., 88, 123852 (abs. ref. 4).

Crystal Data: Cubic. *Point Group:* $4/m\ \overline{3}\ 2/m$. Massive, presumably.

Physical Properties: *Tenacity:* Brittle. Hardness = n.d. VHN = 287–379 (20 g load). D(meas.) = n.d. D(calc.) = 4.9

Optical Properties: Opaque. *Color:* Gray with a brownish tint in reflected light. *Luster:* Metallic.
R: (420) 27.1, (460) 26.7, (500) 26.4, (540) 26.9, (580) 27.0, (620) 26.8, (660) 26.4, (700) 26.0

Cell Data: *Space Group:* $Fd3m$. $a = 5.530(5)$ Z = 1

X-ray Powder Pattern: Erzgebirge, Germany.
3.18 (100), 1.952 (100), 1.671 (50), 1.268 (40), 1.129 (40), 0.978 (40)

Chemistry:

	(1)	(2)
Cu	37.2	37.94
Fe	1.8	
As	14.7	14.91
Se	47.2	47.15
Total	100.9	100.00

(1) Erzgebirge, Germany; by electron microprobe, average of four samples, corresponding to $(Cu_{2.92}Fe_{0.16})_{\Sigma=3.08}As_{0.98}Se_{2.96}$. (2) Cu_3AsSe_3.

Occurrence: In hydrothermal veins.

Association: Clausthalite, berzelianite, umangite, ankerite, calcite.

Distribution: From an undefined locality in the southwestern part of the Erzgebirge, Saxony, Germany.

Name: From the first letters of Moscow Geological Exploration Institute ("Moscow Geol.-Razved Inst."), the laboratory in which the mineral was discovered.

Type Material: A.E. Fersman Mineralogical Museum, Academy of Sciences, Moscow, USSR.

References: (1) Dymkov, Y.M., T.I. Loseva, E.N. Zav'yalov, B.I. Ryzhov, and L.I. Bochek (1982) Mgriite, $(Cu, Fe)_3AsSe_3$, a new mineral. Zap. Vses. Mineral. Obshch., 111, 215–219 (in Russian). (2) (1983) Amer. Mineral., 68, 280–281 (abs. ref. 1).

Crystal Data: Monoclinic. *Point Group:* m. Commonly as thick crystals, as large as 1 cm, tabular on {001}, {100}, or {$\bar{1}$01}; striations ∥ [010] and [0$\bar{1}$1]; massive. *Twinning:* Polysynthetic twinning observed in polished section.

Physical Properties: *Cleavage:* Imperfect on {010}; rare on {100} and {101}. *Fracture:* Subconchoidal. Hardness = 2.5 VHN = n.d. D(meas.) = 5.25 D(calc.) = 5.29

Optical Properties: Nearly opaque; translucent in thin splinters. *Color:* Iron-black to steel-gray; in polished section, white with raspberry-red internal reflections in oil, seldom seen in air; deep cherry-red in transmitted light. *Streak:* Cherry-red. *Luster:* Metallic adamantine.
Optical Class: Biaxial positive (+). *Pleochroism:* In reflected light, distinct, in whites, light gray-whites, and whitish grays. $\beta = > 2.72$ (Li). *Anisotropism:* Strong.
R_1–R_2: (400) 32.2–44.3, (420) 31.8–44.0, (440) 31.4–43.7, (460) 30.7–42.9, (480) 29.8–42.0, (500) 28.8–41.3, (520) 28.0–40.6, (540) 27.4–40.0, (560) 26.8–39.4, (580) 26.3–38.7, (600) 25.8–37.8, (620) 25.3–36.7, (640) 25.0–35.6, (660) 24.6–34.7, (680) 24.3–33.8, (700) 24.0–33.3

Cell Data: *Space Group:* Aa. $a = 13.220$ $b = 4.411$ $c = 12.862$ $\beta = 98°38(0.5)'$ Z = 8

X-ray Powder Pattern: Synthetic.
2.892 (100), 3.440 (80), 2.748 (70), 2.013 (40), 3.186 (30), 3.101 (30), 1.791 (25)

Chemistry:

	(1)	(2)	(3)
Ag	37.06	36.71	36.72
Fe		trace	
Sb	41.13	41.15	41.45
As	0.79		
S	21.50	21.68	21.83
Total	100.48	99.54	100.00

(1) St. Andreasberg, Germany. (2) Příbram, Czechoslovakia. (3) $AgSbS_2$.

Occurrence: In hydrothermal veins of low-temperature origin.

Association: Proustite, pyrargyrite, polybasite, silver, galena, sphalerite, pyrite, quartz, calcite, barite.

Distribution: In small amounts in many mines, rarely a principal ore mineral. In the USA, in Idaho, in the Silver City and Flint districts, Owyhee Co.; at the Kelly, Coyote and Santa Fe mines, in the Randsburg district, San Bernardino Co., California. From Bräunsdorf, near Freiberg, Saxony; and at St. Andreasberg and Clausthal, in the Harz Mountains, Germany. From Příbram, Czechoslovakia. At Hiendelaencina, Guadalajara Province, Spain. In Chile, from Tres Puntas, Atacama, and from Huantajaya, Tarapacá. At Colquechaca, Potosí, and Pulcayo, Huanchaca, Bolivia. From Huancavalica, Peru. In Mexico, from Catorce, San Luis Potosí; and from Sombrerete and Veta Grande, Zacatecas.

Name: From the Greek for *less* and *silver*, as it contains less silver than other red silver sulfosalt minerals.

References: (1) Palache, C., H. Berman, and C. Frondel (1944) Dana's system of mineralogy, (7th edition), v. I, 424–427. (2) Graham, A.R. (1951) Matildite, aramayoite, miargyrite. Amer. Mineral., 36, 436–449. (3) Knowles, C.R. (1964) A redetermination of the structure of miargyrite, $AgSbS_2$. Acta Cryst., 17, 847–851. (4) (1967) NBS Mono. 25, 5, 49–50.

Crystal Data: Cubic. *Point Group:* 23. As minute grains.

Physical Properties: *Fracture:* Conchoidal. *Tenacity:* Brittle. Hardness = 2.5 VHN = 306–317 (25 g load). D(meas.) = ~9.5 D(calc.) = ~10.0

Optical Properties: Opaque. *Color:* Silver-white to grayish white; in polished section, cream-white with a tinge of gray. *Streak:* Black. *Luster:* Metallic.
R: (400) 55.4, (420) 55.8, (440) 56.2, (460) 56.4, (480) 56.6, (500) 57.0, (520) 57.6, (540) 58.2, (560) 58.7, (580) 59.2, (600) 59.8, (620) 60.4, (640) 60.8, (660) 61.2, (680) 61.6, (700) 62.2

Cell Data: *Space Group:* $P2_13$. $a = 6.646$ Z = 4

X-ray Powder Pattern: Synthetic PdBiTe.
2.974 (100), 2.000 (90), 2.715 (80), 1.773 (80), 1.451 (70), 1.234 (70), 0.8668 (70)

Chemistry:

	(1)	(2)	(3)
Pd	24.0	24.1	24.02
Pt	n.d.	1.1	
Ag		0.2	
Ni		0.1	
Bi	46.5	36.6	47.18
Sb	0.4	5.3	
Te	28.8	32.4	28.80
Total	99.7	99.8	100.00

(1) Vermilion mine, Sudbury, Canada; by electron microprobe, resulting in $Pd_{1.00}(Bi_{0.99}Sb_{0.01})_{\Sigma=1.00}Te_{1.00}$. (2) Kambalda, Australia; by electron microprobe, corresponding to $(Pd_{0.96}Pt_{0.02})_{\Sigma=0.98}(Bi_{0.74}Sb_{0.18})_{\Sigma=0.92}Te_{1.08}$. (3) PdBiTe.

Occurrence: A principal palladium mineral in Cu-Ni sulfide deposits.

Association: Froodite, gold, bismuthinite (Vermilion mine, Canada); cubanite, chalcopyrite, hessite, altaite, bismuth, galena, sperrylite (Frood mine, Canada); violarite, chalcopyrite, magnetite, cobaltite, pyrrhotite (Kanichee deposit, Canada); moncheite, hauchecornite, hessite (Levak mine, Canada); testibiopalladite.

Distribution: At numerous mines in the Sudbury area, Ontario; and at the Pipe mine, Manitoba, Canada. In the USA, from the New Rambler mine, Medicine Bow Mountains, Albany Co., Wyoming; and in the Stillwater Complex, Montana. From the USSR, at the Oktyabr mine, Noril'sk region, western Siberia; the Monchegorsk deposit, Kola Peninsula; and the Bissersk placers, in the Ural Mountains. From Hitura, Finland. At Kambalda, Western Australia. In the Rustenburg mine, on the Merensky Reef, Transvaal, South Africa. From Danba, Sichuan Province, China. Known in trace amounts from a few other localities.

Name: For Dr. C.E. Michener, who discovered the mineral.

Type Material: All original material consumed by analysis; Royal Ontario Museum, Toronto, Canada, M31189 and M29438.

References: (1) Hawley, J.E. and L.G. Berry (1958) Michenerite and froodite, palladium bismuth minerals. Can. Mineral., 6, 200–209. (2) Cabri, L.J. and D.C. Harris (1973) Michenerite (PdBiTe) redefined and froodite ($PdBi_2$) confirmed from the Sudbury area. Can. Mineral., 11, 903–912. (3) Genkin, A.D., N.N. Zhuravlev, and E.M. Smirnova (1963) Moncheite and kotulskite—new minerals—and the composition of michenerite. Zap. Vses. Mineral. Obshch., 92, 33–50 (in Russian). (4) (1963) Amer. Mineral., 48, 1184 (abs. ref. 3). (5) Childs, J.D. and S.R. Hall (1973) The crystal structure of michenerite, PdBiTe. Can. Mineral., 12, 61–65. (6) Hudson, D.R., B.W. Robinson, R.B.W. Vigers, and G.A. Travis (1978) Zoned michenerite–testibiopalladite from Kambalda, Western Australia. Can. Mineral., 16, 121–126. (7) Cabri, L.J., Ed. (1981) Platinum group elements: mineralogy, geology, recovery. Can. Inst. Min. & Met., 120–121, 155.

Crystal Data: Orthorhombic. *Point Group:* $2/m\ 2/m\ 2/m$ or $mm2$. As tiny (up to 0.3 mm) grains or irregular masses; as intergrowths with other sulfides.

Physical Properties: Hardness = n.d. VHN = 190–230 (25 g load). D(meas.) = n.d. D(calc.) = 6.06

Optical Properties: Opaque. *Color:* Pale gray to grayish white in reflected light. *Luster:* Metallic. *Anisotropism:* Moderate, from grayish blue to pinkish brown.
R_1–R_2: (480) 29.6–30.3, (548) 30.8–31.5, (589) 31.7–32.6, (657) 32.1–34.2

Cell Data: *Space Group:* *Pbmn*, *Pb2m*, or $Pb2_1m$. $a = 10.854(4)$ $b = 11.985(4)$ $c = 3.871(1)$ Z = 2

X-ray Powder Pattern: Mihara mine, Japan.
3.03 (100), 3.00 (70), 1.935 (70), 2.18 (50), 3.25 (30), 3.11 (30), 2.70 (30)

Chemistry:

	(1)	(2)
Pb	22.72	22.56
Cu	28.24	27.67
Fe	6.05	6.08
Bi	22.75	22.75
S	20.60	20.94
Total	100.36	100.00

(1) Mihara mine, Japan; by electron microprobe, average of six analyses. (2) $PbCu_4FeBiS_6$.

Occurrence: As disseminations in a hedenbergite-garnet-epidote contact metamorphic deposit (Mihara mine, Japan); in a mineralized quartz vein through granite (Imooka mine, Japan).

Association: Wittichenite, bornite, chalcopyrite, galena.

Distribution: At the Honpi deposit, Mihara mine, and the Imooka mine, Okayama Prefecture, Japan. In the Ulsan mine, Kyongsangnam Province, Korea.

Name: For the type locality at the Mihara mine, Japan.

Type Material: The Institute of Mineralogy, Petrology, and Economic Geology, Faculty of Science, Tohoku University, Sendai, Japan.

References: (1) Sugaki, A., H. Shima, and A. Kitakaze (1980) Miharaite, $Cu_4FePbBiS_6$, a new mineral from the Mihara mine, Okayama, Japan. Amer. Mineral., 65, 784–788.

Crystal Data: Hexagonal. *Point Group:* $\bar{3}\ 2/m$. Typically as slender to capillary crystals to several cm, elongated ∥ [0001], in radiating groups of hair-like, interwoven masses; columnar tufted coatings. Single crystals may be helically twisted about [0001]. As cleavable masses up to several cm across. *Twinning:* By pressure on $\{01\bar{1}2\}$.

Physical Properties: *Cleavage:* Perfect on $\{10\bar{1}1\}$ and $\{01\bar{1}2\}$. *Fracture:* Uneven. *Tenacity:* Brittle; capillary crystals elastic. Hardness = 3–3.5 VHN = n.d. D(meas.) = 5.5 D(calc.) = 5.374

Optical Properties: Opaque. *Color:* Pale brass-yellow to bronze-yellow, tarnishes to iridescence. *Streak:* Greenish black. *Luster:* Metallic. *Pleochroism:* Weak in air, appreciably stronger in oil, pale yellow-brown to bright yellow. *Anisotropism:* Strong.

R_1–R_2: (400) 26.7–29.8, (420) 32.0–35.0, (440) 37.3–40.2, (460) 42.0–44.3, (480) 46.6–47.4, (500) 50.0–49.5, (520) 52.8–51.6, (540) 54.8–52.9, (560) 56.1–54.0, (580) 57.2–54.8, (600) 58.0–55.6, (620) 58.8–56.2, (640) 59.5–56.8, (660) 60.1–57.2, (680) 60.6–57.6, (700) 61.0–57.9

Cell Data: *Space Group:* $R\bar{3}m$. $a = 9.607$ $c = 3.143$ Z = 9

X-ray Powder Pattern: Canada.
2.777 (100), 1.8631 (95), 2.513 (65), 4.807 (60), 2.228 (55), 1.8178 (45), 2.946 (40)

Chemistry:

	(1)	(2)
Ni	63.68	64.67
Fe	1.03	
Co	0.21	
Cu	0.00	
S	35.47	35.33
Total	100.39	100.00

(1) Marbridge mine, Canada; by electron microprobe, leading to $(Ni_{0.981}Fe_{0.016}Co_{0.004})_{\Sigma=1.001}S$.
(2) NiS.

Occurrence: Most frequently as a low-temperature mineral, in cavities in limestones and carbonate veins and in barite; as an alteration product of other nickel minerals; also found in sedimentary rocks associated with coal measures and in serpentines.

Association: Gersdorffite, polydymite, nickeline, galena, sphalerite, pyrite, pyrrhotite, calcite, dolomite, siderite, barite, ankerite.

Distribution: Numerous localities world-wide. In the USA, at the Sterling mine, Antwerp, Jefferson Co., New York; and the Gap Nickel mine, Lancaster Co., Pennsylvania. In geodes from Keokuk, Lee Co., Iowa; St. Louis, St. Louis Co., Missouri; Milwaukee, Waukesha Co., Wisconsin; and at Hall's Gap, Lincoln Co., Kentucky. From Kotalahti, Finland, large cleavages. In Germany, at Müsen and Wissen, North Rhine-Westphalia. At Merthyr-Tydfil, Glamorgan, Wales. From Temagami, Ontario; and in large cleavages from the Marbridge mine, Malartic, La Motte Township, Quebec, Canada. At Kambalda, Western Australia.

Name: For William Hallowes Miller (1801–1880), British mineralogist, Cambridge University, Cambridge, England, who first studied crystals of the mineral.

References: (1) Palache, C., H. Berman, and C. Frondel (1944) Dana's system of mineralogy, (7th edition), v. I, 239–241. (2) (1962) NBS Mono. 25, 1, 37. (3) Grice, J.D. and R.B. Ferguson (1974) Crystal structure refinement of millerite (β–NiS). Can. Mineral., 12, 248–252.

Crystal Data: Orthorhombic. *Point Group:* $2/m\ 2/m\ 2/m$. As tiny grains, 0.01 to 0.05 mm.

Physical Properties: Hardness = ~4 VHN = 26 D(meas.) = n.d. D(calc.) = [8.28] for CoAs.

Optical Properties: Opaque. *Color:* Bluish white.
R_1–R_2: n.d.

Cell Data: *Space Group:* n.d. $a = 3.458$ $b = 5.869$ $c = 5.292$ Z = 4.

X-ray Powder Pattern: Synthetic CoAs. (JCPDS 9-94).
1.97 (100), 2.59 (90), 0.957 (70b), 2.55 (60), 1.047 (60), 0.999 (60), 0.927 (60)

Chemistry:

	(1)
Co	35.3 – 41.3
Fe	3.1 – 9.1
Ni	0.0 – 0.4
As	55.9 – 57.5
Total	

(1) Dashkesan deposit, USSR; by electron microprobe, range of analyses of 11 grains, leading to the average composition $(Co_{0.89}Fe_{0.12})_{\Sigma=1.01}As$.

Occurrence: In heavy mineral concentrates.

Association: Nickeline, alloclasite, safflorite, glaucodot, cobaltite, pentlandite, pyrrhotite, chalcopyrite.

Distribution: From the "Far East" Witwatersrand, Transvaal, South Africa. In the Dashkesan deposit, Azerbaijan SSR, USSR.

Name: Presumably named for the Modderfontein mine, South Africa.

Type Material: National Museum of Natural History, Washington, D.C., USA, 161217.

References: (1) Cooper, R.A. (1924) Mineral constituents of Rand concentrates. J. Chem. Met. and Mining Soc. South Africa, 24, 264–265. (2) Makhmudov, A.I. and I.P. Laputina (1977) First occurrence of modderite in the USSR. Zap. Vses. Mineral. Obshch., 106, 347–350 (in Russian). (3) (1978) Amer. Mineral., 63, 600 (abs. ref. 2).

Crystal Data: Triclinic. *Point Group:* n.d. As grains up to 10 x 80 μm.

Physical Properties: Hardness = n.d. VHN = 151–203 (10 g load). D(meas.) = n.d. D(calc.) = [4.86]

Optical Properties: Opaque. *Color:* Gray with a greenish tint in reflected light. *Luster:* Metallic. *Anisotropism:* Distinct but without color effects.

R: (400) — , (420) — , (440) 24.6, (460) 24.7, (480) 25.3, (500) 25.6, (520) 25.4, (540) 25.5, (560) 25.6, (580) 25.7, (600) 25.6, (620) 25.7, (640) 26.0, (660) 25.9, (680) 26.2, (700) 26.4

Cell Data: *Space Group:* n.d. $a = 6.64$ $b = 11.51$ $c = 19.93$ $\alpha = 90°$ $\beta = 109°45'$ $\gamma = 90°$ Z = 12

X-ray Powder Pattern: Kochbulak deposit, USSR.
3.13 (100), 1.920 (70), 2.72 (20), 2.44 (20), 3.66 (10), 3.34 (10), 2.82 (10)

Chemistry:

	(1)	(2)
Cu	37.69	37.16
Sn	35.15	34.71
Sb	0.67	
S	27.91	28.13
Total	101.42	100.00

(1) Kochbulak deposit, USSR; by electron microprobe, average of analyses on seven grains.
(2) Cu_2SnS_3.

Occurrence: Of hydrothermal origin.

Association: Tetrahedrite, famatinite, kuramite, mawsonite, emplectite (Kochbulak deposit, USSR).

Distribution: In the Kochbulak deposit, eastern Uzbekistan, USSR. From the April Fool mine, Delamar, Lincoln Co., Nevada, USA.

Name: For Günter Moh, University of Heidelberg, Germany.

Type Material: n.d.

References: (1) Kovalenker, V.A., V.S. Malov, T.L. Evstigneeva, and L.N. Vyal'sov (1982) Mohite, Cu_2SnS_3, a new sulfide of tin and copper. Zap. Vses. Mineral. Obshch., 111, 110–114 (in Russian). (2) (1983) Amer. Mineral., 68, 281 (abs. ref. 1).

Crystal Data: Hexagonal. *Point Group:* $6/m\ 2/m\ 2/m$ ($2H_1$ polymorph); or $3m$ (3R polymorph). Crystals are commonly tabular, barrel shaped; also as slightly tapered prisms; face development poor; up to 15 cm across. Commonly shows trigonal markings on {0001} parallel to the trace of $\{10\bar{1}1\}$. Foliated, massive, or in scales.

Physical Properties: *Cleavage:* Perfect on {0001}. *Tenacity:* Lamellae flexible, not elastic. Hardness = 1–1.5 VHN = n.d. D(meas.) = 4.62–4.73 D(calc.) = 4.998 Greasy feel.

Optical Properties: Nearly opaque; translucent in thin flakes; transparent in infrared light. *Color:* Lead-gray; very pale yellow to deep reddish brown in transmitted light. *Streak:* Bluish gray. *Luster:* Metallic. *Pleochroism:* Very strong. *Anisotropism:* Very strong.
R_1–R_2: (400) 21.0–55.0, (420) 22.2–54.8, (440) 23.4–54.6, (460) 24.1–53.8, (480) 23.8–52.3, (500) 22.7–49.7, (520) 21.9–47.1, (540) 21.3–45.5, (560) 20.9–44.4, (580) 20.6–44.0, (600) 20.4–44.6, (620) 20.2–45.3, (640) 20.0–45.7, (660) 20.0–45.6, (680) 19.9–45.4, (700) 19.7–44.2

Cell Data: *Space Group:* $P6_3mmc$ ($2H_1$ polymorph), with $a = 3.1604$ $c = 12.295$ Z = 2, and $R3m$ (3R polymorph), with $a = 3.16$ $c = 18.33$ Z = 3

X-ray Powder Pattern: Synthetic ($2H_1$ polymorph).
6.15 (100), 2.277 (45), 1.830 (25), 2.737 (16), 2.049 (14), 1.581 (12), 1.538 (12)

X-ray Powder Pattern: Con mine, Canada (3R polymorph).
6.09 (100), 2.71 (70), 1.581 (70), 1.529 (70), 2.63 (60), 2.344 (60), 2.194 (60)

Chemistry: Nearly pure MoS_2.

Polymorphism & Series: Dimorphous with jordisite; stacking polymorphs $2H_1$ and 3R are known.

Occurrence: In high-temperature hydrothermal veins. In disseminated deposits of the porphyry type, both with and without associated major copper mineralization. Also in contact metamorphic deposits in limestone with calcium silicate minerals as well as in pegmatites, granites and aplites. Rarely in meteorites.

Association: Chalcopyrite, other copper sulfides.

Distribution: Of widespread occurrence; the most abundant molybdenum mineral. Fine crystals occur, in the USA, at the Crown Point mine, Chelan Co., Washington; and at the Frankford quarry, Philadelphia, Pennsylvania. In Canada, in the Temiskaming district, and in Aldfield Township, Quebec. In Norway, from Raade, near Moss, and at Vennesla, near Arendal. In the USSR, in the Adunchilon Mountains, south of Nerchinsk, Transbaikal; and at Miass, Ural Mountains. In Germany, at Altenberg, Saxony. In Morocco, at Azegour, 80 km southwest of Marrakesh. From Kingsgate, New South Wales, Australia. At the Hirase mine, Gifu Prefecture, Honshu, Japan. In the Wolak mine, Danyang, Chungchongnam Province, Korea. The 3R polymorph occurs in the Con mine, Yellowknife, Yukon Territory; and at Mont Saint-Hilaire, Quebec, Canada. From the Yamate mine, Okayama Prefecture, and at Satsuma-Iwo-jima, Kyushu, Japan. In the Slundyanogorsk deposit, Central Ural Mountains, USSR.

Name: A word derived from the Greek *molybdos, lead.*

References: (1) Palache, C., H. Berman, and C. Frondel (1944) Dana's system of mineralogy, (7th edition), v. I, 328–331. (2) (1955) NBS Circ. 539, 5, 47. (3) Traill, R.J. (1963) A rhombohedral polytype of molybdenite. Can. Mineral., 7, 524–526.

Crystal Data: Hexagonal. *Point Group:* $\overline{3}\ 2/m$. Euhedral to subhedral crystals to 1 mm, and as minute grains, usually less than about 0.2 mm.

Physical Properties: *Cleavage:* {0001}. Hardness = n.d. VHN = 73–111, 92 average (10 g load). D(meas.) = n.d. D(calc.) = 9.88

Optical Properties: Opaque. *Color:* Steel-gray; in polished section, white. *Luster:* Metallic. *Pleochroism:* Weak in air, distinct in oil. *Anisotropism:* Strong, except in sections close to {0001}.
R_1–R_2: (400) 50.3–51.6, (420) 51.5–52.6, (440) 52.4–53.5, (460) 53.4–54.4, (480) 54.2–55.1, (500) 54.9–55.8, (520) 55.5–56.4, (540) 56.0–56.9, (560) 56.4–57.3, (580) 56.7–57.6, (600) 57.1–58.0, (620) 57.4–58.4, (640) 57.7–58.7, (660) 58.0–59.0, (680) 58.3–59.2, (700) 58.4–59.4

Cell Data: *Space Group:* $P\overline{3}m1$. $a = 4.049$ $c = 5.288$ Z = 1

X-ray Powder Pattern: Monchegorsk deposit, USSR.
2.93 (100), 2.11 (80), 2.02 (70), 1.462 (70), 1.282 (70), 5.32 (60), 1.664 (7)

Chemistry:

	(1)	(2)	(3)
Pt	26.6	38.3	42.1
Pd	6.93		0.0
Bi	20.93	13.0	1.2
Te	45.55	49.1	56.7
Total	100.01	100.4	100.0

(1) Monchegorsk deposit, USSR; average of microspectrographic analyses of four samples. (2) Impala mine, Transvaal, South Africa; by electron microprobe. (3) Stillwater Complex, Montana, USA; by electron microprobe.

Polymorphism & Series: Forms a series with merenskyite.

Occurrence: As grains in chalcopyrite; also overgrowing kotulskite.

Association: Chalcopyrite, kotulskite, michenerite, cooperite, sperrylite, daomanite, hessite, pentlandite, hauchecornite, millerite, gold, magnetite, sphalerite.

Distribution: At the Monchegorsk deposit, Monche Tundra, Kola Peninsula; and the Oktyabr mine, Noril'sk region, western Siberia, USSR. In South Africa, from several mines along the Merensky Reef, Transvaal. In the Stillwater Complex, Montana; at the New Rambler mine, Medicine Bow Mountains, Albany Co., Wyoming; and in gravels of the San Joaquin River, at Friant, Fresno Co., California, USA. In Canada, at Sudbury, in the Thierry mine, near Pickle Lake, and the Lac des Iles Complex, Ontario. In China, in the Tao and Ma districts - apparently coded names.

Name: For the locality at Monche Tundra in the USSR.

Type Material: n.d.

References: (1) Genkin, A.D., N.N. Zhuravlev, and E.M. Smirnova (1963) Moncheite and kotulskite—new minerals—and the composition of michenerite. Zap. Vses. Mineral. Obshch., 92, 33–50 (in Russian). (2) (1963) Amer. Mineral., 48, 1181 (abs. ref. 1). (3) X-Ray Laboratory, Guiyang Institute of Geochemistry, Academia Sinica (1975) The crystal structures of biteplapalladite and biteplatinite. Geochimica, 184–185 (in Chinese). (4) (1976) Mineral. Abs., 27, 214 (abs. ref. 3). (5) Cabri, L.J., Ed. (1981) Platinum group elements: mineralogy, geology, recovery. Can. Inst. Min. & Met., 121. (6) Volborth, A., M. Tarkian, E.F. Stumpfl, and R.M. Housley (1986) A survey of the Pd– Pt mineralization along the 35-km strike of the J-M reef, Stillwater Complex, Montana. Can. Mineral., 24, 329–346.

Montbrayite $(Au, Sb)_2Te_3$

Crystal Data: Triclinic. *Point Group:* 1. Generally rounded to irregular grains (> 1 mm) and equidimensional masses to about 1 cm in diameter.

Physical Properties: *Cleavage:* $\{1\bar{1}0\}$, $\{0\bar{1}1\}$, $\{1\bar{1}1\}$. *Fracture:* Flat conchoidal. *Tenacity:* Brittle. Hardness = 2.5 VHN = n.d. D(meas.) = 9.94 D(calc.) = [9.91]

Optical Properties: Opaque. *Color:* Yellowish white. *Luster:* Metallic. *Pleochroism:* Very weak. *Anisotropism:* Weak to moderate, light gray, yellow-brown, and blue-gray.
R_1–R_2: n.d.

Cell Data: *Space Group:* $P1$. $a = 12.11$ $b = 13.44$ $c = 10.80$ $\alpha = 104°23'$ $\beta = 97°30'$ $\gamma = 107°56'$ Z = 12

X-ray Powder Pattern: Robb-Montbray mine, Canada.
2.087 (100), 2.983 (80), 2.922 (60), 4.425 (40), 2.119 (40), 7.363 (20), 11.2 (10)

Chemistry:

	(1)	(2)	(3)
Au	47.36	46.8	50.72
Ag	0.37	1.0	
Pb	1.02		
Sb	1.12		
Bi	3.23		
Te	46.66	49.4	49.28
S		2.5	
Total	99.76	99.7	100.00

(1) Robb-Montbray deposit, Canada. (2) Uzbekistan, USSR; by electron microprobe, average analysis. (3) Au_2Te_3.

Occurrence: Found with other tellurides, sulfides, and gold in coarsely crystalline masses which are almost free from gangue (Robb-Montbray mine, Canada); rimming gold (Uzbekistan, USSR).

Association: Calaverite, gold, tellurobismuthite, altaite, petzite, melonite, frohbergite, chalcopyrite, pyrite, sphalerite, chalcocite, covellite, marcasite (Robb-Montbray mine, Canada); altaite, gold (Uzbekistan, USSR).

Distribution: From the Robb-Montbray mine, Montbray Township, Quebec, Canada. At an undefined gold-polymetallic deposit in Uzbekistan, USSR. From the April Fool mine, Delamar, Lincoln Co., Nevada, USA.

Name: For the Robb-Montbray mine, Canada.

Type Material: Royal Ontario Museum, Toronto, Canada, M15815, M19883; Harvard University, Cambridge, Massachusetts, USA, 97681.

References: (1) Peacock M.A. and R.M. Thompson (1946) Montbrayite, a new gold telluride. Amer. Mineral., 31, 515–526. (2) Thompson, R.M. (1949) The telluride minerals and their occurrence in Canada. Amer. Mineral., 34, 342–382. (3) Rucklidge, J. (1969) Frohbergite, montbrayite, and a new Pb–Bi telluride. Can. Mineral., 9, 709–716. (4) Bachechi, F. (1971) Crystal structure of montbrayite. Nature, Phys. Sci., 231, 67–68. (5) Bachechi, F. (1972) Synthesis and stability of montbrayite, Au_2Te_3. Amer. Mineral., 57, 146–154. (6) Ryabeva, E.G., R.P. Badalova, and L.S. Dubakina (1979) Montbrayite – the first discovery in the USSR. Doklady Acad. Nauk SSSR, 246, 463–464 (in Russian). (7) (1979) Chem. Abs., 91, 110305 (abs. ref. 6).

Crystal Data: Tetragonal. *Point Group:* $\overline{4}2m$. Grains up to 1 mm in massive sulfides and as intergrowths with haycockite.

Physical Properties: Hardness = n.d. VHN = 240–246 (100 g load). D(meas.) = 4.36 D(calc.) = 4.37

Optical Properties: Opaque. *Color:* Light yellow in reflected light, tarnishes to pinkish brown and purple. *Luster:* Metallic.
R: (400) 14.3, (420) 16.7, (440) 20.0, (460) 23.9, (480) 27.5, (500) 30.6, (520) 33.2, (540) 35.2, (560) 36.7, (580) 37.8, (600) 38.8, (620) 39.4, (640) 40.0, (660) 40.2, (680) 40.3, (700) 40.4

Cell Data: *Space Group:* $P\overline{4}2m$. $a = 10.585(5)$ $c = 5.383(5)$ Z = 1

X-ray Powder Pattern: Mooihoek Farm, South Africa.
3.07 (100), 1.889 (80), 1.597 (60), 1.083 (60), 1.223 (50), 1.871 (40), 1.323 (40)

Chemistry:

	(1)	(2)	(3)
Cu	35.91	34.08	36.02
Fe	31.88	32.47	31.66
Ni	0.26	0.24	
S	32.44	32.84	32.32
Total	100.49	99.65	100.00

(1) Bushveld Complex, South Africa; by electron microprobe, average of 13 analyses. (2) Duluth Gabbro, Minnesota, USA; by electron microprobe, average of four analyses. (3) $Cu_9Fe_9S_{16}$.

Occurrence: In massive sulfide from a pipe-shaped dunite pegmatite in the Norite Zone of the Bushveld Complex (Mooihoek Farm, South Africa); in troctolite from the basal Duluth Gabbro (Minnesota, USA).

Association: Haycockite, copper, troilite, pentlandite, cubanite, magnetite (Minnesota, USA); haycockite, magnetite, troilite, cuprian pentlandite, mackinawite, sphalerite, moncheite (Mooihoek Farm, South Africa).

Distribution: In the Duluth Gabbro, Minnesota, USA. At the Mooihoek Farm, Lydenburg district, Transvaal, South Africa. From the Talnakh area, Noril'sk region, western Siberia, USSR.

Name: For the Mooihoek Farm locality, in South Africa.

Type Material: Princeton University, Princeton, New Jersey; National Museum of Natural History, Washington, D.C., USA, 124965; National Museum of Canada, Ottawa, 10309; Royal Ontario Museum, Toronto, Canada, M30992; Heidelberg University, Heidelberg, Germany, 2313a.

References: (1) Cabri, L.J. and S.R. Hall (1972) Mooihoekite and haycockite, two new copper–iron sulfides, and their relationship to chalcopyrite and talnakhite. Amer. Mineral., 57, 689–708. (2) Hall, S.R. and J.F. Rowland (1973) The crystal structure of synthetic mooihoekite, $Cu_9Fe_9S_{16}$. Acta Cryst., B29, 2365–2372. (3) Putnis, A. (1978) Talnakhite and mooihoekite: the accessibility of ordered structures in the metal-rich region around chalcopyrite. Can. Mineral., 16, 23–30.

Morozeviczite $(Pb, Fe)_3Ge_{1-x}S_4$ (x = 0.18 to 0.69)

Crystal Data: Cubic. *Point Group:* n.d. Massive with other sulfides.

Physical Properties: Hardness = n.d. VHN = 119–124 (50 g load). D(meas.) = n.d. D(calc.) = 6.62

Optical Properties: Opaque. *Color:* Brownish gray; in reflected light, white with cream-red tint. *Streak:* Dark gray. *Anisotropism:* Distinct.

R_1–R_2: (470) 43.5–44.5, (535) 43.0–44.0, (591) 44.0–45.0, (658) 45.5–46.5

Cell Data: *Space Group:* n.d. $a = 10.61$ Z = 8

X-ray Powder Pattern: Lower Silesia, Poland.
3.08 (10), 2.15 (9), 2.80 (6), 2.047 (6), 1.791 (5), 1.565 (5), 1.467 (5)

Chemistry:

	(1)
Pb	58.6
Fe	8.6
Cu	1.2
Ge	9.0
As	0.8
S	21.0
Total	99.2

(1) Lower Silesia, Poland; by electron microprobe.

Polymorphism & Series: Forms a series with polkovicite.

Occurrence: In epigenetic veinlets and metasomatic replacement zones replacing sandstone and older sulfides, in brecciated sandstones underlying copper-bearing shales.

Association: Marcasite, chalcopyrite, bornite, chalcocite, tennantite, sphalerite, galena.

Distribution: From the Polkovice mine, Lower Silesia, Poland.

Name: For Josef Morozewicz (1865–1941), Professor of Mineralogy, Jagellonian University, Krakow, Poland.

Type Material: Jagellonian University, Krakow, Poland.

References: (1) Haranczyk, C. (1975) Morozeviczite and polkovicite, typochemical minerals of Mesozoic mineralization of the Fore-Sudenten monocline. Rudy Metalle, 20, 288–293 (in Polish). (2) (1981) Amer. Mineral., 66, 437 (abs. ref. 1).

Crystal Data: Cubic. *Point Group:* $4/m\ \overline{3}\ 2/m$. As dodecahedral crystals commonly modified by the cube and trapezohedron; also massive, granular.

Physical Properties: *Cleavage:* {011} and {001} distinct. *Fracture:* Conchoidal. *Tenacity:* Brittle. Hardness = 3.5 VHN = n.d. D(meas.) = 13.48 D(calc.) = 13.5

Optical Properties: Opaque. *Color:* Silver-white. *Luster:* Bright metallic.
R: (400) 60.7, (420) 62.8, (440) 64.9, (460) 66.9, (480) 68.8, (500) 70.7, (520) 72.4, (540) 74.1, (560) 75.8, (580) 77.3, (600) 79.0, (620) 80.5, (640) 82.0, (660) 83.4, (680) 84.8, (700) 86.1

Cell Data: *Space Group:* $Im3m$. $a = 10.1$ Z = 10

X-ray Powder Pattern: Landsberg, Germany.
2.36 (100), 1.365 (70), 1.236 (60), 1.275 (50), 0.941 (50), 0.799 (50), 2.67 (40)

Chemistry:

	(1)	(2)	(3)
Ag	26.48	27.04	26.4
Hg	73.44	72.94	73.6
Total	99.92	99.98	100.0

(1) Sala, Sweden. (2) Landsberg, Germany. (3) Ag_2Hg_3.

Occurrence: Probably of low-temperature hydrothermal origin.

Association: Metacinnabar, cinnabar, mercurian silver, tetrahedrite–tennantite, pyrite, sphalerite, chalcopyrite.

Distribution: At Landsberg, near Obermoschel, Rhineland-Palatinate, Germany. From Sala, Västmanland, Sweden. In the Chalanches mine, near Allemont, Isère, France.

Name: For the locality at Landsberg, near Obermoschel, Germany.

Type Material: n.d.

References: (1) Palache, C., H. Berman, and C. Frondel (1944) Dana's system of mineralogy, (7th edition), v. I, 103–104. (2) Berry, L.G. and R.M. Thompson (1962) X-ray powder data for the ore minerals. Geol. Soc. Amer. Mem. 85, 16–17. (3) Baird, H.W. and F.A. Mueller (1969) Refinement of the crystal structure of γ-phase silver amalgam. J. Biomed. Mat. Res., 375–382.

Crystal Data: Tetragonal. *Point Group:* $\bar{4}2m$, $4mm$, 422, or $4/m\ 2/m\ 2/m$. As small grains (less than 1 μm) and as aggregates up to 0.2 mm.

Physical Properties: *Cleavage:* Imperfect. *Tenacity:* Brittle. Hardness = Soft. VHN = 92.1–123.3, 109.4 average. D(meas.) = n.d. D(calc.) = [3.86]

Optical Properties: Opaque. *Color:* Copper-red to pinchbeck-brown, iridescent; grayish orange-cream in reflected light; a fresh surface oxidizes to a sooty black film. *Luster:* Metallic. *Anisotropism:* Moderate, gray to brownish gray with a bluish tint.
R: (400) — , (420) — , (440) 15.0, (460) 15.6, (480) 16.6, (500) 17.4, (520) 18.2, (540) 19.2, (560) 20.0, (580) 21.0, (600) 21.8, (620) 22.6, (640) 23.3, (660) 24.0, (680) 24.6, (700) 25.4

Cell Data: *Space Group:* $I\bar{4}m2$, $I\bar{4}2m$, $I4mm$, $I422$, or $I4/mmm$. $a = 3.88$ $c = 13.10$ Z = [1]

X-ray Powder Pattern: Murunskii massif, USSR.
6.52 (100), 2.53 (80), 2.90 (60), 1.940 (50), 1.715 (40), 2.10 (30), 3.29 (20)

Chemistry:

	(1)	(2)
K	14.57	17.27
Cu	44.38	42.09
Fe	12.07	12.33
S	28.14	28.31
Total	99.16	100.00

(1) Murunskii massif, USSR; by electron microprobe, corresponding to $K_{1.72}Cu_{3.23}Fe_{1.09}S_{4.05}$. (2) $K_2Cu_3FeS_4$.

Occurrence: In rocks that have undergone intensive potassium metasomatism.

Association: Charoite, acmite, potassium feldspar.

Distribution: From the Murunskii alkalic massif, near Olekminsk, Yakutia, USSR.

Name: For the locality in the Murunskii massif, USSR.

Type Material: A.E. Fersman Mineralogical Museum, Academy of Sciences, Moscow, USSR.

References: (1) Dobrovolskaya, M.G., A.I. Tsepin, T.L. Evstigneeva, L.N. Vyal'sov, and A.O. Zaozerina (1981) Murunskite, $K_2Cu_3FeS_4$, a new sulfide of potassium, copper, and iron. Zap. Vses. Mineral. Obshch., 110, 468–473 (in Russian). (2) (1982) Amer. Mineral., 67, 624 (abs. ref. 1).

Crystal Data: n.d. *Point Group:* n.d. As tabular crystals usually elongated in one direction.

Physical Properties: *Cleavage:* One perfect in zone of elongation. Hardness = 2.5 VHN = n.d. D(meas.) = 5.598 (up to 14% stibnite). D(calc.) = n.d.

Optical Properties: Opaque. *Color:* Brassy yellow, gray-white on fresh fracture surface. *Streak:* Black. *Luster:* Bright metallic.
R_1–R_2: n.d.

Cell Data: *Space Group:* n.d. Z = n.d.

X-ray Powder Pattern: n.d.

Chemistry:

	(1)	(2)	(3)
Ag	26.36	16.69	19.25
Au	22.90	30.03	35.20
Pb	2.58		
Sb		9.75	
Te	46.44	39.14	45.55
S		4.39	
Total	98.28	100.00	100.00

(1) Săcărâmb, Romania. (2) Do.; Sb and S are from stibnite impurity. (3) (Ag, Au)Te with Ag:Au = 1:1.

Occurrence: In close intergrowths with other tellurides, especially krennerite, in epithermal hydrothermal veins, and in ores enriched by secondary processes.

Association: Krennerite, hessite, nagyagite, petzite, pyrargyrite, sylvanite, pyrite, sphalerite, altaite, tetrahedrite–tennantite, alabandite, stibnite.

Distribution: From Săcărâmb (Nagyág), Romania.

Name: For the chemist and crystallographer, Professor Friedrich Wilhelm Muthmann (1861–1913), of Munich, Germany.

Type Material: University of Naples Museum, Naples, Italy.

References: (1) Palache, C., H. Berman, and C. Frondel (1944) Dana's system of mineralogy, (7th edition), v. I, 260–261. (2) Sindeeva, N.D. (1964) Mineralogy and types of deposits of selenium and tellurium, 106–107.

Crystal Data: Tetragonal (?), possibly monoclinic. *Point Group:* n.d. Thin tabular crystals to 1 cm, often bent, with striations on (001) || [100] and [010]; also granular massive. *Twinning:* Crossed twin lamellae exhibited on (001) sections.

Physical Properties: *Cleavage:* Perfect on {010}. *Tenacity:* Flexible, slightly malleable. Hardness = 1.5 VHN = 60–94 (100 g load). D(meas.) = 7.41 D(calc.) = [6.74]

Optical Properties: Opaque. *Color:* Blackish lead-gray; gray-white in polished section. *Streak:* Blackish lead-gray. *Luster:* Metallic, bright on fresh cleavage. *Pleochroism:* Weak in air and oil. *Anisotropism:* Weak but distinct.

R_1–R_2: (400) 43.4–49.0, (420) 42.8–48.0, (440) 42.2–47.0, (460) 41.6–45.9, (480) 40.9–45.0, (500) 40.3–44.2, (520) 39.6–43.5, (540) 38.8–42.8, (560) 38.1–42.1, (580) 37.5–41.5, (600) 37.0–40.9, (620) 36.5–40.3, (640) 36.1–39.8, (660) 35.5–39.2, (680) 34.8–38.6, (700) 34.2–38.0

Cell Data: *Space Group:* n.d. $a = 12.5$ $c = 30.25$ Z = 10

X-ray Powder Pattern: Săcărâmb, Romania.
3.02 (100), 2.81 (60), 1.506 (60), 2.43 (40), 2.08 (30), 1.817 (30), 1.704 (30)

Chemistry:

	(1)	(2)	(3)
Pb	56.81	52.55	58.8
Au	7.51	10.16	9.0
Ag		1.12	0.17
Fe	0.41		
Sb	7.39	7.00	7.8
Te	17.72	18.80	15.1
S	10.76	8.62	10.1
Total	100.60	98.25	100.97

(1) Săcărâmb, Romania. (2) Oroya, Australia. (3) Cripple Creek, Colorado, USA; by electron microprobe, corresponding to $Pb_{5.00}Au_{0.77}(Te_{2.02}Sb_{1.10})_{\Sigma=3.12}S_{5.43}$.

Occurrence: In epithermal hydrothermal veins.

Association: Altaite, arsenic, gold, proustite, rhodochrosite, sphalerite, tetrahedrite.

Distribution: In the USA, at Gold Hill, Boulder Co., and Cripple Creek, Teller Co., Colorado; the Dorleska mine, Coffee Creek district, Trinity Co., California; and the Kings Mountain mine, Gaston Co., North Carolina. In Canada, at the Huronian mine, Moss Township, Ontario; and from the Olive Mabel claim, Gainer Creek, British Columbia. From Săcărâmb (Nagyág) and Baia-de-Arieş (Offenbánya), Romania. At Schellgadener, Austria. From Oroya, Kalgoorlie district, Western Australia. At the Sylvia mine, Tararu Creek, New Zealand. In the Rendaiji mine, Shizuoka Prefecture, Japan. From the Tarua gold field, Vitu Levu, Fiji Islands. At Farallon Negro, Catamarca Province, Argentina.

Name: For the Săcărâmb (Nagyág), Romania locality.

References: (1) Palache, C., H. Berman, and C. Frondel (1944) Dana's system of mineralogy, (7th edition), v. I, 168–169. (2) Berry, L.G. and R.M. Thompson (1962) X-ray powder data for the ore minerals. Geol. Soc. Amer. Mem. 85, 29–30. (3) Stumpfel, E.F. (1970) New electron probe and optical data on gold tellurides. Amer. Mineral., 55, 808–814.

Crystal Data: Orthorhombic, pseudocubic. *Point Group:* 222. Granular massive; infrequently as pseudocubes and plates.

Physical Properties: *Fracture:* Poor, hackly. *Tenacity:* Somewhat sectile. Hardness = 2.5 VHN = 33–84 (25 g load). D(meas.) = 7.0–8.0 D(calc.) = 8.24

Optical Properties: Opaque. *Color:* Grayish iron-black, tarnishes to iridescent brownish. *Streak:* Black. *Luster:* Metallic. *Anisotropism:* Distinct but weak, light gray to dark gray.
R_1–R_2: (400) 36.3–40.7, (420) 37.0–40.6, (440) 37.7–40.5, (460) 38.0–40.2, (480) 38.1–39.6, (500) 37.8–38.9, (520) 37.0–38.1, (540) 35.7–37.3, (560) 34.6–36.7, (580) 33.8–36.2, (600) 33.2–35.8, (620) 32.7–35.5, (640) 32.2–35.0, (660) 31.8–34.7, (680) 31.5–34.3, (700) 31.2–34.8

Cell Data: *Space Group:* $P2_12_12_1$. $a = 4.333$ $b = 7.062$ $c = 7.764$ Z = 4

X-ray Powder Pattern: Tilkerode, Germany.
2.67 (100), 2.57 (100), 2.24 (60), 2.006 (40), 4.15 (20), 2.42 (20), 2.11 (20)

Chemistry:

	(1)	(2)	(3)
Ag	74.8	65.56	73.21
Pb		4.91	
Se	25.0	29.53	26.79
S	1.3		
Total	101.1	100.00	100.00

(1) Silver City, Idaho, USA. (2) Tilkerode, Germany; Pb probably in associated clausthalite. (3) Ag_2Se.

Occurrence: In hydrothermal veins associated with other selenides, quartz, and carbonates.

Association: Clausthalite, aguilarite, acanthite, tiemannite, umangite, eucairite, argentian tetrahedrite, gold, pyrite, chalcopyrite.

Distribution: In the USA, at Republic, Ferry Co., and in the L-D mine, Wenatchee, Chelan Co., Washington; and from the De Lamar mine, Silver City district, Owyhee Co., and the 4th of July mine, Yankee Fork, Custer Co., Idaho. From the Betty claim group, north of Divide Lake, British Columbia; and the Kidd Creek mine, near Timmins, Ontario, Canada. In the Harz, at Tilkerode and St. Andreasberg, Germany. From Săcărâmb (Nagyág), Romania. In the Předbořice deposit, Czechoslovakia. At Kongsberg, Norway. From the Pacajake mine, Colquechaca, Bolivia. In the Sanru mine, Hokkaido, Japan.

Name: For German mineralogist and crystallographer Carl Friedrich Naumann (1797–1873).

References: (1) Palache, C., H. Berman, and C. Frondel (1944) Dana's system of mineralogy, (7th edition), v. I, 179–180. (2) Berry, L.G. and R.M. Thompson (1962) X-ray powder data for the ore minerals. Geol. Soc. Amer. Mem. 85, 34–35. (3) Ramdohr, P. (1969) The ore minerals and their intergrowths, (3rd edition), 474–475. (4) Wiegers, G.A. (1971) The crystal structure of the low temperature form of silver selenide. Amer. Mineral., 56, 1882–1888. (5) Petruk, W., D.R. Owens, J.M. Stewart, and E.J. Murray (1974) Observations on acanthite, aguilarite and naumannite. Can. Mineral., 12, 365–369.

Crystal Data: Cubic. *Point Group:* $\overline{4}3m$. As small rounded grains, less than 1 μm in size.

Physical Properties: *Tenacity:* Brittle. Hardness = n.d. VHN = 286–338 (20 g load). D(meas.) = n.d. D(calc.) = 4.62

Optical Properties: Opaque. *Color:* Pale brown with a pink shadow.
R: (400) — , (420) — , (440) 22.6, (460) 23.8, (480) 25.0, (500) 25.6, (520) 26.0, (540) 26.4, (560) 26.9, (580) 27.4, (600) 28.0, (620) 28.6, (640) 29.3, (660) 29.6, (680) 29.9, (700) 30.0

Cell Data: *Space Group:* $P\overline{4}3n$. $a = 10.73(5)$ Z = 1

X-ray Powder Pattern: Khayragatsch deposit, USSR.
3.09 (10), 1.894 (8), 1.617 (6), 1.230 (5), 1.096 (3), 1.033 (3), 2.68 (2)

Chemistry:

	(1)
Cu	44.18
Fe	4.11
Zn	0.10
V	1.93
Sn	11.40
As	3.03
Sb	4.20
Se	0.32
S	29.86
Total	99.13

(1) Khayragatsch deposit, USSR; by electron microprobe, corresponding to $Cu^{+1}_{18}(Cu, Fe, Zn)^{+2}_8(V, Fe)^{+3}_2(Sn, As, Sb)^{+4}_6S_{32}$.

Occurrence: In ore aggregates within propylitic andesites and dacites (Khayragatsch deposit, USSR).

Association: Tetrahedrite–tennantite, luzonite–famatinite, pyrite, mawsonite, chalcopyrite, emplectite, laitakarite, bismuth, calcite, quartz, barite (Khayragatsch deposit, USSR).

Distribution: From the Khayragatsch deposit, Kuramin Mountains, eastern Uzbekistan, USSR. In the USA, in the Campbell mine, Bisbee, Cochise Co., Arizona.

Name: For I. Y. Nekrasov, Russian mineralogist.

Type Material: A.E. Fersman Mineralogical Museum, Academy of Sciences, Moscow, USSR.

References: (1) Kolavalenker, V.A., T.L. Evstigneeva, V.S. Malov, N.V. Trubkin, A.I. Gorshkov, and V.R. Geinke (1984) Nekrasovite, $Cu_{26}V_2Sn_6S_{32}$, a new mineral of the colusite group. Mineral. Zhurnal, 6–2, 88–97 (in Russian). (2) (1985) Amer. Mineral., 70, 437 (abs. ref. 1).

Crystal Data: Hexagonal. *Point Group:* $\overline{3}\ 2/m$, $3m$, or 32. Irregular grains, up to 2 mm in diameter.

Physical Properties: *Cleavage:* {0001} very perfect. Hardness = n.d. VHN = 60–114, 90 average. D(meas.) = n.d. D(calc.) = 7.85

Optical Properties: Opaque. *Color:* Lead-gray; in reflected light, white with creamy tinge. *Luster:* Metallic. *Anisotropism:* Isotropic on basal sections, distinctly anisotropic in perpendicular sections, with weak color effects from brown to light gray.
R_1–R_2: n.d.

Cell Data: *Space Group:* $P\overline{3}m1$, $P3m1$, or $P321$. $a = 4.197$ $c = 22.80$ Z = 6

X-ray Powder Pattern: Nevskii deposit, USSR.
3.06 (10), 3.59 (4), 2.24 (4), 2.10 (4)

Chemistry:

	(1)	(2)	(3)	(4)
Bi	69.1	69.7	71.2	69.1
Pb	3.6	2.8	1.9	3.5
Ag	0.6	0.6	0.5	0.6
Se	24.7	24.5	24.8	24.6
S	1.8	1.6	1.4	1.7
Total	99.8	99.2	99.8	99.5

(1–4) Nevskii deposit, USSR; by electron microprobe, the average corresponding to $(Bi_{0.92}Pb_{0.04}Ag_{0.02})_{\Sigma=0.98}(Se_{0.86}S_{0.14})_{\Sigma=1.00}$.

Occurrence: In quartz-cassiterite veins.

Association: Wolframite, cassiterite, natanite, laitakarite, guanajuatite.

Distribution: From the Nevskii tin deposit, northeastern USSR.

Name: For the Nevskii deposit in the USSR.

Type Material: A.E. Fersman Mineralogical Museum, Academy of Sciences, Moscow, USSR.

References: (1) Nechelyustov, G.N., N.I. Christyakova, and E.N. Zav'yalov (1984) Nevskite, a new bismuth selenide. Zap. Vses. Mineral. Obshch., 113, 351–355 (in Russian). (2) (1985) Amer. Mineral., 70, 875 (abs. ref. 1).

Crystal Data: Monoclinic. *Point Group:* 2, m or $2/m$. As masses of small intergrown crystals and grains; also as aggregates of prismatic to bladed crystals with stepped surfaces on the broad faces. Some prismatic aggregates are curved.

Physical Properties: *Fracture:* Conchoidal. *Tenacity:* Very brittle. Hardness = 2.5 VHN = n.d. D(meas.) = 7.02 D(calc.) = 7.16

Optical Properties: Opaque. *Color:* Lead-gray, tarnishes yellow then Prussian blue; galena-white in reflected light. *Luster:* Bright metallic. *Anisotropism:* Moderate, light gray, yellow-green, and light reddish brown to gray-black.
R_1–R_2: n.d.

Cell Data: *Space Group:* $C2$, Cm, or $C2/m$. $a = 37.5$ $b = 4.07$ $c = 41.6$ $\beta = 96.8(3)°$ Z = 8

X-ray Powder Pattern: Kitsault, Canada.
3.72 (100), 3.51 (100), 2.92 (100), 2.04 (60), 2.27 (50), 2.08 (40), 1.461 (40)

Chemistry:

	(1)	(2)
Pb	41.76	42.95
Cu	2.84	3.77
Ag	1.52	
Bi	36.62	37.14
S	15.65	16.14
Total	98.39	100.00

(1) Kitsault, Canada. (2) $Pb_7Cu_2Bi_6S_{17}$.

Occurrence: In late-forming crosscutting hydrothermal quartz veins in a quartz-molybdenite stockwork (Lime Creek, Canada); in mesozonal synmetamorphic quartz veins (Johnny Lyon Hills, Arizona, USA).

Association: Pyrite, galena, sphalerite, chalcopyrite, aikinite, cosalite, tetrahedrite, nuffieldite, molybdenite (Lime Creek, Canada); gold, chalcopyrite, nordströmite, quartz (Johnny Lyon Hills, Arizona, USA).

Distribution: In the Lime Creek molybdenum deposit, Kitsault, near Alice Arm, British Columbia, Canada. From the Johnny Lyon Hills, north of Benson, Cochise Co., Arizona, USA.

Name: For Charles Stewart Ney (1918–1975), geologist in charge of early exploration at the Lime Creek deposit.

Type Material: National Museum of Natural History, Washington, D.C., USA, 142527.

References: (1) Drummond, A.D., J. Trotter, R.M. Thompson, and J.A. Gower (1969) Neyite, a new sulphosalt from Alice Arm, British Columbia. Can. Mineral., 10, 90–96. (2) (1970) Amer. Mineral., 55, 1444 (abs. ref. 1).

Crystal Data: Cubic. *Point Group:* $4/m\ \bar{3}\ 2/m$. As euhedral cubic grains or intergrown cubes up to 0.1 mm enclosed in heazlewoodite; also as anhedral "spider-like" irregular masses between heazlewoodite grains; as flakes to 0.75 mm in a placer. *Twinning:* On {111}.

Physical Properties: *Tenacity:* Malleable. Hardness = 4.5 VHN = 172–184 (50 g load). D(meas.) = 8.90 (synthetic). D(calc.) = 8.91

Optical Properties: Opaque. *Color:* Bright silver; white in reflected light, more bluish white than heazlewoodite. *Luster:* Metallic.
R: n.d.

Cell Data: *Space Group:* $Fm3m$. $a = 3.5238$ Z = 4

X-ray Powder Pattern: Synthetic.
2.034 (100), 1.762 (42), 1.246 (21), 1.0624 (20), 0.7880 (15), 0.8084 (14), 1.0172 (7)

Chemistry:

	(1)	(2)
Ni	> 98	96.30
Fe	trace	1.77
Co	trace	0.69
Total	> 98	98.76

(1) Bogota, New Caledonia; by electron microprobe. (2) Jerry River, New Zealand; by electron microprobe.

Occurrence: In serpentinized ultramafic rocks as a result of low-temperature hydrothermal activity.

Association: Heazlewoodite, pyrite, pyrrhotite, pentlandite, godlevskite, millerite.

Distribution: From Bogota, near Canala, New Caledonia. In the Jerry River, in placers derived from the Red Hills Range, southern Westland, New Zealand. From Mount Clifford and Cutmore, near Agnew, Western Australia, and the Nairne pyrite deposit, South Australia. At Kaltenberg, Turtmauntal, Valais, Switzerland. On Grasshopper Mountain, Tulameen, British Columbia, Canada.

Name: From the German *nickel, demon*, from a contraction of *kupfernickel*, or *Devil's copper*, as the mineral was believed to contain copper but yielded none.

Type Material: n.d.

References: (1) Ramdohr, P. (1968) The wide-spread paragenesis of ore minerals originating during serpentinization (with some data on new and insufficiently described minerals). Geol. Rudn. Mestorozhd., 2, 32–43 (in Russian). (2) (1968) Amer. Mineral., 53, 348 (abs. ref. 1). (3) (1953) NBS Circ. 539, 1, 13. (4) Challis, G.A. (1975) Native nickel from the Jerry River, south Westland, New Zealand; an example of natural refining. Mineral. Mag., 40, 247–251. (5) Hudson, D.R. and G.A. Travis (1981) A native nickel-heazlewoodite-ferroan trevorite assemblage from Mount Clifford, Western Australia. Econ. Geol., 76, 1686–1697.

Nickeline NiAs

Crystal Data: Hexagonal. *Point Group:* $6/m\ 2/m\ 2/m$. Commonly in granular aggregates, reniform masses with radial structure and reticulated and arborescent growths. Rarely as distorted, horizontally striated, $\{10\bar{1}1\}$ terminated crystals, to 1 cm. *Twinning:* On $\{10\bar{1}1\}$ producing fourlings; possibly on $\{31\bar{4}1\}$.

Physical Properties: *Fracture:* Conchoidal. *Tenacity:* Brittle. Hardness = 5–5.5 VHN = n.d. D(meas.) = 7.784 D(calc.) = 7.834

Optical Properties: Opaque. *Color:* Pale copper-red, tarnishes gray to blackish; white with strong yellowish pink hue in reflected light. *Streak:* Pale brownish black. *Luster:* Metallic. *Pleochroism:* High, whitish, yellow-pink to light brownish pink. *Anisotropism:* Very strong, light greenish yellow to slate-gray in air.

R_1–R_2: (400) 41.4–48.2, (420) 40.4–47.1, (440) 39.4–46.0, (460) 39.0–45.5, (480) 39.5–46.2, (500) 41.3–47.7, (520) 43.8–49.5, (540) 47.0–51.6, (560) 50.5–53.8, (580) 53.5–56.0, (600) 56.0–57.8, (620) 58.2–59.7, (640) 60.0–60.9, (660) 61.4–62.0, (680) 62.3–62.8, (700) 62.8–63.2

Cell Data: *Space Group:* $P6_3/mcm$. $a = 3.602$ $c = 5.009$ Z = 2

X-ray Powder Pattern: Unknown locality.
2.66 (100), 1.961 (90), 1.811 (80), 1.071 (40), 1.328 (30), 1.033 (30), 0.821 (30)

Chemistry:

	(1)	(2)	(3)
Ni	43.25	40.64	43.93
Co	0.49	2.04	
Fe	0.05	trace	
As	55.10	50.78	56.07
Sb	0.15	4.95	
S	0.15	1.47	
other	0.65		
Total	99.84	99.88	100.00

(1) Hohendahl-Schacht, Eisleben, Germany. (2) Hudson Bay mine, Cobalt, Canada; Sb and S may be in impurities. (3) NiAs.

Occurrence: Partly in high-temperature hydrothermal veins and partly orthomagmatic in peridotite and norite.

Association: Skutterudite–nickel-skutterudite, safflorite, rammelsbergite, maucherite, breithauptite, bismuth, bismuthinite, silver minerals.

Distribution: Numerous minor occurrences. In the USA, at Silver Cliff, Custer Co., and at the Gem mine, Fremont Co., Colorado. In Ontario, Canada, in the Cobalt-Gowganda and Sudbury districts, and at Silver Islet, Thunder Bay district. In Germany, at Eisleben, Thuringia; St. Andreasberg and Mansfeld, Harz Mountains, and at Schneeberg, Saxony. From the Aït Ahmane mine, 10 km east of Bou-Azzer, Morocco. In the Talmessi mine, near Anarak, Iran. At Cochabamba, Cercado, Bolivia.

Name: From the German *kupfernickel*, or *Devil's copper*, as the mineral was believed to contain copper but yielded none.

References: (1) Palache, C., H. Berman, and C. Frondel (1944) Dana's system of mineralogy, (7th edition), v. I, 236–238. (2) Berry, L.G. and R.M. Thompson (1962) X-ray powder data for the ore minerals. Geol. Soc. Amer. Mem. 85, 62.

Crystal Data: Cubic. *Point Group:* $2/m\ \overline{3}$. As euhedral crystals with combinations of {001}, {111}, and rarely {011}; also as reticular skeletal growths, distorted aggregates and granular; {001} frequently warped. *Twinning:* On {112} as sixlings; often complex and distorted.

Physical Properties: *Cleavage:* Distinct on {001} and {111}, in traces on {011}. *Fracture:* Conchoidal to uneven. *Tenacity:* Brittle. Hardness = 5.5–6 VHN = n.d. D(meas.) = 6.5 D(calc.) = [5.07–6.90]

Optical Properties: Opaque. *Color:* Tin-white to silver-gray, sometimes tarnishing gray or iridescent; in reflected light yellowish white, but anisotropic zones can be somewhat reddish. *Streak:* Black. *Luster:* Bright metallic on fresh surface. *Anisotropism:* Weak to distinct in some zones.
R: n.d.

Cell Data: *Space Group:* $Im3$. $a = 8.17$ Z = 8

X-ray Powder Pattern: Timiskaming mine, Cobalt, Canada.
2.61 (100), 1.841 (90), 1.616 (90), 2.20 (80), 1.681 (70), 1.410 (70), 1.213 (70)

Chemistry:

	(1)	(2)	(3)	(4)
Ni	11.35	15.07	28.15	20.71
Co	13.81	15.83		
Fe	1.21	3.69		
As	71.61	63.42	71.85	79.29
Bi		0.86		
S	0.75			
rem.	0.96	0.32		
Total	99.69	99.19	100.00	100.00

(1) Skutterudite; Foster mine, Cobalt, Canada. (2) Skutterudite; Schneeberg, Germany. (3) $NiAs_2$. (4) $NiAs_3$.

Polymorphism & Series: Forms a series with skutterudite.

Occurrence: In hydrothermal veins deposited at moderate temperature.

Association: Arsenopyrite, silver, bismuth, calcite, siderite, barite, quartz.

Distribution: In the USA, at the Trotter mine, Franklin, Sussex Co., New Jersey; and at the Rose mine, Alhambra mine, and Bullards Peak district near Silver City, Grant Co., New Mexico. In Canada, at the O'Brien mine and elsewhere, Cobalt, Ontario. In Germany, at Bieber and Riechelsdorf, Hesse; St. Andreasberg, Harz; and Annaberg and Schneeberg, Saxony. From Horní Slavkov (Schlaggenwald), Czechoslovakia. At Dobšiná (Dobschau), Czechoslovakia. From the Anniviersthal, Switzerland. In the Khovuasinsk deposit, Ural Mountains, Tuva ASSR, USSR.

Name: Nickel, as the nickel member of the series; skutterudite from its locality at Skutterud, Norway.

References: (1) Palache, C., H. Berman, and C. Frondel (1944) Dana's system of mineralogy, (7th edition), v. I, 342–346. (2) Berry, L.G. and R.M. Thompson (1962) X-ray powder data for the ore minerals. Geol. Soc. Amer. Mem. 85, 117–118. (3) Ramdohr, P. (1969) The ore minerals and their intergrowths, (3rd edition), 867–873.

Crystal Data: Hexagonal. *Point Group:* $6/m\ 2/m\ 2/m$. As rounded to anhedral inclusions up to about 75 μm.

Physical Properties: *Tenacity:* Brittle. Hardness = n.d. VHN = 590 (25 g load). D(meas.) = n.d. D(calc.) = 13.4

Optical Properties: Opaque. *Color:* Silver-white; in reflected light, pinkish to blue. *Luster:* Metallic. *Pleochroism:* Strong, from pinkish cream to pale cobalt blue. *Anisotropism:* Very high, bright pinkish cream to very dark blue or black.
R_1–R_2: (400) 52.4–55.0, (420) 51.1–54.9, (440) 49.8–54.8, (460) 47.4–54.4, (480) 46.6–55.0, (500) 46.4–55.4, (520) 45.8–55.9, (540) 45.5–57.3, (560) 45.4–58.9, (580) 45.0–60.0, (600) 44.3–62.0, (620) 42.8–62.9, (640) 41.1–64.4, (660) 39.5–66.2, (680) 37.7–67.2, (700) 36.3–66.7

Cell Data: *Space Group:* $P6_3/mmc$. $a = 4.100(1)$ $c = 5.432(2)$ Z = 2

X-ray Powder Pattern: Synthetic.
2.157 (100), 0.781 (100), 1.203 (90), 2.050 (80), 1.485 (80), 0.7758 (80), 2.971 (70)

Chemistry:

	(1)	(2)	(3)
Pt	61.0	61.1	62.17
Pd		0.38	
Sn	31.4	36.7	37.83
Sb	4.7	1.05	
Bi	2.2	0.35	
Te		0.08	
Total	99.3	99.66	100.00

(1) Insizwa, South Africa; by electron microprobe, corresponding to $Pt_{1.00}(Sn_{0.85}Sb_{0.12}Bi_{0.035})_{\Sigma=1.005}$. (2) Sudbury, Canada; by electron microprobe, corresponding to $(Pt_{0.98}Pd_{0.01})_{\Sigma=0.99}(Sn_{0.97}Sb_{0.03}Bi_{0.01})_{\Sigma=1.01}$. (3) PtSn.

Occurrence: In late forming hydrothermal veins.

Association: Pentlandite, chalcopyrite, parkerite, insizwaite, cubanite (Insizwa, South Africa); stannopalladinite, hessite, platinum and palladium tellurides (Monchegorsk, USSR); froodite (Sudbury, Canada); leadamalgam, chromite, ilmenite, magnetite, gersdorffite, pyrite, chalcopyrite, violarite, millerite, galena, stibnite, argentian gold, sperrylite, iridosmine, platinum, merenskyite, kotulskite (Shiaonanshan, China).

Distribution: From Waterfall Gorge, Insizwa, South Africa. At Monchegorsk, on the Kola Peninsula, and the Talnakh area, Noril'sk region, western Siberia, USSR. From Sudbury, Ontario, Canada. At Fox Gulch, Goodnews Bay, Alaska, USA. In China, at Shiaonanshan, Inner Mongolia Autonomous Region.

Name: For Professor Paul Niggli (1888–1953), of Zurich, Switzerland.

Type Material: National Museum of Natural History, Washington, D.C., USA, 162610.

References: (1) Palache, C., H. Berman, and C. Frondel (1944) Dana's system of mineralogy, (7th edition), v. I, 347. (2) Cabri, L.J. and D.C. Harris (1972) The new mineral insizwaite ($PtBi_2$) and new data on niggliite (PtSn). Mineral. Mag., 38, 794–800. (3) (1973) Amer. Mineral., 58, 805 (abs. ref. 1). (4) Cabri, L.J., Ed. (1981) Platinum group elements: mineralogy, geology, recovery. Can. Inst. Min. & Met., 121–122, 164.

Crystal Data: Cubic. *Point Group:* $4/m\ \overline{3}\ 2/m$. As grains intimately intergrown with kamacite and troilite; grains occasionally contain oriented exsolution lamellae of troilite and minute grains of kamacite.

Physical Properties: Hardness = n.d. VHN = n.d. D(meas.) = n.d. D(calc.) = 3.21–3.68

Optical Properties: Opaque. *Color:* Gray in reflected light. *Luster:* Metallic.
R: n.d.

Cell Data: *Space Group:* $Fm3m$ (probable). $a = 5.17(2)$ Z = 4

X-ray Powder Pattern: Synthetic MgS.
2.601 (100), 1.8388 (60), 1.5010 (16), 1.1630 (14), 1.0617 (10), 3.004 (8), 1.3001 (8)

Chemistry:

	(1)	(2)	(3)	(4)
Mg	10.1	13.2	18.3	23.5
Fe	37.1	35.2	27.0	15.6
Mn	4.02	3.93	6.5	11.6
Ca	3.03	2.55	1.28	0.39
Zn	0.31			
Cr	1.84	1.77	1.66	0.14
S	41.0	42.7	43.4	46.9
Total	97.40	99.35	98.14	98.13

(1) Abee meteorite. (2) Saint-Sauveur meteorite. (3) Indarch meteorite. (4) Kota-Kota meteorite.

Occurrence: In less extensively metamorphosed enstatite chondrite meteorites.

Association: "Nickel-iron" (kamacite), troilite.

Distribution: In the Abee, Saint-Sauveur, Adhi-Kot, Indarch, St. Marks, Qingzhen, and Kota-Kota enstatite chondrite meteorites.

Name: For Harvey Harlow Nininger (1887–1986), of Sedona, Arizona, USA, for his contributions to meteoritics.

Type Material: n.d.

References: (1) Klaus, K. and K.G. Snetsinger (1967) Niningerite: a new meteoric sulfide. Science, 155, 451–453. (2) (1967) Amer. Mineral., 52, 925 (abs. ref. 1). (3) (1957) NBS Mono., 7, 31. (4) Leitch, C.A. and J.V. Smith (1982) Petrography, mineral chemistry and origin of type I enstatite chondrites. Geochim. Cosmochim. Acta, 46, 2083–2097.

Crystal Data: Orthorhombic. *Point Group:* $2/m$ $2/m$ $2/m$. As irregular grains up to 20 μm.

Physical Properties: Hardness = n.d. VHN = 420–513 (5 g load). D(meas.) = n.d. D(calc.) = 8.0

Optical Properties: Opaque. *Color:* White in reflected light. *Luster:* Metallic. *Pleochroism:* Weak. *Anisotropism:* Weak.
R_1–R_2: n.d.

Cell Data: *Space Group:* *Pnnm.* $a = 5.178$ $b = 6.319$ $c = 3.832$ Z = 2

X-ray Powder Pattern: Red Lake area, Canada.
2.761 (100), 2.689 (90), 2.0301 (80), 1.8427 (70), 0.9045 (70), 1.5611 (60), 1.1683 (60)

Chemistry:

	(1)	(2)
Ni	19.15	19.43
Sb	80.43	80.57
Total	99.58	100.00

(1) Red Lake area, Canada; by electron microprobe, average of four grains. (2) $NiSb_2$.

Occurrence: In a high-grade base-metal sulfide deposit in altered mafic rock (Red Lake, Canada); in Pb-Zn-Cu-Ag ore deposits remobilized by hydrothermal solutions from younger granite emplacement (Bergslagen, Sweden).

Association: Chalcopyrite, breithauptite, pyrargyrite, galena, pyrrhotite, tetrahedrite (Red Lake, Canada); costibite, paracostibite, chalcopyrite, pyrrhotite, galena, sphalerite, gersdorffite, ullmannite (Bergslagen, Sweden).

Distribution: From the Red Lake mining division, Kenora district, Ontario, Canada. At the Festivalnoe Cu-Sn mine, Magadan region, Yakutia, USSR. In the Gruvåsen and Getön deposits, Bergslagen, Sweden.

Name: For the composition, NIckel and SB, from the chemical symbol for antimony, *stibium*.

Type Material: National Museum of Canada, Ottawa, Canada.

References: (1) Cabri, L.J., D.C. Harris and J.M. Stewart (1970) Paracostibite (CoSbS) and nisbite ($NiSb_2$), new minerals from the Red Lake area, Ontario, Canada. Can. Mineral., 10, 232–246. (2) (1971) Amer. Mineral., 56, 631 (abs. ref. 1). (3) Zakrzewski, M.A., E.A.J. Burke, and H.W. Nugteren (1980) Cobalt minerals in the Hallëfors area, Bergslagen, Sweden: new occurrences of costibite, paracostibite, nisbite and cobaltian ullmannite. Can. Mineral., 18, 165–171.

Crystal Data: Monoclinic. *Point Group:* $2/m$. Fibrous.

Physical Properties: *Cleavage:* Good, parallel to crystal elongation. Hardness = 2–2.5 VHN = 130–143 (25 g load). D(meas.) = n.d. D(calc.) = 7.12

Optical Properties: Opaque, presumably. *Color:* Lead-gray; white with a gray tint in reflected light. *Anisotropism:* Dark gray to brown.
R_1–R_2: n.d.

Cell Data: *Space Group:* $P2_1/m$. $a = 17.97(8)$ $b = 4.11(2)$ $c = 17.62(8)$ $\beta = 94.3(2)°$ Z = 2

X-ray Powder Pattern: Falun, Sweden.
3.066 (100), 2.242 (70), 3.583 (50), 3.882 (30b), 3.484 (30), 3.427 (30), 3.012 (30)

Chemistry:

	(1)
Pb	21.7
Cu	2.0
Bi	51.9
Se	11.2
S	10.9
Total	97.7

(1) Falun, Sweden; by electron microprobe, corresponding to $Pb_{3.05}Cu_{0.94}Bi_{7.24}(S_{9.89}Se_{4.12})_{\Sigma=14.01}$.

Occurrence: Of hydrothermal origin.

Association: Wittite, friedrichite, bismuthinite (Falun, Sweden); gold, chalcopyrite, neyite, quartz (Johnny Lyon Hills, Arizona, USA).

Distribution: From Falun, Sweden. In the Johnny Lyon Hills, north of Benson, Cochise Co., Arizona, USA.

Name: To honor T. Nordström.

Type Material: Royal Ontario Museum, Toronto, Canada, M12992.

References: (1) Mumme, W.G. (1980) Seleniferous lead–bismuth sulphosalts from Falun, Sweden: weibullite, wittite, and nordströmite. Amer. Mineral., 65, 789–796. (2) Mumme, W.G. (1980) The crystal structure of nordströmite $CuPb_3Bi_7(S, Se)_{14}$, from Falun, Sweden: a member of the junoite homologous series. Can. Mineral., 18, 343–352.

Crystal Data: Monoclinic, pseudotetragonal. *Point Group:* 2, m or $2/m$. As irregular aggregates, with grains up to 3 cm, and as veinlets in arsenic; botryoidal.

Physical Properties: Hardness = 3–3.5 VHN = n.d. D(meas.) = n.d. D(calc.) = 8.01

Optical Properties: Opaque. *Color:* Steel-gray on fresh surface, tarnishes iridescent and then black; white with light cream tint in reflected light. *Streak:* Black. *Luster:* Metallic. *Anisotropism:* Medium, dark blue-gray and light brown-ocher.

R_1–R_2: (400) 43.6–49.0, (420) 45.4–50.5, (440) 47.2–52.0, (460) 48.3–53.0, (480) 49.1–53.9, (500) 49.8–54.6, (520) 50.2–55.2, (540) 50.6–55.6, (560) 50.9–56.0, (580) 51.2–56.3, (600) 51.3–56.6, (620) 51.4–56.8, (640) 51.5–56.9, (660) 51.5–57.0, (680) 51.5–57.1, (700) 51.5–57.2

Cell Data: *Space Group:* $C2$, Cm, or $C2/m$. $a = 16.269(3)$ $b = 11.711(2)$ $c = 10.007(2)$ $\beta = 112.7°$ Z = 4

X-ray Powder Pattern: Černý Důl, Czechoslovakia.
1.877 (100), 1.959 (90), 1.180 (90), 1.998 (80), 1.351 (60), 1.225 (60), 6.41 (50)

Chemistry:

	(1)
Cu	60.30
Ag	4.33
As	35.30
Total	99.93

(1) Černý Důl, Czechoslovakia; by electron microprobe, average of 10 analyses.

Occurrence: In hydrothermal carbonate veins up to 20 cm thick, cutting diopside hornfels lenses in pyroxene gneiss and occasionally in mica schist.

Association: Arsenic, arsenolamprite, koutekite, silver, löllingite, chalcocite, skutterudite, chalcopyrite, bornite, uraninite, calcite.

Distribution: From Černý Důl, Krkonoše (Giant Mountains), Czechoslovakia.

Name: For Jiří Novák, Professor of Mineralogy, Charles University, Prague, Czechoslovakia.

Type Material: n.d.

References: (1) Johan, Z. and J. Hak (1961) Novákite, $(Cu, Ag)_4As_3$, a new mineral. Amer. Mineral., 46, 885–891. (2) Johan, Z. (1985) The Černy Důl deposit (Czechoslovakia): an example of Ni-, Fe-, Ag-, Cu-arsenide mineralization with extremely high activity of arsenic; new data on paxite, novakite and kutinaite. Tschermaks Mineral. Petrog. Mitt., 34, 167–182.

Crystal Data: Hexagonal. *Point Group:* 3. As equant crystals up to 0.3 mm in size. *Twinning:* On {0001}.

Physical Properties: Hardness = n.d. VHN = n.d. D(meas.) = n.d. D(calc.) = 4.3

Optical Properties: Opaque. *Color:* Lead-gray to black. *Luster:* Metallic. R_1–R_2: n.d.

Cell Data: *Space Group:* $R3$. $a = 13.440(1)$ $c = 9.17(1)$ Z = 3

X-ray Powder Pattern: Binntal, Switzerland.
1.019 (100), 1.081 (75), 1.887 (63), 3.127 (61), 1.605 (61), 1.216 (52), 1.085 (20)

Chemistry:

	(1)	(2)	(3)
Cu	31.2	30.9	30.21
Zn	15.9	16.4	15.54
As	22.4	22.8	23.75
S	31.9	32.0	30.50
Total	101.4	102.1	100.00

(1) Binntal, Switzerland; by electron microprobe, corresponding to $Cu_{6.04}Zn_{3.01}As_{3.73}S_{12.70}$. (2) Do.; corresponds to $Cu_{6.02}Zn_{3.11}As_{3.77}S_{12.36}$. (3) $Cu_6Zn_3As_4S_{12}$.

Occurrence: In a hydrothermal deposit in dolomite, noted for a variety of Pb-As-S minerals.

Association: Sphalerite.

Distribution: At the Lengenbach quarry, Binntal, Valais, Switzerland.

Name: For Professor Werner Nowacki (1909–), University of Berne, Switzerland.

Type Material: n.d.

References: (1) Marumo, F. and G. Burri (1965) Nowackiite, a new copper–zinc arsenosulfosalt from Lengenbach (Binnatal, Kanton Wallis). Chimia (Switzerland), 19, 500–501. (2) (1966) Amer. Mineral., 51, 532 (abs. ref. 1). (3) Marumo, F. (1967) The crystal structure of nowackiite, $Cu_6Zn_3As_4S_{12}$. Zeits. Krist., 124, 352–368. (4) Nowacki, W. (1982) Isotypic state of aktashite ($Cu_6Hg_3As_4S_{12}$) and nowackiite ($Cu_6Zn_3As_4S_{12}$). Kristallografiya, 27, 49–50 (in Russian). (5) (1982) Chem. Abs., 96, 107317 (abs. ref. 4).

Crystal Data: Orthorhombic. *Point Group:* $2/m\ 2/m\ 2/m$. As prismatic to acicular crystals, deeply striated and channelled ‖ [001]; as bundles of these unterminated crystals, up to 3 mm long.

Physical Properties: *Cleavage:* Indistinct ⊥ [001]; excellent ‖ [001]. *Fracture:* Uneven to flat conchoidal. *Tenacity:* Very brittle; long needles are quite elastic. Hardness = n.d. VHN = 149–178 D(meas.) = 7.01(7) D(calc.) = 7.21

Optical Properties: Opaque. *Color:* Shiny lead-gray to steel-gray on fresh surface, tarnishes to pale iridescent grayish green to reddish brown; in polished section, pale creamy white. *Streak:* Dark greenish gray to black. *Luster:* Bright metallic. *Anisotropism:* Very weak, from bluish gray to grayish red.

R_1–R_2: n.d.

Cell Data: *Space Group:* *Pbnm.* $a = 14.387(7)$ $b = 21.011(15)$ $c = 4.046(6)$ Z = 4

X-ray Powder Pattern: Lime Creek, Canada.
3.66 (100), 3.54 (100), 4.00 (90), 3.16 (80), 2.54 (70), 1.871 (60), 1.349 (40)

Chemistry:

	(1)
Pb	40.27
Cu	5.88
Bi	37.55
S	16.30
Total	100.00

(1) Lime Creek, Canada; by X-ray spectrography, electron microprobe, colorimetry and atomic absorption spectrophotometry, weighted and manipulated to produce this preferred analysis, corresponding to $Pb_{10.22}Cu_{4.86}Bi_{9.45}S_{26.73}$.

Occurrence: Found in vugs in a small quartz vein in the Lime Creek stock.

Association: Molybdenite, cosalite, aikinite, pyrite, galena, sphalerite.

Distribution: From Lime Creek and Patsy Creek, near Alice Arm, British Columbia, Canada.

Name: For Professor Edward Wilfrid Nuffield (1914–), mineralogist of the University of Toronto, Toronto, Canada.

Type Material: n.d.

References: (1) Kingston, P.W. (1968) Studies of mineral sulphosalts: XXI—Nuffieldite, a new species. Can. Mineral., 9, 439–452. (2) (1969) Amer. Mineral., 54, 574 (abs. ref. 1). (3) Kohatsu, I. and B.J. Wuensch (1973) The crystal structure of nuffieldite, $Pb_2Cu(Pb, Bi)Bi_2S_7$. Zeits. Krist., 138, 343–365.

Crystal Data: Hexagonal. *Point Group:* $\bar{3}\ 2/m$. Tabular hexagonal crystals to 2 mm, in fan-shaped groups 1–2 mm across; also as larger irregular masses up to 4 cm long, sometimes with a palm-like branching structure; as fine lamellae in sphalerite.

Physical Properties: *Cleavage:* Perfect on {0001}. Hardness = n.d. VHN = 103–110 (20 g load). D(meas.) = 4.30(7) D(calc.) = 4.53

Optical Properties: Opaque. *Color:* Copper colored. *Luster:* Metallic. *Pleochroism:* Strong, from reddish orange to pale gray. *Anisotropism:* Very strong, in pale green-gray colors.
R_1–R_2: (400) 15.4–23.8, (420) 14.9–23.8, (440) 14.3–23.8, (460) 14.1–23.6, (480) 14.0–23.7, (500) 14.1–23.7, (520) 14.7–23.7, (540) 15.8–23.7, (560) 17.1–23.8, (580) 18.7–23.9, (600) 20.4–24.0, (620) 22.1–24.7, (640) 24.0–26.1, (660) 25.7–27.9, (680) 27.3–30.0, (700) 28.7–30.2

Cell Data: *Space Group:* $P\bar{3}m1$ (synthetic). $a = 3.782(4)$ $c = 11.187(8)$ Z = 1

X-ray Powder Pattern: Undu mine, Fiji Islands.
3.143 (100), 2.826 (70), 1.891 (60), 1.847 (55), 2.796 (45), 3.273 (30), 1.568 (25)

Chemistry:

	(1)
Cu	56.51
Fe	9.64
Ag	0.09
As	0.04
S	33.51
Total	99.79

(1) Undu mine, Fiji Islands; by electron microprobe, corresponds to $(Cu_{3.37}Fe_{0.66})_{\Sigma=4.03}S_{3.97}$.

Occurrence: As a primary mineral in a Kuroko-type deposit; also as an alteration product of primary chalcopyrite.

Association: Pyrite, covellite, sphalerite.

Distribution: In the Undu mine, Nukundamu, Fiji Islands.

Name: For the Fijian locality at Nukundamu.

Type Material: National Museum of Natural History, Washington, D.C., USA, 148128.

References: (1) Rice, C.M., D. Atkin, J.F.W. Bowles, and A.J. Criddle (1979) Nukundamite, a new mineral, and idaite. Mineral. Mag., 43, 194–200. (2) (1980) Amer. Mineral., 65, 407 (abs. ref. 1). (3) Sugaki, A., H. Shima, A. Kitakaze, and T. Mizota (1981) Hydrothermal synthesis of nukundamite and its crystal structure. Amer. Mineral., 66, 398–402.

Oldhamite (Ca, Mg)S

Crystal Data: Cubic. *Point Group:* $4/m\ \bar{3}\ 2/m$. As single-crystal nodules up to 3 mm (Busti meteorite).

Physical Properties: *Cleavage:* {001}. Hardness = 4 VHN = n.d. D(meas.) = 2.58 D(calc.) = 2.589

Optical Properties: Transparent. *Color:* Pale chestnut-brown, tarnishes strongly on exposure to wet air; in transmitted light, colorless to light brown with many internal reflections.
Optical Class: Isotropic. $n = 2.137$
R: n.d.

Cell Data: *Space Group:* $Fm3m$. $a = 5.6948$ Z = 4

X-ray Powder Pattern: Synthetic CaS.
2.846 (100), 2.013 (70), 1.6439 (21), 1.2737 (20), 1.1627 (14), 1.4238 (10), 0.9491 (8)

Chemistry:

	(1)	(2)	(3)
Ca	53.45	51.0	55.55
Mg	1.49	0.90	
Na		0.041	
Mn		0.170	
Fe	0.32	5.50	
Cu		0.020	
S	44.74	41.6	44.45
Total	100.00	99.231	100.00

(1) Bustee meteorite; recalculated to 100%. (2) Kota-Kota meteorite; by electron microprobe, average of 11 determinations. (3) CaS.

Occurrence: Fills the latest interstices between silicates in enstatite chondrite and achondrite meteorites.

Association: Enstatite, augite, niningerite, osbornite, gypsum, calcite, troilite.

Distribution: In the Bustee, Hvittis, Mayo Belwa, Indarch, Kota-Kota, Adhi-Kot, Norton County, Peña Blanca Spring, Abee, Qingzhen, etc., meteorites.

Name: For Thomas Oldham (1816–1878), Director of the Indian Geological Survey (1850–1876).

Type Material: n.d.

References: (1) Palache, C., H. Berman, and C. Frondel (1944) Dana's system of mineralogy, (7th edition), v. I, 208–209. (2) (1957) NBS Circ. 539, 7, 15. (3) Leitch, C.A. and J.V. Smith (1982) Petrography, mineral chemistry and origin of type I enstatite chondrites. Geochim. Cosmochim. Acta, 46, 2083–2097.

Crystal Data: Orthorhombic. *Point Group:* $2/m\ 2/m\ 2/m$, or $mm2$. Crystals are tabular or elongated || [010], to 20 μm.

Physical Properties: *Cleavage:* Parallel to elongation. *Tenacity:* Brittle. Hardness = n.d. VHN = n.d. D(meas.) = n.d. D(calc.) = 11.20

Optical Properties: Opaque. *Color:* Dull steel-gray; in polished section, white with a yellow tint in air, milky yellow in oil. *Luster:* Metallic. *Pleochroism:* Weak, yellow to grayish yellow. *Anisotropism:* Distinct.

R_1–R_2: n.d.

Cell Data: *Space Group:* $Pnnm$ or $Pnn2$. $a = 5.409$ $b = 6.167$ $c = 3.021$ Z = 2

X-ray Powder Pattern: Danba, China.
2.63 (100), 1.915 (100), 2.67 (80), 4.06 (60), 2.06 (60), 2.01 (60), 1.209 (60)

Chemistry:

	(1)
Os	48.9
Ru	4.0
Ir	0.6
Ni	0.3
Fe	0.2
Co	0.1
As	44.2
Total	98.3

(1) Danba, China; by electron microprobe, average of six analyses, corresponding to $(Os_{0.87}Ru_{0.14}Ir_{0.01})_{\Sigma=1.02}As_{2.00}$.

Polymorphism & Series: Forms a series with anduoite.

Occurrence: Found in a Cu-Ni sulfide deposit associated with an ultramafic body.

Association: Pyrrhotite, pentlandite, chalcopyrite, violarite, cubanite, bornite, sphalerite, galena, linnaeite, magnetite, testibiopalladite, sudburyite, michenerite, sperrylite, kotulskite, gold, argentian gold.

Distribution: From Danba, Sichuan Province, China.

Name: For Omeishan, a well-known mountain in Sichuan Province, China.

Type Material: Museum of Geology, National Bureau of Geology (Beijing ?), China.

References: (1) Ren Yingxin, Hu Qinde, and Xu Jingao (1978) A preliminary study on the new mineral of the platinum group—omeiite, $OsAs_2$. Acta Geol. Sinica, 52, 163–167 (in Chinese with English abs.). (2) (1979) Amer. Mineral., 64, 464 (abs. ref. 1).

Crystal Data: Orthorhombic, pseudotetragonal. *Point Group:* n.d. As irregular grains up to 0.4 mm in size. *Twinning:* Polysynthetic, two orthogonal sets of fine lamellae always observed in polished section.

Physical Properties: Hardness = 4.5–5 VHN = 340 (100 g load). D(meas.) = n.d. D(calc.) = 8.48

Optical Properties: Opaque. *Color:* In polished section, white-yellow with a cream tint. *Anisotropism:* Medium strong, in bluish gray to brownish gray.

R_1–R_2: (400) 38.4–45.1, (420) 39.9–46.5, (440) 41.4–47.9, (460) 42.8–49.2, (480) 43.9–50.3, (500) 44.9–51.2, (520) 45.6–51.7, (540) 46.0–51.8, (560) 46.0–51.6, (580) 46.1–51.3, (600) 46.3–51.2, (620) 46.7–51.4, (640) 47.2–51.6, (660) 47.7–51.8, (680) 48.2–52.0, (700) 48.6–52.2

Cell Data: *Space Group:* n.d. $a = 10.42$ $b = 10.60$ $c = 14.43$ Z = 8

X-ray Powder Pattern: Musonoi mine, Zaire.
2.647 (100), 2.600 (80), 1.847 (80), 2.736 (70d), 2.244 (70d), 1.935 (70), 1.903 (70)

Chemistry:

	(1)	(2)
Pd	44.9	44.1
Cu	17.1	16.7
Se	38.8	39.2
Total	100.8	100.0

(1) Musonoi mine, Zaire; by electron microprobe, corresponds to $(Pd_{4.28}Cu_{2.73})_{\Sigma=7.01}Se_{5.00}$.
(2) Do.; corresponds to $(Pd_{4.17}Cu_{2.64})_{\Sigma=6.81}Se_{5.00}$.

Occurrence: In the zone of oxidation.

Association: Trogtalite, selenian digenite, covellite.

Distribution: From the Musonoi Cu-Co mine, Shaba Province, Zaire.

Name: For M.R. Oosterbosch, for many years involved in the development of the Shaba mines.

Type Material: National School of Mines, Paris, France.

References: (1) Johan, Z., P. Picot, R. Pierrot, and T. Verbeek (1970) L'oosterboschite $(Pd, Cu)_7Se_5$, une nouvelle espèce minérale, et la trogtalite cupro-palladifére de Musonoi (Katanga). Bull. Soc. fr. Minéral., 93, 476–481 (in French with English abs.). (2) (1972) Amer. Mineral., 57, 1553 (abs. ref. 1).

Crystal Data: Hexagonal. *Point Group:* n.d. As inclusions in pentlandite.

Physical Properties: Hardness = n.d. VHN = n.d. D(meas.) = 6.5 D(calc.) = 8.50

Optical Properties: Opaque. *Color:* Rose-bronze, browner than nickeline; orange-white in polished section. *Luster:* Metallic. *Pleochroism:* Low, weaker than nickeline. *Anisotropism:* Noticeable.

R: (400) 42.3, (420) 42.9, (440) 43.6, (460) 44.5, (480) 45.5, (500) 46.7, (520) 47.8, (540) 49.1, (560) 50.5, (580) 51.8, (600) 53.3, (620) 54.7, (640) 56.0, (660) 57.2, (680) 58.2, (700) 59.0

Cell Data: *Space Group:* n.d. $a = 6.815(3)$ $c = 12.498(7)$ Z = 6

X-ray Powder Pattern: Tiébaghi massif, New Caledonia.
1.977 (100), 1.918 (100), 2.110 (40), 1.810 (40), 1.737 (40), 1.380 (40), 1.650 (30)

Chemistry:

	(1)	(2)	(3)	(4)
Ni	57.00	64.41	64.3	65.14
Fe			0.2	
As	31.50	35.59	35.1	34.86
S	1.00		0.1	
SiO_2	4.00			
Fe_2O_3	0.85			
MgO	3.80			
H_2O	1.50			
Total	99.65	100.00	99.7	100.00

(1) Tiébaghi massif, New Caledonia; includes 9.15% antigorite as impurity. (2) Do.; Ni and As recalculated to 100.00%. (3) Nebral, Spain; by electron microprobe. (4) $Ni_{5-x}As_2$, x = 0.23.

Occurrence: As inclusions in pentlandite in serpentinized harzburgite.

Association: Pentlandite, heazlewoodite, millerite, parkerite, maucherite, breithauptite, magnetite.

Distribution: In the Tiébaghi massif, New Caledonia. From Table Mountain and Blow-Me-Down Mountain, Newfoundland, and in the Nipissing mine, Cobalt, Ontario, Canada. In the Ronda massif, at Nebral, Málaga Province, Spain. From Beni Bousera, Morocco. At Polling, Salzburg, Austria. From Vourinos, Greece. In the Vozhmin massif, Karelia, USSR.

Name: For Professor Jean Orcél (1896–), French mineralogist.

Type Material: Museum of Natural History, Paris, France.

References: (1) Caillère, S., J. Avias, and J. Falgueirettes (1959) Découverte en Nouvelle Calédonie d'une minéralisation arsénicale sous forme d'un nouvel arséniure de nickel, Ni_2As. Compt. Rend. Acad. Sci., Paris, 249, 1771–1773. (2) (1960) Amer. Mineral., 45, 753–754 (abs. ref. 1). (3) Caillère, S., J. Avias, and J. Falgueirettes (1961) Sur un nouvel arséniure de nickel (Ni_2As). L'orcélite. Bull. Soc. fr. Minéral., 84, 9–12. (4) Lorand, J.P. and M. Pinet (1984) L'orcelite des peridotites de Beni Bousera (Maroc), Ronda (Espagne), Table Mountain et Blow-Me-Down Mountain (Terre-Neuve) et du Pinde Septentrional (Grèce). Can. Mineral., 22, 553–560 (in French with English abs.).

Oregonite Ni_2FeAs_2

Crystal Data: Hexagonal. *Point Group:* n.d. As fine-grained pebbles, having a smooth brown crust; polygonal grains up to 0.5 mm.

Physical Properties: *Tenacity:* Ductile. Hardness = 5 VHN = n.d. D(meas.) = n.d. D(calc.) = 6.92

Optical Properties: Opaque. *Color:* In polished section, white. *Anisotropism:* Weak.
R: (400) 41.7, (420) 43.0, (440) 44.3, (460) 45.3, (480) 46.1, (500) 46.7, (520) 47.2, (540) 48.0, (560) 48.9, (580) 50.1, (600) 51.3, (620) 52.8, (640) 54.2, (660) 55.4, (680) 56.4, (700) 57.2

Cell Data: *Space Group:* n.d. $a = 6.083$ $c = 7.130$ $Z = 3$

X-ray Powder Pattern: Josephine Creek, Oregon, USA.
2.314 (vs), 2.120 (vs), 1.991 (s), 1.789 (s), 1.757 (s), 1.739 (s), 3.571 (m)

Chemistry: Composition established by X-ray fluorescence, analysis not given.

Occurrence: As water-rolled pebbles in clinochlore and serpentine (Oregon, USA).

Association: Copper, bornite, chalcopyrite, molybdenite, chromite (Oregon, USA); pyrrhotite, magnetite (Cyprus).

Distribution: From Josephine Creek, west of Kerby, Josephine Co., Oregon, USA. In the Alexo mine, near Timmins, Ontario, Canada. From near Skouriotissa, Cyprus.

Name: For the state of Oregon, USA.

Type Material: n.d.

References: (1) Ramdohr, P. and M. Schmitt (1959) Oregonit, ein neues Nickel–Eisenarsenid mit metallartigen Eigenschaften. Neues Jahrb. Mineral., Monatsh., 239–247 (in German).
(2) (1960) Amer. Mineral., 45, 1130 (abs. ref. 1). (3) Ramdohr, P. (1969) The ore minerals and their intergrowths, (3rd edition), 399.

Crystal Data: Hexagonal. *Point Group:* n.d. As grains and laths typically 15 x 150 μm.

Physical Properties: *Cleavage:* Good on {100}. *Fracture:* Conchoidal. Hardness = ~3.5 VHN = n.d. D(meas.) = n.d. D(calc.) = 4.212 Weakly magnetic.

Optical Properties: Opaque. *Color:* Brass-yellow. *Streak:* Black. *Luster:* Metallic. *Pleochroism:* Weak, pale yellow to slightly deeper yellow. *Anisotropism:* Strong, from grayish brown to grayish blue.
R_1–R_2: (470) 34.7, (546) 39.9, (589) 42.8, (650) 46.9

Cell Data: *Space Group:* n.d. $a = 3.695(1)$ $c = 6.16(1)$ Z = 4

X-ray Powder Pattern: Near Orick, California, USA.
3.08 (100), 3.20 (90), 1.85 (70), 2.84 (60), 1.73 (55), 1.583 (30), 2.20 (20)

Chemistry:

	(1)	(2)
Na	0.4	0.4
K	0.2	0.2
Cu	31.7	32.7
Fe	31.0	32.0
S	33.6	34.7
Total	96.9	100.0

(1) Near Orick, California, USA; by electron microprobe, average of six grains; 1.5% to 5.1% O qualitatively determined, not included, presumed to be in H_2O. (2) Do.; recalculated to 100%, leading to the composition $Na_xK_yCu_{0.9}Fe_{1.06}S_2 \cdot zH_2O$, x and y = ~0.3, z = ~0.5.

Occurrence: In a mafic alkaline diatreme, in small pegmatitic clots thought to have crystallized late in the consolidation of the Coyote Peak intrusive.

Association: Djerfisherite, rasvumite, bartonite, erdite, coyoteite, phlogopite, schorlomite, acmite, sodalite, cancrinite, pectolite, natrolite, magnetite, calcite.

Distribution: From Coyote Peak, near Orick, Humboldt Co., California, USA.

Name: For the town of Orick, California, USA, near the locality.

Type Material: National Museum of Natural History, Washington, D.C., USA, 150336.

References: (1) Erd, R.C. and G.K. Czamanske (1983) Orickite and coyoteite, two new sulfide minerals from Coyote Peak, Humboldt Co., California. Amer. Mineral., 68, 245–254.

Crystal Data: Monoclinic. *Point Group:* $2/m$. Usually in foliated columnar or fibrous aggregates, with cleavages as much as 35 cm across; sometimes reniform or botryoidal; also granular or powdery. Occasionally as short prismatic crystals to 3 cm. *Twinning:* On {100}.

Physical Properties: *Cleavage:* {010} perfect, {100} imperfect; cleavage lamellae are flexible. *Tenacity:* Sectile. Hardness = 1.5–2 VHN = n.d. D(meas.) = 3.49 D(calc.) = 3.48

Optical Properties: Transparent. *Color:* Lemon-yellow to golden or brownish yellow. *Streak:* Pale lemon-yellow. *Luster:* Resinous, pearly on cleavage surface.

Optical Class: Biaxial negative (–). *Pleochroism:* In reflected light, strong, white to gray-white with reddish tint; in transmitted light, Y = yellow, Z = greenish yellow. *Orientation:* $X = b$; $Z \wedge c = 2°$. *Dispersion:* $r > v$, strong. $\alpha = 2.4$ (Li). $\beta = 2.81$ (Li). $\gamma = 3.02$ (Li). 2V(meas.) = –76° 2V(calc.) = n.d. *Anisotropism:* Barely observable because of strong internal reflections.

R_1–R_2: (400) 29.8–39.9, (420) 27.6–36.8, (440) 25.4–33.7, (460) 23.6–31.2, (480) 22.3–29.5, (500) 21.4–28.4, (520) 20.9–27.7, (540) 20.5–27.2, (560) 20.1–26.7, (580) 19.7–26.4, (600) 19.4–25.9, (620) 19.1–25.6, (640) 19.0–25.4, (660) 18.8–25.2, (680) 18.7–25.1, (700) 18.6–25.0

Cell Data: *Space Group:* $P2_1/n$. $a = 11.475(5)$ $b = 9.577(4)$ $c = 4.256(2)$ $\beta = 90°41(5)'$ Z = 4

X-ray Powder Pattern: Baia Sprie (Felsőbánya), Romania.
4.85 (100), 4.02 (50), 2.47 (40), 1.755 (40), 3.22 (30), 2.79 (30), 2.72 (30)

Chemistry: Stated to be very near As_2S_3.

Occurrence: In low-temperature hydrothermal veins, hot springs and fumaroles; also commonly as an alteration product of arsenic minerals, especially realgar.

Association: Stibnite, realgar, arsenic, calcite, barite, gypsum.

Distribution: Not uncommon in small amounts, but rare in fine specimens. In the USA, crystallized from Mercur, Tooele Co., Utah; at the Getchell mine, Humboldt Co., and from the White Caps mine, Manhattan, Nye Co., Nevada. In Yugoslavia, from Křeševo, Bosnia, and at Allchar, Macedonia. From Tajov, Czechoslovakia. Fine crystals from the Zarehchuran mine, Takab, Iran. From Jelamerk, Kurdistan, Turkey. In the USSR, from Racha Luyumi, Georgia, and Loukhoumi, Caucasus Mountains. From Shinen, Hunan Province, China. Exceptional specimens from the Quiruvilca mine, La Libertad, Peru.

Name: From the Latin *auripigmentum, golden paint*, in allusion to the color.

References: (1) Palache, C., H. Berman, and C. Frondel (1944) Dana's system of mineralogy, (7th edition), v. I, 266–269. (2) Berry, L.G. and R.M. Thompson (1962) X-ray powder data for the ore minerals. Geol. Soc. Amer. Mem. 85, 80–81. (3) Mullen, D.J.E. and W. Nowacki (1972) Refinement of the crystal structures of realgar, AsS and orpiment, As_2S_3. Zeits. Krist., 136, 48–65.

Crystal Data: Monoclinic. *Point Group:* n.d. As polycrystalline intergrowths.

Physical Properties: Hardness = n.d. VHN = n.d. D(meas.) = n.d. D(calc.) = 8.44

Optical Properties: Opaque. *Color:* Gray in reflected light. *Luster:* Metallic.
Anisotropism: Weak but distinct in air.
R_1–R_2: n.d.

Cell Data: *Space Group:* n.d. $a = 5.933$ $b = 5.916$ $c = 6.009$ $\beta = 112°21(02)'$ Z = 4

X-ray Powder Pattern: Gold Bluff, California, USA.
3.79 (100), 1.892 (100), 1.870 (80), 2.74 (70), 2.78 (60), 2.01 (60), 1.832 (60)

Chemistry:

	(1)	(2)
Os	35.6	37.6
Ru	18.1	10.3
Ir	2.0	0.74
Pd	0.6	
Pt	0.4	
Rh	0.2	
Ni	0.9	3.2
Fe		2.5
Co		2.0
As	30.6	32.6
S	11.5	13.2
Total	99.9	102.14

(1) Gold Bluff, California, USA; by electron microprobe, corresponding to $(Os_{0.48}Ru_{0.46}Ni_{0.04}Ir_{0.03})_{\Sigma=1.01}As_{1.05}S_{0.92}$. (2) Kola Peninsula, USSR; by electron microprobe, corresponding to $(Os_{0.46}Ru_{0.24}Fe_{0.10}Co_{0.08})_{\Sigma=0.88}As_{1.02}S_{0.96}$.

Occurrence: In a platinum-bearing sample of placer sand (Gold Bluff, California, USA).

Association: Irarsite, ruthenarsenite, sperrylite, iridarsenite, osmiridium, anduoite, laurite, ruarsite.

Distribution: From Gold Bluff, Humboldt Co., California, USA. In the USSR, from near Zlatoust, Ural Mountains, and from the Kola Peninsula. From the Witwatersrand, Transvaal, South Africa. At Anduo, Tibet, China. From Vourinos, Greece.

Name: For the content of OSmium and ARSenic.

Type Material: National Museum of Natural History, Washington, D.C., USA, 123218.

References: (1) Snetsinger, K.G. (1972) Osarsite, a new osmium–ruthenium sulfarsenide from California. Amer. Mineral., 57, 1029–1036. (2) Cabri, L.J., Ed. (1981) Platinum group elements: mineralogy, geology, recovery. Can. Inst. Min. & Met., 123, 160.

Crystal Data: Cubic. *Point Group:* $4/m\ \overline{3}\ 2/m$. Crystals tabular on {0001}; as exsolutions in Pt–Fe alloys.

Physical Properties: Hardness = 6–7 VHN = 642–782 (100 g load). D(meas.) = 20.98 D(calc.) = 22.20

Optical Properties: Opaque. *Color:* Tin-white to steel-gray; in polished section, white. *Streak:* Gray. *Luster:* Metallic.
R: (400) — , (420) 68.1, (440) 68.8, (460) 69.3, (480) 69.6, (500) 69.9, (520) 70.2, (540) 70.7, (560) 71.6, (580) 71.9, (600) 72.5, (620) 73.2, (640) 73.4, (660) 73.8, (680) 74.4, (700) 74.4

Cell Data: *Space Group:* $Fm3m$. $a = 3.88$ Z = 4

X-ray Powder Pattern: Nevyansk, USSR. (JCPDS 31-608).
2.23 (100), 1.17 (70), 1.93 (50) 1.37 (40), 1.12 (20)

Chemistry:

	(1)	(2)
Ir	69.95	55.24
Os	17.25	27.32
Rh	11.25	1.51
Ru		5.85
Pt	0.05	10.08
Cu	trace	trace
Fe	trace	trace
Total	98.50	100.00

(1) Borneo. (2) Nizhni Tagil, USSR.

Occurrence: Formed in magmatic segregation deposits and subsequently concentrated into placers during weathering.

Association: Pt–Fe alloys, Os–Ir–Ru alloys, laurite, chromite.

Distribution: In the USSR, from Nevyansk and Nizhni Tagil, Ural Mountains. On the Riam Kanan River, Borneo. In the Sorashigawa placer, Japan. In Canada, from the Atlin, Cariboo, and Spruce Creek districts, British Columbia. From Snow Gulch, Goodnews Bay, Alaska, USA. At Heazlewood, Tasmania, Australia. From the Witwatersrand, Transvaal, South Africa. At Anduo, Tibet, China. From the Tiébaghi massif and Massif du Sud, New Caledonia. At Vourinos, Greece.

Name: From the composition, osmian iridium.

Type Material: n.d.

References: (1) Palache, C., H. Berman, and C. Frondel (1944) Dana's system of mineralogy, (7th edition), v. I, 111–114. (2) Cabri, L.J., Ed. (1981) Platinum group elements: mineralogy, geology, recovery. Can. Inst. Min. & Met., 123–124.

Crystal Data: Hexagonal. *Point Group:* $6/m$ $2/m$ $2/m$. Usually as euhedral prismatic inclusions in Pt–Fe alloys.

Physical Properties: Hardness = n.d. VHN = 1206–1246 (25 g load). D(meas.) = 22.48 D(calc.) = 22.59

Optical Properties: Opaque. *Color:* White with a bluish gray tinge in reflected light. *Pleochroism:* Noticeable. *Anisotropism:* Strong, reddish orange.
R_1–R_2: (400) 66.8–70.9, (420) 66.4–69.6, (440) 66.0–68.3, (460) 64.2–66.0, (480) 60.7–62.7, (500) 58.0–60.0, (520) 56.1–58.2, (540) 54.5–56.6, (560) 53.4–55.3, (580) 52.8–54.1, (600) 52.6–53.0, (620) 53.0–52.1, (640) 54.2–51.2, (660) 56.0–50.5, (680) 58.0–49.7, (700) 60.4–49.0

Cell Data: *Space Group:* $P6_3/mmc$. $a = 2.7341$ $c = 4.3197$ Z = [2]

X-ray Powder Pattern: Synthetic.
2.076 (100), 2.367 (35), 2.160 (35), 1.3668 (20), 1.2300 (20), 1.1551 (20),1.595 (18)

Chemistry:

	(1)	(2)	(3)
Os	94.1	84.8	98.8
Ir	5.4	12.0	0.3
Ru	0.8	2.2	
Pt	0.1	1.7	
Rh		0.5	
Pd		0.1	0.2
Total	100.4	101.3	99.3

(1) Atlin, Canada; by electron microprobe. (2) Joubdo stream, Ethiopia; by electron microprobe. (3) Gusevogorskii pluton, USSR; by electron microprobe, average of four analyses.

Occurrence: With other platinum-group elements and alloys, in ultramafic rocks and placers derived therefrom.

Association: Rutheniridosmine, iridosmine, osmiridium (Ruby Creek, Canada); bowieite, platinum, Ir–Pt alloys, laurite, silicate inclusions (Salmon River, Alaska, USA); isoferroplatinum, cuprorhodsite, malanite, cuproiridsite, iridosmine, laurite, erlichmanite, cooperite, sperrylite, chalcopyrite, bornite (USSR).

Distribution: On Ruby Creek, Atlin, British Columbia, Canada. In the USA, from Fox Gulch and the Salmon River, Goodnews Bay, Alaska. From the Gusevogorskii pluton, Ural Mountains, and the Aldan Shield and Kamchatka, eastern USSR. In Ethiopia, from the Joubdo stream, on the Birbir river. From the Witwatersrand, Transvaal, South Africa. At Anduo, Tibet, China.

Name: From the Greek word for *odor*, in reference to the pungent and irritating odor when heated in air.

Type Material: n.d.

References: (1) Harris, D.C. and L.J. Cabri (1973) The nomenclature of the natural alloys of osmium, iridium and ruthenium based on new compositional data of alloys from world-wide occurrences. Can. Mineral., 12, 104–112. (2) (1975) Amer. Mineral., 60, 946 (abs. ref. 1). (3) Begizov, V.D., L.F. Borisenko, and Y.D. Uskov (1975) Sulfides and natural solid solutions of platinum metals from ultramafic rocks of the Gusevogorskii pluton, Urals. Doklady Acad. Nauk SSSR, 225, 1408–1411. (4) (1955) NBS Circ. 539, 4, 8. (5) Cabri, L.J., Ed. (1981) Platinum group elements: mineralogy, geology, recovery. Can. Inst. Min. & Met., 124–125.

Crystal Data: Orthorhombic. *Point Group:* $2/m$ $2/m$ $2/m$. As small laths. *Twinning:* Commonly twinned.

Physical Properties: Hardness = ~2 VHN = n.d. D(meas.) = 4.87 (synthetic). D(calc.) = 4.806

Optical Properties: Opaque. *Color:* In polished section, gray with orange-brown internal reflections. *Luster:* Metallic. *Pleochroism:* Weak. *Anisotropism:* Strong.
R_1–R_2: n.d.

Cell Data: *Space Group:* *Pnam*. $a = 8.79(3)$ $b = 14.02(5)$ $c = 3.74(1)$ Z = 4

X-ray Powder Pattern: Synthetic.
4.131 (100), 5.495 (75), 2.670 (45), 7.003 (40), 2.747 (35), 3.740 (35), 3.257 (30)

Chemistry: Composition determined by identity of X-ray powder pattern with synthetic material.

Occurrence: In zones of oxidation or secondary enrichment in hydrothermal tin deposits.

Association: Stannite, cassiterite, herzenbergite, berndtite.

Distribution: From the Cerro Rico de Potosí, Bolivia. At the Stiepelmann mine, near Arandis, Namibia. In the Maria Teresa mine, near Huari, between Oruro and Uyuni, Bolivia.

Name: For J. Ottemann, German mineralogist, Heidelberg, Germany.

Type Material: National Museum of Natural History, Washington, D.C., USA, 114486.

References: (1) Moh, G.H. and F. Berndt (1964) Two new natural tin sulfides, Sn_2S_3 and SnS_2. Neues Jahrb. Mineral., Monatsh., 94–95. (2) (1965) Amer. Mineral., 50, 2107 (abs. ref. 1). (3) Moh, G.H. (1966) Das binäre System Zinn–Schwefel und seine Minerale (abs.). Fortschr. Mineral., 42, 211. (4) (1966) Amer. Mineral., 51, 1551 (abs. ref. 3). (5) Mosburg, S., D.R. Ross, P.M. Bethke, and P. Toulmin (1961) X-ray powder data for herzenbergite, teallite, and tin trisulfide. U.S. Geol. Sur. Prof. Paper 424-C, C347–C348. (6) Kniep, R., D. Mootz, U. Severin, and H. Wunderlich (1982) Structure of tin(II)tin(IV) trisulfide, a redetermination. Acta Cryst., 38, 2022-2023.

Crystal Data: Orthorhombic. *Point Group:* $2/m\ 2/m\ 2/m$ or $mm2$. As irregular laths less than 0.1 mm long.

Physical Properties: Hardness = n.d. VHN = n.d. D(meas.) = n.d. D(calc.) = [7.18]

Optical Properties: Opaque. *Color:* In polished section, galena-white. *Pleochroism:* Weak in oil. *Anisotropism:* Distinct to strong, light gray to bluish black.
R_1–R_2: n.d.

Cell Data: *Space Group:* $Bbmm$ or $Bb2_1m$. $a = 13.457(1)$ $b = 44.042(4)$ $c = 4.100(10)$ Z = 4

X-ray Powder Pattern: Old Lout mine, Colorado, USA.
3.43 (100), 2.96 (90), 2.09 (90), 2.04 (70), 1.79 (70), 3.33 (60), 2.85 (60)

Chemistry:

	(1)	(2)
Ag	12.5	12.38
Cu	0.5	
Pb	29.5	31.70
Bi	41.4	39.97
Sb	0.2	15.95
S	16.0	
Total	100.1	100.00

(1) Old Lout mine, Colorado, USA; by electron microprobe, corresponding to $Ag_{3.04}Cu_{0.21}Pb_{3.73}Bi_{5.19}S_{13.07}$. (2) $Ag_3Pb_4Bi_5S_{13}$.

Occurrence: In a hydrothermal sulfide vein (Old Lout mine, Colorado, USA); with topaz and fluorite in a cryolite body (Ivigtut, Greenland); with base-metal sulfides in diopside tactite (Pitiquito, Mexico).

Association: Galena, matildite (Old Lout mine, Colorado, USA); berryite, aikinite, galena. matildite, pyrite, bismuth, gold (Ivigtut, Greenland); sphalerite, galena, chalcopyrite (Pitiquito, Mexico).

Distribution: In the Old Lout mine, Poughkeepsie Gulch, near Ouray, San Juan Co., Colorado, and South Mountain, Owyhee Co., Idaho, USA. From Ivigtut, southern Greenland. At a tungsten prospect, 40 km south of Pitiquito, Sonora, Mexico.

Name: For Ouray, Colorado, USA, near where the mineral was first discovered.

Type Material: Royal Ontario Museum, Toronto, Canada, M4100.

References: (1) Karup-Møller, S. (1977) Mineralogy of some Ag–(Cu)–Pb–Bi sulfide associations. Bull. Geol. Soc. Denmark, 26, 41–68. (2) Makovicky, E. and S. Karup-Møller (1977) Chemistry and crystallography of the lillianite homologous series. Neues Jahrb. Mineral., Abh., 131, 56–82. (3) (1979) Amer. Mineral., 64, 243 (abs. refs. 1 and 2). (4) Makovicky, E. and S. Karup-Møller (1984) Ourayite from Ivigtut, Greenland. Can. Mineral., 22, 565–575.

Owyheeite $Pb_{10-2x}Ag_{3+x}Sb_{11+x}S_{28}$ (x = –0.13 to 0.20)

Crystal Data: Orthorhombic. *Point Group:* $2/m\ 2/m\ 2/m$. Massive to coarsely fibrous, and felted hair-like crystals lining fractures. Needles are striated longitudinally ∥ [001].

Physical Properties: *Cleavage:* {001}. *Tenacity:* Very brittle. Hardness = 2.5 VHN = n.d. D(meas.) = 6.22–6.51 D(calc.) = 6.43

Optical Properties: Opaque. *Color:* Light silvery gray, tarnishes blue or yellow; gray-white in reflected light. *Streak:* Reddish brown. *Luster:* Metallic. *Pleochroism:* Distinct. *Anisotropism:* Strong, yellowish white and gray.

R_1–R_2: (400) 41.5–46.0, (420) 40.9–45.9, (440) 40.3–45.8, (460) 39.7–45.6, (480) 39.2–45.5, (500) 38.8–45.2, (520) 38.4–45.0, (540) 38.0–44.6, (560) 37.7–44.2, (580) 37.4–43.7, (600) 37.1–43.2, (620) 36.7–42.5, (640) 36.3–41.9, (660) 35.8–41.1, (680) 35.3–40.3, (700) 34.7–39.5

Cell Data: *Space Group:* *Pnam.* $a = 22.82$ $b = 27.20$ $c = 8.19$ Z = 8

X-ray Powder Pattern: Poorman mine, Idaho, USA.
3.25 (100), 3.49 (70), 2.84 (60), 2.05 (60), 2.90 (50), 2.23 (50), 3.37 (40)

Chemistry:

	(1)	(2)
Pb	40.77	43.86
Ag	7.40	6.14
Cu	0.75	1.55
Fe	0.46	0.05
Sb	30.61	29.26
S	20.81	19.06
Total	100.80	99.92

(1) Poorman mine, Idaho, USA. (2) Sheba mine, Nevada, USA.

Occurrence: In veins of hydrothermal origin.

Association: Galena, sphalerite, pyrite, arsenopyrite, chalcopyrite, tetrahedrite, pyrargyrite, diaphorite, miargyrite, jamesonite, boulangerite, ramdohrite, andorite, meneghinite.

Distribution: In the USA, at the Poorman mine, Silver City district, Owyhee Co., Idaho; the Banner mine, Boise Co., Idaho; from the Sheba mine, Star City, and from Rochester and the Morey district, Nye Co., Nevada; and the Domingo and Garfield mines, Gunnison Co., Colorado. From the Alma property and Rambler mine, Slocan mining division, British Columbia; and the Tintina silver mines, Yukon Territory, Canada. At the Wongabah mine, Drake and Rivertree mining fields, New South Wales, and at the Meerschaum mine, near Omeo, Victoria, Australia. From Kutná Hora, Czechoslovakia. At Bourneix, Haute-Vienne, France. At Roc-Blanc, Morocco. From near Yecora, Sonora, Mexico. At Rajpura Dariba, Rajasthan, India. Known from a few other localities.

Name: For the locality in Owyhee Co., Idaho, USA.

Type Material: National Museum of Natural History, Washington, D.C., USA, 94054.

References: (1) Robinson, S.C. (1949) Owyheeite. Amer. Mineral., 34, 398–402. (2) Moëlo, Y., N. Mozgova, P. Picot, N. Bortnikov, and Z. Vrubleskaya (1984) Cristallochimie de l'owyheeite: nouvelles données. Tschermaks Mineral. Petrog. Mitt., 32, 271–284 (in French with English abs.). (3) (1985) Amer. Mineral., 70, 440 (abs. ref. 2).

Crystal Data: Monoclinic. *Point Group:* n.d. As irregular grains up to 0.4 mm. *Twinning:* Finely spaced polysynthetic twinning is present.

Physical Properties: *Cleavage:* In one direction. *Tenacity:* Brittle. Hardness = ~2 VHN = 66–87, 77 average. D(meas.) = n.d. D(calc.) = 5.21

Optical Properties: Opaque. *Color:* Dark gray; light pale gray in reflected light, with bright red internal reflections. *Streak:* Gray with a brownish tint. *Luster:* Metallic. *Pleochroism:* Weak. *Anisotropism:* Strong.

R_1–R_2: (400) — , (420) — , (440) 40.9–52.1, (460) 39.8–51.7, (480) 38.7–50.8, (500) 38.0–49.8, (520) 37.5–48.7, (540) 37.2–47.8, (560) 36.9–47.0, (580) 36.7–46.3, (600) 36.5–45.8, (620) 36.5–45.5, (640) 36.4–45.3, (660) 36.2–45.1, (680) 36.2–45.4, (700) 36.0–45.3

Cell Data: *Space Group:* n.d. $a = 5.372(7)$ $b = 3.975(7)$ $c = 11.41(1)$ $\beta = 89.71(15)°$ Z = 2

X-ray Powder Pattern: Kalliosalo deposit, Finland.
2.87 (100), 2.68 (60), 3.90 (40), 3.13 (40), 2.27 (30), 1.750 (30), 2.08 (20)

Chemistry:

	(1)	(2)
Sb	66.9	63.65
As	18.6	19.59
S	15.5	16.76
Total	101.0	100.00

(1) Kalliosalo deposit, Finland; by electron microprobe, average of five analyses, corresponding to $Sb_{2.14}As_{0.97}S_{1.89}$. (2) Sb_2AsS_2.

Occurrence: n.d.

Association: Arsenopyrite, arsenic, löllingite, stibnite.

Distribution: In the Kalliosalo deposit, Seinäjoki region, Finland.

Name: To honor Viekko Pääkkönen, who studied the ore deposits of the type region.

Type Material: A.E. Fersman Mineralogical Museum, Academy of Sciences, Moscow, USSR.

References: (1) Borodaev, Y.S., N.N. Mozgova, N.A. Ozerova, N.S. Bortnikov, P. Oivanen, and V. Iletuinen (1981) Pääkkönenite, Sb_2AsS_2, a new mineral from the Seinäjoki ore region in Finland. Zap. Vses. Mineral. Obshch., 110, 480–487 (in Russian). (2) (1982) Amer. Mineral., 67, 858 (abs. ref. 1). (3) (1982) Mineral. Abs., 33, 169 (abs. ref. 1).

Crystal Data: Monoclinic. *Point Group:* $2/m$. Plate-like fragments, in granular to parallel fibrous aggregates to several mm diameter.

Physical Properties: *Fracture:* Uneven. Hardness = n.d. VHN = n.d. D(meas.) = n.d. D(calc.) = 6.91

Optical Properties: Opaque. *Color:* Steel-gray fresh fractures, turning brown to black; in polished section creamy white. *Anisotropism:* Moderate.
R_1–R_2: n.d.

Cell Data: *Space Group:* $P2_1/m$. $a = 28.44$ $b = 3.95$ $c = 17.55$ $\beta = 106.1°$ Z = 2

X-ray Powder Pattern: Calculated from structure.
3.06 (100), 3.63 (74), 3.21 (61), 2.85 (43), 2.66 (38), 3.18 (34), 2.19 (30)

Chemistry:

	(1)	(2)
Ag	3.26	2.76
Pb	6.71	10.60
Cu	9.45	9.76
Bi	61.40	58.83
S	18.90	18.05
Total	99.72	100.00

(1) Băiţa, Romania; by electron microprobe, average of six analyses on two grains, leading to $Ag_{1.1}Pb_{1.2}Cu_{5.5}Bi_{11.0}S_{22}$. (2) $AgPb_2Cu_6Bi_{11}S_{22}$.

Occurrence: Intimately intergrown with other Pb-Bi sulfosalts.

Association: Hammarite, pekoite, bismuthinite, other Pb-Bi minerals, chalcopyrite, grossular, andradite.

Distribution: From Băiţa (Rézbánya), Romania.

Name: For Dr. K. Paděra, Charles University, Prague, Czechoslovakia, who first worked on the mineral.

Type Material: Mineralogical Department, Charles University, Prague, Czechoslovakia, 11329.

References: (1) Mumme, W.G. and L. Žák (1985) Paděraite, $Cu_{5.9}Ag_{1.3}Pb_{1.6}Bi_{11.2}S_{22}$, a new mineral of the cuprobismutite–hodrushite group. Neues Jahrb. Mineral., Monatsh., 557–567.
(2) Mumme, W.G. (1986) The crystal structure of paderaite, a mineral of the cuprobismutite series. Can. Mineral., 24, 513–521.

Crystal Data: Hexagonal. *Point Group:* n.d. As distinct hexagonal prisms and minute elongated anhedral grains with rectangular outlines, rarely as sinuous grains.

Physical Properties: *Cleavage:* {0001} perfect. *Tenacity:* Brittle. Hardness = n.d. VHN = 470(30) (50 g load). D(meas.) = n.d. D(calc.) = 10.27

Optical Properties: Opaque. *Color:* Steel-gray; grayish white with slight rose tint in reflected light. *Luster:* Metallic. *Pleochroism:* Distinct in oil. *Anisotropism:* Slight, dark gray to brownish gray.

R_1–R_2: (400) — , (420) 46.0–49.2, (440) 47.1–50.1, (460) 48.1–51.1, (480) 49.7–51.9, (500) 50.8–52.8, (520) 52.1–53.5, (540) 53.2–54.5, (560) 54.2–55.4, (580) 55.1–56.4, (600) 56.3–57.4, (620) 57.2–58.3, (640) 58.5–59.3, (660) 59.3–60.1, (680) 59.9–60.6, (700) 60.0–61.0

Cell Data: *Space Group:* n.d. $a = 6.784(5)$ $c = 14.80(1)$ Z = [3]

X-ray Powder Pattern: Talnakh area, USSR.
2.22 (100), 2.50 (40), 1.986 (40), 1.784 (30), 1.187 (30), 2.46 (20), 2.20 (20)

Chemistry:

	(1)
Pd	64.4 – 66.5
Pt	4.0 – 5.3
Au	0.0 – 1.9
Cu	0.0 – 0.5
Pb	2.2 – 7.9
Sn	11.3 – 14.9
As	6.9 – 7.5
Sb	0.8 – 1.9
Bi	0.0 – 1.6
Total	

(1) Talnakh area, USSR; by electron microprobe, ranges of four analyses.

Occurrence: Intergrown with other minerals in massive ores of cubanite, chalcopyrite and talnakhite.

Association: Rustenburgite–atokite, Pt–Fe alloy, polarite, sperrylite, majakite, Au–Ag–Cu alloys.

Distribution: From the Talnakh area, Noril'sk region, western Siberia, USSR.

Name: For the principal elements, PALladium, ARsenic and tin, STANnum.

Type Material: A.E. Fersman Mineralogical Museum, Academy of Sciences, Moscow, USSR.

References: (1) Begizov, V.D., E.M. Zav'yalov, and E.G. Pavlov (1981) Palarstanide, $Pd_8(Sn, As)_3$, a new mineral. Zap. Vses. Mineral. Obshch., 110, 487–492 (in Russian). (2) (1982) Amer. Mineral., 67, 858–859 (abs. ref. 1).

Crystal Data: Cubic. *Point Group:* $4/m\ \overline{3}\ 2/m$. Rarely as octahedra; commonly in grains, sometimes with radial fibrous texture.

Physical Properties: *Tenacity:* Ductile and malleable. Hardness = 4.5–5 VHN = n.d. D(meas.) = 11.9 D(calc.) = 12.04

Optical Properties: Opaque. *Color:* Whitish steel-gray. *Luster:* Metallic.
R: (400) 60.6, (420) 62.0, (440) 63.4, (460) 64.6, (480) 65.6, (500) 66.4, (520) 67.2, (540) 67.8, (560) 68.3, (580) 68.9, (600) 69.5, (620) 70.2, (640) 70.9, (660) 71.7, (680) 72.5, (700) 73.4

Cell Data: *Space Group:* $Fm3m$ (synthetic). $a = 3.8898$ Z = 4

X-ray Powder Pattern: Potaro River, Guyana.
2.259 (100), 1.184 (60), 1.957 (50), 1.387 (40), 1.132 (20), 0.9824 (10), 0.9019 (4B)

Chemistry:

	(1)
Pd	77.2
Hg	22.6
Total	99.8

(1) Potaro River, Guyana; by electron microprobe.

Occurrence: As an oxidation product of palladium-bearing sulfides and as a primary phase in platinum deposits.

Association: Pt–Fe alloys, Pd–Hg minerals, gold, Au–Ag alloy, lead, sobolevskite.

Distribution: In the Department of Chocó, Cauca, Colombia. From Itabira, Minas Gerais, Brazil. In the USSR, in the Ural Mountains. In the Lubin copper mine, Zechstein, Poland. From the Transvaal, South Africa. In Guyana, on the Potaro River.

Name: For the planetoid *Pallas*.

Type Material: n.d.

References: (1) Palache, C., H. Berman, and C. Frondel (1944) Dana's system of mineralogy, (7th edition), v. I, 109–110. (2) Ramdohr, P. (1969) The ore minerals and their intergrowths, (3rd edition), 342. (3) Cabri, L.J., Ed. (1981) Platinum group elements: mineralogy, geology, recovery. Can. Inst. Min. & Met., 125–126, 160.

Crystal Data: Monoclinic. *Point Group:* $2/m$. As small (0.005–0.4 mm) irregular grains in chalcopyrite, long and veinlet-like, vermiform, with curving boundaries.

Physical Properties: *Cleavage:* Perfect in two directions. *Tenacity:* Brittle. Hardness = n.d. VHN = 277–357, 326 average (20 g load). D(meas.) = n.d. D(calc.) = 10.42

Optical Properties: Opaque. *Color:* Steel-gray; grayish white with slight rose tint in polished section. *Luster:* Metallic. *Anisotropism:* Moderate, dark gray with bluish tint to red-brown in air, dark gray to brownish gray with red tint in oil.
R_1–R_2: (400) 44.2–45.1, (420) 46.8–46.8, (440) 48.4–48.8, (460) 49.4–50.3, (480) 50.2–51.6, (500) 51.2–53.1, (520) 51.9–54.1, (540) 52.5–54.8, (560) 53.3–55.4, (580) 53.9–55.8, (600) 54.6–56.3, (620) 55.4–57.1, (640) 56.0–57.6, (660) 56.6–58.1, (680) 57.1–58.6, (700) 57.8–59.2

Cell Data: *Space Group:* $P2/m$. $a = 9.25(1)$ $b = 8.47(2)$ $c = 10.44(2)$ $\beta = 94.0°$ Z = 18

X-ray Powder Pattern: Oktyabr mine, USSR.
2.14 (10), 2.21 (9), 2.60 (7), 1.955 (7), 2.35 (6), 2.31 (6), 2.13 (5)

Chemistry:

	(1)	(2)	(3)	(4)
Pd	67.8	74.0	73.4	73.96
Pt		0.10	0.43	
Au	1.5	0.11	n.d.	
Ag	3.5	n.d.	n.d.	
Ni		n.d.	0.23	
Cu		0.34	n.d.	
Sb			0.04	
As	26.0	24.4	24.2	26.04
Te		1.3	0.54	
Total	98.8	100.25	98.84	100.00

(1) Oktyabr mine, USSR; by electron microprobe, corresponding to $(Pd_{1.87}Ag_{0.09}Au_{0.02})_{\Sigma=1.98}As_{1.02}$. (2) Stillwater Complex, Montana, USA; by electron microprobe, corresponding to $(Pd_{2.01}Cu_{0.02})_{\Sigma=2.03}(As_{0.94}Te_{0.03})_{\Sigma=0.97}$. (3) Lac des Iles Complex, Canada; by electron microprobe, leading to $(Pd_{2.02}Pt_{0.01}Ni_{0.01})_{\Sigma=2.04}(As_{0.95}Te_{0.01})_{\Sigma=0.96}$. (4) Pd_2As.

Occurrence: As minute grains in chalcopyrite that occurs in veins along the footwall of a Cu-Ni deposit (Talnakh area, USSR); from heavy-mineral concentrates (Stillwater Complex, Montana, USA).

Association: Chalcopyrite, sperrylite, gold, kotulskite.

Distribution: In the Oktyabr mine, Talnakh area, Noril'sk district, western Siberia, USSR. In the USA, from the Banded and Upper Zones of the Stillwater Complex, Montana. In Canada, at the Lac des Iles Complex, Ontario.

Name: For its chemical composition.

Type Material: A.E. Fersman Mineralogical Museum, Academy of Sciences, Moscow, USSR.

References: (1) Begizov, V.D., V.I. Meschankina, and L.S. Dubakina (1974) Palladoarsenide, Pd_2As, a new natural palladium arsenide from the copper–nickel deposits of the Oktyabr deposits. Zap. Vses. Mineral. Obshch., 103, 104–107 (in Russian). (2) (1975) Amer. Mineral., 60, 162 (abs. ref. 1). (3) Cabri, L.J., J.H.G. Laflamme, J.M. Stewart, J.F. Rowland, and T.T. Chen (1975) New data on some palladium arsenides and antimonides. Can. Mineral., 13, 321–335.

Crystal Data: Orthorhombic. *Point Group:* $2/m\ 2/m\ 2/m$ or $mm2$. As irregular grains, approximately 80 x 95 to 135 x 165 μm.

Physical Properties: Hardness = n.d. VHN = 429–483, 450 average (25 g load). D(meas.) = n.d. D(calc.) = 10.8

Optical Properties: Opaque. *Color:* In polished section, cream colored. *Luster:* Metallic. *Anisotropism:* Weak to distinct, in air from gray to extinction; under oil varying from gray to brownish gray or brown at extinction.

R_1–R_2: (470) 53.6–54.5, (546) 52.1–53.0, (589) 53.6–54.6, (650) 55.8–56.1

Cell Data: *Space Group:* $Pmcn$ or $P2_1cn$. $a = 7.504(4)$ $b = 18.884(1)$ $c = 6.841(7)$ Z = 20

X-ray Powder Pattern: Stillwater Complex, Montana, USA.
2.224 (100), 2.505 (90), 2.089 (60), 2.596 (40), 1.880 (40), 2.380 (30), 1.211 (30)

Chemistry:

	(1)
Pd	67.3
As	19.1
Bi	13.4
Total	99.8

(1) Stillwater Complex, Montana, USA; by electron microprobe, corresponding to $Pd_{1.99}(As_{0.81}Bi_{0.20})_{\Sigma=1.01}$.

Occurrence: From heavy mineral concentrates (Stillwater Complex, Montana, USA).

Association: Palladoarsenide, calcite, an undetermined (Pd,Te,Bi) mineral.

Distribution: In the Banded and Upper Zones, Stillwater Complex, Montana, USA.

Name: For the composition, palladium, bismuth, and arsenic.

Type Material: Royal Ontario Museum, Toronto, Canada, M34218; National Museum of Natural History, Washington, D.C., USA, 135407.

References: (1) Cabri, L.J., T.T. Chen, J.W. Stewart, and J.H.G. Laflamme (1976) Two new palladium–arsenic–bismuth minerals from the Stillwater Complex, Montana. Can. Mineral., 14, 410–413. (2) Cabri, L.J., Ed. (1981) Platinum group elements: mineralogy, geology, recovery. Can. Inst. Min. & Met., 126–127.

Crystal Data: Cubic. *Point Group:* $4/m\ \overline{3}\ 2/m$. As grains up to 0.5 mm.

Physical Properties: Hardness = n.d. VHN = 390–437, 414 average (100 g load). D(meas.) = 8.30 (synthetic). D(calc.) = 8.15

Optical Properties: Opaque. *Color:* In reflected light, white in air, light gray in oil.
R: (400) 42.4, (420) 42.8, (440) 43.3, (460) 43.8, (480) 44.3, (500) 44.8, (520) 45.4, (540) 45.7, (560) 45.8, (580) 45.7, (600) 45.6, (620) 45.6, (640) 45.6, (660) 45.7, (680) 45.8, (700) 46.2

Cell Data: *Space Group:* $Pm3m$. $a = 10.635$ Z = 2

X-ray Powder Pattern: Itabira, Brazil.
1.887 (vvs), 2.832 (vs), 2.571 (s), 2.430 (s), 2.040 (s), 1.723 (s)

Chemistry:

	(1)	(2)
Pd	55.77	60.43
Cu	3.99	
Hg	1.66	
Se	38.59	39.57
Total	100.01	100.00

(1) Itabira, Brazil; by electron microprobe, corresponding to $(Pd_{15.47}Cu_{1.85}Hg_{0.24})_{\Sigma=17.56}Se_{14.43}$.
(2) $Pd_{17}Se_{15}$.

Occurrence: Found sparingly in residual concentrates from gold washings.

Association: Arsenopalladinite, isomertieite, atheneite.

Distribution: From Itabira, Minas Gerais, Brazil.

Name: For the chemical composition, PALLADium and SElenium.

Type Material: British Museum (Natural History), London, England.

References: (1) Davis, R.J., A.M. Clark, and A.J. Criddle (1977) Palladseïte, a new mineral from Itabira, Minas Gerais, Brazil. Mineral. Mag., 41, 123. (2) (1977) Amer. Mineral., 62, 1059 (abs. ref. 1).

Paolovite Pd_2Sn

Crystal Data: Orthorhombic. *Point Group:* $2/m$ $2/m$ $2/m$. As irregular grains. *Twinning:* Polysynthetic.

Physical Properties: Hardness = n.d. VHN = 380 (50 g load). D(meas.) = n.d. D(calc.) = 11.08

Optical Properties: Opaque. *Color:* Lilac-rose. *Luster:* Metallic. *Pleochroism:* Dark lilac-rose to pale rose.
R_1–R_2: (400) — , (420) 44.4–46.5, (440) 44.6–47.2, (460) 45.0–48.2, (480) 45.5–49.1, (500) 45.9–50.3, (520) 46.7–51.6, (540) 47.7–52.8, (560) 49.0–54.1, (580) 50.9–55.4, (600) 52.7–56.7, (620) 55.3–57.7, (640) 57.7–59.2, (660) 59.8–59.5, (680) 61.5–60.3, (700) 62.7–60.6

Cell Data: *Space Group:* *Pbnm.* $a = 8.11(1)$ $b = 5.662(6)$ $c = 4.324(2)$ Z = 4

X-ray Powder Pattern: Oktyabr mine, USSR.
2.28 (100), 2.16 (70), 1.955 (50), 2.36 (40), 1.397 (40), 1.315 (40), 1.120 (40)

Chemistry:

	(1)	(2)	(3)
Pd	64.8	64.3	64.19
Pt	2.5		
Sn	35.5	35.0	35.81
Sb	0.3		
Bi	0.2		
Total	103.3	99.3	100.00

(1) Oktyabr mine, USSR; by electron microprobe, corresponding to $(Pd_{2.02}Pt_{0.04})_{\Sigma=2.06}Sn_{1.00}$. (2) Western Platinum mine, South Africa; by electron microprobe. (3) Pd_2Sn.

Occurrence: In Cu-Ni sulfide ores; in cubanite-chalcopyrite, cubanite-talnakhite, and cubanite-mooihoekite assemblages (Oktyabr mine, USSR).

Association: Cubanite, chalcopyrite, talnakhite, mooihoekite, magnetite, sperrylite, sobolevskite, polarite, atokite, silver, palladium, bismuth.

Distribution: From the Oktyabr mine, Talnakh area, Noril'sk region, western Siberia, USSR. At the Atok and Western Platinum mines, Witwatersrand, Transvaal, South Africa.

Name: For the chemical composition, PAlladium and *olovo*, *tin* (in Russian).

Type Material: A.E. Fersman Mineralogical Museum, Academy of Sciences, Moscow, USSR.

References: (1) Genkin, A.D., T.L. Evstigneeva, L.N. Vyal'sov, I.P. Laputina, and N.V. Groneva (1974) Paolovite, Pd_2Sn, a new mineral from copper–nickel sulfide ores. Geol. Rudn. Mestorozhd., 16, 98–103 (in Russian). (2) (1974) Amer. Mineral., 59, 1331–1332 (abs. ref. 1).

Crystal Data: Orthorhombic. *Point Group:* $2/m\ 2/m\ 2/m$. As irregular and subhedral grains up to 130 μm.

Physical Properties: Hardness = n.d. VHN = 654–763 (100 g load). D(meas.) = 6.9 D(calc.) = 6.97

Optical Properties: Opaque. *Color:* White with a faint grayish tinge. *Luster:* Metallic. *Pleochroism:* Weak. *Anisotropism:* Weak, faint pink to pale buff.

R_1–R_2: (400) 47.8–47.9, (420) 48.0–48.5, (440) 48.2–48.9, (460) 48.2–49.2, (480) 48.2–49.4, (500) 48.4–49.6, (520) 48.7–49.6, (540) 48.9–49.4, (560) 49.0–49.3, (580) 49.2–49.2, (600) 49.5–49.3, (620) 49.7–49.6, (640) 49.9–49.9, (660) 50.2–50.3, (680) 50.3–50.6, (700) 50.5–50.9

Cell Data: *Space Group:* *Pbca.* $a = 5.842$ $b = 5.951$ $c = 11.666$ Z = 8

X-ray Powder Pattern: Mulcahy Township, Canada.
2.555 (100), 2.033 (90), 0.9593 (80), 2.877 (70), 1.7619 (70), 0.9581 (70), 5.83 (60)

Chemistry:

	(1)	(2)	(3)
Co	26.4	25.0	27.70
Fe	0.4	0.4	
Ni	0.8	1.0	
Sb	56.8	57.6	57.23
As		2.1	
S	14.9	14.5	15.07
Total	99.3	100.6	100.00

(1) Mulcahy Township, Canada; by electron microprobe, average of five analyses. (2) Wheal Cock, Cornwall, England; by electron microprobe. (3) CoSbS.

Polymorphism & Series: Dimorphous with costibite.

Occurrence: In drill core from a massive base-metal sulfide deposit in a carbonatized chlorite-anthophyllite schist that is most likely an altered mafic rock, in a sequence of meta-volcanics and meta-sediments (Mulcahy Township, Canada); in Pb-Zn-Cu-Ag ore deposits remobilized by hydrothermal solutions from younger granite emplacement (Bergslagen, Sweden).

Association: Sphalerite, chalcopyrite, galena, pyrargyrite, pyrrhotite, antimonial silver (Mulcahy Township, Canada); costibite, nisbite, chalcopyrite, pyrrhotite, galena, sphalerite, gersdorffite, ullmannite (Bergslagen, Sweden).

Distribution: In drill core from Mulcahy Township, Kenora district, Red Lake mining division, Ontario, Canada. From Wheal Cock, Botallack, St. Just, Cornwall, England. In the Gruvåsen and Getön deposits, Bergslagen, Sweden.

Name: For the chemical composition and probable structural relation to pararammelsbergite.

Type Material: National Museum of Canada, Ottawa; Royal Ontario Museum, Toronto, Canada.

References: (1) Cabri, L.J., D.C. Harris, and J.M. Stewart (1970) Paracostibite (CoSbS) and nisbite ($NiSb_2$), new minerals from the Red Lake Area, Ontario, Canada. Can. Mineral., 10, 232–246. (2) (1971) Amer. Mineral., 56, 631 (abs. ref. 1). (3) Rowland, J.F., E.J. Gabe, and S.R. Hall (1975) The crystal structures of costibite (CoSbS) and paracostibite (CoSbS). Can. Mineral., 13, 188–196.

Crystal Data: Monoclinic. *Point Group:* 2. As aggregates of equant, polygonal grains and short (0.5 mm) stubby prisms that occasionally are striated parallel to their length; prisms occasionally curved, sometimes in nests of curved, grooved, pseudohexagonal plates. *Twinning:* Polysynthetically twinned on {010} as parallel lamellae 25 μm wide, on $\{\bar{2}01\}$ as short tapering lamellae and on $\{\bar{1}10\}$ and {001}.

Physical Properties: *Cleavage:* Parting {010} perfect, several others less perfect, paralleling the twinning composition planes. *Tenacity:* Brittle. Hardness = n.d. VHN = 118 (100 g load). D(meas.) = 6.52 D(calc.) = 6.44

Optical Properties: Opaque. *Color:* Bright white. *Streak:* Black. *Luster:* Metallic. *Pleochroism:* Distinct, from yellowish white with a slightly pinkish tint, to slightly gray-white in air; in oil, strong, from pale pink-white-gray with a very faint blue-green tint. *Anisotropism:* Strong, in vivid pinks, pale orange, yellow, pale greenish blue and pale green.
R_1–R_2: n.d.

Cell Data: *Space Group:* $C2$. $a = 7.252(1)$ $b = 4.172(4)$ $c = 4.431(2)$ $\beta = 123°8.4(1.4)'$ Z = 1

X-ray Powder Pattern: Broken Hill, Australia.
3.06 (100), 2.09 (70), 2.21 (60), 3.72 (40), 1.730 (40), 1.521 (40), 1.392 (40)

Chemistry:

	(1)
Sb	82.9
As	18.6
Total	101.5

(1) Broken Hill, Australia; by electron microprobe.

Occurrence: Replacing calcite.

Association: Antimonian löllingite, stibarsen, calcite.

Distribution: From the Consols mine, Broken Hill, New South Wales, Australia.

Name: From the Greek for *unexpected alloy.*

Type Material: National Museum of Natural History, Washington, D.C., USA, R419.

References: (1) Leonard, B.F., C.W. Mead, and J.J. Finney (1971) Paradocrasite, $Sb_2(Sb,As)_2$, a new mineral. Amer. Mineral., 56, 1127–1146.

Crystal Data: Hexagonal. *Point Group:* $\bar{3}\ 2/m$. Intimately intergrown with guanajuatite.

Physical Properties: *Cleavage:* Perfect on {0001}. Hardness = 2 VHN = 27–50 (10 g load). D(meas.) = 6.2–7.0 D(calc.) = [7.704]

Optical Properties: Opaque. *Color:* Yellow-white, often tarnished lead-gray. *Luster:* Metallic. *Pleochroism:* Distinct in air. *Anisotropism:* Distinct.
R_1–R_2: (400) 47.0–51.7, (420) 48.1–52.6, (440) 49.2–53.5, (460) 50.0–54.2, (480) 50.5–54.6, (500) 50.9–55.0, (520) 51.1–55.1, (540) 51.0–55.0, (560) 50.8–54.6, (580) 50.5–54.2, (600) 50.0–53.7, (620) 49.5–53.0, (640) 48.8–52.2, (660) 48.2–51.5, (680) 47.4–50.6, (700) 46.5–49.8

Cell Data: *Space Group:* $R\bar{3}m$. $a = 4.133$ (synthetic). $c = 28.62$ (synthetic). Z = 3

X-ray Powder Pattern: Synthetic Bi_2Se_3.
3.03 (100), 2.23 (60), 1.404 (40), 4.80 (30), 2.07 (30), 1.907 (30), 1.320 (30)

Chemistry: Identity of natural and synthetic material established by similarity of X-ray powder patterns.

Polymorphism & Series: Dimorphous with guanajuatite.

Occurrence: Intergrown with guanajuatite in contact metamorphic as well as in hydrothermal veins.

Association: Guanajuatite, bismuthinite, ferroselite.

Distribution: In the Santa Catarina and Leon mines, Guanajuato, Mexico. From Falun, Sweden.

Name: From the supposed relationship to guanajuatite.

Type Material: n.d.

References: (1) Ramdohr, P. (1948) Los especes mineralogicas guanajuatite y paraguanajuatite. Comite Direct. Invest. Recursos Minerales Mexico, Bol. 20, 1–15. (2) (1949) Amer. Mineral., 34, 619 (abs. ref. 1). (3) Godovikov, A.A. and V.A. Klyakhin (1966) Guanajuatite and paraguanajuatite. Akad. Nauk SSSR, Sibirsk. Otdel., Geol. Geofiz., 7, 67–76 (in Russian). (4) (1967) Amer. Mineral., 52, 1588 (abs. ref. 3). (5) Berry, L.G. and R.M. Thompson (1962) X-ray powder data for the ore minerals. Geol. Soc. Amer. Mem. 85, 28. (6) Sindeeva, N.D. (1964) Mineralogy and types of deposits of selenium and tellurium, 71–74. (7) Ramdohr, P. (1969) The ore minerals and their intergrowths, (3rd edition), 702–703.

Crystal Data: Orthorhombic or lower. *Point Group:* n.d. Crystals columnar with imperfect, rounded faces and without good terminations, up to 8 x 2 mm in size.

Physical Properties: Hardness = n.d. VHN = n.d. D(meas.) = 5.482 D(calc.) = n.d.

Optical Properties: Opaque. *Color:* Gray. *Luster:* Metallic.
R_1–R_2: n.d.

Cell Data: *Space Group:* n.d. Z = n.d.

X-ray Powder Pattern: Herja, Romania.
4.21 (80-90), 3.78 (50), 4.67 (30), 2.49 (20), 2.23 (20), 2.02 (20), 3.29 (10)

Chemistry:

	(1)	(2)
Pb	39.81	40.15
Fe	2.98	2.71
Sb	34.74	35.39
S	21.96	21.75
insol.	0.13	
Total	99.62	100.00

(1) Herja, Romania. (2) $Pb_4FeSb_6S_{14}$.

Polymorphism & Series: Dimorphous with jamesonite.

Occurrence: In a hydrothermal deposit, later formed than other associated sulfides.

Association: Galena, pyrrhotite, chalcopyrite, tetrahedrite.

Distribution: At Herja (Kisbánya), Romania.

Name: For the paramorphous relationship to jamesonite.

Type Material: n.d.

References: (1) Zsivny, V. and I. v. Náray-Szabó (1947) Parajamesonit, ein neues Mineral von Kisbánya. Schweiz. Mineral. Petrog. Mitt., 27, 183–189 (in German). (2) (1949) Amer. Mineral., 34, 133 (abs. ref. 1).

Crystal Data: Monoclinic. *Point Group:* $2/m$. As terminated crystals up to 3 mm, prismatic along [001], with prominent {100}, {110}, and {100}.

Physical Properties: Hardness = 2.5–3 VHN = n.d. D(meas.) = 5.07 D(calc.) = 5.04

Optical Properties: Opaque. *Color:* Black; white with a creamy tint in reflected light, with rare purple-red internal reflections. *Luster:* Semi-metallic. *Pleochroism:* Weak in air, grayish white to brownish gray in oil. *Anisotropism:* Distinct, brownish gray to bluish gray.
R_1–R_2: (400) 31.7–42.8, (420) 31.3–42.2, (440) 30.9–41.6, (460) 30.3–41.1, (480) 29.9–40.6, (500) 29.5–40.3, (520) 29.3–40.0, (540) 29.1–39.7, (560) 28.8–39.2, (580) 28.6–38.5, (600) 28.2–37.9, (620) 27.9–37.1, (640) 27.3–36.1, (660) 26.7–35.1, (680) 26.0–34.1, (700) 26.6–34.2

Cell Data: *Space Group:* $P2_1/n$. $a = 8.098(5)$ $b = 19.415(12)$ $c = 9.059(6)$
$\beta = 91.96(8)°$ Z = 4

X-ray Powder Pattern: Allchar, Yugoslavia.
3.493 (100), 2.832 (100), 4.15 (90), 3.696 (90), 2.913 (90), 2.356 (90), 3.599 (70)

Chemistry:

	(1)	(2)	(3)
Tl	19.3	19.4	20.2
Sb	50.8	51.8	50.8
As	5.3	5.0	3.7
S	24.7	25.4	24.5
Total	100.1	101.6	99.2

(1–3) Allchar, Yugoslavia; by electron microprobe, the average corresponding to $Tl_{1.01}(Sb_{4.36}As_{0.64})_{\Sigma=5.00}S_{8.01}$.

Occurrence: Of hydrothermal origin, in cavities in realgar (Allchar, Yugoslavia).

Association: Realgar (Allchar, Yugoslavia); avicennite (Lookout Pass, Utah, USA).

Distribution: From Allchar, Macedonia, Yugoslavia. At Jas Roux, Hautes-Alpes, France. In the Hemlo gold deposit, Thunder Bay district, Ontario, Canada. From near Lookout Pass, Tooele Co., Utah, USA.

Name: For its relation to pierrotite.

Type Material: National School of Mines, Paris, France.

References: (1) Johan, Z., P. Picot, J. Hak, and M. Kvaček (1975) La parapierrotite, un nouveau minéral thallifère d'Allchar (Yougoslavie). Tschermaks Mineral. Petrog. Mitt., 22, 200–210 (in French with English abs.). (2) (1976) Amer. Mineral., 61, 504 (abs. ref. 1). (3) Engle, P. (1980) Die Kristallstruktur von synthetischem Parapierrotit, $TlSb_5S_8$. Zeits. Krist., 151, 203–216 (in German with English abs.).

Crystal Data: Orthorhombic. *Point Group:* $2/m\ 2/m\ 2/m$. Crystals 1–2 mm in size, tabular on {001}, in rounded grains, also massive and in dendrites.

Physical Properties: *Cleavage:* {001} perfect and easy. Hardness = ~5 VHN = 681–830 (100 g load). D(meas.) = 7.12 D(calc.) = 7.24

Optical Properties: Opaque. *Color:* Tin-white. *Luster:* Metallic. *Anisotropism:* Strong, in russet-brown, brown, yellow, gray.
R_1–R_2: (400) 59.1–58.7, (420) 59.3–58.8, (440) 59.5–58.9, (460) 59.7–59.2, (480) 59.7–59.5, (500) 59.5–59.6, (520) 59.2–59.8, (540) 59.9–59.9, (560) 58.5–60.0, (580) 58.2–60.0, (600) 58.0–60.1, (620) 57.8–60.2, (640) 57.8–60.3, (660) 57.7–60.6, (680) 57.8–60.8, (700) 58.0–61.2

Cell Data: *Space Group: Pcba.* $a = 5.753$ $b = 5.799$ $c = 11.407$ Z = 8

X-ray Powder Pattern: Synthetic. (JCPDS 18-876).
2.559 (100), 2.521 (95), 2.371 (65), 2.337 (65), 1.741 (65), 2.855 (50), 2.827 (50)

Chemistry:

	(1)	(2)	(3)
Ni	28.1	25.79	28.15
Co	0.4	1.26	
Cu		0.77	
As	68.5	71.23	71.85
S	2.6	0.02	
Total	99.6	99.07	100.00

(1) Moose Horn mine, Canada. (2) Černý Důl, Czechoslovakia; by electron microprobe, giving $(Ni_{0.923}Co_{0.045}Cu_{0.025})_{\Sigma=0.993}(As_{1.999}S_{0.001})_{\Sigma=2.000}$. (3) $NiAs_2$.

Polymorphism & Series: Trimorphous with rammelsbergite and krutovite.

Occurrence: In hydrothermal veins bearing Ni-Co mineralization.

Association: Nickeline, skutterudite, cobaltite, löllingite, gersdorffite, rammelsbergite.

Distribution: In Canada, at the Moose Horn mine, Elk Lake, Gowganda; several mines in Cobalt; and the Keeley mine, South Lorrain, all in Ontario; at the D uranium deposit, Saskatchewan; and on the Camsell River, six km south of Conjurer Bay, Northwest Territories. At Franklin, Sussex Co., New Jersey, USA. From Dobšiná (Dobschau) and Černý Důl, Krkonoše (Giant Mountains), Czechoslovakia. In the Shorbulaksk mercury deposit, Azerbaijan SSR, USSR. At Bou Azzer, Morocco. In the Talmessi and Meskani mines, Anarak district, Iran.

Name: For its chemical similarity to rammelsbergite.

Type Material: Royal Ontario Museum, Toronto, Canada, M12411, M11772, M14242.

References: (1) Palache, C., H. Berman, and C. Frondel (1944) Dana's system of mineralogy, (7th edition), v. I, 310–311. (2) Fleet, M.E. (1972) The crystal structure of pararammelsbergite. Amer. Mineral., 57, 1–9. (3) Johan, Z. (1985) The Černy Důl deposit (Czechoslovakia): an example of Ni-, Fe-, Ag-, Cu-arsenide mineralization with extremely high activity of arsenic; new data on paxite, novakite and kutinaite. Tschermaks Mineral. Petrog. Mitt., 34, 167–182.

Crystal Data: Monoclinic. *Point Group:* m or $2/m$. As fine powder, occasionally as granular (< 0.02 mm) aggregates.

Physical Properties: *Fracture:* Uneven. *Tenacity:* Brittle. Hardness = 1–1.5 VHN = n.d. D(meas.) = 3.52(5) D(calc.) = 3.499

Optical Properties: Translucent when coarsely granular. *Color:* Bright yellow when powdery, to orange-yellow and orange-brown when granular, with internal reflections from gold to orange-red. *Streak:* Bright yellow. *Luster:* Vitreous to resinous.
Optical Class: n.d. *Pleochroism:* High. $n = > 2.02$ *Anisotropism:* Distinct.
R_1–R_2: n.d.

Cell Data: *Space Group:* Pc or $P2/c$. $a = 9.929(4)$ $b = 9.691(6)$ $c = 8.503(3)$ $\beta = 97.06(2)°$ Z = 16

X-ray Powder Pattern: Mount Washington, Canada.
5.14 (100), 5.56 (91), 3.75 (78), 2.795 (71), 3.025 (51), 3.299 (50), 3.105 (33)

Chemistry:

	(1)	(2)
As	69.81	70.0
S	29.97	30.0
Total	99.78	100.0

(1) Mount Washington, Canada; by electron microprobe, average of two analyses. (2) AsS.

Polymorphism & Series: Dimorphous with realgar.

Occurrence: As an alteration product of realgar in stibnite-bearing quartz veins.

Association: Realgar, stibnite, tetrahedrite, arsenopyrite, duranusite, arsenic, arsenolite, sulfur, lepidocrocite, pyrite.

Distribution: In Canada, in British Columbia, at Mount Washington, Comox district, on Vancouver Island; at Siwash Creek in the Kamloops district; and from the Gray Rock property, head of Traux Creek, in the Bridge River Area, Lillooet district; from Ontario, in the Hemlo gold deposit, Thunder Bay district. From the Golconda mine, Humboldt Co., Nevada, USA. At the Lengenbach quarry, Binntal, Valais, Switzerland. Probably to be found at a number of other arsenic-rich localities.

Name: In allusion to its chemical identity with realgar.

Type Material: Canadian Geological Survey, Ottawa, Canada, 61566, 61567.

References: (1) Roberts, A.C., H.G. Ansell, and M. Bonardi (1980) Pararealgar, a new polymorph of AsS, from British Columbia. Can. Mineral., 18, 525–527. (2) (1981) Amer. Mineral., 66, 1277 (abs. ref. 1).

Paraschachnerite $Ag_{1.2}Hg_{0.8}$

Crystal Data: Orthorhombic, pseudohexagonal. *Point Group:* $2/m\ 2/m\ 2/m$ or $mm2$. As crystals up to 1 cm in length but usually much smaller. *Twinning:* Always twinned in a complex fashion; twin plane {110}; trillings common.

Physical Properties: Hardness = > 3.5 VHN = n.d. D(meas.) = n.d. D(calc.) = 12.98

Optical Properties: Opaque. *Color:* In polished section, creamy white. *Luster:* Metallic. *Pleochroism:* Brownish rose-white to creamy white. *Anisotropism:* Distinct.
R_1–R_2: n.d.

Cell Data: *Space Group:* $Cmcm$ (probable) or $Cmc2$, $Cmc2_1$. $a = 2.961$ $b = 5.13$ $c = 4.83$ Z = 2

X-ray Powder Pattern: Landsberg, Germany.
2.267 (100), 2.404 (60), 1.263 (60), 1.361 (50), 1.481 (40), 0.8310 (40), 2.564 (30)

Chemistry:

	(1)	(2)
Ag	47.85	44.65
Hg	52.15	55.35
Total	100.00	100.00

(1) Analysis of a synthetic product formed by heating reactants; shown to be equivalent to paraschachnerite by X-ray and optical examination; corresponds to $Ag_{1.26}Hg_{0.74}$. (2) $Ag_{1.2}Hg_{0.8}$.

Occurrence: Found in the oxidized zone, formed by the alteration of moschellandsbergite.

Association: Schachnerite, mercurian silver, acanthite, cinnabar, "limonite", ankerite (Landsberg, Germany); mercurian silver, cinnabar, many secondary copper minerals (Kremikova deposit, Bulgaria); schachnerite, sphalerite, chalcopyrite, pyrite, pyrrhotite, gudmundite, cubanite, ilmenite, galena (Sala, Sweden).

Distribution: From the Vertraun Gott mercury mine, at Landsberg, near Obermoschel, Rhineland-Palatinate, Germany. In the Kremikova deposit, Sofia district, Bulgaria. From Sala, Västmanland, Sweden.

Name: For the relation to schachnerite.

Type Material: Technical University, Berlin, Germany; National Museum of Natural History, Washington, D.C., USA, 145618, 150256.

References: (1) Seeliger, E. and A. Mücke (1972) Para-schachnerite, $Ag_{1.2}Hg_{0.8}$, und Schachnerite, $Ag_{1.1}Hg_{0.9}$, von Landsberg bei Obermoschel, Pfalz. Neues Jahrb. Mineral., Abh., 117, 1–18 (in German with English abs.). (2) (1973) Amer. Mineral., 58, 347 (abs. ref. 1).
(3) Zakrzewski, M.A. and E.A.J. Burke (1987) Schachnerite, paraschachnerite and silver amalgam from the Sala mine, Sweden. Mineral. Mag., 51, 318–321.

$Ni_3(Bi,Pb)_2S_2$ **Parkerite**

Crystal Data: Monoclinic. *Point Group:* $2/m$. As grains (< 1 mm), the rhombic shape of which is controlled by intersecting parting and cleavage planes. *Twinning:* Lamellar twinning on {111}.

Physical Properties: *Cleavage:* {001} perfect; parting on {111}. *Fracture:* Uneven. *Tenacity:* Brittle. Hardness = 3 VHN = n.d. D(meas.) = 8.50 D(calc.) = 8.53

Optical Properties: Opaque. *Color:* Bright bronze, tarnishing darker and duller; light cream in polished section. *Streak:* Black, shining. *Luster:* Metallic. *Pleochroism:* Perceptible. *Anisotropism:* Strong, greenish gray to yellowish brown.

R_1–R_2: (400) 44.1–46.2, (420) 44.6–47.1, (440) 45.1–48.0, (460) 45.5–48.8, (480) 45.7–49.5, (500) 45.8–50.0, (520) 45.8–50.3, (540) 45.8–50.5, (560) 46.0–50.7, (580) 46.3–51.0, (600) 46.8–51.5, (620) 47.4–51.9, (640) 48.0–52.3, (660) 48.5–52.8, (680) 49.1–53.3, (700) 49.6–53.8

Cell Data: *Space Group:* $C2/m$ (probable). $a = 11.066(1)$ $b = 8.085(1)$ $c = 7.965(1)$ $\beta = 134.0°$ Z = 4

X-ray Powder Pattern: Synthetic $Ni_3Bi_2S_2$.
2.836 (100), 2.864 (65), 2.336 (55), 3.98 (30), 4.04 (25), 1.989 (25), 1.648 (25)

Chemistry:

	(1)	(2)	(3)	(4)
Ni	30.3	26.8	29.7	26.76
Bi	60.1	63.6	50.8	63.50
Pb		trace	9.5	
S	9.1	9.2	9.6	9.74
Total	99.5	99.6	99.6	100.00

(1) Langis mine, Canada; by electron microprobe. (2) Sudbury, Canada. (3) Insizwa, South Africa; by electron microprobe. (4) $Ni_3Bi_2S_2$.

Occurrence: As grains in other hydrothermal sulfide and arsenide minerals.

Association: Galena, bismuth, bismuthinite, tetradymite, hessite, cubanite, maucherite, nickeline, sperrylite, gold, chalcopyrite, pyrrhotite, pentlandite, siegenite, bravoite.

Distribution: From Waterfall Gorge, Insizwa, East Griqualand, South Africa. In Canada, from Sudbury, and in the Langis mine, Casey Township, Cobalt-Gowganda area, Ontario; in the Gaspe copper mine, Quebec; and the Gros Cap area, near Great Slave Lake, Northwest Territories. In the Noril'sk region, western Siberia; Karik'yavr, Kola Peninsula; and the Allarechensk region, Murmansk, USSR. In the Zinkwand mine, Schladminger, Styria, Austria. From Nebral, Málaga Province, Spain. In the Rakha copper deposit, Bihar, India.

Name: After Professor Robert Lüling Parker (1893–1973), Zurich, Switzerland.

Type Material: n.d.

References: (1) Michener, C.E. and M.A. Peacock (1943) Parkerite ($Ni_3Bi_2S_2$) from Sudbury, Ontario: redefinition of the species. Amer. Mineral., 28, 343–355. (2) Peacock, M.A. and J. McAndrew (1950) On parkerite and shandite and the crystal structure of $Ni_3Pb_2S_2$. Amer. Mineral., 35, 425–439. (3) Petruk, W., D.C. Harris, and J.M. Stewart (1969) Langisite, a new mineral, and the rare minerals cobalt pentlandite, siegenite, parkerite and bravoite from the Langis mine, Cobalt–Gowganda area, Ontario. Can. Mineral., 9, 597–616. (4) Fleet, M.E. (1973) The crystal structure of parkerite ($Ni_3Bi_2S_2$). Amer. Mineral., 58, 435–439. (5) Brower, W.S., H.S. Parker, and R.S. Roth (1974) Reexamination of synthetic parkerite and shandite. Amer. Mineral., 59, 296–301.

Patronite VS_4(?)

Crystal Data: Monoclinic. *Point Group:* $2/m$. Massive, and as aggregates of columnar crystals.

Physical Properties: *Cleavage:* Pronounced columnar cleavage. Hardness = Very low. VHN = n.d. D(meas.) = 2.82(1) D(calc.) = 2.834

Optical Properties: Opaque. *Color:* Lead-gray on fresh surface, gray-black on old; dark gray in reflected light. *Luster:* Metallic. *Pleochroism:* Strong.
R_1–R_2: (400) 20.2–29.9, (420) 20.0–30.0, (440) 19.8–30.1, (460) 19.8–30.8, (480) 20.0–31.9, (500) 20.1–32.6, (520) 20.1–32.9, (540) 20.0–32.8, (560) 19.8–32.4, (580) 19.6–31.4, (600) 19.1–30.5, (620) 18.9–29.6, (640) 18.8–29.6, (660) 19.0–29.8, (680) 19.2–30.3, (700) 19.5–31.0

Cell Data: *Space Group:* $I2/c$. $a = 6.775(5)$ $b = 10.42(1)$ $c = 12.11(1)$ $\beta = 100.8(2.0)°$ Z = 8

X-ray Powder Pattern: Synthetic VS_4.
5.604 (100), 5.181 (65), 2.473 (30), 2.216 (30), 2.047 (25), 3.151 (20), 2.962 (20)

Chemistry: Composition established by comparison of X-ray patterns with synthetic material.

Occurrence: As interstitial filling in the core of a porous 2.5 m layer of admixed vanadium-bearing minerals. These vanadian materials are in fissures that cut red shales and that were probably filled by a remobilized asphaltite deposit.

Association: Sulfur, bravoite, quartz, vanadian lignite, natural coke.

Distribution: A major ore mineral in what was the world's richest vanadium deposit, at Minasragra, near Cerro de Pasco, Peru.

Name: After Antenor Rizo-Patrón, Peruvian engineer, discoverer of the Peruvian occurrence.

Type Material: n.d.

References: (1) Palache, C., H. Berman, and C. Frondel (1944) Dana's system of mineralogy, (7th edition), v. I, 347. (2) Tudo, J. (1965) Sur l'etude du sulfate de vanadyle et de sa réduction par l'hydrogène sulfuré: les sulfures de vanadium. Rev. Chim. Minérale, 2, 53–117 (in French). (3) Allmann, R., I. Baumann, A. Kutoglu, H. Rösch, and E. Hellner (1963) Die Kristallstruktur des Patronits $V(S_2)_2$. Naturwiss., 51, 263–264 (in German). (4) Kutoglu, A. and R. Allmann (1972) Strukturverfeinerung des Patronits, $V(S_2)_2$. Neues Jahrb. Mineral., Monatsh., 339–345 (in German with English abs.).

Crystal Data: Monoclinic. *Point Group:* $2/m$. Massive, and as tiny bladed crystals elongated parallel to [010].

Physical Properties: *Cleavage:* Indistinct. Hardness = 2 VHN = 188 (50 g load). D(meas.) = 6.8 (synthetic). D(calc.) = [6.80]

Optical Properties: Opaque. *Color:* Lead-gray to tin-white; in polished section, white. *Luster:* Metallic. *Pleochroism:* Strong, gray-pink-white and blue-gray-white in oil. *Anisotropism:* Strong, in pale to intense blue and light tan to brown.
R_1–R_2: (400) 46.3–50.7, (420) 46.1–50.5, (440) 45.9–50.3, (460) 45.6–50.0, (480) 45.3–49.8, (500) 45.0–49.5, (520) 44.5–49.0, (540) 44.0–48.4, (560) 43.4–47.8, (580) 42.9–47.3, (600) 42.5–47.0, (620) 42.2–46.7, (640) 42.0–46.4, (660) 41.9–46.2, (680) 41.7–46.0, (700) 41.6–45.8

Cell Data: *Space Group:* $C2/m$. $a = 13.35$ $b = 4.03$ $c = 16.34$ $\beta = 94.5°$ Z = 4

X-ray Powder Pattern: Synthetic $AgBi_3S_5$. Easily confused with benjaminite.
2.858 (100), 3.59 (90), 3.38 (80), 3.46 (40), 5.45 (35), 2.259 (35), 4.08 (30)

Chemistry:

	(1)
Ag	11.1
Cu	1.7
Pb	3.5
Bi	64.5
S	18.3
Total	99.1

(1) Porvenir mine, Bolivia; by electron microprobe, corresponding to $(Ag_{0.93}Cu_{0.24})_{\Sigma=1.17}(Bi_{2.80}Pb_{0.15})_{\Sigma=2.95}S_{5.17}$.

Occurrence: In hydrothermal vein deposits.

Association: Cupropavonite, bismuthinite, chalcopyrite, aikinite, hodrushite.

Distribution: From the Porvenir mine, Cerro Bonete, about 75 km south-southwest of Esmoraca, Potosí, Bolivia. In the USA, in Colorado, in the Silver Bell mine, Red Mountain, Ouray Co.; and at the Old Lout, Gladiator and Alaska mines, Poughkeepsie Gulch, near Ouray, San Juan Co. In Nevada, at Manhattan, Nye Co., and Pioche, Lincoln Co.; and in the Apache Hills mine, southeast of Hachita, Grant Co., New Mexico. In Canada, in the Keeley mine, South Lorrain Township, Ontario. From Portugal, at Panasqueira, Beira Baixa Province.

Name: From the Latin *pavo*, *peacock*, honoring Professor Martin Alfred Peacock (1898–1950), Canadian mineralogist.

Type Material: Harvard University, Cambridge, Massachusetts, USA; University of Toronto, Toronto, Canada.

References: (1) Nuffield, E.W. (1954) Studies of mineral sulpho-salts: XVIII—pavonite, a new mineral. Amer. Mineral., 39, 409–415. (2) Van Hook, H.J. (1960) The ternary system Ag_2S–Bi_2S_3–PbS. Econ. Geol., 55, 759–788. (3) Karup-Møller, S. (1972) New data on pavonite, gustavite and some related sulfosalt minerals. Neues Jahrb. Mineral., Abh., 117, 19–38. (4) Harris, D.C. and T.T. Chen (1975) Studies of type pavonite material. Can. Mineral., 13, 408–410. (5) Makovicky, E., W.G. Mumme, and J.A. Watts (1977) The crystal structure of synthetic pavonite, $AgBi_3S_5$, and the definition of the pavonite homologous series. Can. Mineral., 15, 339–348. (6) Karup-Møller, S. and E. Makovicky (1979) On pavonite, cupropavonite, benjaminite, and "oversubstituted" gustavite. Bull. Minéral., 102, 351–367. (7) (1980) Amer. Mineral., 65, 206 (abs. ref. 6).

Crystal Data: Monoclinic. *Point Group:* $2/m$. Intergrown with novákite, koutekite and arsenic. *Twinning:* Polysynthetic along [010].

Physical Properties: *Cleavage:* Perfect in one direction. Hardness = 3.5–4 VHN = 146 (25 g load). D(meas.) = 5.4 D(calc.) = 5.97

Optical Properties: Opaque. *Color:* Light steel-gray on fresh surface; in polished section, light grayish white. *Streak:* Black. *Luster:* Metallic, tarnishing black. *Anisotropism:* Very strong, in light gray-green and dark brown with violet tint.

R_1–R_2: (400) 46.4–48.2, (420) 45.8–48.4, (440) 45.2–48.6, (460) 44.8–48.4, (480) 44.6–48.0, (500) 44.6–48.5, (520) 44.5–49.2, (540) 44.2–49.5, (560) 43.8–49.4, (580) 43.5–49.1, (600) 43.4–48.9, (620) 43.5–48.8, (640) 43.7–48.6, (660) 43.8–48.3, (680) 43.7–47.9, (700) 43.3–47.6

Cell Data: *Space Group:* $P2_1/c$. $a = 5.839(2)$ $b = 5.111(2)$ $c = 8.084(3)$ $\beta = 99.7°$ Z = 10

X-ray Powder Pattern: Černý Důl, Czechoslovakia.
3.164 (100), 3.633 (80), 2.772 (70), 2.618 (70), 1.795 (70), 3.825 (60), 1.882 (60)

Chemistry:

	(1)	(2)
Cu	30.35	29.77
As	69.91	70.23
Total	100.26	100.00

(1) Černý Důl, Czechoslovakia; by electron microprobe, average of 10 analyses. (2) $CuAs_2$.

Occurrence: In hydrothermal calcite veins cutting diopside hornfels lenses, in pyroxene gneisses and mica schists. Probably a late-stage reaction product formed at the expense of novákite and arsenic below 130°C (Černý Důl, Czechoslovakia).

Association: Novákite, koutekite, arsenic, arsenolamprite, silver, löllingite, nickeline, chalcocite, skutterudite, bornite, chalcopyrite, tiemannite, clausthalite, uraninite, hematite, fluorite (Černy Důl, Czechoslovakia); lautite, kutinaite (Niederbeerbach, Germany); domeykite, algodonite, koutekite (Mohawk, Michigan, USA).

Distribution: From Černý Důl, Krkonoše (Giant Mountains), Czechoslovakia. At Niederbeerbach, Odenwald, Germany. From Mohawk, Keeweenaw Co., Michigan, USA.

Name: From the Latin *pax*, *peace*.

Type Material: National Museum of Natural History, Washington, D.C., USA, 162605.

References: (1) Johan, Z. (1961) Paxite—Cu_2As_3, a new copper arsenide from Černý Důl in the Giant Mts. (Krkonoše). Acta Univer. Carolinae, Geologica, 77–88 (1962) (in Czech with English abs.). (2) (1962) Amer. Mineral., 47, 1484–1485 (abs. ref. 1). (3) Johan, Z. (1985) The Černy Důl deposit (Czechoslovakia): an example of Ni-, Fe-, Ag-, Cu-arsenide mineralization with extremely high activity of arsenic; new data on paxite, novakite and kutinaite. Tschermaks Mineral. Petrog. Mitt., 34, 167–182.

Crystal Data: Monoclinic. *Point Group:* $2/m$. As short, tabular pseudohexagonal prisms with bevelled edges, showing triangular striations on {001}; as rosettes of such crystals, to 2 cm.

Physical Properties: *Fracture:* Conchoidal to irregular. Hardness = 3 VHN = 130–142 (100 g load). D(meas.) = 6.15(2) D(calc.) = 6.07

Optical Properties: Opaque, translucent in very thin splinters. *Color:* Dull black; in polished section, white, with very dark red internal reflections. *Streak:* Black. *Luster:* Metallic. *Pleochroism:* Very weak in air, fair in oil. *Anisotropism:* Moderate, often dark violet.
R_1–R_2: (400) 34.8–34.3, (420) 34.2–34.2, (440) 33.4–34.1, (460) 32.5–33.8, (480) 31.5–33.2, (500) 30.6–32.8, (520) 29.8–32.5, (540) 29.3–32.3, (560) 29.0–32.0, (580) 29.0–31.7, (600) 29.1–31.2, (620) 29.1–30.7, (640) 28.9–30.2, (660) 28.6–29.8, (680) 28.0–29.3, (700) 27.4–29.1.

Cell Data: *Space Group:* $C2/m$. $a = 12.64$ $b = 7.29$ $c = 11.90$ $\beta = 90.0°$ Z = 2

X-ray Powder Pattern: Aspen, Colorado, USA.
2.97 (100), 2.80 (90), 2.47 (60), 2.30 (60), 1.827 (60), 2.34 (50), 1.992 (50)

Chemistry:

	(1)	(2)	(3)
Ag	56.90	63.54	77.46
Cu	14.85	10.70	
Zn	2.81		
Fe		0.60	
As	7.01	7.29	6.72
Sb	0.30	0.43	
S	18.13	17.07	15.82
Total	100.00	99.63	100.00

(1) Molly Gibson mine, Colorado, USA; recalculated to 100% after deducting 12.81% impurities. (2) Arqueros, Chile. (3) $Ag_{16}As_2S_{11}$.

Polymorphism & Series: Forms a series with polybasite.

Occurrence: In hydrothermal deposits formed at low to moderate temperatures.

Association: Acanthite, silver, proustite, quartz, barite, calcite.

Distribution: In the USA, in Colorado, at the Molly Gibson mine, Aspen, Pitkin Co., and Rico, Dolores Co.; in Utah, in the Tintic district, Juab Co., at Eureka and elsewhere. In Montana, at Neihart, Cascade Co.; in the Flathead mine, Niarada, Flathead Co.; and at the Drumlummon mine, Marysville, Lewis and Clark Co. In the Lakeview district, Bonner Co., Idaho. In the Morrison mine, and the Ross mine, Holtyre, Ontario, and the Husky mine, Yukon Territory, Canada. At Baňská Štiavnica (Schemnitz), Czechoslovakia. From the Clara mine, Schwartzwald, Germany. In Spain, from Francoli, Tarragona Province. At Sark's Hope mine, Sark, Channel Islands. In Mexico, at the Veta Rica mine, Sierra Mojada, Coahuila. From Arqueros, Chile. At Seikoshi, Shizuoka Prefecture, Japan.

Name: For Richard Pearce (1837–1927), chemist and metallurgist, Denver, Colorado, USA.

Type Material: n.d.

References: (1) Palache, C., H. Berman, and C. Frondel (1944) Dana's system of mineralogy, (7th edition), v. I, 353–355. (2) Peacock, M.A. and L.G. Berry (1947) Studies of mineral sulfo-salts: XIII—polybasite and pearceite. Mineral. Mag., 28, 1–13. (3) Frondel, C. (1963) Isodimorphism of the polybasite and pearceite series. Amer. Mineral., 48, 565–572. (4) Hall, H.T. (1967) The pearceite and polybasite series. Amer. Mineral., 52, 1311–1321.

Pekoite $PbCuBi_{11}(S, Se)_{18}$

Crystal Data: Orthorhombic. *Point Group:* $mm2$. Crystals up to 1–2 mm.

Physical Properties: *Cleavage:* {010} good. Hardness = n.d. VHN = n.d. D(meas.) = n.d. D(calc.) = 6.8

Optical Properties: Opaque. *Color:* Lead-gray; white to cream in reflected light. *Streak:* Lead-gray. *Luster:* Metallic.
R_1–R_2: n.d.

Cell Data: *Space Group:* $P2_1am$. $a = 11.472(2)$ $b = 33.744(6)$ $c = 4.016(1)$ Z = 2

X-ray Powder Pattern: Juno mine, Australia.
3.140 (100), 3.025 (40), 3.622 (30), 1.971 (30), 3.564 (25), 3.593 (20), 2.554 (20)

Chemistry:

	(1)
Pb	5.8
Cu	1.6
Ag	0.2
Fe	0.1
Bi	76.1
Se	6.9
S	14.6
Total	105.3

(1) Juno mine, Australia; by electron microprobe.

Occurrence: In magnetite pipes that cut felsic sediments and pyroclastics; the pipes are probably hydrothermal replacements associated with volcanism (Juno mine, Australia).

Association: Gladite, junoite, emplectite, magnetite, chalcopyrite (Juno mine, Australia); paděraite, hammarite, chalcopyrite, grossular, andradite (Băiţa, Romania).

Distribution: From the Juno mine, Tennant Creek, Northern Territory, Australia. Near Narechen, southern Rhodope Mountains, Bulgaria. In the Kochbulak deposit, eastern Uzbekistan, USSR. From Băiţa (Rézbánya), Romania. In the Tanco pegmatite, Bernic Lake, Manitoba, Canada. In the USA, from the Germania Consolidated mine, Fruitland, Stevens Co., Washington; and the Comstock mine, Dos Cabezas, Cochise Co., Arizona.

Name: For the Peko mine, Tennant Creek, Australia.

Type Material: Geology Department, University of New England, Australia, R27788.

References: (1) Mumme, W.G., E. Welin, and B.J. Wuensch (1976) Crystal chemistry and proposed nomenclature for sulfosalts intermediate in the system bismuthinite–aikinite (Bi_2S_3–$CuPbBiS_3$). Amer. Mineral., 61, 15–20. (2) Mumme, W.G. and J.A. Watts (1976) Pekoite, $CuPbBi_{11}S_{18}$, a new member of the bismuthinite–aikinite mineral series: its crystal structure and relationship with naturally- and synthetically-formed members. Can. Mineral., 14, 322–333. (3) Large, R.R. and W.G. Mumme (1975) Junoite, "wittite", and related seleniferous bismuth sulfosalts from Juno mine, Northern Territory, Australia. Econ. Geol., 70, 369–383. (4) Mumme, W.G. and J.A. Watts (1976) Additional physical, optical and x-ray data for pekoite. Can. Mineral., 14, 578.

Crystal Data: Cubic. *Point Group:* $2/m\ \overline{3}$. As reniform masses with radiating columnar structure.

Physical Properties: *Cleavage:* Perfect on {001}; distinct on {011}. *Fracture:* Subconchoidal. *Tenacity:* Brittle. Hardness = 2.5–3 VHN = 500–583 (100 g load). D(meas.) = 6.58–6.74 D(calc.) = 6.7

Optical Properties: Opaque. *Color:* Lead-gray; in polished section, creamy grayish white. *Streak:* Black. *Luster:* Metallic, tarnishes rapidly.

R: (400) 39.0, (420) 39.2, (440) 39.4, (460) 39.6, (480) 40.0, (500) 40.3, (520) 40.7, (540) 41.1, (560) 41.5, (580) 42.0, (600) 42.5, (620) 43.0, (640) 43.3, (660) 43.4, (680) 43.7, (700) 44.2

Cell Data: *Space Group:* $P3a$. $a = 5.991$ Z = 4

X-ray Powder Pattern: Colquechaca, Bolivia.
2.68 (100), 2.45 (100), 1.806 (90), 1.599 (40), 3.02 (30), 1.662 (30), 1.151 (30)

Chemistry:

	(1)	(2)	(3)	(4)
Ni	8.89	9.88	12.5	22.9
Co	1.10	0.71	9.2	3.2
Cu	4.50	3.72	5.7	2.1
Ag	2.13	5.00		
Hg	1.45	4.12		
Pb	13.72	10.88		
Fe	trace	0.72		
S	0.39	0.52		
Se	67.01	66.59	72.3	71.7
Total	99.19	102.14	99.7	99.9

(1) Colquechaca, Bolivia.; Pb content probably due to impurities. (2) Do. (3) Hope's Nose, England; by electron microprobe. (4) Colquechaca, Bolivia; by electron microprobe.

Occurrence: In hydrothermal veins.

Association: Naumannite, clausthalite, tiemannite, pyrite, chalcopyrite (Pacajake mine, Bolivia); clausthalite, sederholmite, trüstedtite (Kuusamo, Finland).

Distribution: From the Pacajake mine, Hiaco, 30 km east-northeast of Colquechaca, Potosí, Bolivia. At Shinkolobwe, Shaba Province, Zaire. From Hope's Nose, Torquay, Devon, England. In Germany, in the Harz, at Tilkerode. At Kuusamo, northeastern Finland.

Name: For Dr. Richard Alexander Fullerton Penrose, Jr. (1863–1931), American economic geologist, of Philadelphia, Pennsylvania, USA.

Type Material: British Museum (Natural History), London, England, 1926, 1; Harvard University, Cambridge, Massachusetts, USA, 87472.

References: (1) Palache, C., H. Berman, and C. Frondel (1944) Dana's system of mineralogy, (7th edition), v. I, 294–296. (2) Early, J.W. (1950) Description and synthesis of the selenide minerals. Can. Mineral., 35, 337–364. (3) Ramdohr, P. (1969) The ore minerals and their intergrowths, (3rd edition), 806–807.

Pentlandite $(Fe,Ni)_9S_8$

Crystal Data: Cubic. *Point Group:* $4/m\ \bar{3}\ 2/m$. Massive, usually in granular aggregates; sometimes in crystals as large as 10 cm, imbedded in other sulfides.

Physical Properties: *Cleavage:* None; parting on {111}. *Fracture:* Conchoidal. *Tenacity:* Brittle. Hardness = 3.5–4 VHN = 268–285 (100 g load). D(meas.) = 4.6–5.0 D(calc.) = 4.956

Optical Properties: Opaque. *Color:* Light bronze-yellow; reddish brown when argentian. *Streak:* Light bronze-brown. *Luster:* Metallic.

R: (400) 30.9, (420) 34.8, (440) 38.7, (460) 41.8, (480) 44.2, (500) 46.3, (520) 47.9, (540) 49.2, (560) 50.4, (580) 51.4, (600) 52.3, (620) 53.1, (640) 53.8, (660) 54.5, (680) 55.0, (700) 55.6

Cell Data: *Space Group:* $Fm3m$. $a = 9.928(1)$ Z = 4

X-ray Powder Pattern: Worthington mine, Sudbury, Canada.
1.775 (100), 3.028 (80), 1.931 (50), 2.896 (40), 5.780 (30), 2.301 (30), 1.307 (20)

Chemistry:

	(1)	(2)	(3)	(4)
Fe	30.68	25.6	43.1	37.9
Ni	34.48	30.8	19.3	19.7
Co	1.28	11.0	4.1	0.1
Ag				10.2
Cu				0.3
S	32.74	32.4	33.6	31.5
insol.	0.56			
Total	99.74	99.8	100.1	99.7

(1) Worthington mine, Sudbury, Canada. (2) Mt. Colin, Australia; by electron microprobe. (3) Mt. Dun area, New Zealand; by electron microprobe. (4) Outokumpu, Finland; by electron microprobe.

Polymorphism & Series: Forms a series with cobalt pentlandite.

Occurrence: Nearly always intimately associated with pyrrhotite; found in ultramafic rocks and perhaps derived from them by magmatic segregation. Rarely in meteorites.

Association: Pyrrhotite, chalcopyrite, cubanite, mackinawite, magnetite.

Distribution: The most important ore of nickel, occurring at many localities. In Canada, from Sudbury, Ontario. At Outokumpu, Vuonos, Miihkali and Hietajärvi, in east Finland; in west Finland at the Hitura nickel deposit; and in north Finland at Rarrua. In Norway, at Espedalen, near Lillehammar, Opland. At Wheal Jane, Truro, Cornwall, England. In South Africa, in the Rustenburg district, Transvaal. In New Zealand, in the Mount Dun area, Nelson Province. In Australia, in northwest Queensland, at Lime Creek, and at Mount Colin; Coolgardie and Kambalda, Western Australia; and from the Lord Brassey mine, northwestern Tasmania.

Name: For Joseph Barclay Pentland (1797–1873), Irish natural scientist, who first noted the mineral.

References: (1) Palache, C., H. Berman, and C. Frondel (1944) Dana's system of mineralogy, (7th edition), v. I, 242–243. (2) Hall, S.R. and J.M. Stewart (1973) The crystal structure of argentian pentlandite $(Fe,Ni)_8AgS_8$, compared with the refined structure of pentlandite $(Fe,Ni)_9S_8$. Can. Mineral., 12, 169–177. (3) Riley, J.F. (1977) The pentlandite group $(Fe,Ni,Co)_9S_8$: new data and an appraisal of structure-composition relationships. Mineral. Mag., 41, 345–349.

Crystal Data: Hexagonal. *Point Group:* 622. Intergrowths of elongate or platy deposits to 7 μm in size.

Physical Properties: Hardness = n.d. VHN = n.d. D(meas.) = n.d. D(calc.) = 8.35

Optical Properties: Opaque. *Color:* In reflected light, grayish white. *Anisotropism:* Evident, in colors of creamy yellow and greenish gray.
R_1–R_2: n.d.

Cell Data: *Space Group:* $P6_322$. $a = 13.779$ $c = 16.980$ Z = 18

X-ray Powder Pattern: Penzhina River, USSR.
2.59 (10), 2.71 (9), 2.14 (9), 2.11(9), 1.989 (6), 3.38 (5), 1.784 (5)

Chemistry:

	(1)
Ag	51.3
Au	24.3
Cu	3.1
Se	7.1
S	13.8
Total	99.6

(1) Penzhina River, USSR; by electron microprobe, corresponding to $(Ag_{3.65}Cu_{0.32})_{\Sigma=3.97}Au_{0.97}(S_{3.31}Se_{0.69})_{\Sigma=4.00}$.

Occurrence: In a near surface Au-Ag deposit.

Association: Gold, aguilarite, chalcopyrite, galena.

Distribution: From an undefined locality near the Penzhina River, eastern USSR.

Name: For the Penzhina River, near the Kamchatka Peninsula, USSR.

Type Material: A.E. Fersman Mineralogical Museum, Academy of Sciences, Moscow, USSR.

References: (1) Bochek, L.I., S.M. Sandomirskaya, N.G. Chuvikina, and V.P. Khvorostov (1984) A new selenium-containing sulfide of silver, gold, and copper—penzhinite $(Ag, Cu)_4Au(S, Se)_4$. Zap. Vses. Mineral. Obshch., 113, 356–360 (in Russian). (2) (1985) Amer. Mineral., 70, 875–876 (abs. ref. 1).

Crystal Data: Tetragonal, pseudocubic. *Point Group:* $\bar{4}2m$. As microscopic grains containing hakite inclusions. *Twinning:* Observed in polished section.

Physical Properties: Hardness = 4–4.5 VHN = 234 (50 g load). D(meas.) = n.d. D(calc.) = 5.82

Optical Properties: Opaque. *Color:* In polished section, light brownish pink. *Luster:* Metallic. *Anisotropism:* Strong, with bright colors in rose and green.
R_1–R_2: (400) 30.6–30.9, (420) 29.8–31.0, (440) 29.0–31.1, (460) 28.2–30.5, (480) 27.6–29.4, (500) 27.0–28.2, (520) 26.6–27.1, (540) 26.2–26.3, (560) 25.6–25.8, (580) 25.0–25.5, (600) 24.5–25.6, (620) 24.5–25.9, (640) 24.7–26.5, (660) 25.3–27.4, (680) 26.2–28.4, (700) 27.6–29.3

Cell Data: *Space Group:* $I\bar{4}2m$ (probable). $a = 5.631(2)$ $c = 11.230(5)$ Z = 2

X-ray Powder Pattern: Předbořice, Czechoslovakia.
3.251 (100), 1.503 (100d), 1.986 (90), 1.697 (80), 1.148 (70), 1.290 (60), 1.408 (50)

Chemistry:

	(1)	(2)
Cu	30.7	30.3
Sb	17.4	19.4
As	1.5	
Se	50.4	50.3
Total	100.0	100.0

(1) Předbořice, Czechoslovakia; by electron microprobe, leading to $Cu_{3.01}(Sb_{0.89}As_{0.12})_{\Sigma=1.01}Se_{3.98}$. (2) Cu_3SbSe_4.

Occurrence: A product of epithermal hydrothermal mineralization.

Association: Berzelianite, hakite, eskebornite, umangite, naumannite, hematite, clausthalite, ferroselite, klockmannite, chalcopyrite, pyrite, gold, tetrahedrite, uraninite, goethite, calcite.

Distribution: From Předbořice, Czechoslovakia.

Name: For François Permingeat (1917–1988), mineralogist, University of Toulouse, France.

Type Material: Charles University, Prague, Czechoslovakia; National School of Mines, Paris, France.

References: (1) Johan, Z., P. Picot, R. Pierrot, and M. Kvaček (1971) La permingeatite Cu_3SbSe_4, un nouveau minéral du groupe de la luzonite. Bull. Soc. fr. Minéral., 94, 162–165. (2) (1972) Amer. Mineral., 57, 1554–1555 (abs. ref. 1).

Crystal Data: Orthorhombic. *Point Group:* $2/m\ 2/m\ 2/m$. Fibrous tufted aggregates of crystals up to 0.02 mm wide and 0.07 mm long, prismatic on [001], flattened on {100}, sometimes with hollow terminations. *Twinning:* Observed as contact twins on $\{0hl\}$.

Physical Properties: *Cleavage:* Perfect on {100}. *Fracture:* Irregular. *Tenacity:* Brittle. Hardness = Soft. VHN = n.d. D(meas.) = n.d. D(calc.) = 6.60–6.92

Optical Properties: Transparent. *Color:* Bright red. *Streak:* Reddish orange. *Luster:* Vitreous to adamantine.
Optical Class: Biaxial positive (+). *Pleochroism:* Very intense, X = dark brownish red, Y = yellow, Z = brownish yellow. *Orientation:* $X = c$, $Y = b$, $Z = a$. *Dispersion:* $r > v$, very strong.
$\alpha = 2.3$ $\beta = 2.4$ γ = n.d. 2V(meas.) = ~70° 2V(calc.) = n.d.
R_1–R_2: n.d.

Cell Data: *Space Group:* $P2_12_12$. $a = 17.43(2)$ $b = 12.24(2)$ $c = 4.35(1)$ Z = 2

X-ray Powder Pattern: Coppin Pool, Australia.
2.982 (100), 2.724 (40), 3.948 (30), 2.629 (20), 2.442 (20), 2.141 (20), 2.071 (20)

Chemistry:

	(1)	(2)	(3)
Hg	53.08	51.95	57.68
Ag	22.86	23.54	19.43
S	6.67	9.07	8.79
I	10.32	10.80	6.45
Cl	2.89	2.83	3.56
Br	3.67	0.96	2.11
Total	99.49	99.15	98.02

(1) Cap Garonne, France; by electron microprobe, average of five determinations on two crystals, leading to $Hg_{5.04}Ag_{4.03}S_{3.96}(Cl_{1.55}I_{1.55}Br_{0.87})_{\Sigma=3.97}$. (2) Broken Hill, Australia; by electron microprobe, average of seven analyses, corresponding to $Hg_{5.00}Ag_{4.20}S_{5.45}(I_{1.60}Cl_{1.55}Br_{0.25})_{\Sigma=3.40}$. (3) Coppin Pool, Australia; by electron microprobe, average of 32 analyses on two specimens, leading to $Hg_{5.18}Ag_{4.75}S_{6.00}(Cl_{2.40}I_{1.81}Br_{0.98})_{\Sigma=5.19}$.

Occurrence: As an alteration of tennantite in sandstones and conglomerates (Cap-Garonne, France); in a quartz vein carrying oxidized galena (Coppin Pool, Australia).

Association: Mercurian and argentian tennantite, secondary copper minerals (Cap-Garonne, France); iodargyrite, gold, kaolinite (Broken Hill, Australia); anglesite, cerussite, phosgenite, covellite, pyromorphite, cinnabar (Coppin Pool, Australia).

Distribution: From Cap-Garonne, Var, France. At Broken Hill, New South Wales, and Coppin Pool, Pilbara district, Western Australia.

Name: For Professor Pierre Perroud (1943–), Voltaire College, Geneva, Switzerland, for his work on Cap-Garonne minerals.

Type Material: Natural History Museum, Geneva, Switzerland; Western Australian Government Chemical Laboratories, Perth, Western Australia; Museum of Victoria, Melbourne, Australia.

References: (1) Sarp, H., W.D. Birch, P.F. Hlava, A. Pring, D.K.B. Sewell, and E.H. Nickel (1987) Perroudite, a new sulfide-halide of Hg and Ag from Cap-Garonne, Var, France, and from Broken Hill, New South Wales, and Coppin Pool, Western Australia. Amer. Mineral., 72, 1251–1256. (2) Mumme, W.G. and E.H. Nickel (1987) Crystal structure and crystal chemistry of perroudite: a mineral from Coppin Pool, Western Australia. Amer. Mineral., 72, 1257–1262.

Crystal Data: Orthorhombic. *Point Group:* $2/m$ $2/m$ $2/m$ or $mm2$. Crystals are tabular and up to a few tenths of one mm in size.

Physical Properties: *Cleavage:* Imperfect, parallel to flattening of crystals. Hardness = ~3 VHN = 102 (25 g load). D(meas.) = n.d. D(calc.) = 7.707

Optical Properties: Opaque. *Color:* In polished section, cream colored. *Luster:* Strong metallic. *Anisotropism:* Weak, in dark green and violet.

R_1–R_2: (400) 44.0–44.4, (420) 44.1–44.8, (440) 44.2–45.2, (460) 44.3–45.4, (480) 44.4–45.7, (500) 44.7–46.1, (520) 45.1–46.3, (540) 45.2–46.5, (560) 45.2–46.4, (580) 45.1–46.2, (600) 45.1–46.0, (620) 45.0–45.9, (640) 44.9–45.8, (660) 44.9–45.7, (680) 44.8–45.6, (700) 44.8–45.5

Cell Data: *Space Group:* *Pnam* or $Pna2_1$. a = 16.176(5) b = 14.684(5) c = 4.331(3) Z = 4

X-ray Powder Pattern: Petrovice deposit, Czechoslovakia.
3.120 (100), 2.961 (100), 3.546 (80), 3.186 (80), 3.621 (70), 2.720 (50), 2.262 (50)

Chemistry:

	(1)	(2)	(3)
Pb	16.7	17.3	17.24
Hg	17.4	17.9	16.69
Cu	16.0	15.3	15.85
Bi	16.6	17.4	17.38
Se	34.4	32.6	32.84
Total	101.1	100.5	100.00

(1) Petrovice deposit, Czechoslovakia; by electron microprobe, corresponding to $Pb_{0.95}Hg_{1.02}Cu_{2.97}Bi_{0.93}Se_{5.13}$. (2) Do.; corresponding to $Pb_{1.01}Hg_{1.08}Cu_{2.91}Bi_{1.01}Se_{4.99}$. (3) $PbHgCu_3BiSe_5$.

Occurrence: Found in hydrothermal dolomite-calcite veins with other selenides.

Association: Berzelianite, eucairite, crookesite, tyrrellite, ferroselite, bukovite, krutaite, athabascaite, umangite, eskebornite, calcite, dolomite.

Distribution: In the Petrovice deposit, Czechoslovakia.

Name: For its occurrence in the Petrovice deposit, Czechoslovakia.

Type Material: National School of Mines, Paris, France.

References: (1) Johan, Z., M. Kvaček, and P. Picot (1976) La petrovicite, $Cu_3HgPbBiSe_5$, un nouveau minéral. Bull. Soc. fr. Minéral., 99, 310–313. (2) (1977) Amer. Mineral., 62, 594–595 (abs. ref. 1).

Crystal Data: Monoclinic. *Point Group:* $2/m$, 2, or m. As fine-grained rims, to 20 μm, on gold.

Physical Properties: *Tenacity:* Brittle. Hardness = 2–2.5 VHN = 39.9–47.8
D(meas.) = n.d. D(calc.) = 9.5

Optical Properties: Opaque. *Color:* Dark gray to black. *Streak:* Dark gray. *Luster:* Dull metallic.
R_1–R_2: n.d.

Cell Data: *Space Group:* $P2/m$, $P2$, or Pm. $a = 4.943(9)$ $b = 6.670(9)$ $c = 7.221(9)$
$\beta = 95.68(7)^\circ$ Z = 4

X-ray Powder Pattern: Synthetic AgAuS.
2.77 (10), 2.63 (5), 2.39 (4), 2.25 (4), 7.16 (3), 3.96 (3)

Chemistry:

	(1)
Au	58.6
Ag	31.0
S	9.54
Se	1.35
Total	100.5

(1) Maikain deposit, USSR; by electron microprobe, average of seven analyses, corresponding to $Au_{0.99}Ag_{0.96}(S_{0.99}Se_{0.06})_{\Sigma=1.05}$.

Occurrence: At a depth of about 60-65 m in a gold deposit.

Association: Gold, chlorargyrite.

Distribution: In the Maikain gold deposit, central Kazakhstan, USSR.

Name: For Nina Petrovskaya, Russian mineralogist.

Type Material: Central Siberian Geological Museum, Novosibirsk, USSR.

References: (1) Nesterenko, G.V., A.I. Kuznetsova, N.A. Pal'chik, and Y.G. Lavrent'ev (1984) Petrovskaite, AuAg(S, Se), a new selenium-containing sulfide of gold and silver. Zap. Vses. Mineral. Obshch., 113, 602–607 (in Russian). (2) (1985) Amer. Mineral., 70, 1331 (abs. ref. 1).

Crystal Data: Cubic. *Point Group:* 432. Massive, fine granular to compact and as irregular shaped blebs to 2 mm.

Physical Properties: *Cleavage:* {001}. *Fracture:* Subconchoidal. *Tenacity:* Slightly sectile to brittle. Hardness = 2.5–3 VHN = 48 (10 g load). D(meas.) = 8.7–9.4 D(calc.) = 8.74

Optical Properties: Opaque. *Color:* Bright steel-gray to iron-gray to iron-black, often tarnished from bronze-yellow to sooty black; in reflected light, grayish white with a pale bluish tint. *Luster:* Metallic. *Anisotropism:* Noticeable in part.

R: (400) 45.4, (420) 43.6, (440) 41.8, (460) 40.7, (480) 39.8, (500) 39.2, (520) 38.6, (540) 38.2, (560) 37.9, (580) 37.6, (600) 37.4, (620) 37.3, (640) 37.1, (660) 37.0, (680) 36.9, (700) 36.9

Cell Data: *Space Group:* $I4_132$. $a = 10.385(4)$ Z = 8

X-ray Powder Pattern: Botés, Romania.
2.77 (100), 2.12 (80), 2.03 (70), 2.44 (60), 2.32 (60), 7.31 (50), 1.893 (50)

Chemistry:

	(1)	(2)	(3)
Ag	41.37	41.87	41.71
Au	23.42	25.16	25.39
Te	33.00	33.21	32.90
Hg	2.26		
Cu	0.16		
Total	100.21	100.24	100.00

(1) Kalgoorlie, Western Australia. (2) Mother Lode district, California, USA. (3) Ag_3AuTe_2.

Occurrence: With other tellurides in vein-controlled gold deposits.

Association: Gold, hessite, sylvanite, krennerite, calaverite, altaite, montbrayite, melonite, frohbergite, tetradymite, rickardite, vulcanite, pyrite.

Distribution: Noted in small amounts at a number of localities other than those listed here. In the USA, at Gold Hill, Boulder Co., Lake City, Hinsdale Co., and Leadville, Lake Co., Colorado; from California, in the Golden Rule and Norwegian mines, Tuttletown, Tuolumne Co., in the Stanislaus and Melones mines, Carson Hill, Calaveras Co., and in other mines along the Mother Lode. In Canada, at the Hollinger mine, Timmins, and the Lake Shore mine, Kirkland Lake, Ontario; in Quebec, at the Robb-Montbray mine, Montbray Township, the Noranda mine at Rouyn, the Horne mine at Noranda, and many other localities. From Botés, Săcărâmb (Nagyág), and Baia Sprie (Felsőbánya), Romania. At Kalgoorlie, Western Australia. In the Kuramin Mountains, eastern Uzbekistan, the Byn'govsk Au-Te deposit, central Ural Mountains, and the Zhana-Tyube deposit, northern Kazakhstan, USSR.

Name: After W. Petz, who first analyzed the mineral.

Type Material: Harvard University, Cambridge, Massachusetts, 99348; National Museum of Natural History, Washington, D.C., USA, R9556.

References: (1) Palache, C., H. Berman, and C. Frondel (1944) Dana's system of mineralogy, (7th edition), v. I, 186–187. (2) Frueh, A.J., Jr. (1959) The crystallography of petzite, Ag_3AuTe_2. Amer. Mineral., 44, 693–701. (3) Chamid, S., E.A. Pobedimskaya, E.M. Spiridonov, and N.V. Belov (1978) Refinement of the structure of petzite $AuAg_3Te_2$. Kristallografiya, 23, 483–486 (in Russian). (4) (1979) Mineral. Abs., 30, 353 (abs. ref. 3).

Crystal Data: Orthorhombic. *Point Group:* 222, $mm2$, or $2/m\ 2/m\ 2/m$. Crystals are typically pseudohexagonal plates 0.5 mm or less in size. *Twinning:* Penetration twins on {120} mimic hexagonal symmetry.

Physical Properties: Hardness = n.d. VHN = ~41 (15 g load). D(meas.) = n.d. D(calc.) = 5.20

Optical Properties: Opaque. *Color:* In polished section, creamy white. *Luster:* Metallic. *Pleochroism:* Strong. *Anisotropism:* Strong, in gray-violet tints.

R_1–R_2: (400) 23.5–28.7, (420) 24.5–30.3, (440) 25.5–31.9, (460) 25.8–32.4, (480) 25.4–32.0, (500) 24.6–31.3, (520) 24.3–31.0, (540) 24.2–31.0, (560) 24.4–31.4, (580) 24.6–32.2, (600) 25.2–33.1, (620) 25.7–33.8, (640) 26.5–34.7, (660) 27.3–35.6, (680) 28.2–36.6, (700) 29.2–37.6

Cell Data: *Space Group:* $C222$, $Cmm2$, $Amm2$, or $Cmmm$. $a = 5.40(2)$ $b = 10.72(5)$ $c = 9.04(4)$ Z = 4

X-ray Powder Pattern: Allchar, Yugoslavia.
2.912 (100), 4.26 (90), 3.80 (70), 3.33 (70), 5.40 (50), 4.53 (50), 2.556 (50)

Chemistry:

	(1)	(2)	(3)
Tl	46.8	47.8	49.57
Pb	2.8	1.9	
Fe	28.6	28.2	27.09
S	24.1	23.9	23.34
Total	102.3	101.8	100.00

(1) Allchar, Yugoslavia; by electron microprobe, corresponding to $(Tl_{0.91}Pb_{0.05})_{\Sigma=0.96}Fe_{2.04}S_{3.00}$. (2) Do.; $(Tl_{0.93}Pb_{0.04})_{\Sigma=0.97}Fe_{2.03}S_{3.00}$. (3) $TlFe_2S_3$.

Occurrence: In a hydrothermal ore deposit with other arsenic minerals.

Association: Realgar, lorandite, raguinite, pyrite.

Distribution: From Allchar, Macedonia, Yugoslavia.

Name: For Paul Picot, mineralogist with the B.R.G.M., Orléans, France.

Type Material: National School of Mines, Paris, France.

References: (1) Johan, Z., R. Pierrot, H.-J. Schubnel, and F. Permingeat (1970) La picotpaulite, $TlFe_2S_3$, une nouvelle espèce minérale. Bull. Soc. fr. Minéral., 93, 545–549 (in French with English abs.). (2) (1972) Amer. Mineral., 57, 1909 (abs. ref. 1).

Crystal Data: Orthorhombic. *Point Group:* $mm2$. As polycrystalline aggregates.

Physical Properties: Hardness = n.d. VHN = n.d. D(meas.) = 4.97 D(calc.) = 4.75

Optical Properties: Translucent. *Color:* Gray-black; in polished section, white, with red internal reflections; dark red in thin section. *Luster:* Metallic, lively. *Anisotropism:* Moderate. R_1–R_2: (400) 35.9–39.6, (420) 35.2–38.8, (440) 34.5–38.0, (460) 33.7–37.2, (480) 33.1–36.6, (500) 32.5–36.0, (520) 31.9–35.5, (540) 31.3–35.0, (560) 30.7–34.5, (580) 30.2–34.0, (600) 29.7–33.4, (620) 29.1–32.8, (640) 28.6–32.1, (660) 28.0–31.5, (680) 27.4–30.9, (700) 26.7–30.3

Cell Data: *Space Group:* $Pna2_1$. $a = 38.746(8)$ $b = 8.816(2)$ $c = 7.989(2)$ Z = 4

X-ray Powder Pattern: Jas-Roux, France.
3.59 (100), 3.49 (90), 2.70 (90), 3.63 (80), 2.84 (80), 2.347 (80), 2.52 (60)

Chemistry:

	(1)	(2)
Tl	20.95	20.94
Sb	37.65	37.43
As	15.54	15.35
S	26.57	26.28
Total	100.70	100.00

(1) Jas-Roux, France; by electron microprobe, average of seven analyses. (2) $Tl_2Sb_6As_4S_{16}$.

Occurrence: In quartz veins.

Association: Stibnite, realgar, pyrite, an unnamed amorphous mineral of composition $Tl(As,Sb)_{10}S_{16}$.

Distribution: From Jas-Roux, Hautes-Alpes, France.

Name: For Roland Pierrot, Head of Mineralogy, B.R.G.M., Orléans, France.

Type Material: National School of Mines, Paris, France.

References: (1) Guillemin, C., Z. Johan, C. Laforêt, and P. Picot (1970) La pierrotite, $Tl_2(Sb,As)_{10}S_{17}$, une nouvelle espèce minérale. Bull. Soc. fr. Minéral., 93, 66–71 (in French with English abs.). (2) (1972) Amer. Mineral., 57, 1909–1910 (abs. ref. 1). (3) Engel, P., M. Gostojić, and W. Nowacki (1983) The crystal structure of pierrotite, $Tl_2(As,Sb)_{10}S_{16}$. Zeits. Krist., 165, 209–215. (4) (1985) Amer. Mineral., 70, 220 (abs. ref. 3).

Crystal Data: Hexagonal. *Point Group:* $\overline{3}\ 2/m$. Interlayered with thin (several μm thick) layers of hessite.

Physical Properties: Hardness = Soft. VHN = n.d. D(meas.) = n.d. D(calc.) = [8.4]

Optical Properties: Opaque. *Color:* Tin-white to light steel-gray. *Luster:* Metallic. *Anisotropism:* Noticeable.
R_1–R_2: n.d.

Cell Data: *Space Group:* $R\overline{3}m$. $a = 4.446(2)$ $c = 41.94(2)$ Z = 3

X-ray Powder Pattern: Börzsöny, Hungary.
3.25 (vs), 2.36 (s), 2.22 (s), 1.998 (s), 1.833 (s), 1.485 (s), 4.68 (m)

Chemistry:

	(1)	(2)
Bi	64.7 – 66.0	68.59
Ag	0.0 – 0.3	
Pb	0.9 – 1.3	
Fe	trace	
Te	30.6 – 31.2	31.41
S	trace – 0.1	
Total		100.00

(1) Börzsöny, Hungary; by electron microprobe, range of analyses, the average of which gives $(Bi_{3.87}Pb_{0.07}Ag_{0.01})_{\Sigma=3.95}(Te_{3.01}S_{0.04})_{\Sigma=3.05}$. (2) Bi_4Te_3.

Occurrence: Of hydrothermal origin.

Association: Hessite (Börzsöny, Hungary); tsumoite (Sylvanite, New Mexico, USA); argentian pentlandite, hessite, argentian tetrahedrite (Koronuda deposit, Greece).

Distribution: From Börzsöny, near Gran, Hungary. At Sylvanite, Hidalgo Co., New Mexico, USA. In the Koronuda Au-Cu deposit, central Macedonia, Greece.

Name: For the type locality, Börzsöny (Deutsch-Pilsen), Hungary.

Type Material: University of Tokyo Museum, Tokyo, Japan.

References: (1) Palache, C., H. Berman, and C. Frondel (1944) Dana's system of mineralogy, (7th edition), v. I, 167. (2) Ozawa, T. and H. Shimazaki (1982) Pilsenite redefined and wehrlite discredited. Proc. Japan Acad., 58, Series B, 291–294 (in English). (3) (1984) Amer. Mineral., 69, 215 (abs. ref. 1). (4) Ramdohr, P. (1969) The ore minerals and their intergrowths, (3rd edition), 438–439.

Crystal Data: Tetragonal. *Point Group:* $\overline{4}2m$ or $\overline{4}$. As irregular grains intergrown with other minerals. *Twinning:* Polysynthetic.

Physical Properties: Hardness = n.d. VHN = 170–251 (25 g load). D(meas.) = n.d. D(calc.) = 4.822

Optical Properties: Opaque. *Color:* Brownish gray in reflected light, with red internal reflections. *Luster:* Metallic. *Anisotropism:* Strong, from brick-red to light green.
R_1–R_2: (400) — , (420) 24.6–21.7, (440) 24.2–21.8, (460) 24.1–21.9, (480) 23.9–21.9, (500) 23.8–21.8, (520) 23.9–21.9, (540) 24.1–22.2, (560) 24.3–22.5, (580) 24.2–22.7, (600) 24.1–22.4, (620) 23.8–21.8, (640) 23.2–21.3, (660) 22.5–20.9, (680) 22.2–20.5, (700) 21.6–20.1

Cell Data: *Space Group:* $I\overline{4}2m$ or $I\overline{4}$. $a = 5.786(4)$ $c = 10.829(6)$ Z = 2

X-ray Powder Pattern: Pirquitas deposit, Argentina.
3.267 (100), 1.976 (80), 1.735 (80), 2.049 (60), 2.901 (40), 1.289 (40), 1.165 (40)

Chemistry:

	(1)	(2)
Ag	39.72	40.85
Cu	0.06	
Zn	11.40	12.38
Fe	1.31	
Sn	23.12	22.48
S	24.42	24.29
Total	100.03	100.00

(1) Pirquitas deposit, Argentina; by electron microprobe, corresponding to $(Ag_{1.93}Cu_{0.01})_{\Sigma=1.94}(Zn_{0.92}Fe_{0.12})_{\Sigma=1.04}Sn_{1.02}S_{4.00}$. (2) Ag_2ZnSnS_4.

Polymorphism & Series: Forms a series with hocartite.

Occurrence: As hydrothermal mineralization in veins associated with subvolcanic environments, similar to the classic Bolivian tin and silver deposits.

Association: Hocartite, pyrite, marcasite, wurtzite, franckeite, miargyrite, aramayoite, chalcostibite, stannite, kësterite, rhodostannite, cassiterite.

Distribution: In the Pirquitas deposit, Rinconada Department, Jujuy Province, Argentina. From the Toyoha mine, Hokkaido, Japan.

Name: For the type locality at the Pirquitas deposit in Argentina.

Type Material: National School of Mines, Paris, France.

References: (1) Johan, Z. and P. Picot (1982) La pirquitasite, Ag_2ZnSnS_4, un nouveau membre du groupe de la stannite. Bull. Minéral., 105, 229–235 (in French with English abs.). (2) (1983) Amer. Mineral., 68, 1249 (abs. ref. 1).

Crystal Data: Monoclinic. *Point Group:* $2/m$. Thick crystals to 1 cm, tabular on {001}, with numerous well developed prisms and pinacoids; occasionally prismatic along [20$\bar{1}$]; striated [110]. Also massive, granular to compact.

Physical Properties: *Cleavage:* Very good on {112}. *Fracture:* Conchoidal to uneven. *Tenacity:* Brittle. Hardness = 2.5 VHN = n.d. D(meas.) = 5.54 D(calc.) = 5.55

Optical Properties: Opaque. *Color:* Blackish lead-gray; pale white in reflected light. *Streak:* Blackish lead-gray, deep red-brown. *Luster:* Metallic. *Pleochroism:* Noticeable in some orientations. *Anisotropism:* Distinct in oil.

R_1–R_2: (400) 33.6–43.4, (420) 33.8–43.5, (440) 34.0–43.6, (460) 34.2–43.6, (480) 34.3–43.5, (500) 34.3–43.3, (520) 34.2–43.0, (540) 34.0–42.5, (560) 33.6–41.9, (580) 33.2–41.0, (600) 32.7–40.2, (620) 32.0–39.2, (640) 31.2–38.3, (660) 30.3–37.4, (680) 29.5–36.6, (700) 28.9–35.8

Cell Data: *Space Group:* $C2/c$. $a = 13.4857(8)$ $b = 11.8656(4)$ $c = 19.9834(7)$ $\beta = 107.168(4)°$ Z = 4

X-ray Powder Pattern: Wolfsberg, Germany.
3.21 (100), 3.26 (90), 2.911 (90), 3.87 (80), 3.77 (70), 2.622 (60), 3.61 (50)

Chemistry:

	(1)	(2)	(3)
Pb	41.24	40.28	40.55
Ag		0.18	
Sb	37.35	38.30	38.12
S	21.10	21.43	21.33
Total	99.69	100.19	100.00

(1) Wolfsberg, Germany. (2) Oruro, Bolivia. (3) $Pb_5Sb_8S_{17}$.

Occurrence: In hydrothermal vein deposits.

Association: Cassiterite, franckeite, andorite, semseyite, geocronite, boulangerite, galena, pyrite.

Distribution: In Germany, from Wolfsberg, in the Harz Mountains, and Arnsberg, North Rhine-Westphalia. In France, at Leyraux, Cantal; Chazelles, Haute-Loire; Les Cougnasses and Riou Beyrou, Hautes-Alpes; and Bournac, Montagne Noire. At Rujevac, Yugoslavia. From the Azatec deposit, Caucasus Mountains, and the Bal'Kumeisk deposit, Yakutia, USSR. In Canada, from the Porter claim, Carbon Hill, Wheaton River district, Yukon Territory. From the Purisima vein, San José mine, Oruro, Bolivia.

Name: From the Greek for *oblique*, in reference to the crystal morphology.

References: (1) Palache, C., H. Berman, and C. Frondel (1944) Dana's system of mineralogy, (7th edition), v. I, 464–465. (2) Nuffield, E.W. and M.A. Peacock (1945) Studies of mineral sulpho-salts: VIII — plagionite and semseyite. Univ. Toronto Studies, Geol. Series, 49, 17–39. (3) Jambor, J.L. (1969) Sulphosalts of the plagionite group. Mineral. Mag., 37, 442–446. (4) Cho, S.-A. and B.J. Wuensch (1970) Crystal chemistry of the plagionite group. Nature, 225, 444–445. (5) Cho, S.-A. and B.J. Wuensch (1974) The crystal structure of plagionite, $Pb_5Sb_8S_{17}$, the second member in the homologous series $Pb_{3+2n}Sb_8Sb_{15+2n}$. Zeits. Krist., 139, 351–378.

Crystal Data: Cubic. *Point Group:* $2/m\ \overline{3}$. As subhedral grains, the largest being a triangular crystal about 1.1 mm on the side.

Physical Properties: Hardness = n.d. VHN = 1379–1584, 1486 average (50 g load). D(meas.) = 8.0 D(calc.) = 8.375

Optical Properties: Opaque. *Color:* In polished section, gray. *Luster:* Metallic.
R: (400) — , (420) 47.7, (440) 47.4, (460) 47.3, (480) 47.3, (500) 47.1, (520) 47.5, (540) 48.0, (560) 48.2, (580) 48.7, (600) 49.0, (620) 49.9, (640) 48.9, (660) 49.5, (680) 49.8, (700) 49.6

Cell Data: *Space Group:* $Pa3$. $a = 5.790(1)$ Z = 4

X-ray Powder Pattern: Onverwacht mine, South Africa (all reflections broad).
1.746 (100), 2.896 (90), 3.345 (80), 1.114 (70), 2.047 (60), 2.590 (50), 2.364 (50)

Chemistry:

	(1)	(2)	(3)
Pt	26.9	30.4	22.4
Rh	12.8	10.8	8.0
Ru	11.4	9.1	6.4
Ir	3.6	5.9	17.5
Os	0.58	0.10	2.7
As	31.7	33.2	31.2
S	13.0	10.7	10.7
Total	99.98	100.2	98.9

(1) Onverwacht mine, South Africa; by electron microprobe, corresponding to $(Pt_{0.34}Rh_{0.30}Ru_{0.28}Ir_{0.05})_{\Sigma=0.97}As_{1.03}S_{0.99}$. (2) Do.; leading to $(Pt_{0.40}Rh_{0.27}Ru_{0.23}Ir_{0.08})_{\Sigma=0.98}As_{1.15}S_{0.87}$. (3) Western Platinum mine, South Africa; by electron microprobe.

Occurrence: In replacement pegmatite deposits.

Association: Stibiopalladinite, ruthenarsenite, mertieite-II, genkinite, Pt–Fe–Cu–Ni alloys, bornite, chromite.

Distribution: Found in the Onverwacht, Driekop, Union, and Western Platinum mines, Transvaal, South Africa. At Fox Gulch, Goodnews Bay, Alaska, USA.

Name: For the composition.

Type Material: Royal Ontario Museum, Toronto; National Museum of Canada, Ottawa, Canada; National Museum of Natural History, Washington, D.C., USA, 136485; A.E. Fersman Mineralogical Museum, Academy of Sciences, Moscow, USSR.

References: (1) Cabri, L.J., J.H.G. Laflamme, and J.M. Stewart (1977) Platinum-group minerals from Onverwacht. II. Platarsite, a new sulfarsenide of platinum. Can. Mineral., 15, 385–388. (2) Szymański, J.T. (1979) The crystal structure of platarsite, $Pt(As, S)_2$, and a comparison with sperrylite, $PtAs_2$. Can. Mineral., 17, 117–123. (3) (1979) Amer. Mineral., 64, 657 (abs. ref. 2). (4) Cabri, L.J., Ed. (1981) Platinum group elements: mineralogy, geology, recovery. Can. Inst. Min. & Met., 128–129.

Crystal Data: Cubic. *Point Group:* $4/m\ \overline{3}\ 2/m$. As cubic crystals, commonly in rounded or angular grains; also found as microscopic ($\sim$20 μm across) grains in Pt–Fe alloy. *Twinning:* Polysynthetic on {111}.

Physical Properties: *Cleavage:* Indistinct on {001}. *Fracture:* Hackly. Hardness = 6–7 VHN = n.d. D(meas.) = 22.65 D(calc.) = 22.66

Optical Properties: Opaque. *Color:* Silver-white with a yellow tinge; in polished section, white. *Luster:* Metallic.
R: n.d.

Cell Data: *Space Group:* $Fm3m$. $a = 3.8312(5)$ Z = 4

X-ray Powder Pattern: n.d.

Chemistry:

	(1)	(2)
Ir	76.85	72.3
Pt	19.64	14.7
Os		5.3
Pd	0.89	
Ru		4.5
Fe		1.9
Cu	1.78	0.33
Ni		0.14
Total	99.16	99.17

(1) Nizhni Tagil, USSR. (2) Tulameen River area, Canada; by electron microprobe, corresponding to $Ir_{0.67}Pt_{0.13}Os_{0.05}Ru_{0.08}Fe_{0.06}Cu_{0.01}$.

Occurrence: Found in placer deposits derived from platinum-bearing dunites.

Association: Pt–Fe alloys, tulameenite.

Distribution: In the USA, from placers in Siskiyou Co., California, and at Fox Gulch, Goodnews Bay, Alaska. In Canada, on the south bank of the Tulameen River, between Princeton and Coalmont, British Columbia. In the USSR, at Nizhni Tagil, Ural Mountains; also from the Inagli massif, southern Yakutia. In Burma, at Ava, near Mandalay.

Name: For the composition, with iridium comprising 50–80 atomic percent of platinum + iridium.

Type Material: n.d.

References: (1) Palache, C., H. Berman, and C. Frondel (1944) Dana's system of mineralogy, (7th edition), v. I, 110–111. (2) Cabri, L.J. and M.H. Hey (1974) Platiniridium—confirmation as a valid mineral species. Can. Mineral., 12, 299–308.

Platinum Pt

Crystal Data: Cubic. *Point Group:* $4/m\ \overline{3}\ 2/m$. Cubic crystals, often distorted; usually as grains or scales, sometimes as nuggets or lumps up to 9 kg in weight. *Twinning:* On {111}.

Physical Properties: *Fracture:* Hackly. *Tenacity:* Malleable and ductile. Hardness = 4–4.5 VHN = 297–339 (100 g load). D(meas.) = 14–19 D(calc.) = 21.472 Nonmagnetic to distinctly magnetic when rich in iron.

Optical Properties: Opaque. *Color:* Whitish steel-gray to dark gray; in polished section, white. *Luster:* Metallic.
R: (400) 57.0, (420) 58.0, (440) 59.0, (460) 60.0, (480) 61.0, (500) 61.8, (520) 62.7, (540) 63.5, (560) 64.2, (580) 64.8, (600) 65.6, (620) 66.3, (640) 66.8, (660) 67.4, (680) 68.0, (700) 68.5

Cell Data: *Space Group:* $Fm3m$. $a = 3.9231$ Z = 4

X-ray Powder Pattern: Synthetic.
2.265 (100), 1.9616 (53), 1.1826 (33), 1.3873 (31), 0.8008 (29), 0.9000 (22), 0.8773 (20)

Chemistry:

	(1)	(2)	(3)
Pt	86.20	79.48	92.2
Ir	0.85	0.82	
Ir–Os	0.95	1.41	
Rh	1.40	0.75	1.2
Pd	0.50	0.49	1.3
Au	1.00	0.49	
Cu	0.60		0.5
Fe	7.80	16.50	5.3
gangue	0.95		
Total	100.25	99.94	100.5

(1) Chocó, Colombia. (2) Birbir River, Ethiopia. (3) Nizhni Tagil, USSR; by electron microprobe.

Occurrence: Chiefly in placer deposits, or in mafic and ultramafic igneous rocks; rarely found in hydrothermal quartz veins or contact metamorphic deposits.

Association: Pt–Fe alloys, chalcopyrite, chromite, magnetite.

Distribution: From many deposits world-wide. In the USA, from Platinum Creek, Goodnews Bay, Alaska; in California, in a number of placers, as in Trinity Co.; and at Oroville, Butte Co. In Oregon, at Cape Blanco, Port Orford, Curry Co. In Canada, at Rivière-du-Loup and Rivière des Plantes, Beauce Co., Quebec; in British Columbia, in the Kamloops district, on the Fraser and Tranquille rivers, and in the Similkameen district, on Granite, Cedar, and Olivine Creeks, tributaries to the Tulameen River; in Alberta, near Edmonton. From near Papayan, in the Department of Chocó, Cauca, Colombia. In the USSR, the Ural Mountains, in a large district surrounding Nizhni Tagil. In South Africa, at a number of deposits along the Merensky Reef of the Bushveld Complex, Transvaal, South Africa.

Name: From the Spanish *platina*, diminutive of *plata, silver.*

References: (1) Palache, C., H. Berman, and C. Frondel (1944) Dana's system of mineralogy, (7th edition), v. I, 106–109. (2) (1953) NBS Circ. 539, 1, 31. (3) Cabri, L.J. and C.E. Feather (1975) Platinum–iron alloys; nomenclature based on a study of natural and synthetic alloys. Can. Mineral., 13, 117–126. (4) Criddle, A.J. and C.J. Stanley, Eds. (1986) The quantitative data file for ore minerals. British Museum (Natural History), London, England, 285.

Crystal Data: Hexagonal. *Point Group:* $\bar{3}$. As thin plates or foliae.

Physical Properties: *Cleavage:* Good on {0001}; fair on $\{10\bar{1}1\}$. Hardness = 2–3 VHN = n.d. D(meas.) = 7.98 D(calc.) = 7.45

Optical Properties: Opaque. *Color:* Iron-black to dark steel-gray. *Streak:* Black, shining. *Luster:* Metallic. *Anisotropism:* Strong.
R_1–R_2: n.d.

Cell Data: *Space Group:* n.d. $a = 8.49$ $c = 20.8$ Z = 2

X-ray Powder Pattern: n.d.

Chemistry:

	(1)
Pb	26.45
Bi	50.22
Se	19.20
S	4.13
Total	100.00

(1) Falun, Sweden; recalculated after deduction of Cu 0.32% and Fe 0.30% as chalcopyrite and insoluble 0.36%.

Occurrence: In hydrothermal veins.

Association: Chalcopyrite, quartz.

Distribution: At Falun, Sweden.

Name: From the Greek *to broaden*, in allusion to its platy habit.

Type Material: n.d.

References: (1) Palache, C., H. Berman, and C. Frondel (1944) Dana's system of mineralogy, (7th edition), v. I, 474. (2) Sindeeva, N.D. (1964) Mineralogy and types of deposits of selenium and tellurium, 82–83. (3) Ramdohr, P. and H. Strunz (1978) Klockmanns Lehrbuch der Mineralogie, 434 (in German).

Crystal Data: Monoclinic. *Point Group:* 2, m, or $2/m$. Tabular crystals heavily striated parallel to elongation. *Twinning:* Exhibits very fine twin lamellae.

Physical Properties: *Cleavage:* {100} perfect. Hardness = n.d. VHN = 154 (50 g load). D(meas.) = n.d. D(calc.) = 5.72

Optical Properties: Opaque. *Color:* Lead-gray to black. *Streak:* Black. *Luster:* Metallic. *Pleochroism:* Strong, white to brownish gray.
R_1–R_2: n.d.

Cell Data: *Space Group:* $P2$, Pm, $P2/m$, $P2_1$ or $P2_1/m$. $a = 45.4$ $b = 8.29$ $c = 21.3$ $\beta = 92°30(30)'$ Z = 4

X-ray Powder Pattern: Madoc, Canada.
3.39 (100), 3.32 (100), 2.785 (70), 2.086 (60), 3.98 (40), 3.49 (40), 2.97 (40)

Chemistry:

	(1)	(2)	(3)
Pb	51.0	50.7	48.15
Sb	28.0	26.3	31.83
As	2.4	3.3	
S	18.8	21.3	20.02
Total	100.2	101.6	100.00

(1) Madoc, Canada; by electron microprobe. (2) Novoye, USSR; by electron microprobe. (3) $Pb_{16}Sb_{18}S_{43}$.

Occurrence: As masses and stringers through dolomitic and calcitic marbles. At the edges of other sulfosalt minerals, and extending into them along microscopic veinlets.

Association: Boulangerite, jamesonite, antimonian baumhauerite, zinkenite, semseyite, geocronite, robinsonite (Madoc, Canada); sphalerite, pyrite, galena, sorbyite, twinnite, guettardite, baumhauerite, realgar, orpiment, cinnabar, fluorite, quartz (Novoye, USSR).

Distribution: From near Madoc, Ontario, Canada. At Novoye, Khaidarkan, Kirgizia, USSR.

Name: In honor of John Playfair (1748–1819), Professor of Natural Philosophy, Edinburgh, Scotland.

Type Material: National Museum of Canada, Ottawa, Canada.

References: (1) Jambor, J.L. (1967) New lead sulfantimonides from Madoc, Ontario. Part 2 — mineral descriptions. Can. Mineral., 9, 191–213. (2) (1968) Amer. Mineral., 53, 1424 (abs. ref. 1). (3) Mozgova, N.N., N.S. Bortnikov, Y.S. Borodaev, and A.I. Tzépine (1982) Sur la non-stoechiométrie des sulfosels antimonieux arséniques de plomb. Bull. Minéral., 105, 3–10 (in French with English abs.).

Crystal Data: Hexagonal. *Point Group:* 6*mm* (synthetic). As tiny aggregates of minute grains, up to 0.7 mm.

Physical Properties: Hardness = ~5 VHN = 407–441 (50 g load). D(meas.) = n.d. D(calc.) = 12.36

Optical Properties: Opaque. *Color:* Bright white with faint rose tint in reflected light. *Pleochroism:* Noticeable in air. *Anisotropism:* Strong, orange-brown to dark brown.
R_1–R_2: (400) — , (420) — , (440) 43.5–46.6, (460) 44.6–48.2, (480) 45.8–49.5, (500) 47.5–50.8, (520) 49.0–52.6, (540) 50.7–54.4, (560) 52.4–55.8, (580) 53.9–57.1, (600) 55.2–58.0, (620) 56.6–59.2, (640) 58.3–60.8, (660) 60.0–62.6, (680) 61.4–64.4, (700) 62.8–66.1

Cell Data: *Space Group:* $P6_3mc$ (synthetic). $a = 4.470$ $c = 5.719$ Z = 1

X-ray Powder Pattern: Talnakh area, USSR.
2.30 (10), 2.23 (10), 1.207 (6), 3.20 (5), 1.302 (5), 1.760 (4), 1.602 (4)

Chemistry:

	(1)	(2)
Pd	42.35	43.51
Pb	55.12	56.49
Ag	2.07	
Cu	0.22	
Sn	0.47	
Bi	0.80	
Sb	0.08	
Total	101.11	100.00

(1) Talnakh area, USSR; by electron microprobe, average of four analyses. (2) Pd_3Pb_2.

Occurrence: In Ni-Cu ores (Talnakh area, USSR).

Association: Cubanite, talnakhite, polarite, stannopalladinite, silver, sphalerite, galena (Talnakh area, USSR).

Distribution: In the Mayak mine, Talnakh area, Noril'sk region, western Siberia, USSR. From the Stillwater Complex, Montana, USA.

Name: For the chemical composition.

Type Material: n.d.

References: (1) Genkin, A.D., T.L. Evstigneeva, L.N. Vyal'sov, I.P. Laputina, and N.V. Troneva (1970) Plumbopalladinite, a new mineral from copper–nickel ores. Geol. Rudn. Mestorozhd, 5, 63–68 (in Russian). (2) (1971) Amer. Mineral., 56, 1121 (abs. ref. 1).

Polarite $Pd(Bi, Pb)$

Crystal Data: Orthorhombic. *Point Group:* $mm2$. As grains up to 0.3 mm.

Physical Properties: Hardness = n.d. VHN = 168–232 (50 g load). D(meas.) = n.d. D(calc.) = 12.51

Optical Properties: Opaque. *Color:* White with yellowish tint in reflected light. *Luster:* Metallic. *Anisotropism:* Slight gray to pale brown in oil.
R: (400) — , (420) — , (440) 55.4, (460) 56.8, (480) 55.9, (500) 56.4, (520) 58.2, (540) 59.2, (560) 59.6, (580) 59.6, (600) 59.9, (620) 59.5, (640) 60.9, (660) 61.2, (680) 60.9, (700) 61.8

Cell Data: *Space Group:* $Ccm2_1$. $a = 7.191$ $b = 8.693$ $c = 10.681$ Z = 16

X-ray Powder Pattern: Talnakh area, USSR.
2.65 (100), 2.16 (90), 2.25 (50), 1.638 (50), 2.50 (30), 1.400 (30), 1.220 (30)

Chemistry:

	(1)	(2)	(3)
Pd	32.8	33.1	35.0
Pt			2.0
Pb	34.0	29.0	
Bi	33.4	36.4	59.4
Te			5.2
Total	100.2	98.5	101.6

(1–2) Talnakh area, USSR; by electron microprobe, averages. (3) Union mine, South Africa; by electron microprobe, giving $(Pd_{1.00}Pt_{0.03})_{\Sigma=1.03}(Bi_{0.85}Te_{0.12})_{\Sigma=0.97}$.

Occurrence: In hydrothermal Cu-Ni-Fe veins.

Association: Chalcopyrite, talnakhite, cubanite, stannopalladinite, paolovite, sobolevskite, sperrylite, nickeloan platinum, sphalerite, silver.

Distribution: From the Talnakh area, Noril'sk region, Polar Ural Mountains, western Siberia, USSR. In the Union mine, on the Merensky Reef, Bushveld Complex, Transvaal, South Africa. At Fox Gulch, Goodnews Bay, Alaska, USA.

Name: For its occurrence in the Polar Ural Mountains, USSR.

Type Material: National Museum of Canada, Ottawa, Canada.

References: (1) Genkin, A.D., T.L. Evstigneeva, N.V. Troneva, and L.N. Vyal'sov (1969) Polarite, Pd(Pb, Bi) a new mineral from copper–nickel sulfide ores. Zap. Vses. Mineral. Obshch., 98, 708–715 (in Russian). (2) (1970) Amer. Mineral., 55, 1810 (abs. ref. 1). (3) Cabri, L.J. and R.J. Traill (1966) New palladium minerals from Noril'sk, western Siberia. Can. Mineral., 8, 541–550. (4) (1967) Amer. Mineral., 52, 1579–1580 (abs. ref. 3). (5) Cabri, L.J., Ed. (1981) Platinum group elements: mineralogy, geology, recovery. Can. Inst. Min. & Met., 130–131. (6) Tarkian, M. (1987) Compositional variations and reflectance of the common platinum-group minerals. Mineral. Petrol., 36, 169–190.

Crystal Data: Tetragonal. *Point Group:* $4/m$ or $4/m\ 2/m\ 2/m$. As microscopic grains and stubby prisms and dipyramids less than 25 μm long. *Twinning:* Knee-shaped twins with {605} as composition plane; contact twins on {*hkl*} less common; twinning on {*h0l*} may also be present, developing trains of prisms and discontinuous lamellae.

Physical Properties: *Cleavage:* Rare. Hardness = n.d. VHN = 262 (25 g load). D(meas.) = n.d. D(calc.) = 4.23–5.63

Optical Properties: Nearly opaque; translucent in thin section. *Color:* Black; gray in polished section, with intense red internal reflections. *Luster:* Resinous to adamantine. *Pleochroism:* Very slightly brownish gray to slightly lavender-gray. *Anisotropism:* Moderate in air, strong in oil.
R_1–R_2: n.d.

Cell Data: *Space Group:* $P4/n$, $P4_2/n$, $P4/nbm$, $P4/nmm$, $P4_2/nnm$, or $P4_2/ncm$. $a = 8.71$ $c = 14.74$ Z = 24 – 32(?)

X-ray Powder Pattern: B and B deposit, Idaho, USA.
3.08 (vs), 1.888 (s), 3.16 (ms), 1.608 (ms), 3.60 (m), 1.222 (m), 1.086 (m)

Chemistry:

	(1)
Zn	49.1
Hg	25.8
Fe	0.5
S	26.7
Total	102.1

(1) B and B deposit, Idaho; by electron microprobe, mean analyses of 15 grains, corresponding to $(Zn_{0.87}Hg_{0.15}Fe_{0.01})_{\Sigma=1.03}S_{0.97}$.

Occurrence: As part of a replacement deposit of stibnite (B and B deposit, Idaho, USA).

Association: Stibnite, cinnabar, mercurian sphalerite, zincian metacinnabar (B and B deposit, Idaho, USA); realgar (Getchell mine, Nevada, USA).

Distribution: In the B and B deposit, Big Creek district, Valley Co., Idaho, and in the Getchell mine, Humboldt Co., Nevada, USA.

Name: For Clyde Polhemus Ross, American economic geologist.

Type Material: National Museum of Natural History, Washington, D.C., USA, 145549.

References: (1) Leonard, B.F., G.A. Desborough, and C.W. Mead (1978) Polhemusite, a new Hg–Zn sulfide from Idaho. Amer. Mineral., 63, 1153–1161.

Polkovicite $(Fe, Pb)_3(Ge, Fe)_{1-x}S_4$ (x = 0.18 to 0.69)

Crystal Data: Cubic. *Point Group:* n.d. Massive with other sulfides.

Physical Properties: Hardness = n.d. VHN = 119–124 (50 g load). D(meas.) = n.d. D(calc.) = n.d.

Optical Properties: Opaque. *Color:* Brownish gray; in reflected light, white with cream-red tint. *Streak:* Dark gray. *Anisotropism:* Distinct.
R_1–R_2: (470) 43.5–44.5, (535) 43.0–44.0, (591) 44.0–45.0, (658) 45.5–46.5

Cell Data: *Space Group:* n.d. Z = n.d.

X-ray Powder Pattern: n.d.

Chemistry:

	(1)
Fe	29.3
Pb	14.6
Cu	3.6
Ge	4.5
As	1.8
S	34.2
Total	88.0

(1) Lower Silesia, Poland; by electron microprobe.

Polymorphism & Series: Forms a series with morozeviczite.

Occurrence: In epigenetic veinlets and metasomatic replacement zones replacing sandstone and older sulfides, in brecciated sandstones underlying copper-bearing shales.

Association: Marcasite, chalcopyrite, bornite, chalcocite, tennantite, sphalerite, galena.

Distribution: From the Polkovice mine, Lower Silesia, Poland.

Name: For the Polkovice mine, Poland.

Type Material: Jagellonian University, Kraków, Poland.

References: (1) Haranczyk, C. (1975) Morozeviczite and polkovicite, typochemical minerals of Mesozoic mineralization of the Fore-Sudenten monocline. Rudy Metalle, 20, 288–293 (in Polish). (2) (1981) Amer. Mineral., 66, 437 (abs. ref. 1).

Crystal Data: Monoclinic. *Point Group:* $2/m$. Crystals are pseudohexagonal, tabular on {001}, to 6 cm; triangular patterns formed by striae on {001}; also massive. *Twinning:* {110} as twin plane, repeated.

Physical Properties: *Cleavage:* Imperfect on {001}. *Fracture:* Uneven. Hardness = 2–3 VHN = n.d. D(meas.) = 6.1 D(calc.) = 6.36

Optical Properties: Nearly opaque; in thin splinters, translucent and dark red in transmitted light. *Color:* Iron-black; in polished section, gray-white with occasional red internal reflections. *Streak:* Black. *Luster:* Metallic.

Optical Class: Biaxial negative (–). *Pleochroism:* Weak. *Orientation:* $X = c$, $Y = a$. $n = > 2.72$ (Li). 2V(meas.) = 22° 2V(calc.) = n.d. *Anisotropism:* Moderate.

R_1–R_2: (400) 33.9–32.5, (420) 33.6–32.8, (440) 33.4–33.0, (460) 33.0–33.9, (480) 32.4–34.0, (500) 31.9–33.8, (520) 31.3–33.4, (540) 30.8–32.8, (560) 30.5–32.2, (580) 30.2–31.7, (600) 29.8–31.1, (620) 29.3–30.5, (640) 29.1–29.3, (660) 27.3–29.1, (680) 27.2–28.4, (700) 26.7–27.8

Cell Data: *Space Group:* $C2/m$. $a = 26.17$ $b = 15.11$ $c = 23.89$ $\beta = 90°00'$ Z = 16

X-ray Powder Pattern: Keeley mine, South Lorrain, Ontario, Canada
3.00 (100), 3.19 (90), 2.88 (80), 2.53 (60), 1.892 (60), 2.70 (50), 2.42 (40)

Chemistry:

	(1)	(2)
Ag	68.90	67.95
Cu	5.21	6.07
Sb	8.85	5.15
As	1.07	3.88
S	15.33	16.37
rem.	0.09	0.76
Total	99.45	100.18

(1) Arizpe, Mexico; remainder is Fe. (2) Quespisoza, Peru; remainder is Pb.

Polymorphism & Series: Forms a series with pearceite.

Occurrence: In silver veins of low to moderate temperature of formation.

Association: Pyrargyrite, tetrahedrite, stephanite, other silver sulfosalts, acanthite, gold, quartz, calcite, dolomite, barite.

Distribution: Common in small amounts, occasionally a major ore mineral; only rarely as fine specimens. In the USA, in Colorado, especially in the Molly Gibson and Smuggler mines, Aspen, Pitkin Co.; the Ouray district, Ouray Co.; the Leadville district, Lake Co.; and the Red Mountain district, San Juan Co. In Nevada, at Tonopah and Goldfield, Esmeralda Co.; in the Comstock Lode, Storey Co. In Germany, from Freiberg, Saxony, and St. Andreasberg, Harz. From Banská-Hodruša, near Baňská Štiavnica (Schemnitz), and Příbram, Czechoslovakia. At Tres Puntas, Atacama, Chile. In Mexico, at many localities, especially in the Las Chiapas mine, Arizpe, Sonora; and at Guanajuato.

Name: From the Greek for *many* and *base*, in allusion to the many metallic bases present.

References: (1) Palache, C., H. Berman, and C. Frondel (1944) Dana's system of mineralogy, (7th edition), v. I, 351–353. (2) Frondel, C. (1963) Isodimorphism of the polybasite and pearceite series. Amer. Mineral., 48, 565–572. (3) Berry, L.G. and R.M. Thompson (1962) X-ray powder data for the ore minerals. Geol. Soc. Amer. Mem. 85, 119–120. (4) Hall, H.T. (1967) The pearceite and polybasite series. Amer. Mineral., 52, 1311–1321.

Crystal Data: Cubic. *Point Group:* $4/m\ \overline{3}\ 2/m$. Crystals are dominantly octahedral, also massive, granular to compact. *Twinning:* On {111}.

Physical Properties: *Cleavage:* Imperfect on {001}; reported on {111}. *Fracture:* Subconchoidal to uneven. Hardness = 4.5–5.5 VHN = 379–427 (100 g load). D(meas.) = 4.5–4.8 D(calc.) = 4.83

Optical Properties: Opaque. *Color:* Light gray to steel-gray, tarnishing to copper-red. *Luster:* Metallic, brilliant on fresh surface.

R: (400) 41.4, (420) 42.0, (440) 42.6, (460) 43.3, (480) 43.8, (500) 44.4, (520) 44.9, (540) 45.4, (560) 46.2, (580) 47.1, (600) 48.1, (620) 49.2, (640) 50.5, (660) 51.7, (680) 53.1, (700) 54.5

Cell Data: *Space Group:* $Fd3m$. $a = 9.405$ Z = [8]

X-ray Powder Pattern: Siegen, Germany.
2.87 (100), 1.678 (80), 2.37 (60), 1.825 (50), 0.994 (50), 1.060 (40), 3.36 (30)

Chemistry:

	(1)	(2)	(3)
Ni	54.30	55.2	57.86
Fe	3.98	3.1	
Co	0.63	0.8	
S	41.09	41.2	42.14
Total	100.00	100.3	100.00

(1) Grunau mine, Germany; recalculated to 100% after deducting 5% gersdorffite and ullmannite. (2) Madziwa mine, Zimbabwe; by electron microprobe. (3) $NiNi_2S_4$.

Polymorphism & Series: Forms a series with linnaeite.

Occurrence: Found in hydrothermal veins.

Association: Chalcopyrite, pyrrhotite, pyrite, millerite, gersdorffite, ullmannite, sphalerite, galena, bismuthinite, quartz, siderite.

Distribution: In Germany, in the Grünau mine, Daaden, near Siegen, North Rhine-Westphalia. From Saint Marina, Khaskovo district, Bulgaria. At Kunratice and Rozany, Czechoslovakia. From Novo-Aidyrlinsk, Southern Ural Mountains, and the Noril'sk region, western Siberia, USSR. In the USA, from Hamilton, Hancock Co., Illinois; and in the Miliken (Sweetwater) mine, Reynolds Co., Missouri. In the Madziwa (Dry Nickel) mine, Bindura; and at Shamva, Zimbabwe. From the Kalgoorlie district, Western Australia.

Name: From the Greek for *many* and *twin*, as the mineral is observed in twinned forms.

References: (1) Palache, C., H. Berman, and C. Frondel (1944) Dana's system of mineralogy, (7th edition), v. I, 262–265. (2) Berry, L.G. and R.M. Thompson (1962) X-ray powder data for the ore minerals. Geol. Soc. Amer. Mem. 85, 78. (3) Ramdohr, P. (1969) The ore minerals and their intergrowths, (3rd edition), 686–691.

Crystal Data: Tetragonal. *Point Group:* $4/m\ 2/m\ 2/m$. As small grains or nuggets, up to 10 mm, exhibiting octahedral points and having indistinct, slightly divergent columnar or fibrous structure.

Physical Properties: *Tenacity:* Brittle. Hardness = 3.5 VHN = n.d. D(meas.) = 14.88 D(calc.) = 15.09

Optical Properties: Opaque. *Color:* Silver-white; in polished section, cream to creamish white. *Streak:* Silver-white. *Luster:* Metallic, bright. *Anisotropism:* Noted on inclusions in the fine-granular, apparently isotropic principal component; brown with an orange tinge.

R: (400) — , (420) — , (440) 57.5, (460) 58.1, (480) 58.4, (500) 59.0, (520) 60.0, (540) 61.0, (560) 61.7, (580) 62.4, (600) 63.0, (620) 63.8, (640) 64.5, (660) 65.4, (680) 66.0, (700) 66.8

Cell Data: *Space Group:* $P4/mmm$. $a = 3.026$ $c = 3.702$ Z = 1

X-ray Powder Pattern: Potaro River area, Guyana.
2.33 (100), 1.269 (70), 0.849 (50), 0.818 (50), 1.395 (40), 0.957 (20), 0.924 (20)

Chemistry:

	(1)	(2)
Pd	34.5	34.66
Hg	64.3	65.34
Cu	0.5	
Total	99.3	100.00

(1) Potaro River area, Guyana; by electron microprobe. (2) PdHg.

Occurrence: In diamond washings (Potaro River area, Guyana).

Association: Platinum, palladium, Pd–Hg compounds.

Distribution: In Guyana, in the Potaro river area and at Amu Creek, Essequibo River. At Morro de Pilar, Minas Gerais, Brazil. At Fox Gulch, Goodnews Bay, Alaska, USA.

Name: For the locality on the Potaro River, Guyana.

Type Material: National Museum of Natural History, Washington, D.C., USA, 95350.

References: (1) Palache, C., H. Berman, and C. Frondel (1944) Dana's system of mineralogy, (7th edition), v. I, 105. (2) Terada, K. and F.W. Cagle, Jr. (1960) The crystal structure of potarite (PdHg) with some comments on allopalladium. Amer. Mineral., 45, 1093–1097. (3) Cabri, L.J., Ed. (1981) Platinum group elements: mineralogy, geology, recovery. Can. Inst. Min. & Met., 131, 156.

Potosiite $Pb_6Sn_2FeSb_2S_{14}$

Crystal Data: Triclinic, pseudotetragonal and pseudohexagonal. *Point Group:* 1 or $\bar{1}$. In crystals up to 10 μm long, in felted masses. *Cleavage:* Perfect ∥ (001); less perfect ∥ (010).

Physical Properties: Hardness = n.d. VHN = 71.7 (50 g load). D(meas.) = n.d. D(calc.) = 6.20

Optical Properties: Opaque. *Color:* White in reflected light. *Luster:* Metallic. *Pleochroism:* Faint, from bluish gray to yellowish gray. *Anisotropism:* Moderate, bluish gray to grayish pale yellow.
R_1–R_2: (486) 36.2–36.9, (551) 34.9–35.35, (589) 34.4–35.1, (656) 33.9–34.5

Cell Data: *Space Group:* $P1$ or $P\bar{1}$, with two incommensurate cells, one pseudotetragonal with: $a = 5.915(10)$ $b = 5.938(13)$ $c = 17.239(17)$ $\alpha = 91.63(28)°$ $\beta = 91.02(25)°$ $\gamma = 90.84(21)°$ and the other pseudohexagonal with: $a = 6.253(7)$ $b = 3.734(5)$ $c = 17.229(19)$ $\alpha = 90.80(19)°$ $\beta = 91.71(16)°$ $\gamma = 90.18(14)°$ Z = n.d.

X-ray Powder Pattern: Andacaba deposit, Bolivia.
2.876 (100), 3.45 (90), 4.32 (30), 2.936 (10), 2.067 (10), 2.159 (9), 1.920 (8)

Chemistry:

	(1)	(2)	(3)
Pb	55.23	55.3	55.78
Ag	0.21	0.3	
Sn	11.57	10.7	10.65
In		0.5	
Fe	2.32	2.4	2.51
Sb	10.58	10.6	10.92
S	19.80	20.4	20.14
Total	99.71	100.2	100.00

(1) Andacaba deposit, Bolivia; by electron microprobe, average of seven analyses, corresponding to $(Pb_{6.00}, Ag_{0.04})_{\Sigma=6.04}Sn_{2.20}Fe_{0.94}Sb_{1.96}S_{13.91}$. (2) Herb claim, Canada; by electron microprobe, average of six analyses, corresponding to $(Pb_{5.91}Ag_{0.07}In_{0.09})_{\Sigma=6.07}Sn_{1.99}Fe_{0.95}Sb_{1.93}S_{14.07}$. (3) $Pb_6Sn_2FeSb_2S_{14}$.

Occurrence: On layered sulfide ore in a complex xenothermal-type hydrothermal tin deposit associated with subvolcanic granitic intrusive bodies (Andacaba deposit, Bolivia); in hydrothermal veins cutting rhyolite intrusions into highly kaolinized granite (Herb claim, Canada).

Association: Galena, sphalerite, semseyite, cerussite, cassiterite, quartz (Andacaba deposit, Bolivia); galena, pyrite, sphalerite, arsenopyrite, quartz (Herb claim, Canada).

Distribution: In the Andacaba deposit, Potosí, Bolivia. From the Herb claim, Turnagain River area, Cassiar district, British Columbia, Canada.

Name: For the type locality in Potosí, Bolivia.

Type Material: n.d.

References: (1) Wolf, M., H.-J. Hunger, and K. Bewilogua (1981) Potosiite, ein neues Mineral der Kylindrite-Franckeite-Gruppe. Freiberger Forschungshefte, 364, 113–133 (in German).
(2) (1983) Amer. Mineral., 68, 1249–1250 (abs. ref. 1). (3) Kissin, S.A. and D.R. Owens (1986) The properties and modulated structure of potosiite from the Cassiar district, British Columbia. Can. Mineral., 24, 45–50.

Crystal Data: Hexagonal. *Point Group:* $\bar{3}\ 2/m$. In euhedral to subhedral lath-like crystals 30-40 μm, rarely up to 80 μm long, 8–10 μm thick.

Physical Properties: *Cleavage:* Perfect {0001}. Hardness = n.d. VHN = 74–122, 99 average (25 g load). D(meas.) = n.d. D(calc.) = 7.85

Optical Properties: Opaque. *Color:* In polished section, creamy white. *Luster:* Metallic. *Pleochroism:* Maximum parallel to elongation, creamy. *Anisotropism:* Strong, in light grayish blue.

R_1–R_2: (400) — , (420) 46.0–50.3, (440) 46.6–51.4, (460) 47.2–52.2, (480) 47.6–52.7, (500) 48.0–53.0, (520) 48.3–53.2, (540) 48.5–53.3, (560) 48.6–53.3, (580) 48.8–53.3, (600) 48.8–53.3, (620) 48.9–53.2, (640) 49.0–53.2, (660) 49.0–53.1, (680) 49.1–53.1, (700) 49.1–53.0

Cell Data: *Space Group:* $R\bar{3}m$. $a = 4.252$ $c = 40.095$ Z = 3

X-ray Powder Pattern: Oldřichov, Czechoslovakia.
3.093 (100), 2.127 (80), 2.251 (70b), 1.752 (60), 1.352 (60), 4.048 (50), 2.727 (50)

Chemistry:

	(1)	(2)
Pb	22.95	18.22 – 19.07
Bi	40.29	42.23 – 43.21
Cu	0.61	
Se	15.48	13.39 – 14.17
Te	17.47	22.71 – 23.55
S	2.10	0.74 – 1.21
Total	98.90	

(1) Oldřichov, Czechoslovakia; by electron microprobe, corresponding to $Pb_{1.11}Bi_{1.94}Cu_{0.10}Se_{1.97}(Te_{1.37}S_{0.66})_{\Sigma=2.03}$. (2) Otish Mountains deposit, Canada; by electron microprobe, ranges of six grains.

Occurrence: In hydrothermal dolomite-calcite-quartz veins in the western contact of the Bor granite massif with gneisses and amphibolites, associated with a selenide-sulfide-telluride aggregate and uraninite (Oldřichov, Czechoslovakia); in a vein-type uranium deposit with other tellurides and selenides (Otish Mountains deposit, Canada).

Association: Selenian rucklidgeite, several incompletely characterized new species (Oldřichov, Czechoslovakia); součekite, wittichenite (Otish Mountains deposit, Canada).

Distribution: From Oldřichov, near Tachov, Czechoslovakia. In the Otish Mountains uranium deposit, Quebec, Canada.

Name: For Professor Z. Pouba, economic geologist, Charles University, Prague, Czechoslovakia.

Type Material: Department of Geology, Charles University, Prague, Czechoslovakia.

References: (1) Čech, F. and I. Vavřín (1978) Poubaite, $PbBi_2(Se,Te,S)_4$, a new mineral. Neues Jahrb. Mineral., Monatsh., 9–19. (2) (1978) Amer. Mineral., 63, 1283 (abs. ref. 1). (3) Johan, Z., P. Picot, and F. Ruhlmann (1987) The ore mineralogy of the Otish Mountains uranium deposit, Quebec: skippenite, Bi_2Se_2Te, and watkinsonite, $Cu_2PbBi_4(Se,S)_8$, two new mineral species. Can. Mineral., 25, 625–638.

Proudite $Cu_{0-1}Pb_{7.5}Bi_{9.3-9.7}(S,Se)_{22}$

Crystal Data: Monoclinic. *Point Group:* $2/m$. As elongate to acicular grains or irregular laths.

Physical Properties: *Cleavage:* One good cleavage parallel to grain elongation, a second at an apparent angle of 40° to the first. Hardness = n.d. VHN = 38–87 (50 g load). D(meas.) = n.d. D(calc.) = 7.08

Optical Properties: Opaque. *Color:* Silver-gray; in polished section, creamy white. *Pleochroism:* Strong, from cream-white to white. *Anisotropism:* Strong, from cream-gray to tan. R_1–R_2: n.d.

Cell Data: *Space Group:* $C2/m$. $a = 31.96(1)$ $b = 4.12(1)$ $c = 36.69(3)$ $\beta = 109.52(3)°$ Z = 4

X-ray Powder Pattern: Juno mine, Australia.
2.960 (100), 2.059 (86), 3.494 (65), 2.066 (60), 3.834 (48), 3.224 (48), 3.447 (42)

Chemistry:

	(1)	(2)
Cu	0.5	1.38
Pb	33.7	33.78
Bi	43.3	42.39
Se	14.0	12.02
S	9.9	10.43
Total	101.4	100.00

(1) Juno mine, Australia; by electron microprobe, average of 11 analyses. (2) "ideal" proudite – $CuPb_{7.5}Bi_{9.33}S_{15}Se_7$.

Occurrence: With large, presumably hydrothermal magnetite bodies.

Association: Gold, junoite, selenian heyrovskýite, krupkaite, magnetite.

Distribution: In Australia, at the Juno mine, Tennant Creek, Northern Territory. From Janos, Chihuahua, Mexico.

Name: For J.S. Proud, a Director of the Peko-Wallsend mining company, developers of the Tennant Creek deposits.

Type Material: n.d.

References: (1) Mumme, W.G. (1976) Proudite from Tennant Creek, Northern Territory, Australia: its crystal structure and relationship to weibullite and wittite. Amer. Mineral., 61, 839–852. (2) Large, R.R. and W.G. Mumme (1975) Junoite, "wittite", and related seleniferous bismuth sulfosalts from Juno mine, Northern Territory, Australia. Econ. Geol., 70, 369–383.

Crystal Data: Hexagonal. *Point Group:* $\bar{3}\ 2/m$. Crystals prismatic, to 8 cm; commonly rhombohedral with dominant $\{01\bar{1}2\}$ or $\{10\bar{1}1\}$; also scalenohedral with prominent $\{12\bar{3}1\}$; massive, compact. *Twinning:* On $\{10\bar{1}4\}$ to produce trillings; also common on $\{10\bar{1}1\}$ and on $\{0001\}$, $\{01\bar{1}2\}$.

Physical Properties: *Cleavage:* Distinct on $\{10\bar{1}1\}$. *Fracture:* Conchoidal to uneven. *Tenacity:* Brittle. Hardness = 2–2.5 VHN = 70–105 (25 g load). D(meas.) = 5.57 D(calc.) = 5.625

Optical Properties: Translucent, darkens with exposure to light. *Color:* Scarlet-vermilion. *Streak:* Vermilion. *Luster:* Adamantine.
Optical Class: Uniaxial negative (–). *Pleochroism:* Moderate, cochineal-red to blood-red. $\omega = 3.0877$ (Na). $\epsilon = 2.7924$ (Na). *Anisotropism:* Strong.
R_1–R_2: (400) 36.9–39.6, (420) 36.8–39.5, (440) 36.7–39.4, (460) 35.8–38.2, (480) 34.0–36.8, (500) 32.5–35.0, (520) 31.2–33.5, (540) 30.0–32.3, (560) 29.0–31.2, (580) 28.2–30.3, (600) 27.5–29.6, (620) 26.9–29.0, (640) 26.3–28.5, (660) 25.9–28.2, (680) 25.4–27.9, (700) 25.0–27.6

Cell Data: *Space Group:* *R3c*. $a = 10.79$ $c = 8.69$ Z = 6

X-ray Powder Pattern: Cobalt, Canada.
2.76 (10), 3.28 (8), 3.18 (8), 2.56 (8), 2.48 (8), 1.929 (2), 1.672 (2)

Chemistry:

	(1)	(2)	(3)
Ag	64.12	64.65	65.42
Sb	0.08	trace	
As	15.90	15.25	15.14
S	19.28	20.18	19.44
rem.	0.75	0.70	
Total	100.13	100.78	100.00

(1) Cobalt, Canada; remainder Fe 0.25%, Co(Ni) 0.12%, insoluble 0.38%. (2) Veta Rica mine, Coahuila, Mexico; remainder Cu. (3) Ag_3AsS_3.

Polymorphism & Series: Dimorphous with xanthoconite.

Occurrence: A late-forming mineral in hydrothermal deposits, in the oxidized and enriched zone, associated with other silver minerals and sulfides.

Association: Silver, arsenic, xanthoconite, stephanite, acanthite, tetrahedrite, chlorargyrite.

Distribution: Occurs at many localities, but only occasionally in fine crystals or as an important ore mineral. In the USA, in Colorado, at Red Mountain, San Juan Co., and at Georgetown, Clear Creek Co.; in Idaho, at the Poorman mine, Silver district, Owyhee Co. In Canada, at the Keeley mine, Cobalt, Ontario. From Germany, at the Himmelsfürst mine, Erbisdorf, near Freiberg; and at Niederschlema, Saxony. In Romania, at Săcărâmb (Nagyág). At Jáchymov (Joachimsthal) and Přibram, Czechoslovakia. From Sainte-Marie-aux-Mines, Haut-Rhin, France. At Sarrabus, Sardinia, Italy. In Chile, in Atacama, at Chañarcillo, in exceptional crystals. From Mexico, at Batopilas, Chihuahua, and Sombrerete, Zacatecas.

Name: For Joseph Louis Proust (1754-1826), celebrated French chemist.

References: (1) Palache, C., H. Berman, and C. Frondel (1944) Dana's system of mineralogy, (7th edition), v. I, 366–369. (2) Berry, L.G. and R.M. Thompson (1962) X-ray powder data for the ore minerals. Geol. Soc. Amer. Mem. 85, 124. (3) Toulmin, P. (1963) Proustite–pyrargyrite solid solutions. Amer. Mineral., 48, 725–736. (4) Engel, P. and W. Nowacki (1966) Die Verfeinerung der Kristallstruktur von Proustit, Ag_3AsS_3, und Pyrargyrit, Ag_3SbS_3. Neues Jahrb. Mineral., Monatsh., 181–184 (in German).

Crystal Data: Cubic. *Point Group:* n.d. Rarely as crystals; in coarse-grained aggregates, up to 1–2 cm, and as small grains, 0.1 mm, finely intergrown with mooihoekite. *Twinning:* Polysynthetic.

Physical Properties: Hardness = n.d. VHN = 263 (50 g load). D(meas.) = n.d. D(calc.) = n.d.

Optical Properties: Opaque. *Color:* Light yellow in reflected light, similar to mooihoekite and chalcopyrite. *Anisotropism:* Nickelian variety distinctly anisotropic.
R: (400) — , (420) — , (440) — , (460) 24.4, (480) 25.1, (500) 30.0, (520) 31.8, (540) 38.0, (560) 40.0, (580) 41.2, (600) 41.2, (620) 40.9, (640) 41.9, (660) 41.6, (680) 41.2, (700) 41.0

Cell Data: *Space Group:* n.d. $a = 5.30$ Z = n.d.

X-ray Powder Pattern: Oktyabr mine, USSR.
3.05 (10), 1.873 (9), 1.596 (6), 1.081 (5), 1.216 (3), 2.65 (2), 1.326 (2)

Chemistry:

	(1)	(2)
Cu	35.68	32.99
Fe	31.22	32.11
Ni	0.51	1.63
S	32.49	33.14
Total	99.90	99.87

(1) Oktyabr mine, USSR; by electron microprobe, average of seven samples, corresponding to $Cu_{18}(Fe,Ni)_{18}S_{32}$. (2) Do.; average of twelve samples, corresponding to $Cu_{16}(Fe,Ni)_{19}S_{32}$.

Occurrence: In massive mooihoekite ores.

Association: Mooihoekite, talnakhite, cubanite, pentlandite, magnetite, galena, sphalerite, platinum group minerals, silver, alabandite, valleriite, mackinawite, manganoan shadlunite, djerfisherite.

Distribution: In the Oktyabr mine, Talnakh area, Noril'sk region, western Siberia, USSR.

Name: For Putoran Mountain, in the northwestern part of the Siberian platform, USSR.

Type Material: A.E. Fersman Mineralogical Museum, Academy of Sciences, Moscow, USSR.

References: (1) Filimonova, A.A., T.L. Evstigneeva, and I.P. Laputina (1980) Putoranite and nickel-bearing putoranite, new minerals of the chalcopyrite group. Zap. Vses. Mineral. Obshch., 109, 335–341 (in Russian). (2) (1981) Amer. Mineral., 66, 638–639 (abs. ref. 1).

Crystal Data: Hexagonal. *Point Group:* $3m$. Crystals commonly prismatic ∥ [0001], showing hemimorphism, often with prominent development of rhombohedra; also steep scalenohedra with $\{05\bar{5}1\}$, to 6 cm. Massive, granular. *Twinning:* On $\{10\bar{1}4\}$ as twins of complex aggregates of individuals and as lamellar twins; less commonly on $\{10\bar{1}1\}$ and about $[11\bar{2}0]$; rarely on $\{01\bar{1}2\}$.

Physical Properties: *Cleavage:* Distinct on $\{10\bar{1}1\}$; very imperfect on $\{01\bar{1}2\}$. *Fracture:* Conchoidal to uneven. *Tenacity:* Brittle. Hardness = 2.5 VHN = n.d. D(meas.) = 5.82 D(calc.) = 5.855

Optical Properties: Translucent, darkens with exposure to light. *Color:* Deep red. *Streak:* Purplish red. *Luster:* Adamantine.

Optical Class: Uniaxial negative (–). *Pleochroism:* Distinct in air. $\omega = 3.084$ (Li). $\epsilon = 2.881$ (Li). *Anisotropism:* Strong, in yellow-white and gray-blue.

R_1–R_2: (400) 35.0–41.0, (420) 34.8–40.8, (440) 34.6–40.6, (460) 34.0–40.2, (480) 32.8–39.5, (500) 31.0–37.4, (520) 29.6–35.4, (540) 28.2–34.0, (560) 27.2–32.7, (580) 26.4–31.7, (600) 25.6–30.9, (620) 25.0–30.2, (640) 24.4–29.6, (660) 23.8–29.0, (680) 23.4–28.5, (700) 22.9–28.1

Cell Data: *Space Group:* $R3c$. $a = 11.047$ $c = 8.719$ Z = 6

X-ray Powder Pattern: Beaverdell, British Columbia, Canada.
2.79 (10) 3.21 (8), 3.35 (5), 2.58 (5), 2.54 (5), 1.965 (2), 1.870 (2)

Chemistry:

	(1)	(2)	(3)
Ag	59.82	60.17	59.76
Sb	22.00	21.64	22.48
As	0.08	0.52	
S	17.82	17.65	17.76
Total	99.72	99.98	100.00

(1) Săcărâmb (Nagyág), Romania. (2) Freiberg, Germany. (3) Ag_3SbS_3.

Polymorphism & Series: Dimorphous with pyrostilpnite.

Occurrence: Formed in hydrothermal veins as a primary late-stage, low-temperature mineral; also formed by secondary processes.

Association: Silver, acanthite, tetrahedrite, other silver sulfosalts, calcite, dolomite, quartz.

Distribution: An important ore of silver, not uncommon in oxidized silver deposits, but only occasionally in fine specimens. In the USA, in Nevada, at the Comstock Lode, Storey Co., and in the Reese River district, Lander Co.; in Idaho, at the Poorman mine, Owyhee Co. In Canada, at Cobalt, Ontario. At Jáchymov (Joachimsthal) and Příbram, Czechoslovakia. In Germany, in the Harz, at St. Andreasberg; and at the Himmelsfürst mine, Erbisdorf, near Freiberg, Saxony. In Bolivia, at Colquechaca, Potosí. At Chañarcillo, Atacama, Chile. From San Cristobal, Peru. From many localities in Mexico, especially at Fresnillo, Zacatecas, and Guanajuato.

Name: From the Greek for *fire* and *silver*, in allusion to its composition and color.

References: (1) Palache, C., H. Berman, and C. Frondel (1944) Dana's system of mineralogy, (7th edition), v. I, 362–366. (2) Berry, L.G. and R.M. Thompson (1962) X-ray powder data for the ore minerals. Geol. Soc. Amer. Mem. 85, 123. (3) Ramdohr, P. (1969) The ore minerals and their intergrowths, (3rd edition), 774–777. (4) Engel, P. and W. Nowacki (1966) Die Verfeinerung der Kristallstruktur von Proustit, Ag_3AsS_3, und Pyrargyrit, Ag_3SbS_3. Neues Jahrb. Mineral., Monatsh., 181–184 (in German).

Crystal Data: Cubic. *Point Group:* $2/m\ \overline{3}$. Commonly cubic, pyritohedral, octahedral, and combinations of these and other forms, to 25 cm or more. Striated conforming to pyritohedral symmetry. Sometimes elongated to acicular; frequently granular, globular, stalactitic. *Twinning:* Twin axis [001] and twin plane {011}, penetration and contact twins.

Physical Properties: *Cleavage:* Indistinct on {001}; partings on {011} and {111}, indistinct. *Fracture:* Conchoidal to uneven. *Tenacity:* Brittle. Hardness = 6–6.5 VHN = 1505–1520 (100 g load). D(meas.) = 5.018 D(calc.) = 5.013 Paramagnetic; a semiconductor.

Optical Properties: Opaque. *Color:* Pale brass-yellow, tarnishes darker and iridescent; in polished section creamy white. *Streak:* Greenish black to brownish black. *Luster:* Metallic, splendent. *Anisotropism:* Sometimes.
R: (400) 38.2, (420) 40.5, (440) 42.8, (460) 45.5, (480) 48.5, (500) 51.0, (520) 52.6, (540) 53.8, (560) 54.6, (580) 55.0, (600) 55.2, (620) 55.5, (640) 56.0, (660) 56.4, (680) 56.8, (700) 57.0

Cell Data: *Space Group:* *Pa*3. *a* = 5.4179(11) Z = 4

X-ray Powder Pattern: Synthetic.
1.6332 (100), 2.7088 (85), 2.4281 (65), 2.2118 (50), 1.9155 (40), 3.128 (35), 1.4479 (25)

Chemistry:

	(1)	(2)	(3)	(4)
Fe	46.49	29.30	33.32	46.55
Ni		16.69	0.19	
Co		trace	13.90	
S	53.49	53.40	52.45	53.45
Total	99.98	99.39	99.86	100.00

(1) Elba, Italy; remainder 0.04% SiO_2. (2) Mill Close mine, Derbyshire, England. (3) Gladhammar, Sweden. (4) FeS_2.

Polymorphism & Series: Dimorphous with marcasite; forms a series with cattierite.

Occurrence: Formed under a wide variety of conditions. In hydrothermal veins as very large bodies, as magmatic segregations, as an accessory mineral in igneous rocks, in pegmatites; in contact metamorphic deposits, also in metamorphic rocks; as diagenetic replacements in sedimentary rocks.

Association: Pyrrhotite, marcasite, galena, sphalerite, arsenopyrite, chalcopyrite, many other sulfides and sulfosalts, hematite, fluorite, quartz, barite, calcite.

Distribution: The most abundant and widespread sulfide. Only a few localities for large or fine crystals can be mentioned. In Peru, from many districts, with exceptional crystals from the Quiruvilca mine, La Libertad, and Huanzala, Huanaco. At Rio Marina, in Elba, and Traversella, Piedmont, Italy. At Ambasaguas and Navajun, Logroño Province, Spain, sculptural groups of crystals. From Aktchitao, Kazakhstan, USSR. In the USA, at the Ibex mine, Leadville, Lake Co., Colorado; in Illinois, as "suns" at Sparta, Randolph Co.; from the Santo Niño mine, near Duquesne, Santa Cruz Co., Arizona. In Pennsylvania, from the French Creek mines, Chester Co., and in the Carleton talc mine, Chester, Windsor Co., Vermont. From Butte, Silver Bow Co., Montana; at the Spruce Claim, King Co., Washington; as "bars" from the Buick mine, Bixby, Iron Co., Missouri.

Name: From the Greek for *fire*, as sparks may be struck from it.

References: (1) Palache, C., H. Berman, and C. Frondel (1944) Dana's system of mineralogy, (7th edition), v. I, 282–290. (2) (1955) NBS Circ. 539, 5, 29. (3) Brostigen, G. and A. Kjekshus (1969) Redetermined crystal structure of FeS_2 (pyrite). Acta Chem. Scand., 23, 2186–2188.

Crystal Data: Monoclinic. *Point Group:* $2/m$. Crystals tabular {010} giving flat rhombic forms; also laths by elongation ‖ [001]; as sub-parallel sheaf-like aggregates. *Twinning:* On {100} with (100) as composition plane.

Physical Properties: *Fracture:* Conchoidal. *Tenacity:* Somewhat flexible in thin plates. Hardness = 2 VHN = n.d. D(meas.) = 5.94 D(calc.) = 5.97

Optical Properties: Transparent. *Color:* Hyacinth-red; lemon-yellow by transmitted light. *Streak:* Orange-yellow. *Luster:* Adamantine.
Optical Class: Biaxial positive (+). *Orientation:* $Y = b$; $X \wedge c = 8–11°$. α = Very high. β = Very high. γ = Very high.
$R_1–R_2$: n.d.

Cell Data: *Space Group:* $P2_1/c$. $a = 6.84$ $b = 15.84$ $c = 6.24$ $\beta = 117°09'$ Z = 4

X-ray Powder Pattern: Příbram, Czechoslovakia.
2.85 (100), 2.65 (50), 2.42 (50), 1.895 (50b), 1.887 (50b), 1.824 (20b), 1.813 (20b)

Chemistry:

	(1)	(2)
Ag	59.44	59.76
Sb	22.30	22.48
S	18.11	17.76
Total	99.85	100.00

(1) St. Andreasberg, Germany. (2) Ag_3SbS_3.

Polymorphism & Series: Dimorphous with pyrargyrite.

Occurrence: Found in low-temperature hydrothermal veins as a late stage mineral.

Association: Pyrargyrite, stephanite, acanthite, other silver minerals.

Distribution: In the USA, from the Silver City district, Owyhee Co., Idaho; at the Bulldog mine, Creede, Mineral Co., Colorado; and Randsburg, San Bernardino Co., California. In Canada, from Cobalt, Ontario. In Germany, at St. Andreasberg, Harz Mountains. In Czechoslovakia, at Příbram and Trebsko. From Hiendelaencina, Guadalajara Province, Spain. In Australia, from Broken Hill, New South Wales; and the Long Tunnel, at Heazlewood, Tasmania. In Chile, from Chañarcillo, Atacama. From the Kushikino mine, Kagoshima Prefecture, Japan. In small amounts from other localities.

Name: From the Greek for *fire* and *shining*, in allusion to its color and luster.

References: (1) Palache, C., H. Berman, and C. Frondel (1944) Dana's system of mineralogy, (7th edition), v. I, 369–370. (2) Peacock, M.A. (1950) Studies of mineral sulpho-salts: XV. xanthoconite and pyrostilpnite. Mineral. Mag., 29, 346–358. (3) Kutoglu, A. (1968) Die Struktur des Pyrostilpnits (Feuerblende) Ag_3SbS_3. Neues Jahrb. Mineral., Monatsh., 145–160 (in German with English abs.). (4) Ramdohr, P. (1969) The ore minerals and their intergrowths, (3rd edition), 777–778.

Pyrrhotite $Fe_{1-x}S$ (x = 0 to 0.17)

Crystal Data: Monoclinic, pseudohexagonal. *Point Group:* $2/m$. Crystals commonly tabular or platy on {0001} to 15 cm; steep pyramidal faces and short pyramidal; as rosettes showing nearly parallel aggregation on {0001}; usually massive, granular. *Twinning:* On $\{10\overline{1}2\}$.

Physical Properties: *Cleavage:* None; distinct {0001} parting. *Fracture:* Uneven to subconchoidal. Hardness = 3.5–4.5 VHN = 373–409 (100 g load). D(meas.) = 4.58–4.65 D(calc.) = 4.69 Magnetic, varying in intensity inversely with iron content.

Optical Properties: Opaque. *Color:* Bronze-yellow to pinchbeck-brown; tarnishes quickly, occasionally to iridescence. *Streak:* Dark grayish black. *Luster:* Metallic. *Pleochroism:* Weak. *Anisotropism:* Strong.

R_1–R_2: (400) 27.9–31.0, (420) 28.6–32.2, (440) 29.4–33.6, (460) 30.3–34.8, (480) 31.4–36.2, (500) 32.4–37.6, (520) 33.4–38.6, (540) 34.5–39.6, (560) 35.5–40.4, (580) 36.5–41.2, (600) 37.4–42.0, (620) 38.3–42.6, (640) 39.1–43.0, (660) 39.9–43.5, (680) 40.7–43.9, (700) 41.4–44.1

Cell Data: *Space Group:* Depends on polytype. $a = 6.865$ $b = 11.9$ $c = 22.72$ $\beta = 90°5'$ Z = 64

X-ray Powder Pattern: Santa Eulalia, Mexico.
2.08 (10), 2.65 (6), 1.728 (5), 3.00 (4), 1.328 (4), 1.105 (4), 1.052 (3)

Chemistry:

	(1)	(2)	(3)
Fe	60.18	59.83	61.57
S	39.82	39.55	38.53
rem.		0.55	
Total	100.00	99.93	100.10

(1) Homestake mine, Lead, Lawrence Co., South Dakota, USA; Fe:S = 0.87. (2) Kongsberg, Norway; average of two analyses, Fe:S = 0.87. (3) Setregruben, Østfold, Norway; Fe:S = 0.92.

Polymorphism & Series: Numerous polymorphs, with 1-C, 3-C, 4-C, 5-C, 6-C, 7-C, and 11-C known, and a high-temperature hexagonal polymorph which may remain stable at ordinary temperatures.

Occurrence: Mainly in mafic igneous rocks, frequently as magmatic segregations; also in pegmatites, and in high-temperature hydrothermal and replacement veins, and in sedimentary and metamorphic rocks; in iron meteorites (troilite).

Association: Pyrite, marcasite, chalcopyrite, pentlandite, many other sulfides, magnetite, calcite, dolomite.

Distribution: Massive material occurs at many localities. Well crystallized from Herja (Kisbánya), Romania. From the Stari-Trg mine, Trepča, Yugoslavia. In Val Passiria, Trentino, Italy. In Germany, in the Harz Mountains, at St. Andreasberg. From the Bristenstock tunnel, Canton Uri, Switzerland. In Italy, at Bottino, near Servezza, Tuscany. In Sweden, at Falun, Kopparberg. Large crystals from the Morro Velho gold mine, near Nova Lima, Minas Gerais, Brazil. At the Potosí and San Antonio mines, Santa Eulalia, Chihuahua, Mexico.

Name: From the Greek for *redness*, in allusion to its color.

References: (1) Palache, C., H. Berman, and C. Frondel (1944) Dana's system of mineralogy, (7th edition), v. I, 231–235. (2) Berry, L.G. and R.M. Thompson (1962) X-ray powder data for the ore minerals. Geol. Soc. Amer. Mem. 85, 60. (3) Deer, W.A., R.A. Howie and J. Zussman (1963) Rock forming minerals. v. 5, non-silicates, 145–157.

Crystal Data: Tetragonal. *Point Group:* $4/m$, 4, $\bar{4}$, 422, $4mm$, $\bar{4}2m$, or $4/m\ 2/m\ 2/m$. Flame-like aggregates and veinlets up to 20 μm across, in rims and intergrowths with other tellurides and sulfides.

Physical Properties: Hardness = n.d. VHN = n.d. D(meas.) = n.d. D(calc.) = 8.89

Optical Properties: Opaque. *Color:* Rose-brown in reflected light. *Anisotropism:* Distinct.
R_1–R_2: 23.7–24.3 (400), 24.7–25.2 (420), 25.6–26.1 (440), 26.6–27.1 (460), 27.6–28.2 (480), 28.6–29.2 (500), 29.6–30.1 (520), 30.3–31.0 (540), 31.1–31.8 (560), 31.8–32.4 (580), 32.4–32.9 (600), 32.7–33.3 (620), 33.0–33.5 (640), 33.2–33.6 (660), 33.2–33.6 (680), 33.2–33.6 (700)

Cell Data: *Space Group:* $P4/m$, $P4$, $P\bar{4}$, $P422$, $P4mm$, $P\bar{4}2m$, $P4/mmm$ or $P\bar{4}m2$.
$a = 5.71(5)$ $c = 3.77(5)$ Z = 1

X-ray Powder Pattern: Champion Reef mine, India. Electron diffraction.
3.16 (100), 3.78 (60), 1.92 (50), 1.78 (50), 2.73 (40), 2.29 (40), 1.59 (40)

Chemistry:

	(1)	(2)
Pb	29.17	31.35
Te	59.80	57.92
S	0.77	
Cl	9.90	10.73
Total	99.64	100.00

(1) Champion Reef mine, India; by electron microprobe, leading to $Pb_{0.94}Te_{3.06}(Cl_{1.84}S_{0.16})_{\Sigma=2.00}$.
(2) $PbTe_3Cl_2$.

Occurrence: Of primary origin, in a hydrothermal gold-quartz vein containing sulfides and tellurides.

Association: Galena, altaite, kolarite, cotunnite, volynskite, hessite, pyrrhotite, chalcopyrite, ullmannite, hawleyite, cadmian tetrahedrite, cadmian sphalerite.

Distribution: From the Champion Reef mine, Kolar Gold Fields, Karnataka, India.

Name: For B.P. Radhakrishna, Indian geologist.

Type Material: Institute of Geology of Ore Deposits, Petrography, Mineralogy and Geochemistry; A.E. Fersman Mineralogical Museum, Academy of Sciences, Moscow, USSR.

References: (1) Genkin, A.D., Y.G. Safonov, V.N. Vasudev, B. Krishna Rao, V.A. Boronikhin, L.N. Vyalsov, A.I. Gorshkov, and A.V. Mokhov (1985) Kolarite $PbTeCl_2$ and radhakrishnaite $PbTe_3(Cl,S)_2$, new mineral species from the Kolar gold deposit, India. Can. Mineral., 23, 501–506. (2) (1986) Amer. Mineral., 71, 1545–1546 (abs. ref. 1).

Crystal Data: Orthorhombic. *Point Group:* n.d. As pseudohexagonal plates pseudomorphic after an unknown mineral, consisting of fibers intimately mixed with pyrite.

Physical Properties: Hardness = n.d. VHN = n.d. D(meas.) = 6.4(2) D(calc.) = 6.29

Optical Properties: Opaque. *Color:* Brilliant bronze; in polished section, creamy grayish white perpendicular to the fibers, and rose colored parallel. *Luster:* Metallic. *Pleochroism:* Strong, grayish white to rose. *Anisotropism:* Strong, with orange dominant.
R_1–R_2: (400) 23.7–28.0, (420) 24.1–30.4, (440) 24.5–32.8, (460) 24.8–33.6, (480) 25.0–31.8, (500) 25.1–29.8, (520) 25.2–30.2, (540) 25.4–32.1, (560) 25.6–34.3, (580) 25.7–36.3, (600) 25.8–38.0, (620) 26.0–39.5, (640) 26.3–40.7, (660) 26.6–41.5, (680) 26.9–42.0, (700) 27.2–42.4

Cell Data: *Space Group:* n.d. $a = 12.40(5)$ $b = 10.44(5)$ $c = 5.26(5)$ Z = 8

X-ray Powder Pattern: Allchar, Yugoslavia.
2.89 (vvs), 4.17 (s), 3.35 (s), 2.64 (m), 2.35 (m), 4.70 (w), 6.03 (vw)

Chemistry:

	(1)	(2)	(3)
Tl	62.50	61.3	63.01
Fe	17.10	19.7	17.22
S	19.50	20.2	19.77
Total	99.10	101.2	100.00

(1–2) Allchar, Yugoslavia; by electron microprobe. (3) $TlFeS_2$.

Occurrence: Formed in a hydrothermal deposit.

Association: Pyrite, lorandite, orpiment, realgar, vrbaite.

Distribution: From Allchar, Macedonia, Yugoslavia.

Name: For Professor E. Raguin, National School of Mines, Paris, France.

Type Material: National School of Mines, Paris, France.

References: (1) Laurent, Y., P. Picot, R. Pierrot, and T. Ivanov (1969) La raguinite, $TlFeS_2$, une nouvelle espèce minérale et le problème de l'allcharite. Bull. Soc. fr. Minéral., 92, 38–48 (in French with English abs.). (2) (1969) Amer. Mineral., 54, 1495 (abs. ref. 1). (3) Johan, Z., P. Picot, and R. Pierrot (1969) Nouvelles données sur la raguinite. Bull. Soc. fr. Minéral., 92, 237 (in French). (4) (1969) Amer. Mineral., 54, 1741 (abs. ref. 3).

Crystal Data: Monoclinic. *Point Group:* $2/m$. Crystals are long prismatic or thick lance-shaped, from 0.5–1 cm. *Twinning:* Lamellar twinning on (010).

Physical Properties: *Fracture:* Uneven. *Tenacity:* Brittle. Hardness = 2 VHN = 206 (20 g load). D(meas.) = 5.43 D(calc.) = [5.64]

Optical Properties: Opaque. *Color:* Gray-black; in polished section, white. *Streak:* Gray-black. *Luster:* Metallic. *Pleochroism:* Very weak. *Anisotropism:* Moderate.
R_1–R_2: (400) 39.3–43.1, (420) 38.8–42.8, (440) 38.3–42.5, (460) 37.8–42.2, (480) 37.4–42.0, (500) 36.9–41.6, (520) 36.5–41.2, (540) 36.0–40.9, (560) 35.7–40.5, (580) 35.3–40.2, (600) 34.8–39.7, (620) 34.4–39.2, (640) 34.0–38.7, (660) 33.4–38.0, (680) 32.8–37.3, (700) 32.0–36.4

Cell Data: *Space Group:* $P2_1/n$. $a = 19.24$ $b = 13.08$ $c = 8.73$ $\beta = 90.28°$ Z = [2]

X-ray Powder Pattern: Chocaya mine, Bolivia.
3.32 (100), 2.94 (60), 2.78 (50), 2.21 (50), 3.48 (30), 3.04 (30), 3.82 (20)

Chemistry:

	(1)	(2)	(3)	(4)
Ag	8.96	8.79	9.6	8.80
Pb	33.84	34.46	35.7	33.82
Fe		0.21		
Cd		0.60		
In		0.20		
Sb	34.91	34.40	36.1	36.44
S	21.14	20.41	19.6	20.94
Total	98.85	99.07	100.8	100.00

(1–2) Chocaya mine, Bolivia; by electron microprobe. (3) Do.; by electron microprobe, average of ten analyses. (4) $Ag_3Pb_6Sb_{11}S_{24}$.

Occurrence: Found in fine-grained quartz in a vein of hydrothermal origin (Chocaya mine, Bolivia).

Association: Pyrite, stannite, andorite, jamesonite, sphalerite, quartz (Chocaya mine, Bolivia); andorite (Bear Basin, Washington, USA).

Distribution: In Bolivia, from the Colorado Ag-Sn vein, Chocaya mine, Potosí, and from Tatasi. In the USA, at the Round Valley tungsten mine, Bishop Creek area, Inyo Co., California; and at Bear Basin, King Co., Washington.

Name: For Professor Paul Ramdohr (1890–1985), German mineralogist.

Type Material: National Museum of Natural History, Washington, D.C., USA, R6595.

References: (1) Palache, C., H. Berman, and C. Frondel (1944) Dana's system of mineralogy, (7th edition), v. I, 450–451. (2) Donnay, J.D.H. and G. Donnay (1954) Syntaxic intergrowths in the andorite series. Amer. Mineral., 39, 161–171. (3) Ramdohr, P. (1969) The ore minerals and their intergrowths, (3rd edition), 731–733. (4) Borodaev, Y.S., O.L. Sveshnikova, and N.N. Mozgova (1971) The inhomogeneity of ramdohrite. Doklady Acad. Nauk SSSR, 199, 1138–1141 (in Russian). (5) (1972) Amer. Mineral., 57, 1560 (abs. ref. 4). (6) Makovicky, E. and W.G. Mumme (1983) The crystal structure of ramdohrite, $Pb_6Sb_{11}Ag_3S_{24}$, and its implications for the andorite group and zinckenite. Neues Jahrb. Mineral., Abh., 147, 58–79. (7) Moëlo, Y., E. Makovicky, and S. Karup-Møller (1984) New data on the minerals of the andorite series. Neues Jahrb. Mineral., Monatsh., 175–182. (8) (1985) Amer. Mineral., 70, 219–220 (abs. ref. 7).

Crystal Data: Orthorhombic. *Point Group:* $2/m\ 2/m\ 2/m$. As prismatic crystals, elongated [010], about 1 mm long; massive, granular, radial, fibrous. *Twinning:* On {101}.

Physical Properties: *Cleavage:* {101} distinct. Hardness = 5.5–6 VHN = 630–758 (100 g load). D(meas.) = 7.1(1) D(calc.) = 7.091

Optical Properties: Opaque. *Color:* Tin-white with a faint pinkish hue; in polished section, pure white. *Streak:* Grayish black. *Luster:* Metallic. *Pleochroism:* Weak, in yellow to pinkish hue and bluish white. *Anisotropism:* Strong, especially in blues.

R_1–R_2: (400) 53.1–52.2, (420) 54.2–54.2, (440) 55.3–56.2, (460) 56.0–57.6, (480) 56.5–58.4, (500) 56.6–58.7, (520) 56.3–58.6, (540) 56.0–58.2, (560) 55.8–58.0, (580) 55.8–57.8, (600) 55.8–57.8, (620) 55.7–57.6, (640) 55.6–57.3, (660) 55.6–57.0, (680) 55.5–56.9, (700) 55.4–56.8

Cell Data: *Space Group:* *Pnnm.* $a = 4.759$ $b = 5.797$ $c = 3.539$ Z = 2

X-ray Powder Pattern: Synthetic. Cannot be distinguished from ferroselite.
2.552 (100), 2.476 (90), 2.843 (65), 1.870 (55), 1.696 (30), 1.771 (25), 1.790 (20)

Chemistry:

	(1)	(2)	(3)
Ni	27.84	28.6	28.15
Co	1.80	0.5	
Fe	trace	0.1	
As	67.32	71.0	71.85
Sb	0.83		
S	2.03	0.4	
Total	99.82	100.6	100.00

(1) University mine, Cobalt, Canada. (2) Coniston mines, England; by electron microprobe. (3) $NiAs_2$.

Polymorphism & Series: Trimorphous with pararammelsbergite and krutovite.

Occurrence: In hydrothermal veins formed at moderate temperatures with other Ni and Co minerals.

Association: Skutterudite, safflorite, löllingite, nickeline, bismuth, silver, algodonite, domeykite, uraninite.

Distribution: In the USA, at the Mohawk mine, Keweenaw Co., Michigan. In Canada, at the Eldorado mine, Great Bear Lake, Northwest Territories; in various mines at South Lorrain and Cobalt, Ontario, and other scattered localities. In Germany, at Schneeberg, Saxony, and Eisleben and Mansfeld, Halle district. From the Lölling-Hüttenberg district, Carinthia, Austria. In France, at Sainte-Marie-aux-Mines, Haut-Rhin. From the Coniston mines, Cumbria, England. In the Leadhills-Wanlockhead district, Scotland. At the Sedmochislenitsi mine, Vratsa district, western part of the Stara Planina (Balkan Mountains), Bulgaria. From Lainijaur, Västerbotten, Sweden. Known from a number of other occurrences world-wide.

Name: For Karl Friedrich Rammelsberg (1813–1899), mineral chemist of Berlin, Germany.

Type Material: n.d.

References: (1) Palache, C., H. Berman, and C. Frondel (1944) Dana's system of mineralogy, (7th edition), v. I, 309–310. (2) (1960) NBS Circ. 539, 10, 42. (3) Ramdohr, P. (1969) The ore minerals and their intergrowths, (3rd edition), 839–845. (4) Williams, S.A. (1963) Crystals of rammelsbergite and algodonite. Amer. Mineral., 48, 421–422.

Crystal Data: Orthorhombic. *Point Group:* $2/m\ 2/m\ 2/m$. In grains up to 2 mm in size, which are aggregates of very fine, often curved needles; massive crusts.

Physical Properties: *Cleavage:* Perfect on {110}. Hardness = n.d. VHN = 243–433 D(meas.) = 3.1 D(calc.) = 3.029 Slightly magnetic.

Optical Properties: Opaque. *Color:* Steel-gray, becoming iridescent purple or bronze, to dull black with oxidation. *Luster:* Metallic. *Pleochroism:* Strong. *Anisotropism:* Strong.
R_1–R_2: (400) 15.9–23.6, (420) 16.2–23.9, (440) 16.5–24.2, (460) 16.7–23.6, (480) 16.9–23.3, (500) 17.1–23.3, (520) 17.4–23.5, (540) 17.7–23.8, (560) 18.1–24.2, (580) 18.6–24.7, (600) 19.2–25.3, (620) 19.8–25.8, (640) 20.2–26.3, (660) 20.5–26.7, (680) 20.6–27.0, (700) 20.7–27.0

Cell Data: *Space Group:* *Cmcm.* $a = 9.049(6)$ $b = 11.019(7)$ $c = 5.431(4)$ Z = 4

X-ray Powder Pattern: Coyote Peak, California, USA.
5.513 (100), 6.98 (62), 3.403 (43), 3.492 (17), 1.778 (14), 2.935 (12), 2.905 (12)

Chemistry:

	(1)	(2)	(3)	(4)
K	10.60	16.3	16.1	15.8
Na	0.39			
Fe	45.50	45.2	44.8	45.2
Mg	0.50			
Ca	0.00			
Cu	trace			
S	42.40	38.0	37.3	38.9
Total	99.39	99.5	98.2	99.9

(1–2) Khibina massif, USSR; by electron microprobe. (3) Coyote Peak, California, USA; by electron microprobe. (4) KFe_2S_3.

Occurrence: In pegmatites (USSR); in mafic alkalic diatremes (Coyote Peak, California, USA); in sodalite syenite xenoliths (Mont Saint-Hilaire, Canada).

Association: Djerfisherite, pyrrhotite, cubanite, potassium feldspar, nepheline, acmite (USSR); pyrrhotite, nepheline, acmite, analcime, alkali feldspar (Point of Rocks, New Mexico, USA); ussingite, villiaumite, siderenkite, sodalite (Mont Saint-Hilaire, Canada).

Distribution: In the USSR, in the Rasvumchorr and Kikusvumchorr apatite deposits, Khibina massif, and in the Lovozero massif, Kola Peninsula. In the USA, at Coyote Peak, Humboldt Co., California, and from Point of Rocks, Colfax Co., New Mexico. From Mont Saint-Hilaire, Quebec, Canada.

Name: For the Rasvumchorr deposit in the USSR.

Type Material: A.E. Fersman Mineralogical Museum, Academy of Sciences, Moscow, USSR.

References: (1) Sokolova, M.N., M.G. Dobrovol'skaya, N.I. Organova, M.E. Kazalova, and A.L. Dmitrik (1970) A sulfide of iron and potassium, the new mineral rasvumite. Zap. Vses. Mineral. Obshch., 99, 712–720 (in Russian). (2) (1971) Amer. Mineral., 56, 1121–1122 (abs. ref. 1). (3) Czamanske, G.R., R.C. Erd, M.N. Sokolova, M.G. Dobrovol'skaya, and M.T. Dmitrieva (1979) New data for rasvumite and djerfisherite. Amer. Mineral., 64, 776–778. (4) Clark, J.R. and G.E. Brown, Jr. (1980) Crystal structure of rasvumite, KFe_2S_3. Amer. Mineral., 65, 477–482.

Crystal Data: Monoclinic. *Point Group:* $2/m$. Crystals prismatic to short prismatic and striated parallel to [001]. *Twinning:* Polysynthetic on {100} producing crystals of pseudo-orthorhombic habit; commonly seen in polished section.

Physical Properties: *Cleavage:* Perfect on {100}; parting on {010}. *Fracture:* Subconchoidal. Hardness = 3 VHN = 161 D(meas.) = 5.37(4) D(calc.) = 5.31

Optical Properties: Not fully opaque. *Color:* Lead-gray, may tarnish to iridescence; in polished section, white with deep red internal reflections. *Streak:* Chocolate-brown. *Luster:* Metallic. *Pleochroism:* Strong. *Anisotropism:* Intense, olive-green or yellow and bluish violet.

R_1–R_2: (400) 38.6–42.9, (420) 37.8–42.2, (440) 37.0–41.5, (460) 36.2–41.0, (480) 35.5–40.5, (500) 34.9–40.0, (520) 34.3–39.6, (540) 33.8–39.2, (560) 33.3–38.6, (580) 32.8–38.0, (600) 32.2–37.1, (620) 31.6–36.2, (640) 30.8–35.4, (660) 30.0–34.7, (680) 29.2–34.1, (700) 28.4–33.6

Cell Data: *Space Group:* $P2_1/n$. $a = 25.16$ $b = 7.94$ $c = 8.47$ $\beta = 100°28'$ Z = 4

X-ray Powder Pattern: Binntal, Switzerland.
2.75 (100), 3.60 (80), 3.39 (70), 2.87 (70), 4.19 (60), 2.97 (60), 2.22 (50)

Chemistry:

	(1)	(2)	(3)
Pb	51.51	52.43	41.2
Tl			3.6
Fe		0.33	
As	24.62	21.96	27.0
Sb		0.43	
S	23.41	24.11	28.0
Total	99.54	99.26	99.8

(1–2) Binntal, Switzerland. (3) Do.; by electron microprobe.

Occurrence: In crystalline dolomite with other Pb-As-S minerals.

Association: Liveingite, baumhauerite, sartorite, hutchinsonite.

Distribution: From the Lengenbach quarry, Binntal, Valais, Switzerland.

Name: For Gerhard von Rath (1830–1888), Professor of Mineralogy, Bonn, Germany.

References: (1) Palache, C., H. Berman, and C. Frondel (1944) Dana's system of mineralogy, (7th edition), v. I, 455–457. (2) Berry, L.G. and R.M. Thompson (1962) X-ray powder data for the ore minerals. Geol. Soc. Amer. Mem. 85, 152–153. (3) Ramdohr, P. (1969) The ore minerals and their intergrowths, (3rd edition), 741–744. (4) Marumo, F. and W. Nowacki (1965) The crystal structure of rathite-I. Zeits. Krist., 122, 433–456. (5) Engel, P. and W. Nowacki (1970) Die Kristallstruktur von Rathite-II $[As_{25}S_{56}|Pb^{VII}_{6.5}Pb^{IX}_{12}]$. Zeits. Krist., 131, 356–375 (in German).

Crystal Data: Monoclinic. *Point Group:* n.d. As tabular grains (30 μm) and as patches (0.5 mm) in other minerals.

Physical Properties: Hardness = n.d. VHN = n.d. D(meas.) = n.d. D(calc.) = 6.13

Optical Properties: Opaque. *Color:* Lead-gray; white with greenish and bluish tints in reflected light. *Streak:* Lead-gray *Luster:* Metallic. *Pleochroism:* Greenish to greenish blue. *Anisotropism:* Perceptible, from dark blue to dark reddish brown.

R_1–R_2: (400) 38.3–41.8, (420) 38.0–41.3, (440) 37.8–40.9, (460) 37.6–40.6, (480) 37.6–40.4, (500) 37.5–40.1, (520) 37.5–39.9, (540) 37.5–39.6, (560) 37.4–39.4, (580) 37.2–38.9, (600) 37.0–38.4, (620) 36.7–37.9, (640) 36.3–37.5, (660) 35.9–37.3, (680) 35.6–37.2, (700) 35.3–37.1

Cell Data: *Space Group:* n.d. $a = 13.60(2)$ $b = 11.96(3)$ $c = 24.49(5)$ $\beta = 103.94(12)°$ Z = [4]

X-ray Powder Pattern: Rajpura-Dariba, India.
3.37 (100), 3.26 (90), 2.98 (50), 3.90 (30), 3.74 (30), 2.06 (30), 2.88 (20)

Chemistry:

	(1)	(2)
Ag	4.54	4.49
Tl	2.04	2.83
Pb	47.06	45.98
Cu	0.03	
Sb	27.42	27.02
S	19.59	19.68
Total	100.68	100.00

(1) Rajpura-Dariba, India; by electron microprobe, average of four samples, giving $(Ag, Tl)_{1.8}Pb_{5.3}(Pb_{2.4}Sb_{7.6})_{\Sigma=10.0}S_{20.9}$. (2) $(Ag, Tl)_2Pb_8Sb_8S_{21}$ with Ag:Tl = 3:1.

Occurrence: In Precambrian polymetallic massive-sulfide deposits interbedded with kyanite-graphite schists, diopside-bearing calcsilicates, and meta-cherts.

Association: Galena, meneghinite, owyheeite.

Distribution: From Rajpura-Dariba, Udaipur district, Rajasthan, India.

Name: To honor Professor Santosh K. Ray of President College, Calcutta, India.

Type Material: Indian Institute of Technology, Kharagpur, India; Institute of Mineralogy and Geochemistry of Rare Elements, Moscow, USSR.

References: (1) Basu, K., N.S. Bortinikov, A. Mookherjee, N.N. Mozgova, A.I. Tsepin, and L.N. Vyal'sov (1983) Rare minerals from Rajpura-Dariba, Rajasthan, India. IV: A new Pb–Ag–Tl–Sb sulfosalt, rayite. Neues Jahrb. Mineral., Monatsh., 296–304. (2) (1984) Amer. Mineral., 69, 211 (abs. ref. 1).

Crystal Data: Monoclinic. *Point Group:* $2/m$. Crystals prismatic, to 7 cm, and striated $\parallel$ [001]; more commonly massive, coarse to fine granular, or as incrustations. *Twinning:* As contact twins on {100}.

Physical Properties: *Cleavage:* Good on {010}; less so on {$\bar{1}$01}, {100}, {120} and {110}. *Tenacity:* Sectile, also slightly brittle. Hardness = 1.5–2 VHN = n.d. D(meas.) = 3.56 D(calc.) = 3.59 Disintegrates on long exposure to light to a powder composed of arsenolite and orpiment.

Optical Properties: Transparent when fresh. *Color:* Red to orange-yellow; in polished section, gray-white, with abundant yellow to red internal reflections. *Streak:* Orange-red to red. *Luster:* Resinous to greasy.
Optical Class: Biaxial negative (–). *Pleochroism:* Nearly colorless to pale golden yellow. *Orientation:* $X \wedge c = -11°$ *Dispersion:* $r > v$, very strong, inclined. $\alpha = 2.538$ $\beta = 2.684$ $\gamma = 2.704$ 2V(meas.) = 40°34′ 2V(calc.) = n.d. *Anisotropism:* Strong, in polished section.
R: (400) 23.3, (420) 23.0, (440) 22.7, (460) 22.3, (480) 21.8, (500) 21.2, (520) 20.6, (540) 20.1, (560) 19.7, (580) 19.4, (600) 19.1, (620) 19.0, (640) 18.8, (660) 18.6, (680) 18.5, (700) 18.4

Cell Data: *Space Group:* $P2_1/n$. $a = 9.325(3)$ $b = 13.571(5)$ $c = 6.587(3)$ $\beta = 106°23(5)'$ Z = 16

X-ray Powder Pattern: Allchar, Yugoslavia.
5.40 (100), 3.19 (90), 2.94 (80), 2.73 (80), 1.859 (60), 2.49 (50), 2.14 (50)

Chemistry:

	(1)	(2)
As	69.54	70.0
S	30.29	30.0
rem.	0.11	
Total	99.94	100.0

(1) Binntal, Switzerland; remainder SiO_2. (2) AsS.

Polymorphism & Series: Dimorphous with pararealgar.

Occurrence: Most commonly as a low-temperature hydrothermal vein mineral associated with As-Sb minerals; also as volcanic sublimation and hot spring deposits; found in carbonate and clay sedimentary rocks.

Association: Orpiment, arsenolite, other arsenic minerals, calcite, barite.

Distribution: Only localities for finely crystallized material are given. In the USA, at the Getchell mine, Humboldt Co., Nevada; in Tooele Co., Utah, at Mercur; in Washington, in the Monte Cristo district, Snohomish Co., and in the Green River Gorge, Franklin, King Co. In Romania, at Baia Sprie (Felsőbánya), Cavnic (Kapnik), and Săcărâmb (Nagyág). In Czechoslovakia, at Jáchymov (Joachimsthal). In Yugoslavia, at Allchar, Macedonia. In Germany, at Schneeberg, Saxony, and St. Andreasberg, in the Harz. At the Lengenbach quarry, Binntal, Valais, Switzerland. At Men Kule, Yakutia, USSR.

Name: From the Arabic for *powder of the mine.*

References: (1) Palache, C., H. Berman, and C. Frondel (1944) Dana's system of mineralogy, (7th edition), v. I, 255–258. (2) Berry, L.G. and R.M. Thompson (1962) X-ray powder data for the ore minerals. Geol. Soc. Amer. Mem. 85, 69–70. (3) Ito, T., N. Morimoto, and R. Sadanaga (1952) The crystal structure of realgar. Acta Cryst., 5, 775–782. (4) Mullen, D.J.E. and W. Nowacki (1972) Refinement of the crystal structures of realgar, AsS and orpiment, As_2S_3. Zeits. Krist., 136, 48–65.

Crystal Data: Monoclinic. *Point Group:* $2/m$. Crystals exhibit the forms {100}, {001} and {111}; all faces in the zone [010] are striated.

Physical Properties: Hardness = n.d. VHN = n.d. D(meas.) = 4.81 D(calc.) = 4.90

Optical Properties: Opaque. *Color:* Dark gray. *Streak:* Brownish red. *Luster:* Sub-metallic. R_1–R_2: n.d.

Cell Data: *Space Group:* $P2_1/c$. $a = 17.441(5)$ $b = 7.363(2)$ $c = 32.052(7)$ $\beta = 105.03(7)°$ Z = 4

X-ray Powder Pattern: n.d.

Chemistry:

	(1)	(2)
Tl	32.76	34.81
Sb	22.88	20.75
As	20.46	20.41
S	24.33	24.03
Total	100.43	100.00

(1) Allchar, Yugoslavia; by electron microprobe. (2) $Tl_5Sb_5As_8S_{22}$.

Occurrence: In a hydrothermal As-Tl deposit.

Association: Simonite, $TlHgAsS_3$ (christite or routhierite).

Distribution: From Allchar, Macedonia, Yugoslavia.

Name: Derivation not stated.

Type Material: n.d.

References: (1) Balić-Žunić, T., S. Šćavničar, and P. Engel (1982) The crystal structure of rebulite, $Tl_5Sb_5As_8S_{22}$. Zeits. Krist., 160, 109–125. (2) (1983) Amer. Mineral., 68, 644 (abs. ref. 1).

Reniérite $(Cu, Zn)_{11}(Ge, As)_2Fe_4S_{16}$

Crystal Data: Tetragonal, pseudocubic. *Point Group:* $\bar{4}2m$. As disseminated granular aggregates, irregular patches, and as regular lamellae replacing germanite grains; also as small (to 3 mm) equant crystals. *Twinning:* Polysynthetic.

Physical Properties: Hardness = 3.5 VHN = 340–363, 351 average (25 g load). D(meas.) = 4.38 D(calc.) = 4.40 Magnetic.

Optical Properties: Opaque. *Color:* Bronze; in polished section, orange-brown. *Luster:* Metallic. *Pleochroism:* Orange to brown or bronze with violet tint. *Anisotropism:* Strong, yellow to reddish orange.
R_1–R_2: (400) 19.3–20.2, (420) 18.8–19.5, (440) 18.3–18.8, (460) 18.3–18.3, (480) 18.9–19.5, (500) 20.2–21.4, (520) 22.2–23.5, (540) 24.3–25.7, (560) 26.2–27.8, (580) 28.0–29.6, (600) 29.5–31.2, (620) 30.8–32.5, (640) 31.8–33.6, (660) 32.8–34.5, (680) 33.7–35.2, (700) 34.4–36.0

Cell Data: *Space Group:* $P\bar{4}2m$. $a = 10.622(1)$ $c = 10.551(1)$ Z = 2

X-ray Powder Pattern: Kipushi, Zaire.
3.06 (100), 1.87 (80), 1.595 (60), 2.65 (30), 1.214 (30), 4.31 (20), 1.083 (20)

Chemistry:

	(1)	(2)
Cu	41.63	41.6
Fe	13.73	13.0
Zn	3.53	2.7
Pb	trace	
Ge	7.75	7.3
Ga		0.5
As	0.87	2.0
S	31.51	32.6
Total	99.02	99.7

(1) Kipushi, Zaire. (2) Tsumeb, Namibia; by electron microprobe.

Occurrence: In a few dolomite-hosted Cu-Pb-Zn deposits, and in hydrothermal polymetallic deposits.

Association: Germanite, galena, tennantite, enargite, digenite, bornite, chalcopyrite, sphalerite.

Distribution: At Tsumeb, Namibia. From M'Passa, 150 km west of Brazzaville, Niari Province, Congo Republic. In Zaire, at Kipushi, Shaba Province. From the Ruby Creek deposit, southern Brooks Range, Alaska; and at the Inexco # 1 mine, Jamestown, Boulder Co., Colorado, USA. In Bulgaria, in the Chelopech deposit, Sofia; and the Radka deposit, Pazardzhik. From the San Giovanni mine, Iglesiente, Sardinia, Italy. In the USSR, at Vaygach, Arkhangel'sk; Akhtala, Armenia SSR; the Urup deposit, Caucasus Mountains, and other less well defined localities.

Name: For Armand Reniér, Belgian geologist and Director of the Belgian Geological Survey.

Type Material: n.d.

References: (1) Vaes, J.F. (1948) La reniérite (anciennement appelée "bornite orange") un sulfure germanifère provenant de la Mine Prince-Léopold, Kipushi (Congo Belge). Ann. (Bull.) Soc. Géol. Belgique, 72, 20–32 (in French). (2) (1950) Amer. Mineral., 35, 136 (abs. ref. 1). (3) Murdoch, J. (1953) X-ray investigation of colusite, germanite and reniérite. Amer. Mineral., 38, 794–801. (4) Springer, G. (1969) Microanalytical investigations into germanite, renierite, briartite, and gallite. Neues Jahrb. Mineral., Monatsh., 435–441. (5) Bernstein, L.R. (1986) Renierite, $Cu_{10}ZnGe_2Fe_4S_{16}$—$Cu_{11}GeAsFe_4S_{16}$: a coupled solid solution series. Amer. Mineral., 71, 210–221.

Crystal Data: Cubic. *Point Group:* $4/m\ \overline{3}\ 2/m$ (synthetic). As subhedral grains.

Physical Properties: Hardness = n.d. VHN = 165 (30 g load). D(meas.) = n.d. D(calc.) = [16.5]

Optical Properties: Opaque. *Color:* In polished section, bright white. *Luster:* Metallic. R: (470) 75.2, (546) 72.6, (589) 73.3, (650) 75.7

Cell Data: *Space Group:* $Fm3m$ (synthetic). $a = 3.856(1)$ Z = [4]

X-ray Powder Pattern: Stillwater Complex, Montana, USA.
0.7874 (100), 0.8623 (80), 0.8847 (70), 2.227 (60), 1.362 (50), 1.162 (50), 1.927 (30)

Chemistry:

	(1)
Pt	59.6
Rh	41.7
Total	101.3

(1) Stillwater Complex, Montana, USA; by electron microprobe, leading to $Rh_{0.57}Pt_{0.43}$.

Occurrence: A single grain was found in heavy mineral concentrates (Stillwater Complex, Montana, USA).

Association: Platinum, Pt–Fe alloy, gold, moncheite, kotulskite, merenskyite, cooperite, braggite, vysotskite, sperrylite, pyrite, chalcopyrite, pyrrhotite, chromite, magnetite, marcasite, violarite, graphite (Stillwater Complex, Montana, USA).

Distribution: From the Stillwater Complex, Montana, and at Fox Gulch, Goodnews Bay, Alaska, USA.

Name: From the Greek *rhodon, rose.*

Type Material: Royal Ontario Museum, Toronto, Canada.

References: (1) Cabri, L.J. and J.H.G. Laflamme (1974) Rhodium, platinum, and gold alloys from the Stillwater Complex. Can. Mineral., 12, 399–403. (2) (1976) Amer. Mineral., 61, 340 (abs. ref. 1). (3) Cabri, L.J., Ed. (1981) Platinum group elements: mineralogy, geology, recovery. Can. Inst. Min. & Met., 132–133.

Rhodostannite $Cu_2FeSn_3S_8$

Crystal Data: Tetragonal. *Point Group:* $4/m$. Very fine-grained massive, slightly porous.

Physical Properties: Hardness = n.d. VHN = 243–266 D(meas.) = n.d. D(calc.) = 4.79

Optical Properties: Opaque. *Color:* Reddish. *Luster:* Metallic. *Anisotropism:* Distinct, from bluish gray to dark brown.

R_1–R_2: (400) 29.1–30.4, (420) 28.7–29.9, (440) 28.3–29.4, (460) 28.2–29.3, (480) 28.2–29.4, (500) 28.3–29.8, (520) 28.5–30.2, (540) 28.8–30.4, (560) 28.9–30.6, (580) 28.9–30.8, (600) 28.9–30.8, (620) 28.9–30.8, (640) 28.8–30.7, (660) 28.8–30.7, (680) 28.7–30.5, (700) 28.6–30.3

Cell Data: *Space Group:* $I4_1/a$ (synthetic). $a = 7.305(2)$ $c = 10.330(5)$ Z = 2

X-ray Powder Pattern: Vila Apacheta, Bolivia.
3.12 (100), 5.93 (60), 2.58 (50), 6.09 (40), 1.819 (30), 1.837 (20), 3.64 (16)

Chemistry:

	(1)	(2)	(3)
Cu	16.0	15.9	15.97
Fe	6.8	6.5	7.02
Sn	45.5	45.4	44.76
S	31.3	32.8	32.25
Total	99.6	100.6	100.00

(1) Vila Apacheta, Bolivia; by electron microprobe, corresponding to $Cu_{2.03}Fe_{0.98}Sn_{3.10}S_{7.89}$. (2) Do.; corresponding to $Cu_{1.98}Fe_{0.92}Sn_{3.02}S_{8.08}$. (3) $Cu_2FeSn_3S_8$.

Occurrence: As a replacement (alteration) product of stannite.

Association: Stannite, pyrite.

Distribution: At Vila Apacheta, Bolivia. From the Pirquitas deposit, Rinconada Department, Jujuy Province, Argentina. In the Dean mine, southeast of Battle Mountain, Lander Co., Nevada, USA.

Name: From the Greek for *rose*, in allusion to its color; and for its affinity to stannite.

Type Material: n.d.

References: (1) Springer, G. (1968) Electronprobe analyses of stannite and related tin minerals. Mineral. Mag., 36, 1045–1051. (2) (1969) Amer. Mineral., 54, 1218 (abs. ref. 1). (3) Wang, N. (1975) Investigation of synthetic rhodostannite, $Cu_2FeSn_3S_8$. Neues Jahrb. Mineral., Monatsh., 166–171. (4) Jumas, J.C., E. Philippot, and M. Maurin (1979) Structure du rhodostannite synthétique. Acta Cryst., B35, 2195–2197 (in French with English abs.).

Crystal Data: Hexagonal. *Point Group:* $\overline{3}\ 2/m$. As grains to 20 x 40 μm in a platinum nugget.

Physical Properties: Hardness = n.d. VHN = n.d. D(meas.) = n.d. D(calc.) = 9.74

Optical Properties: Opaque. *Color:* White (?) in reflected light. *Pleochroism:* Notable, from cream-pink to grayish blue. *Anisotropism:* Strong.
R_1–R_2: n.d.

Cell Data: *Space Group:* $R\overline{3}m$. $a = 5.73$ $c = 14.00$ Z = 3

X-ray Powder Pattern: Ural Mountains, USSR.
2.86 (10), 2.33 (6), 2.01 (6), 4.02 (3), 2.44 (3), 1.807 (3), 4.64 (2)

Chemistry:

	(1)	(2)
Rh	40.27	39.21
Ir	0.27	
Pt	0.24	
Pb	51.48	52.64
S	8.05	8.15
Total	100.31	100.00

(1) Ural Mountains, USSR; by electron microprobe. (2) $Pb_2Rh_3S_2$.

Occurrence: In a platinum nugget.

Association: Platinum, iridosmine, osmiridium.

Distribution: From an undefined location in the Ural Mountains, USSR.

Name: For the composition.

Type Material: n.d.

References: (1) Genkin, A.D., L.N. Vyal'sov, T.L. Evstigneeva, I.P. Laputina, and G.V. Basova (1983) Rhodplumsite, $Rh_3Pb_2S_2$, a new sulfide of rhodium and lead. Mineral. Zhurnal, 5, 87–91 (in Russian). (2) (1984) Mineral. Abs., 35, 88 (abs. ref. 1).

Crystal Data: Orthorhombic. *Point Group:* n.d. Massive, somewhat porous.

Physical Properties: *Tenacity:* Brittle. Hardness = 3.5 VHN = 72–85 (50 g load). D(meas.) = 7.54 D(calc.) = 7.467

Optical Properties: Opaque. *Color:* Purple-red on fresh surface, tarnishes rapidly. *Luster:* Metallic. *Pleochroism:* Strong, carmine to violet-gray. *Anisotropism:* Strong, exceptional, in fiery orange colors.

R_1–R_2: (400) 29.4–37.4, (420) 26.8–34.8, (440) 24.2–32.2, (460) 21.2–29.1, (480) 17.8–25.8, (500) 14.9–22.8, (520) 13.3–20.3, (540) 13.0–18.4, (560) 15.6–18.0, (580) 20.2–18.4, (600) 26.5–19.8, (620) 33.2–21.3, (640) 39.4–24.8, (660) 45.0–28.0, (680) 49.4–31.8, (700) 53.8–35.6

Cell Data: *Space Group:* n.d. $a = 4.0032(14)$ $b = 19.893(7)$ $c = 12.220(4)$ Z = [4]

X-ray Powder Pattern: Good Hope Mine, Colorado, USA.
2.07 (100), 3.36 (60), 2.55 (40), 1.988 (40), 1.158 (30), 2.82 (20), 1.706 (20)

Chemistry:

	(1)	(2)
Cu	40.0	41.08
Te	60.9	58.92
Total	100.9	100.00

(1) Good Hope mine, Colorado, USA; by electron microprobe. (2) Cu_7Te_5.

Occurrence: A late-stage mineral of low-temperature hydrothermal origin.

Association: Vulcanite, pyrite, bornite, tellurium, petzite, sylvanite, berthierite.

Distribution: In the USA, in Colorado, at the Good Hope mine, Vulcan, Gunnison Co.; also found at the Empress Josephine mine, Bonanza district, Saguache Co.; in Arizona, from the Junction mine, Bisbee; and at Tombstone, Cochise Co. In the Horne mine, Noranda, Quebec, Canada. In San Salvador, from the Sebastian mine. In Japan, from the Teine mine, Hokkaido, and the Kawazu mine, Shizuoka Prefecture. From Kalgoorlie, Western Australia. In South Africa, at Ookiep, Namaqualand. From the Byn'govsk Au-Te deposit, Central Ural Mountains, and the Koshmansaisk deposits, Chatkal Ridge, in the basin of the Arbulak River, USSR. Other occurrences are known.

Name: For Thomas Arthur Rickard (1864-1953), mining engineer and Editor of the Engineering and Mining Journal, New York and London.

Type Material: National Museum of Natural History, Washington, D.C., USA, 114590.

References: (1) Palache, C., H. Berman, and C. Frondel (1944) Dana's system of mineralogy, (7th edition), v. I, 198–199. (2) Forman, S.A. and M.A. Peacock (1949) Crystal structure of rickardite, $Cu_{4-x}Te_2$. Amer. Mineral., 34, 441–451. (3) Ramdohr, P. (1969) The ore minerals and their intergrowths, (3rd edition), 416–417. (4) Mizota, T., K. Koto, and N. Morimoto (1973) Crystallography and composition of synthetic rickardite. Mineral. J. (Japan), 7, 252–261.

Crystal Data: Triclinic. *Point Group:* $\bar{1}$. Crystals are slender, prismatic and striated ∥ [001]; massive, fibrous to compact.

Physical Properties: *Fracture:* Irregular. *Tenacity:* Brittle. Hardness = 2.5–3 VHN = n.d. D(meas.) = 5.64–5.75 D(calc.) = 5.74

Optical Properties: Opaque. *Color:* Bluish lead-gray. *Luster:* Metallic. *Anisotropism:* Strong, in blue-gray, cream-white and brown-gray.
R_1–R_2: n.d.

Cell Data: *Space Group:* $P\bar{1}$. $a = 16.519(2)$ $b = 17.641(2)$ $c = 3.971(1)$ $\alpha = 96.12(2)°$ $\beta = 96.32(2)°$ $\gamma = 91.15(1)°$ Z = 2

X-ray Powder Pattern: Red Bird mine, Nevada, USA.
3.39 (100), 4.04 (80), 3.92 (80), 3.03 (80), 2.75 (80), 2.67 (80), 3.79 (60)

Chemistry:

	(1)	(2)	(3)
Pb	42.6	41.5	41.94
Sb	35.5	20.0	36.97
Bi		22.0	
S	20.9	18.0	21.09
Total	99.0	101.5	100.00

(1) Red Bird mine, Nevada, USA; by electron microprobe. (2) Salmo, Canada. (3) $Pb_4Sb_6S_{13}$.

Occurrence: A primary hydrothermal mineral (Red Bird mine, Nevada, USA).

Association: Pyrite, sphalerite, stibnite, boulangerite, zinkenite, quartz, calcite, plumboan aragonite.

Distribution: At the Red Bird mercury mine, about 37 road km east of Lovelock, Pershing Co.; and in the Silver Coin mine, Golconda Summit, Humboldt Co., Nevada, USA. From Madoc, Ontario; and from the Dodger tungsten mine, Salmo, British Columbia, Canada. At Vall de Ribes, eastern Pyrenees, Spain. From Male Zelezne, Czechoslovakia. At Rujevac, Yugoslavia.

Name: For Dr. Stephen Clive Robinson (1911–), Geological Survey of Canada, who first synthesized the mineral at Queen's University, Kingston, Ontario, Canada.

Type Material: National Museum of Natural History, Washington, D.C., USA, 106568.

References: (1) Berry, L.G., J.J. Fahey, and E.H. Bailey (1952) Robinsonite, a new lead antimony sulfide. Amer. Mineral., 37, 438–446. (2) Jambor, J.L. and A.G. Plant (1975) The composition of the lead sulphantimonide, robinsonite. Can. Mineral., 13, 415–417. (3) Jambor, J.L. and G.R. Lachance (1968) Bismuthian robinsonite. Can. Mineral., 9, 426–428. (4) Ayora, C. and S. Gali (1981) Additional data on robinsonite. Can. Mineral., 19, 415–417. (5) Jambor, J.L. and D.R. Owens (1982) Re-examination of robinsonite from Vall de Ribes, Spain. Can. Mineral., 20, 97–100. (6) Petrova, I.V., L.N. Kaplunnik, N.S. Bortnikov, E.A. Pobedimskaya, and N.V. Belov (1978) Crystal structure of synthetic robinsonite. Doklady Acad. Nauk SSSR, 241, 88–90 (in Russian). (7) (1978) Chem. Abs., 89, 98177 (abs. ref. 6).

Crystal Data: Orthorhombic, pseudotetragonal. *Point Group:* 222, $mm2$, or $2/m\ 2/m\ 2/m$. In patchy aggregates about 0.5 cm in size; rarely as tiny euhedral and subhedral crystals in the size range 0.05 to 0.5 mm.

Physical Properties: *Tenacity:* Malleable. Hardness = n.d. VHN = 94 (25 g load). D(meas.) = n.d. D(calc.) = 7.752

Optical Properties: Opaque. *Color:* In reflected light, yellowish cream to dull bluish gray. *Luster:* Metallic. *Pleochroism:* Distinct to strong. *Anisotropism:* Very strong, from yellowish cream to black, with dark red-purple colors present near extinction.

R_1–R_2: (400) 27.9–36.2, (420) 27.7–36.4, (440) 27.5–36.6, (460) 26.8–36.7, (480) 25.9–36.5, (500) 25.2–36.1, (520) 24.5–36.0, (540) 23.9–35.4, (560) 23.6–35.2, (580) 23.4–35.0, (600) 23.2–34.6, (620) 23.3–34.4, (640) 23.1–34.0, (660) 22.9–33.9, (680) 22.6–33.7, (700) 22.3–33.7

Cell Data: *Space Group:* $P222$, $Pmm2$, or $Pmmm$. $a = 7.602(2)$ $b = 3.801(2)$ $c = 20.986(8)$ Z = 2

X-ray Powder Pattern: Ilímaussaq Intrusion, Greenland.
3.078 (100), 2.393 (100), 3.800 (90), 1.902 (90), 2.623 (50), 2.605 (50), 2.577 (50)

Chemistry:

	(1)	(2)	(3)
Tl	26.6	27.7	28.87
Cu	18.6	17.8	17.20
Pb	2.0	1.6	
Fe	0.3	0.3	
Sb	41.0	43.5	44.87
S	9.3	9.6	9.06
Total	97.8	100.5	100.00

(1) Ilímaussaq Intrusion, Greenland; by electron microprobe, average of analyses of seven grains, corresponding to $Tl_{0.97}(Cu_{4.79}Fe_{0.07})_{\Sigma=4.86}Pb_{0.04}Sb_{1.13}S_{2.15}$. (2) Do.; average of analyses of eight grains, corresponding to $Tl_{0.97}(Cu_{4.89}Fe_{0.06})_{\Sigma=4.95}Pb_{0.04}Sb_{1.05}S_{2.14}$. (3) $TlCu_5SbS_2$.

Occurrence: In veins cutting an alkalic intrusive.

Association: Chalcocite, löllingite, cuprite, antimonian silver, analcime, sodalite.

Distribution: From the Ilímaussaq Intrusion, southern Greenland.

Name: For John Rose-Hansen, University of Copenhagen, Denmark.

Type Material: University of Copenhagen Geological Museum, Copenhagen, Denmark.

References: (1) Karup-Møller, S. (1978) Primary and secondary ore minerals associated with cuprostibite. Bull. Grønlands Geol. Undersøgelse, 126, 23–45. (2) (1980) Amer. Mineral., 65, 208–209 (abs. ref. 1). (3) Makovicky, E., Z. Johan, and S. Karup-Møeller (1980) New data on bukovite, thalcusite, chalcothallite and rohaite. Neues Jahrb. Mineral., Abh., 138, 122-146.

Crystal Data: Tetragonal. *Point Group:* n.d. As tiny (0.2 x 0.3 mm) inclusions in bornite. *Twinning:* Polysynthetic.

Physical Properties: Hardness = n.d. VHN = 241(5) (100 g load). D(meas.) = n.d. D(calc.) = [4.78]

Optical Properties: Opaque. *Color:* In polished section, gray with a slight bluish tint. *Luster:* Metallic. *Anisotropism:* Very weak.

R_1–R_2: (400) 27.1–27.2, (420) 25.6–26.0, (440) 24.1–24.8, (460) 23.4–24.0, (480) 23.0–23.4, (500) 22.6–23.0, (520) 22.5–22.7, (540) 22.2–22.5, (560) 22.2–22.4, (580) 22.1–22.4, (600) 22.1–22.4, (620) 22.2–22.5, (640) 22.3–22.6, (660) 22.4–22.7, (680) 22.6–22.9, (700) 22.8–23.1

Cell Data: *Space Group:* n.d. $a = 5.51$ $c = 11.05$ Z = 4

X-ray Powder Pattern: Charrier mine, France.
3.19 (vvs), 1.95 (vs), 1.66 (s), 2.76 (mw), 1.268 (vw), 1.127 (vvw), 3.08 (f)

Chemistry:

	(1)	(2)	(3)
Cu	26.8	28.4	26.20
Fe		0.6	
In	47.8	45.9	47.35
S	27.3	26.2	26.45
Total	101.9	101.1	100.00

(1) Charrier, France; by electron microprobe. (2) Geevor mine, England; by electron microprobe. (3) $CuInS_2$.

Occurrence: In association with copper sulfides in high-temperature Sn-W-Bi-Mo hydrothermal veins in highly metamorphosed rocks (Charrier, France); as a late-stage mineral in a skarn Fe-W ore pipe (Ulsan mine, Korea); in magnetite-bearing massive chalcopyrite ore (Akenobe mine, Japan).

Association: Chalcopyrite, wittichenite, chalcocite, covellite, bornite, sphalerite (Charrier, France); bornite, sphalerite, chalcopyrite, tetrahedrite, löllingite, arsenopyrite, bismuth (Ulsan mine, Korea).

Distribution: In France, from Charrier, Allier; Lautaret Pass, Hautes-Alpes; La Telhaie, Morbihan; and Vaultry, Haute-Vienne. In the Geevor mine, Pendeen, Penzance, Cornwall, England. From Listulli, Telemark, and Långban, Värmland, Sweden. At the Akenobe and Ikuno mines, Hyogo Prefecture, Japan. In the Ulsan mine, Kyongsangnam Province, Korea. At the Tosham tin prospect, Bhiwani district, Haryana, India. In Canada, from Mount Pleasant, New Brunswick.

Name: For Professor Maurice Roques, University of Clairmont-Ferrand, France.

Type Material: National School of Mines, Paris, France.

References: (1) Picot, P. and R. Pierrot (1963) La roquésite, premier minéral d'indium: $CuInS_2$. Bull. Soc. fr. Minéral., 86, 7–14. (2) (1963) Amer. Mineral., 48, 1178–1179 (abs. ref. 1). (3) Vlasov, K.A., Ed. (1966) Mineralogy of rare elements, v. II, 585–586. (4) Imai, N. and S. Choi (1984) The first Korean occurrence of roquesite. Mineral. J. (Japan), 12, 162–172. (5) Criddle, A.J. and C.J. Stanley, Eds. (1986) The quantitative data file for ore minerals. British Museum (Natural History), London, England, 314.

Crystal Data: Monoclinic. *Point Group:* $2/m$. As minute (0.5 mm) equidimensional crystals; thick tabular on {010}; short prismatic on {110}; rarely acicular [001]; also dipyramidal with {111} and {$\bar{1}$11}. *Twinning:* On {101}, the twinning lamellae parallel to the edge (010).

Physical Properties: Hardness = Low. VHN = n.d. D(meas.) = n.d. D(calc.) = [2.02]

Optical Properties: Transparent to translucent. *Color:* Colorless to pale yellow with a greenish tinge. *Luster:* Adamantine.
Optical Class: Biaxial negative (–). *Orientation:* Axial plane || {010}; $X \wedge c = 1.25°$
Dispersion: High. n = High. 2V(meas.) = Large. 2V(calc.) = n.d.
R_1–R_2: n.d.

Cell Data: *Space Group:* $P2/n$. $a = 8.50$ $b = 13.16$ $c = 9.29$ $\beta = 124°49'$ Z = [32]

X-ray Powder Pattern: n.d.

Chemistry: Identity established by morphological similarity with synthetic material.

Polymorphism & Series: Dimorphous with sulfur to which it changes in time.

Occurrence: Found in fumaroles (Vulcano, Lipari Islands); in hollow limonite nodules which occur in a thin clay stratum (Havírna, Czechoslovakia).

Association: n.d.

Distribution: From Vulcano, Lipari Islands, Italy. At Visky and Havírna, near Letovice, Czechoslovakia. In the USA, from Point Rincon, Ventura Co., California.

Name: For Professor Vojtěch Rosický, former Director of the Mineralogical and Petrological Institute of Masaryk University, Brno, Czechoslovakia.

Type Material: n.d.

References: (1) Palache, C., H. Berman, and C. Frondel (1944) Dana's system of mineralogy, (7th edition), v. I, 145–146. (2) Sekanina, J. (1931) Rosickyit, die natürliche γ–Schwefelmodifikation. Zeits. Krist., 80, 174–189 (in German). (3) (1932) Amer. Mineral., 17, 251 (abs. ref. 2).

Crystal Data: Tetragonal. *Point Group:* 4 (possible). As anhedral grains and as veinlets. *Twinning:* Fine polysynthetic lamellae seen in polished section.

Physical Properties: *Cleavage:* Two perpendicular. Hardness = n.d. VHN = 148 (25 g load). D(meas.) = n.d. D(calc.) = 5.83

Optical Properties: Opaque. *Color:* Violet-red; in polished section, bluish white, with strong internal reflections. *Luster:* Metallic. *Pleochroism:* Weak. *Anisotropism:* Weak.
R_1–R_2: (400) 31.1–31.8, (420) 30.9–31.6, (440) 30.7–31.4, (460) 30.4–31.2, (480) 30.0–30.8, (500) 29.5–30.4, (520) 28.9–30.0, (540) 28.2–29.3, (560) 27.2–28.5, (580) 26.3–27.5, (600) 25.6–26.8, (620) 25.2–26.2, (640) 25.0–26.0, (660) 25.0–25.7, (680) 24.8–25.5, (700) 24.6–25.4

Cell Data: *Space Group:* $I4$ (possible). $a = 9.977(2)$ $c = 11.290(2)$ Z = 8

X-ray Powder Pattern: Jas Roux, France.
4.146 (100), 2.989 (100), 2.495 (90), 3.525 (80), 1.870 (70), 1.763 (70), 1.827 (60)

Chemistry:

	(1)	(2)	(3)
Tl	20.4	19.7	35.48
Cu	3.9	3.8	
Ag	3.8	4.2	
Hg	34.7	34.4	34.82
Zn	2.0	2.1	
As	13.2	13.2	13.00
Sb	2.6	2.9	
S	19.6	19.6	16.70
Total	100.2	99.9	100.00

(1) Jas Roux, France; by electron microprobe, corresponding to $(Tl_{0.50}Cu_{0.30}Ag_{0.18})_{\Sigma=0.98}(Hg_{0.86}Zn_{0.15})_{\Sigma=1.01}(As_{0.88}Sb_{0.10})_{\Sigma=0.98}S_{3.03}$. (2) Do.; giving $(Tl_{0.48}Cu_{0.30}Ag_{0.19})_{\Sigma=0.97}(Hg_{0.85}Zn_{0.16})_{\Sigma=1.01}(As_{0.87}Sb_{0.12})_{\Sigma=0.99}S_{3.03}$. (3) $TlHgAsS_3$.

Polymorphism & Series: Dimorphous with christite.

Occurrence: In hydrothermal deposits in dolomite (Jas Roux, France).

Association: Realgar, stibnite, pierrotite, sphalerite, pyrite, smithite.

Distribution: At Jas Roux, 10 km east of Chapelle-en-Valgaudemar, Hautes-Alpes, France. From the Hemlo gold deposit, Thunder Bay district, Ontario, Canada.

Name: For Pierre Routhier, Professor of Economic Geology, University of Paris, France.

Type Material: National School of Mines, Paris, France.

References: (1) Johan, Z., J. Mantienne, and P. Picot (1974) La routhiérite, $TlHgAsS_3$, et la laffittite, $AgHgAsS_3$, deux nouvelles espèces minérales. Bull. Soc. fr. Minéral., 97, 48–53 (in French with English abs.). (2) (1975) Amer. Mineral., 60, 947 (abs. ref. 1).

Crystal Data: Monoclinic, pseudohexagonal. *Point Group:* $2/m$, m, or 2. As small single crystals up to 0.01 mm, epitaxially intergrown with djurleite; in powdery mixtures with djurleite.

Physical Properties: *Cleavage:* Poor on {100}. Hardness = n.d. VHN = 83 (50 g load). D(meas.) = n.d. D(calc.) = n.d.

Optical Properties: Opaque. *Color:* Blue-black; in reflected light, white to off-white with a blue tinge. *Anisotropism:* Very weak.
R_1–R_2: n.d.

Cell Data: *Space Group:* $C2/m$, Cm, or $C2$. $a = 53.79$ $b = 30.90$ $c = 13.36$ $\beta = 90.0°$ Z = 512

X-ray Powder Pattern: Olympic Dam, Australia.
1.933 (100), 2.375 (90), 1.857 (80), 2.86 (70), 2.630 (50), 1.673 (25), 1.626 (20)

Chemistry:

	(1)	(2)	(3)
Cu	77.25	79.47	78.10
Fe	0.77	0.07	
S	22.19	21.94	21.90
Total	100.21	101.48	100.00

(1) Olympic Dam, Australia; by electron microprobe, corresponding to $(Cu_{1.76}Fe_{0.02})_{\Sigma=1.78}S_{1.00}$. (2) Do.; corresponding to $Cu_{1.83}S_{1.00}$. (3) Cu_9S_5.

Occurrence: As a low-temperature alteration product of djurleite, in a complex hydrothermal deposit (Olympic Dam, Australia); in a porphyry copper deposit (El Teniente mine, Chile).

Association: Djurleite, bornite, pyrite, chalcopyrite, hematite (Olympic Dam, Australia); djurleite (El Teniente mine, Chile).

Distribution: From the Olympic Dam Cu-Au-U deposit, Roxby Downs Station, South Australia. In the El Teniente mine, 67 km west of Rancagua, Chile.

Name: For its occurrence at Roxby Downs, South Australia.

Type Material: Museum of Victoria, Melbourne, Australia.

References: (1) Mumme, W.G., G.J. Sparrow, and G.S. Walker (1988) Roxbyite, a new copper sulphide mineral from the Olympic Dam deposit, Roxby Downs, South Australia. Mineral. Mag., 52, 323-330. (2) (1989) Amer. Mineral., 74, 947 (abs. ref. 1).

Crystal Data: Monoclinic. *Point Group:* n.d. As irregular grains or as aggregates, 100–150 μm in dimension, having rough textured surfaces.

Physical Properties: *Tenacity:* Brittle. Hardness = n.d. VHN = 893 (50 g load) low % Os grains; 734 (50 g load) for 14% Os. D(meas.) = n.d. D(calc.) = n.d.

Optical Properties: Opaque. *Color:* Lead-gray to dark lead-gray; grayish white in polished section. *Streak:* Grayish black. *Luster:* Metallic. *Pleochroism:* Pale yellowish white to light grayish yellow-white. *Anisotropism:* Distinct, bluish gray to light reddish.
R_1–R_2: (480) 43.4, (546) 44.1, (589) 43.8, (656) 45.1

Cell Data: *Space Group:* n.d. $a = 5.931$ $b = 5.915$ $c = 6.003$ $\beta = 112°27'$ Z = n.d.

X-ray Powder Pattern: Tibet, China.
2.77 (100), 1.870 (90), 1.695 (90), 1.890 (70), 1.660 (60), 3.79 (50), 1.580 (50)

Chemistry:

	(1)	(2)
Ru	42.45	48.58
Os	5.94	
Ir	1.67	
Pt	0.07	
Cu	0.00	
As	36.25	36.01
Sb	0.00	
S	14.08	15.41
Total	100.36	100.00

(1) Tibet, China; by electron microprobe, average of eight analyses of seven grains, corresponding to $(Ru_{0.913}Os_{0.048}Ir_{0.018}Pt_{0.001})_{\Sigma=0.980}As_{1.050}S_{0.934}$. (2) RuAsS.

Occurrence: Found in heavy mineral concentrates and in chromium ore derived from an Alpine-type ultramafic.

Association: Iridosmine, rutheniridosmine, ruthenosmiridium, Pt–Ir alloy, Ru–Fe alloy, osmium, sperrylite, laurite, chrome spinel, pyrite, pyrrhotite, löllingite, magnetite, chalcopyrite, molybdenite, galena, millerite.

Distribution: From Anduo, northern Tibet, China.

Name: For the chemical composition RUthenium, ARsenic, and Sulfur.

Type Material: Museum of Geology, (Beijing ?), China.

References: (1) Tsu-hsiang Yu and Hsueh-tsi Chou (1979) Ruarsite, a new mineral. Ko'Hsueh Tu'ng Pao (Science Bulletin) 24, 310–316 (in Chinese). (2) (1980) Amer. Mineral., 65, 1068–1069 (abs. ref. 1). (3) Cabri, L.J., Ed. (1981) Platinum group elements: mineralogy, geology, recovery. Can. Inst. Min. & Met., 133.

Crystal Data: Hexagonal. *Point Group:* $\bar{3}\,2/m$. As foliated aggregates (up to 9 x 13 mm) intergrown with gold, and as smaller (up to 0.5 mm) grains.

Physical Properties: *Cleavage:* Perfect on {0001}; average on rhombohedron. *Tenacity:* Flexible; brittle. Hardness = n.d. VHN = 51.6–62.9 (10 g load). D(meas.) = 7.739 D(calc.) = 8.06

Optical Properties: Opaque. *Color:* Silver-white; in polished section, white with very weak rose tint. *Streak:* Lead-gray. *Luster:* Metallic. *Pleochroism:* Very weak in air, noticeable in oil. *Anisotropism:* Distinct, reddish brown to bluish gray.

R_1–R_2: (400) 62.9–62.1, (420) 62.9–62.0, (440) 63.0–61.8, (460) 63.2–61.8, (480) 63.7–61.8, (500) 64.4–61.9, (520) 65.0–62.0, (540) 65.3–62.1, (560) 65.5–62.2, (580) 65.8–62.2, (600) 66.1–62.3, (620) 66.4–62.4, (640) 66.5–62.6, (660) 66.6–62.8, (680) 66.7–63.0, (700) 66.6–63.2

Cell Data: *Space Group:* $R\bar{3}m$. $a = 4.422$ $c = 41.49$ Z = 3

X-ray Powder Pattern: Kochkar, USSR.
3.22 (100), 2.34 (90), 1.473 (60), 1.976 (50), 2.21 (40), 1.822 (40), 1.607 (40)

Chemistry:

	(1)	(2)	(3)	(4)
Bi	37.4	38.1	39.6	39.9
Pb	15.4	13.6	9.4	13.3
Ag	1.2	1.1	1.9	
Sb	1.0	2.9	[2.9]	
Te	43.9	44.2	44.9	45.5
Total	98.9	99.9	98.7	98.7

(1) Kochkar deposit, USSR; by electron microprobe, corresponding to $(Bi_{2.08}Pb_{0.88}Ag_{0.13})_{\Sigma=3.09}Te_{4.00}$. (2) Zod deposit, USSR; by electron microprobe, corresponding to $(Bi_{2.11}Pb_{0.78}Sb_{0.10}Ag_{0.12})_{\Sigma=3.11}Te_{4.00}$. (3) Do.; corresponding to $(Bi_{2.15}Pb_{0.52}Sb_{0.27}Ag_{0.20})_{\Sigma=3.14}Te_{4.00}$. (4) Ashley deposit, Ontario, Canada; by electron microprobe, corresponding to $Bi_{2.15}Pb_{0.72}Te_{4.00}$.

Occurrence: Of hydrothermal origin.

Association: Gold, arsenopyrite, boulangerite, dolomite (USSR); tellurobismuthite, volynskite, calaverite, hessite (Ashley deposit, Canada); altaite, galena, volynskite, hessite, melonite, chalcopyrite, michenerite, hawleyite (Kambalda, Australia).

Distribution: From the Zod and Kochkar gold deposits, Ural Mountains, USSR. In the Campbell mine, Bisbee, Cochise Co., Arizona, and the Hesperus mine, La Plata Co., Colorado, USA. From the Ashley deposit, Bannockburn Township, Ontario; and in the Robb-Montbray mine, Montbray Township, Quebec, Canada. At Oldřichov, near Tachov, Czechoslovakia. From the Lunnon Shoot, Kambalda nickel deposit, Western Australia.

Name: For Dr. J.C. Rucklidge, University of Toronto, Canada, who first noted a mineral of analogous composition in the Robb-Montbray deposit, Ontario, Canada.

Type Material: A.E. Fersman Mineralogical Museum, Academy of Sciences, Moscow, USSR.

References: (1) Zav'yalov, E.N. and V.D. Begizov (1977) Rucklidgeite, $(Bi,Pb)_3Te_4$, a new mineral from the Zod and Kochkar gold ore deposits. Zap. Vses. Mineral. Obshch., 106, 62–68 (in Russian). (2) (1978) Amer. Mineral., 63, 599 (abs. ref. 1). (3) Harris, D.C., W.D. Sinclair, and R.I. Thorpe (1983) Telluride minerals from the Ashley deposit, Bannockburn Township, Ontario. Can. Mineral., 21, 137–143.

Crystal Data: Cubic. *Point Group:* $4/m\ \overline{3}\ 2/m$. As small grains less than 100 μm in diameter, rarely showing {001}, and as drop-like inclusions.

Physical Properties: Hardness = n.d. VHN = 365 (25 g load). D(meas.) = n.d. D(calc.) = 15.08

Optical Properties: Opaque. *Color:* In polished section, light cream. *Luster:* Metallic. *Pleochroism:* None under the microscope but slight through a photometer. *Anisotropism:* Slight, due to strain.

R_1–R_2: (480) 56.2, (546) 59.8, (589) 60.4, (656) 61.65

Cell Data: *Space Group:* $Fm3m$ (probable) or $Pm3m$. $a = 3.991$ Z = 4

X-ray Powder Pattern: Rustenburg mine, South Africa.
2.295 (100), 1.202 (100), 1.408 (90), 0.9153 (90), 0.8145 (90), 1.992 (80),0.8922 (80)

Chemistry:

	(1)
Pt	53.44
Pd	28.28
Sn	17.70
Total	99.42

(1) Rustenburg and Atok mines, South Africa; by electron microprobe, corresponding to $(Pt, Pd)_3Sn_{0.83}$.

Occurrence: As very sparse grains in concentrates.

Association: Unspecified platinum tellurides (Merensky Reef, South Africa); moncheite, pyrrhotite, pentlandite (Stillwater Complex, Montana, USA).

Distribution: In the Atok and Rustenburg mines, Merensky Reef, Bushveld Complex, Transvaal, South Africa. From the Upper Banded Zone of the Stillwater Complex, Montana, USA. At Noril'sk, western Siberia, USSR.

Name: For its occurrence in the Rustenburg mine, South Africa.

Type Material: n.d.

References: (1) Mihálik, S.A., S.A. Hiemstra, and J.P.R. de Villiers (1975) Rustenburgite and atokite, two new platinum-group minerals from the Merensky Reef, Bushveld Igneous Complex. Can. Mineral., 13, 146–150. (2) (1976) Amer. Mineral., 61, 340 (abs. ref. 1).

Crystal Data: Orthorhombic. *Point Group:* $2/m\ 2/m\ 2/m$. As irregular inclusions up to 100 μm in diameter in a matrix of rutheniridosmine.

Physical Properties: Hardness = n.d. VHN = 743–933 (100 g load). D(meas.) = n.d. D(calc.) = 10.0 for $Ru_{0.89}Ni_{0.11}As$.

Optical Properties: Opaque. *Color:* In polished section, pale orange-brown to brownish gray. *Pleochroism:* Distinct. *Anisotropism:* Strong, orange-brown to light steel-gray.
R_1–R_2: (470) 46.1–48.8, (546) 47.5–49.5, (589) 49.3–50.9, (650) 51.1–52.4

Cell Data: *Space Group:* *Pnma* (synthetic RuAs). $a = 5.628$ $b = 3.239$ $c = 6.184$ Z = 4

X-ray Powder Pattern: Papua New Guinea.
2.061 (100), 2.696 (70), 2.124 (50), 1.780 (40), 1.750 (40), 1.343 (40), 1.302 (40)

Chemistry:

	(1)	(2)
Ru	44.6	29.1
Ir	4.0	
Os	2.1	
Pt		1.5
Rh	2.3	27.4
Pd	1.8	0.59
Ni	4.4	0.69
As	39.4	41.1
Sb		1.4
Total	98.6	101.78

(1) Papua New Guinea; by electron microprobe, corresponding to $(Ru_{0.79}Ni_{0.13}Ir_{0.04}Rh_{0.04}Pd_{0.03}Os_{0.02})_{\Sigma=1.05}As_{0.95}$. (2) Onverwacht mine, South Africa; by electron microprobe, corresponding to $(Ru_{0.51}Rh_{0.47}Ni_{0.02}Pt_{0.01}Pd_{0.01})_{\Sigma=1.02}(As_{0.96}Sb_{0.02})_{\Sigma=0.98}$.

Occurrence: As inclusions in Os–Ir–Ru alloys (Papua New Guinea), and in Alpine-type ultramafics.

Association: Rutheniridosmine, irarsite, iridarsenite.

Distribution: From Papua New Guinea. At Anduo, Tibet, China. In the Onverwacht mine, Witwatersrand, Transvaal, South Africa.

Name: For the composition.

Type Material: National Museum of Canada, Ottawa, Canada.

References: (1) Harris, D.C. (1974) Ruthenarsenite and iridarsenite, two new minerals from the Territory of Papua and New Guinea and associated irarsite, laurite and cubic iron-bearing platinum. Can. Mineral., 12, 280–284. (2) (1976) Amer. Mineral., 61, 177 (abs. ref. 1). (3) Cabri, L.J., Ed. (1981) Platinum group elements: mineralogy, geology, recovery. Can. Inst. Min. & Met., 134, 154.

Crystal Data: Hexagonal. *Point Group:* $6/m\ 2/m\ 2/m$. Euhedral laths and grains, and as inclusions in Pt–Fe alloys.

Physical Properties: Hardness = n.d. VHN = 907–1018 (100 g load). D(meas.) = n.d. D(calc.) = [21.4]

Optical Properties: Opaque. *Color:* In polished section, white with a bluish gray tinge. *Luster:* Metallic. *Pleochroism:* Weak to absent. *Anisotropism:* Weak to moderate, sometimes with wavy extinction, colors tinged with orange.
R_1–R_2: (487) 72.0, (556) 72.0, (589) 67.8, (650) 68.0

Cell Data: *Space Group:* $P6_3mmc$. $a = 2.726(2)$ $c = 4.326(3)$ Z = [2]

X-ray Powder Pattern: Papua New Guinea.
2.071 (100), 2.363 (80), 1.364 (60), 1.153 (50), 0.8737 (50B), 1.139 (40), 1.594 (40)

Chemistry:

	(1)	(2)
Os	48.9	54.0
Ir	44.7	29.4
Ru	5.8	7.9
Pt	1.1	2.9
Pd		0.6
Rh	0.59	3.8
Fe	0.30	0.1
Ni	0.03	
Total	101.42	98.7

(1) Heazlewood, Tasmania, Australia; by electron microprobe. (2) Atlin, Canada; by electron microprobe.

Occurrence: Found in placer sands, generally derived from ultramafic rocks.

Association: Other Os–Ir–Ru alloys, sperrylite, hollingworthite, iridarsenite, ruthenarsenite, michenerite, laurite, geversite, moncheite, chromite.

Distribution: In Canada, from several properties around Atlin, and from the Spruce Creek and Tulameen River areas, British Columbia. From Papua New Guinea. Around Zlatoust, Ural Mountains; in the Konder massif, Aldan Shield, Yakutia; and from several other less well defined localities, USSR. In South Africa, in the Witwatersrand. From the Heazlewood and Adamsfield districts, Tasmania, Australia. From Vourinos, Greece. In the Massif du Sud, New Caledonia.

Name: For the composition.

Type Material: n.d.

References: (1) Harris, D.C. and L.J. Cabri (1973) The nomenclature of the natural alloys of osmium, iridium and ruthenium based on new compositional data of alloys from world-wide occurrences. Can. Mineral., 12, 104–112. (2) (1975) Amer. Mineral., 60, 946 (abs. ref. 1). (3) Cabri, L.J., Ed. (1981) Platinum group elements: mineralogy, geology, recovery. Can. Inst. Min. & Met., 134–135.

Crystal Data: Hexagonal. *Point Group:* $6/m\ 2/m\ 2/m$. As a tabular crystal 35 x 7 μm, partly included in platy rutheniridosmine, and in platinum.

Physical Properties: Hardness = n.d. VHN = n.d. D(meas.) = n.d. D(calc.) = 12.438

Optical Properties: Opaque. *Color:* White with light creamy tint. *Anisotropism:* Weak. R_1–R_2: n.d.

Cell Data: *Space Group:* $P6_3/mmc$ (synthetic). $a = 2.7058$ $c = 4.2819$ Z = 2

X-ray Powder Pattern: Synthetic.
2.056 (100), 2.343 (40), 2.142 (35), 1.5808 (25), 1.3530 (25), 1.2189 (25), 1.1434 (25)

Chemistry:

	(1)
Ru	64.43
Ir	14.62
Pt	9.14
Rh	7.05
Os	5.29
Pd	0.49
Fe	0.21
Ni	trace
Cu	trace
Total	101.23

(1) Uryu River, Japan; by electron microprobe, corresponding to $Ru_{1.47}Ir_{0.18}Rh_{0.16}Pt_{0.11}Os_{0.06}Pd_{0.01}Fe_{0.01}$; proportions are Ru:Ir:Rh = 81.5:9.7:8.8.

Occurrence: Associated with rutheniridosmine in a platinum grain.

Association: Rutheniridosmine, platinum.

Distribution: In gravels of the Uryu River, near Horokanai, Hokkaido, Japan.

Name: From the Latin *Ruthenia*, for *Ukraine* or *Russia*, as the element was found associated with platinum in the Ural Mountain placers, USSR.

Type Material: University of Kagoshima, Kagoshima, Japan.

References: (1) Urashima, Y., T. Wakabayashi, T. Masaki, and Y. Terasaki (1974) Ruthenium, a new mineral from Horokanai, Hokkaido, Japan. Mineral. J. (Japan), 7, 438–444. (2) (1976) Amer. Mineral., 61, 177 (abs. ref. 1). (3) (1959) NBS Circ. 539, 4, 5.

Crystal Data: Cubic. *Point Group:* $4/m\ \overline{3}\ 2/m$. As tiny composite grains intergrown with rutheniridosmine.

Physical Properties: *Cleavage:* Perfect. Hardness = n.d. VHN = n.d. D(meas.) = 18.97 D(calc.) = [19.7]

Optical Properties: Opaque. *Color:* Tin-white, with a pinkish or bluish tinge.
Luster: Metallic, bright.
R: n.d.

Cell Data: *Space Group:* $Fm3m$ a = 3.8394 (synthetic Ir). Z = [4]

X-ray Powder Pattern: n.d.

Chemistry:

	(1)	(2)
Ir	39.018	59.8
Os	38.885	25.4
Ru	21.080	5.3
Pt		9.2
Pd		0.05
Rh	0.986	1.0
Fe		0.40
Ni		0.12
Total	99.969	101.27

(1) Uryu River, Hokkaido, Japan. (2) Heazlewood, Tasmania, Australia; by electron microprobe.

Occurrence: As grains in iridosmine (Uryu River, Hokkaido, Japan), and in ultramafic Alpine-type intrusives.

Association: Osmium, iridosmine, osmiridium, iridium, rutheniridosmine, sperrylite, isoferroplatinum, hollingworthite, iridarsenite.

Distribution: In gravels of the Uryu River, near Horokanai, Hokkaido, Japan. In South Africa, in the Witwatersrand. At Heazlewood, Tasmania, Australia. From the Ural Mountains, USSR. At Tiébaghi, New Caledonia.

Name: For the composition, *ruthenian osmiridium.*

Type Material: n.d.

References: (1) Shin'ichi, A. (1936) A new mineral "ruthenosmiridium." Sci. Repts. Tohoku Imp. Univ., 1st series, Honda Anniv. Vol. (Oct. 1936), 527–546. (2) (1940) Amer. Mineral., 25, 440 (abs. ref. 1). (3) Harris, D.C. and L.J. Cabri (1973) The nomenclature of the natural alloys of osmium, iridium and ruthenium based on new compositional data of alloys from world-wide occurrences. Can. Mineral., 12, 104–112. (4) (1975) Amer. Mineral., 60, 946 (abs. ref. 3). (5) Feather, C.E. (1976) Mineralogy of platinum-group minerals in the Witwatersrand, South Africa. Econ. Geol., 71, 1399–1428. (6) Cabri, L.J., Ed. (1981) Platinum group elements: mineralogy, geology, recovery. Can. Inst. Min. & Met., 135, 161.

Crystal Data: Orthorhombic. *Point Group:* n.d. As radiating polycrystalline aggregates.

Physical Properties: Hardness = n.d. VHN = 54 (15 g load). D(meas.) = n.d. D(calc.) = 6.78

Optical Properties: Opaque. *Color:* In polished section, bluish gray with slight cream tint. *Luster:* Metallic. *Pleochroism:* Distinct, in light brown to light blue. *Anisotropism:* Strong, with undulatory extinction, colors in light gray-blue to dark yellow-brown.

R_1–R_2: (400) 31.1–32.1, (420) 31.0–33.2, (440) 30.9–34.3, (460) 30.6–33.8, (480) 30.1–33.0, (500) 29.5–32.2, (520) 28.7–31.5, (540) 27.9–30.6, (560) 27.0–29.6, (580) 26.1–28.4, (600) 25.3–27.2, (620) 24.5–26.0, (640) 23.8–24.8, (660) 23.0–23.8, (680) 22.2–22.7, (700) 21.3–21.6

Cell Data: *Space Group:* n.d. $a = 3.986(4)$ $b = 5.624(4)$ $c = 9.778(8)$ Z = 1

X-ray Powder Pattern: Bukov deposit, Czechoslovakia.
3.089 (100), 2.706 (70), 1.991 (70), 2.445 (60), 1.847 (60), 3.987 (50), 2.525 (50)

Chemistry:

	(1)	(2)
Cu	42.71	42.29
Tl	22.44	22.67
Se	34.57	35.04
Total	99.72	100.00

(1) Bukov deposit, Czechoslovakia; by electron microprobe, average of five analyses, corresponding to $Cu_{6.06}Tl_{0.99}Se_{3.95}$. (2) Cu_6TlSe_4.

Occurrence: In calcite veins.

Association: Crookesite, berzelianite, umangite, bukovite.

Distribution: In the Bukov deposit, Rožná district, Czechoslovakia.

Name: For Germain Sabatier, former Director of Research, C.N.R.S., Orléans, France.

Type Material: National School of Mines, Paris, France.

References: (1) Johan, Z., M. Kvaček, and P. Picot (1978) La sabatierite, un nouveau séléniure de cuivre et de thallium. Bull. Minéral., 101, 557–560 (in French with English abs.). (2) (1979) Amer. Mineral., 64, 1331–1332 (abs. ref. 1). (3) Berger, R.A. (1987) Crookesite and sabatierite in a new light – a crystallographer's comment. Zeits. Krist., 181, 241–249.

Crystal Data: Orthorhombic. *Point Group:* $2/m\ 2/m\ 2/m$. Crystals prismatic [010], also with prominent {101} and {310}; commonly massive with a radiated fibrous structure. *Twinning:* As fivelings with {011} as twin plane; also forming cruciform penetration twins, as with arsenopyrite, with twin plane {101}.

Physical Properties: *Cleavage:* Distinct on {100}. *Fracture:* Uneven to conchoidal. *Tenacity:* Brittle. Hardness = 4.5–5 VHN = 792–882 (100 g load). D(meas.) = 7.2 D(calc.) = 7.471

Optical Properties: Opaque. *Color:* Tin-white, readily tarnishes to dark gray; in polished section, white. *Streak:* Grayish black. *Luster:* Metallic. *Pleochroism:* Weak. *Anisotropism:* Strong.
R_1–R_2: (400) 55.8–52.6, (420) 55.8–53.0, (440) 55.8–53.4, (460) 55.6–53.8, (480) 55.2–54.1, (500) 54.7–54.3, (520) 54.2–54.4, (540) 53.5–54.5, (560) 52.8–54.5, (580) 52.2–54.5, (600) 51.8–54.4, (620) 51.3–54.3, (640) 51.0–54.2, (660) 50.7–54.1, (680) 50.5–53.9, (700) 50.3–53.8

Cell Data: *Space Group:* *Pnnm.* $a = 5.173$ $b = 5.954$ $c = 2.999$ Z = 2

X-ray Powder Pattern: Cobalt, Canada. (JCPDS 23-88).
2.379 (100), 2.572 (80), 2.597 (55), 1.862 (45), 1.849 (20), 1.650 (20), 1.636 (20)

Chemistry:

	(1)	(2)	(3)
Co	18.58	12.99	28.23
Fe	9.51	15.28	
Ni	0.00	0.20	
Cu	0.62	0.33	
As	70.36	71.13	71.77
S	0.90	0.68	
Total	99.97	100.61	100.00

(1) Schneeberg, Germany. (2) Nordmarken, Sweden. (3) $CoAs_2$.

Polymorphism & Series: Dimorphous with clinosafflorite.

Occurrence: In hydrothermal veins of moderate temperature and pressure.

Association: Skutterudite, rammelsbergite, nickeline, silver, bismuth, löllingite.

Distribution: In the USA, from the Quartzburg district, Grant Co., Oregon. At Cobalt and South Lorrain, Ontario; and at Great Bear Lake, Saskatchewan, Canada. In Germany, at Schneeberg and Annaberg, Saxony; Bieber and Mackenheim, Hesse; St. Andreasberg, in the Harz Mountains; and at Wittichen, Black Forest. From Sweden, at Tunaberg, Södermanland; and at Nordmark, Wermland. From Burguillos de Cerro, Badajoz Province, Spain. At Sarrabus and Gonnosfanadiga, Sardinia, Italy. In Australia, at Broken Hill, New South Wales. A number of other less prominent localities are known.

Name: From *safflower*, in allusion to its use as a pigment.

References: (1) Palache, C., H. Berman, and C. Frondel (1944) Dana's system of mineralogy, (7th edition), v. I, 307–309. (2) Radcliffe, D. and L.G. Berry (1968) The safflorite–loellingite solid solution series. Amer. Mineral., 53, 1856–1881.

Sakharovaite $(Pb, Fe)(Bi, Sb)_2S_4$

Crystal Data: Monoclinic (?). *Point Group:* n.d. As radiating aggregates of fine capillary crystals up to 1 cm long and a fraction of a mm in cross-section.

Physical Properties: *Cleavage:* Perfect in one direction. Hardness = Low. VHN = n.d. D(meas.) = n.d. D(calc.) = n.d.

Optical Properties: Opaque. *Color:* Lead-gray; in polished section, pure white. *Luster:* Metallic. *Pleochroism:* Birefringent. *Anisotropism:* Strong, with reddish internal reflections seen only in oil.
R_1–R_2: n.d.

Cell Data: *Space Group:* n.d. Z = n.d.

X-ray Powder Pattern: Ustarasaisk deposit, USSR.
3.412 (100), 2.721 (60), 2.031 (60b), 3.093 (40), 2.811 (40), 2.299 (40), 2.244 (40)

Chemistry:

	(1)
Pb	32.25
Fe	1.39
Cu	0.30
Bi	30.50
Sb	16.50
S	17.62
insol.	1.59
Total	100.15

(1) Ustarasaisk deposit, USSR.

Occurrence: In carbonate veinlets cutting arsenopyrite ore.

Association: Antimony, realgar, cinnabar.

Distribution: In the Ustarasaisk deposit, western Tyan Shan Mountains, USSR. In the Spissko-Gemer deposits, Czechoslovakia.

Name: For M.S. Sakharova, Soviet mineralogist.

Type Material: n.d.

References: (1) Sakharova, M.S. (1955) Bismuth sulfosalts of the Ustarasaisk deposits. Trudy Mineralog. Muzeya Akad. Nauk. SSSR, 7, 112–116 (in Russian). (2) (1956) Amer. Mineral., 41, 814 (abs. ref. 1). (3) Kostov, I. (1959) Bismuth jamesonite or sakharovaite—a new mineral species. Trudy Mineralog. Muzeya Akad. Nauk. SSSR, 10, 148–149 (in Russian). (4) (1960) Amer. Mineral., 45, 1134 (abs. ref. 3).

Crystal Data: Cubic. *Point Group:* 432, $\bar{4}3m$, or $4/m\ \bar{3}\ 2/m$. Forms in exsolution texture with stannite, up to 0.5 x 0.03 mm.

Physical Properties: Hardness = 4 VHN = 243–282, 265 average (100 g load). D(meas.) = n.d. D(calc.) = 4.45

Optical Properties: Opaque. *Color:* Greenish steel-gray; in polished section, purplish olive-gray with a red tint. *Streak:* Lead-gray with olive tint. *Luster:* Metallic. *Anisotropism:* Nearly isotropic.

R_1–R_2: (400) 22.5, (420) 22.5, (440) 22.5, (460) 22.6, (480) 22.6, (500) 22.6, (520) 22.6, (540) 22.7, (560) 22.6, (580) 22.5, (600) 22.5, (620) 22.6, (640) 22.6, (660) 22.6, (680) 22.6, (700) 22.4

Cell Data: *Space Group:* $P432$, $P\bar{4}3m$, or $Pm3m$. $a = 5.4563(24)$ Z = 4

X-ray Powder Pattern: Ikuno mine, Japan.
3.15 (100), 1.927 (40), 1.650 (20), 2.73 (10), 5.47 (6), 3.85 (6), 2.44 (6)

Chemistry:

	(1)	(2)
Cu	23	21
Zn	10	14
Fe	9	5
Ag	4	3.5
In	17	23
Sn	9	4
S	31	30
Total	103	100.5

(1) Ikuno mine, Japan; by electron microprobe; corresponding to $(Cu_{1.6}Zn_{0.7}Fe_{0.7}Ag_{0.2}In_{0.7}Sn_{0.3})_{\Sigma=4.2}S_{4.2}$. (2) Do.; corresponding to $(Cu_{1.4}Zn_{0.9}Fe_{0.4}Ag_{0.1}In_{0.9}Sn_{0.1})_{\Sigma=3.8}S_{4.0}$.

Occurrence: In a banded hydrothermal vein.

Association: Stannite, sphalerite, chalcopyrite, cassiterite, matildite, cobaltian arsenopyrite, quartz, calcite.

Distribution: From the Ikuno mine, Hyogo Prefecture, Japan.

Name: For Dr. Kin-ichi Sakurai, a Japanese amateur mineralogist.

Type Material: National Museum of Natural History, Washington, D.C., 120592; Harvard University, Cambridge, Massachusetts, USA, 108788.

References: (1) Kato, A. (1965) Sakuraiite, a new mineral. Chigaku Kenkyu [Earth Science Studies], Sakurai volume, 1–5 (in Japanese). (2) (1968) Amer. Mineral., 53, 1421 (abs. ref. 1). (3) Shimizu, M., A. Kato, and T. Shiozawa (1986) Sakuraiite: chemical composition and extent of (Zn, Fe)In–for–CuSn substitution. Can. Mineral., 24, 405–409. (4) Kissin, S.A. and D.R. Owens (1986) The crystallography of sakuraiite. Can. Mineral., 24, 679–683.

Crystal Data: Monoclinic. *Point Group:* $2/m$. Crystals prismatic ∥ [001], to 1 cm; striated ∥ [001].

Physical Properties: *Fracture:* Conchoidal. *Tenacity:* Brittle. Hardness = 2.5 VHN = 187–212 (100 g load). D(meas.) = 5.51 D(calc.) = 5.50

Optical Properties: Nearly opaque; translucent in thin splinters. *Color:* Steel-black; in polished section, bluish white with deep red internal reflections; deep red to brown in transmitted light. *Streak:* Dark red. *Luster:* Metallic. *Pleochroism:* Weak. *Anisotropism:* Weak.

R_1–R_2: (400) 36.2–37.6, (420) 35.5–36.6, (440) 34.8–35.6, (460) 34.0–34.7, (480) 33.0–34.0, (500) 32.0–33.2, (520) 31.0–32.2, (540) 30.0–31.2, (560) 29.2–30.0, (580) 28.5–29.2, (600) 27.8–28.5, (620) 27.2–27.8, (640) 26.7–27.4, (660) 26.2–26.9, (680) 25.8–26.5, (700) 25.6–26.1

Cell Data: *Space Group:* $P2_1/n$. $a = 10.362(4)$ $b = 8.101(3)$ $c = 6.647(3)$ $\beta = 92°38(4)'$ Z = 2

X-ray Powder Pattern: Samson vein, St. Andreasberg, Germany.
3.20 (100), 3.01 (90), 2.57 (60), 2.86 (50), 2.43 (50), 2.51 (40), 6.46 (30)

Chemistry:

	(1)	(2)
Ag	45.95	46.79
Mn	5.86	5.96
Cu	0.18	
Fe	0.22	
Sb	26.33	26.40
S	20.55	20.85
rem.	0.87	
Total	99.96	100.00

(1) Samson vein, St. Andreasberg, Germany; remainder is $CaCO_3$ 0.41%, $MgCO_3$ 0.46%.
(2) $Ag_4MnSb_2S_6$.

Occurrence: In a hydrothermal vein (Samson vein, St. Andreasberg, Germany).

Association: Pyrargyrite, galena, dyscrasite, tetrahedrite, pyrolusite, quartz, calcite, apophyllite.

Distribution: In the Samson vein, St. Andreasberg, Harz Mountains, Germany. From the Brady Lake property of the Silver Miller mine, near Cobalt, Ontario, Canada.

Name: For the occurrence in the Samson vein, St. Andreasberg, Germany.

Type Material: n.d.

References: (1) Palache, C., H. Berman, and C. Frondel (1944) Dana's system of mineralogy, (7th edition), v. I, 393–395. (2) Berry, L.G. and R.M. Thompson (1962) X-ray powder data for the ore minerals. Geol. Soc. Amer. Mem. 85, 128–129. (3) Hrušková, J. and V. Syneček (1969) The crystal structure of samsonite, $2Ag_2S.MnS.Sb_2S_3$. Acta Cryst., B25, 1004–1006. (3) Edenharter, A. and W. Nowacki (1974) Verfeinerung der Kristallstruktur von Samsonit, $(SbS_3)_2Ag_2^{III}Ag_2^{IV}Mn^{VI}$. Zeits. Krist., 140, 87–99 (in German with English abs.).

Crystal Data: Monoclinic. *Point Group:* $2/m$. Crystals tabular and prismatic ∥ [010], up to 1 mm in length.

Physical Properties: *Tenacity:* Somewhat sectile. Hardness = n.d. VHN = 272 (20 g load). D(meas.) = 4.8 D(calc.) = 4.99

Optical Properties: Translucent. *Color:* Carmine-red; gray-white in polished section with remarkable reddish orange internal reflections. *Streak:* Orange. *Luster:* Resinous.
Optical Class: Biaxial (–) (synthetic). *Pleochroism:* Discernible in reflected light, from brownish to purplish tint; in transmitted light reddish yellow ∥ Y; brownish red ∥ Z'. 2V(meas.) = ~90° 2V(calc.) = n.d.
R_1–R_2: (486) 20.1–25.8, (546) 19.7–24.7, (589) 19.1–23.5, (656) 14.0–20.0

Cell Data: *Space Group:* $C2/c$. $a = 25.37(2)$ $b = 5.654(1)$ $c = 16.87(1)$ $\beta = 117.58(4)°$ Z = 4

X-ray Powder Pattern: Sarabau mine, Malaysia.
3.215 (100), 2.817 (88), 3.466 (78), 3.182 (60), 3.164 (50), 4.227 (42), 2.583 (40)

Chemistry:

	(1)	(2)
Ca	2.43	2.49
Sb	74.89	75.62
S	11.91	11.95
O	9.61	9.94
Total	98.84	100.00

(1) Sarabau mine, Malaysia; by electron microprobe, leading to $Ca_{0.99}Sb_{10.00}O_{9.76}S_{6.04}$.
(2) $CaSb_{10}O_{10}S_6$.

Occurrence: In a hydrothermal mineral deposit.

Association: Stibnite, senarmontite, wollastonite, calcite, quartz.

Distribution: In the Sarabau mine, about 40 km southwest of Kuching, Sarawak, Malaysia.

Name: For the Sarabau mine, Malaysia.

Type Material: n.d.

References: (1) Nakai, I., H. Adachi, S. Matsubara, A. Kato, K. Masutomi, T. Fujiwara, and K. Nagashima (1978) Sarabauite, a new oxide sulfide mineral from the Sarabau mine, Sarawak, Malaysia. Amer. Mineral., 63, 715–719. (2) Nakai, I., K. Nagashima, K. Koto, and N. Morimoto (1978) Crystal chemistry of oxide-chalcogenide. I. The crystal structure of sarabauite $CaSb_{10}O_{10}S_6$. Acta Cryst., B34, 3569–3572.

Crystal Data: Monoclinic, pseudo-orthorhombic. *Point Group:* $2/m$. Prismatic, to 10 cm long, and deeply striated ∥ [010]; rounded cavernous crystal terminations; parallel and subparallel groupings common due to twinning in part. *Twinning:* Often repeated on {100} to produce lamination of crystals; lamellae seen in polished section are sometimes bent.

Physical Properties: *Cleavage:* Fair on {100}. *Fracture:* Conchoidal. *Tenacity:* Extremely brittle. Hardness = 3 VHN = 196 D(meas.) = 5.10 D(calc.) = [5.13]

Optical Properties: Opaque. *Color:* Dark lead-gray; in polished section, pure white, rarely with deep red internal reflections. *Streak:* Chocolate-brown. *Luster:* Metallic. *Pleochroism:* Rarely visible, only in oil. *Anisotropism:* Weak.

R_1–R_2: (400) 38.6–43.6, (420) 38.0–42.7, (440) 37.4–41.8, (460) 36.8–41.0, (480) 36.4–40.2, (500) 35.8–39.7, (520) 35.2–39.2, (540) 34.6–38.6, (560) 33.8–37.8, (580) 33.0–37.0, (600) 32.2–36.0, (620) 31.5–35.1, (640) 30.8–34.2, (660) 30.2–33.4, (680) 29.6–32.8, (700) 29.2–32.2

Cell Data: *Space Group:* $P2_1/n$ (pseudocell). $a = 19.62$ $b = 7.89$ $c = 4.19$ $\beta = 90°$ Z = 4

X-ray Powder Pattern: Binntal, Switzerland.
3.49 (100), 2.76 (90), 2.96 (80), 2.33 (60), 3.87 (50), 2.64 (50), 4.15 (40)

Chemistry:

	(1)	(2)
Pb	43.63	42.70
As	30.46	30.87
S	25.51	26.43
Total	99.60	100.00

(1) Binntal, Switzerland; average of three analyses. (2) $PbAs_2S_4$.

Occurrence: Found in a hydrothermal deposit in dolomite (Binntal, Switzerland).

Association: Tennantite, pyrite, dufrénoysite, realgar (Binntal, Switzerland).

Distribution: At the Lengenbach quarry, Binntal, Valais, Switzerland. From the Zuni mine, San Juan Co., Colorado, USA. In the Pitone marble quarry, near Seravezza, Apuane Alps, Tuscany, Italy.

Name: For Sartorius von Waltershausen (1809–1876), Professor at the University of Göttingen, Germany, who first announced the species.

Type Material: n.d.

References: (1) Palache, C., H. Berman, and C. Frondel (1944) Dana's system of mineralogy, (7th edition), v. I, 478–481. (2) Berry, L.G. and R.M. Thompson (1962) X-ray powder data for the ore minerals. Geol. Soc. Amer. Mem. 85, 166–167. (3) Iitake, Y. and W. Nowacki (1961) A refinement of the pseudo crystal structure of scleroclase $PbAs_2S_4$. Acta Cryst., 14, 1291–1292.

Crystal Data: Hexagonal. *Point Group:* $6/m\ 2/m\ 2/m$. As crystals up to 1 cm in length, but most much smaller.

Physical Properties: Hardness = Low. VHN = n.d. D(meas.) = n.d. D(calc.) = 13.52

Optical Properties: Opaque. *Color:* In polished section, gray. *Luster:* Metallic. *Anisotropism:* Very weak.
R_1–R_2: n.d.

Cell Data: *Space Group:* $P6_3/mmc$. $a = 2.978$ $c = 4.842$ Z = 2

X-ray Powder Pattern: Landsberg, Germany.
2.273 (100), 0.8595 (60), 2.420 (50), 1.268 (50), 0.9538 (50), 1.489 (40), 0.9373 (40)

Chemistry: Microprobe analyses give low totals because of high absorption; results cluster about $Ag_{1.12}Hg_{0.98}$.

Occurrence: Found in the zone of oxidation, formed by the alteration of moschellandsbergite.

Association: Paraschachnerite, mercurian silver, acanthite, cinnabar, ankerite, "limonite" (Landsberg, Germany); paraschachnerite, mercurian silver, sphalerite, pyrite (Sala, Sweden).

Distribution: In the Vertraun Gott mercury mine at Landsberg, near Obermoschel, Rhineland-Palatinate, Germany. At Sala, Västmanland, Sweden.

Name: For Professor Doris Schachner, ore mineralogist, Institute for Mineralogy and Ore Deposits, Rhine Westphalian Technical School, Aachen, Germany.

Type Material: Technical University, Berlin, Germany; National Museum of Natural History, Washington, D.C., USA, 150256.

References: (1) Seeliger, E. and A. Mücke (1972) Para-schachnerite, $Ag_{1.2}Hg_{0.8}$, und Schachnerite, $Ag_{1.1}Hg_{0.9}$, vom Landsberg bei Obermoschel, Pfalz. Neues Jahrb. Mineral., Abh., 117, 1–18 (in German with English abs.). (2) (1973) Amer. Mineral., 58, 347 (abs. ref. 1).
(3) Zakrzewski, M.A. and E.A.J. Burke (1987) Schachnerite, paraschachnerite and silver amalgam from the Sala mine, Sweden. Mineral. Mag., 51, 318–321.

Crystal Data: Orthorhombic. *Point Group:* $2/m$ $2/m$ $2/m$ or $mm2$. As lath-shaped grains, also massive and fine-granular.

Physical Properties: *Cleavage:* Noticeable along grain length; transverse and longitudinal partings. Hardness = ~2 VHN = 150–206 D(meas.) = n.d. D(calc.) = 7.58

Optical Properties: Opaque. *Color:* Lead-gray to iron-black; in polished section, white. *Luster:* Metallic. *Pleochroism:* Weak in air, weak to distinct in oil. *Anisotropism:* Weak in air, distinct to strong in oil, in gray to black.

R_1–R_2: (400) 42.6–46.6, (420) 43.0–47.3, (440) 43.4–48.0, (460) 43.6–48.2, (480) 43.5–48.2, (500) 43.2–48.0, (520) 42.9–47.7, (540) 42.4–47.2, (560) 42.0–46.7, (580) 41.7–46.2, (600) 41.4–46.0, (620) 41.2–45.7, (640) 41.1–45.6, (660) 41.0–45.5, (680) 40.9–45.3, (700) 40.9–45.1

Cell Data: *Space Group:* $Bbmm$, $Bb2_1m$, or $Bbm2$. $a = 13.448(2)$ $b = 44.386(1)$ $c = 4.022(12)$ Z = 8

X-ray Powder Pattern: Treasure Vault mine, Colorado, USA.
2.035 (100), 3.207 (90), 2.923 (90), 2.878 (70), 2.873 (70), 2.786 (70), 2.382 (70)

Chemistry:

	(1)	(2)	(3)	(4)
Ag	9.1	12.5	9.51	8.97
Cu	0.1	0.8		
Pb	29.0	28.4	18.27	34.47
Bi	45.9	42.6	55.26	40.56
S	16.7	16.6	16.96	16.00
Total	100.8	100.9	100.00	100.00

(1) Treasure Vault mine, Colorado, USA; by electron microprobe. (2) Old Lout mine, Colorado, USA; by electron microprobe. (3) $Ag_3Pb_3Bi_9S_{18}$. (4) $Ag_3Pb_6Bi_7S_{18}$.

Occurrence: Disseminated through quartz.

Association: Galena.

Distribution: In the USA, in Colorado, from several mines on the Treasure Vault and parallel lodes, Geneva (Montezuma) district, Park and Summit Cos.; the Magnolia district, Boulder Co.; Lake City, Hinsdale Co.; and the Old Lout mine, Poughkeepsie Gulch, near Ouray, San Juan Co. From Darwin, Inyo Co., California. In Scotland, from Corrie Buie, Meal nan Oighreag, Perthshire. In the Zambaraks deposit, eastern Karamazar, USSR.

Name: For J.H.L. Schirmer, former Superintendent of the U.S. Mint, Denver, Colorado, USA.

References: (1) Palache, C., H. Berman, and C. Frondel (1944) Dana's system of mineralogy, (7th edition), v. I, 424. (2) Karup-Møller, S. (1973) New data on schirmerite. Can. Mineral., 11, 952–957. (3) Ramdohr, P. (1969) The ore minerals and their intergrowths, (3rd edition), 736–737. (4) Makovicky, E. and S. Karup-Møller (1977) Chemistry and crystallography of the lillianite homologous series. Neues Jahrb. Mineral., Abh., 131, 56–82. (5) (1979) Amer. Mineral., 64, 243 (abs. ref. 4).

Crystal Data: Hexagonal. *Point Group:* $3m$, $\bar{3}\ 2/m$, or 32. As thin bands a few μm wide in caswellsilverite, also as individual grains up to 250 μm.

Physical Properties: *Cleavage:* Basal, perfect (synthetic); distinct parting on $\{h0l\}$. Hardness = n.d. VHN = 41.8–90.8 on {0001} (15 g load) (synthetic). D(meas.) = 2.70 (synthetic). D(calc.) = 2.74

Optical Properties: Opaque. *Color:* In reflected light, gray in air, bluish gray in oil. *Luster:* Submetallic (synthetic). *Pleochroism:* Distinct, white with yellowish tint to light gray with bluish or greenish tint. *Anisotropism:* Strong.

R_1–R_2: (400) — , (420) 15.4–17.0, (440) 16.0–17.4, (460) 16.6–19.3, (480) 16.5–19.3, (500) 16.4–19.4, (520) 16.2–19.3, (540) 16.0–19.3, (560) 15.8–19.3, (580) 15.5–19.3, (600) 15.2–19.2, (620) 15.2–19.1, (640) 15.0–19.0, (660) 15.1–19.1, (680) 15.1–19.4, (700) 14.8–19.0

Cell Data: *Space Group:* $R3m$, $R\bar{3}m$, or $R32$. $a = 3.32(1)$ $c = 26.6(1)$ Z = 3

X-ray Powder Pattern: Synthetic.
8.85 (vsb), 2.81 (mb), 2.53 (mb), 1.66 (mb), 4.43 (wb), 2.21 (vvwb)

Chemistry:

	(1)	(2)	(3)
Na	5.10	4.95	4.89
Cr	36.3	36.2	36.87
Ti	0.17		
Mn	0.17		
S	45.5	44.9	45.47
H_2O	14.3	13.9	12.77
Total	101.5	100.0	100.00

(1) Norton County meteorite; by electron microprobe, average values. (2) Synthetic schöllhornite. (3) $Na_{0.3}CrS_2 \cdot H_2O$.

Occurrence: In an enstatite achondrite meteorite with other chromium-rich minerals.

Association: Caswellsilverite, daubréelite, titanoan troilite, ferromagnesian alabandite, oldhamite, kamacite, perryite.

Distribution: In the Norton County enstatite achondrite meteorite.

Name: For Professor Robert Schöllhorn, Inorganic Chemical Institute, Münster University, Germany.

Type Material: Institute of Meteoritics and Department of Geology, University of New Mexico, Albuquerque, New Mexico, USA.

References: (1) Okada, A., K. Keil, B.F. Leonard, and I.D. Hutcheon (1985) Schöllhornite, $Na_{0.3}(H_2O)_1[CrS_2]$, a new mineral in the Norton County enstatite achondrite. Amer. Mineral., 70, 638–643.

Crystal Data: Hexagonal. *Point Group:* $6/m\ 2/m\ 2/m$. As grains in clausthalite.

Physical Properties: Hardness = n.d. VHN = n.d. D(meas.) = n.d. D(calc.) = 7.06

Optical Properties: Opaque. *Color:* In polished section, yellow to orange-yellow. *Pleochroism:* Distinct, in shades of yellow. *Anisotropism:* Strong to distinct, in pinkish to greenish colors.

R_1–R_2: n.d.

Cell Data: *Space Group:* $P6_3/mmc$. $a = 3.624$ $c = 5.288$ Z = 2

X-ray Powder Pattern: Kuusamo, Finland.
2.70 (100), 2.015 (80), 1.806 (60), 1.535 (40), 1.50 (40), 1.348 (30), 1.155 (30)

Chemistry:

	(1)	(2)
Ni	36.8	42.65
Co	1.9	
Se	61.3	57.35
Total	100.0	100.00

(1) Kuusamo, Finland; by X-ray fluorescence analysis. (2) NiSe.

Occurrence: In calcite veins, in sills of albitite diabase in schist, associated with low-grade uranium mineralization.

Association: Wilkmanite, penroseite, clausthalite, calcite.

Distribution: From Kuusamo, northeastern Finland.

Name: For J.J. Sederholm, former Director of the Geological Survey of Finland.

Type Material: n.d.

References: (1) Vuorelainen, Y., A. Huhma, and A. Häkli (1964) Sederholmite, wilkmanite, kullerudite, mäkinenite, and trüstedtite, five new nickel selenide minerals. Compt. Rendus Soc. Géol. Finlande, 36, 113–125. (2) (1965) Amer. Mineral., 50, 519–520 (abs. ref. 1).

Crystal Data: Orthorhombic or pseudo-orthorhombic. *Point Group:* n.d. Massive.

Physical Properties: Hardness = n.d. VHN = 332 D(meas.) = n.d. D(calc.) = 7.938

Optical Properties: Opaque. *Color:* In polished section, light gray with a rose tint. *Luster:* Metallic. *Anisotropism:* Distinct, brown to blue.
R_1–R_2: n.d.

Cell Data: *Space Group:* n.d. $a = 3.190(1)$ $b = 5.81(1)$ $c = 6.490(15)$ Z = 2

X-ray Powder Pattern: Seinäjoki, Finland.
2.81 (100), 2.59 (90), 2.03 (80), 1.790 (60), 1.212 (5), 1.174 (30), 1.626 (20)

Chemistry:

	(1)	(2)
Fe	14.7 – 16.0	15.4
Ni	3.4 – 5.4	4.0
Co	0.3 – 0.6	0.6
Sb	71.6 – 75.4	73.5
As	5.7 – 7.8	6.9
Te	1.0 – 1.6	1.2
Total		101.6

(1–2) Seinäjoki, Finland; range and average of six analyses, corresponding to $(Fe_{0.78}Ni_{0.19}Co_{0.03})_{\Sigma=1.00}(Sb_{1.71}As_{0.26}Te_{0.03})_{\Sigma=2.00}$.

Occurrence: In an antimony deposit, replacing antimony (Seinäjoki, Finland).

Association: Antimony, antimonian westerveldite, breithauptite, altaite (Seinäjoki, Finland).

Distribution: From Seinäjoki, Vaasa, Finland. In the Ilímaussaq Intrusion, southern Greenland.

Name: For the occurrence at Seinäjoki, Finland.

Type Material: A.E. Fersman Mineralogical Museum, Academy of Sciences, Moscow, USSR.

References: (1) Mozgova, N.N., Y.S. Borodaev, N.A. Ozerova, V. Paakkonen, O.L. Sveshnikova, V.S. Balitskii, and B.A. Dorogovin (1976) Seinäjokite, $(Fe_{0.8}Ni_{0.2})(Sb_{1.7}As_{0.3})$ and antimonian westerveldite from Seinäjoki, Finland. Zap. Vses. Mineral. Obshch., 105, 617–630 (in Russian). (2) (1977) Amer. Mineral., 62, 1059 (abs. ref. 1). (3) Mozgova, N.N., Y.S. Borodaev, N.A. Ozerova, and V. Paakkonen (1977) New minerals of the group of iron antimonides and arsenides from the Seinäjoki deposit, Finland. Bull. Geol. Soc. Finland, 49, 47–52 (in English). (4) (1978) Chem. Abs., 88, 76550 (abs. ref. 3).

Crystal Data: Hexagonal. *Point Group:* 32. Forms acicular crystals, to 2 cm in length, commonly hollow and tube-like; as sheet-like clusters and felted masses; also as glassy droplets.

Physical Properties: *Cleavage:* Good on $\{01\bar{1}2\}$. *Tenacity:* Flexible. Hardness = 2 VHN = n.d. D(meas.) = 4.80 D(calc.) = 4.809

Optical Properties: Opaque; transparent in very thin splinters. *Color:* Gray to purple-gray; in polished section, white in air, darker white to grayish brown in oil; red in transmitted light. *Streak:* Red. *Luster:* Metallic.

Optical Class: Uniaxial positive (+). *Pleochroism:* Distinct, cream-white toward brown. $\omega = 3.0$ $\epsilon = 4.04$ *Anisotropism:* Very strong.

R_1–R_2: (400) 28.4–38.4, (420) 27.4–37.8, (440) 26.4–37.2, (460) 25.5–36.3, (480) 24.8–35.4, (500) 24.3–34.7, (520) 24.0–34.4, (540) 24.1–34.1, (560) 24.4–34.0, (580) 24.8–33.9, (600) 25.4–33.8, (620) 25.8–33.5, (640) 26.0–33.3, (660) 25.8–32.9, (680) 25.2–32.4, (700) 24.6–31.9

Cell Data: *Space Group:* $P3_121$ or $P3_221$. $a = 4.3662$ $c = 4.9536$ Z = 3

X-ray Powder Pattern: Synthetic.
3.00 (100), 3.78 (55), 2.072 (35), 1.998(20), 1.766 (20), 2.184 (16), 1.650 (10)

Chemistry: Very pure selenium, with a trace of S.

Occurrence: Formed at relatively low temperatures by sublimation of fumarolic vapors, and from oxidation of selenium-bearing organic compounds in sandstone U-V deposits; from burning coal.

Association: Pyrite, ferroselite, zippeite, metatyuyamunite, metarossite, montroseite, corvusite.

Distribution: In the USA, at the United Verde mine, Jerome, Yavapai Co., Arizona; at a number of localities in U-V deposits of the Colorado Plateau, as at the Peanut mine, Bull Canyon, Montrose Co., Colorado, and the Parco No. 23 mine, Thompsons district, Grand Co., Utah. From the Road Hog No. 1A mine, Black Hills, Fall River Co., South Dakota; in the Mopung Hills, Churchill Co., Nevada; and in uranium mines around Grants, McKinley Co., New Mexico. In Czechoslovakia, at Kladno. In the USSR, at the Kul'Yurt Tau pyrite deposit, and the in Zhivetski horizon, Tuva ASSR. From the Pacajake mine, Colquechaca, Bolivia.

Name: From the Greek, *selene, the moon*, in allusion to its similarity to tellurium, named for *the earth*, Latin *tellus.*

References: (1) Palache, C., H. Berman, and C. Frondel (1944) Dana's system of mineralogy, (7th edition), v. I, 136–137. (2) (1955) NBS Circ. 539, 5, 54. (3) Thompson, M.E., C. Roach, and W. Braddock (1956) New occurrences of native selenium. Amer. Mineral., 41, 156–157. (4) Ramdohr, P. (1969) The ore minerals and their intergrowths, (3rd edition), 382–383. (5) Sindeeva, N.D. (1964) Mineralogy and types of deposits of selenium and tellurium, 42–44.

Crystal Data: Orthorhombic. *Point Group:* $2/m\ 2/m\ 2/m$. Platy crystals, in irregular aggregates up to 0.08 mm, filling interstices.

Physical Properties: Hardness = 2.5–3 VHN = 95–116 (20 g load). D(meas.) = n.d. D(calc.) = 7.5

Optical Properties: Opaque. *Color:* In polished section, grayish white with olive tint. *Anisotropism:* Brownish gray to gray.
R_1–R_2: (420) 33.3–34.8, (520) 33.6–36.0, (580) 32.6–34.8, (640) 31.0–33.0, (700) 30.4–32.3

Cell Data: *Space Group:* $P2_12_12_1$. $a = 7.86$ $b = 11.84$ $c = 8.92$ Z = 4

X-ray Powder Pattern: Chukotka, USSR.
2.96 (10), 2.64 (9), 2.28 (9), 2.23 (9), 1.918 (9b), 1.888 (9b)

Chemistry:

	(1)
Ag	55.80
Sb	12.63
Se	29.04
S	1.44
Total	98.91

(1) Chukotka, USSR; by electron microprobe, corresponding to $Ag_{5.00}Sb_{1.00}(Se_{3.56}S_{0.44})_{\Sigma=4.00}$.

Occurrence: In quartz and orthoclase-quartz veins with other silver minerals.

Association: Miaragyrite, clausthalite, Au–Ag alloy, orthoclase, quartz.

Distribution: At an undefined locality in central Chukotka, USSR.

Name: For the composition, and similarity to stephanite.

Type Material: C.N.I.G.R.I. Institute; A.E. Fersman Mineralogical Museum, Academy of Sciences, Moscow, USSR.

References: (1) Botova, M.M., S.M. Sandomirskaya, and N.G. Tschuvikina (1985) Selenostephanite $Ag_5Sb(Se,S)_4$ – a new mineral. Zap. Vses. Mineral. Obshch., 114, 627–630 (in Russian). (2) (1987) Amer. Mineral., 72, 225 (abs. ref. 1).

Crystal Data: Orthorhombic. *Point Group:* $mm2$. Crystals equant, short prismatic [001], tabular on {001}, to 20 mm; multiple striations common on {110}. *Twinning:* {110} very common; polysynthetic twinning observed in polished section.

Physical Properties: *Cleavage:* Very poor on {001}, {100}, and {010}. *Fracture:* Conchoidal. *Tenacity:* Brittle. Hardness = 2.5–3 VHN = 168–181 (100 g load). D(meas.) = 5.38 D(calc.) = 5.41

Optical Properties: Opaque. *Color:* Dark lead-gray to black; in polished section, rose-white. *Streak:* Chocolate-brown to purple-black. *Luster:* Metallic. *Anisotropism:* Strong.
R_1–R_2: (400) 34.8–33.6, (420) 34.6–33.6, (440) 34.4–33.6, (460) 34.2–33.7, (480) 33.8–33.8, (500) 33.4–33.8, (520) 32.9–33.7, (540) 32.4–33.6, (560) 32.0–33.4, (580) 31.6–33.1, (600) 31.4–32.7, (620) 31.0–32.2, (640) 30.6–31.5, (660) 30.2–30.9, (680) 29.7–30.1, (700) 29.2–29.4

Cell Data: *Space Group:* $Pn2_1m$. $a = 8.076(2)$ $b = 8.737(5)$ $c = 7.634(3)$ Z = 4

X-ray Powder Pattern: Binntal, Switzerland.
2.72 (100), 3.85 (80), 1.749 (70), 5.72 (60), 2.56 (50), 1.914 (50), 1.414 (50)

Chemistry:

	(1)	(2)
Pb	46.34	46.89
Cu	13.09	14.38
Ag	0.11	
Zn	0.27	
Fe	0.06	
As	16.88	16.96
Sb	0.64	
S	21.73	21.77
Total	99.12	100.00

(1) Binntal, Switzerland. (2) $PbCuAsS_3$.

Polymorphism & Series: Forms a series with bournonite.

Occurrence: Found in cavities in dolomite (Binntal, Switzerland).

Association: Tennantite, sphalerite, pyrite, dufrénoysite, rathite, baumhauerite (Binntal, Switzerland).

Distribution: In the USA, from Bingham, Tooele Co., Utah; at Butte, Silver Bow Co., Montana; and in the Balmat-Edwards mine, Balmat, St. Lawrence Co., New York. From the Whisky Creek group of mines, near Woodcock, British Columbia, and in the Hemlo gold deposit, Thunder Bay district, Ontario, Canada. In the Lengenbach quarry, Binntal, Valais, Switzerland. From Tsumeb, Namibia. At Noche Buena, Zacatecas, Mexico. At Cerro de Pasco, Peru. From the Desierto mine, near Iqueque, Chile.

Name: For Gustav Seligmann (1849–1920), a mineral collector of Koblenz, Germany.

Type Material: n.d.

References: (1) Palache, C., H. Berman, and C. Frondel (1944) Dana's system of mineralogy, (7th edition), v. I, 411–412. (2) Berry, L.G. and R.M. Thompson (1962) X-ray powder data for the ore minerals. Geol. Soc. Amer. Mem. 85, 172. (3) Leineweber, G. (1956) Struktur-analyse des Bournonits und Seligmannits mit Hilfe der Superpositions-Methoden. Zeits. Krist., 108, 161–184 (in German). (4) Takéuchi, Y. and N. Haga (1969) On the crystal structure of seligmannite, $PbCuAsS_3$, and related minerals. Zeits. Krist., 130, 254–260. (5) Edenharter, A., W. Nowacki, and Y. Takéuchi (1970) I. Verfeinerung der Kristallstruktur von Bournonit $[(SbS_3)_2|Cu_2^{IV}Pb^{VII}Pb^{VIII}]$ und von Seligmannit $[(AsS_3)_2|Cu_2^{IV}Pb^{VII}Pb^{VIII}]$. Zeits. Krist., 131, 397–417 (in German).

Crystal Data: Monoclinic. *Point Group:* $2/m$. Crystals both tabular {001} and prismatic [010]; elongated [$\bar{4}$01] and [$\bar{2}$01]; often twisted and composite; also rosette-like groups of crystals, measured in cm.

Physical Properties: *Cleavage:* Perfect on {112}. *Tenacity:* Brittle. Hardness = 2.5 VHN = n.d. D(meas.) = 6.03 D(calc.) = 6.12

Optical Properties: Opaque. *Color:* Gray to black; in polished section, white with a light green tint. *Streak:* Black. *Luster:* Metallic, tarnishes to dull. *Pleochroism:* Very weak, in yellow green to darker, grayish green. *Anisotropism:* Strong.

R_1–R_2: (400) 39.2–47.7, (420) 38.3–46.6, (440) 37.4–45.5, (460) 36.9–44.8, (480) 36.5–44.2, (500) 36.3–43.8, (520) 36.1–43.6, (540) 36.0–43.3, (560) 35.8–42.9, (580) 35.3–42.3, (600) 34.8–41.6, (620) 34.1–40.7, (640) 33.4–39.8, (660) 32.5–38.7, (680) 31.5–37.6, (700) 30.5–36.8

Cell Data: *Space Group:* $C2/c$. a = 13.603(3) b = 11.936(8) c = 24.435(7) β = 106.047(10)° Z = 4

X-ray Powder Pattern: Herja, Romania.
3.260 (100), 3.81 (90), 2.949 (90), 2.857 (80), 3.351 (70), 3.88 (60), 2.152 (50)

Chemistry:

	(1)	(2)	(3)
Pb	54.27	51.88	53.10
Ag	0.25	0.56	
Cu		0.11	
Fe		trace	
Sb	26.17	27.20	27.73
S	18.99	19.73	19.17
insol.	0.01	1.15	
Total	99.69	100.63	100.00

(1–2) Herja, Romania. (3) $Pb_9Sb_8S_{21}$.

Occurrence: In hydrothermal veins formed at moderate temperature.

Association: Sorbyite, jamesonite, bournonite, zinkenite, guettardite, jordanite, diaphorite, sphalerite, galena, siderite.

Distribution: In Romania, at Baia Sprie (Felsőbánya), Herja (Kisbánya), and Rodna. In Germany, at Wolfsberg, in the Harz Mountains, and from the Caspari mine, at Arnsberg, North Rhine-Westphalia. From the Les Anglais mine, Massaic, Cantal; Montlucan, and Bournac, Montagne Noire; and elsewhere in France. In Scotland, at the Louisa mine, Glendinning, Eskdale, Dumfriesshire. In Bolivia, at the San José mine, Oruro. From Herminia, in the Julcani district, Peru. At the Brobdignag prospect, near Silverton, San Juan Co., Colorado, USA. From near Madoc, Ontario, Canada. In small amounts from other occurrences.

Name: For Andor von Semsey (1833–1923), a Hungarian nobleman much interested in minerals.

References: (1) Palache, C., H. Berman, and C. Frondel (1944) Dana's system of mineralogy, (7th edition), v. I, 466–468. (2) Jambor, J.L. (1969) Sulphosalts of the plagionite group. Mineral. Mag., 37, 442–446. (3) Ramdohr, P. (1969) The ore minerals and their intergrowths, (3rd edition), 753–754. (4) Kohatsu, J.J. and B.J. Wuensch (1974) Semseyite ($Pb_9Sb_8S_{21}$) and the crystal chemistry of the plagionite group, $Pb_{3+2n}Sb_8S_{15+2n}$ (abs.) Amer. Mineral., 59, 1127–1138.

Shadlunite $(Pb, Cd)(Fe, Cu)_8S_8$

Crystal Data: Cubic. *Point Group:* $4/m\ \overline{3}\ 2/m$. Massive, as irregular grains from a few μm to 0.4 mm in size; also in veinlets less than 0.1 mm, mainly in cubanite. *Twinning:* Polysynthetic, seen in polished section.

Physical Properties: Hardness = n.d. VHN = 210 (20 g load). D(meas.) = n.d. D(calc.) = 4.72

Optical Properties: Opaque. *Color:* In polished section, grayish yellow. *Luster:* Metallic. *Pleochroism:* Weak. *Anisotropism:* Weak.

R: (400) — , (420) — , (440) 18.0, (460) 19.5, (480) 20.8, (500) 22.0, (520) 23.4, (540) 24.5, (560) 25.8, (580) 26.9, (600) 27.7, (620) 28.7, (640) 29.3, (660) 30.2, (680) 30.8, (700) 31.2

Cell Data: *Space Group:* $Fm3m$. $a = 10.91$ Z = 4

X-ray Powder Pattern: Majak mine, USSR.
3.29 (100), 1.925 (90), 3.84 (40), 2.11 (40), 5.42 (30), 3.16 (20), 1.666 (20)

Chemistry:

	(1)
Cu	27.5
Fe	24.1
Pb	16.6
Cd	3.9
S	27.4
Total	99.5

(1) Majak mine, USSR; by electron microprobe, leading to $(Fe, Cu)_8(Pb_{0.7}Cd_{0.2})_{\Sigma=0.9}S_8$.

Occurrence: As tiny grains and veinlets cutting Cu-Ni ores.

Association: Manganese-shadlunite, cubanite, talnakhite.

Distribution: At the Majak and Oktyabr mines, Talnakh area, Noril'sk region, western Siberia, USSR.

Name: For Soviet mineralogist Tatanya Shadlun, researcher on ore minerals.

Type Material: A.E. Fersman Mineralogical Museum, Academy of Sciences, Moscow, USSR.

References: (1) Evstigneeva, T.L., A.D. Genkin, N.V. Troneva, A.A. Filimonova, and A.I. Tsepin (1973) Shadlunite, a new sulfide of copper, iron, lead, manganese and cadmium from copper–nickel ores. Zap. Vses. Mineral. Obshch., 102, 63–74 (in Russian). (2) (1973) Amer. Mineral., 58, 1114 (abs. ref. 1).

Crystal Data: Hexagonal, pseudocubic. *Point Group:* $\overline{3}\ 2/m$. As minute grains embedded in serpentine; as rims on other sulfides.

Physical Properties: *Cleavage:* Rhombohedral. Hardness = 4 VHN = n.d. D(meas.) = 8.72 D(calc.) = 8.87

Optical Properties: Opaque. *Color:* In polished section, cream-white. *Luster:* Metallic. *Pleochroism:* Distinct. *Anisotropism:* Strong, in gray-blue and yellow-brown.
R_1–R_2: n.d.

Cell Data: *Space Group:* $R\overline{3}m$. $a = 5.591(1)$ $c = 13.579(1)$ Z = 3

X-ray Powder Pattern: Synthetic.
2.794 (100), 2.779 (98), 1.971 (86), 3.943 (84), 2.281 (81), 1.610 (42), 1.759 (31)

Chemistry:

	(1)	(2)
Pb	61.64	63.30
Cu	0.00	
Zn	0.01	
Ni	28.40	26.90
Fe	0.43	
Co	0.14	
Mn	0.08	
Bi	0.00	
S	9.58	9.80
Total	100.29	100.00

(1) Isua Belt, Greenland; by electron microprobe. (2) $Pb_2Ni_3S_2$.

Occurrence: In serpentine (Trial Harbour, Australia and Isua Belt, Greenland); in segregations in an iron-rich lens in basalt (Kitdlît, Greenland).

Association: Heazlewoodite, pentlandite, sphalerite, chromite, magnetite (Trial Harbour, Australia); galena, altaite, lead, iron, troilite (Kitdlît, Greenland); heazlewoodite, pentlandite (Isua Belt, Greenland).

Distribution: In Australia, at Trial Harbour, Tasmania, and at Nullagine, Western Australia. From near Kitdlît, Disko Island, and the Isua Belt, about 150 km northeast of Godthåb, western Greenland.

Name: For the Scottish petrologist, Professor Samuel James Shand (1882–1957), of Columbia University, New York, USA.

Type Material: n.d.

References: (1) Peacock, M.A. and J. McAndrew (1950) On parkerite and shandite and the crystal structure of $Ni_3Pb_2S_2$. Amer. Mineral., 35, 425–439. (2) Brower, W.S., H.S. Parker, and R.S. Roth (1974) Reexamination of synthetic parkerite and shandite. Amer. Mineral., 59, 296–301. (3) Clauss, A., M. Warasteh, and K. Weber (1978) Kristallchemische Untersuchung der Mischungsreihe $Ni_3Pb_2S_2$–$Ni_3Pb_2Se_2$ sowie eine Bemerkung zur Shandit-Struktur. Neues Jahrb. Mineral., Monatsh., 256–268 (in German with English abs.). (4) Dymek, R.F. (1987) Shandite, $Ni_3Pb_2S_2$, in a serpentinized metadunite from the Isua Supracrustal Belt, west Greenland. Can. Mineral., 25, 245–249. (5) Ramdohr, P. (1969) The ore minerals and their intergrowths, (3rd edition), 404–405.

Siegenite $(Ni, Co)_3S_4$

Crystal Data: Cubic. *Point Group:* $4/m\ \overline{3}\ 2/m$. Crystals dominantly octahedral, to 1 cm; also massive, granular to compact. *Twinning:* On {111}; may be polysynthetic.

Physical Properties: *Cleavage:* Imperfect on {001}. *Fracture:* Subconchoidal to uneven. Hardness = 4.5–5.5 VHN = 459–540 (100 g load). D(meas.) = 4.5–4.8 D(calc.) = 4.83

Optical Properties: Opaque. *Color:* Light gray to steel-gray; tarnishes easily to copper-red or violet-gray. *Luster:* Metallic, brilliant on fresh surface.
R: (400) 44.3, (420) 44.5, (440) 44.7, (460) 45.0, (480) 45.2, (500) 45.5, (520) 45.8, (540) 46.2, (560) 46.8, (580) 47.5, (600) 48.2, (620) 49.1, (640) 50.0, (660) 51.0, (680) 52.1, (700) 53.1

Cell Data: *Space Group:* $Fd3m$. $a = 9.41$ Z = 8

X-ray Powder Pattern: Mine la Motte, Missouri, USA.
2.86 (100), 1.670 (80), 2.36 (70), 1.815 (60), 0.988 (50), 3.34 (40), 1.228 (40)

Chemistry:

	(1)	(2)	(3)
Ni	31.18	31.24	23.5
Co	26.08	20.36	33.0
Fe	0.62	3.22	
Cu		3.16	
S	42.63	42.43	44.0
insol.	0.16		
Total	100.67	100.41	100.5

(1) Littfeld, Germany. (2) Mine la Motte, Missouri, USA. (3) Langis mine, Canada; by electron microprobe, corresponding to $(Co_{1.63}Ni_{1.17})_{\Sigma=2.80}S_{4.00}$.

Occurrence: In hydrothermal veins with other Cu-Ni-Fe sulfides.

Association: Chalcopyrite, pyrrhotite, galena, sphalerite, pyrite, millerite, gersdorffite, ullmannite.

Distribution: In the USA, in the Mine la Motte, Madison Co.; the Buick mine, at Bixby, and the Miliken (Sweetwater) mine, Reynolds Co., Missouri. In Canada, at the Langis mine, Cobalt-Gowganda area, Ontario. From near Brestovsko, central Bosnian Mountains, Yugoslavia. In Germany, in the Siegen district, North Rhine-Westphalia, at Littfeld. At Kladno, Czechoslovakia. From Blackcraig, Kirkcudbrightshire, Scotland. In the Shinkolobwe mine, Shaba Province, Zaire. At Kilembe, Uganda. In the Kamaishi mine, Iwate Prefecture, and the Yokozuru mine, north Kyushu, Japan. From Kalgoorlie, Western Australia. Known from a few other localities.

Name: For the occurrence in the Siegen district, Germany.

Type Material: n.d.

References: (1) Palache, C., H. Berman, and C. Frondel (1944) Dana's system of mineralogy, (7th edition), v. I, 262–265. (2) Petruk, W., D.C. Harris, and J.M. Stewart (1968) Langisite, a new mineral, and the rare minerals cobalt pentlandite, siegenite, parkerite and bravoite from the Langis mine, Cobalt–Gowganda, Ontario. Can. Mineral., 9, 597–616. (3) Berry, L.G. and R.M. Thompson (1962) X-ray powder data for the ore minerals. Geol. Soc. Amer. Mem. 85, 76–77.

Crystal Data: Cubic. *Point Group:* $4/m\ \bar{3}\ 2/m$. Crystals commonly cubic, octahedral or dodecahedral, to 2 cm; in parallel groups; more commonly as elongated, wiry, arborescent or reticulated forms. Massive, in scales, sheets, and as coatings. *Twinning:* On {111} as simple pairs and repeated in aggregates radiating along [111] axes.

Physical Properties: *Fracture:* Hackly. *Tenacity:* Ductile, malleable. Hardness = 2.5–3 VHN = 61–65 (100 g load). D(meas.) = 10.1–11.1 (10.5 when pure). D(calc.) = 10.497

Optical Properties: Opaque. *Color:* Silver-white, tarnishes gray to black; in polished section, brilliant silver-white. *Streak:* Silver-white. *Luster:* Metallic.
R: (400) 69.8, (420) 71.0, (440) 73.2, (460) 75.0, (480) 76.8, (500) 78.2, (520) 79.5, (540) 80.8, (560) 81.7, (580) 82.5, (600) 83.2, (620) 84.0, (640) 84.6, (660) 85.2, (680) 86.0, (700) 86.5

Cell Data: *Space Group:* $Fm3m$. $a = 4.0862$ Z = 4

X-ray Powder Pattern: Synthetic.
2.359 (100), 2.044 (40), 1.231 (26), 1.445 (25), 0.9375 (15), 1.8341 (13), 0.9137 (12)

Chemistry:

	(1)
Ag	98.450
Au	0.004
Cu	0.011
Fe	0.024
Hg	1.130
Sb	0.581
Total	100.200

(1) Kongsberg, Norway.

Polymorphism & Series: Forms a series with gold; the cubic form is 3C; hexagonal stacking polymorphs 2H and 4H are known.

Occurrence: A primary hydrothermal mineral, also formed by secondary processes, especially in the oxidized portions of mineral deposits.

Association: Acanthite, chlorargyrite, embolite, silver sulfosalts, copper.

Distribution: Numerous localities even for fine specimens. Well crystallized examples are from, in the USA, the Keweenaw Peninsula, Houghton and Keweenaw Cos., Michigan; Aspen, Pitkin Co., and Creede, Mineral Co., Colorado; and in Arizona, in the Silver King mine, Pinal Co. In Canada, in large amounts from Cobalt; and in the Thunder Bay district, at Silver Islet, on the north shore of Lake Superior, Ontario. Exceptionally developed at Kongsberg, Buskerud, Norway. In Germany, near Freiberg and Marienberg, Saxony, and at St. Andreasberg, in the Harz Mountains. From Příbram, Czechoslovakia. In Italy, from Monte Narba, Sarrabus, Sardinia. In Australia, at Broken Hill, New South Wales. At Chañarcillo, Atacama, Chile. Important production from Mexico, in many states; finely crystallized from Batopilas, Chihuahua; masses over 1500 kg from Arizonac, Sonora.

Name: From an Old English word for the metal *soelfer*, related to the German *silber* and the Dutch *zilver*; the chemical symbol from the Latin *argentum*.

References: (1) Palache, C., H. Berman, and C. Frondel (1944) Dana's system of mineralogy, (7th edition), v. I, 96–99. (2) (1953) NBS Circ. 539, 1, 23. (3) Novgorodova, M.I., A.I. Gorshkov, and A.V. Mokhov (1979) Native silver and its new structural modifications. Zap. Vses. Mineral. Obshch., 108, 552–563 (in Russian). (4) (1980) Amer. Mineral., 65, 1069 (abs. ref. 3).

Crystal Data: Monoclinic. *Point Group:* $2/m$. As irregular crystals, 0.1–0.2 mm across.

Physical Properties: Hardness = n.d. VHN = n.d. D(meas.) = n.d. D(calc.) = 5.036

Optical Properties: Opaque. *Color:* Red.
R_1–R_2: n.d.

Cell Data: *Space Group:* $P2_1/n$. $a = 5.948(2)$ $b = 11.404(6)$ $c = 15.979(5)$
$\beta = 90.15(1)°$ Z = 4

X-ray Powder Pattern: n.d.

Chemistry:

	(1)	(2)
Tl	24.00	24.86
Hg	23.80	24.40
As	25.55	27.34
Sb	1.68	0.00
S	24.97	23.40
Total	100.00	100.00

(1) Allchar, Yugoslavia; by electron microprobe. (2) $TlHgAs_3S_6$.

Occurrence: As inclusions in rebulite.

Association: Rebulite, christite or routhierite.

Distribution: From Allchar, Macedonia, Yugoslavia.

Name: Derivation not stated.

Type Material: n.d.

References: (1) Engel, P., W. Nowacki, T. Balić-Žunić, and S. Šćavnićar (1982) The crystal structure of simonite, $TlHgAs_3S_6$. Zeits. Krist., 161, 159–166. (2) (1984) Amer. Mineral., 69, 211 (abs. ref. 1).

Crystal Data: Triclinic, pseudocubic. *Point Group:* 1. Flattened to columnar crystals, usually corroded, to about 12 mm. *Twinning:* Crystals are twinned complexly; principal laws are twin planes {010} and {110}.

Physical Properties: *Fracture:* Irregular. *Tenacity:* Brittle. Hardness = ~4 VHN = 373 D(meas.) = 5.2(3) D(calc.) = [4.46] (synthetic).

Optical Properties: Opaque. *Color:* Steel-gray. *Streak:* Gray-black. *Luster:* Metallic. R_1–R_2: n.d.

Cell Data: *Space Group:* $P1$. $a = 9.064(8)$ $b = 9.830(8)$ $c = 9.078(8)$ $\alpha = 90°00(20)'$ $\beta = 109°30(20)'$ $\gamma = 107°48(20)'$ Z = 2

X-ray Powder Pattern: Binntal, Switzerland.
3.02 (100), 1.852 (80), 1.581 (70), 1.205 (30), 3.34 (20), 1.556 (20), 1.312 (20)

Chemistry:

	(1)	(2)	(3)
Cu	41.3	39.1	39.33
As	29.2	29.7	30.93
S	29.8	28.7	29.74
Total	100.3	97.5	100.00

(1–2) Binntal, Switzerland; by electron microprobe, corresponding to $Cu_{1.4}As_{0.9}S_{2.1}$, content in the pseudocubic cell. (3) Synthetic $Cu_6As_4S_9$.

Occurrence: On sulfides in crystalline dolomite.

Association: Tennantite, galena, sphalerite.

Distribution: From the Lengenbach quarry, Binntal, Valais, Switzerland.

Name: For Rudolph von Sinner (1890–1960), President of the Commission of the Natural History Museum, Bern, Switzerland.

Type Material: n.d.

References: (1) Nowacki, W., F. Marumo, and Y. Takéuchi (1964) Untersuchungen an Sulfiden aus dem Binnatal (Kt. Wallis, Schweiz.). Schweiz. Mineral. Petrog. Mitt., 44, 5–9. (2) (1975) Mineral. Abs., 17, 74 (abs. ref. 1). (3) Marumo, F. and W. Nowacki (1964) The crystal structure of lautite and sinnerite, a new mineral from the Lengenbach Quarry. Schweiz. Mineral. Petrog. Mitt., 44, 439–454. (4) (1965) Amer. Mineral., 50, 1504 (abs. ref. 3). (5) Makovicky, E. and B.J. Skinner (1972) Studies of the sulfosalts of copper. II. The crystallography and composition of sinnerite, $Cu_6As_4S_9$. Amer. Mineral., 57, 824–834. (6) Makovicky, E. and B.J. Skinner (1975) Studies of the sulfosalts of copper. IV. Structure and twinning of sinnerite, $Cu_6As_4S_9$. Amer. Mineral., 60, 998–1012.

Crystal Data: Monoclinic. *Point Group:* $2/m$. As small irregular grains (ca. 0.1 mm) embedded in analcime-natrolite. *Twinning:* Microscopic to submicroscopic about {001}.

Physical Properties: Hardness = n.d. VHN = 148–166 (50 g load). D(meas.) = n.d. D(calc.) = 5.10

Optical Properties: Opaque; transparent in very thin sections. *Color:* In polished section, light bluish gray, with deep red internal reflections; very deep red in transmitted light. *Luster:* Metallic. *Pleochroism:* Weak in air, weak to distinct in oil. *Anisotropism:* Distinct in air, distinct to strong in oil, from pale purple to grayish yellow.

R_1–R_2: n.d.

Cell Data: *Space Group:* $P2_1/c$. $a = 7.81$ $b = 10.25$ $c = 13.27$ $\beta = 90°21'$ Z = 8

X-ray Powder Pattern: Synthetic.
2.830 (100), 2.628 (90), 2.619 (90), 3.911 (80), 3.208 (70), 3.192 (70), 3.048 (70)

Chemistry:

	(1)	(2)
Cu	46.08	46.66
Ag	2.01	
Sb	29.06	29.80
S	22.79	23.54
Total	99.94	100.00

(1) Ilímaussaq Intrusion, Greenland; by electron microprobe, average of 15 analyses, corresponding to $Cu_{3.0}Ag_{0.1}Sb_{1.0}S_{2.9}$. (2) Cu_3SbS_3.

Occurrence: Found in a complex of analcime-natrolite veins cutting naujaites (Ilímaussaq Intrusion, Greenland).

Association: Senarmontite, valentinite, tetrahedrite, antimony, chalcostibite, löllingite, galena, natrolite, analcime, ussingite, sodalite, feldspars, acmite, arfvedsonite, steenstrupine-(Ce) (Ilímaussaq Intrusion, Greenland).

Distribution: From the Ilímaussaq Intrusion, southern Greenland. From near Belmont, Nye Co., Nevada, USA. At Kosice, Czechoslovakia.

Name: For Professor Brian J. Skinner, Yale University, New Haven, Connecticut, USA.

Type Material: n.d.

References: (1) Karup-Møller, S. and E. Makovicky (1974) Skinnerite, Cu_3SbS_3, a new sulfosalt from the Ilímaussaq alkaline intrusion, South Greenland. Amer. Mineral., 59, 889–895.

Crystal Data: Hexagonal. *Point Group:* $\overline{3}\ 2/m$, $3m$ or 32. Crystals flattened ∥ {0001}, to 1 mm, strongly deformed to produce lamellae, in massive aggregates.

Physical Properties: *Cleavage:* Perfect {0001}. Hardness = n.d. VHN = 52–74, 63 average (25 g load). D(meas.) = n.d. D(calc.) = 7.94

Optical Properties: Opaque. *Color:* Steel-gray; white with a yellow tint in polished section. *Streak:* Black. *Luster:* Metallic. *Anisotropism:* Moderate, blue to bluish gray.

R_1–R_2: (400) — , (420) 41.0–43.0, (440) 44.6–46.1, (460) 46.5–47.8, (480) 47.7–48.8, (500) 48.3–49.5, (520) 48.7–50.0, (540) 49.0–50.3, (560) 49.1–50.4, (580) 49.1–50.4, (600) 49.2–50.5, (620) 49.3–50.6, (640) 49.4–50.8, (660) 49.4–50.8, (680) 49.4–50.8, (700) 49.4–50.8

Cell Data: *Space Group:* $R\overline{3}m$, $R3m$, or $R32$. $a = 4.183(4)$ $c = 29.12(8)$ Z = 3

X-ray Powder Pattern: Otish Mountains deposit, Canada.
3.074 (10), 2.090 (8), 2.267 (7), 4.85 (6), 3.584 (6), 9.71 (5), 2.133 (5)

Chemistry:

	(1)	(2)
Bi	59.67 – 61.62	59.41
Cu	0.00 – 0.47	
Pb	0.00 – 0.44	
Se	21.79 – 22.77	22.45
Te	15.41 – 16.02	18.14
S	0.51 – 0.75	
Total		100.00

(1) Otish Mountains deposit, Canada; by electron microprobe, ranges of six analyses, the average corresponding to $(Bi_{2.05}Cu_{0.04}Pb_{0.01})_{\Sigma=2.10}Se_{2.00}(Te_{0.87}S_{0.13})_{\Sigma=1.00}$. (2) Bi_2Se_2Te.

Occurrence: In a vein-type uranium deposit with other tellurides and selenides.

Association: Watkinsonite, součekite, clausthalite, chalcopyrite, Au–Ag alloy.

Distribution: From the Otish Mountains uranium deposit, Quebec, Canada.

Name: For Professor George Skippen, Carleton University, Ottawa, Canada.

Type Material: n.d.

References: (1) Johan, Z., P. Picot, and F. Ruhlmann (1987) The ore mineralogy of the Otish Mountains uranium deposit, Quebec: skippenite, Bi_2Se_2Te, and watkinsonite, $Cu_2PbBi_4(Se, S)_8$, two new mineral species. Can. Mineral., 25, 625–637. (2) (1989) Amer. Mineral., 74, 947 (abs. ref. 1).

Skutterudite $CoAs_{2-3}$

Crystal Data: Cubic. *Point Group:* $2/m\ \bar{3}$. Crystals are cubes, octahedra, dodecahedra, to 9 cm; rarely prismatic; in skeletal growth forms, distorted aggregates; also massive, granular and dense. *Twinning:* On {112} as sixlings and complex shapes; also reported on {011}.

Physical Properties: *Cleavage:* Distinct on {001} and {111}; in traces on {011}. *Fracture:* Conchoidal to uneven. Hardness = 5.5–6 VHN = 810–915 (100 g load). D(meas.) = 6.5 D(calc.) = 6.821

Optical Properties: Opaque. *Color:* Tin-white to silver-gray, tarnishes gray or iridescent; in polished section, gray, creamy or golden white. *Streak:* Black. *Luster:* Metallic.
R: (400) 59.1, (420) 57.8, (440) 56.5, (460) 55.6, (480) 55.0, (500) 54.5, (520) 54.2, (540) 54.0, (560) 53.8, (580) 53.6, (600) 53.5, (620) 53.2, (640) 53.0, (660) 52.9, (680) 52.8, (700) 52.7

Cell Data: *Space Group:* $Im3$. $a = 8.204$ Z = 8

X-ray Powder Pattern: Synthetic $CoAs_3$.
2.592 (100), 2.193 (35), 1.835 (35), 1.609 (30), 1.675 (20), 0.9537 (20), 3.35 (18)

Chemistry:

	(1)	(2)	(3)	(4)
Co	19.70	19.0	28.23	20.77
Fe	2.80	2.0		
Ni		1.8		
As	76.41	75.7	71.77	79.23
S	1.03	2.1		
Total	99.94	100.6	100.00	100.00

(1) Skutterud, Norway. (2) Cobalt, Canada; by electron microprobe, leading to $(Co_{0.85}Fe_{0.10}Ni_{0.23})_{\Sigma=1.18}(As_{2.68}S_{0.17})_{\Sigma=2.85}$. (3) $CoAs_2$. (4) $CoAs_3$.

Polymorphism & Series: Forms a series with nickel-skutterudite.

Occurrence: Typically found in moderate to high-temperature hydrothermal veins with other Ni-Co minerals.

Association: Nickeline, cobaltite, arsenopyrite, silver, silver sulfosalts, bismuth, calcite, siderite, barite, quartz.

Distribution: From many localities as an accessory mineral, only occasionally as an important ore or in fine specimens. In the USA, in Connecticut, at Chatham, Middlesex Co.; and at the Mine la Motte, Madison Co., Missouri. In Canada, in large amounts at Cobalt, and at Sudbury, South Lorrain, and Gowganda, Ontario. From Norway, at Skutterud, in Modum. In Germany, at Schneeberg. From Austria, at Lölling. In Spain, in the valley of Gistain, Huesca Province. From France, at Riomanou, Hautes-Pyrénées and at Mount Chalanches, Isère. At Talnotry, Kirkcudbrightshire, Scotland. Well crystallized from the Aghbar and Irhtem mines, near Bou Azzer, Morocco.

Name: For the Norwegian locality at Skutterud.

References: (1) Palache, C., H. Berman, and C. Frondel (1944) Dana's system of mineralogy, (7th edition), v. I, 342–346. (2) Roseboom, E.H., Jr. (1962) Skutterudites $(Co, Ni, Fe)As_{3-x}$: composition and cell dimensions. Amer. Mineral., 47, 310–327. (3) Radcliffe, D. (1971) Structural formula and composition of skutterudite. Can. Mineral., 9, 559–563. (4) (1960) NBS Circ. 539, 10, 21. (5) Mandel, N. and J. Donohue (1971) The refinement of the crystal structure of skutterudite, $CoAs_3$. Acta Cryst., B27, 2288–2289. (6) Pauling, L. (1978) Covalent chemical bonding of transition metals in pyrite, cobaltite, skutterudite, millerite and related minerals. Can. Mineral., 16, 447–452.

Crystal Data: Monoclinic. *Point Group:* $2/m$. Crystals commonly equant, also tabular on {100}, to 3 mm; pseudohexagonal.

Physical Properties: *Cleavage:* Perfect on {100}. *Fracture:* Conchoidal. *Tenacity:* Brittle. Hardness = 1.5–2 VHN = n.d. D(meas.) = 4.88 D(calc.) = 4.926

Optical Properties: Opaque; translucent in thin splinters. *Color:* Light red, becoming orange-red on exposure to light. *Streak:* Vermilion. *Luster:* Adamantine.

Optical Class: Biaxial negative (–). *Pleochroism:* Slight in transmitted light. *Orientation:* $Y = b$, $Z \wedge c = 6.5°$ *Dispersion:* Strong. $n = 3.27(9)$ (Na). 2V(meas.) = ~65° 2V(calc.) = n.d.

R_1–R_2: (400) 40.7–44.0, (420) 40.0–43.2, (440) 39.3–42.4, (460) 38.7–41.5, (480) 37.9–40.4, (500) 37.0–39.2, (520) 36.0–38.0, (540) 35.0–36.6, (560) 33.8–35.4, (580) 32.7–34.7, (600) 32.0–34.0, (620) 31.4–33.4, (640) 31.0–32.9, (660) 30.6–32.4, (680) 30.4–31.8, (700) 30.1–31.4

Cell Data: *Space Group:* $A2/a$. $a = 17.23$ $b = 7.78$ $c = 15.19$ $\beta = 101°12'$ Z = 24

X-ray Powder Pattern: Synthetic.
2.82 (100), 3.21 (80), 2.72 (60), 1.953 (50), 1.701 (40), 1.661 (40), 1.608 (40)

Chemistry:

	(1)	(2)
Ag	43.9	43.69
As	28.9	30.34
Sb	0.4	
S	26.0	25.97
Total	99.2	100.00

(1) Binntal, Switzerland. (2) $AgAsS_2$.

Polymorphism & Series: Dimorphous with trechmannite.

Occurrence: Of hydrothermal origin with other Ag-As sulfides.

Association: Sphalerite, pyrite, realgar, orpiment, lead arsenic sulfosalts (Binntal, Switzerland); chabournéite, pierrotite, parapierrotite, stibnite, pyrite, sphalerite, twinnite, zinkenite, madocite, andorite, laffittite, routhierite, aktashite, wakabayashilite, realgar, orpiment (Jas Roux, France); proustite, gersdorffite, argentian miargyrite (Silvermines, Ireland).

Distribution: At the Lengenbach quarry, Binntal, Valais, Switzerland. From the Jas Roux deposit, Hautes-Alpes, France. At Wiesloch, Black Forest, Germany. From Silvermines, Co. Tipperary, Ireland.

Name: For George Frederick Herbert Smith (1872–1953), crystallographer of the British Museum (Natural History), London, England.

Type Material: n.d.

References: (1) Palache, C., H. Berman, and C. Frondel (1944) Dana's system of mineralogy, (7th edition), v. I, 430–432. (2) Berry, L.G. and R.M. Thompson (1962) X-ray powder data for the ore minerals. Geol. Soc. Amer. Mem. 85, 142–143. (3) Hellner, E. and H. Burzlaff (1964) Die Struktur des Smithits $AgAsS_2$. Naturwiss., 51, 35–36 (in German).

Crystal Data: Hexagonal. *Point Group:* $\bar{3}\ 2/m$. As very thin hexagonal flakes, 0.05 to 2 mm across.

Physical Properties: *Cleavage:* Perfect basal. *Fracture:* Subconchoidal. *Tenacity:* Flexible and elastic lamellae. Hardness = n.d. VHN = 388 D(meas.) = ~4.32 D(calc.) = ~4.32 Strongly ferromagnetic, crystals attracted edge-on.

Optical Properties: Opaque. *Color:* Jet-black with a brown tinge against a white background, light bronze-yellow viewed on the mirror-like base; in polished section, pinkish cream. *Streak:* Dark gray. *Luster:* Metallic; splendent on basal section. *Pleochroism:* Strong, from grayish yellow to reddish brown. *Anisotropism:* Strong, in yellow and blue-gray.
R_1–R_2: n.d.

Cell Data: *Space Group:* $R\bar{3}m$. $a = 3.47(1)$ $c = 34.4(1)$ Z = 1

X-ray Powder Pattern: Bloomington, Indiana, USA.
1.732 (100), 1.897 (80), 1.979 (70), 11.5 (60), 3.00 (60), 2.56 (60), 2.26 (60)

Chemistry:

	(1)
Fe	56.98
Ni	1.04
S	40.92
Total	98.94

(1) Bloomington, Indiana, USA; by electron microprobe, average of seven analyses.

Occurrence: Formed as a low-temperature oxidation product of the strongly magnetic monoclinic phase of pyrrhotite, as inclusions in calcite crystals in quartz geodes; also in hydrothermal veins.

Association: Pyrite, pyrrhotite, marcasite, magnetite, sphalerite, galena, chalcopyrite, calcite.

Distribution: In the USA, in Monroe Co., Indiana, north of Bloomington, and near Unionville; from Boron, Kern Co., California; and from near Helena, Boulder Co., Montana. In Canada, from the Silverfields mine, at Cobalt, and the Nicopor mine, north of Schreiber, Ontario; in the Bird River mines, Lac du Bonnet area, Manitoba; and elsewhere. In the USSR, where it occurs in the Kimmerian sedimentary iron ores of the Kerch Peninsula. In the Kamaishi mine, Iwate Prefecture, Japan. In the Lengenbach quarry, Binntal, Valais, and Trimbach, near Olten, Switzerland. In Germany, from Hartenstein, Saxony, and the Clara mine, Black Forest.

Name: For Professor Charles Henry Smyth, Jr. (1866–1937), economic geologist and petrologist of Princeton University, Princeton, New Jersey, USA.

Type Material: National Museum of Natural History, Washington, D.C., 112704; Harvard University, Cambridge, Massachusetts, USA, 106149.

References: (1) Erd, R.C., H.T. Evans, Jr., and D.H. Richter (1957) Smythite, a new iron sulfide and associated pyrrhotite from Indiana. Amer. Mineral., 42, 309–333. (2) Taylor, L.A. and K.L. Williams (1972) Smythite, $(Fe,Ni)_9S_{11}$ — a redefinition. Amer. Mineral., 57, 1571–1577. (3) Taylor, L.A. (1970) Smythite, $Fe_{3+x}S_4$, and associated minerals from the Silverfields mine, Cobalt, Ontario. Amer. Mineral., 55, 1650–1658.

Crystal Data: Hexagonal (probable). *Point Group:* n.d. As tiny (less than 0.1 mm) grains in Cu-Fe sulfides or gold.

Physical Properties: Hardness = n.d. VHN = 236 (50 g load). D(meas.) = n.d. D(calc.) = 11.88

Optical Properties: Opaque. *Color:* In polished section, grayish white with creamy tint. *Anisotropism:* Strong, in reddish lilac, brown, and pale cream shades.
R_1–R_2: (400) — , (420) — , (440) — , (460) — , (480) 46.7–52.2, (500) 48.3–53.6, (520) 50.4–56.1, (540) 52.4–57.7, (560) 54.3–59.7, (580) 55.5–61.0, (600) 56.5–62.0, (620) 57.8–63.5, (640) 59.3–65.0, (660) 60.2–66.2, (680) 60.5–67.5, (700) 60.8–68.3

Cell Data: *Space Group:* n.d. $a = 4.23$ $c = 5.69$ Z = 2

X-ray Powder Pattern: Oktyabr mine, USSR.
3.07 (100), 2.26 (100), 2.11 (90), 1.690 (50), 1.182 (50), 1.74 (40), 1.342 (40)

Chemistry:

	(1)	(2)	(3)
Pd	34.4	32.0 – 34.0	33.74
Pt		0.8 – 1.5	
Ag		0.5 – 1.0	
Bi	66.2	65.0 – 66.5	66.26
Total	100.6		100.00

(1) Oktyabr mine, USSR; by electron microprobe, corresponding to $Pd_{1.01}Bi$. (2) Lubin mine, Poland: by electron microprobe, range of values, the average corresponding to $(Pb_{0.98}Pt_{0.02}Ag_{0.02})_{\Sigma=1.02}Bi_{1.00}$. (3) PdBi.

Occurrence: In massive and disseminated mooihoekite-chalcopyrite, troilite-pyrrhotite, and chalcopyrite-cubanite ores (Oktyabr mine, USSR); in a copper deposit in organic-rich limestones (Lubin mine, Poland).

Association: Paolovite, sperrylite, silver, polarite, chalcopyrite, cubanite, mooihoekite, troilite, pyrrhotite (Oktyabr mine, USSR); platinian gold, palladium (Lubin mine, Poland).

Distribution: From the Oktyabr mine, Talnakh area, Noril'sk region, western Siberia, USSR. In the Lubin copper mine, Zechstein, Poland.

Name: For Petr Grigorevich Sobolevski (1781–1841), Russian metallurgist, who studied the platinum deposits of the Ural Mountains.

Type Material: A.E. Fersman Mineralogical Museum, Academy of Sciences, Moscow, USSR.

References: (1) Evstigneeva, T.L., A.D. Genkin, and V.A. Kovalenker (1975) A new bismuthide of palladium, sobolevskite, and the nomenclature of minerals of the system PdBi–PdTe–PdSb. Zap. Vses. Mineral. Obshch., 104, 568–579 (in Russian). (2) (1976) Amer. Mineral., 61, 1054 (abs. ref. 1). (3) Kucha, H. (1981) Precious metal alloys and organic matter in the Zechstein copper deposits, Poland. Tschermaks Mineral. Petrog. Mitt., 28, 1–16.

Crystal Data: Orthorhombic. *Point Group:* n.d. Massive, presumably.

Physical Properties: Hardness = n.d. VHN = 134–209 (10 g load). D(meas.) = n.d. D(calc.) = 9.948

Optical Properties: Opaque. *Color:* Light gray with a brownish tint in reflected light. *Luster:* Metallic. *Anisotropism:* Weak, yellowish red to bluish.

R_1–R_2: (400) 34.9–39.2, (420) 35.6–39.2, (440) 37.2–39.5, (460) 39.1–40.0, (480) 41.4–40.6, (500) 42.6–41.0, (520) 43.1–41.3, (540) 43.1–41.6, (560) 43.3–42.2, (580) 43.7–43.1, (600) 44.1–44.0, (620) 44.3–44.9, (640) 44.4–45.6, (660) 44.8–46.4, (680) 45.2–47.3, (700) 45.6–48.1

Cell Data: *Space Group:* n.d. $a = 9.645$ $b = 7.906$ $c = 11.040$ Z = 4

X-ray Powder Pattern: Sopcha massif, USSR.
3.33 (100), 1.805 (70b), 2.56 (60), 2.70 (50), 2.30 (50b), 4.12 (40), 2.15 (40)

Chemistry:

	(1)	(2)	(3)
Ag	32.62	33.56	34.22
Pd	25.26	23.92	25.31
Fe	0.80	2.13	
Cu	0.09	0.00	
Ni	0.03	0.00	
Bi	0.17		
Te	41.32	42.13	40.47
Total	100.29	101.74	100.00

(1) Sopcha massif, USSR; by electron microprobe, corresponding to $(Ag_{3.78}Fe_{0.18}Cu_{0.02}Ni_{0.01})_{\Sigma=3.99}Pd_{2.96}(Te_{4.04}Bi_{0.01})_{\Sigma=4.05}$. (2) Do.; corresponding to $(Ag_{3.79}Fe_{0.21})_{\Sigma=4.00}(Pd_{2.74}Fe_{0.25})_{\Sigma=2.99}Te_{4.02}$. (3) $Ag_4Pd_3Te_4$.

Occurrence: In veinlets cutting chalcopyrite.

Association: Merenskyite, kotulskite, chalcopyrite, mackinawite.

Distribution: From the Sopcha massif, Monchegorsk pluton, Kola Peninsula, USSR. In Canada, in Ontario, in the Roby zone, Lac des Iles Complex; and the Levack West mine, Sudbury.

Name: For the type locality, the Sopcha massif, USSR.

Type Material: Mineralogical Museum, Kola Branch, Academy of Sciences of the USSR, Apatite, USSR.

References: (1) Orsoev, D.A., S.A. Rezhenova, and A.N. Bodanova (1982) Sopcheite, $Ag_4Pd_3Te_4$, a new mineral from copper–nickel ores of the Monchegorsk pluton. Zap. Vses. Mineral. Obshch., 111, 114–117 (in Russian). (2) (1983) Amer. Mineral., 68, 472 (abs. ref. 1). (3) Dunning, G.R., J.H.G. Laflamme, and A.J. Criddle (1984) Sopcheite: a second Canadian occurrence, from the Lac-des-Iles Complex, Ontario. Can. Mineral., 22, 233–237.

Crystal Data: Monoclinic. *Point Group:* $2/m$. Crystals are equant to thin tabular, elongated [010], and commonly heavily striated ∥ [010]. *Twinning:* Irregular lamellae seen in polished section.

Physical Properties: *Cleavage:* Perfect on {001}. Hardness = n.d. VHN = 175 (50 g load). D(meas.) = n.d. D(calc.) = 5.52

Optical Properties: Opaque. *Color:* Lead-gray; in polished section, white. *Streak:* Black. *Luster:* Metallic. *Pleochroism:* Strong. *Anisotropism:* Strong.
R_1–R_2: n.d.

Cell Data: *Space Group:* $C2/m$. $a = 45.1$ $b = 4.14$ $c = 26.5$ $\beta = 113°25'$ Z = 2

X-ray Powder Pattern: Madoc, Canada.
3.44 (100), 3.38 (90), 4.13 (60), 2.96 (60), 2.099 (50), 4.02 (40), 3.04 (40)

Chemistry:

	(1)	(2)	(3)
Pb	46	47	46.9
Sb	25	25.5	25.8
As	6	5	5.7
S	21	21.5	22.1
Total	98	99.0	100.5

(1–2) Madoc, Canada; by electron microprobe. (3) Novoye, USSR; by electron microprobe.

Occurrence: Of hydrothermal origin, in a deposit in marble (Madoc, Canada); in a hydrothermal deposit in limestone, replacing pyrite and sphalerite, with other Pb-As-Sb sulfides (Novoye, USSR).

Association: Boulangerite, jamesonite, antimonian baumhauerite, zinkenite, semseyite, geocronite, robinsonite (Madoc, Canada); sphalerite, pyrite, galena, playfairite, twinnite, guettardite, baumhauerite, realgar, orpiment, cinnabar, fluorite, quartz (Novoye, USSR).

Distribution: In Canada, from near Madoc, Ontario. In the USA, from near Candelaria, Mineral Co., Nevada. From Novoye, Khaidarkan, Kirgizia, USSR.

Name: For Henry Clifton Sorby (1826–1908), English chemist and the founder of metallography.

Type Material: n.d.

References: (1) Jambor, J.L. (1967) New lead sulfantimonides from Madoc, Ontario, Part 2, mineral descriptions. Can. Mineral., 9, 191–207. (2) (1968) Amer. Mineral., 53, 1425 (abs. ref. 1). (3) Mozgova, N.N., N.S. Bortnikov, Y.S. Borodaev, and A.I. Tzépine (1982) Sur la non-stoechiométrie des sulfosels antimonieux arséniques de plomb. Bull. Minéral., 105, 3–10 (in French with English abs.).

Crystal Data: Orthorhombic, probable. *Point Group:* $mm2$, by analogy to bournonite. As anhedral grains up to 0.01 mm across, typically filling interstices between lath-shaped crystals of poubaite; also as veinlets about 0.5 mm thick. *Twinning:* Very fine polysynthetic.

Physical Properties: Hardness = n.d. VHN = 179 (25 g load). D(meas.) = n.d. D(calc.) = 7.60

Optical Properties: Opaque. *Color:* Lead-gray; in polished section, creamy with a brownish tint. *Luster:* Metallic. *Pleochroism:* Distinct in oil, in creamy brown to light gray with a bluish tint. *Anisotropism:* Medium strong, in light brown to dark blue-gray.
R_1–R_2: n.d.

Cell Data: *Space Group:* $Pn2_1m$, by analogy to bournonite. $a = 8.153(3)$ $b = 8.498(4)$ $c = 8.080(3)$ Z = 4

X-ray Powder Pattern: Oldřichov, Czechoslovakia.
2.757 (100), 2.717 (100), 4.040 (80), 4.249 (60), 2.019 (60), 3.774 (50), 1.833 (50)

Chemistry:

	(1)	(2)	(3)
Pb	33.86	32.68	31.06 – 32.30
Cu	9.76	9.93	9.68 – 10.13
Bi	31.71	32.41	32.79 – 33.84
Se	14.55	14.55	15.94 – 16.93
Te	0.79	0.51	0.92 – 1.25
S	8.83	9.25	7.77 – 7.91
Total	99.50	99.33	

(1) Oldřichov, Czechoslovakia; by electron microprobe, corresponds to $Pb_{1.05}Cu_{0.99}Bi_{0.98}(S_{1.77}Se_{1.19}Te_{0.04})_{\Sigma=3.00}$. (2) Do.; by electron microprobe. (3) Otish Mountains deposit, Canada; by electron microprobe, ranges of seven grains; the average analysis corresponds to $Pb_{1.00}Cu_{1.03}Bi_{1.04}(S_{1.60}Se_{1.35}Te_{0.05})_{\Sigma=3.00}$.

Occurrence: In hydrothermal quartz-carbonate veins (Oldřichov, Czechoslovakia); in a vein-type uranium deposit with other tellurides and selenides (Otish Mountains deposit, Canada).

Association: Poubaite, galena, clausthalite, selenian-sulfurian rucklidgeite, uraninite (Oldřichov, Czechoslovakia); watkinsonite, poubaite, wittichenite (Otish Mountains deposit, Canada).

Distribution: From Oldřichov, near Tachov, Czechoslovakia. In the Otish Mountains uranium deposit, Quebec, Canada.

Name: For Frantisek Souček, Department of Mineralogy, Charles University, Prague, Czechoslovakia.

Type Material: Department of Mineralogy, Charles University, Prague, Czechoslovakia.

References: (1) Čech, F. and I. Vavřín (1979) Součekte, $CuPbBi(S, Se)_3$, a new mineral of the bournonite group. Neues Jahrb. Mineral., Monatsh., 289–295. (2) (1980) Amer. Mineral., 65, 209 (abs. ref. 1). (3) Johan, Z., P. Picot, and F. Ruhlmann (1987) The ore mineralogy of the Otish Mountains uranium deposit, Quebec: skippenite, Bi_2Se_2Te, and watkinsonite, $Cu_2PbBi_4(Se, S)_8$, two new mineral species. Can. Mineral., 25, 625–638.

Crystal Data: Cubic. *Point Group:* $2/m\ \overline{3}$. Usually well crystallized as cubes and cubo-octahedrons, from microcrystals to as large as 2.4 cm; sometimes highly modified with rounded edges and corners; as intergrowths with Pt–Fe alloys.

Physical Properties: *Cleavage:* Indistinct on {001}. *Fracture:* Conchoidal. *Tenacity:* Brittle. Hardness = 6–7 VHN = 960–1277 ⊥ (100); 960–1277 ⊥ (111) (100 g load). D(meas.) = 10.58 D(calc.) = 10.78

Optical Properties: Opaque. *Color:* Tin-white. *Streak:* Black. *Luster:* Metallic. R: (400) 59.4, (420) 58.2, (440) 57.0, (460) 56.2, (480) 55.7, (500) 55.4, (520) 55.2, (540) 55.2, (560) 55.2, (580) 55.2, (600) 54.9, (620) 54.4, (640) 53.7, (660) 53.2, (680) 53.0, (700) 53.4

Cell Data: *Space Group:* $Pa3$. $a = 5.967$ Z = 4

X-ray Powder Pattern: Vermilion mine, Sudbury, Canada.
1.801 (100), 0.777 (90), 1.148 (70), 2.98 (60), 0.798 (60), 2.11 (50), 3.43 (40)

Chemistry:

	(1)	(2)	(3)
Pt	56.8	54.83	56.56
Rh	n.d.	1.66	
Sb	0.08		
As	42.8	39.89	43.44
rem.		2.90	
Total	99.68	99.28	100.00

(1) Vermilion mine, Sudbury, Canada; by electron microprobe, corresponding to $Pt_{1.00}(As_{1.962}Sb_{0.003})_{\Sigma=1.965}$. (2) Tweefontein Farm, South Africa. (3) $PtAs_2$.

Occurrence: The most widespread platinum mineral, occurring in every type of deposit.

Association: Pyrrhotite, pentlandite, chalcopyrite, violarite, cubanite, bornite, sphalerite, galena, linnaeite, magnetite, minor testibiopalladite, sudburyite, omeiite, gold, argentian gold (Danba, China); leadamalgam, chromite, ilmenite, magnetite, gersdorffite, pyrite, chalcopyrite, violarite, millerite, galena, stibnite, argentian gold, niggliite, iridosmine, platinum, merenskyite, kotulskite (Shiaonanshan, China); also cooperite, laurite.

Distribution: In the Bushveld Complex, Transvaal, South Africa, fine crystals from the Tweefontein Farm, near Potgietersrust; also at the Atok, Onverwacht, and Rustenburg mines. From Antamponbato, Madagascar. In the USA, at the New Rambler mine, Medicine Bow Mountains, Albany Co., Wyoming; and the Stillwater Complex, Montana. In Canada, at the Vermilion, Victoria, and Frood mines, Algoma district, near Sudbury, Ontario. In China, from Danba, Sichuan Province; and Shiaonanshan, Inner Mongolia Autonomous Region. In the USSR, at Nikolaevsky, Amur, Siberia; in the Konder massif, Aldan Shield, Yakutia, and elsewhere. At Rometölväs Hill, in the Koillismaa Complex, and from the Hitura Cu-Ni deposit, western Finland.

Name: For Francis L. Sperry (?–1906), chemist of Sudbury, Ontario, Canada, who first found the mineral.

Type Material: Royal Ontario Museum, Toronto, Canada, topotypic.

References: (1) Palache, C., H. Berman, and C. Frondel (1944) Dana's system of mineralogy, (7th edition), v. I, 292–293. (2) Berry, L.G. and R.M. Thompson (1962) X-ray powder data for the ore minerals. Geol. Soc. Amer. Mem. 85, 89. (3) Ramdohr, P. (1969) The ore minerals and their intergrowths, (3rd edition), 809–811. (4) Cabri, L.J., Ed. (1981) Platinum group elements: mineralogy, geology, recovery. Can. Inst. Min. & Met., 136–138.

Crystal Data: Cubic. *Point Group:* $\bar{4}3m$. Crystals tetrahedral, dodecahedral, frequently complex and distorted, curved and conical faces common, to 30 cm. Also fibrous, botryoidal, stalactitic, cleavable, coarse to fine granular, massive. *Twinning:* Twin axis [111], twin plane {111}, simple contact twins or complex lamellar forms.

Physical Properties: *Cleavage:* Perfect on {011}. *Fracture:* Conchoidal. *Tenacity:* Brittle. Hardness = 3.5–4 VHN = 208–224 (100 g load). D(meas.) = 3.9–4.1 D(calc.) = 4.096 Pyroelectric, sometimes triboluminescent, fluorescent.

Optical Properties: Transparent to translucent, opaque when iron-rich. *Color:* Highly variable, ranging from colorless to dark brown, gray, black, commonly brown, yellow, red, green. *Streak:* Brownish to light yellow and white. *Luster:* Resinous to adamantine.

Optical Class: Isotropic. $n = 2.369$ (Na) (ZnS). *Anisotropism:* May show strain-induced birefringence.

R: (400) 18.4, (420) 18.2, (440) 18.0, (460) 17.7, (480) 17.4, (500) 17.1, (520) 16.8, (540) 16.6, (560) 16.5, (580) 16.4, (600) 16.3, (620) 16.2, (640) 16.2, (660) 16.1, (680) 16.0, (700) 15.9

Cell Data: *Space Group:* $F\bar{4}3m$. $a = 5.4060$ Z = 4

X-ray Powder Pattern: Synthetic ZnS.
3.123 (100), 1.912 (51), 1.561 (30), 2.705 (10), 1.240 (9), 1.1034 (9), 1.351 (6)

Chemistry:

	(1)	(2)	(3)	(4)
Zn	66.98	57.38	44.67	67.10
Fe	0.15	7.99	18.25	
Cd		1.23	0.28	
Mn			2.66	
S	32.78	32.99	33.57	32.90
Total	99.91	99.59	99.43	100.00

(1) Sonora, Mexico. (2) Gadoni, Sardinia, Italy. (3) St. Christoph mine, Isère, France. (4) ZnS.

Polymorphism & Series: Trimorphous with matraite and wurtzite.

Occurrence: Formed under a wide variety of hydrothermal conditions, low- to high-temperature; in coal, limestone and other sedimentary deposits.

Association: Galena, chalcopyrite, marcasite, pyrite, fluorite, barite, quartz, many other hydrothermal minerals.

Distribution: The most important ore of zinc. Only a few localities for the finest crystallized examples can be given. In the USA, from the Tri-State district of the Mississippi Valley; near Baxter Springs, Cherokee Co., Kansas; Joplin, Jasper Co., Missouri and Picher, Ottawa Co., Oklahoma. From near Carthage, Smith Co., Tennessee. In the Eagle mine, Gilman, Eagle Co., Colorado. At Horní Slavkov (Schlaggenwald) and Příbram, Czechoslovakia. From Rodna, Romania. In Germany, from Freiberg, Saxony, and Neudorf, Harz Mountains. Colorless crystals from the Lengenbach quarry, Binntal, Valais, Switzerland. In England, from Alston Moor, Cumbria. Transparent crystals from Picos de Europa, Santander, Santander Province, Spain. In Mexico, from Santa Eulalia, Chihuahua, and Cananea, Sonora.

Name: From the Greek for *treacherous*, the mineral sometimes being mistaken for galena, but yielding no lead.

References: (1) Palache, C., H. Berman, and C. Frondel (1944) Dana's system of mineralogy, (7th edition), v. I, 210–215.

Crystal Data: Hexagonal. *Point Group:* $\overline{3}\ 2/m$, $3m$, or 32. Massive.

Physical Properties: *Cleavage:* {0001}. Hardness = n.d. VHN = 63–93 (15 g load). D(meas.) = n.d. D(calc.) = 5.13

Optical Properties: Opaque. *Color:* Blue. *Luster:* Metallic. *Pleochroism:* Blue to bluish white; formerly called "blaubleibender Covellit" (blue-remaining covellite) for this reason. *Anisotropism:* Orange.

R_1–R_2: (400) 26.1–33.4, (420) 25.8–32.9, (440) 25.1–32.3, (460) 24.2–31.4, (480) 23.0–30.5, (500) 21.8–29.5, (520) 20.5–28.5, (540) 19.0–27.4, (560) 17.6–26.4, (580) 16.3–25.4, (600) 14.9–24.4, (620) 13.6–23.8, (640) 12.4–23.8, (660) 11.2–24.3, (680) 10.0–24.6, (700) 9.06–24.3

Cell Data: *Space Group:* $P\overline{3}m1$, $P3m1$, or $P321$. $a = 22.962$ $c = 41.429$ Z = 18

X-ray Powder Pattern: Spionkop Creek, Canada.
1.910 (100), 3.076 (85), 2.777 (30), 1.820 (30), 2.297 (25), 3.681 (20), 2.849 (20)

Chemistry:

	(1)	(2)	(3)	(4)
Cu	67.1	70.8	67.2	73.41
Fe			7.8	
Zn			0.7	
S	25.8	27.2	24.5	26.59
Total	92.9	98.0	100.2	100.00

(1–2) Yarrow and Spionkop Creeks, Canada; by electron microprobe. (3) Eretria, Greece; by electron microprobe. (4) $Cu_{39}S_{28}$.

Occurrence: As weathering-produced lamellar replacements of anilite and djurleite in stratabound red-bed copper deposits (Spionkop Creek, Canada); in a serpentine-hosted magnetite-chromite deposit (Eretria, Greece).

Association: Anilite, djurleite, yarrowite, tennantite (Yarrow and Spionkop Creeks, Canada); geerite, chalcopyrite, cobalt pentlandite, magnetite, chromite, andradite, chlorite, diopside (Eretria, Greece).

Distribution: In the Upper Grinnell Formation, Spionkop Creek and Yarrow Creek areas of southwestern Alberta, Canada. From the Campbell mine, Bisbee, Cochise Co., Arizona; and Dekalb Township, St. Lawrence Co., New York, USA. From near Eretria, Greece. At Schulenberg, near Clausthal, Harz Mountains, and in the Clara mine, Black Forest, Germany.

Name: For the locality at Spionkop Creek, Canada.

Type Material: Canadian Geological Survey, Ottawa; and Queen's University, Kingston, Ontario, Canada; Harvard University, Cambridge, Massachusetts, USA, 122290.

References: (1) Globe, R.J. (1980) Copper sulfides from Alberta: yarrowite, Cu_9S_8, and spionkopite, $Cu_{39}S_{28}$. Can. Mineral., 18, 511–518. (2) (1981) Amer. Mineral., 66, 1279 (abs. ref. 1). (3) Economou, M.I. (1981) A second occurrence of the copper sulfides geerite and spionkopite in Eretria area, central Greece. Neues Jahrb. Mineral., Monatsh., 489–494.

Crystal Data: Tetragonal. *Point Group:* $\bar{4}2m$. Rarely as crystals, to 3 cm, with a pseudo-octahedral habit due to twinning; also massive, granular, and disseminated. *Twinning:* As penetration twins on {102} and with [112] as twin axis and {112} as composition plane. Polysynthetic lamellae seen in polished section.

Physical Properties: *Cleavage:* Indistinct on {110} and {001}. *Fracture:* Uneven. Hardness = 4 VHN = 216–265 (25 g load). D(meas.) = 4.3–4.5 D(calc.) = 4.490

Optical Properties: Opaque. *Color:* Steel-gray to iron-black, sometimes tarnishes blue; in polished section gray with olive-green tinge. *Streak:* Blackish. *Luster:* Metallic. *Pleochroism:* Indistinct in air, distinct in oil. *Anisotropism:* Distinct, in violet and slate-green. R_1–R_2: (400) 18.5–20.4, (420) 20.0–22.0, (440) 21.5–23.6, (460) 23.0–24.9, (480) 24.3–26.0, (500) 25.5–27.1, (520) 26.5–28.0, (540) 27.4–28.6, (560) 28.0–29.0, (580) 28.6–29.4, (600) 28.8–29.5, (620) 29.0–29.5, (640) 28.8–29.2, (660) 28.2–28.6, (680) 27.6–27.8, (700) 26.4–26.8

Cell Data: *Space Group:* $I\bar{4}2m$. $a = 5.4432$ $c = 10.7299$ Z = 2

X-ray Powder Pattern: Synthetic.
3.11 (100), 1.908 (80), 1.640 (60), 1.623 (60), 2.427 (50), 1.784 (50), 1.240 (50)

Chemistry:

	(1)	(2)	(3)	(4)
Cu	29.6	29.24	29.2	29.58
Fe	10.8	13.95	12.4	12.99
Zn	2.2	0.08	0.8	
Cd	0.11			
Sn	27.7	27.14	28.1	27.61
In	0.3			
S	29.7	28.88	30.3	29.82
insol.		0.51		
Total	100.41	99.80	100.8	100.00

(1) Oruro, Bolivia; by electron microprobe, leading to $Cu_{2.01}Fe_{0.84}Zn_{0.14}Sn_{1.01}S_{4.00}$. (2) Chocaya, Bolivia. (3) Animas, Bolivia; by electron microprobe. (4) Cu_2FeSnS_4.

Occurrence: In tin-bearing vein deposits of hydrothermal origin.

Association: Chalcopyrite, sphalerite, tetrahedrite, arsenopyrite, pyrite, cassiterite, wolframite.

Distribution: From Czechoslovakia, at Cínvald (Zinnwald). In Cornwall, England, at a number of mines, sometimes constituting an ore mineral. From the Noril'sk region, western Siberia, USSR. In Bolivia, in the Itos and San José mines, Oruro; at Llallagua; Chocaya, Potosí; and in the Salvadora vein at Uncia. In Australia, from Broken Hill, New South Wales, and in Tasmania, at Zeehan. In the USA, from mines in the Black Hills, Pennington Co., South Dakota. From the Brunswick tin mines, 56 km southwest of Fredericton, New Brunswick, Canada. Noted in small amounts from a number of other localities world-wide.

Name: From the Latin *stannum, tin.*

References: (1) Palache, C., H. Berman, and C. Frondel (1944) Dana's system of mineralogy, (7th edition), v. I, 224–226. (2) Berry, L.G. and R.M. Thompson (1962) X-ray powder data for the ore minerals. Geol. Soc. Amer. Mem. 85, 51–52. (3) Ramdohr, P. (1969) The ore minerals and their intergrowths, (3rd edition), 542–553. (4) Springer, G. (1968) Electronprobe analysis of stannite and related tin minerals. Mineral. Mag., 36, 1045–1051. (5) Kissin, S.A. and D.R. Owens (1979) New data on stannite and related tin sulfide minerals. Can. Mineral., 17, 125–135.

Crystal Data: Orthorhombic. *Point Group:* 222. Massive, in veinlets to 1 mm.

Physical Properties: *Fracture:* Uneven to subconchoidal. Hardness = ~4 VHN = n.d. D(meas.) = n.d. D(calc.) = [4.68]

Optical Properties: Opaque. *Color:* Brass-brown; in polished section, pinkish brown. *Streak:* Dark brown-gray. *Luster:* Metallic. *Pleochroism:* Distinct, light salmon-brown to brown. *Anisotropism:* Strong, dark orange-red to yellow-gray.
R_1–R_2: (400) 19.5–20.4, (420) 19.4–20.8, (440) 19.3–21.2, (460) 19.6–21.7, (480) 20.3–22.4, (500) 21.2–23.3, (520) 22.2–24.2, (540) 23.4–25.2, (560) 24.6–26.3, (580) 25.9–27.5, (600) 27.1–28.7, (620) 28.3–30.0, (640) 29.5–31.2, (660) 30.6–32.3, (680) 31.7–33.4, (700) 32.6–34.2

Cell Data: *Space Group:* $I222$. a = 10.767 b = 5.411 c = 16.118 Z = [2]

X-ray Powder Pattern: Konjo mine, Japan.
3.11 (100), 1.906 (70), 1.621 (20b), 2.70 (16), 4.83 (10), 5.40 (5), 4.13 (4)

Chemistry:

	(1)	(2)	(3)
Cu	37.2	38.2	38.0
Ag	0.1		
Fe	12.5	11.9	11.1
Zn	1.2	0.8	4.1
Sn	16.5	18.7	15.6
S	31.2	29.9	29.2
Total	98.7	99.5	98.0

(1) Konjo mine, Japan; by electron microprobe, corresponding to $(Cu_{4.83}Ag_{0.01})_{\Sigma=4.84}(Fe_{1.96}Zn_{0.15})_{\Sigma=2.11}Sn_{1.15}S_{8.03}$. (2) Vila Apacheta, Bolivia; by electron microprobe. (3) Tingha mine, Australia.

Occurrence: In hydrothermal Cu-Sn deposits.

Association: Chalcopyrite, galena, tetrahedrite, stannite, cassiterite, siderite, quartz (Konjo mine, Japan); hemusite, enargite, luzonite, colusite, reniérite, tennantite, chalcopyrite, pyrite (Chelopech, Bulgaria).

Distribution: In Japan, at the Konjo mine, Okayama Prefecture; the Fukoku mine, Kyoto Prefecture; and in Hyogo Prefecture, at the Ikuno mine, Kanagase; Akenobe mine, Yabu; and elsewhere. From Bolivia, at Vila Apacheta. In Australia, at Tingha, New South Wales. From Colquijirca, Hunin, Peru. At Chelopech, Bulgaria. From St. Michael's Mount, Marazion, Cornwall, England. At Vaultry, Haute-Vienne; Montebras, Creuse; and Chizeuil, Saône-et-Loire, France. From Långban, Värmland, Sweden. In the Campbell mine, Bisbee, Cochise Co., Arizona, USA. From the Maggie copper deposit, British Columbia, Canada.

Name: For its physical and chemical similarity to stannite.

Type Material: Geological Institute, University of Tokyo, Japan.

References: (1) Kato, A. (1969) Stannoidite, $Cu_5(Fe,Zn)_2SnS_8$, a new stannite-like mineral from the Konjo mine, Okayama Prefecture, Japan. Bull. Nat. Sci. Mus. Tokyo, 12, 165–172. (2) (1969) Amer. Mineral., 54, 1495 (abs. ref. 1). (3) Springer, G. (1968) Electronprobe analyses of stannite and related tin minerals. Mineral. Mag., 36, 1045–1051. (4) Yamanaka, T. and A. Kato (1976) Mössbauer effect study of ^{57}Fe and ^{119}Sn in stannite, stannoidite, and mawsonite. Amer. Mineral., 61, 260–265. (5) Shimizu, M. and N. Shikazono (1987) Stannoidite-bearing tin ore: Mineralogy, texture and physicochemical environment of formation. Can. Mineral., 25, 229–236.

Stannopalladinite $(Pd, Cu)_3Sn_2(?)$

Crystal Data: Hexagonal. *Point Group:* n.d. As elongated and rounded cubic crystals up to 0.1 mm, intergrown with niggliite. *Twinning:* Occasionally exhibited polysynthetically.

Physical Properties: Hardness = n.d. VHN = 220–228; 387–452. D(meas.) = n.d. D(calc.) = n.d.

Optical Properties: Opaque. *Color:* Brown-rose, pale pink. *Anisotropism:* Strong, from lilac-red to gray-blue.

R_1–R_2: (460) 46.2–48.5, (500) 50.3–51.5, (540) 53.0–54.0, (580) 54.0–55.5, (620) 56.0–57.8, (660) 57.0–60.0, (700) 57.0–61.0

Cell Data: *Space Group:* n.d. $a = 4.40$ $c = 5.66$ Z = [1]

X-ray Powder Pattern: Synthetic Pd_3Sn_2. (JCPDS 4-801).
2.27 (100), 2.20 (100), 1.58 (70), 1.28 (70), 1.19 (70), 0.834 (70), 0.830 (70)

Chemistry:

	(1)	(2)
Pd	40. – 45.	58
Pt	15. – 20.	2
Fe	0.3 – 2.3	
Sn	28. – 33.	38
Cu	5. – 12.	
Ni	0.1 – 0.7	
insol.	0.25 – 2.5	
Total		98

(1) Noril'sk region (?), USSR; by electron microprobe, ranges of analyses. (2) Monchegorsk, USSR; by electron microprobe, average of analyses.

Occurrence: In sulfide Cu-Ni ores (Noril'sk region (?), USSR).

Association: Chalcopyrite, Pt–Fe alloy (Noril'sk region (?), USSR); niggliite, hessite, tellurides of Pt and Pd (Monchegorsk, USSR).

Distribution: Originally described from a locality thought to be in the Noril'sk region, and since described from the Taimyr mine, Talnakh area, of that region in western Siberia; the species later redefined from Monchegorsk, Kola Peninsula; also at the Ugol'nri Ruch' placer, all in the USSR.

Name: For the composition.

Type Material: n.d.

References: (1) Maslenitzky, I.N., P.V. Faleev, and E.V. Iskyul (1947) Tin-bearing minerals of the platinum group in sulfide copper–nickel ores. Doklady Acad. Nauk SSSR, 58, 1137–1140 (in Russian). (2) (1949) Mineral. Abs., 10, 453 (abs. ref. 1). (3) Chernyaev, I.A. and O.E. Yushko-Zakharova (1970) Diagnosis of micro-inclusions of minerals of the platinum group by means of the JXA-3A X-ray spectroscopic analyzer. In: Physical properties of rare-metal minerals and methods for their study. Izdat. "Nauka", 80–81 (in Russian). (4) (1971) Amer. Mineral., 56, 360–361 (abs. ref. 3). (5) Cabri, L.J., Ed. (1981) Platinum group elements: mineralogy, geology, recovery. Can. Inst. Min. & Met., 139.

Crystal Data: Orthorhombic. *Point Group:* $mm2$. Crystals short prismatic to tabular [001], to 6 cm across; also elongate [100], to 5 cm; {110} striated parallel to [1$\bar{1}$4]; massive, compact and disseminated. *Twinning:* On {110}, frequently repeated to form pseudohexagonal groups; less common on {130}, {100} and {010}, with composition plane {001}.

Physical Properties: *Cleavage:* Imperfect on {010} and {021}. *Fracture:* Subconchoidal. *Tenacity:* Brittle. Hardness = 2–2.5 VHN = n.d. D(meas.) = 6.26 D(calc.) = 6.28

Optical Properties: Opaque. *Color:* Iron-black; in polished section, white-gray. *Streak:* Iron-black. *Luster:* Metallic. *Pleochroism:* Very weak, white to pink-white. *Anisotropism:* Strong, in vivid colors.

R_1–R_2: (400) 31.0–31.8, (420) 30.8–31.6, (440) 30.6–31.4, (460) 30.2–31.1, (480) 29.7–30.8, (500) 29.0–30.3, (520) 28.2–29.7, (540) 27.6–29.0, (560) 27.0–28.5, (580) 26.5–28.1, (600) 26.2–28.0, (620) 26.1–27.9, (640) 26.0–27.8, (660) 26.0–27.6, (680) 26.2–27.4, (700) 26.6–27.0

Cell Data: *Space Group:* $Cmc2_1$. $a = 7.837(3)$ $b = 12.467(6)$ $c = 8.538(2)$ Z = 4

X-ray Powder Pattern: Freiberg, Germany.
3.08 (100), 2.58 (90), 2.89 (60), 2.13 (50), 2.19 (40), 1.834 (40), 3.56 (30)

Chemistry:

	(1)	(2)	(3)
Ag	68.65	68.21	68.33
Sb	15.22	15.86	15.42
S	16.02	15.95	16.25
Total	99.89	100.02	100.00

(1) Copiapó, Chile; traces of As and Cu. (2) Cornwall, England; trace of Fe. (3) Ag_5SbS_4.

Occurrence: A late-stage mineral in hydrothermal silver deposits.

Association: Proustite, acanthite, silver, tetrahedrite, galena, sphalerite, pyrite.

Distribution: From many silver mining localities, usually in small amounts, and only rarely in fine specimens. In the USA, in Nevada, as an important ore mineral in the Comstock Lode, Virginia City, Storey Co. In Canada, in the Cobalt district, Ontario; and at United Keno Hill Mines, Yukon Territory. In Czechoslovakia, at Hodruša, Jáchymov (Joachimsthal), Baňská Štiavnica (Schemnitz), and Příbram. In Germany, at Freiberg, Schneeberg, and Marienberg, Saxony; and as exceptional crystals from St. Andreasberg, Harz Mountains. In Italy, fine twins from Monte Narba, near Sarrabus, Sardinia. In England, as twinned crystals from Wheal Boys, Endellion, Cornwall. From Chañarcillo, Atacama, Chile. In Bolivia, at Colquechaca. From San Cristobal, Peru. In Mexico, large crystals from Arizpe, Sonora; and at Guanajuato.

Name: For the Archduke Victor Stephan (1817–1867), former Mining Director of Austria.

References: (1) Palache, C., H. Berman, and C. Frondel (1944) Dana's system of mineralogy, (7th edition), v. I, 358–361. (2) Berry, L.G. and R.M. Thompson (1962) X-ray powder data for the ore minerals. Geol. Soc. Amer. Mem. 85, 122–123. (3) Ribárt, B. and W. Nowacki (1970) Die Kristallstruktur von Stephanit, $[SbS_3|S|Ag_5^{III}]$. Acta Cryst., B26, 201–207 (in German).

Crystal Data: Orthorhombic, pseudohexagonal. *Point Group:* $2/m$ $2/m$ $2/m$. Crystals as thin pseudohexagonal basal plates to 0.6 cm, striated on {001} || [100], also || [010]; forming rosettes and fan-like aggregates. *Twinning:* On {130}, lamellar.

Physical Properties: *Cleavage:* Perfect and easy on {001}. *Tenacity:* Thin lamellae are flexible, foil-like. Hardness = 1–1.5 VHN = 31–44 (50 g load). D(meas.) = 4.101–4.215 D(calc.) = 4.275

Optical Properties: Opaque. *Color:* Pinchbeck-brown; sometimes tarnishes violet-blue on selected faces; in reflected light, light brown. *Streak:* Black. *Luster:* Metallic, brilliant on {001}. *Pleochroism:* Distinct, in browns. *Anisotropism:* Strong, bluish and reddish to lilac, reddish brown to light green with purplish pink shades.

R_1–R_2: (400) — , (420) 27.2–31.0, (440) 28.4–32.3, (460) 29.4–33.5, (480) 30.4–34.7, (500) 31.1–35.5, (520) 31.6–36.3, (540) 32.3–37.2, (560) 32.9–37.6, (580) 33.4–37.9, (600) 34.0–38.1, (620) 34.5–38.3, (640) 34.9–38.4, (660) 35.1–38.5, (680) 35.2–38.4, (700) 35.2–38.3

Cell Data: *Space Group:* *Ccmb.* $a = 6.615(2)$ $b = 11.639(4)$ $c = 12.693(4)$ Z = 8

X-ray Powder Pattern: Měděnec, Czechoslovakia.
5.780 (10), 2.785 (7), 3.226 (6), 2.616 (5), 4.281 (3), 1.938 (3), 1.893 (3)

Chemistry:

	(1)	(2)
Ag	35.27	34.17
Fe	35.97	35.37
S	29.10	30.46
Total	100.34	100.00

(1) Jáchymov, Czechoslovakia. (2) $AgFe_2S_3$.

Polymorphism & Series: Dimorphous with argentopyrite.

Occurrence: In hydrothermal veins with other silver sulfosalt minerals and in Co-Ni-Ag assemblages.

Association: Stephanite, acanthite, proustite, pyrargyrite, argentopyrite, xanthoconite, pyrite, galena, sphalerite, dolomite, calcite, quartz.

Distribution: In Czechoslovakia, at Jáchymov (Joachimsthal), Příbram, and in the Krušné hory Mountains at Měděnec. In Germany, at Johanngeorgenstadt, Freiberg, Marienberg, and Schneeberg, Saxony, and St. Andreasberg, Harz Mountains. From the Ruen Pb-Zn deposit, Osogovo region, Bulgaria. In the USA, in Arizona, in the Leroy mine, Dos Cabezas Mountains, Cochise Co. In the Highland Bell mine, Beaverdell, British Columbia, Canada. From the Kamikita mine, Aomori Prefecture, Japan.

Name: For Count Casper Maria Sternberg (1761–1838), of the National Museum in Prague.

References: (1) Palache, C., H. Berman, and C. Frondel (1944) Dana's system of mineralogy, (7th edition), v. I, 246–248. (2) Ramdohr, P. (1969) The ore minerals and their intergrowths, (3rd edition), 628–630. (3) Šrein, V., T. Řídkošil, P. Kašpar, and J. Šourek (1986) Argentopyrite and sternbergite from polymetallic veins of the skarn deposit Měděnec, Krušné hory Mts., Czechoslovakia. Neues Jahrb. Mineral., Abh., 154, 207–222. (4) Pertlik, F. (1987) Crystal structure of sternbergite. Neues Jahrb. Mineral., Monatsh., 458–464.

Crystal Data: Orthorhombic. *Point Group:* $mm2$ or $2/m$ $2/m$ $2/m$. Plumose and as bundles of fibers elongate || [001], also as anhedral grains. *Twinning:* Very fine twin lamellae occasionally seen in polished section.

Physical Properties: *Cleavage:* Perfect, parallel to [001]. Hardness = ~3.5 VHN = n.d. D(meas.) = n.d. D(calc.) = 5.91

Optical Properties: Opaque. *Color:* Black; in polished section, white. *Streak:* Black. *Luster:* Metallic. *Pleochroism:* Strong, from gray to white.
R_1–R_2: n.d.

Cell Data: *Space Group:* $Pba2$ or $Pbam$ (pseudocell). $a = 28.4(5)$ $b = 42.6(6)$ $c = 8.20(5)$ Z = 4

X-ray Powder Pattern: Madoc, Canada.
3.26 (100), 3.68 (90), 2.836 (70), 3.54 (60), 2.965 (60), 4.14 (50), 3.94 (50)

Chemistry:

	(1)	(2)	(3)
Ag	< 0.5	3.1 – 3.8	3.39
Cu	trace	0.75 – 0.97	0.86
Pb	44.5	44.6 – 45.9	45.43
Sb	21	20.5 – 26.6	22.63
As	5.5	2.9 – 8.3	6.18
S	21.5	20.3 – 21.1	20.76
Total	92.5		99.25

(1) Madoc, Canada; by electron microprobe. (2) Do.; ranges of 12 grains. (3) Do.; average of 12 grains, corresponding to $(Ag, Cu)_{2.04}Pb_{10.08}(Sb, As)_{12.24}S_{29.64}$.

Occurrence: Of hydrothermal origin, in marble.

Association: Veenite.

Distribution: From near Madoc, Ontario, Canada.

Name: For T. Sterry Hunt (1826–1892), first mineralogist with the Geological Survey of Canada.

Type Material: National Museum of Natural History, Washington, D.C., USA, 160258.

References: (1) Jambor, J.L. (1967) New lead sulfantimonides from Madoc, Ontario. Part 2 – mineral descriptions. Can. Mineral., 9, 191–213. (2) (1968) Amer. Mineral., 53, 1423 (abs. ref. 1). (3) Jambor, J.L., J.H.G. Laflamme, and D.A. Walker (1982) A re-examination of the Madoc sulfosalts. Mineral. Record, 13, 93–100.

Crystal Data: Hexagonal. *Point Group:* $\overline{3}\ 2/m$. As indistinct crystals; more commonly reniform or mammillary; also curved lamellar, fine granular. Commonly graphically intergrown with arsenic or antimony in exsolution texture.

Physical Properties: *Cleavage:* Perfect in one direction. Hardness = 3–4 VHN = n.d. D(meas.) = 5.8–6.2 D(calc.) = 6.307

Optical Properties: Opaque. *Color:* Tin-white or reddish gray, tarnishes gray. *Streak:* Gray. *Luster:* Metallic, sometimes splendent, sometimes dull.

R_1–R_2: (400) 64.8–67.8, (420) 65.0–68.0, (440) 65.2–68.2, (460) 64.8–67.9, (480) 64.0–67.2, (500) 63.0–66.8, (520) 62.3–66.6, (540) 61.6–66.3, (560) 61.2–66.0, (580) 60.8–65.9, (600) 60.5–65.8, (620) 60.3–65.7, (640) 60.1–65.8, (660) 60.0–66.2, (680) 59.8–66.3, (700) 59.6–66.4

Cell Data: *Space Group:* $R\overline{3}m$. $a = 4.045$ $c = 10.961$ Z = 6

X-ray Powder Pattern: Varuträsk, Sweden.
2.92 (100), 2.01 (70), 2.13 (60), 1.661 (40), 1.282 (40), 3.60 (30), 1.467 (20)

Chemistry:

	(1)	(2)
Sb	61.5	61.90
As	35.0	38.10
Bi	0.02	
Fe	0.85	
S	0.20	
insol.	2.20	
Total	99.77	100.00

(1) Varuträsk, Sweden. (2) SbAs.

Occurrence: Most commonly in hydrothermal veins, but also in pegmatites.

Association: Arsenic, arsenolite, antimony, kermesite, stibnite, stibiconite, cervantite, sphalerite, siderite, calcite, quartz.

Distribution: In the USA, in the Ophir mine, Comstock Lode, Virginia City, Storey Co., Nevada. In Canada, at Atlin and Alder Island, British Columbia, and at Bernic Lake, Manitoba. At Marienberg, Saxony; and St. Andreasberg, Harz Mountains, Germany. In France, from near Allemont, Isère. In Czechoslovakia, at Příbram. At Hüttenberg, Carinthia, Austria. From Valtellina, Italy. At Varuträsk, Sweden. From Broken Hill, New South Wales, Australia. At the Moctezuma mine, Sonora, Mexico.

Name: For the composition, intermediate between antimony, STIBium, and ARSENic.

References: (1) Wretblad, P.E. (1941) Minerals of the Varuträsk pegmatite, XX. Die allemontite und das system As–Sb. Geol. För. Förh. Stockholm, 63, 19–48. (2) (1941) Amer. Mineral., 26, 456 (abs. ref. 1). (3) Palache, C., H. Berman, and C. Frondel (1944) Dana's system of mineralogy, (7th edition), v. I, 130–132 ("allemontite"). (4) Berry, L.G. and R.M. Thompson (1962) X-ray powder data for the ore minerals. Geol. Soc. Amer. Mem. 85, 19.

Crystal Data: Hexagonal. *Point Group:* $6/m\ 2/m\ 2/m$, $6mm$, or $\overline{6}m2$. As hexagonal thin to thick tabular crystals, up to about 200 μm. *Twinning:* Ubiquitous.

Physical Properties: Hardness = n.d. VHN = 610(7) (100 g load). D(meas.) = n.d. D(calc.) = 10.8

Optical Properties: Opaque. *Color:* In polished section, yellowish white. *Pleochroism:* Weak, in pale yellow to yellowish white, also sometimes faint greenish, pinkish, or lavender. *Anisotropism:* Distinct, colors usually lacking but some grains are purplish gray to greenish gray. R_1–R_2: (400) 39.9–40.6, (420) 42.2–43.4, (440) 44.5–46.2, (460) 46.6–48.5, (480) 48.4–50.4, (500) 50.0–52.0, (520) 51.4–53.2, (540) 52.5–54.2, (560) 53.5–54.8, (580) 54.4–55.2, (600) 55.4–55.6, (620) 56.4–56.0, (640) 57.4–56.6, (660) 58.6–57.1, (680) 59.7–57.6, (700) 61.0–58.3

Cell Data: *Space Group:* $P6_3/mmc$, $P6_3mc$, or $P\overline{6}2c$. $a = 7.598(2)$ $c = 28.112(9)$ Z = 12

X-ray Powder Pattern: Tweefontein Farm, South Africa.
2.236 (100), 2.194 (100), 1.279 (50), 1.228 (50), 1.183 (50), 1.576 (40), 1.267 (30)

Chemistry:

	(1)	(2)	(3)
Pd	68.0	67.8	68.60
Cu	1.6	1.9	
Sn	0.2		
Sb	30.5	31.2	31.40
As	0.2		
Total	100.5	100.9	100.00

(1) Tweefontein Farm, South Africa; by electron microprobe, corresponding to $(Pd_{4.87}Cu_{0.19})_{\Sigma=5.06}Sb_{1.91}$. (2) Goodnews Bay, Alaska, USA; by electron microprobe, average of analyses of seven grains, corresponding to $(Pd_{4.83}Cu_{0.23})_{\Sigma=5.06}Sb_{1.94}$. (3) Pd_5Sb_2.

Occurrence: An uncommon constituent of platinum deposits.

Association: Braggite, cooperite, mertieite-II, sperrylite, Pt–Fe–Cu–Ni alloys, genkinite, platarsite, chromite, chalcopyrite, pentlandite, pyrrhotite, geversite, gold, violarite.

Distribution: In the USA, at Goodnews Bay, Alaska. In Transvaal, South Africa, from Tweefontein Farm, Potgietersrust district, and from the Elephant Winze, near Rietfontein; at the Onverwacht and Driekop mines. From the Department of Chocó, Cauca, Colombia. In the USSR, from the Morozova mine, Noril'sk region, western Siberia, and at Zlatoust, Ural Mountains. From the Lac des Iles Complex, Ontario, Canada.

Name: For the composition.

Type Material: British Museum (Natural History), London, England, 1930,950-2; Royal Ontario Museum, Toronto, Canada, M34229; National Museum of Natural History, Washington, D.C., USA, 135408; all probably topotypic.

References: (1) Cabri, L.J. and T.T. Chen (1976) Stibiopalladinite from the type locality. Amer. Mineral., 61, 1249–1254. (2) Desborough, G.A., J.J. Finney, and B.F. Leonard (1973) Mertieite, a new palladium mineral from Goodnews Bay, Alaska. Amer. Mineral., 58, 1–10. (3) Ramdohr, P. (1969) The ore minerals and their intergrowths, (3rd edition), 413–416. (4) Cabri, L.J., Ed. (1981) Platinum group elements: mineralogy, geology, recovery. Can. Inst. Min. & Met., 139–140.

Crystal Data: Orthorhombic. *Point Group:* $2/m\ 2/m\ 2/m$. Slender to stout, complexly terminated crystals, elongate [001], as long as 0.6 m; bent crystals not uncommon, occasionally twisted. In radiating and confused groups of acicular crystals; also columnar, granular, or very fine masses. *Twinning:* Rare; twin planes {130}, {120}, and perhaps {310}.

Physical Properties: *Cleavage:* Perfect and easy on {010}; imperfect on {100} and {110}. *Fracture:* Subconchoidal. *Tenacity:* Highly flexible but not elastic; slightly sectile. Hardness = 2 VHN = 71–86, (010) section (100 g load). D(meas.) = 4.63 D(calc.) = 4.625

Optical Properties: Opaque. *Color:* Lead-gray, tarnishing blackish or iridescent; in polished section, white. *Streak:* Lead-gray. *Luster:* Metallic, splendent on cleavage surfaces. *Anisotropism:* Strong.

R_1–R_2: (400) 34.2–55.2, (420) 33.6–54.5, (440) 33.0–53.8, (460) 32.8–52.8, (480) 32.8–51.8, (500) 32.8–50.7, (520) 32.8–49.5, (540) 32.6–48.2, (560) 31.8–46.8, (580) 31.2–45.5, (600) 30.7–44.4, (620) 30.6–43.5, (640) 30.8–42.6, (660) 31.2–41.8, (680) 31.0–41.2, (700) 30.6–40.5

Cell Data: *Space Group:* *Pbnm.* $a = 11.229$ $b = 11.310$ $c = 3.8389$ Z = 4

X-ray Powder Pattern: Synthetic.
2.764 (100), 3.053 (95), 3.556 (70), 3.573 (65), 5.052 (55), 2.680 (50), 2.525 (45)

Chemistry:

	(1)	(2)
Sb	71.45	71.69
S	28.42	28.31
Total	99.87	100.00

(1) Wolfsberg, Germany. (2) Sb_2S_3.

Polymorphism & Series: Dimorphous with metastibnite.

Occurrence: Of hydrothermal origin, formed in veins through a wide range of temperatures.

Association: Realgar, orpiment, cinnabar, galena, lead sulfantimonides, pyrite, marcasite, arsenopyrite, cervantite, stibiconite, calcite, ankerite, barite, chalcedonic quartz.

Distribution: The most important ore of antimony, although large deposits are rare. The following localities have produced outstanding crystallized material. In the USA, at the White Caps mine, Manhattan, Nye Co., Nevada. From Oruro, Bolivia. In Germany, at Wolfsberg, in the Harz Mountains, and near Arnsberg, North Rhine-Westphalia. At Kremnica (Kremnitz), Baňská Štiavnica (Schemnitz), and Příbram, Czechoslovakia. At Baia Sprie (Felsőbánya) and Herja (Kisbánya), Romania. In France, at Massaic, Cantal; and at La Lucette, Mayenne. In Borneo, at the Kusa mine, near Bau, Sarawak Province. From Thames, New Zealand. At Bahar-Lou, near Hamadan, Iran. Magnificent groups of crystals from the Ichinokawa mine, Ehime Prefecture (Iyo Province), Shikoku Island, Japan. In China, from Xikuangshan, Hunan Province, economically very important.

Name: From the Greek *stibi*, then Latin *stibium*, an old name for the mineral.

References: (1) Palache, C., H. Berman, and C. Frondel (1944) Dana's system of mineralogy, (7th edition), v. I, 270–275. (2) (1955) NBS Circ. 539, 5, 6. (3) Ramdohr, P. (1969) The ore minerals and their intergrowths, (3rd edition), 692–697. (4) Bayliss, P. and W. Nowacki (1972) Refinement of the crystal structure of stibnite Sb_2S_3. Zeits. Krist., 135, 308–315.

Crystal Data: Cubic. *Point Group:* $\bar{4}3m$. As microscopic angular inclusions. *Twinning:* Twin lamellae noted in polished section.

Physical Properties: Hardness = ~5 VHN = n.d. D(meas.) = 5.42 D(calc.) = 5.267

Optical Properties: Opaque to translucent. *Color:* Gray; in polished section, resembles tetrahedrite without the olive-brown or greenish blue hues.
Optical Class: Isotropic. $n = \sim 2.5$
R: n.d.

Cell Data: *Space Group:* $F\bar{4}3m$. $a = 5.667$ Z = [4]

X-ray Powder Pattern: Synthetic.
3.273 (100), 2.003 (70), 1.707 (44), 1.1561 (15), 1.299 (13), 1.416 (9), 1.0901 (8)

Chemistry:

	(1)	(2)
Zn	40.30	45.29
Hg	7.04	
Se	54.95	54.71
Total	102.29	100.00

(1) Santa Brigida mine, Argentina; by electron microprobe. (2) ZnSe.

Occurrence: Included in linnaeite (Shinkolobwe mine, Zaire); intermixed with other selenides (Santa Brigida mine, Argentina).

Association: Pyrite, linnaeite, clausthalite, selenian vaesite, molybdenite, dolomite (Shinkolobwe mine, Zaire); tiemannite, clausthalite, eucairite, umangite, klockmannite (Santa Brigida mine, Argentina).

Distribution: From Shinkolobwe, Shaba Province, Zaire. In the Santa Brigida mine, La Rioja Province, Argentina.

Name: For the distinguished German geologist Hans Stille (1876–1966).

Type Material: n.d.

References: (1) Ramdohr, P. (1956) Stilleit, ein neues Mineral, natürliches Zinkselenid, von Shinkolobwe. Geotektonisches Symposium zu Ehren von Hans Stille, 481–483 (in German). (2) (1957) Amer. Mineral., 42, 584 (abs. ref. 1). (3) (1959) NBS Circ. 539, 3, 23. (4) DeMontreuil, L.A. (1974) Occurrence of mercurian stilleite. Bol. Soc. Geol. Peru, 44, 28–41 (in Spanish). (5) (1975) Chem. Abs., 82, 61925 (abs. ref. 4).

Crystal Data: Hexagonal. *Point Group:* $\bar{3}$ or 3. As small anhedral grains 40 x 75 to 120 x 265 μm. *Twinning:* Observed in 1 of 13 grains examined in polished section.

Physical Properties: Hardness = n.d. VHN = 384 (50 g load). D(meas.) = n.d. D(calc.) = 10.96

Optical Properties: Opaque. *Color:* In polished section, light creamy gray. *Luster:* Metallic. *Anisotropism:* Weak in air, dark gray to brownish gray; in oil, distinct in brownish black with a blue to yellow-brown tinge.

R_1–R_2: (470) 51.6–52.7, (546) 52.5–53.2, (589) 53.1–53.7, (650) 54.4–55.0

Cell Data: *Space Group:* $P\bar{3}$ or $P3$. $a = 7.399$ $c = 10.311$ Z = 3

X-ray Powder Pattern: Stillwater Complex, Montana, USA.
2.115 (100), 2.355 (80), 1.351 (50), 0.8858 (40), 2.700 (30), 1.991 (30b), 2.521 (20)

Chemistry:

	(1)	(2)
Pd	79.0	79.11
As	21.2	20.89
Total	100.2	100.00

(1) Stillwater Complex, Montana, USA; by electron microprobe, elements sought for but not detected, Au, Ag, Cu, Pt, Ni, Hg, Sb, Te, Sn. (2) Pd_8As_3.

Occurrence: In an ultramafic igneous body.

Association: Gold, palladoarsenide, sperrylite, braggite, chalcopyrite, pentlandite, pyrrhotite.

Distribution: From the Banded and Upper zones of the Stillwater Complex, Montana, USA. From the Roby zone, Lac des Iles Complex, Ontario, Canada.

Name: For the Stillwater Complex geological rock unit, Montana, USA.

Type Material: Royal Ontario Museum, Toronto, Canada; National Museum of Natural History, Washington, D.C., USA, 132500.

References: (1) Cabri, L.J., J.L.G. Laflamme, J.M. Stewart, J.F. Rowland, and Tzong R. Chen (1975) New data on some palladium arsenides and antimonides. Can. Mineral., 13, 321–335.
(2) (1977) Amer. Mineral., 62, 1060 (abs. ref. 1).

Crystal Data: Cubic. *Point Group:* n.d. As well-formed cubic crystals 0.02–0.15 mm in size, and as aggregates in tin.

Physical Properties: *Tenacity:* Malleable. Hardness = > 2 VHN = 103–127, 115 average. D(meas.) = n.d. D(calc.) = 5.59

Optical Properties: Opaque. *Color:* Light gray; in polished section, creamy white.
R: (460) 78.0, (480) 79.3, (500) 60.5, (520) 81.2, (540) 81.6, (560) 81.5, (580) 81.3, (600) 81.5, (620) 81.8, (640) 82.0, (660) 82.5

Cell Data: *Space Group:* n.d. $a = 4.15$ Z = 1

X-ray Powder Pattern: Elkiaidan River, USSR.
3.09 (10), 2.19(10), 1.374 (8), 1.253(7), 1.022 (6), 1.779 (4), 1.537 (4)

Chemistry:

	(1)	(2)	(3)
Sn	56.9	55.55	49.36
Cu		0.12	
Sb	43.1	44.33	50.64
Total	100.0	100.00	100.00

(1) Elkiaidan River, USSR; by electron microprobe. Contains small amounts of metallic tin, so the analyzed ratio of Sn:Sb = 1.35:1 is considered to be 1:1. (2) Rio Tamaná, Colombia; by electron microprobe, recalculated from 93.84% to 100%. (3) SnSb.

Occurrence: In concentrates from placer deposits from a region of Silurian shaly-sandy sediments (Elkiaidan River, USSR); in concentrates from precious metal placers (Rio Tamaná, Colombia).

Association: Tin, zircon, "leucoxene", rutile, apatite, barite, celestine, scheelite, cinnabar.

Distribution: From the Elkiaidan River, eastern North Nuratin Range, Uzbekistan, USSR. In the Rio Tamaná, Colombia.

Name: For the composition, from Greek STIbium, *antimony*, and STAnnum, *tin*.

Type Material: n.d.

References: (1) Nikolaeva, E.P., V.A. Grigorenko, S.D. Gatarkina, and P.E. Tsypkina (1970) New natural intermetallic compounds of tin, antimony and copper. Zap. Vses. Mineral. Obshch., 99, 68–70 (in Russian). (2) (1971) Amer. Mineral., 56, 358 (abs. ref. 1). (3) Rose, D. (1981) New data for stistaite and antimony-bearing ν–Cu_6Sn_5 from Rio Tamaná, Colombia. Neues Jahrb. Mineral., Monatsh., 117–126.

Crystal Data: Orthorhombic. *Point Group:* $2/m\ 2/m\ 2/m$. Crystals are prismatic, pseudohexagonal about [010]; usually massive, compact. *Twinning:* Common on {101}.

Physical Properties: *Fracture:* Subconchoidal. *Tenacity:* Brittle. Hardness = 2.5–3 VHN = 70–72 (25 g load). D(meas.) = 6.2–6.3 D(calc.) = 6.33

Optical Properties: Opaque. *Color:* Dark steel-gray, tarnishes bluish; in polished section, dull grayish white. *Streak:* Steel-gray. *Luster:* Metallic. *Pleochroism:* Weak in air, strong in oil, grayish brown to whitish gray, with a faint bluish gray tinge or creamy with a pink tint. *Anisotropism:* Strong, vivid blue and deep violet.

R_1–R_2: (400) 32.9–40.0, (420) 32.2–40.7, (440) 31.5–41.4, (460) 30.4–38.6, (480) 29.2–36.0, (500) 28.4–33.8, (520) 28.0–32.1, (540) 27.7–31.0, (560) 27.5–30.2, (580) 27.4–29.7, (600) 27.3–29.4, (620) 27.3–29.1, (640) 27.3–28.9, (660) 27.4–28.8, (680) 27.6–28.9, (700) 27.8–29.0

Cell Data: *Space Group:* *Bbmm.* $a = 6.62$ $b = 7.94$ $c = 4.06$ Z = 4

X-ray Powder Pattern: Silver King mine, Arizona, USA.
2.61 (100), 3.30 (60), 2.02 (50), 1.985 (50), 1.884 (50), 1.740 (40), 1.421 (40)

Chemistry:

	(1)	(2)	(3)
Ag	53.31	52.10	53.01
Cu	31.00	32.14	31.24
Fe	trace		
S	16.02	15.26	15.75
Total	100.33	99.50	100.00

(1) Foster mine, Cobalt, Canada. (2) Guarisamey, Mexico; average of three analyses. (3) AgCuS.

Occurrence: In hydrothermal veins, formed most commonly by secondary processes, although it also forms as a primary mineral.

Association: Freibergite, bornite, chalcopyrite, galena, other sulfides.

Distribution: As an accessory mineral in a number of deposits, although only rarely in good crystals. At Zmeyewskaja-Goro, near Zmyeinogorsk, Altai Mountains, Siberia, USSR. From Rudelstadt and Kupferberg, Silesia, Poland. At Příbram and Vrančice, Czechoslovakia. In Australia, in Tasmania, at Mt. Lyell. In Chile, at Santiago, at San Lorenzo in Aconcagua, and at Copiapó and Tarapacá. In the USA, in Arizona, an important ore mineral at the Silver King and Magma mines, Superior, Gila Co. In Colorado, at many mines in the state; crystallized from Aspen, Pitkin Co.; the Red Mountain district, San Juan Co.; the American Sisters mine, Clear Creek Co., and elsewhere. From Butte, Silver Bow Co., Montana. In Canada, from Cobalt and Gowganda, Timiskaming district, Ontario; and at the Silver King mine, south of Nelson, British Columbia.

Name: For Friedrich Stromeyer (1776–1835), chemist of Göttingen, Germany, who first analyzed the mineral.

References: (1) Palache, C., H. Berman, and C. Frondel (1944) Dana's system of mineralogy, (7th edition), v. I, 190–191. (2) Berry, L.G. and R.M. Thompson (1962) X-ray powder data for the ore minerals. Geol. Soc. Amer. Mem. 85, 36–37. (3) Frueh, A.J., Jr. (1955) The crystal structure of stromeyerite, AgCuS : a possible defect structure. Zeits. Krist., 106, 299–307. (4) Suhr, N. (1955) The Ag_2S–Cu_2S system. Econ. Geol., 50, 347–350. (5) Ramdohr, P. (1969) The ore minerals and their intergrowths, (3rd edition), 477–480.

Crystal Data: Hexagonal. *Point Group:* n.d. In microscopic masses up to a few tenths of a mm, intergrown with geversite.

Physical Properties: Hardness = n.d. VHN = 385 (50 g load). D(meas.) = n.d. D(calc.) = 13.52

Optical Properties: Opaque. *Color:* In polished section, cream. *Luster:* Metallic. *Pleochroism:* Perceptible. *Anisotropism:* Strong, in brownish yellow.
R_1–R_2: (400) 42.7–46.6, (420) 46.0–49.4, (440) 49.3–52.2, (460) 52.1–54.9, (480) 54.5–57.3, (500) 56.6–59.4, (520) 58.1–61.3, (540) 59.3–63.0, (560) 60.2–64.4, (580) 61.2–65.6, (600) 62.1–66.6, (620) 62.9–67.5, (640) 63.6–68.5, (660) 64.4–69.2, (680) 64.9–70.0, (700) 65.2–70.6

Cell Data: *Space Group:* n.d. $a = 4.175$ $c = 5.504$ Z = 2

X-ray Powder Pattern: Driekop mine, South Africa.
3.027 (100), 2.192 (100), 2.088 (80), 3.618 (60), 1.512 (50), 1.720 (40), 1.224 (40)

Chemistry:

	(1)
Pt	57.0
Sb	26.1
Bi	16.3
Total	99.4

(1) Driekop mine, South Africa; by electron microprobe, corresponding to $Pt_{1.00}(Sb_{0.73}Bi_{0.27})_{\Sigma=1.00}$.

Occurrence: In platinum concentrates from an ultramafic pipe deposit (Driekop mine, South Africa).

Association: Geversite, Pt–Fe alloys (Driekop mine, South Africa).

Distribution: In the Driekop mine, Transvaal, South Africa. From near Nizhni Tagil, Ural Mountains, USSR.

Name: For Professor E.F. Stumpfl, University of Hamburg, Germany, who first described the mineral.

Type Material: Bureau de Recherches Géologiques et Minières (B.R.G.M.), Orléans, France.

References: (1) Johan, Z. and P. Picot (1972) La stumpflite, Pt(Sb, Bi), un nouveau minéral. Bull. Soc. fr. Minéral., 95, 610–613 (in French). (2) (1974) Amer. Mineral., 59, 211 (abs. ref. 1).

Stützite $Ag_{5-x}Te_3$ (x = 0.24 to 0.36)

Crystal Data: Hexagonal. *Point Group:* $6/m\ 2/m\ 2/m$. Hexagonal, equant, and highly modified; commonly massive, compact, granular.

Physical Properties: *Fracture:* Subconchoidal. *Tenacity:* Brittle. Hardness = 3.5 VHN = 73–90 (25 g load). D(meas.) = 8.00 D(calc.) = 8.18

Optical Properties: Opaque. *Color:* Dark lead-gray, tarnishes rapidly to a dark bronze to iridescence; in polished section, light gray. *Luster:* Metallic. *Anisotropism:* Moderate, in gray reddish brown-blue.

R_1–R_2: (400) 41.4–43.8, (420) 41.4–43.5, (440) 41.4–43.2, (460) 41.3–42.8, (480) 41.2–42.4, (500) 41.1–41.9, (520) 40.9–41.3, (540) 40.8–40.7, (560) 40.6–40.0, (580) 40.6–39.4, (600) 40.6–38.9, (620) 40.6–38.3, (640) 40.7–37.8, (660) 41.0–37.3, (680) 41.2–36.8, (700) 41.4–36.4

Cell Data: *Space Group:* $C6/mmm$. $a = 13.38$ $c = 8.45$ Z = 7

X-ray Powder Pattern: May Day mine, Colorado, USA.
2.16 (100b), 2.55 (80b), 3.03 (70), 2.62 (70), 3.56 (60), 3.52 (60), 2.11 (60)

Chemistry:

	(1)	(2)
Ag	57.1	58.49
Te	42.6	41.51
Total	99.7	100.00

(1) Red Cloud mine, Colorado, USA; by electron microprobe, corresponding to $Ag_{4.76}Te_3$.
(2) Ag_5Te_3.

Occurrence: As replacement masses in hydrothermal deposits associated with other tellurides and sulfides.

Association: Sylvanite, hessite, altaite, petzite, empressite, tellurium, galena, sphalerite, colusite, tetrahedrite–tennantite, pyrite.

Distribution: In the USA, in Colorado, at the May Day mine, La Plata district, La Plata Co.; the Golden Fleece mines, Lake City, Hinsdale Co.; at the Empress Josephine mine, Kerber Creek district, Saguache Co.; at Buckeye Gulch, near Leadville, Lake Co.; and at the Red Cloud mine, Gold Hill district, Boulder Co. From the Campbell mine, Bisbee, Cochise Co., Arizona. On Temagami Island, Lake Temagami, Ontario; and at Lindquist Lake, Quebec, Canada. From Glava, Värmland, Sweden. Probably from Săcărâmb (Nagyág), Romania. In the Kochbulak deposit, eastern Uzbekistan, USSR. In the Kawazu mine, Shizuoka Prefecture, Japan. From Moctezuma, Sonora, Mexico.

Name: For Andreas Stütz (1747–1806), mineralogist of Vienna, Austria.

Type Material: Vienna University, Vienna, Austria, 5808; Harvard University, Cambridge, Massachusetts, USA, 108098.

References: (1) Thompson, R.M., M.A. Peacock, J.F. Rowland, and L.G. Berry (1951) Empressite and "stuetzite". Amer. Mineral., 36, 458–469. (2) Honea, R. (1964) Empressite and stuetzite redefined. Amer. Mineral., 49, 325–338. (3) Cabri, L.J. (1965) Discussion of "empressite and stuetzite redefined" by R.M. Honea. Amer. Mineral., 50, 795–801. (4) Stumpfl, E.F. and J. Rucklidge (1968) New data on natural phases in the system Ag–Te. Amer. Mineral., 53, 1513–1522.

Crystal Data: Hexagonal. *Point Group:* $6/m\ 2/m\ 2/m$. As small inclusions, often elongated, (18 x 100 μm or less), most frequently in cobaltite or maucherite.

Physical Properties: Hardness = n.d. VHN = 281 and 311 on two grains (25 g load). D(meas.) = 9.37 (PdSb) D(calc.) = 9.41 (PdSb)

Optical Properties: Opaque. *Color:* In polished section, white with a yellow tint in air; in oil, pale yellow. *Luster:* Metallic. *Anisotropism:* Weak to moderate, in light grayish yellow and dark grayish brown.

R_1–R_2: (400) 49.1–45.0, (420) 50.9–46.9, (440) 52.9–49.2, (460) 54.7–51.5, (480) 56.6–53.8, (500) 58.6–55.8, (520) 60.7–57.7, (540) 62.8–59.3, (560) 64.4–60.4, (580) 65.7–61.4, (600) 66.9–62.2, (620) 67.8–62.8, (640) 68.6–63.4, (660) 69.3–63.6, (680) 70.0–63.9, (700) 70.5–64.2

Cell Data: *Space Group:* $P6_3/mmc$. $a = 4.083$ $c = 5.602$ Z = 2

X-ray Powder Pattern: Synthetic PdSb.
2.18 (100), 0.7790 (100), 0.8237 (90), 1.202 (80), 2.03 (70), 0.9006 (70), 1.489 (60)

Chemistry:

	(1)	(2)
Pd	36.1	46.0
Ni	7.6	
Sb	55.5	53.3
Bi	0.77	
As	0.71	< 0.01
Te	0.79	
Total	101.47	99.3

(1) Copper Cliff mine, Sudbury, Canada; by electron microprobe, leading to $(Pd_{0.72}Ni_{0.28})_{\Sigma=1.00}(Sb_{0.97}Bi_{0.01}As_{0.02}Te_{0.01})_{\Sigma=1.01}$. (2) Witwatersrand, South Africa; by electron microprobe, leading to $Pd_{0.99}Sb_{1.01}$.

Occurrence: As tiny inclusions, most often in cobaltite or maucherite, in mechanical concentrates of platinum-group minerals (Sudbury, Canada).

Association: Cobaltite, maucherite, gersdorffite, breithauptite, chalcopyrite, nickeline, galena (Sudbury, Canada); pyrrhotite, pentlandite, chalcopyrite, violarite, cubanite, bornite, sphalerite, galena, linnaeite, magnetite, minor testibiopalladite, sperrylite, omeiite, gold (Danba, China).

Distribution: At the Copper Cliff and Frood mines, Sudbury, Ontario, Canada. From Danba, Sichuan Province, China. In the Witwatersrand, South Africa.

Name: For the localities at Sudbury, Canada.

Type Material: n.d.

References: (1) Cabri, L.J. and J.H.G. Laflamme (1974) Sudburyite, a new palladium–antimony mineral from Sudbury, Ontario. Can. Mineral., 12, 275–279. (2) (1976) Amer. Mineral., 61, 178 (abs. ref. 1). (3) Fu, Ping-Chiu, Yu-Hua Kung, and Liu Chang (1979) Crystal structure of sudburyite. Ti Ch'iu Hua Hseuh, 72–75 (in Chinese). (4) (1979) Chem. Abs., 90, 213618 (abs. ref. 3). (5) Cabri, L.J., Ed. (1981) Platinum group elements: mineralogy, geology, recovery. Can. Inst. Min. & Met., 141, 158.

Crystal Data: Orthorhombic. *Point Group:* $2/m$ $2/m$ $2/m$. Crystals dipyramidal on {111}, thick tabular and disphenoidal, to 20 cm; also massive, reniform, and forming stalactites; as a powder. *Twinning:* On {101}, {011}, {110}, rare.

Physical Properties: *Cleavage:* Imperfect on {001}, {110}, {111}; parting on {111}. *Fracture:* Conchoidal to uneven. *Tenacity:* Rather brittle to somewhat sectile. Hardness = 1.5–2.5 VHN = n.d. D(meas.) = 2.07 D(calc.) = 2.076

Optical Properties: Transparent to translucent. *Color:* Sulfur-yellow to honey-yellow, yellowish brown, greenish, reddish or yellowish gray; may be black from included organic matter. *Streak:* White. *Luster:* Resinous to greasy.

Optical Class: Biaxial positive (+). *Pleochroism:* Distinct. *Orientation:* $X = a$; $Y = b$; $Z = c$. *Dispersion:* $r < v$. $\alpha = 1.9579$ (Na). $\beta = 2.0377$ (Na). $\gamma = 2.2452$ (Na). 2V(meas.) = 68°58′ 2V(calc.) = n.d.

R_1–R_2: n.d.

Cell Data: *Space Group:* $Fddd$. $a = 10.468$ $b = 12.870$ $c = 24.49$ Z = 128

X-ray Powder Pattern: Synthetic.
3.90 (100), 3.24 (60), 5.75 (50), 7.76 (40), 3.48 (40), 3.12 (40), 2.12 (40)

Chemistry: Sulfur, sometimes with Se and Te.

Polymorphism & Series: Dimorphous with rosickýite.

Occurrence: A sublimation product at volcanic fumaroles; a product of the activity of biological micro-organisms; as a result of low oxidation potential and highly acidic chemical reactions in mineral deposits; formed by the decomposition of sulfides, especially pyrite, during mine fires; found in sedimentary rocks.

Association: Gypsum, anhydrite, halite, aragonite, calcite, celestine.

Distribution: In the USA, large deposits occur in salt domes, as in Louisiana, especially in the area of Lake Charles, Chalcasieu Parish, and in Texas near Freeport, Brazoria Co. At Sulfur Mountain, in Yellowstone Park, Wyoming. In California, at the Sulfur Bank mercury mine, on Clear Creek, Lake Co. Crystals from quarries at Maybee and Scofield, Monroe Co., Michigan. On Sicily, at Cianciana, Agrigento, and Racalmuto, the source of exceptionally large and fine crystals; from many other places in Italy, notably at Solfatara di Pozzuoli, near Naples; at Perticara, near Rimini, Marche; and at Carrara, Tuscany. Large crystals from Spain, at Conil, near Cádiz, Cádiz Province. In Baja California, Mexico, at San Felipe. Numerous other localities are known, the occurrence often inconspicuous.

Name: From the Latin *sulfur.*

References: (1) Palache, C., H. Berman, and C. Frondel (1944) Dana's system of mineralogy, (7th edition), v. I, 140–144. (2) (1960) NBS Circ. 539, 9, 54.

Crystal Data: Hexagonal. *Point Group:* $\overline{3}\ 2/m$. Massive and as aggregates.

Physical Properties: *Cleavage:* Very perfect in one direction. *Tenacity:* Brittle. Hardness = Very soft. VHN = 63.9–66.2 (5 g load). D(meas.) = n.d. D(calc.) = 8.13

Optical Properties: Opaque. *Color:* Grayish white. *Luster:* Metallic.
R_1–R_2: (400) — , (420) 51.8–53.8, (440) 51.8–54.1, (460) 52.0–54.7, (480) 52.3–55.3, (500) 52.7–56.0, (520) 53.2–56.7, (540) 53.6–57.3, (560) 53.8–57.8, (580) 53.9–58.0, (600) 53.8–58.0, (620) 53.7–58.0, (640) 53.6–58.0, (660) 53.6–57.9, (680) 53.6–57.9, (700) 53.6–57.8

Cell Data: *Space Group:* $P\overline{3}m1$. $a = 4.316$ $c = 23.43$ Z = 2

X-ray Powder Pattern: Magadan region or Egerlyakh deposit, USSR.
3.16 (100), 2.32 (60), 2.16 (50), 1.779 (40) 1.367 (40), 4.67 (30), 1.960 (30)

Chemistry:

	(1)	(2)	(3)
Bi	70.1	67.7	68.58
Te	27.0	28.6	27.91
Se	0.5		
S	3.2	3.4	3.51
Total	100.8	99.7	100.00

(1) Magadan region or Egerlyakh deposit, USSR; by electron microprobe, corresponding to $Bi_{3.08}Te_{1.94}(S_{0.92}Se_{0.06})_{\Sigma=0.98}$. (2) Do.; corresponding to $Bi_{2.97}Te_{2.06}S_{0.97}$. (3) Bi_3Te_2S.

Occurrence: Of hydrothermal origin, with other bismuth tellurides.

Association: Tsumoite, joséite-B.

Distribution: From the Magadan region, and the Egerlyakh deposit, Yakutia, USSR.

Name: In allusion to the chemical relationship to tsumoite.

Type Material: n.d.

References: (1) Zav'yalov, E.N. and V.D. Begizov (1982) Sulphotsumoite, Bi_3Te_2S, a new bismuth mineral. Zap. Vses. Mineral. Obshch., 111, 316–320 (in Russian). (2) (1983) Amer. Mineral., 68, 1250 (abs. ref. 1).

Crystal Data: Cubic. *Point Group:* $\bar{4}3m$. Crystals cubic, to 2.5 cm, but most commonly massive.

Physical Properties: *Cleavage:* Perfect on {001}. Hardness = 3.5 VHN = 135–157 (100 g load). D(meas.) = 3.86–4.00 D(calc.) = 3.94

Optical Properties: Opaque. *Color:* Bronze-gold; yellow-white in polished section. *Streak:* Black. *Luster:* Metallic, tarnishes to dull. *Anisotropism:* Sometimes weak.
R: (400) 30.8, (420) 28.5, (440) 26.2, (460) 25.9, (480) 27.6, (500) 30.5, (520) 32.1, (540) 31.0, (560) 30.2, (580) 31.0, (600) 31.4, (620) 33.5, (640) 34.6, (660) 33.4, (680) 32.0, (700) 30.6

Cell Data: *Space Group:* $P\bar{4}3m$. $a = 5.3912(7)$ Z = 1

X-ray Powder Pattern: Mercur district, Utah, USA.
5.44 (100), 1.910 (80), 3.12 (50), 2.40 (40), 1.634 (40), 1.101 (40), 1.804 (30)

Chemistry:

	(1)	(2)	(3)
Cu	47.97	51.6	51.55
V	12.15	14.3	13.77
S	31.66	34.4	34.68
rem.	7.49		
Total	99.27	100.3	100.00

(1) Burra-Burra, South Australia; average of two analyses, remainder 1.04% $(Fe, Al)_2O_3$ and 6.45% gangue. (2) Mercur, Utah, USA; by electron microprobe. (3) Cu_3VS_4.

Polymorphism & Series: Forms a series with arsenosulvanite.

Occurrence: In hydrothermal copper deposits which contain vanadium, as a primary sulfide.

Association: Chalcopyrite, chalcocite, digenite, covellite, chrysocolla, malachite, azurite, atacamite, vésigniéite, mottramite, gypsum (Burra-Burra, Australia); yushkinite, cadmian sphalerite, fluorite (Pay-Khoy, USSR).

Distribution: In the USA, from the Thorpe Hills, and near Mercur, Tooele Co., Utah. In the Edelweiss mine, Burra-Burra district, South Australia. From Tsumeb, Namibia. At Pay-Khoy, in the middle stream of the Silova-Yakha River, USSR. From near Carrara, Tuscany, Italy. At Bor, Serbia, Yugoslavia. From Ponte Castiola, Corsica, France.

Name: For the composition: SULfur and VANadium.

Type Material: n.d.

References: (1) Palache, C., H. Berman, and C. Frondel (1944) Dana's system of mineralogy, (7th edition), v. I, 384–385. (2) Berry, L.G. and R.M. Thompson (1962) X-ray powder data for the ore minerals. Geol. Soc. Amer. Mem. 85, 57. (3) Trojer, F.J. (1966) Refinement of the structure of sulvanite. Amer. Mineral., 51, 890–894. (4) Ramdohr, P. (1969) The ore minerals and their intergrowths, (3rd edition), 566–567.

Crystal Data: Monoclinic. *Point Group:* $2/m$. Crystals to 1 cm, short prismatic [001] or [010], thick tabular on {100} and {010}; may be skeletal, bladed, imperfectly columnar, granular. *Twinning:* Common on {100} as contact, lamellar, or penetration twins, resembling cuneiform characters.

Physical Properties: *Cleavage:* Perfect on {010}. *Fracture:* Uneven. *Tenacity:* Brittle. Hardness = 1.5–2 VHN = 154–172 (100 g load). D(meas.) = 8.16 D(calc.) = 8.161

Optical Properties: Opaque. *Color:* Steel-gray to silver-white inclining to yellow; in polished section, creamy white. *Streak:* Steel-gray to silver-white. *Luster:* Metallic, brilliant. *Pleochroism:* Strong, cream-white to leather-brown. *Anisotropism:* Very strong, in grayish brown, grayish white, and yellowish and bluish tints.
R_1–R_2: (400) 38.6–52.6, (420) 41.4–54.8, (440) 44.2–57.0, (460) 46.4–58.5, (480) 47.9–59.6, (500) 49.0–60.3, (520) 50.0–60.8, (540) 50.6–61.1, (560) 50.8–61.4, (580) 51.0–61.6, (600) 51.3–61.7, (620) 51.6–61.9, (640) 52.0–62.0, (660) 52.6–62.2, (680) 53.5–62.5, (700) 54.8–63.1

Cell Data: *Space Group:* $P2/c$. $a = 8.95(1)$ $b = 4.478(5)$ $c = 14.62(2)$ $\beta = 145.35(5)°$ Z = 2

X-ray Powder Pattern: Fiji Islands.
3.05 (100), 2.15 (50), 2.25 (30), 1.989 (30), 2.98 (20), 1.797 (20), 1.523 (20)

Chemistry:

	(1)	(2)	(3)
Au	25.45	29.85	24.19
Ag	13.94	9.18	13.22
Cu		0.15	
Ni		0.10	
Te	60.61	60.45	62.59
Total	100.00	99.73	100.00

(1) Cripple Creek, Colorado, USA. (2) Kalgoorlie, Western Australia. (3) $(Au, Ag)_2Te_4$ with Au:Ag = 1:1.

Occurrence: Most commonly in low-temperature hydrothermal veins; also in moderate and high-temperature deposits, usually among the last minerals formed.

Association: Gold, calaverite, krennerite, altaite, hessite, petzite, acanthite, pyrite, galena, sphalerite, chalcopyrite, quartz, chalcedony, fluorite.

Distribution: A mineral present in minor amounts in many Au-Ag deposits, but only rarely in rich specimens or of economic importance. In the USA, in California, in the Melones and Stanislaus mines, Carson Hill, Calaveras Co.; in Colorado, in abundance at Cripple Creek, Teller Co., and important in the Magnolia, Gold Hill and Sunshine districts, Boulder Co., and elsewhere; in the Cornucopia district, Baker Co., Oregon. In Canada, in the Dome mine, Porcupine, Ontario. At Facebánya, Săcărâmb (Nagyág), and Baia-de-Arieş (Offenbánya), Romania. At Kalgoorlie, Western Australia. From Glava, Värmland, Sweden. At the Arakara goldfields, Guyana. From the Emperor mine, Vatukoula, Viti Levu, Fiji Islands.

Name: From TranSYLVANia, for the part of Romania where it was first found, and in allusion to the element tellurium (*sylvanium*, a name early proposed for the element), which the mineral contains.

References: (1) Palache, C., H. Berman, and C. Frondel (1944) Dana's system of mineralogy, (7th edition), v. I, 338–341. (2) Berry, L.G. and R.M. Thompson (1962) X-ray powder data for the ore minerals. Geol. Soc. Amer. Mem. 85, 148. (3) Ramdohr, P. (1969) The ore minerals and their intergrowths, (3rd edition), 424–426. (4) Pertlik, F. (1984) Kristallchemie natürlicher Telluride. I: Verfeinerung der Kristallstruktur des Sylvanites, $AuAgTe_4$. Tschermaks Mineral. Petrog. Mitt., 33, 203–212 (in German with English abs.).

Taenite (Ni, Fe)

Crystal Data: Cubic. *Point Group:* $4/m\ \overline{3}\ 2/m$. As small plates and grains, fine scales and large ellipsoidal masses, to 40 kg (terrestrial); in intergrowths with, or narrow selvages around, kamacite in meteorites.

Physical Properties: *Tenacity:* Malleable, flexible. Hardness = 5–5.5 VHN = n.d. D(meas.) = 7.8–8.22 D(calc.) = 8.29 (ordered). Strongly magnetic.

Optical Properties: Opaque. *Color:* Silver-white to grayish white. *Luster:* Metallic. R: n.d.

Cell Data: *Space Group:* $Fm3m$. $a = 7.146$ Z = 32

X-ray Powder Pattern: Linville ataxite meteorite - ordered structure.
3.340 (100), 2.879 (80), 2.526 (80), 4.239 (60), 2.279 (10), 2.187 (10), 2.070 (10)

Chemistry:

	(1)	(2)	(3)
Fe	68.13	74.78	25.24
Ni	30.85	24.32	74.17
Co	0.69	0.33	0.46
Cu	0.33		
P			0.04
S			0.09
C		0.50	
Total	100.00	99.93	100.00

(1) Cañon Diablo meteorite. (2) Welland meteorite. (3) Josephine Co., Oregon, USA (terrestrial).

Occurrence: In massive serpentine bodies, as particles in placer sands, and as loose, detached masses which give few clues as to their ultimate origin, which was most likely hydrothermal, igneous, or metamorphic. Important in meteorites.

Association: Kamacite, graphite, cohenite, moissanite, schreibersite, troilite, daubréelite, oldhamite, other meteorite minerals.

Distribution: In the USA, in Josephine and Jackson Counties, Oregon; and in California, near South Fork, Smith River, Del Norte Co. In Canada, on the Fraser River, Lillooet district, British Columbia. Taenite is found in all octahedrite meteorites which exhibit Widmannstätten structures as well as in some nickel-rich ataxites.

Name: From the Greek for *band* or *strip*, in allusion to its platy structure.

References: (1) Palache, C., H. Berman, and C. Frondel (1944) Dana's system of mineralogy, (7th edition), v. I, 117–119 ("nickel-iron"). (2) Ramsden, A.R. and E.N. Cameron (1966) Kamacite and taenite superstructures and a metastable tetragonal phase in iron meteorites. Amer. Mineral., 51, 37–55. (3) Ramdohr, P. (1969) The ore minerals and their intergrowths, (3rd edition), 360–361.

Crystal Data: Orthorhombic. *Point Group:* n.d. As rounded grains (0.3–0.5 mm) and as veinlets. *Twinning:* Most grains show polysynthetic twinning.

Physical Properties: Hardness = n.d. VHN = 480(25) (50 g load). D(meas.) = n.d. D(calc.) = n.d.

Optical Properties: Opaque. *Color:* Bronze-gray; light gray with a rose tint in reflected light. *Luster:* Metallic. *Pleochroism:* Distinct, from light gray with a rose tint to a creamy tint. *Anisotropism:* Colors from dark gray with a blue tint to yellowish gray.

R_1–R_2: (400) 33.0–37.1, (430) 37.8–41.2, (460) 39.6–42.3, (490) 42.3–45.2, (520) 44.0–47.8, (550) 45.5–49.6, (580) 47.4–51.6, (610) 49.7–54.0, (640) 51.4–56.5, (670) 53.0–59.2, (700) 54.0–61.9

Cell Data: *Space Group:* n.d. a = 16.11(2) or 12.57(2) b = 11.27(1) or 13.40(2) c = 8.64(1) or 17.09(2) Z = n.d.

X-ray Powder Pattern: Talnakh area, USSR.
2.155 (100), 2.29 (55), 2.365 (40), 1.44 (40)

Chemistry:

	(1)
Pd	46.9 – 54.0
Cu	8.6 – 10.3
Pt	6.3 – 15.4
Pb	0.0 – 2.8
Sn	18.7 – 24.7
Sb	1.8 – 7.4
Total	

(1) Talnakh area, USSR; by electron microprobe, range of analyses.

Occurrence: As grains and veinlets near the contact between sulfide and rock-forming minerals in gabbro-dolerites.

Association: Argentian gold, cuprian gold, polarite, sperrylite, sobolevskite, galena, sphalerite.

Distribution: From the Talnakh area, Noril'sk region, Taimyr Peninsula, western Siberia, USSR.

Name: For the type locality on the Taimyr Peninsula, USSR.

Type Material: Mineralogical Museum of the Moscow Geological Prospecting Institute; A.E. Fersman Mineralogical Museum, Academy of Sciences, Moscow, USSR.

References: (1) Begizov, V.D., E.N. Zav'yalov, and E.G. Palov (1982) New data on taimyrite, $(Pd, Cu, Pt)_3Sn$, from copper–nickel ores of the Talnakh deposit. Zap. Vses. Mineral. Obshch., 111, 78–83 (in Russian). (2) (1983) Amer. Mineral., 68, 1252 (abs. ref. 1).

Crystal Data: Cubic. *Point Group:* $\bar{4}3m$. Seen in polished section as narrow laths, up to several mm in length.

Physical Properties: Hardness = n.d. VHN = 261–277 (50 g load). D(meas.) = 4.29 D(calc.) = [4.32]

Optical Properties: Opaque. *Color:* Chalcopyrite-yellow, tarnishes rapidly to hues of pink and brown.

R: (400) 13.0, (420) 17.1, (440) 21.2, (460) 26.2, (480) 30.6, (500) 34.1, (520) 37.0, (540) 39.4, (560) 40.8, (580) 41.7, (600) 42.3, (620) 42.8, (640) 43.0, (660) 43.2, (680) 43.3, (700) 43.4

Cell Data: *Space Group:* $I\bar{4}3m$ (probable). $a = 10.59(1)$ Z = [2]

X-ray Powder Pattern: Talnakh area, USSR.
3.06 (100), 1.873 (90), 1.599 (70), 1.079 (60), 2.64 (50), 1.213 (50), 1.019 (50)

Chemistry:

	(1)
Cu	37.15
Fe	29.1
Ni	0.75
S	33.31
Total	100.31

(1) Talnakh area, USSR; by electron microprobe, average of five analyses, corresponds to $Cu_{17.9}(Fe_{15.9}Ni_{0.4})_{\Sigma=16.3}S_{31.8}$.

Occurrence: In concentrations of up to 70% in hydrothermal vein ores (Talnakh area, USSR); in the layered ultramafic Stillwater Complex (Nye, Montana, USA).

Association: Chalcopyrite, cubanite, djerfisherite, pentlandite, valleriite or mackinawite, magnetite, Ag–Au alloy, Pd–Pb alloy (Talnakh area, USSR).

Distribution: From the Talnakh area, Noril'sk region, western Siberia, USSR. From Nye, Sweetwater Co., Montana, USA.

Name: For the Talnakh area, USSR, in which it occurs.

Type Material: National Museum of Canada, Ottawa; Royal Ontario Museum, Toronto, Canada; National Museum of Natural History, Washington, D.C., USA, 121944.

References: (1) Bud'ko, I.A. and E.A. Kulagov (1968) The new mineral talnakhite, a cubic variety of chalcopyrite. Zap. Vses. Mineral. Obshch., 97, 63 (in Russian). (2) (1970) Amer. Mineral., 55, 2135 (abs. ref. 1). (3) Cabri, L.J. and D.C. Harris (1971) New compositional data for talnakhite, $Cu_{18}(Fe,Ni)_{16}S_{32}$. Econ. Geol., 66, 673–675. (4) (1971) Amer. Mineral., 56, 2159 (abs. ref. 3). (5) Cabri, L.J. (1972) Mooihoekite and haycockite, two new copper–iron sulfides and their relationship to chalcopyrite and talnakhite. Amer. Mineral., 57, 689–708. (6) Hall, S.R. and E.J. Gabe (1972) The crystal structure of talnakhite, $Cu_{18}Fe_{16}S_{32}$. Amer. Mineral., 57, 368–380. (7) Putnis, A. (1978) Talnakhite and mooihoekite: The accessibility of ordered structures in the metal-rich region around chalcopyrite. Can. Mineral., 16, 23–30.

Crystal Data: Orthorhombic. *Point Group:* $2/m\ 2/m\ 2/m$. Crystals thin tabular on {001} with nearly square section; frequently with lateral faces striated ∥ their intersection with {001}, and with striations on {001} ∥ [110]. More usually as massive aggregates of thin, graphite-like folia having irregular edges; warped or bent crystals common. *Twinning:* Seen only in polished section.

Physical Properties: *Cleavage:* Perfect on {001}. *Tenacity:* Flexible; somewhat malleable. Hardness = 1.5 VHN = n.d. D(meas.) = 6.36 D(calc.) = 6.567

Optical Properties: Opaque. *Color:* Grayish black, may tarnish dull or iridescent; in polished section, white with slight yellow tint. *Streak:* Black. *Luster:* Metallic. *Pleochroism:* Weak, in white with light golden tint, and pure white. *Anisotropism:* Distinct to rather high.
R_1–R_2: (400) 41.9–48.2, (420) 42.6–48.9, (440) 43.3–49.6, (460) 43.7–49.6, (480) 43.8–49.4, (500) 43.8–48.7, (520) 43.6–48.0, (540) 43.4–47.2, (560) 43.1–46.6, (580) 42.7–46.0, (600) 42.2–45.4, (620) 41.6–44.9, (640) 41.1–44.4, (660) 40.6–44.0, (680) 40.1–43.6, (700) 39.6–43.2

Cell Data: *Space Group:* *Pbnm.* $a = 4.266(3)$ $b = 11.419(7)$ $c = 4.090(2)$ Z = 2

X-ray Powder Pattern: Montserrat, Bolivia.
2.84 (100), 3.41 (60), 1.419 (50), 3.27 (40), 2.33 (40), 2.03 (40), 1.090 (40)

Chemistry:

	(1)	(2)
Pb	52.09	53.13
Fe	0.17	
Sn	30.55	30.43
S	16.91	16.44
Total	99.72	100.00

(1) Santa Rosa mine, Bolivia; recalculated after deduction of 1.08% Zn as ZnS. (2) $PbSnS_2$.

Occurrence: A hydrothermal mineral found in tin veins, sometimes as an important ore mineral.

Association: Cassiterite, stannite, franckeite, cylindrite, galena, sphalerite, wurtzite, pyrite.

Distribution: Found in many Bolivian tin veins, including those of the Ichucollo, Santa Rosa, and El Salvador mines, Montserrat, and the Porvenir mine, Huanuni, Oruro; Mount Cerillos, near Carguaicollo in the Cordillera de los Frailes, Potosí; the Lipey Huaico mine, Ocuri; the San Alfredo mine, Colquiri; and the Aliada mine, Colquechaca. At Ivigtut, southern Greenland. From Sinantscha, Sichota-Alin, and Smirnowsk, Transbaikalia, USSR. At the Wallah Wallah mine, Rye Park, New South Wales, Australia.

Name: For Jethro Justinian Harris Teall (1849–1924), Director of the Geological Survey of Great Britain and Ireland.

Type Material: n.d.

References: (1) Palache, C., H. Berman, and C. Frondel (1944) Dana's system of mineralogy, (7th edition), v. I, 439–441. (2) Berry, L.G. and R.M. Thompson (1962) X-ray powder data for the ore minerals. Geol. Soc. Amer. Mem. 85, 75. (3) Ramdohr, P. (1969) The ore minerals and their intergrowths, (3rd edition), 660–662. (4) Mosburg, S., D.R. Ross, P.M. Bethke, and P. Toulmin (1961) X-ray powder data for herzenbergite, teallite, and tin trisulfide. U.S. Geol. Sur. Prof. Paper 424-C, C347–C348.

Telargpalite $(Pd, Ag)_3Te(?)$

Crystal Data: Cubic. *Point Group:* n.d. As tiny, rounded, often elongate grains dispersed and intergrown with sulfide minerals.

Physical Properties: Hardness = n.d. VHN = 62 (10 g load). D(meas.) = n.d. D(calc.) = 7.378

Optical Properties: Opaque. *Color:* In polished section, light gray with lilac tint.
R: (400) — , (420) — , (440) 44.7, (460) 45.3, (480) 46.6, (500) 47.5, (520) 49.1, (540) 50.1, (560) 50.9, (580) 51.7, (600) 52.4, (620) 53.7, (640) 54.7, (660) 55.6, (680) 56.7, (700) 57.2

Cell Data: *Space Group:* n.d. $a = 12.60(2)$ Z = 16

X-ray Powder Pattern: Oktyabr mine, USSR.
2.42 (100), 2.10 (50), 3.05 (40), 2.74 (30), 1.475 (30), 2.22 (20), 1.8221 (20)

Chemistry:

	(1)	(2)
Pd	39.0	41.4
Ag	30.8	29.8
Pb	6.7	
Bi	3.7	2.1
Te	19.8	27.0
Se	0.6	
Total	100.6	100.3

(1) Oktyabr mine, USSR; by electron microprobe, average of nine samples, corresponding to $(Pd_{2.25}Ag_{1.75}Pb_{0.20}Bi_{0.11})_{\Sigma=4.31}(Te_{0.95}Se_{0.05})_{\Sigma=1.00}$. (2) Lukkulaisvaara massif, USSR; by electron microprobe, corresponding to $(Pd_{1.76}Ag_{1.25})_{\Sigma=3.01}(Te_{0.95}Bi_{0.05})_{\Sigma=1.00}$.

Occurrence: In metasomatically altered Cu-Ni ores (Talnakh area, USSR); in the layered ultramafic Stillwater Complex (Nye, Montana, USA).

Association: Chalcopyrite, bornite, millerite, braggite, silver, kotulskite, clausthalite (Talnakh area, USSR).

Distribution: In the Oktyabr mine, Talnakh area, Noril'sk region, western Siberia, and the Lukkulaisvaara massif, USSR. From Nye, Sweetwater Co., Montana, USA.

Name: For the constituent chemical elements, TELlurium, ARGentum *silver*, and PALladium.

Type Material: A.E. Fersman Mineralogical Museum, Academy of Sciences, Moscow, USSR.

References: (1) Kovalenker, V.A., A.D. Genkin, T.L. Evstigneeva, and I.P. Laputina (1974) Telargpalite, a new mineral of palladium, silver, and tellurium, from the copper–nickel ores of the Oktyabr deposit. Zap. Vses. Mineral. Obshch., 103, 595–600 (in Russian). (2) (1975) Amer. Mineral., 60, 489 (abs. ref. 1). (3) Begizov, V.D. and E.V. Batashev (1978) Platinum minerals of the Lukkulaisvaara massif. Doklady Acad. Nauk SSSR, 243, 1265–1268 (in Russian). (4) (1981) Amer. Mineral., 66, 1103 (abs. ref. 3). (5) Cabri, L.J., Ed. (1981) Platinum group elements: mineralogy, geology, recovery. Can. Inst. Min. & Met., 142.

Crystal Data: Hexagonal. *Point Group:* $\overline{3}\ 2/m$. As lath-shaped crystals up to 175 x 350 μm, but more frequently as laths 40 μm wide and less, in altaite. *Twinning:* In polished section, twinning perpendicular to the elongation axis of the lath-shaped crystals is commonly seen.

Physical Properties: Hardness = n.d. VHN = 49.8 (25 g load). D(meas.) = n.d. D(calc.) = 6.511

Optical Properties: Opaque. *Color:* In polished section, pinkish cream. *Luster:* Metallic. *Pleochroism:* Weak, from pink to cream. *Anisotropism:* Moderate, in pink and dark gray.
R_1–R_2: (400) 54.5–56.8, (420) 57.0–58.6, (440) 59.5–60.4, (460) 61.0–61.5, (480) 62.3–62.4, (500) 63.1–63.1, (520) 63.7–63.7, (540) 64.3–64.1, (560) 64.8–64.4, (580) 65.0–64.8, (600) 65.1–64.9, (620) 65.3–65.1, (640) 65.4–65.2, (660) 65.4–65.2, (680) 65.3–65.1, (700) 65.0–65.0

Cell Data: *Space Group:* $R\overline{3}m$. $a = 4.258$ $c = 30.516$ Z = 3

X-ray Powder Pattern: Synthetic.
3.157 (100), 2.349 (35), 2.130 (25), 1.766 (10), 1.578 (8), 1.470 (8), 1.3597 (8)

Chemistry:

	(1)	(2)	(3)
Sb	37.5	37.1	38.88
Pb		1.3	
Bi	0.3		
Te	61.8	61.5	61.12
Total	99.6	99.9	100.00

(1) Mattagami mine, Canada; by electron microprobe, average of analyses of eight grains, leading to $Sb_{1.91}Te_{3.00}$. (2) Kobetsuzawa mine, Japan; by electron microprobe, giving $(Sb_{1.96}Pb_{0.04})_{\Sigma=2.00}Te_{3.00}$. (3) Sb_2Te_3.

Polymorphism & Series: Forms a series with tellurobismuthite.

Occurrence: In a zinc-rich deposit of the stratiform massive sulfide type of volcanogenic origin in Archean volcanics (Mattagami Lake mine, Canada).

Association: Altaite, pyrrhotite, chalcopyrite (Mattagami Lake mine, Canada); altaite, tellurium (Kobetsuzawa mine, Japan).

Distribution: In Canada, in the Mattagami Lake mine, Galinee Township, about 8 km southwest of the town of Matagami, Quebec. From the Kobetsuzawa mine, Sapporo, Japan.

Name: For the composition.

Type Material: Canadian Geological Survey, Ottawa; Royal Ontario Museum, Toronto, Canada.

References: (1) Thorpe, R.I. and D.C. Harris (1973) Mattagamite and tellurantimony, two new telluride minerals from Mattagami Lake mine, Matagami area, Quebec. Can. Mineral., 12, 55–60. (2) (1974) Amer. Mineral., 59, 383 (abs. ref. 1). (3) (1964) NBS Mono. 25, 3, 8. (4) Nakata, M., J. Chung, H. Honma, and K. Sakurai (1985) On tellurantimony from the Kobetsuzawa mine, Sapporo, Japan. J. Mineral. Soc. Japan, 17, 79–83 (in Japanese with English abs.).

Crystal Data: Hexagonal. *Point Group:* 32. Crystals usually tiny, but occasionally up to 3.5 cm, prismatic and acicular ∥ [0001] with (+) and (–) rhombohedra equally developed; also massive, columnar to fine granular.

Physical Properties: *Cleavage:* Perfect on $\{10\bar{1}0\}$; imperfect on {0001}. *Tenacity:* Brittle, somewhat. Hardness = 2–2.5 VHN = 48–66 (10 g load). D(meas.) = 6.1–6.3 D(calc.) = 6.225

Optical Properties: Opaque. *Color:* Tin-white; in polished section, white. *Streak:* Gray. *Luster:* Metallic. *Pleochroism:* Feeble. *Anisotropism:* Fairly strong.

R_1–R_2: (400) 58.0–69.8, (420) 57.8–69.5, (440) 57.6–69.2, (460) 57.4–69.0, (480) 57.3–68.9, (500) 57.2–68.8, (520) 57.1–68.8, (540) 57.1–68.7, (560) 57.0–68.5, (580) 56.6–68.2, (600) 56.0–67.6, (620) 55.4–67.0, (640) 54.5–66.2, (660) 53.6–65.4, (680) 52.7–64.5, (700) 51.6–63.5

Cell Data: *Space Group:* $P3_121$ or $P3_221$. $a = 4.4572$ $c = 5.9290$ Z = 3

X-ray Powder Pattern: Synthetic.
3.230 (100), 2.351 (37), 2.228 (31), 3.86 (20), 1.835 (20), 1.479 (13), 1.616 (12)

Chemistry:

	(1)	(2)
Te	94.28	99.45
Se		0.40
Fe		0.11
Bi	3.38	
S	2.01	
Total	99.67	99.96

(1) Grant Co., New Mexico, USA; analysis recalculated deducting 7.70% insolubles. (2) Gunnison Co., Colorado, USA.

Occurrence: Found in hydrothermal veins; may be of both primary and secondary origin.

Association: Gold, sylvanite, empressite, altaite, pyrite, galena, alabandite, barite, quartz, carbonates.

Distribution: In the USA, at many localities in Colorado, including Cripple Creek, Teller Co.; in the Magnolia, Gold Hill, Ballarat, and Central districts, Boulder Co.; and at the Vulcan mine, Gunnison Co. In Nevada, at Delamar, Lincoln Co. In Romania, at Zlatna and Baia-de-Arieş (Offenbánya). At Kalgoorlie, Western Australia. From Balia, Turkey. In the Teine mine, northwest of Sapporo, Hokkaido; and in the Kawazu mine, near Sizewka, Honshu, Japan. From the Emperor mine, Vatukoula, Viti Levu, Fiji Islands. At the Moctezuma mine, Sonora, Mexico.

Name: Latin name for *the earth, tellus.*

References: (1) Palache, C., H. Berman, and C. Frondel (1944) Dana's system of mineralogy, (7th edition), v. I, 138–139. (2) (1953) NBS Circ. 539, 1, 26. (3) Berry, L.G. and R.M. Thompson (1962) X-ray powder data for the ore minerals. Geol. Soc. Amer. Mem. 85, 21–22.

Crystal Data: Hexagonal. *Point Group:* $\overline{3}\ 2/m$. As irregular plates and foliated masses.

Physical Properties: *Cleavage:* Perfect on {0001}. *Tenacity:* Laminae flexible but not elastic; slightly sectile. Hardness = 1.5–2 VHN = 51 (25 g load). D(meas.) = 7.815 D(calc.) = 7.857

Optical Properties: Opaque. *Color:* Pale lead-gray; in polished section, white. *Streak:* Pale lead-gray. *Luster:* Metallic, splendent on fresh cleavages. *Anisotropism:* Weak.
R_1–R_2: (400) 58.6–61.3, (420) 59.0–61.4, (440) 59.4–61.5, (460) 59.6–61.5, (480) 59.6–61.4, (500) 59.5–61.5, (520) 59.5–61.7, (540) 59.6–62.0, (560) 60.0–62.6, (580) 60.3–63.2, (600) 60.8–63.8, (620) 61.4–64.3, (640) 62.0–64.8, (660) 62.5–65.2, (680) 63.0–65.5, (700) 63.4–65.8

Cell Data: *Space Group:* $R\overline{3}m$. $a = 4.3852$ $c = 30.483$ Z = 3

X-ray Powder Pattern: Synthetic.
3.222 (100), 2.376 (25), 2.192 (25), 5.078 (8), 1.812 (8), 1.4901 (8), 2.031 (6)

Chemistry:

	(1)	(2)	(3)
Bi	53.07	51.99	52.20
Te	48.19	47.89	47.80
S		0.12	
Total	101.26	100.00	100.00

(1) Tellurium mine, Virginia, USA. (2) Oya, Japan. (3) Bi_2Te_3.

Polymorphism & Series: Forms a series with tellurantimony.

Occurrence: Typically formed in hydrothermal gold-quartz veins of low sulfur content.

Association: Gold, bismuth, other gold tellurides, tetradymite, altaite, chalcopyrite, pyrrhotite.

Distribution: In the USA, in the Tellurium mine, Fluvanna Co., Virginia; and at Field's vein, Dahlonega, Lumpkin Co., Georgia. In New Mexico, from the Little Mildred mine, near Hachita, Grant Co. In Colorado, from near Whitehorn, Fremont Co.; at the Little Gerald and Hamilton mines, Sierra Blanca, San Louis Co.; and at Mount Chipeta, Chaffee Co. In Canada, a number of minor occurrences, as at the Hunter mine, Khutze Inlet, near Swanson Bay; at the Lucky Jim mine, Quadra Island, and at the Ashloo mine, near Squamish, Howe Sound, British Columbia; from the Robb-Montbray mine, Montbray Township, and the Horne mine, Noranda, Quebec. In Japan, in the Oya mine, Miyagi Prefecture; the Ojiri mine, Iwate Prefecture; the Suwa mine, Ibaragi Prefecture; and the Kiura mine, Oita Prefecture. In Sweden, at Boliden, Västerbotten. In Wales, from the Clogau-St. Davids mine, Mangtall Mountains, Dollgelly, and the Vigra mine, Llanaker, Merionshire.

Name: For the composition.

References: (1) Palache, C., H. Berman, and C. Frondel (1944) Dana's system of mineralogy, (7th edition), v. I, 160–161. (2) (1964) NBS Mono. 25, 3, 16.

Crystal Data: Tetragonal. *Point Group:* $4/m\ 2/m\ 2/m$. As irregular grains up to 150 μm in diameter.

Physical Properties: Hardness = n.d. VHN = 812–825 (50 g load). D(meas.) = n.d. D(calc.) = 6.50

Optical Properties: Opaque. *Color:* Bronze-yellow.
R_1–R_2: (470) 41.2–44.8, (546) 43.9–47.7, (589) 45.6–49.4, (650) 48.2–51.9

Cell Data: *Space Group:* $P4/mmm$. $a = 14.64$ $c = 10.87$ Z = 8

X-ray Powder Pattern: Strathcona mine, Sudbury, Canada.
2.80 (100), 2.314 (60), 2.405 (50), 4.35 (40), 3.66 (40), 3.28 (40), 1.868 (40)

Chemistry:

	(1)	(2)
Ni	44.1	47.12
Fe	0.9	
Co	0.9	
Bi	22.4	18.63
As	0.0	
Sb	0.0	
Te	8.5	11.38
S	21.9	22.87
Total	98.7	100.00

(1) Strathcona mine, Sudbury, Canada; by electron microprobe. (2) Ni_9BiTeS_8.

Occurrence: In hydrothermal Ni-Co-Cu veins.

Association: Millerite, chalcopyrite.

Distribution: At the Strathcona mine, Sudbury, Ontario, Canada.

Name: Alludes to its chemical relation to the hauchecornite group.

Type Material: Royal Ontario Museum, Toronto, Canada, M30942.

References: (1) Gait, R.L. and D.C. Harris (1980) Arsenohauchecornite and tellurohauchecornite: new minerals in the hauchecornite group. Mineral. Mag., 43, 877–888. (2) (1981) Amer. Mineral., 66, 436 (abs. ref. 1). (3) Gait, R.L. and D.C. Harris (1972) Hauchecornite—antimonian, arsenian and tellurian varieties. Can. Mineral., 11, 819–825.

Crystal Data: Monoclinic. *Point Group:* $2/m$. As tiny (27 x 27 to 55 x 100 μm) grains.

Physical Properties: Hardness = n.d. VHN = 376–399, 388 average, and 296–308, 302 average on two grains (15 g load). D(meas.) = 10.25 (synthetic). D(calc.) = 10.62

Optical Properties: Opaque. *Color:* In polished section, cream with a yellowish tint. *Luster:* Metallic. *Pleochroism:* None in air, very faintly discernible in oil. *Anisotropism:* Moderate to strong in air, strong in oil, in gray to light brownish gray.
R: (400) — , (420) 44.6, (440) 45.3, (460) 45.9, (480) 46.7, (500) 47.8, (520) 49.0, (540) 50.0, (560) 51.3, (580) 52.6, (600) 53.9, (620) 55.2, (640) 56.5, (660) 57.8, (680) 58.9, (700) 60.1

Cell Data: *Space Group:* $P2_1/c$. $a = 7.456(6)$ $b = 13.936(5)$ $c = 8.842(9)$ $\beta = 91.94°$ Z = 4

X-ray Powder Pattern: Stillwater Complex, Montana, USA.
2.237 (100), 1.305 (80b), 2.543 (70), 2.134 (60), 2.094 (60), 1.961 (60), 2.196 (50b)

Chemistry:

	(1)	(2)
Pd	65.4	65.23
Sn	0.07	
As	0.33	
Bi	0.15	
Sb	0.12	
Te	34.1	34.77
Total	100.17	100.00

(1) Stillwater Complex, Montana, USA; by electron microprobe, yielding $Pd_{8.99}(Te_{3.91}As_{0.07}Bi_{0.01}Sn_{0.01}Sb_{0.01})_{\Sigma=4.01}$. (2) Pd_9Te_4.

Occurrence: As tiny grains in mineral concentrates.

Association: Keithconnite, magnetite, kotulskite, merenskyite, moncheite, braggite, vysotskite.

Distribution: From the Banded and Upper Zones of the Stillwater Complex, Montana, USA.

Name: For the composition.

Type Material: National Museum of Natural History, Washington, D.C., USA, 144958; Canadian Geological Survey, Ottawa; Royal Ontario Museum, Toronto, Canada.

References: (1) Cabri, L.J., J.R. Rowland, J.H.G. Laflamme, and J. M. Stewart (1979) Keithconnite, telluropalladinite and other Pd–Pt tellurides from the Stillwater Complex, Montana. Can. Mineral., 17, 589–594. (2) Matkovic, P. and K. Schubert (1978) Kristallstruktur von Pd_9Te_4. Jour. Less Common Metals, 58, 39–46.

Crystal Data: Orthorhombic. *Point Group:* n.d. As small (115 μm or less) rounded to irregular inclusions in chalcopyrite.

Physical Properties: Hardness = n.d. VHN = 92 (25 g load). D(meas.) = 9.5 (synthetic). D(calc.) = 9.45

Optical Properties: Opaque. *Color:* In polished section, white with a gray tinge. *Luster:* Metallic. *Anisotropism:* Weak in air, stronger in oil, in pale gray to dark gray.
R_1–R_2: (470) 51.8–52.8, (546) 52.9–53.9, (589) 54.2–55.0, (650) 57.1–57.7

Cell Data: *Space Group:* n.d. $a = 11.608$ $b = 12.186$ $c = 6.793$ Z = 6

X-ray Powder Pattern: Synthetic.
2.912 (100), 2.187 (90), 1.959 (70), 1.661 (50), 1.624 (50), 1.462 (50), 1.155 (50)

Chemistry:

	(1)	(2)
Pd	34.9	34.5
Pt		1.0
Hg	22.1	22.0
Bi	n.d.	0.13
Te	42.1	42.1
Total	99.1	99.73

(1) Temagami Island, Canada; by electron microprobe, leading to $Pd_{2.99}Hg_{1.00}Te_{3.01}$.
(2) Stillwater Complex, Montana, USA; by electron microprobe, giving $(Pd_{2.95}Pt_{0.05})_{\Sigma=3.00}Hg_{1.00}Te_{3.00}$.

Occurrence: Cogenetic with moderately high temperature invasive chalcopyrite magma (Temagami Island, Canada).

Association: Merenskyite, hessite, chalcopyrite, stützite.

Distribution: At Temagami Island, Lake Temagami, Temiskaming mining division, Ontario, Canada. In the USA, from the Stillwater Complex, Montana; and the New Rambler mine, Medicine Bow Mountains, Albany Co., Wyoming.

Name: For the Temagami deposit in Canada, in which the mineral was first found.

Type Material: Royal Ontario Museum, Toronto, Canada.

References: (1) Cabri, L.J., J.H.G. Laflamme, and J.M. Stewart (1973) Temagamite, a new palladium–mercury telluride from the Temagami copper deposit, Ontario, Canada. Can. Mineral., 12, 193–198. (2) Cabri, L.J., Ed. (1981) Platinum group elements: mineralogy, geology, recovery. Can. Inst. Min. & Met., 143, 157.

Crystal Data: Cubic. *Point Group:* $\bar{4}3m$. Crystals are tetrahedral, as large as 15 cm; sometimes as groups of parallel crystals; massive, coarse or fine granular to compact. *Twinning:* Twin axis [111], twin plane {111}, as contact and penetration twins, often repeated.

Physical Properties: *Fracture:* Subconchoidal to uneven. *Tenacity:* Somewhat brittle. Hardness = 3–4.5 VHN = 294–380 (100 g load). D(meas.) = 4.62 D(calc.) = 4.61

Optical Properties: Opaque, except in very thin splinters. *Color:* Flint-gray to iron-black; in polished section, gray inclining to black to brown to cherry-red (high As and low Fe); cherry-red in transmitted light. *Luster:* Metallic, often splendent.
Optical Class: Isotropic. $n = > 2.72$ (Li)
R: (400) 31.4, (420) 30.6, (440) 29.8, (460) 29.2, (480) 29.0, (500) 29.0, (520) 28.8, (540) 28.5, (560) 28.0, (580) 27.3, (600) 26.4, (620) 25.6, (640) 24.8, (660) 24.2, (680) 23.8, (700) 23.4

Cell Data: *Space Group:* $I\bar{4}3m$. $a = 10.19$ Z = 2

X-ray Powder Pattern: Binntal, Switzerland.
2.94 (100), 1.801 (80), 1.535 (50), 2.55 (30), 1.169 (30), 1.041 (30), 2.40 (20)

Chemistry:

	(1)	(2)	(3)
Cu	42.05	35.72	51.57
Ag	0.04	13.65	
Zn	6.09	6.90	
Pb		0.86	
Fe	1.48	0.42	
As	12.57	17.18	20.26
Sb	10.87	0.13	
S	27.12	25.04	28.17
Total	100.22	99.90	100.00

(1) San Lorenzo mine, Santiago, Chile. (2) Molly Gibson mine, Colorado, USA. (3) $Cu_{12}As_4S_{13}$.

Polymorphism & Series: Forms series with tetrahedrite and argentotennantite.

Occurrence: Found in hydrothermal veins and contact metamorphic deposits.

Association: Cu-Pb-Zn-Ag sulfides and sulfosalts, pyrite, calcite, dolomite, siderite, barite, fluorite, quartz.

Distribution: From numerous localities; rarer than tetrahedrite, and not often well crystallized. In the USA, in Colorado, at many localities, including the Freeland mine, Idaho Springs, Clear Creek Co.; in the Central City district, Gilpin Co.; at the Molly Gibson and other mines, Aspen, Pitkin Co.; and in the Red Mountain district of San Juan Co. At Butte, Silver Bow Co., Montana. From England, in Cornwall, at Wheal Jewel in Gwennap, Wheal Unity in Gwinear, the East Relistian mine at Carn Brea, and at Roskear and Dolcoath, Camborne. In Germany, from Freiberg, Saxony. In Poland, at Kupferberg-Rudelstadt, Silesia. In Switzerland, at the Lengenbach quarry, Binntal, Valais. At Tsumeb, Namibia, remarkable large crystals. From Morococha and Quiruvilca, Peru. At Concepción del Oro, Zacatecas, Mexico.

Name: For Smithson Tennant (1761–1815), English chemist.

References: (1) Palache, C., H. Berman, and C. Frondel (1944) Dana's system of mineralogy, (7th edition), v. I, 374–384. (2) Berry, L.G. and R.M. Thompson (1962) X-ray powder data for the ore minerals. Geol. Soc. Amer. Mem. 85, 55–56. (3) Weunsch, B.J., Y. Takéuchi, and W. Nowacki (1966) Refinement of the crystal structure of binnite, $Cu_{12}As_4S_{13}$. Zeits. Krist., 123, 1–20.

Crystal Data: Cubic. *Point Group:* 23. As irregular to short prismatic grains (0.07 to 0.10 mm) in sulfide minerals.

Physical Properties: *Cleavage:* Imperfect in two directions. *Tenacity:* Brittle. Hardness = 3.5–4 VHN = 165 (10 g load). D(meas.) = n.d. D(calc.) = 8.991

Optical Properties: Opaque. *Color:* Bright steel-gray with light brown tint; in polished section, bright white with faint blue tint. *Luster:* Metallic, often tarnished yellow-brown.
R: (400) — , (420) — , (440) — , (460) — , (480) 54.4, (500) 54.2, (520) 54.0, (540) 53.9, (560) 54.0, (580) 54.9, (600) 54.0, (620) 54.0, (640) 54.0, (660) 54.7, (680) — , (700) —

Cell Data: *Space Group:* $P2_13$. $a = 6.569(12)$. Z = 4

X-ray Powder Pattern: Locality "Y", China.
2.940 (100), 1.983 (90), 2.680 (80), 1.755 (70), 1.066 (70), 1.267 (60), 1.162 (60)

Chemistry:

	(1)	(2)	(3)
Pd	25	26	26.6
Pt			0.5
Ag			0.2
Ni	1		0.1
Sb	20	21	19.8
Bi	19	17	18.1
Te	35	36	34.4
Total	100	100	99.7

(1) Locality "Y", China; by electron microprobe, corresponding to $(Pd_{0.89}Ni_{0.06})_{\Sigma=0.95}(Sb_{0.63}Bi_{0.34})_{\Sigma=0.97}Te_{1.00}$. (2) Do.; corresponding to $Pd_{0.91}(Sb_{0.65}Bi_{0.30})_{\Sigma=0.95}Te_{1.05}$. (3) Kambalda, Western Australia; by electron microprobe, corresponding to $(Pd_{0.97}Pt_{0.01}Ni_{0.01}Ag_{0.01})_{\Sigma=1.00}(Sb_{0.63}Bi_{0.33})_{\Sigma=0.96}Te_{1.04}$.

Occurrence: In a Cu-Ni sulfide deposit in a serpentine body intruded into a Permian formation of metamorphic rocks (Locality "Y", China).

Association: Michenerite, gersdorffite, cobaltite, pyrrhotite, chalcopyrite, pentlandite (Locality "Y", China); pyrrhotite, pentlandite, chalcopyrite, violarite, cubanite, bornite, sphalerite, galena, linnaeite, magnetite, sudburyite, sperrylite, omeiite, gold (Danba, China).

Distribution: From Locality "Y" (a code name), in southwestern China; also reported from the "W" deposit, in northeastern China. From Danba, Sichuan Province, China. At Kambalda, Western Australia.

Name: For the composition.

Type Material: n.d.

References: (1) Platinum Metal Mineral Research Group, Microprobe Analysis Laboratory, X-ray Powder Laboratory, and Mineral Dressing Laboratory, Kweiyang Institute of Geochemistry, Academia Sinica (1974) Tellurostibnide of palladium and nickel and other new minerals and varieties of platinum metals. Geochimica, 3, 169–181 (in Chinese with English abs.). (2) (1976) Amer. Mineral., 61, 182 (abs. ref. 1). (3) Cabri, L.J., Ed. (1981) Platinum group elements: mineralogy, geology, recovery. Can. Inst. Min. & Met., 143–144.

Crystal Data: Tetragonal. *Point Group:* $4/m\ 2/m\ 2/m$. As irregular grains (600 x 350 x 50 μm), often striated.

Physical Properties: *Tenacity:* Malleable. Hardness = 1.6 VHN = 294 (20 g load). D(meas.) = n.d. D(calc.) = 14.67

Optical Properties: Opaque. *Color:* Golden yellow; copper-red with a yellow tint in reflected light. *Luster:* Metallic. *Anisotropism:* Weak, with colors from gray to light gray.
R: (405) 44.6, (436) 46.8, (480) 50.3, (526) 55.8, (546) 61.2, (578) 74.4, (589) 76.6, (622) 83.6, (644) 84.9, (656) 85.4, (664) 86.6, (700) 91.3

Cell Data: *Space Group:* $C4/mmm$. $a = 3.98$ $c = 3.72$ Z = 2

X-ray Powder Pattern: Sardala, China.
2.24 (100), 1.195 (100), 0.797 (90b), 0.877 (70b), 1.125 (60b), 0.832 (60), 1.99 (50)

Chemistry:

	(1)	(2)
Au	75.18	75.61
Cu	23.74	24.39
Total	98.92	100.00

(1) Sardala, China; by electron microprobe, corresponding to $Cu_{0.99}Au_{1.01}$. (2) AuCu.

Occurrence: In mafic to ultramafic rocks that also contain platinum group elements.

Association: Pyrrhotite, pyrite, chalcopyrite, gold, silver, platinum group minerals, magnetite, chromite, tremolite, diopside, serpentine, chlorite, epidote, apatite, zircon.

Distribution: From Sardala, Xinjiang Autonomous Region, China.

Name: In allusion to the composition and symmetry.

Type Material: Geological Museum, Ministry of Geology, Beijing, China.

References: (1) Chen Keqiao, Yu Tinggao, Zhang Yongge, and Peng Zhizhong (1982) Tetraauricupride, CuAu, discovered in China. Scientia Geologica Sinica, 111–116 (in Chinese with English abs.). (2) (1983) Amer. Mineral., 68, 1250–1251 (abs. ref. 1).

Tetradymite Bi_2Te_2S

Crystal Data: Hexagonal. *Point Group:* $\overline{3}\ 2/m$. Crystals steep pyramidal, resembling hexagonal prisms but rarely distinct; commonly granular, massive to foliated, also bladed. *Twinning:* Fourlings with twin plane $\{01\overline{1}8\}$, also $\{01\overline{1}5\}$.

Physical Properties: *Cleavage:* Perfect on {0001}. *Tenacity:* Laminae flexible but not elastic; slightly sectile. Hardness = 1.5–2 VHN = 30–44 (25 g load). D(meas.) = 7.3 D(calc.) = 7.271

Optical Properties: Opaque. *Color:* Pale steel-gray, tarnishes dull or iridescent; in polished section, white. *Streak:* Pale steel-gray. *Luster:* Metallic, splendent on fresh surfaces. *Anisotropism:* Weak.

R_1–R_2: (400) 50.0–52.7, (420) 50.8–53.6, (440) 51.6–54.5, (460) 52.0–55.0, (480) 52.4–55.4, (500) 52.7–55.8, (520) 53.0–56.2, (540) 53.2–56.4, (560) 53.3–56.5, (580) 53.4–56.6, (600) 53.5–56.7, (620) 53.5–56.8, (640) 53.4–56.9, (660) 53.4–57.0, (680) 53.4–57.1, (700) 53.5–57.2

Cell Data: *Space Group:* $R\overline{3}m$. $a = 4.2381$ $c = 29.589$ Z = 3

X-ray Powder Pattern: Paonia, Delta Co., Colorado, USA.
3.10 (100), 2.292 (100), 2.111 (75), 1.965 (75), 1.929 (75), 1.638 (75b), 1.348 (75)

Chemistry:

	(1)	(2)	(3)	(4)
Bi	59.12	62.23	60.4	59.27
Te	35.94	33.25	34.5	36.19
Se			0.9	
S	4.75	4.50	4.0	4.54
Total	99.81	99.98	99.8	100.00

(1) Ciclova, Romania. (2) Near Bradshaw City, Yavapai Co., Arizona, USA. (3) Žubkov, Czechoslovakia; by electron microprobe. (4) Bi_2Te_2S.

Occurrence: Found in hydrothermal gold-quartz veins of moderate to high temperature of formation; also found in contact metamorphic deposits.

Association: Gold, bismuth, hessite, petzite, calaverite, matildite, altaite, pyrite, pyrrhotite, galena, sphalerite, arsenopyrite, hematite.

Distribution: Occurs in small amounts at numerous localities world-wide; those following are only a sampling. In the USA, at Trail Creek, Blaine Co., Idaho; in New Mexico, in the Sylvanite district, Hidalgo Co.; and in Virginia, at the Whitehall mines, Spotsylvania Co. In Canada, at many localities, as near Liddle Creek, West Kootenay, and at the White Elephant mine, near Vernon, British Columbia; Red Lake at Bigstone Bay, Lake of the Woods, Ontario; in Quebec, at the McWatters mine, Rouyn Township, and at the Eureka mine, Abitibi Co. In Romania, at Băiţa (Rezbánya), also Ciclova, Moraviţa (Moravicza), and Oraviţa (Oravicza). In Czechoslovakia, at Žubkov (Schubkau), as fine twinned crystals. In Norway, at Narverud and Seljord, Telemark. At Boliden, Västerbotten, Sweden.

Name: From the Greek for *fourfold*, an allusion to the twin crystals.

References: (1) Palache, C., H. Berman, and C. Frondel (1944) Dana's system of mineralogy, (7th edition), v. I, 161–164. (2) Berry, L.G. and R.M. Thompson (1962) X-ray powder data for the ore minerals. Geol. Soc. Amer. Mem. 85, 26. (3) Glatz, A.C. (1967) The Bi_2Te_3–Bi_2S_3 system and the synthesis of the mineral tetradymite. Amer. Mineral., 52, 161–170. (4) Criddle, A.J. and C.J. Stanley, Eds. (1986) The quantitative data file for ore minerals. British Museum (Natural History), London, England, 375.

Crystal Data: Tetragonal. *Point Group:* n.d. As tiny grains embedded in other Pt–Fe alloys.

Physical Properties: Hardness = n.d. VHN = n.d. D(meas.) = n.d. D(calc.) = [15.7] Ferromagnetic.

Optical Properties: Opaque. *Color:* White in reflected light. *Luster:* Metallic. *Anisotropism:* Weak, in oil.
R_1–R_2: (470) 61.3, (546) 61.3, (589) 62.2, (650) 63.0

Cell Data: *Space Group:* n.d. $a = 3.850(5)$ $c = 3.693(6)$ Z = [2]

X-ray Powder Pattern: Mooihoek, South Africa.
2.191 (100), 1.920 (70b), 1.342 (60b), 1.152 (50b), 1.099 (30b), 2.721 (20), 1.709 (20)

Chemistry:

	(1)	(2)	(3)
Pt	78.0	77.4	77.74
Rh		0.5	
Ir	0.32		
Fe	14.9	22.0	22.26
Cu	6.1		
Ni	0.50		
Sb	0.40		
Total	100.22	99.9	100.00

(1) Mooihoek, South Africa; by electron microprobe, corresponding to $(Pt_{1.03}Ir_{0.01})_{\Sigma=1.04}(Fe_{0.69}Cu_{0.24}Ni_{0.02}Sb_{0.01})_{\Sigma=0.96}$. (2) Sorashigawa placers, Japan; by electron microprobe, corresponding to $Pt_{0.50}Rh_{<0.01}Fe_{0.50}$. (3) PtFe.

Occurrence: In a hortonolite dunite (Mooihoek, South Africa); in Uralian ultramafics and placers derived therefrom.

Association: Geversite, irarsite, antimonian sperrylite, chalcopyrite, pentlandite, stannopalladinite, chromite, magnetite.

Distribution: In the Mooihoek and Onverwacht mines, Transvaal, South Africa. At the Taimyr mine, Noril'sk region, western Siberia, and placers around Nizhni Tagil, Ural Mountains, USSR. In Canada, in the Tulameen area, British Columbia. At Nye, Sweetwater Co., Montana, USA. From Yubdo, Ethiopia. In the Sorashigawa placers, Japan.

Name: For composition and crystallization in the tetragonal system.

Type Material: National Museum of Natural History, Washington, D.C., USA, 136552.

References: (1) Cabri, L.J. and C.E. Feather (1975) Platinum–iron alloys: A nomenclature based on a study of natural and synthetic alloys. Can. Mineral., 13, 117–126. (2) (1976) Amer. Mineral., 61, 341 (abs. ref. 1). (3) Cabri, L.J., Ed. (1981) Platinum group elements: mineralogy, geology, recovery. Can. Inst. Min. & Met., 144–145. (4) Tarkian, M. (1987) Compositional variations and reflectance of the common platinum-group minerals. Mineral. Petrol., 36, 169–190.

Crystal Data: Cubic. *Point Group:* $\bar{4}3m$. Crystals are tetrahedral, to as much as 15 cm; sometimes as groups of parallel crystals; massive, coarse or fine and granular to compact. *Twinning:* On {111} around [111] as twin axis; contact and penetration twins, often repeated.

Physical Properties: *Fracture:* Subconchoidal. *Tenacity:* Somewhat brittle. Hardness = 3–4.5 VHN = 312–351 (100 g load). D(meas.) = 4.97 D(calc.) = 4.99

Optical Properties: Opaque, except in very thin splinters. *Color:* Flint-gray to iron-black to dull black; in polished section gray inclining to olive-brown; cherry-red in transmitted light. *Streak:* Black to brown. *Luster:* Metallic, often splendent.
Optical Class: Isotropic. $n = > 2.72$ (Li).
R: (400) 30.3, (420) 30.9, (440) 31.5, (460) 31.8, (480) 31.8, (500) 32.0, (520) 32.4, (540) 32.8, (560) 33.0, (580) 33.0, (600) 32.5, (620) 31.6, (640) 31.0, (660) 30.8, (680) 30.6, (700) 30.4

Cell Data: *Space Group:* $I\bar{4}3m$. $a = 10.23–10.55$ Z = 2

X-ray Powder Pattern: Neuglück mine, Wittichen, Black Forest, Germany.
3.00 (100), 1.831 (60), 1.563 (30), 2.61 (20), 1.056 (20), 3.69 (10), 2.46 (10)

Chemistry:

	(1)	(2)	(3)
Cu	45.39	38.95	45.77
Ag		0.02	
Pb	0.11		
Zn		2.21	
Fe	1.32	4.77	
Sb	28.85	27.00	29.22
As	trace	1.40	
S	24.48	25.66	25.01
Total	100.15	100.01	100.00

(1) Bourg d'Oisans, France. (2) Hornachuelas, Spain. (3) $Cu_{12}Sb_4S_{13}$.

Polymorphism & Series: Forms series with tennantite and freibergite.

Occurrence: Usually found in hydrothermal veins or contact metamorphic deposits of low to moderate temperature of formation.

Association: Chalcopyrite, galena, sphalerite, pyrite, bornite, acanthite, calcite, dolomite, siderite, barite, fluorite, quartz.

Distribution: One of the most common of the sulfosalts. Only a few localities which have produced well crystallized material can be mentioned. In the USA, in the Park City district, Summit Co., Utah. In Germany, at Freiberg, Saxony; in the Harz Mountains, at Clausthal; and at Horhausen. From near Brixlegg, Tyrol, Austria. At Botés, near Zlatna, and Cavnic (Kapnik), Romania. From France, at Irazein, Arriége, France, a few exceptionally large crystals; and at Sainte-Marie-aux-Mines, Haut-Rhin. In Algeria, near Tenés and Mouzaía. From Cornwall, England, at the Herodsfoot mine. In Bolivia, at the San José mine, Oruro. From Peru, at Casapalca, Junin. From Noche Buena and Concepción del Oro, Zacatecas, Mexico.

Name: In allusion to the tetrahedral crystal shape.

References: (1) Palache, C., H. Berman, and C. Frondel (1944) Dana's system of mineralogy, (7th edition), v. I, 374–384. (2) Berry, L.G. and R.M. Thompson (1962) X-ray powder data for the ore minerals. Geol. Soc. Amer. Mem. 85, 53–54. (3) Wuensch, B.J., Y. Takéuchi, and W. Nowacki (1966) Refinement of the crystal structure of binnite, $Cu_{12}As_4S_{13}$. Zeits. Krist., 123, 1–20. (4) Johnson, N.E., J.R. Craig, and J.D. Rimstidt (1988) Crystal chemistry of tetrahedrite. Amer. Mineral., 73, 389–397.

Crystal Data: Tetragonal. *Point Group:* $4/m\ 2/m\ 2/m$. As grains (to 50 μm) and as rims (to 20 μm) composed of irregular crystals (to 15 μm) around taenite.

Physical Properties: Hardness = n.d. VHN = 170–200 (25 g load). D(meas.) = n.d. D(calc.) = n.d.

Optical Properties: Opaque. *Color:* Creamy in reflected light. *Luster:* Metallic. *Anisotropism:* Distinct on well-polished surfaces, bluish green to brownish orange.
R_1–R_2: n.d.

Cell Data: *Space Group:* $P4/mmm$. $a = 2.533(2)$ $c = 3.582(2)$ Z = 1

X-ray Powder Pattern: Linville Ni-rich ataxite meteorite.
3.40 (100), 2.879 (80), 2.526 (80), 4.239 (60), 2.279 (10), 2.187 (10), 2.070 (10)

Chemistry:

	(1)	(2)	(3)
Fe	44.00 – 52.00	49.00	48.75
Ni	48.00 – 57.00	51.00	51.25
Cu	0.11 – 0.36	0.20	
Co	< 0.02 – 2.00	0.08	
P	< 0.01	< 0.01	
Total		100.28	100.00

(1) From 18 meteorites; by electron microprobe, range of analyses. (2) Average of same analyses. (3) FeNi.

Occurrence: In slowly cooled meteorites, by the ordering of Fe and Ni atoms in taenite. It is most abundant in mesosiderites and chondrites.

Association: Kamacite, troilite, taenite.

Distribution: Widely distributed in chondrite, mesosiderite, iron and pallas types of meteorites.

Name: In allusion to the symmetry of the mineral, and the genetic link to taenite.

Type Material: National Museum of Natural History, Washington, D.C., USA, meteorite collection 1025.

References: (1) Ramsden, A.R. and E.N. Cameron (1966) Kamacite and taenite superstructures and a metastable tetragonal phase in iron meteorites. Amer. Mineral., 51, 37–55. (2) Clark, R.S., Jr. and E.R.D. Scott (1980) Tetrataenite – ordered FeNi, a new mineral in meteorites. Amer. Mineral., 65, 624–630. (3) Albertson, J.F., G.B. Jensen, and J.M. Knudsen (1978) Structure of taenite in two iron meteorites. Nature, 273, 453–454.

Crystal Data: Tetragonal. *Point Group:* $\bar{4}2m$, $4mm$, 422, or $4/m\ 2/m\ 2/m$. Bladed crystals to 1 mm, most commonly as grains 10–15 μm in diameter.

Physical Properties: *Cleavage:* Parallel to grain elongation. *Tenacity:* Brittle. Hardness = n.d. VHN = 88 (10 g load). D(meas.) = 6.15 (synthetic). D(calc.) = 6.54

Optical Properties: Opaque. *Color:* Bronzy-black; in reflected light, rose-gray to pale lilac. *Luster:* Metallic. *Pleochroism:* Weak, in pale gray with brownish lilac tint to dark gray. *Anisotropism:* Moderate, light yellowish to very dark gray.
R_1–R_2: (400) — , (420) — , (440) 29.3–30.7, (460) 30.4–31.3, (480) 29.6–30.1, (500) 29.8–30.8, (520) 29.7–30.8, (540) 28.4–29.9, (560) 27.9–29.8, (580) 27.9–30.3, (600) 28.4–30.3, (620) 29.2–32.1, (640) 29.7–33.0, (660) 30.6–33.7, (680) 30.7–34.5, (700) 31.7–35.9

Cell Data: *Space Group:* $I\bar{4}md$, $I\bar{4}2m$, $I4mm$, $I422$, or $I4/mmm$. a = 3.882(5) c = 13.25(2) Z = 1

X-ray Powder Pattern: Talnakh area, USSR.
2.91 (100), 2.53 (100), 3.73 (80), 1.941 (50), 1.717 (50), 3.31 (40), 2.18 (40)

Chemistry:

	(1)	(2)	(3)
K		2.30	0.4
Tl	52.2	42.34	42.1
Cu	22.6	26.55	29.4
Fe	9.1	7.43	9.3
Pb		0.24	
Cr		0.17	
Sb		0.23	
As		0.54	
S	16.3	18.17	19.1
Total	100.2	97.97	100.3

(1) Talnakh area, USSR; by electron microprobe, corresponding to $Tl_{2.01}(Cu_{2.79}Fe_{1.28})_{\Sigma=4.07}S_{4.00}$. (2) Murun massif, USSR; by electron microprobe, corresponding to $(Tl_{0.94}K_{0.42})_{\Sigma=1.36}(Cu_{2.995}Fe_{0.94})_{\Sigma=3.935}S_{4.00}$. (3) Ilímaussaq, Greenland; by electron microprobe, corresponding to $(Tl_{1.41}K_{0.07})_{\Sigma=1.48}(Cu_{3.1}Fe_{1.11})_{\Sigma=4.21}S_{4.00}$.

Occurrence: In base-metal sulfide deposits and in high-alkali massifs.

Association: Platinum and palladium minerals, altaite, galena, sphalerite, djerfisherite (Talnakh area, USSR); chalcopyrite, chalcocite, bornite, idaite (Murun massif, USSR); gudmundite, cuprostibite (Ilímaussaq, Greenland).

Distribution: From the Talnakh area, Noril'sk region, western Siberia; and the Murun massif, USSR. From the Ilímaussaq Intrusion, southern Greenland. At Rajpura-Dariba, Rajasthan, India. From Mont Saint-Hilaire, Quebec, Canada.

Name: For the constituents THALlium; CUprum, *copper*; and Sulfur.

Type Material: A.E. Fersman Mineralogical Museum, Academy of Sciences, Moscow, USSR.

References: (1) Kovalenker, V.A., I.P. Laputina, T.L. Evstigneeva, and V.M. Izoitko (1976) Thalcusite, $Cu_{3-x}Tl_2Fe_{1+x}S_4$, a new thallium sulfide from copper–nickel ores of the Talnakh deposits. Zap. Vses. Mineral. Obshch., 105, 202–206 (in Russian). (2) (1977) Amer. Mineral., 62, 396 (abs. ref. 1). (3) Kovalenker, V.A., I.P. Laputina, Y.E. Semenov, and T.L. Evstigneeva (1978) Potassium-bearing thalcusite from the Ilímaussaq pluton and new data on chalcothallite. Doklady Acad. Nauk SSSR, 239, 1203–1206 (in Russian). (4) Dobrovol'skaya, M.G., V.S. Malov, V.P. Rogova, and L.N. Vyal'sov (1982) New occurrence of potassium-containing thalcusite in charoite rocks of the Murun massif. Doklady Acad. Nauk SSSR, 267, 1214–1217 (in Russian). (5) (1983) Chem. Abs., 98, 129389 (abs. ref. 4).

Crystal Data: Cubic. *Point Group:* n.d. Aggregates (0.1–0.3 mm) of grains to 5 μm.

Physical Properties: *Tenacity:* Brittle. Hardness = n.d. VHN = 130–164 (10 g load). D(meas.) = n.d. D(calc.) = 5.26

Optical Properties: Opaque. *Color:* Brown in reflected light, dark brown in oil.
R: (400) 24.2, (440) 23.6, (480) 23.6, (520) 24.3, (560) 25.6, (600) 27.3, (640) 28.7, (700) 30.3

Cell Data: *Space Group:* n.d. $a = 10.29(2)$ Z = 1

X-ray Powder Pattern: Oktyabr mine, USSR.
2.96 (10), 3.42 (9), 3.24 (7), 1.810 (7), 2.35 (6), 4.16 (5), 1.965 (4)

Chemistry:

	(1)	(2)
Tl	33.4	26.1
K	0.03	1.51
Fe	29.4	31.1
Ni	10.3	10.1
Cu	1.74	2.11
S	24.8	26.1
Cl	0.84	1.01
Total	100.51	98.03

(1) Oktyabr mine, USSR; by electron microprobe, average of 15 grains of 2 samples, corresponding to $(Tl_{5.58}K_{0.03})_{\Sigma=5.61}(Fe_{17.96}Ni_{5.99}Cu_{0.93})_{\Sigma=24.88}S_{26.39}Cl_{0.81}$. (2) Do.; the border zone of a zoned grain, corresponding to $(Tl_{4.15}K_{1.52})_{\Sigma=5.67}(Fe_{18.01}Ni_{5.51}Cu_{1.07})_{\Sigma=24.59}S_{26.33}Cl_{0.92}$.

Occurrence: In pentlandite-galena-chalcopyrite ores, localized at the contact of chalcopyrite and galena and included in pentlandite.

Association: Pentlandite, galena, chalcopyrite.

Distribution: In the Oktyabr mine, Talnakh area, Noril'sk region, western Siberia, USSR.

Name: For the principal constituents, THALlium, FErrum for *iron*, NI for *nickel*, and Sulfur.

Type Material: Leningrad Mining Institute, Leningrad, USSR.

References: (1) Rudashevskii, N.S., A.M. Karpenov, G.S. Shipova, N.N. Shishkin, and V.A. Ryabkin (1979) Thalfenisite, the thallium analog of djerfisherite. Zap. Vses. Mineral. Obshch., 108, 696–701 (in Russian). (2) (1981) Amer. Mineral., 66, 219 (abs. ref. 1).

Crystal Data: Cubic. *Point Group:* $\overline{4}3m$. Crystals tetrahedral, to 0.5 cm; {111} is usually dull, {$\overline{1}$11} is bright; frequently striated ∥ [1$\overline{1}$0]. Commonly massive, compact, granular. *Twinning:* Frequent with [111] as twin axis.

Physical Properties: *Fracture:* Uneven to conchoidal. *Tenacity:* Brittle. Hardness = 2.5 VHN = 22–26 (5 g load). D(meas.) = 8.19–8.47 D(calc.) = 8.239

Optical Properties: Opaque. *Color:* Steel-gray to brownish lead-gray; in polished section, grayish white. *Streak:* Nearly black. *Luster:* Metallic.
R: (400) 33.4, (420) 33.6, (440) 33.8, (460) 33.2, (480) 32.2, (500) 31.3, (520) 30.5, (540) 29.8, (560) 29.3, (580) 29.0, (600) 28.7, (620) 28.5, (640) 28.3, (660) 28.2, (680) 28.0, (700) 28.0

Cell Data: *Space Group:* $F\overline{4}3m$. a = 6.085 (synthetic). Z = 4

X-ray Powder Pattern: Synthetic.
3.51 (100), 2.151 (50), 1.835 (30), 3.04 (16), 1.396 (10), 1.2424 (8), 1.521 (6)

Chemistry:

	(1)	(2)	(3)
Hg	75.15	69.84	71.75
Pb	0.12		
Cd		0.34	
Se	24.88	29.19	28.25
S	0.20	0.37	
insol.		0.06	
Total	100.35	99.80	100.00

(1) Clausthal, Germany. (2) Utah, USA. (3) HgSe.

Occurrence: Found in hydrothermal veins, usually with other selenides and calcite.

Association: Barite, manganese oxides, calcite.

Distribution: In the USA, near Marysvale, Piute Co., Utah. At Nicholson Bay, Lake Athabasca, Saskatchewan, Canada. In Germany, at Tilkerode, Lerbach, Zorge, and Clausthal, in the Harz Mountains, and at Niederschlema, Saxony. From Hope's Nose, Torquay, Devon, England. At Předbořice, Czechoslovakia. In Bolivia, at the Pacajake deposit, near Colquechaca, Potosí. In Argentina, in the Sierra de Umango and in the Santa Brigida mine, La Rioja Province. At El Sharana, Northern Territories, Australia. In Mexico, from Rio Blanco, Pinal de Arvole, Artiaga district, and the Cadereyta district, Queretaro, Mexico.

Name: For Johann Carl Wilhelm Ferdinand Tiemann (1848–1899), chemist of Berlin, Germany, who discovered the mineral.

References: (1) Palache, C., H. Berman, and C. Frondel (1944) Dana's system of mineralogy, (7th edition), v. I, 217–218. (2) Early, J.W. (1950) Description and synthesis of the selenide minerals. Amer. Mineral., 35, 337–364. (3) (1957) NBS Circ. 539, 7, 35. (4) Sindeeva, N.D. (1964) Mineralogy and types of deposits of selenium and tellurium, 55–57. (5) Ramdohr, P. (1969) The ore minerals and their intergrowths, (3rd edition), 519–520.

Crystal Data: Tetragonal. *Point Group:* $4/m\ 2/m\ 2/m$. Natural crystals are not known; in irregular, rounded grains or aggregates of grains from 0.1 to 1.5 mm in size.

Physical Properties: *Fracture:* Hackly. *Tenacity:* Ductile, malleable. Hardness = 2 VHN = 7–9 (10 g load). D(meas.) = 7.31 D(calc.) = 7.286

Optical Properties: Opaque. *Color:* Tin-white. *Luster:* Metallic. *Anisotropism:* Moderate. R_1–R_2: (400) 76.3–72.7, (420) 79.4–74.2, (440) 81.9–75.8, (460) 84.0–77.6, (480) 85.5–79.2, (500) 86.7–80.5, (520) 87.5–81.4, (540) 88.1–81.9, (560) 88.2–82.0, (580) 88.0–81.9, (600) 87.7–81.5, (620) 87.2–81.1, (640) 86.7–80.7, (660) 86.3–80.4, (680) 85.8–80.1, (700) 85.4–79.8

Cell Data: *Space Group:* $I4_1/amd$. $a = 5.831$ $c = 3.182$ Z = 4

X-ray Powder Pattern: Synthetic.
2.915 (100), 2.793 (90), 2.017 (74), 2.062 (34), 1.484 (23), 1.442 (20), 1.205 (20)

Chemistry: Tin.

Occurrence: In placer sands (Oban, Australia); in calcite as discrete grains (Beaverlodge, Canada).

Association: Platinum, iridosmine, gold, copper, cassiterite, corundum.

Distribution: In the Nesbitt LaBine uranium mines, Beaverlodge area, Saskatchewan, Canada. On the Aberfoyle and Sam Rivers, near Oban, New South Wales, Australia. At the Elkiaidan River, eastern North Nuratin Range, Uzbekistan, USSR. From the Badiko district, Bauchi, Nigeria. From the Rio Tamaná, the Department of Chocó, Cauca, Colombia. In the Ilímaussaq Intrusion, southern Greenland.

Name: A word of Old English origin, related to the Dutch *tin* and German *zinn*; the chemical symbol from the Latin *stannum*.

References: (1) Palache, C., H. Berman, and C. Frondel (1944) Dana's system of mineralogy, (7th edition), v. I, 126–127. (2) (1953) NBS Circ. 539, 11, 24. (3) Ramdohr, P. (1969) The ore minerals and their intergrowths, (3rd edition), 364–365.

Crystal Data: Orthorhombic. *Point Group:* $2/m\ 2/m\ 2/m$. As small masses up to 2 mm in diameter and as veinlets; also as blades which form parallel aggregates up to 2 mm x 0.5 mm.

Physical Properties: *Cleavage:* Distinct on {010}. Hardness = n.d. VHN = n.d. D(meas.) = 5.48 D(calc.) = 5.51

Optical Properties: Opaque. *Color:* Lead-gray. *Streak:* Black. *Luster:* Metallic. R_1–R_2: n.d.

Cell Data: *Space Group:* *Pnnm*. $a = 22.30$ $b = 34.00$ $c = 4.04$ Z = [1]

X-ray Powder Pattern: Tintina mine, Canada.
3.40 (100), 3.51 (80), 2.71 (70), 2.022 (60), 2.87 (50), 3.96 (40), 3.27 (40)

Chemistry:

	(1)	(2)
Pb	40.97	42.3
Cu	2.21	n.d.
Ag	0.12	n.d.
Sb	34.96	17.6
Bi		23.0
S	21.33	20.0
Total	99.59	102.9

(1) Tintina mine, Canada; by electron microprobe, corresponding to $Pb_{10.6}Cu_{1.8}Ag_{0.2}Sb_{15.3}S_{34.5}$.
(2) Rossland, Canada; by electron microprobe, corresponding to $Pb_{5.0}(Bi_{2.7}Sb_{3.5})_{\Sigma=6.2}S_{15.3}$.

Polymorphism & Series: Forms a series with kobellite.

Occurrence: As small masses and veinlets in malachite-stained quartz (Tintina mines, Canada).

Association: Jamesonite, argentian tetrahedrite, galena, sphalerite, pyrite, arsenopyrite, marcasite, bournonite, owyheeite, pyrrhotite, chalcopyrite (Tintina mines, Canada); joséite, kobellite (Boliden, Sweden).

Distribution: In Canada, found at the Tintina silver mines, Yukon Territory; a bismuthian variety occurs at the Deer Park mine, Rossland, British Columbia. From Boliden, Västerbotten, Sweden.

Name: For the Tintina mines, Canada.

Type Material: Geological Survey of Canada, Ottawa (Tintina material); Royal Ontario Museum, Toronto, Canada (Rossland material).

References: (1) Harris, D.C., J.L. Jambor, G.R. Lachance, and R.I. Thorpe (1968) Tintinaite, the antimony analogue of kobellite. Can. Mineral., 9, 371–382. (2) (1969) Amer. Mineral., 54, 573 (abs. ref. 1). (3) Moëlo, Y., J.L. Jambor, and D.C. Harris (1984) Tintinaïte et sulfosels associés de Tintina (Yukon): la cristallochimie de la série de la kobellite. Can. Mineral., 22, 219–226 (in French with English abs.). (4) (1985) Amer. Mineral., 70, 441 (abs. ref. 3).

Crystal Data: Monoclinic. *Point Group:* 2, *m*, or $2/m$. As radiating aggregates of cylindrical acicular crystals, sometimes hollow, to 1 cm; as paper-thin coatings, felted masses; and as clots up to 10 cm.

Physical Properties: *Cleavage:* {001}. Hardness = n.d. VHN = 15–49 (5 g load). D(meas.) = 2.96 D(calc.) = 3.03

Optical Properties: Opaque. *Color:* Bronze-black; yellowish brown in reflected light. *Luster:* Moderately metallic. *Pleochroism:* Strong, from yellowish brown to yellowish gray. *Anisotropism:* Strong, pinkish-cream to gray.
R_1–R_2: (486) 10.0–13.3, (546) 10.2–14.9, (589) 10.6–15.8, (644) 10.8–17.6

Cell Data: *Space Group:* $P2$, Pm, or $P2/m$. $a = 5.42$ $b = 15.77$ $c = 10.74$ $\beta = 95°$ Z = 6

X-ray Powder Pattern: Voronezh region, USSR.
5.34 (10), 10.68 (9), 1.845 (7b), 2.20 (6b), 2.04 (5b), 1.537 (5vb)

Chemistry:

	(1)
Fe	40.25
FeO	6.56
MgO	16.87
S	23.00
H_2O	11.25
rem.	1.60
Total	99.53

(1) Voronezh region, USSR; by electron microprobe, remainder Al_2O_3, deducted as gibbsite, $Al(OH)_3$, giving $2(Fe_{0.88}S) \cdot 1.67[(Mg_{0.71}Fe_{0.29})_{\Sigma=1.00}(OH)_2]$ by analogy to valleriite.

Occurrence: In sulfide-rich serpentinized ultramafic bodies; in some carbonaceous chondrite meteorites.

Association: Lizardite (Voronezh region, USSR); pyrrhotite, pentlandite, chalcopyrite, djerfisherite, ilmenite (Yakutia, USSR); forsterite, perovskite, clinohumite, spinel, hydromagnesite, apatite, mica, graphite, pyrrhotite, calcite (Bancroft, Canada); vesuvianite, tilleyite, titaniferous andradite, calcite (Kamaishi mine, Japan).

Distribution: From the Lower Mamon and Staromelovatskii intrusives, Voronezh region; and in the Mir, Udachnaya, and Obnazhennaya diamond pipes, Yakutia, USSR. In the Grace mine, Morgantown, Berks Co., and the Cornwall open pit, Lebanon Co., Pennsylvania, USA. In the Cross and Maxwell quarries, near Wakefield; in the Jeffrey mine, Asbestos; and the Amos area, Quebec; in the Muskox Intrusion, Northwest Territories; and near Bancroft, Ontario, Canada. In the Kamaishi mine, Iwate Prefecture, Japan. At Mt. Keith, 15 km north of Betheno, Western Australia. In the Jacupiranga mine, São Paulo, Brazil. From the Lizard ultramafic body, Cornwall, England.

Name: For Mitrofan Stepanovich Tochilin, Professor at Voronezh University, USSR.

Type Material: A.E. Fersman Mineralogical Museum, Academy of Sciences, Moscow, USSR.

References: (1) Organova, N.I., A.D. Genkin, V.A. Drits, S.P. Molotkov, O.V. Kuz'mina, and A.L. Dmitrik (1971) Tochilinite, a new sulfide-hydroxide of iron and magnesium. Zap. Vses. Mineral. Obshch., 100, 477–487 (in Russian). (2) (1972) Amer. Mineral., 57, 1552 (abs. ref. 1). (3) Muramatsu, Y. and M. Nambu (1980) Tochilinite and cuprian tochilinite from the Kamaishi mine, Iwate Prefecture, Japan. Ganseki Kobutsu Kosho Gakkaishi, 75, 377-384 (in English). (4) (1982) Chem. Abs., 97, 58731 (abs. ref. 3). (5) Zolensky, M.E. and I.D.R. Mackinnon (1986) Microstructures of cylindrical tochilinites. Amer. Mineral., 71, 1201–1209.

Crystal Data: Cubic. *Point Group:* n.d. As aggregates of corroded, anhedral grains (to 18 μm) that define a skeletal isometric outline (up to 50 x 72 μm).

Physical Properties: *Fracture:* Conchoidal. *Tenacity:* Brittle. Hardness = Very hard. VHN = 1431–1703 (10 g load). D(meas.) = n.d. D(calc.) = 10.50

Optical Properties: Opaque. *Color:* Steel-gray; gray with a pale brown tint in reflected light. *Luster:* Metallic.

R: (442) 36.7, (468) 39.0, (484) 39.9, (525) 42.0, (554) 42.5, (586) 42.9, (621) 43.5, (666) 43.8, (699) 44.0

Cell Data: *Space Group:* n.d. $a = 6.027(3)$ Z = 4

X-ray Powder Pattern: Tolovka River, USSR.
1.813 (100), 2.99 (90), 1.146 (90), 1.065 (90), 2.126 (80), 1.005(80), 1.233 (70)

Chemistry:

	(1)	(2)	(3)
Ir	55.60	55.00	55.55
Pt	0.25	0.69	
Os	0.12	0.49	
Ni	0.06	0.06	
Sb	35.00	34.70	35.19
S	9.22	9.20	9.26
Total	100.25	100.14	100.00

(1–2) Tolovka River, USSR; by electron microprobe, the average of which leads to $(Ir_{0.993}Pt_{0.009}Os_{0.006})_{\Sigma=1.008}Sb_{0.993}S_{0.993}$. (3) IrSbS.

Occurrence: In placers derived from an Alpine-type gabbro massif (Ust'-Bel'skii massif, USSR).

Association: Os–Ir alloys.

Distribution: From the Ust'-Bel'skii massif, Tolovka River, northeastern USSR. In the Similkameen River, British Columbia, Canada. At Fox Gulch, Goodnews Bay, Alaska, USA.

Name: For the type locality near the Tolovka River, USSR.

Type Material: The Mining Museum, Leningrad Mining Institute, Leningrad, USSR.

References: (1) Razin, L.V., N.S. Rudashevskii, and G.A. Sidorenko (1981) Tolovkite, IrSbS, a new sulfoantimonide of iridium from northeastern USSR. Zap. Vses. Mineral. Obshch., 110, 474–480 (in Russian). (2) (1982) Amer. Mineral., 67, 1076–1077 (abs. ref. 1).

Crystal Data: Monoclinic. *Point Group:* $2/m$, 2, or m. As tiny grains.

Physical Properties: Hardness = n.d. VHN = n.d. D(meas.) = n.d. D(calc.) = 7.25

Optical Properties: Opaque. *Color:* In polished section, white. *Anisotropism:* Distinct to strong, in light gray to steel bluish black.
R_1–R_2: n.d.

Cell Data: *Space Group:* $B2/m$, $B2$, or Bm. a = 13.349(10) b = 26.538(20) c = 4.092(7) β = 92.77(7)° Z = 4

X-ray Powder Pattern: Treasury (sic) mine, Colorado, USA.
3.49 (100), 3.22 (80), 1.989 (60), 1.955 (60), 3.63 (50), 2.93 (50), 2.86 (50)

Chemistry:

	(1)	(2)
Ag	12.7	12.26
Pb	19.6	20.18
Bi	50.5	50.90
S	16.4	16.66
Total	99.2	100.00

(1) Treasury (sic) mine, Colorado, USA; by electron microprobe, corresponding to $Ag_{1.82}Pb_{1.46}Bi_{3.73}S_{7.89}$. (2) $Ag_7Pb_6Bi_{15}S_{32}$.

Occurrence: In hydrothermal vein material.

Association: A fine-grained decomposition product of treasurite having very similar optical properties (Treasure Vault mine, Colorado, USA).

Distribution: From the Treasure Vault (misnamed Treasury) mine, Geneva district, Park and Summit Cos., Colorado; and from a prospect, 10 km southwest of Tyrone, Grant Co., New Mexico, USA. In the Kochbulak deposit, eastern Uzbekistan, USSR.

Name: For the Treasure Vault lode, Colorado, USA, in which it occurs.

Type Material: National Museum of Natural History, Washington, D.C., USA, R9714.

References: (1) Karup-Møller, S. (1977) Mineralogy of some Ag–(Cu)–Pb–Bi sulfide associations. Bull. Geol. Soc. Denmark, 26, 41–68. (2) Makovicky, E. and S. Karup-Møller (1977) Chemistry and crystallography of the lillianite homologeous series. Neues Jahrb. Mineral., Abh., 131, 56–82. (3) (1979) Amer. Mineral., 64, 243 (abs. refs. 1 and 2).

Crystal Data: Hexagonal. *Point Group:* $\bar{3}$. Crystals short prismatic; equant, also irregular.

Physical Properties: *Cleavage:* Good on $\{10\bar{1}1\}$, distinct on $\{0001\}$. *Fracture:* Conchoidal. *Tenacity:* Brittle. Hardness = 1.5–2 VHN = 97–100 (25 g load). D(meas.) = n.d. D(calc.) = 4.77 (synthetic).

Optical Properties: Opaque except in thin pieces. *Color:* Scarlet-vermilion; in polished section, white having a bluish cast, with orange-red internal reflections. *Streak:* Scarlet-vermilion. *Luster:* Adamantine.

Optical Class: Uniaxial negative (–). *Pleochroism:* In transmitted light, faint, O = pale reddish; E = clear, near colorless. $n = > 2.61$ (Li). *Anisotropism:* Moderate (synthetic).

R_1–R_2: (400) 37.6–26.2, (420) 36.9–25.7, (440) 35.5–24.9, (460) 33.8–23.9, (480) 32.6–23.0, (500) 31.6–23.4, (520) 30.7–23.7, (540) 29.9–23.4, (560) 29.1–22.7, (580) 28.5–22.1, (600) 28.0–21.5, (620) 27.6–21.0, (640) 27.2–20.6, (660) 26.9–20.3, (680) 26.4–20.0, (700) 26.2–19.9

Cell Data: *Space Group:* $R\bar{3}$. $a = 13.98$ $c = 9.12$ Z = 18

X-ray Powder Pattern: Binntal, Switzerland. (JCPDS 16-700).
2.702 (100), 3.15 (80), 1.887 (80), 1.937 (70), 7.0 (60), 4.26 (60), 3.64 (60)

Chemistry:

	(1)	(2)
Ag	43.9	43.69
As	30.8	30.34
S	26.1	25.97
Total	100.8	100.00

(1) Binntal, Switzerland; by electron microprobe. (2) $AgAsS_2$.

Polymorphism & Series: Dimorphous with smithite.

Occurrence: Of hydrothermal origin, in dolomite.

Association: Seligmannite, tennantite, pyrite, chromian muscovite.

Distribution: From the Lengenbach quarry, Binntal, Valais, Switzerland. At Niederbeerbach, Odenwald, Germany.

Name: For Dr. Charles O. Trechmann (1851–1917), English crystallographer.

Type Material: n.d.

References: (1) Palache, C., H. Berman, and C. Frondel (1944) Dana's system of mineralogy, (7th edition), v. I, 432–433. (2) Roland, G.W. (1968) Synthetic trechmannite. Amer. Mineral., 53, 1208–1214. (3) Matsumo, T. and W. Nowacki (1969) The crystal structure of trechmannite, $AgAsS_2$. Zeits. Krist., 129, 163–177.

Crystal Data: Cubic. *Point Group:* $2/m\ \overline{3}$. Intergrown with clausthalite as irregular grains.

Physical Properties: Hardness = ~7 VHN = n.d. D(meas.) = n.d. D(calc.) = 7.12

Optical Properties: Opaque. *Color:* Rose-violet. *Luster:* Metallic.
R: (400) 42.5, (420) 42.2, (440) 41.9, (460) 41.4, (480) 41.0, (500) 40.8, (520) 41.1, (540) 41.3, (560) 42.0, (580) 42.6, (600) 43.3, (620) 43.9, (640) 44.5, (660) 45.2, (680) 45.6, (700) 45.8

Cell Data: *Space Group:* *Pa*3. $a = 5.87(2)$ Z = 4

X-ray Powder Pattern: Musonoi mine, Zaire.
2.64 (100), 2.419 (90), 1.788 (80), 2.95 (70), 1.585 (70), 1.644 (50), 1.295 (50)

Chemistry:

	(1)	(2)
Co	17.09	27.18
Cu	9.02	
Pd	4.67	
Se	69.77	72.82
Total	100.55	100.00

(1) Musonoi mine, Zaire; by electron microprobe, corresponding to a cuproan-palladian variety $(Co_{0.65}Cu_{0.32}Pd_{0.09})_{\Sigma=1.06}Se_{2.00}$. (2) $CoSe_2$.

Polymorphism & Series: Dimorphous with hastite.

Occurrence: Intergrown with clausthalite.

Association: Clausthalite, hastite, bornhardtite, selenium, gold (Trogtal quarry, Germany).

Distribution: At the Trogtal quarry, near Lautenthal, Harz, Germany. From the Musonoi mine, Shaba Province, Zaire.

Name: For the occurrence at the Trogtal quarry, Germany.

Type Material: n.d.

References: (1) Ramdohr, P. and M. Schmitt (1955) Vier neue natürliche Kobaltselenide vom Steinbruch Trogtal bei Lautenthal im Harz. Neues Jahrb. Mineral., Monatsh., 133–142 (in German). (2) (1956) Amer. Mineral., 41, 164–165 (abs. ref. 1). (3) Johan, Z., P. Picot, R. Pierrot, and T. Verbeek (1970) L'oosterboschite $(Pd, Cu)_7Se_5$, une nouvelle espèce minérale et la trogtalite cupro-palladifère de Musonoï (Katanga). Bull. Soc. fr. Minéral., 93, 476–481 (in French with English abs.).

Crystal Data: Hexagonal. *Point Group:* $6/m\ 2/m\ 2/m$. Massive, granular; nodular.

Physical Properties: Hardness = 3.5–4.5 VHN = 250(3) D(meas.) = 4.67–4.79 D(calc.) = 4.85

Optical Properties: Opaque. *Color:* Light grayish brown, tarnishes rapidly on exposure. *Streak:* Dark grayish brown. *Luster:* Metallic. *Anisotropism:* Strong.
R_1–R_2: n.d.

Cell Data: *Space Group:* $P6_3/mmc$. $a = 5.958$ $c = 11.74$ Z = 12

X-ray Powder Pattern: Del Norte Co., California, USA.
2.09 (100), 2.66 (60), 1.719 (50), 2.98 (40), 1.331 (40), 1.119 (40), 1.923 (30)

Chemistry:

	(1)	(2)	(3)
Fe	62.70	63.0	63.53
S	35.40	35.0	36.47
Total	98.10	98.0	100.00

(1) Del Norte Co., California, USA. (2) Cranbourne meteorite. (3) FeS.

Occurrence: Found in serpentine (Del Norte Co., California, USA); with Fe-Cu-Ni sulfides in a layered ultramafic intrusive (Sally Malay deposit, Australia); and as nodules in meteorites.

Association: Pyrrhotite, pentlandite, mackinawite, cubanite, valleriite, chalcopyrite, pyrite (Wannaway deposit, Australia); daubréelite, chromite, sphalerite, graphite, various phosphates and silicates (meteorites).

Distribution: From the Alta mine, Del Norte Co., California, USA. In the Wannaway Fe-Ni-Cu deposit, and at the Sally Malay Cu-Ni deposit, 120 km north of Halls Creek, Western Australia. From Disco Island and the Ilímaussaq Intrusion, southern Greenland. In many meteorites.

Name: After Dominico Troili, who described, in 1766, a meteorite which fell in Albareto, near Modena, Italy, which contained the species.

References: (1) Palache, C., H. Berman, and C. Frondel (1944) Dana's system of mineralogy, (7th edition), v. I, 231–234. (2) Berry, L.G. and R.M. Thompson (1962) X-ray powder data for the ore minerals. Geol. Soc. Amer. Mem. 85, 61. (3) Ramdohr, P. (1969) The ore minerals and their intergrowths, (3rd edition), 582–601. (4) Evans, H.T., Jr. (1970) Lunar troilite: crystallography. Science, 20, 621–623. (5) Buchwald, V.F. (1977) The mineralogy of iron meteorites. Phil. Trans. Royal Soc. London, A. 286, 453–491.

Crystal Data: Cubic. *Point Group:* $4/m\ \overline{3}\ 2/m$. As minute euhedral crystals imbedded in clausthalite.

Physical Properties: Hardness = ~2.5 VHN = n.d. D(meas.) = n.d. D(calc.) = 6.62

Optical Properties: Opaque. *Color:* Yellow. *Luster:* Metallic.
R: (400) 42.2, (420) 43.6, (440) 45.0, (460) 46.4, (480) 47.5, (500) 48.5, (520) 49.4, (540) 50.3, (560) 51.2, (580) 52.0, (600) 52.6, (620) 53.1, (640) 53.8, (660) 54.3, (680) 54.9, (700) 55.4

Cell Data: *Space Group:* $Fd3m$, probable. $a = 9.94$ Z = 8

X-ray Powder Pattern: Kuusamo, northeast Finland.
2.48 (100), 1.755 (100), 3.00 (80), 2.87 (80), 1.905 (60), 5.75 (40), 3.52 (40)

Chemistry:

	(1)	(2)
Ni	29.5	35.80
Co	6.4	
Cu	trace	
Se	64.4	64.20
S	trace	
Total	100.3	100.00

(1) Kuusamo, Finland; by X-ray fluorescence analysis. (2) Ni_3Se_4.

Polymorphism & Series: Dimorphous with wilkmanite.

Occurrence: In calcite- and uranium-bearing veins in sills of albite diabase in schist.

Association: Clausthalite, penroseite, sederholmite, wilkmanite, kullerudite.

Distribution: From Kuusamo, northeastern Finland.

Name: For O. Trüstedt, whose work on prospecting methods lead to the discovery of the Outokumpu ore deposit, Finland.

Type Material: n.d.

References: (1) Vuorelainen, Y., A. Huhma, and A. Häkli (1964) Sederholmite, wilkmanite, kullerudite, mäkinenite, and trüstedtite, five new nickel selenide minerals. Compt. Rendus Soc. Géol. Finlande, 36, 113–125. (2) (1965) Amer. Mineral., 50, 519–520 (abs. ref. 1).

Tsumoite BiTe

Crystal Data: Hexagonal. *Point Group:* $\bar{3}\ 2/m$. As irregular aggregates, up to 1 cm, which include tsumoite tablets a few mm across in their centers.

Physical Properties: *Cleavage:* Perfect on {0001}. Hardness = n.d. VHN = 51–90 (15 g load). D(meas.) = 8.16(5) D(calc.) = 8.23

Optical Properties: Opaque. *Color:* Silver-white; in polished section, white with a creamy tint. *Streak:* Steel-gray. *Luster:* Metallic. *Pleochroism:* Very weak. *Anisotropism:* Moderate. R_1–R_2: n.d.

Cell Data: *Space Group:* $P\bar{3}m1$. $a = 4.422(2)$ $c = 24.05(2)$ Z = 6

X-ray Powder Pattern: Tsumo mine, Japan.
3.23 (vs), 2.36 (s), 2.21 (s), 1.825 (s), 1.487 (s), 2.01 (m), 1.617 (m)

Chemistry:

	(1)	(2)
Bi	61.1	62.09
Pb	1.0	
Te	37.6	37.91
Total	99.7	100.00

(1) Tsumo mine, Japan; by electron microprobe, corresponding to $(Bi_{0.99}Pb_{0.02})_{\Sigma=1.01}Te_{1.00}$.
(2) BiTe.

Occurrence: As small irregular aggregates in skarns associated with pyrometasomatic Cu-Zn-Pb deposits (Tsumo mine, Japan).

Association: Tetradymite, cosalite, bismuthinite, galena (Magadan region, USSR); pilsenite (Sylvanite, New Mexico, USA).

Distribution: From Japan, in the Tsumo mine, about 50 km northwest of Hiroshima City, Akita Prefecture. At an undefined locality in the Magadan region, Yakutia, and from Tyrong Auz, Caucasus Mountains, USSR. From Sylvanite, Hidalgo Co., New Mexico, USA.

Name: For the Tsumo mine, Japan, in which it was first found.

Type Material: University Museum, University of Tokyo, Tokyo, Japan.

References: (1) Shimazaki, H. and T. Ozawa (1978) Tsumoite, BiTe, a new mineral from the Tsumo mine, Japan. Amer. Mineral., 63, 1162–1165.

Crystal Data: Tetragonal. *Point Group:* $4/m\ 2/m\ 2/m$. As tiny grains, and as rims and irregular grains partly replacing millerite.

Physical Properties: *Fracture:* Conchoidal. *Tenacity:* Brittle. Hardness = n.d. VHN = 718 (20 g load). D(meas.) = n.d. D(calc.) = 6.15

Optical Properties: Opaque. *Color:* Pale brass-yellow; in polished section, brownish yellow. *Luster:* Metallic. *Pleochroism:* Very weak to absent. *Anisotropism:* Very strong to distinct, deep brown to grayish blue.
R_1–R_2: (400) 37.5–38.2, (420) 38.0–39.4, (440) 38.6–41.0, (460) 39.5–42.5, (480) 40.7–44.0, (500) 42.3–45.6, (520) 44.0–47.0, (540) 45.4–48.0, (560) 46.6–48.8, (580) 47.6–49.4, (600) 48.4–50.1, (620) 49.2–50.7, (640) 49.9–51.4, (660) 50.6–52.0, (680) 51.2–52.6, (700) 51.8–53.1

Cell Data: *Space Group:* $P4/mmm$. $a = 7.174(6)$ $c = 5.402(7)$ Z = 1

X-ray Powder Pattern: Witwatersrand, South Africa.
2.76 (100), 2.38 (80), 2.28 (80b), 1.850 (80), 4.33 (70), 1.793 (70), 3.21 (60)

Chemistry:

	(1)	(2)	(3)	(4)
Ni	47.34	47.8	46.8	51.38
Co	1.06		4.51	
Fe	3.61	3.75	0.10	
As	0.86	1.34	1.74	
Sb	21.62	21.87	21.8	23.68
Bi	1.84	1.02		
Te	0.30			
S	25.19	25.13	24.9	24.94
Total	101.81	100.91	99.85	100.00

(1) Kanowna, Australia; by electron microprobe, estimates of a range of measurements, leading to $(Ni_{8.21}Fe_{0.66}Co_{0.18})_{\Sigma=9.05}(Sb_{0.89}Bi_{0.09}Te_{0.02})_{\Sigma=1.00}(Sb_{0.92}As_{0.12})_{\Sigma=1.04}S_{8.00}$. (2) Witwatersrand, South Africa; by electron microprobe, average of a range of measurements, corresponding to $(Ni_{8.31}Fe_{0.69})_{\Sigma=9.00}(Sb_{0.95}Bi_{0.05})_{\Sigma=1.00}(Sb_{0.88}As_{0.18})_{\Sigma=1.06}S_{8.00}$. (3) Vozhmin massif, USSR; by electron microprobe. (4) $Ni_9Sb_2S_8$.

Occurrence: In a hydrothermal deposit (Kanowna, Australia); in a heavy mineral concentrate (Witwatersrand, South Africa); in serpentinites (Vozhmin massif, USSR).

Association: Millerite, pyrite, chalcopyrite, gersdorffite, pentlandite, magnetite, polydymite (Kanowna, Australia); gold, dyscrasite, michenerite, geversite, tetrahedrite, stibnite, sudburyite, stibiopalladinite (Klerkdorp, South Africa); vozhminite, heazlewoodite, magnetite, geversite, copper (Vozhmin massif, USSR).

Distribution: In Australia, at Kanowna, near Kalgoorlie, and the Whim Creek deposit, Western Australia; and from the Central Balstrup lease, Zeehan, Tasmania. In South Africa, in the Witwatersrand at the Vaal Reefs mine, Klerkdorp. In the Vozhmin massif, northeastern Karelia, USSR. At Rocheservières, Vendeé, France.

Name: For Dr. Karel Tuček, Curator of Minerals in the National Museum, Prague, Czechoslovakia.

Type Material: National Museum of Natural History, Washington, D.C., USA, 146920, 146921

References: (1) Just, J. and C.E. Feather (1978) Tučekite, a new mineral. Mineral. Mag., 42, M21–M22. (2) (1979) Amer. Mineral., 64, 465 (abs. ref. 1). (3) Rudashevskii, N.S., N.I. Shumskaya, A.G. Tutov, S.N. Soshnikova, A.B. Lobanova, and G.N. Goncharov (1979) First occurrence of tucekite in the USSR. Doklady Acad. Nauk SSSR, 249, 181–185 (in Russian). (4) (1980) Chem. Abs., 92, 79585 (abs. ref. 3).

Tulameenite Pt_2FeCu

Crystal Data: Tetragonal. *Point Group:* n.d. As rounded to irregular areas up to about 400 μm in diameter, as free grains, or as grains with complex inclusions, associated with Pt–Fe alloys.

Physical Properties: Hardness = n.d. VHN = 420–456, 442 average (50 g load). D(meas.) = 14.9 (synthetic). D(calc.) = 15.62 Distinctly ferromagnetic.

Optical Properties: Opaque. *Color:* White in reflected light. *Luster:* Metallic. *Anisotropism:* Very weak.

R_1–R_2: (470) 61.0–65.3, (546) 60.0–66.5, (589) 61.5–65.5, (650) 61.1–64.9

Cell Data: *Space Group:* n.d. $a = 3.891(2)$ $c = 3.577(2)$ Z = 2

X-ray Powder Pattern: Tulameen River, Canada.
2.179 (100), 1.163 (80), 1.093 (80), 1.946 (70), 1.016 (60), 1.317 (50), 2.753 (40)

Chemistry:

	(1)	(2)	(3)
Pt	73.98	76.7	76.57
Ir	1.99		
Fe	10.38	10.6	10.96
Cu	13.13	7.0	12.47
Ni	n.d.	3.8	
Sb	n.d.	2.1	
Total	99.48	100.2	100.00

(1) Similkameen River area, Canada; by electron microprobe, corresponding to $(Pt_{1.94}Ir_{0.06})_{\Sigma=2.00}Fe_{1.06}Cu_{0.94}$. (2) Tulameen River area, Canada; by electron microprobe, corresponding to $Pt_{2.04}Fe_{0.98}(Cu_{0.56}Ni_{0.54}Sb_{0.08})_{\Sigma=1.18}$. (3) Pt_2FeCu.

Occurrence: In placers (Canada); in Uralian ultramfics (USSR).

Association: Pt–Fe alloys, geversite, chalcopyrite, chromite, magnetite.

Distribution: From placers in the Tulameen and Similkameen River areas, British Columbia, Canada. In the Stillwater Complex, Montana, USA. From Guma Water, Sierra Leone. At Yubdo, Ethiopia. From Nizhni Tagil, Ural Mountains, USSR.

Name: For the Tulameen River, Canada, from the vicinity of which the mineral was first noted.

Type Material: National Museum of Natural History, Washington, D.C., USA, 128460.

References: (1) Cabri, L.J., D.R. Owens, and J.H.G. Laflamme (1973) Tulameenite, a new platinum–iron–copper mineral from placers in the Tulameen River area, British Columbia. Can. Mineral., 12, 21–25. (2) (1974) Amer. Mineral., 59, 383 (abs. ref. 1). (3) Cabri, L.J., Ed. (1981) Platinum group elements: mineralogy, geology, recovery. Can. Inst. Min. & Met., 145.

Crystal Data: Hexagonal. *Point Group:* $6/m\ 2/m\ 2/m$ (2H polymorph); $3m$ (3R polymorph). Massive, fine scaly or feathery aggregates.

Physical Properties: *Cleavage:* On {0001}. *Tenacity:* Sectile. Hardness = 2.5 VHN = n.d. D(meas.) = 7.4 D(calc.) = 7.732 Soils the fingers.

Optical Properties: Opaque. *Color:* Dark lead-gray; in polished section, pure white. *Luster:* Metallic. *Pleochroism:* Very high. *Anisotropism:* Strong, in striking colors including pink and deep blue.

R_1–R_2: (400) 24.4–53.4, (420) 22.6–48.0, (440) 20.8–48.6, (460) 19.9–40.5, (480) 19.3–39.0, (500) 18.8–37.8, (520) 18.5–36.8, (540) 18.2–36.0, (560) 18.0–35.3, (580) 17.8–35.0, (600) 17.8–34.9, (620) 17.9–34.9, (640) 18.0–34.8, (660) 18.1–34.7, (680) 18.0–34.5, (700) 17.9–34.2

Cell Data: *Space Group:* $P6_3/mmc$ (2H polymorph), with: $a = 3.154$ $c = 12.362$ Z = 2, or $R3m$ (3R polymorph), with: $a = 3.1492$ $c = 18.434$ Z = 3

X-ray Powder Pattern: Synthetic (2H polymorph).
6.18 (100), 2.2772 (35), 2.731 (25), 2.667 (25), 1.8335 (18), 1.5783 (16), 3.089 (14)

X-ray Powder Pattern: Emma mine, Utah, USA; 3R polymorph. (JCPDS 35-651).
6.13 (100), 1.528 (70), 2.674 (60), 1.571 (50), 2.037 (40), 1.100 (40), 3.04 (35)

Chemistry:

	(1)	(2)
W	73.8	74.14
S	25.5	25.86
Total	99.3	100.00

(1) Angokitsk deposit, USSR; by electron microprobe. (2) WS_2.

Polymorphism & Series: Both 2H and 3R polymorphs occur naturally.

Occurrence: In a deposit replacing limestone (Emma mine, Utah, USA); replacing scheelite (Angokitsk deposit, USSR).

Association: Pyrite, sphalerite, galena, tetrahedrite, wolframite, scheelite, quartz.

Distribution: In the Emma mine, Little Cottonwood Canyon, Salt Lake Co., Utah, USA (2H and 3R polymorphs). From Tsumeb, Namibia. Near Chase, British Columbia, and in the Kidd Creek mine, near Timmins, Ontario, Canada. At Crevola d'Ossola, Piedmont, Italy. In the Angokitsk tungsten deposit, Buryat ASSR, and in the Tamvatnei deposit, Kamchatka, USSR.

Name: For the composition.

Type Material: National Museum of Natural History, Washington, D.C., USA, 94490.

References: (1) Palache, C., H. Berman, and C. Frondel (1944) Dana's system of mineralogy, (7th edition), v. I, 331–332. (2) (1959) NBS Circ. 539, 8, 65. (3) Ramdohr, P. (1969) The ore minerals and their intergrowths, (3rd edition), 866. (4) Gait, R.I. and J.A. Mandarino (1970) Polytypes of tungstenite. Can. Mineral., 10, 729–730. (5) Getmanskaya, T.I., E.G. Ryabeva, G.A. Sidorenko, and K.V. Yurkina (1979) Tungstenite — a new discovery in the USSR. Doklady Acad. Nauk SSSR, 247, 194–198 (in Russian). (6) (1979) Chem. Abs., 91, 196073 (abs. ref. 5).

Crystal Data: Monoclinic. *Point Group:* n.d. As granular aggregates, without well formed crystals.

Physical Properties: *Cleavage:* Perfect in one direction. *Fracture:* Uneven and conchoidal. Hardness = < 3 VHN = 172 D(meas.) = 7.380(5) D(calc.) = 7.74

Optical Properties: Translucent. *Color:* Lead-gray; whitish in reflected light, with strong dark red internal reflections; in transmitted light, cherry-red. *Streak:* Nearly black, with a dark reddish tint. *Luster:* Adamantine. *Pleochroism:* White to grayish white, weak lilac to a distinctly greenish tint. *Anisotropism:* Distinct, dark olive to dark violet.
R_1–R_2: n.d.

Cell Data: *Space Group:* n.d. $a = 11.51(4)$ $b = 4.39(2)$ $c = 14.62(6)$ $\beta = 92.14°$ Z = 1

X-ray Powder Pattern: Gomi deposit, USSR.
3.49 (100), 2.92 (100), 2.89 (100), 2.080 (100), 3.29 (70), 2.031 (35), 3.19 (30)

Chemistry:

	(1)	(2)	(3)
Hg	64.60	64.94	66.1
Cu			0.15
Sb	14.32	14.12	12.9
As	7.60	7.75	8.54
S	12.83	12.71	10.3
Total	99.35	99.52	97.99

(1) Gomi deposit, USSR; by electron microprobe, corresponding to $Hg_{12}(Sb_{4.38}As_{3.78})_{\Sigma=8.16}S_{14.92}$. (2) Do.; corresponds to $Hg_{12}(Sb_{4.30}As_{3.84})_{\Sigma=8.14}S_{14.69}$. (3) Tyute deposit, USSR; by electron microprobe.

Occurrence: In a hydrothermal As-Sb-Hg deposit (Gomi deposit, USSR).

Association: Cinnabar, metacinnabar, realgar, dickite (Gomi deposit, USSR).

Distribution: In the Gomi deposit, Caucasus Mountains, Georgian SSR, and the Tyute mercury deposit, Altai Mountains, USSR. From the Hemlo gold deposit, Thunder Bay district, Ontario, Canada. In the Getchell mine, Humboldt Co., Nevada.

Name: For A.A. Tvalchrelidze, founder of the Georgian Mineralogical-Petrographic School, USSR.

Type Material: A.E. Fersman Mineralogical Museum, Academy of Sciences, Moscow, USSR; National Museum of Natural History, Washington, D.C., USA, 143827.

References: (1) Gruzdev, V.S., N.M. Mchedlishvili, G.A. Terekhova, Z.Y. Tsertsvadze, N.M. Chernitsova, and N.G. Shumkova (1975) Tvalchrelidzeite, $Hg_{12}(Sb,As)_8S_{15}$, a new mineral from the Gomi arsenic–antimony–mercury deposit, Caucasus. Doklady Acad. Nauk SSSR, 225, 911–913 (in Russian). (2) (1977) Amer. Mineral., 62, 174 (abs. ref. 1). (3) Vasil'ev, V.I. (1979) The second discovery of tvalchrelidzeite $Hg_{12}(Sb,As)_8S_{15}$ in ores of mercury deposits. Geol. Geofiz., 9, 159–162 (in Russian with English abs.). (4) (1980) Chem. Abs., 92, 25693 (abs. ref. 3).

Crystal Data: Triclinic probable, pseudo-orthorhombic. *Point Group:* $2/m\ 2/m\ 2/m$, apparent. As grains rarely up to 1.5 mm in diameter. *Twinning:* Polysynthetic, the trace of the composition plane being parallel to {100}.

Physical Properties: *Cleavage:* Perfect on {100}. *Tenacity:* Brittle. Hardness = n.d. VHN = 147 (50 g load). D(meas.) = n.d. D(calc.) = 5.323 (for Sb:As=3:2).

Optical Properties: Opaque. *Color:* Black; white in polished section. *Streak:* Black, with a slightly brownish tint. *Luster:* Metallic. *Pleochroism:* Strong, displaying twin lamellae.
R_1–R_2: (400) 39.1–45.2, (420) 38.2–44.7, (440) 37.3–44.2, (460) 36.6–43.7, (480) 36.2–43.2, (500) 35.8–42.9, (520) 35.5–42.5, (540) 35.0–42.0, (560) 34.4–41.4, (580) 34.0–40.8, (600) 33.5–40.0, (620) 33.2–39.4, (640) 32.8–38.7, (660) 32.3–38.1, (680) 31.7–37.5, (700) 31.0–37.0

Cell Data: *Space Group:* *Pnmn*, pseudocell. $a = 19.6(2)$ $b = 7.99(5)$ $c = 8.60(5)$ $\alpha = 90°$ $\beta = 90°$ $\gamma = 90°$ Z = 8

X-ray Powder Pattern: Madoc, Canada.
3.51 (100), 2.344 (80), 2.78 (70), 4.18 (50), 3.91 (50), 2.689 (50), 2.645 (50)

Chemistry:

	(1)	(2)	(3)
Pb	41	39.3	38.7
Sb	28	28.1	24.8
As	11	8.9	12.0
S	23	23.7	23.9
Total	103	100.0	99.4

(1) Madoc, Canada; by electron microprobe, leading to $Pb(Sb_{1.162}As_{0.742})_{\Sigma=1.904}S_{3.625}$. (2) Do. (3) Novoye, USSR; by electron microprobe, corresponding to $Pb(Sb_{1.06}As_{0.84})_{\Sigma=1.90}S_{3.89}$.

Polymorphism & Series: Dimorphous with guettardite.

Occurrence: In marble with other lead sulfantimonides (Madoc, Canada).

Association: Chabournéite, pierrotite, parapierrotite, stibnite, pyrite, sphalerite, zinkenite, madocite, andorite, smithite, laffittite, routhierite, aktashite, wakabayashilite, realgar, orpiment (Jas Roux, France); sphalerite, pyrite, galena, playfairite, sorbyite, guettardite, baumhauerite, realgar, orpiment, cinnabar, fluorite, quartz (Novoye, USSR).

Distribution: From near Madoc, and in the Hemlo gold deposit, Thunder Bay district, Ontario, Canada. In France, from Jas Roux, Hautes-Alpes. At Rujevac, Yugoslavia. From Novoye, Khaidarkan, Kirgizia, USSR.

Name: For Robert Mitchell Thompson (1918–1967), Canadian mineralogist, University of British Columbia. Thompson is "son of Thomas"; the latter is Aramaic for "a twin". The name is doubly appropriate in alluding to the polysynthetic twinning present in the mineral.

Type Material: National Museum of Canada, Ottawa, Canada.

References: (1) Jambor, J.L. (1967) New lead sulfantimonides from Madoc, Canada. Part 2 — Mineral descriptions. Can. Mineral., 9, 191–213. (2) (1968) Amer. Mineral., 53, 1424 (abs. ref. 1). (3) Jambor, J.L., J.H.G. Laflamme, and D.A. Walker (1982) A re-examination of the Madoc sulfosalts. Mineral. Record, 13, 93–100. (4) Mozgova, N.N., N.S. Bortnikov, Y.S. Borodaev, and A.I. Tzépine (1982) Sur la non-stoechiométrie des sulfosels antimonieux arséniques de plomb. Bull. Minéral., 105, 3–10 (in French with English abs.).

Tyrrellite $(Cu, Co, Ni)_3Se_4$

Crystal Data: Cubic. *Point Group:* $4/m\ \overline{3}\ 2/m$. Rounded grains and subhedral cubes.

Physical Properties: *Cleavage:* {001}, poor. *Fracture:* Conchoidal. *Tenacity:* Brittle. Hardness = ~3.5 VHN = 343–368 (100 g load). D(meas.) = n.d. D(calc.) = 6.6(2)

Optical Properties: Opaque. *Color:* Light bronze; light brassy bronze in reflected light. *Streak:* Black. *Luster:* Metallic.

R: (400) 43.2, (420) 44.2, (440) 45.2, (460) 45.8, (480) 46.0, (500) 46.7, (520) 47.0, (540) 47.3, (560) 47.5, (580) 47.8, (600) 48.0, (620) 48.4, (640) 48.8, (660) 49.3, (680) 49.7, (700) 50.2

Cell Data: *Space Group:* $Fm3m$. $a = 10.005(4)$ Z = 8

X-ray Powder Pattern: Goldfields district, Canada.
1.769 (100), 2.501 (90), 2.886 (70), 3.016 (60), 1.926 (60), 5.780 (40), 3.537 (40)

Chemistry:

	(1)	(2)
Cu	12.7	13.7
Co	17.7	11.6
Ni	6.9	12.0
Se	62.4	62.0
Total	99.7	99.3

(1) Beaverlodge district (sic), Canada; by electron microprobe. (2) Hope's Nose, England; by electron microprobe.

Occurrence: With other selenides, as the youngest hydrothermal replacements and open space fillings in sheared Precambrian rocks, which also contain uraninite deposits (Goldfields district, Canada).

Association: Umangite, klockmannite, clausthalite, pyrite, hematite, chalcopyrite, chalcomenite (Ato Bay, Canada); berzelianite, eucairite, crookesite, ferroselite, bukovite, krutaite, athabascaite, calcite, dolomite (Petrovice deposit, Czechoslovakia).

Distribution: At the western part of the Eagle claims, and also at the head of Ato Bay, Beaverlodge Lake, Goldfields district, Saskatchewan, Canada. From Bukov, and in the Petrovice deposit, Czechoslovakia. At Hope's Nose, Torquay, Devon, England.

Name: For Joseph Burr Tyrell (1858–1957), first geologist to visit the area of discovery in Canada.

Type Material: n.d.

References: (1) Robinson, S.C. and E.J. Brooker (1952) A cobalt–nickel–copper selenide from the Goldfields district, Saskatchewan. Amer. Mineral., 37, 542–544. (2) Sindeeva, N.D. (1964) Mineralogy and types of deposits of selenium and tellurium, 77–78. (3) Criddle, A.J. and C.J. Stanley, Eds. (1986) The quantitative data file for ore minerals. British Museum (Natural History), London, England, 391.

Crystal Data: Monoclinic likely, pseudo-orthorhombic. *Point Group:* $2/m$ $2/m$ $2/m$, 222, or $mm2$. Subhedral crystals, anhedral grains, to 200 μm. *Twinning:* Polysynthetic along one direction.

Physical Properties: Hardness = n.d. VHN = 168 (100 g load). D(meas.) = n.d. D(calc.) = 5.61

Optical Properties: Opaque. *Color:* Gray. *Luster:* Metallic.
R_1–R_2: (400) 36.2–44.3, (420) 36.3–44.2, (440) 35.8–44.2, (460) 35.2–43.8, (480) 35.0–43.9, (500) 34.7–43.7, (520) 34.5–43.4, (540) 34.1–43.3, (560) 33.8–43.0, (580) 33.5–42.6, (600) 33.2–42.4, (620) 33.0–42.0, (640) 32.7–41.8, (660) 32.3–41.2, (680) 31.8–40.6, (700) 31.2–39.9

Cell Data: *Space Group:* $Pmmm$, $P222$, or $Pmm2$. $a = 12.67$ $b = 19.32$ $c = 4.38$ Z = 2

X-ray Powder Pattern: Uchuc-Chacua, Peru.
3.30 (100), 2.90 (80), 3.80 (30), 3.49 (30), 2.75 (30), 2.08 (30), 2.29 (10)

Chemistry:

	(1)	(2)
Ag	5.9	6.07
Pb	34.8	34.96
Mn	2.8	3.09
Fe	0.2	
Sb	34.4	34.24
Se	0.3	
S	21.1	21.64
Total	99.5	100.00

(1) Uchuc-Chacua, Peru; by electron microprobe, corresponding to $Ag_{0.98}Pb_{3.04}(Mn_{0.91}Fe_{0.06})_{\Sigma=0.97}Sb_{5.09}(S_{11.93}Se_{0.07})_{\Sigma=12.00}$. (2) $AgPb_3MnSb_5S_{12}$.

Occurrence: In a telescoped hydrothermal deposit.

Association: Alabandite, galena, benavidesite, sphalerite, pyrite, pyrrhotite, arsenopyrite.

Distribution: From Uchuc-Chacua, Cajatambo Province, Peru.

Name: For the Uchuc-Chacua deposit in Peru.

Type Material: National School of Mines, Paris, France.

References: (1) Moëlo, Y., E. Oudin, P. Picot, and R. Caye (1984) L'uchucchacuaïte, $AgMnPb_3Sb_5S_{12}$, une nouvelle espèce minérale de la sèrie de l'andorite. Bull. Minéral., 107, 597–604 (in French with English abs.). (2) (1985) Amer. Mineral., 70, 1332–1333 (abs. ref. 1).

Ullmannite $NiSbS$

Crystal Data: Cubic. *Point Group:* 23. As cubes and, less frequently, octahedra, pyritohedra, tetrahedra, to 3 cm. Cube faces striated by [110], twin boundaries of enantiomorphs. *Twinning:* Forms penetration twins about [110] with {001} the approximate composition plane. Re-entrants develop on cube edges.

Physical Properties: *Cleavage:* Perfect on {001}. *Fracture:* Uneven. *Tenacity:* Brittle. Hardness = 5–5.5 VHN = 592–627 (100 g load). D(meas.) = 6.65 D(calc.) = 6.793

Optical Properties: Opaque. *Color:* Steel-gray to silver-white; white in reflected light. *Streak:* Grayish black. *Luster:* Metallic. *Anisotropism:* Occasionally individuals are weakly anisotropic revealing a fine lamellar structure.
R: (400) 52.0, (420) 51.0, (440) 50.0, (460) 49.0, (480) 48.2, (500) 47.4, (520) 46.7, (540) 46.1, (560) 45.7, (580) 45.5, (600) 45.5, (620) 45.6, (640) 46.0, (660) 46.4, (680) 47.0, (700) 47.6

Cell Data: *Space Group:* $P2_13$ ($P1$ for arsenian ullmannite). a = 5.88–5.93 Z = 4

X-ray Powder Pattern: Salchendorf, Germany.
2.64 (100), 1.774 (70), 2.40 (60), 1.573 (50), 1.092 (50), 0.810 (50), 0.802 (50)

Chemistry:

	(1)	(2)	(3)	(4)
Ni	28.91	27.3	23.3	27.62
Co	1.13	0.8	3.8	
Fe	0.40			
Sb	42.93	52.8	58.9	57.29
As	10.28	3.5	0.4	
Bi	0.68	1.0		
S	16.22	15.1	14.8	15.09
Total	100.55	100.5	101.2	100.00

(1) Gosenbach, Germany. (2) Petersbach, Germany; by electron microprobe. (3) Broken Hill, Australia; by electron microprobe. (4) NiSbS.

Polymorphism & Series: Forms a series with willyamite.

Occurrence: With nickel minerals in hydrothermal veins.

Association: Nickeline, gersdorffite, pentlandite, chalcopyrite, pyrrhotite, galena, tetrahedrite, dyscrasite.

Distribution: From Broken Hill, New South Wales, Australia. In Germany, at the Friedrich mine, near Wissen; the Petersbach mine near Eichelhardt, at Gosenbach, and Salchendorf, in Siegerland; and at Neudorf, near Harzgerode, in the Harz Mountains. At Lölling, Carinthia, Austria. In France, at Ar, near Eaux-Bonnes, Basses-Pyrénées. From Monte Narba and Masaloni, Sarrabus, Sardinia, Italy. At the Settlingstones mine, Fourstones, Northumberland, and at New Brancepeth colliery, Durham, England. At the Mina Esperanza, Salta Province, Argentina. In Canada, at the Kerr Addison mine, Timiskaming district, Ontario; and from the Nicholson mine, near Goldfields, Saskatchewan. Additional minor localities are known.

Name: For Johan Christoph Ullmann (1771–1821), German chemist and mineralogist, who first discovered the mineral.

References: (1) Palache, C., H. Berman, and C. Frondel (1944) Dana's system of mineralogy, (7th edition), v. I, 301–302. (2) Berry, L.G. and R.M. Thompson (1962) X-ray powder data for the ore minerals. Geol. Soc. Amer. Mem. 85, 95–96. (3) Bayliss, P. (1986) Subdivision of the pyrite group, and a chemical and X-ray diffraction investigation of ullmannite. Can. Mineral., 24, 27–33.

Crystal Data: Tetragonal. *Point Group:* $\overline{4}2m$. Massive, as disseminated small grains; individual grains commonly exhibit rectangular outlines. *Twinning:* Sometimes lamellar.

Physical Properties: *Cleavage:* Poor, rectangular. *Fracture:* Uneven, subconchoidal. *Tenacity:* Brittle. Hardness = 3 VHN = 88–100 (100 g load). D(meas.) = 6.44–6.49 D(calc.) = 6.590

Optical Properties: Opaque. *Color:* Bluish black with reddish cast, weathers to a dull iridescent purple. *Streak:* Black. *Pleochroism:* Pronounced, reddish purple to grayish blue. *Anisotropism:* Very strong, fiery orange to straw-yellow with a rose tint.

R_1–R_2: (400) 25.6–24.4, (420) 23.3–23.0, (440) 21.0–21.6, (460) 19.0–20.3, (480) 17.3–19.1, (500) 15.8–18.1, (520) 14.6–17.2, (540) 13.6–16.5, (560) 13.2–16.1, (580) 13.5–16.0, (600) 15.0–16.1, (620) 17.5–16.4, (640) 20.7–16.8, (660) 24.2–17.4, (680) 27.6–18.1, (700) 30.6–18.8

Cell Data: *Space Group:* $P\overline{4}2_1m$. $a = 6.4024$ $c = 4.2786$ Z = 2

X-ray Powder Pattern: Sierra de Umango, Argentina.
3.559 (100), 1.829 (90), 1.778 (80), 3.108 (70), 2.258 (70), 3.202 (60), 1.908 (50)

Chemistry:

	(1)	(2)
Cu	54.35	54.69
Ag	0.55	
Se	45.10	45.31
Total	100.00	100.00

(1) Sierra de Umango, Argentina. (2) Cu_3Se_2.

Occurrence: With other selenides in hydrothermal veins.

Association: Clausthalite, tiemannite, berzelianite, guanajuatite, eucairite, hessite, naumannite, chalcomenite, chalcopyrite, cobaltite, pyrite, malachite, calcite.

Distribution: In Canada, From Lodge Bay on Lake Athabasca, to Ato Bay on Beaverlodge Lake, Saskatchewan, and on Christopher Island, Baker Lake, Northwest Territories. In Argentina, at Sierra de Umango, Sierra de Sanagasta, and the Santa Brigida mine, La Rioja Province; and at Sierra Cacheuta, Mendoza Province. At Skrikerum, Sweden. In Germany, in the Harz, at St. Andreasberg, Tilkerode, Zorge, Lerbach and Clausthal. From Chaméane, Puy-de-Dôme, France. At Kletno, Poland. From the Bukov and Habrí mines, near Tisnova, Předbořice, and Petrovice, Czechoslovakia. At Shinkolobwe, Shaba Province, Zaire. From Tsumeb, Namibia. At Kalgoorlie, Western Australia. In the Frederik VII's mine, near Julianehåb, southern Greenland.

Name: For the Sierra de Umango, Argentina, place of first discovery.

References: (1) Palache, C., H. Berman, and C. Frondel (1944) Dana's system of mineralogy, (7th edition), v. I, 194–195. (2) Berry, L.G. and R.M. Thompson (1962) X-ray powder data for the ore minerals. Geol. Soc. Amer. Mem. 85, 43–44. (3) Earley, J.W. (1950) Description and synthesis of the selenide minerals. Amer. Mineral., 35, 337–364. (4) Morimoto, N. and K. Koto (1966) Crystal structure of umangite, Cu_3Se_2. Science, 152, 345. (5) Heyding, R.D. and R.M. Murray (1976) The crystal structures of $Cu_{1.8}Se$, Cu_3Se_2, α– and β–CuSe, $CuSe_2$, and $CuSe_2$II. Can. J. Chem., 54, 841–848.

Urvantsevite $Pd(Bi, Pb)_2$

Crystal Data: Hexagonal. *Point Group:* n.d. In irregular polymineralic intergrowths up to 4 mm.

Physical Properties: *Cleavage:* Perfect in one direction. Hardness = n.d. VHN = 47–68 (10 g load). D(meas.) = n.d. D(calc.) = 9.66

Optical Properties: Opaque. *Color:* Grayish white in reflected light. *Pleochroism:* Weak to none. *Anisotropism:* Slight.

R_1–R_2: (400) — , (420) — , (440) 50.5–48.7, (460) 54.9–53.2, (480) 53.8–54.6, (500) 53.3–54.1, (520) 53.7–54.8, (540) 54.0–55.2, (560) 54.9–56.2, (580) 55.2–56.7, (600) 56.1–57.0, (620) 56.8–58.3, (640) 57.8–59.2, (660) 58.6–60.1, (680) 59.6–60.9, (700) 59.6–61.3

Cell Data: *Space Group:* n.d. $a = 13.82$ $c = 6.53$ Z = 12

X-ray Powder Pattern: Majak mine, USSR.
2.643 (100), 2.372 (80), 1.420 (70), 1.111 (70), 1.166 (60), 2.043 (50), 1.685 (50)

Chemistry:

	(1)
Pd	20.5
Bi	64.6
Pb	15.3
Total	100.4

(1) Majak mine, USSR; by electron microprobe, corresponding to $Pd_{1.00}(Bi_{1.61}Pb_{0.38})_{\Sigma=1.99}$.

Occurrence: In massive zoned Cu-Ni sulfide ores (Talnakh area, USSR).

Association: Froodite, sobolevskite, paolovite, silver, altaite, galena, pentlandite, cubanite, chalcopyrite (Talnakh area, USSR).

Distribution: In the Majak mine, Talnakh area, Noril'sk region, western Siberia, USSR. From Tilkerode, Harz Mountains, Germany.

Name: For Professor Nikolai Nikolaevich Urvantsev (1893–), of Leningrad, USSR, who discovered the Noril'sk deposits.

Type Material: Mining Museum, Leningrad Mining Institute, Leningrad, USSR.

References: (1) Rudashevskii, N.S., V.N. Makarov, E.M. Mededeva, V.V. Ballakh, Y.I. Permyakov, G.A. Mitenkov, A.M. Karpenkov, I.A. Bud'ko, and N.N. Shishkin (1976) Urvantsevite, $Pd(Bi, Pb)_2$, a new mineral in the system Pd–Bi–Pb. Zap. Vses. Mineral. Obshch., 105, 704–709 (in Russian). (2) (1977) Amer. Mineral., 62, 1260–1261 (abs. ref. 1). (3) Cabri, L.J., Ed. (1981) Platinum group elements: mineralogy, geology, recovery. Can. Inst. Min. & Met., 145–146.

Crystal Data: n.d. *Point Group:* n.d. Prismatic crystals; may be bent or twisted.

Physical Properties: *Cleavage:* One perfect; another present. Hardness = 2.5 VHN = n.d. D(meas.) = n.d. D(calc.) = n.d.

Optical Properties: Opaque. *Color:* Silver-gray to gray; pure white in reflected light. *Luster:* Metallic. *Anisotropism:* Strong.

R_1–R_2: n.d.

Cell Data: *Space Group:* n.d. Z = n.d.

X-ray Powder Pattern: Ustarasaisk deposits, USSR.
3.53 (100), 1.102 (100), 1.057 (100), 3.08 (70), 2.508 (70), 1.915 (70), 1.732 (70)

Chemistry:

	(1)	(2)
Pb	10.51	11.35
Cu	0.30	0.74
Fe	0.60	1.40
Bi	65.33	64.90
Sb	2.96	1.87
As		0.15
S	17.25	17.25
insol.	0.34	0.54
Total	97.29	98.20

(1–2) Ustarasaisk deposits, USSR.

Occurrence: In quartz-bismuthinite veins.

Association: Bismuthinite, kobellite, bismuthian jamesonite, quartz.

Distribution: From the Ustarasaisk deposits, western Tyan-Shan, Siberia, USSR.

Name: For the locality of first discovery, Ustarasaisk, USSR.

Type Material: n.d.

References: (1) Sakharova, M.S. (1955) Bismuth sulfosalts of the Ustarasaisk deposits. Trudy Mineralog. Muzeya Akad. Nauk SSSR, 7, 112–126. (2) (1956) Amer. Mineral., 41, 814 (abs. ref. 1).

Crystal Data: Tetragonal. *Point Group:* 422 or 4. As small (up to 100 μm) blebs and rims.

Physical Properties: *Tenacity:* More brittle than acanthite. Hardness = ~2 VHN = n.d. D(meas.) = n.d. D(calc.) = 8.45

Optical Properties: Opaque. *Color:* In polished section, gray-white. *Pleochroism:* Distinct in air, gray-white to gray-white with a brownish tint. *Anisotropism:* Strong, without distinct colors. R_1–R_2: n.d.

Cell Data: *Space Group:* $P4_122$ or $P4_1$. $a = 9.75$ $c = 9.85$ Z = 8

X-ray Powder Pattern: Tambang Sawah, Indonesia.
2.731 (100), 2.609 (90), 6.98 (80), 2.124 (80), 4.38 (60), 3.09 (60), 2.809 (60)

Chemistry:

	(1)	(2)	(3)	(4)
Ag	56.1	57.1	53.2	55.34
Au	29.8	32.7	35.1	33.69
Cu	2.2			
Se	trace			
Te	trace			
S	11.2	10.3	11.7	10.97
Total	99.3	100.1	100.0	100.00

(1) Comstock Lode, Nevada, USA; by electron microprobe. (2) Smeinogorski, USSR; by electron microprobe. (3) Tambang Sawah, Indonesia; by electron microprobe. (4) Ag_3AuS_2.

Occurrence: In low-temperature hydrothermal Au-Ag quartz veins.

Association: Acanthite, Au–Ag alloy, quartz.

Distribution: At Tambang Sawah, Benkoelen district, Sumatra, Indonesia. From the Comstock Lode, Virginia City, Storey Co., Nevada, USA. At Smeinogorski (Schlangenberg), Altai Mountains, USSR.

Name: For Willem Uytenbogaardt, Professor of Geology, Technical University, Delft, The Netherlands.

Type Material: Free University of Amsterdam, The Netherlands; National Museum of Natural History, Washington, D.C., USA, 105328, B239.

References: (1) Barton, M.D., C. Kieft, E.A.J. Burke, and I.S. Oen (1978) Uytenbogaardtite, a new silver–gold sulfide. Can. Mineral., 16, 651–657. (2) (1980) Amer. Mineral., 65, 209 (abs. ref. 1).

Crystal Data: Monoclinic. *Point Group:* $2/m$. As prismatic crystals up to 0.5 mm, elongated [001], with the forms {110}, {001}, {101} present. The {110} faces exhibit fine striations ∥ $[1\bar{1}1]$. *Twinning:* Cruciform twins with twin axis [100] and a growth plane near (011).

Physical Properties: *Cleavage:* On {001}. Hardness = 1.5 VHN = 66–71 (8 g load). D(meas.) = 3.385 D(calc.) = [3.37]

Optical Properties: Transparent. *Color:* Yellow; in reflected light, grayish white with bright yellow internal reflections. *Luster:* Pearly to slightly greasy on a broken surface.
Optical Class: Biaxial positive (+). $\alpha' = 2.38(1)$ $\gamma' = 2.68(1)$
R_1–R_2: (400) 20.1–18.8, (425) 20.0–18.8, (450) 19.9–18.7, (475) 19.7–18.3, (500) 19.2–18.1, (525) 19.8–19.0, (550) 20.0–19.4, (575) 19.9–19.4, (600) 19.7–19.3, (625) 19.4–19.2, (650) 19.3–19.1, (675) 19.2–19.0, (700) 19.0–18.8

Cell Data: *Space Group:* $P2_1/m$. $a = 7.98(1)$ $b = 8.10(1)$ $c = 7.09(1)$ $\beta = 100.14(3)°$ Z = 2

X-ray Powder Pattern: Uzon caldera, USSR.
5.81 (100), 3.602 (80), 2.905 (80), 5.31 (60), 3.100 (60), 2.820 (60)

Chemistry:

	(1)	(2)
As	64.65	65.15
S	34.09	34.85
Total	98.74	100.00

(1) Uzon caldera, USSR; by electron microprobe, average of three grains, giving $As_{4.03}S_{4.97}$.
(2) As_4S_5.

Occurrence: As intergrowths with realgar in the pore spaces of tuffaceous sediments 10-30 cm deep in a caldera.

Association: Realgar, cubic α–arsenic sulfide.

Distribution: From the Uzon caldera, Kamchatka, USSR.

Name: For the type locality, Uzon caldera, USSR.

Type Material: A.E. Fersman Mineralogical Museum, Academy of Sciences, Moscow, USSR.

References: (1) Popova, V.I. and V.O. Polyakov (1985) Uzonite As_4S_5 — a new sulphide of arsenic from Kamchatka. Zap. Vses. Mineral. Obshch., 114, 369–373 (in Russian). (2) (1986) Amer. Mineral., 71, 1280 (abs. ref. 1).

Crystal Data: Cubic. *Point Group:* $2/m\ \overline{3}$. As octahedra and cubes to 1 cm, and massive.

Physical Properties: *Cleavage:* {001}. Hardness = ~3.5 VHN = 743–837 (100 g load). D(meas.) = n.d. D(calc.) = 4.45

Optical Properties: Opaque. *Color:* Black; pale violet-gray on polished surface. *Luster:* Metallic to submetallic.

R: (400) 35.0, (420) 34.3, (440) 33.6, (460) 33.0, (480) 32.5, (500) 32.0, (520) 31.8, (540) 31.6, (560) 31.5, (580) 31.5, (600) 31.6, (620) 31.8, (640) 32.0, (660) 32.4, (680) 33.0, (700) 33.6

Cell Data: *Space Group:* *Pa*3. $a = 5.66787(8)$ Z = 4

X-ray Powder Pattern: Kasompi mine, Zaire.
2.83 (100), 1.707 (80), 1.091 (60), 2.00 (50), 2.54 (40), 2.32 (40), 1.003 (40)

Chemistry:

	(1)	(2)	(3)
Ni	41.24	47.9	47.79
Co	3.41	1.1	
Fe	2.20	2.6	
S	53.15	48.4	52.21
Total	100.00	100.0	100.00

(1) Kasompi mine, Zaire. (2) Kalgoorlie, Australia; by electron microprobe, giving $(Ni_{0.93}Fe_{0.05}Co_{0.02})_{\Sigma=1.00}S_{2.00}$. (3) NiS_2.

Polymorphism & Series: Forms a series with cattierite.

Occurrence: Disseminated through dolomite (Kasompi mine, Zaire); as a secondary alteration product of arsenic-deficient nickel-skutterudite (Germany).

Association: Arsenic-deficient nickel-skutterudite, pyrite, polydymite, uraninite.

Distribution: From the Shinkolobwe mine, and in the Kasompi mine, 70 km west-southwest of Kambove, Shaba Province, Zaire. At Schneeberg, Nentershausen, and Iserlohn, Germany. From Saint Marina, Khaskovo district, Bulgaria. In the Agyatagsk mercury deposit, Kel'badzhar, Azerbaijan SSR, USSR. At the Cármenes district, Léon Province, Spain. From the Miliken (Sweetwater) mine, Reynolds Co., Missouri, and the Orphan mine, in the Grand Canyon, Coconino Co., Arizona, USA. From Mina San Santiago, La Rioja Province, Argentina. Near the Scotia mine, Kalgoorlie, Western Australia.

Name: For Mr. Johannes Vaes, mineralogist for the Union Minière du Haut Katanga.

Type Material: Columbia University, New York, New York, USA (?).

References: (1) Kerr, P.F. (1945) Cattierite and vaesite: new Co–Ni minerals from the Belgian Congo. Amer. Mineral., 30, 483–497. (2) Berry, L.G. and R.M. Thompson (1962) X-ray powder data for the ore minerals. Geol. Soc. Amer. Mem. 85, 93–94. (3) Ramdohr, P. (1969) The ore minerals and their intergrowths, (3rd edition), 815–816. (4) Ostwald, J. (1980) Notes on a Co–Ni disulphide and a Co–Ni–Fe thiospinel from the Kalgoorlie district, Western Australia. Mineral. Mag., 43, 950–951.

$4(Fe, Cu)S \cdot 3(Mg, Al)(OH)_2$ **Valleriite**

Crystal Data: Hexagonal *Point Group:* $\overline{3}\ 2/m$ or $3m$. Massive, as bean-sized nodules, tiny splinters, very thin flakes, and thin crusts.

Physical Properties: *Cleavage:* {0001}, excellent. Hardness = Very low, less than graphite. VHN = n.d. D(meas.) = 3.14 D(calc.) = 3.21

Optical Properties: Opaque. *Color:* Bronze-yellow, resembling pyrrhotite. *Luster:* Moderately metallic. *Pleochroism:* Strong, pale yellow to deep brown. *Anisotropism:* Strong, golden yellow.
R_1–R_2: (400) 11.0–8.6, (420) 10.8–9.6, (440) 10.6–10.6, (460) 10.4–11.6, (480) 10.4–12.7, (500) 10.3–13.8, (520) 10.2–15.0, (540) 10.2–16.2, (560) 10.2–17.2, (580) 10.2–18.1, (600) 10.2–19.0, (620) 10.2–19.7, (640) 10.2–20.4, (660) 10.3–21.1, (680) 10.3–21.7, (700) 10.4–22.3

Cell Data: *Space Group:* $R\overline{3}m$ or $R3m$. $a = 64.46$ $c = 34.10$ Z = [390]

X-ray Powder Pattern: Loolekop, South Africa.
11.39 (100), 5.71 (100), 3.27 (60), 3.80 (50), 3.23 (50), 2.846 (50), 1.885 (50)

Chemistry:

	(1)	(2)	(3)
Cu	17.7	17.6	19.8
Fe	26.3	21.2	20.0
S	22.5	21.4	21.6
Al_2O_3	5.1	8.1	8.5
MgO	10.6	16.2	16.0
CaO	0.3	1.7	1.3
K_2O	0.3		
Na_2O	0.6		
H_2O	10.8	12.2	10.8
SiO_2 + insol.		1.8	3.3
Total	94.2	100.2	101.3

(1) Kopparberg, Sweden. (2) Kaveltorp, Sweden. (3) Loolekop, South Africa.

Occurrence: As an alteration product of chalcopyrite in chromitites and dunites (Cyprus); in copper-bearing carbonatites, replacing magnetite (Palabora, South Africa); in Cu-Ni sulfide-bearing serpentinized ultramafic rocks.

Association: Chalcopyrite, cubanite, magnetite, serpentine.

Distribution: An inconspicuous mineral, now recognized from a number of localities in addition to those listed here. From the Aurora mine, Ljusnarsberg, and Kaveltorp, Kopparberg, Sweden. In South Africa, in the Transvaal, from Loolekop, Palabora, and the Mooihoek and Onverwacht pipes. From near Pefkos, Cyprus. In Canada, at the Little Chief mine, near Whitehorse, Yukon Territory; the Marbridge mine, Malartic, Quebec; at Sudbury, Ontario. In the USA, at the Elizabeth mine, South Strafford, Strafford Co., Vermont; the Pima mine, near Tucson, Pima Co., and the Christmas mine, Gila Co., Arizona; and at the Continental mine, Fierro, Grant Co., New Mexico. In the Talnakh area, Noril'sk region, western Siberia, USSR. From Wannaway, Western Australia.

Name: For Göran Wallerius (Vallerius) (1683–1742), a Swedish mining geologist.

References: (1) Evans, H.T., Jr., C. Milton, E.C.T. Chao, I. Adler, C. Mead, B. Ingram, and R.A. Berner (1964) Valleriite and the new iron sulfide, mackinawite. U.S. Geol. Sur. Prof. Paper 475-D, D64–D69. (2) Evans, H.T., Jr. and R. Allmann (1968) The crystal structure and crystal chemistry of valleriite. Zeits. Krist., 127, 73–93. (3) Harris, D.C. and D.J. Vaughan (1972) Two fibrous iron sulfides and valleriite from Cyprus with new data on valleriite. Amer. Mineral., 57, 1037–1052. (4) Ramdohr, P. (1969) The ore minerals and their intergrowths, (3rd edition), 683–692.

Crystal Data: Orthorhombic. *Point Group:* $mm2$ or $2/m\ 2/m\ 2/m$. Masses up to 2 cm in diameter or as anhedral grains; also as prisms elongated along [100] but flattened on {010}. The prism zone is almost equant. Faces are grooved and roughened. *Twinning:* Lamellar.

Physical Properties: *Fracture:* Conchoidal. *Tenacity:* Brittle. Hardness = n.d. VHN = 164 (50 g load). D(meas.) = 5.92 D(calc.) = 5.96

Optical Properties: Opaque. *Color:* Steel-gray; white in reflected light. *Streak:* Black, faint brownish tinge against white background. *Pleochroism:* Weak; white to pale pinkish gray. *Anisotropism:* Moderate.
R_1–R_2: n.d.

Cell Data: *Space Group:* $P2_1cn$ or $Pmcn$. $a = 8.44$ $b = 26.2$ $c = 7.90$ Z = 8

X-ray Powder Pattern: Madoc, Canada.
3.81 (100), 3.03 (90), 3.42 (80), 3.26 (80), 2.76 (70), 3.23 (50), 2.93 (50)

Chemistry:

	(1)	(2)	(3)
Pb	53.5	53.	51.
Sb	19.5	19.5	20.
As	6.0	7.5	7.
S	22.0	19.2	21.5
Total	101.0	99.2	99.5

(1–3) Madoc, Canada; by electron microprobe.

Occurrence: As small masses, stringers, and disseminated grains in marbles developed in a sequence of Precambrian metasediments near a contact with plutonic granitic gneiss (Madoc, Canada).

Association: Gratonite, calcite, boulangerite, other lead antimonides, sphalerite, pyrite, chalcopyrite, arsenopyrite, galena (Madoc, Canada).

Distribution: From near Madoc, Ontario, Canada. At Huachocolpa, Huancavelica, Peru.

Name: For R.W. van der Veen, eminent metallographer.

Type Material: National Museum of Canada, Ottawa, Canada.

References: (1) Jambor, J.L. (1967) New lead sulfantimonides from Madoc, Ontario, Part I. Can. Mineral., 9, 7–24. (2) (1968) Amer. Mineral., 53, 1422 (abs. ref. 1).

Crystal Data: Monoclinic. *Point Group:* $2/m$ or m. Lamellar with an average grain size of 0.5 mm. *Twinning:* On {010}; twin lamellae parallel to [001] often observed in thin section.

Physical Properties: Hardness = n.d. VHN = 185 (50 g load). D(meas.) = n.d. D(calc.) = 6.94

Optical Properties: Opaque. *Color:* In polished section, galena-white. *Pleochroism:* Absent in air, absent to weak in oil. *Anisotropism:* Distinct to strong, in light gray to steel bluish black. R_1–R_2: n.d.

Cell Data: *Space Group:* $P2/a$ or Pa; $B2/m$ or Bm (subcell). $a = 13.603$ $b = 25.248$ $c = 4.112(4)$ $\beta = 95.55(3)°$ Z = 1

X-ray Powder Pattern: Ivigtut, Greenland.
3.40 (100), 1.754 (90), 2.06 (80), 2.91 (50), 3.62 (40), 3.05 (40), 2.10 (40)

Chemistry:

	(1)	(2)
Ag	8.9	9.18
Cu	0.2	
Pb	27.7	28.21
Bi	47.6	46.24
S	16.5	16.37
Total	100.9	100.00

(1) Ivigtut, Greenland; by electron microprobe, corresponding to $Ag_{1.21}Cu_{0.05}Pb_{1.96}Bi_{3.33}S_{7.53}$.
(2) $Ag_5Pb_8Bi_{13}S_{30}$.

Occurrence: Associated with cosalite and as lamellae in galena in the gustavite-galena paragenesis (Ivigtut, Greenland).

Association: Cosalite, galena (Ivigtut, Greenland).

Distribution: From Ivigtut, southern Greenland. In the USA, from Gabbs, Nye Co., Nevada; South Mountain, Owyhee Co., Idaho; and in the Apache Hills, southeast of Hachita, Grant Co., New Mexico. In the Kochbulak deposit, eastern Uzbekistan, USSR.

Name: For the Vikings, early settlers of Greenland.

Type Material: National Museum of Natural History, Washington, D.C., USA, 136172.

References: (1) Karup-Møller, S. (1977) Mineralogy of some Ag–(Cu)–Pb–Bi sulfide associations. Bull. Geol. Soc. Denmark, 26, 41–68. (2) Makovicky, E. and S. Karup-Møller (1977) Chemistry and crystallography of the lillianite homologous series. Neues Jahrb. Mineral., Abh., 131, 56–82. (3) (1979) Amer. Mineral., 64, 243 (abs. refs. 1 and 2).

Crystal Data: Cubic. *Point Group:* $2/m\ \overline{3}$. As cubes and octahedra, with rounded cube faces; also as nodules with radial fibrous structure. *Twinning:* Noted, but not described.

Physical Properties: *Cleavage:* Perfect cubic. *Fracture:* Uneven. Hardness = 4.5 VHN = 535–710 (spherical aggregates); 440–520 (euhedral crystals) (20 g load). D(meas.) = 4.4–4.5 D(calc.) = [4.81]

Optical Properties: Opaque. *Color:* Iron-black; light blue-gray to violet-gray in reflected light. *Streak:* Sooty black. *Luster:* Dull metallic.

R: (400) 26.9, (420) 26.7, (440) 26.5, (460) 25.8, (480) 25.1, (500) 24.5, (520) 23.8, (540) 23.4, (560) 23.2, (580) 23.4, (600) 23.9, (620) 24.8, (640) 26.0, (660) 27.6, (680) 29.4, (700) 31.4

Cell Data: *Space Group:* $Pa3$. $a = 5.6944(3)$ Z = 4

X-ray Powder Pattern: Providencia mine, Spain.
2.852 (100), 1.7174 (40), 2.548 (30), 2.325 (25), 2.014 (25), 3.289 (15), 1.0959 (15)

Chemistry:

	(1)	(2)
Cu	24.0	24.1
Ni	11.8	17.3
Co	4.0	5.7
Fe	5.3	2.2
Se	0.06	1.5
S	54.0	50.2
Total	99.2	101.0

(1) Providencia mine, Spain. (2) Do.; by electron microprobe.

Occurrence: As subhedral disseminated grains in black bituminous dolostone and in white coarse-grained recrystallized dolostone or as nodular aggregates in vuggy dolomite veinlets (Providencia mine, Spain).

Association: Pyrite, cuprian bravoite, chalcopyrite, dolomite, quartz.

Distribution: From the Providencia mine, near Villamanin, Cármenes district, León Province, Spain. In the Lubin mine, near Legnica, Lower Silesia, Poland.

Name: For the town of Villamanin, Spain.

Type Material: n.d.

References: (1) Schoeller, W.R. and A.R. Powell (1920) Villamaninite, a new mineral. Mineral. Mag., 19, 14–18. (2) Hey, M.H. (1962) A new analysis of villamaninite. Mineral. Mag., 33, 169–170. (3) Ypma, P.J.M., H.J. Evers, and C.F. Woensdregt (1968) Mineralogy and geology of the Providencia mine (León, Spain), type-locality of villamaninite. Neues Jahrb. Mineral., Monatsh., 174–191. (4) Bayliss, P. (1977) X-ray powder data for villamaninite. Mineral. Mag., 41, 545.

Crystal Data: Tetragonal, pseudocubic. *Point Group:* $4/m\ 2/m\ 2/m$, $4mm$, or 422. As grains up to 1 mm. *Twinning:* Simple and poor polysynthetic twins.

Physical Properties: *Fracture:* Conchoidal. *Tenacity:* Brittle. Hardness = n.d. VHN = 280 (25 g load). D(meas.) = n.d. D(calc.) = 4.29

Optical Properties: Opaque. *Color:* Orange. *Luster:* Metallic. *Anisotropism:* Weak, purplish blue to greenish brown-yellow.

R_1–R_2: (420) 19.4–19.4, (460) 21.8–20.6, (500) 23.9–22.8, (540) 27.7–26.5, (580) 30.7–29.5, (620) 33.6–32.1, (660) 35.9–34.3, (700) 37.8–36.0

Cell Data: *Space Group:* $P4/mmm$, $P4mm$, $P422$, or $P4_122$. $a = 10.697(6)$ $c = 10.697(6)$ Z = 2

X-ray Powder Pattern: Chizeuil, France.
3.088 (100), 1.895 (90), 1.614 (70), 2.676 (50), 1.227 (40), 1.091 (40), 4.37 (30)

Chemistry:

	(1)	(2)
Cu	40.90	42.6
Fe	14.63	13.2
Zn		0.7
Sn	7.33	8.3
As	3.43	4.5
Sb	1.60	
S	31.85	31.3
Total	99.74	100.6

(1) Chizeuil, France; by electron microprobe, average of six analyses, yielding $Cu_{10.19}Fe_{4.15}Sn_{0.98}(As_{0.73}Sb_{0.21})_{\Sigma=0.94}S_{15.74}$. (2) Maggie deposit, Canada; by electron microprobe, giving $Cu_{10.60}(Fe^{+3}_{3.00}Fe^{+2}_{0.74}Zn_{0.17})_{\Sigma=3.91}Sn_{1.11}As_{0.95}S_{15.44}$.

Occurrence: In a porphyry copper deposit with other tin-bearing sulfides (Maggie deposit, Canada).

Association: Pyrite, chalcopyrite, colusite, stannite, stannoidite, mawsonite, bornite, enargite, tetrahedrite–tennantite, quartz, barite.

Distribution: From Chizeuil, Saône-et-Loire, France. At Huaron, Peru. In the Maggie copper deposit, 15 km north of Ashcroft, British Columbia, Canada.

Name: To honor Professor Henri Vincienne (1898–1965), who first called attention to the mineral.

Type Material: National School of Mines, Paris, France.

References: (1) Cesbron, F., R. Girauld, P. Picot, and F. Pillard (1985) La vinciennite $Cu_{10}Fe_4Sn(As,Sb)S_{16}$, une nouvelle espèce minérale. Etude paragénétique du gîte type de Chizeuil, Saône-et-Loire. Bull. Minéral., 108, 447–456 (in French with English abs.). (2) (1986) Amer. Mineral., 71, 1280–1281 (abs. ref. 1). (3) Jambor, J. and D.R. Owens (1987) Vinciennite in the Maggie porphyry copper deposit, British Columbia. Can. Mineral., 25, 227–228.

Crystal Data: Cubic. *Point Group:* $4/m\ \overline{3}\ 2/m$. As nodules up to 0.5 cm and massive.

Physical Properties: *Cleavage:* Perfect on {001}. *Tenacity:* Brittle. Hardness = 4.5–5.5 VHN = 455–493 (100 g load). D(meas.) = n.d. D(calc.) = 4.79

Optical Properties: Opaque. *Color:* Violet-gray; distinctly violet in reflected light. *Luster:* Metallic.
R: (400) 39.0, (420) 39.6, (440) 40.2, (460) 40.6, (480) 41.0, (500) 41.4, (520) 41.9, (540) 42.5, (560) 43.1, (580) 43.8, (600) 44.3, (620) 44.8, (640) 45.4, (660) 45.8, (680) 46.2, (700) 46.6

Cell Data: *Space Group:* $Fd3m$. $a = 9.51$ Z = 8

X-ray Powder Pattern: Vermilion mine, Sudbury, Canada.
2.85 (100), 1.674 (80), 1.820 (60), 2.36 (50), 1.059 (50), 1.183 (40), 1.115 (40)

Chemistry:

	(1)	(2)	(3)
Fe	17.01	19.33	18.52
Ni	38.68	33.94	38.94
Co	1.05	2.50	
Cu	1.12	1.05	
S	41.68	42.17	42.54
insol.	0.40	1.31	
Total	99.94	100.30	100.00

(1) Vermilion mine, Sudbury, Canada; contains trace chalcopyrite. (2) Friday mine, Julian, California, USA; contains trace chalcopyrite. (3) $FeNi_2S_4$.

Occurrence: Of hydrothermal origin, with other sulfides.

Association: Pyrrhotite, millerite, chalcopyrite, pentlandite.

Distribution: In the USA, from the Friday mine, Julian, San Diego Co., California; the Key West mine, Clark Co., Nevada; the Lick Fork deposit, Floyd Co., Virginia; and the Gap Nickel mine, Lancaster Co., Pennsylvania. In Canada, from the Vermilion, Levack and Worthington mines, Sudbury, Ontario; at the Marbridge mine, Malartic, Quebec; in the Rottenstone mine, Saskatchewan; and from a number of other minor occurrences. At the Madziwa (Dry Nickel) mine, Bindura, Zimbabwe. From the Praborna mine, Aosta Valley, St. Marcel, Italy. At Kambalda and Kalgoorlie, Western Australia, and near Mount Colin, Queensland, Australia. Known from a few other localities.

Name: From the Latin for *violet*, its color on a polished surface.

References: (1) Palache, C., H. Berman, and C. Frondel (1944) Dana's system of mineralogy, (7th edition), v. I, 262–265. (2) Berry, L.G. and R.M. Thompson (1962) X-ray powder data for the ore minerals. Geol. Soc. Amer. Mem. 85, 77–78.

Crystal Data: Orthorhombic. *Point Group:* n.d. Massive (?).

Physical Properties: *Cleavage:* One perfect, two imperfect. Hardness = n.d. VHN = 42–66 (25 g load). D(meas.) = n.d. D(calc.) = n.d.

Optical Properties: Opaque. *Color:* Pale purplish in reflected light. *Pleochroism:* Very slight in oil. *Anisotropism:* Very weak, pale pink to gray.

R_1–R_2: (400) 56.4–58.1, (420) 56.0–57.7, (440) 55.6–57.3, (460) 55.2–56.9, (480) 54.6–56.3, (500) 53.8–55.5, (520) 53.2–54.9, (540) 53.0–54.6, (560) 53.4–54.7, (580) 54.0–54.9, (600) 54.6–55.1, (620) 55.0–55.5, (640) 55.3–55.6, (660) 55.3–55.8, (680) 55.3–55.8, (700) 55.4–55.8

Cell Data: *Space Group:* n.d. Z = n.d.

X-ray Powder Pattern: Armenia, USSR.
3.09 (100), 3.21 (80), 2.21 (50), 2.33 (30), 2.15 (30), 1.82 (30), 1.61 (20)

Chemistry:

	(1)	(2)	(3)
Ag	17.7	18.0	18.86
Cu	0.1		
Bi	38.8	36.6	36.53
Te	45.0	45.2	44.61
Total	101.6	99.8	100.00

(1) Campbell mine, Bisbee, Arizona, USA; by electron microprobe. (2) Ashley deposit, Ontario, Canada; by electron microprobe. (3) $AgBiTe_2$.

Occurrence: As complex intergrowths with other tellurides in gold ores.

Association: Tellurobismuthite, hessite, altaite, calaverite, melonite, petzite.

Distribution: In the USSR, from an undefined locality in Armenia; in the Zhana-Tyube Au-Te deposit, Kazakhstan; and the Kochbulak deposit, eastern Uzbekistan. In the Campbell mine, Bisbee, Cochise Co., Arizona, USA. From the Ashley deposit, Bannockburn Township, Ontario, Canada. At the Yokozuru mine, north Kyushu, Japan. In the Champion Reef mine, Kolar Gold Fields, Karnataka, India. From Kambalda, Western Australia. At Glava, Värmland, Sweden. From Ivigtut, southern Greenland.

Name: For I.S. Volynskii (1900–1962), former Director of the Mineragraphic Laboratory, Institute of Mineralogy and Geochemistry of Rare Elements, Moscow, USSR.

Type Material: n.d.

References: (1) Bezsmertnaya, M.S. and L.N. Soboleva (1965) Volynskite, a new telluride of bismuth and silver. Akad. Nauk SSSR, Eksperimental'no Metod. Issled. Rudnykh Mineralov, 129–141. (2) (1966) Amer. Mineral., 51, 531 (abs. ref. 1). (3) Bezsmertnaya, M.S. and L.N. Soboleva (1963) A new telluride of bismuth and silver, established by the newest micromethods. Trudy Inst. Mineralog., Geokhim., Kristallokhim Redkikh. Elementov, 18, 70–84 (in Russian). (4) (1964) Amer. Mineral., 49, 818 (abs. ref. 3). (5) Harris, D.C., W.D. Sinclair, and R.I. Thorpe (1983) Telluride minerals from the Ashley deposit, Bannockburn Township, Ontario. Can. Mineral., 21, 137–143.

Vozhminite $(Ni, Co)_4(As, Sb)S_2$

Crystal Data: Hexagonal. *Point Group:* n.d. Massive (?).

Physical Properties: *Cleavage:* One, distinct. Hardness = n.d. VHN = 240–300, 270 average, to 376–480, 436 average (100 g load), depending on orientation. D(meas.) = n.d. D(calc.) = 6.2

Optical Properties: Opaque. *Color:* Yellowish with a brown tint; in reflected light, rose-orange. *Streak:* Black. *Luster:* Metallic.

R_1–R_2: (400) — , (420) 35.0–41.2, (440) 37.0–43.0, (460) 39.2–45.0, (480) 41.7–47.0, (500) 44.0–48.6, (520) 45.9–49.8, (540) 47.5–50.9, (560) 48.9–51.9, (580) 50.2–52.7, (600) 51.1–53.3, (620) 52.1–54.2, (640) 52.8–55.0, (660) 53.4–55.8, (680) 53.7–56.6, (700) 54.1–57.1

Cell Data: *Space Group:* n.d. $a = 17.46(4)$ $c = 7.20(1)$ Z = 18

X-ray Powder Pattern: Vozhmin massif, USSR.
8.7 (10), 1.776 (10b), 3.07 (9), 2.111 (9), 2.303 (7), 2.717 (6)

Chemistry:

	(1)
Ni	52.7
Co	5.56
Fe	0.05
As	13.1
Sb	11.3
S	16.8
Total	99.51

(1) Vozhmin massif, USSR; by electron microprobe, average of 22 points on 2 samples, corresponding to $(Ni_{3.43}Co_{0.36})_{\Sigma=3.79}(As_{0.67}Sb_{0.35})_{\Sigma=1.02}S_{2.00}$.

Occurrence: In heazlewoodite ore in serpentinites.

Association: Heazlewoodite, tučekite, magnetite, geversite, copper.

Distribution: From the Vozhmin massif, northeast Karelia, USSR.

Name: For its occurrence in the Vozhmin massif, USSR.

Type Material: Mining Museum, Leningrad Mining Institute, Leningrad, USSR.

References: (1) Rudashevskii, N.S., Y.P. Men'shikov, A.A. Lentsi, N.I. Shumskaya, A.B. Lobanova, G.N. Goncharov, and A.G. Tutov (1982) Vozhminite, $(Ni, Co)_4(As, Sb)S_2$, a new mineral. Zap. Vses. Mineral. Obshch., 111, 480–485 (in Russian). (2) (1983) Amer. Mineral., 68, 645 (abs. ref. 1).

Crystal Data: Orthorhombic. *Point Group:* $2/m\ 2/m\ 2/m$. Crystals small, tabular {010} or pyramidal {111}.

Physical Properties: *Cleavage:* {010}, good. *Fracture:* Uneven to conchoidal. *Tenacity:* Brittle. Hardness = 3.5 VHN = n.d. D(meas.) = 5.30(3) D(calc.) = [5.40]

Optical Properties: Opaque. *Color:* Dark gray-black with a bluish tint; in polished section, bluish white with red internal reflections; dark red in translucent thin splinters. *Streak:* Light red with a yellow tone. *Luster:* Submetallic to metallic.

Optical Class: Biaxial (+). *Dispersion:* $r > v$. $n = > 2.72$

R_1–R_2: (400) 45.8–48.6, (420) 42.0–44.6, (440) 38.2–40.6, (460) 35.7–38.2, (480) 33.8–36.7, (500) 32.2–35.3, (520) 30.8–34.0, (540) 29.6–32.5, (560) 28.6–31.1, (580) 28.0–30.0, (600) 27.6–29.2, (620) 27.3–28.8, (640) 27.1–28.8, (660) 26.8–28.6, (680) 26.6–28.4, (700) 26.4–28.0

Cell Data: *Space Group:* $C2ca$. $a = 13.399$ $b = 23.389$ $c = 11.287$ Z = 4

X-ray Powder Pattern: Allchar, Yugoslavia.
4.04 (100), 3.33 (80), 2.57 (80), 5.18 (60), 3.15 (60), 2.29 (60), 4.31 (50)

Chemistry:

	(1)	(2)
Tl	28.8	28.15
Hg	20.5	20.73
Sb	8.2	8.39
As	20.5	20.64
S	22.1	22.09
Total	100.1	100.00

(1) Allchar, Yugoslavia; by electron microprobe. (2) $Tl_4Hg_3Sb_2As_8S_{20}$.

Occurrence: In a hydrothermal deposit with other As-Tl sulfide minerals.

Association: Realgar, orpiment.

Distribution: From Allchar, Macedonia, Yugoslavia.

Name: For Karl Vrba (1845–1922), Czech mineralogist.

Type Material: National Museum of Natural History, Washington, D.C., USA, R939.

References: (1) Palache, C., H. Berman, and C. Frondel (1944) Dana's system of mineralogy, (7th edition), v. I, 484–485. (2) Caye, R., P. Picot, R. Pierrot, and F. Permingeat (1967) Nouvelles données sur la vrbaïte, sa teneur en mercure. Bull. Soc. fr. Minéral., 90, 185–191 (in French). (3) (1968) Amer. Mineral., 53, 351 (abs. ref. 2). (4) Ohmasa, M. and W. Nowacki (1971) The crystal structure of vrbaite, $Hg_3Tl_4As_8Sb_2S_{20}$. Zeits. Krist., 134, 360–380.

Crystal Data: Orthorhombic. *Point Group:* $2/m\ 2/m\ 2/m$. Minute (0.1 to 1.1 mm) prismatic or blade-like grains, massively encrusting and cementing rock fragments. *Twinning:* Very thin twin lamellae at approximately 45° to the prominent cleavage in some grains.

Physical Properties: *Cleavage:* One direction parallel to the length of the lath-shaped grains; another direction, less prominent, perpendicular to the first. *Tenacity:* Sectile. Hardness = 1–2 VHN = 34–40 (50 g load). D(meas.) = n.d. D(calc.) = 7.1

Optical Properties: Opaque. *Color:* Light bronze to yellow-bronze. *Luster:* Metallic. *Pleochroism:* Very strong, bright yellow to blue-gray. *Anisotropism:* Very strong, brilliant yellow-white, grayish yellow-white, yellow-orange to gray.

R_1–R_2: (400) 25.6–30.7, (420) 24.4–30.0, (440) 23.0–30.5, (460) 21.3–37.0, (480) 19.7–47.9, (500) 18.1–56.8, (520) 16.6–63.4, (540) 15.2–67.8, (560) 13.8–70.8, (580) 12.7–73.1, (600) 11.8–74.9, (620) 11.3–76.4, (640) 11.2–77.4, (660) 11.6–78.2, (680) 12.9–78.9, (700) 15.0–79.3

Cell Data: *Space Group:* *Pmnm.* $a = 4.09$ $b = 6.95$ $c = 3.15$ Z = 2

X-ray Powder Pattern: Good Hope mine, Colorado, USA.
2.03 (100), 2.86 (70), 3.52 (60), 6.94 (40), 3.47 (30), 2.65 (30), 2.32 (30)

Chemistry:

	(1)	(2)
Cu	32.5	33.24
Te	67.8	66.76
Total	100.3	100.00

(1) Good Hope mine, Colorado, USA; by electron microprobe. (2) CuTe.

Occurrence: Intermixed with other tellurides.

Association: Rickardite, tellurium (Good Hope mine, Colorado); rickardite, petzite, sylvanite (Byn'govsk deposit, USSR).

Distribution: In the Good Hope mine, Vulcan, Gunnison Co., Colorado, USA. From the Byn'govsk Au-Te deposit, Central Ural Mountains, USSR. In the Iriki mine, Kagoshima Prefecture, Japan. From the Jabal Sayid deposit, Saudi Arabia.

Name: After the locality, Vulcan, Colorado, USA.

Type Material: National Museum of Natural History, Washington, D.C., USA, R933, 85136.

References: (1) Cameron, E.N. and I.M. Threadgold (1961) Vulcanite, a new copper telluride from Colorado, with notes on certain associated minerals. Amer. Mineral., 46, 258–268.
(2) Anderko, K. and K. Schubert (1954) Untersuchungen im System Kupfer–Tellur. Zeits. für Metallkunde, 4, 371–378 (in German).

Crystal Data: Tetragonal. *Point Group:* $4/m$. As minute irregular grains, and rarely as prismatic crystals up to 0.07 mm.

Physical Properties: Hardness = n.d. VHN = 715–864, 806 average (100 g load). D(meas.) = n.d. D(calc.) = 6.705

Optical Properties: Opaque. *Color:* Silvery; in reflected light, creamy white. *Luster:* Metallic, strong. *Anisotropism:* Weak to moderate, dark brown to dark blue-gray.
R_1–R_2: (400) 41.2–42.0, (420) 42.2–43.0, (440) 43.0–43.9, (460) 43.7–44.5, (480) 44.2–45.0, (500) 44.5–45.4, (520) 44.9–45.6, (540) 45.1–45.8, (560) 45.4–46.1, (580) 45.6–46.3, (600) 45.8–46.5, (620) 46.0–46.6, (640) 46.0–46.6, (660) 46.0–46.5, (680) 45.7–46.3, (700) 45.4–46.2

Cell Data: *Space Group:* $P4_2/m$. $a = 6.368(3)$ $c = 6.562(3)$ Z = 8

X-ray Powder Pattern: Lac des Iles Complex, Canada. Easily confused with braggite.
2.846 (100), 2.914 (90), 2.612 (80), 2.650 (70), 1.857 (70), 1.142 (70), 2.150 (60)

Chemistry:

	(1)	(2)	(3)
Pd	57.6	67.6	60.8
Pt	4.4	2.2	13.5
Ni	16.6	5.9	4.1
S	21.4	24.1	22.1
Total	100.0	99.8	100.5

(1) Noril'sk region, USSR; by electron microprobe. (2–3) Stillwater Complex, Montana, USA; by electron microprobe.

Polymorphism & Series: Forms a series with braggite.

Occurrence: In andesine diabases (Noril'sk region, USSR) and in ultramafic layered intrusives elsewhere. As disseminations that probably are crystallized products, at magmatic temperatures, of a residual immiscible sulfide melt.

Association: Chalcopyrite, millerite, nickelian pyrite, linnaeite, cooperite (Noril'sk region, USSR); pentlandite, pyrrhotite, chalcopyrite, cubanite, nickelian mackinawite, gold, braggite, cooperite, moncheite, isoferroplatinum, kotulskite, keithconnite, palladian tulameenite (Stillwater Complex, Montana, USA); hongshiite, cooperite, sperrylite, isomertieite, magnetite, bornite, polydymite, diopside, actinolite, epidote (Yen deposit, China).

Distribution: From the Noril'sk region, western Siberia, USSR. In the Upper and Banded portions of the Stillwater Complex, Montana; from near Moapa, Clark Co., Nevada; and from the Yuba River, Nevada Co., California, USA. At the Lac des Iles Complex, Ontario, Canada. In the Yen deposit (a code name), China. From the Konttijärvi Intrusion, northern Finland.

Name: For N.K. Vysotskii, Soviet geologist, who first found platinum at Noril'sk, USSR.

Type Material: A.E. Fersman Mineralogical Museum, Academy of Sciences, Moscow, USSR.

References: (1) Genkin, A.D. and O.E. Zvyagintsev (1962) Vysotskite, a new sulfide of palladium and nickel. Zap. Vses. Mineral. Obshch., 91, 718–725 (in Russian). (2) (1963) Amer. Mineral., 48, 708 (abs. ref. 1). (3) Cabri, L.J., J.H.G. Laflamme, J.M. Stewart, K. Turner, and B.J. Skinner (1978) On cooperite, braggite, and vysotskite. Amer. Mineral., 63, 832–839. (4) Cabri, L.J., Ed. (1981) Platinum group elements: mineralogy, geology, recovery. Can. Inst. Min. & Met., 146. (5) Criddle, A.J. and C.J. Stanley (1985) Characteristic optical data for cooperite, braggite and vysotskite. Can. Mineral., 23, 149–162.

Crystal Data: Cubic. *Point Group:* n.d. As euhedral grains up to 7 x 4 μm in diameter, but usually below 2 μm, many showing the cube and octahedron as common forms.

Physical Properties: Hardness = 4.5 VHN = 255 D(meas.) = n.d. D(calc.) = 8.32 Highly magnetic.

Optical Properties: Opaque. *Color:* In polished section, white. *Luster:* Metallic.
R: n.d.

Cell Data: *Space Group:* n.d. $a = 2.86$ Z = [1]

X-ray Powder Pattern: n.d.

Chemistry:

	(1)	(2)	(3)
Co	48.8	50.2	51.34
Fe	49.8	49.7	48.66
Ni	0.5	0.4	
Total	99.1	100.3	100.00

(1) Wairau Valley, New Zealand; by electron microprobe. (2) Muskox Intrusion, Canada; by electron microprobe. (3) CoFe.

Occurrence: Found in dominantly lizardite serpentine at the western contact of an ultramafic intrusion; thought to have formed under low-sulfur reducing conditions during the serpentinization process (Wairau Valley, New Zealand).

Association: Chromite, magnetite, awaruite, copper (Wairau Valley, New Zealand).

Distribution: From the Wairau Valley, South Island, New Zealand. In the Muskox Intrusion, Northwest Territories, Canada.

Name: For the Wairau Valley locality in New Zealand.

Type Material: National Museum of Natural History, Washington, D.C., USA, 137192.

References: (1) Challis, G.A. and J.V.P. Long (1964) Wairauite—a new cobalt–iron mineral. Mineral. Mag., 33, 942–948. (2) (1965) Amer. Mineral., 50, 520 (abs. ref. 1).

Crystal Data: Monoclinic, pseudohexagonal. *Point Group:* 2 or $2/m$. As fibers elongated $\parallel$ [010], up to 2 cm long, or as prisms.

Physical Properties: *Cleavage:* {100}, {010}, $\{10\bar{1}\}$ perfect. *Tenacity:* Flexible. Hardness = ~1.5 VHN = n.d. D(meas.) = 3.96 D(calc.) = 4.06

Optical Properties: Translucent. *Color:* Golden to lemon-yellow, with golden yellow internal reflections. *Streak:* Orange-yellow. *Luster:* Silky (for fibrous aggregates) to resinous. *Anisotropism:* Weak. *Pleochroism:* Strong.

R: (400) 28.4, (420) 27.8, (440) 27.2, (460) 26.0, (480) 24.5, (500) 23.3, (520) 22.5, (540) 22.0, (560) 21.8, (580) 21.6, (600) 21.6, (620) 21.5, (640) 21.4, (660) 21.3, (680) 21.2, (700) 21.2

Cell Data: *Space Group:* $P2_1$ or $P2_1/m$; $P6_3/mmc$ (pseudocell). $a = 29.128$ $b = 6.480$ $c = 29.128$ $\beta = 120.0°$ Z = [8]

X-ray Powder Pattern: White Caps mine, Nevada, USA.
6.28 (100), 3.488 (80), 4.78 (70), 3.239 (40), 3.078 (40), 2.423 (40), 1.590 (40)

Chemistry:

	(1)	(2)
As	52.3	54.5
Sb	8.3	5.7
S	39.0	39.5
Total	99.6	99.7

(1–2) White Caps mine, Nevada, USA; by electron microprobe.

Occurrence: As fibers in druses of quartz or embedded in calcite.

Association: Realgar, orpiment, stibnite, pyrite (Nishinomaki mine, Japan); realgar, orpiment, calcite (White Caps mine, Nevada, USA); chabournéite, pierrotite, parapierrotite, stibnite, pyrite, sphalerite, twinnite, zinkenite, madocite, andorite, smithite, laffittite, routhierite, aktashite, realgar, orpiment (Jas Roux, France).

Distribution: In the Nishinomaki mine, Gumma Prefecture, Japan. From the White Caps mine, Manhattan, Nye Co., Nevada, USA. At Jas Roux, Hautes-Alpes, France. In the Gal-Khaya deposit, Yakutia, and at Khaidarkan, Kirgizia, USSR. From the Shuiluo arsenic deposit, Zhuang district, Guangxi Autonomous Region, China.

Name: For Yaichiro Wakabayashi (1874–1943), mineralogist for the Mitsubishi Mining Company, Japan.

Type Material: National Science Museum and the Sakurai Museum, Tokyo, Japan; National Museum of Natural History, Washington, D.C., USA, C252, 98012, 94600.

References: (1) Kato, A., K.I. Sakurai, and K. Ohsumi (1970) An introduction to Japanese minerals, In: Geol. Survey Japan, 1970, 92–93. (2) (1972) Amer. Mineral., 57, 1311 (abs. ref. 1). (3) Scott, J.D. and W. Nowacki (1975) New data on wakabayashilite. Can. Mineral., 13, 418–419.

Wallisite $PbTl(Cu,Ag)As_2S_5$

Crystal Data: Triclinic. *Point Group:* $\bar{1}$ or 1. As small crystals to 1 mm, usually massive.

Physical Properties: *Cleavage:* Pronounced on {001}. Hardness = n.d. VHN = n.d. D(meas.) = n.d. D(calc.) = 5.71

Optical Properties: Opaque. *Color:* Lead-gray. *Luster:* Metallic.
R_1–R_2: n.d.

Cell Data: *Space Group:* $P\bar{1}$ or $P1$. $a = 9.215$ $b = 8.524$ $c = 7.980$ $\alpha = 55°59(6)'$ $\beta = 62°30(6)'$ $\gamma = 69°24(6)'$ Z = 2

X-ray Powder Pattern: Binntal, Switzerland.
3.339 (100), 2.834 (50), 2.879 (30), 2.667 (30), 4.555 (25), 4.233 (25), 2.985 (25)

Chemistry:

	(1)	(2)
Pb	25.2	26.38
Tl	26.	26.03
Cu	6.9	8.09
Ag	2.7	
As	20.5	19.08
S	19.	20.42
Total	100.3	100.00

(1) Binntal, Switzerland; by electron microprobe. (2) $PbTlCuAs_2S_5$.

Occurrence: Overgrowing other lead sulfosalts.

Association: Dufrénoysite, rathite, pyrite.

Distribution: In Switzerland, at the Lengenbach quarry, Binntal, Valais.

Name: For Wallis, the German name for the Swiss Canton in which the Lengenbach quarry is located.

Type Material: n.d.

References: (1) Nowacki, W. (1965) Über einige Mineralfunde aus dem Lengenbach (Binnatal, Kt. Wallis). Eclogae Geol. Helveticae, 58, 403–406 (in German). (2) (1966) Amer. Mineral., 51, 532 (abs. ref. 1). (3) Nowacki, W., G. Burri, P. Engel, and F. Marumo (1965) Über einige Mineralstufen aus dem Lengenbach (Binnatal) II. Neues Jahrb. Mineral., Monatsh., 43–48 (in German). (4) (1969) Amer. Mineral., 54, 1497 (abs. ref. 3). (5) Takéuchi, Y., M. Ohmasa, and W. Nowacki (1968) The crystal structure of wallisite, $PbTlCuAs_2S_5$, the Cu analogue of hatchite, $PbTlAgAs_2S_5$. Zeits. Krist., 127, 349–365.

Crystal Data: Monoclinic. *Point Group:* $2/m$, 2, or m. As irregular grains to 3 mm, intergrown within aggregates of skippenite.

Physical Properties: *Fracture:* Conchoidal. Hardness = n.d. VHN = 155–186, 166 average (25 g load). D(meas.) = n.d. D(calc.) = 7.82

Optical Properties: Opaque. *Color:* Black on a fresh fracture; white with a bluish tint in reflected light. *Luster:* Metallic. *Anisotropism:* Moderate, in blue-gray to dark grayish brown.
R_1–R_2: (400) — , (420) 46.0–46.6, (440) 46.7–47.2, (460) 47.3–47.6, (480) 47.7–48.0, (500) 47.9–48.3, (520) 48.1–48.5, (540) 48.0–48.6, (560) 47.8–48.7, (580) 47.8–48.8, (600) 47.9–48.8, (620) 47.9–48.9, (640) 48.0–48.9, (660) 48.0–49.0, (680) 47.9–49.0, (700) 47.9–49.0

Cell Data: *Space Group:* $P2/m$, $P2$, or Pm. $a = 12.921(3)$ $b = 3.997(1)$ $c = 14.989(3)$ $\beta = 109.2(2)°$ Z = 2

X-ray Powder Pattern: Otish Mountains deposit, Canada.
2.976 (10), 2.929 (10), 3.573 (9b), 2.407 (7), 2.140 (7b), 2.065 (7b), 1.484 (7)

Chemistry:

	(1)
Cu	8.06 – 9.16
Pb	14.03 – 16.39
Bi	42.69 – 46.44
Se	27.44 – 28.73
Te	0.19 – 0.53
S	3.03 – 3.47
Total	

(1) Otish Mountains deposit, Canada; by electron microprobe, ranges of 11 grains, the average of which corresponds to $Cu_{2.36}Pb_{1.26}Bi_{3.70}(Se_{6.21}S_{1.74}Te_{0.05})_{\Sigma=8.00}$.

Occurrence: In a vein-type uranium deposit with other tellurides and selenides.

Association: Skippenite, souček ite, clausthalite, chalcopyrite, Au–Ag alloy.

Distribution: From the Otish Mountains uranium deposit, Quebec, Canada.

Name: For Professor David H. Watkinson, Carleton University, Ottawa, Canada.

Type Material: n.d.

References: (1) Johan, Z., P. Picot, and F. Ruhlmann (1987) The ore mineralogy of the Otish Mountains uranium deposit, Quebec: skippenite, Bi_2Se_2Te, and watkinsonite, $Cu_2PbBi_4(Se,S)_8$, two new mineral species. Can. Mineral., 25, 625–637. (2) (1989) Amer. Mineral., 74, 948 (abs. ref. 1).

Crystal Data: Orthorhombic. *Point Group:* $2/m\ 2/m\ 2/m$. Massive prismatic to fibrous or foliated habit; also as indistinct prismatic crystals.

Physical Properties: *Tenacity:* Very brittle, flexible. Hardness = 2–2.5 VHN = 136–156 (100 g load). D(meas.) = n.d. D(calc.) = [6.73]

Optical Properties: Opaque. *Color:* Steel-gray; white in reflected light. *Luster:* Metallic. *Pleochroism:* Weak. *Anisotropism:* Strong in reflected light, dark gray to red-brown.
R_1–R_2: (400) 43.6–51.5, (420) 43.0–51.0, (440) 42.4–50.5, (460) 41.8–49.3, (480) 41.4–48.4, (500) 40.9–47.8, (520) 40.5–47.4, (540) 40.0–46.6, (560) 39.5–45.9, (580) 38.9–45.1, (600) 38.4–44.4, (620) 38.0–43.9, (640) 37.5–43.3, (660) 37.0–42.8, (680) 36.7–42.5, (700) 36.4–42.2

Cell Data: *Space Group:* *Pnma.* $a = 53.68(9)$ $b = 4.11(1)$ $c = 15.40(2)$ Z = 4

X-ray Powder Pattern: Falun, Sweden.
3.847 (100), 3.080 (90), 3.268 (80), 3.019 (70), 2.811 (70), 2.278 (60), 2.138 (60)

Chemistry:

	(1)	(2)	(3)
Pb	29.7	31.5	28.3
Ag		0.9	1.0
Cu		0.3	
Bi	46.6	46.5	48.0
As	0.8		
Se	15.3	11.8	12.8
S	9.6	10.2	10.3
Total	102.0	101.2	100.4

(1–3) Falun, Sweden; by electron microprobe.

Occurrence: Intimately intergrown with laitakarite and bismuthinite in a hornblende rock.

Association: Laitakarite, bismuthinite, gold, chalcopyrite, bismuth, pyrrhotite, quartz.

Distribution: From Falun, Sweden.

Name: For Mats Weibull (1856–1923), who first described the mineral.

References: (1) Palache, C., H. Berman, and C. Frondel (1944) Dana's system of mineralogy, (7th edition), v. I, 473–474. (2) Karup-Møller, S. (1970) Weibullite, laitakarite and bismuthinite from Falun, Sweden. Geol. Foren Förb., 92, 181–187. (3) (1971) Amer. Mineral., 56, 639 (abs. ref. 2). (4) Johan, Z. and P. Picot (1976) Definition nouvelle de la weibullite et de la wittite. Compte Rendus Acad. Sci. Paris, Series D, 282, 137–139. (5) (1977) Amer. Mineral., 62, 397 (abs. ref. 4). (6) Mumme, W.G. (1980) Weibullite $Ag_{0.32}Pb_{5.09}Bi_{8.55}Se_{6.08}S_{11.92}$ from Falun, Sweden: a higher homologue of galenobismutite. Can. Mineral., 18, 1–12. (7) Mumme, W.G. (1980) Seleniferous lead–bismuth sulphosalts from Falun, Sweden: weibullite, wittite, and nordströmite. Amer. Mineral., 65, 789–796.

Crystal Data: Hexagonal. *Point Group:* $6/m\ 2/m\ 2/m$. Grains to 30 μm, in irregular aggregates from 0.05 to 0.4 mm.

Physical Properties: *Tenacity:* Ductile, malleable. Hardness = 2.4 VHN = 50.5 D(meas.) = n.d. D(calc.) = 18.17

Optical Properties: Opaque. *Color:* Pale yellow. *Luster:* Metallic. *Pleochroism:* Weak. *Anisotropism:* Weak.
R_1–R_2: (480) 63.75, (534) 76.30, (589) 81.03, (656) 68.6

Cell Data: *Space Group:* $P6_3/mmc$, by analogy with Au_3Hg. $a = 2.9265$ $c = 4.8178$ Z = 2

X-ray Powder Pattern: Poshan district, China.
2.243 (100), 0.9396 (90), 1.3593 (80), 1.2509 (80), 0.9954 (70), 1.4609 (60), 1.2255 (60)

Chemistry:

	(1)
Au	56.91
Ag	3.17
Hg	39.92
Total	100.00

Poshan district, China; by electron microprobe, corresponding to $(Au_{2.89}Ag_{0.29})_{\Sigma=3.18}Hg_{1.99}$.

Occurrence: In the silicified zone of the silver-rich part of an Au-Ag deposit in biotite granulite.

Association: Gold, silver, acanthite, pyrite, galena, sphalerite, pyrrhotite, scheelite.

Distribution: From the Poshan mining district, Tongbai, Henan Province, China.

Name: Derivation not stated.

Type Material: National Geological Museum, Beijing, China.

References: (1) Li Yuheng, Ouyang Shan, and Tian Peixue (1984) Weishanite — a new gold-bearing mineral. Acta Mineral. Sinica, 4, 102–105 (in Chinese with English abs.). (2) (1988) Amer. Mineral., 73, 196 (abs. ref. 1).

Crystal Data: Triclinic. *Point Group:* 1. As grains of irregular shape, up to 0.5 mm; also prismatic or tabular with striations parallel to the long dimension.

Physical Properties: *Cleavage:* One perfect, two excellent, one good. Hardness = 1.5 VHN = n.d. D(meas.) = 5.79 D(calc.) = 6.1

Optical Properties: Opaque. *Color:* Steel-gray; creamy white in reflected light. *Streak:* Dark gray. *Luster:* Bright metallic. *Pleochroism:* Weak. *Anisotropism:* Strong, blue-green, blue-black, orange-brown, gray.

R_1–R_2: (470) 36.0–38.4, (546) 35.0–36.8, (589) 34.5–35.8, (650) 32.1–33.8

Cell Data: *Space Group:* $P1$, based on morphology. $a = 11.8$ $b = 6.4$ $c = 6.1$ $\alpha = 109.9°$ $\beta = 81.8°$ $\gamma = 105.4°$ Z = 4

X-ray Powder Pattern: Synthetic; strong preferred orientation.
2.973 (100), 3.650 (8), 3.583 (7), 2.839 (7), 2.718 (6), 2.351 (6), 3.752 (5)

Chemistry:

	(1)	(2)
Tl	52.7	52.37
Sb	31.2	31.20
S	16.4	16.43
Total	100.2	100.00

(1) Carlin gold deposit, Nevada, USA; by electron microprobe, average of four analyses.
(2) $TlSbS_2$.

Occurrence: In late stage hydrothermal veins that brecciate silicified dolomitic carbonate rocks. The assemblage appears to fill open space between breccia fragments and to have been formed during the boiling stage at the end of the hydrothermal episode.

Association: Stibnite, quartz.

Distribution: From the east ore zone, Carlin gold mine, Eureka Co., Nevada, USA.

Name: For Dr. Byron G. Weissberg, Chemistry Division, D.S.I.R., New Zealand.

Type Material: Department of Geology, Stanford University, Palo Alto, California, Epithermal Minerals Collection; National Museum of Natural History, Washington, D.C., USA, 144274.

References: (1) Dickson, F.W. and A.S. Radtke (1978) Weissbergite, $TlSbS_2$, a new mineral from the Carlin gold deposit, Nevada. Amer. Mineral., 63, 720–724.

Crystal Data: Hexagonal. *Point Group:* $6/m\ 2/m\ 2/m$. Massive, as small lenses up to 2.5 cm across.

Physical Properties: *Fracture:* Uneven. Hardness = 3 VHN = 145–158 (100 g load). D(meas.) = ~6 D(calc.) = [5.83]

Optical Properties: Opaque. *Color:* Dark bluish black, tarnishing to deep black. *Streak:* Black. *Luster:* Shiny metallic. *Pleochroism:* Distinct. *Anisotropism:* Distinct.
R_1–R_2: (400) 40.3–41.6, (420) 39.8–41.3, (440) 38.9–40.6, (460) 37.9–39.6, (480) 36.9–38.4, (500) 35.9–37.1, (520) 34.8–35.9, (540) 33.7–34.9, (560) 32.6–34.0, (580) 31.5–33.1, (600) 30.5–32.5, (620) 29.6–32.0, (640) 28.9–31.6, (660) 28.1–31.2, (680) 27.3–30.8, (700) 26.4–30.2

Cell Data: *Space Group:* $P6/mmm$. $a = 12.54$ $c = 21.71$ Z = [15]

X-ray Powder Pattern: Synthetic.
2.09 (100), 3.61 (70), 2.01 (70), 3.24 (60), 1.80 (50), 2.55 (30), 2.17 (30)

Chemistry:

	(1)	(2)
Cu	45.84	45.35
Te	53.97	54.65
Total	99.81	100.00

(1) Good Hope mine, Colorado, USA; average of two analyses. (2) Cu_5Te_3.

Occurrence: In hydrothermal veins with other tellurides.

Association: Pyrite, tellurium, sylvanite, petzite, rickardite, sulfur (Colorado, USA); gold, calaverite, krennerite (Kalgoorlie, Australia).

Distribution: In the USA, from the Good Hope and Mammoth Chimney mines, Vulcan, Gunnison Co., Colorado; near Winston, Sierra Co., New Mexico. In Japan, from the Teine mine, Hokkaido. From the Kalgurli gold mines, Kalgoorlie, Western Australia.

Name: For Louis Weiss, owner of the Good Hope mine and discoverer of rickardite.

Type Material: National Museum of Natural History, Washington, D.C., USA, 95782.

References: (1) Palache, C., H. Berman, and C. Frondel (1944) Dana's system of mineralogy, (7th edition), v. I, 199. (2) Thompson, R.M. (1949) The telluride minerals and their occurrence in Canada. Amer. Mineral., 34, 342–382. (3) Patzak, I. (1956) Über die Struktur und die Lage der Phasen im System Kupfer–Tellur. Zeits. für Metallkunde, 47, 418–420 (in German). (4) Ramdohr, P. (1969) The ore minerals and their intergrowths, (3rd edition), 417–418.

Westerveldite $(Fe,Ni)As$

Crystal Data: Orthorhombic. *Point Group:* $2/m\ 2/m\ 2/m$. As irregularly shaped blebs and stringers up to 25 μm.

Physical Properties: Hardness = > 5 VHN = 707–798 (25 g load). D(meas.) = n.d. D(calc.) = 8.13

Optical Properties: Opaque. *Color:* Brownish white to gray. *Pleochroism:* Distinct. *Anisotropism:* Distinct in air, orange brown and bluish in oil.

R_1–R_2: (400) 46.9–48.6, (420) 46.3–48.5, (440) 45.7–48.4, (460) 44.9–48.0, (480) 44.3–48.0, (500) 43.7–47.8, (520) 43.4–47.9, (540) 43.4–47.9, (560) 43.4–48.0, (580) 43.5–48.0, (600) 43.7–48.1, (620) 44.1–48.3, (640) 44.4–48.6, (660) 44.9–48.9, (680) 45.6–49.1, (700) 46.2–49.4

Cell Data: *Space Group:* *Pmcn.* $a = 3.46(1)$ $b = 5.97(1)$ $c = 5.33(1)$ Z = 4

X-ray Powder Pattern: Synthetic FeAs.
2.588 (100), 1.996 (60), 2.635 (55), 2.076 (35), 2.019 (35), 1.725 (30), 2.116 (18)

Chemistry:

	(1)	(2)
Fe	26.6	30.1
Ni	17.4	13.9
Co	0.6	0.5
As	55.5	55.1
Sb		0.5
Total	100.1	100.1

(1) La Gallega mine, Spain; by electron microprobe, average of eight determinations on one grain, corresponding to $(Fe_{0.640}Ni_{0.400}Co_{0.015})_{\Sigma=1.055}As_{1.000}$. (2) Birchtree mine, Canada; by electron microprobe, average of 12 determinations, corresponding to $(Fe_{0.73}Ni_{0.32}Co_{0.01})_{\Sigma=1.06}As_{1.00}$.

Occurrence: As minute inclusions in maucherite which occurs in chromite-nickeline ores. The chromite ores and associated cordierite rock are schlieren veins through serpentinized ultramafic rocks (La Gallega mine, Spain).

Association: Maucherite, nickeline, cobaltite, nickeloan löllingite, Fe-Co-rich gersdorffite, rammelsbergite, antimony, serpentine.

Distribution: At the La Gallega mine, 3 km west of Ojén, Málaga Province, Spain. From the Ilímaussaq Intrusion, southern Greenland. From Seinäjoki, Finland. At the Birchtree mine, near Thompson, Manitoba, Canada.

Name: For Dr. Jan Westerveld (1905–1962), Professor of Geology and Mineralogy, University of Amsterdam, The Netherlands.

Type Material: Geological Institute of the University of Amsterdam; Institute of Earth Sciences of the Free University of Amsterdam, The Netherlands.

References: (1) Oen, I.S., E.A.J. Burke, C. Kieft, and A.B. Westerhof (1972) Westerveldite (Fe, Ni, Co)As, a new mineral from La Gallega, Spain. Amer. Mineral., 57, 354–363. (2) (1962) NBS Mono. 25, 1, 19. (3) Sizgoric, M.B. and C.M. Duesing (1973) Westerveldite, a Canadian occurrence. Can. Mineral., 12, 137–138.

Crystal Data: Monoclinic. *Point Group:* $2/m$. Massive (?).

Physical Properties: Hardness = n.d. VHN = n.d. D(meas.) = n.d. D(calc.) = 6.96

Optical Properties: Opaque. *Color:* Pale grayish yellow. *Pleochroism:* Distinct, pale yellow to grayish yellow. *Anisotropism:* Strong, pink to yellowish green.
R_1–R_2: n.d.

Cell Data: *Space Group:* $I2/m$. $a = 6.22$ $b = 3.63$ $c = 10.52$ $\beta = 90.53°$ Z = 2

X-ray Powder Pattern: Kuusamo, Finland.
2.70 (100), 2.02 (100), 1.800 (100), 2.00 (80), 1.815 (80), 1.532 (60), 1.497 (60)

Chemistry:

	(1)	(2)
Ni	33.7	35.80
Co	1.0	
Cu	trace	
Fe	trace	
Se	65.3	64.20
Total	100.0	100.00

(1) Kuusamo, Finland; by electron microprobe. (2) Ni_3Se_4.

Polymorphism & Series: Dimorphous with trüstedtite.

Occurrence: As a primary phase and as an alteration product of sederholmite, in albite diabase sills that cut a schist formation, associated with low-grade uranium mineralization.

Association: Sederholmite, penroseite, selenium, ferroselite, selenian vaesite, cattierite, calcite.

Distribution: From Kuusamo, northeastern Finland.

Name: For W.W. Wilkman, geologist.

Type Material: n.d.

References: (1) Vuorelainen, Y., A. Huhma, and A. Häkli (1964) Sederholmite, wilkmanite, kullerudite, mäkinenite, and trüstedtite, five new nickel selenide minerals. Compt. Rendus Soc. Géol. Finlande, 36, 113–125. (2) (1965) 50, 519 (abs. ref. 1).

Willyamite $(Co,Ni)SbS$

Crystal Data: Monoclinic or triclinic, pseudocubic. *Point Group:* n.d. Crystals several mm on a side, showing zonal growth patterns.

Physical Properties: *Fracture:* Uneven. *Tenacity:* Brittle. Hardness = 5–5.5
VHN = Slightly higher than ullmannite. D(meas.) = 6.76(3) D(calc.) = n.d.

Optical Properties: Opaque. *Color:* Steel-gray. *Luster:* Metallic.
R_1–R_2: n.d.

Cell Data: *Space Group:* n.d. $a = 5.86$ Z = 4

X-ray Powder Pattern: Broken Hill, Australia. (JCPDS 26-1106).
2.621 (100), 2.390 (70), 1.767 (60), 1.565 (50), 1.625 (40), 1.278 (40), 1.087 (40)

Chemistry:

	(1)	(2)	(3)
Co	13.88	20.6	23.2
Ni	13.41	6.9	3.8
Fe	trace	0.2	0.4
Sb	56.78	55.9	54.7
As		0.7	1.6
S	15.78	15.1	14.6
Total	99.85	99.4	98.3

(1) Broken Hill, Australia. (2–3) Do.; by electron microprobe.

Polymorphism & Series: Forms a series with ullmannite.

Occurrence: In calcite-siderite veins.

Association: Dyscrasite, calcite, siderite.

Distribution: In the Consols lode, Broken Hill, Willyama Township, New South Wales, Australia.

Name: For the locality in Willyama Township, New South Wales, Australia.

Type Material: National Museum of Natural History, Washington, D.C., USA, R849.

References: (1) Palache, C., H. Berman, and C. Frondel (1944) Dana's system of mineralogy, (7th edition), v. I, 301–302. (2) Cabri, L.J., D.C. Harris, J.M. Stewart, and J.F. Rowland (1970) Willyamite redefined. Proc. Australasian Inst. Mining Met., 233, 95–100. (3) (1971) Amer. Mineral., 56, 361 (abs. ref. 2).

Crystal Data: Orthorhombic. *Point Group:* 222. Usually massive. Crystals rare, prismatic parallel to [001]; prisms show striae and are often blocky in aspect.

Physical Properties: *Fracture:* Conchoidal. *Tenacity:* Brittle. Hardness = 2–3
VHN = 170–187 (100 g load). D(meas.) = 6.01, 6.19 (synthetic). D(calc.) = 6.19

Optical Properties: Opaque. *Color:* Steel-gray to tin-white, tarnishing pale lead-gray or brass-yellow; white to creamy white to grayish white in reflected light. *Streak:* Black. *Luster:* Metallic. *Anisotropism:* Weak, shades of dark brown.

R_1–R_2: (400) 35.0–36.9, (420) 34.1–35.8, (440) 33.2–34.7, (460) 32.7–34.0, (480) 32.6–34.1, (500) 32.9–34.7, (520) 33.2–35.0, (540) 33.5–35.2, (560) 33.6–35.3, (580) 33.5–35.3, (600) 33.2–35.2, (620) 33.0–35.2, (640) 32.8–35.2, (660) 32.4–35.0, (680) 31.7–34.6, (700) 30.8–34.0

Cell Data: *Space Group:* $P2_12_12_1$. $a = 7.723$ $b = 10.395$ $c = 6.716$ Z = 4

X-ray Powder Pattern: Wittichen, Germany.
2.85 (100), 3.08 (80), 4.55 (40), 2.66 (40), 3.83 (30), 3.19 (30), 2.39 (30)

Chemistry:

	(1)	(2)	(3)	(4)
Cu	37.79	38.0	37.7	38.46
Bi	42.56	42.8	43.4	42.15
S	19.13	18.3	18.7	19.39
Total	99.48	99.1	99.8	100.00

(1) Daniel mine, Wittichen, Germany; by electron microprobe. (2) Seathwaite Tarn, England; by electron microprobe. (3) Wittichen, Germany; by electron microprobe. (4) Cu_3BiS_3.

Occurrence: In hydrothermal veins with other bismuth minerals (Wittichen, Germany); with Cu-Fe sulfides (Seathwaite Tarn, England); with secondary uranium minerals and selenides of Cu, Pb, and Bi (Kletno, Poland).

Association: Bornite, chalcocite, chalcopyrite, djurleite, digenite, tennantite, pyrite, stromeyerite, bismuth, emplectite, rammelsbergite, calcite, aragonite, fluorite, barite.

Distribution: In the USA, at Butte, Silver Bow Co., Montana; in the Fairfax quarry, Centreville, Fairfax Co., Virginia; and at Bisbee, Cochise Co., Arizona. In Canada, in the Maid of Erin mine, Rainy Hollow district, British Columbia; and at Cobalt, Ontario. In Germany, in the Schapbachtal, and at the Neuglück, Daniel, and King David mines, near Wittichen, Black Forest. In the Sedmochislenitsi mine, Vratsa district, Bulgaria. At Colquijirca, Peru. From Seathwaite Tarn, northwest of Coniston, Cumbria, England. At Tjøstulflaten, Telemark, Norway. In the Kletno mine, Sudetes Mountains, Poland. Additional localities are known.

Name: After the type locality, Wittichen, Germany.

Type Material: Royal Ontario Museum, Toronto, Canada, M23304; National Museum of Natural History, Washington, D.C., C770; Harvard University, Cambridge, Massachusetts, USA, 82408, 82409.

References: (1) Nuffield, E.W. (1947) Studies of mineral sulpho-salts: XI—wittichenite (klaprothite). Econ. Geol., 42, 147–160. (2) Criddle, A.J. and C.J. Stanley (1979) New data on wittichenite. Mineral. Mag., 43, 109–113. (3) Kocman, V. and E.W. Nuffield (1973) The crystal structure of wittichenite, Cu_3BiS_3. Acta Cryst., B29, 2528–2535. (4) Palache, C., H. Berman, and C. Frondel (1944) Dana's system of mineralogy, (7th edition), v. I, 373 (wittichenite) and 418–419 (klaprothite).

Crystal Data: Monoclinic. *Point Group:* $2/m$. Massive, striations parallel [010].

Physical Properties: *Cleavage:* Good, giving platy appearance. Hardness = 2–2.5 VHN = 45–63 (25 g load). D(meas.) = 7.12 D(calc.) = [6.91]

Optical Properties: Opaque. *Color:* Lead-gray; cream-white in reflected light. *Streak:* Black. *Luster:* Metallic.

R_1–R_2: (400) 43.2–45.3, (420) 42.8–44.9, (440) 42.4–44.5, (460) 41.9–44.2, (480) 41.5–43.8, (500) 41.1–43.3, (520) 40.5–42.8, (540) 40.0–42.3, (560) 39.4–41.8, (580) 38.9–41.3, (600) 38.4–40.9, (620) 38.0–40.6, (640) 37.7–40.4, (660) 37.4–40.2, (680) 37.3–39.8, (700) 37.2–39.7

Cell Data: *Space Group:* $P2/m$ subcell: $a = 4.19(1)$ $b = 4.08(1)$ $c = 15.56(4)$ $\beta = 101.35(16)°$ Z = [1] *Space Group:* $C2/m$ subcell: $a = 7.21(2)$ $b = 4.08(1)$ $c = 15.50(5)$ $\beta = 98.75(15)°$ Z = [1]

X-ray Powder Pattern: Falun, Sweden.
3.004 (s), 2.887 (sb), 2.041 (s), 3.434 (ms), 3.848 (m), 3.607 (m), 3.529 (m)

Chemistry:

	(1)	(2)
Pb	33.2	35.3
Bi	45.5	43.9
Se	9.7	7.7
S	11.8	12.6
Total	100.2	99.5

(1) Falun, Sweden; average of two analyses. (2) Do.; by electron microprobe.

Occurrence: In an amphibolite rock (Falun, Sweden); in garnet tactite (Middlemarch Canyon, Arizona, USA).

Association: Nordströmite, friedrichite, magnetite, pyrite, cordierite, quartz (Falun, Sweden).

Distribution: From Falun, Sweden. At Middlemarch Canyon, Cochise Co., Arizona, USA.

Name: For T. Witt, Swedish mining engineer.

Type Material: Swedish Museum of Natural History, Stockholm, Sweden; Royal Ontario Museum, Toronto, Canada, M12992; National School of Mines, Paris, France.

References: (1) Palache, C., H. Berman, and C. Frondel (1944) Dana's system of mineralogy, (7th edition), v. I, 451. (2) Mumme, W.G. (1976) Proudite from Tennant Creek, Northern Territory, Australia: its crystal structure and relationship with weibullite and wittite. Amer. Mineral., 61, 839–852. (3) Mumme, W.G. (1980) Seleniferous lead–bismuth sulfosalts from Falun, Sweden: weibullite, wittite and nordströmite. Amer. Mineral., 65, 789–796.

Crystal Data: Hexagonal. *Point Group:* $6mm$. Crystals to 1.5 cm, commonly hemimorphic pyramidal $\{50\bar{5}2\}$ and $\{10\bar{1}1\}$; also short prismatic to tabular $\{0001\}$; usually striated horizontally on $\{10\bar{1}0\}$ and $\{10\bar{1}1\}$. The polytypes show steepening of the pyramid as the repeat distance increases along $\{0001\}$. As concentrically banded crusts, fibrous or columnar.

Physical Properties: *Cleavage:* Easy on $\{11\bar{2}0\}$; difficult on $\{0001\}$. *Tenacity:* Brittle. Hardness = 3.5–4 VHN = n.d. D(meas.) = 4.09 D(calc.) = 4.10

Optical Properties: Translucent. *Color:* Deep reddish brown to dark brown to brown-black; yellow to dark brown internal reflections common in reflected light. *Streak:* Brown. *Luster:* Resinous, brilliant submetallic on crystal faces.

Optical Class: Uniaxial positive (+). $\epsilon = 2.378$ $\omega = 2.356$

R: (400) 19.2, (420) 18.7, (440) 18.2, (460) 17.8, (480) 17.4, (500) 17.1, (520) 16.9, (540) 16.6, (560) 16.4, (580) 16.3, (600) 16.2, (620) 16.1, (640) 16.1, (660) 16.1, (680) 16.0, (700) 15.8.

Cell Data: *Space Group:* $P6_3mc$ (synthetic, 2H polymorph). $a = 3.820$ $c = 6.260$ Z = 2

X-ray Powder Pattern: Synthetic (2H polymorph).
3.309 (100), 3.128 (86), 2.925 (84), 1.911 (74), 1.764 (52), 1.630 (45), 2.273 (29)

Chemistry:

	(1)	(2)
Zn	62.64	67.10
Fe	2.43	
Cd	1.84	
Pb	0.41	
S	32.10	32.90
Total	99.42	100.00

(1) Příbram, Czechoslovakia. (2) ZnS.

Polymorphism & Series: Trimorphous with matraite and sphalerite. Polymorphs 2H, 4H, 6H, 8H, 10H, 15R, 18R, and 21R are known.

Occurrence: Of hydrothermal origin in veins with other sulfides. Also along shrinkage fractures in clay-ironstone concretions, of low-temperature origin.

Association: Sphalerite, pyrite, chalcopyrite, barite, marcasite.

Distribution: Numerous localities; a few of those providing good crystals or especially rich material follow. In the USA, from near the Thomaston Dam, Litchfield Co., Connecticut; at Butte, Silver Bow Co., Montana; at Frisco, Beaver Co., Utah; and from the Joplin district, Jasper Co., Missouri. From Rachelshausen, near Gladenbach, Hesse, Germany. At Liskeard, Cornwall, England. From Mežica (Mies), Yugoslavia. At Příbram, Czechoslovakia. From Romania, at Baia Sprie (Felsőbánya). At Quispisiza, near Castro Virreyna, Peru. In a number of mines in the Huanuni district, Oruro, and Chocaya, Potosí, Bolivia.

Name: For Charles Adolphe Wurtz (1817–1884), French chemist, of Paris, France.

References: (1) Palache, C., H. Berman, and C. Frondel (1944) Dana's system of mineralogy, (7th edition), v. I, 226–228. (2) Frondel, C. and C. Palache (1950) Three new polymorphs of zinc sulfide. Amer. Mineral., 35, 29–42. (3) (1953) NBS Circ. 539, 2, 14.

Crystal Data: Monoclinic. *Point Group:* $2/m$. Tabular on {001}, frequently producing flat rhombs, also as laths elongated along [010], to 0.5 cm; rarely pyramidal. Reniform and as hemispherical radial aggregates. *Twinning:* On {001} producing pseudo-orthorhombic twins.

Physical Properties: *Cleavage:* Distinct on {001}. *Fracture:* Subconchoidal. *Tenacity:* Brittle. Hardness = 2–3 VHN = 71–93 (25 g load). D(meas.) = 5.54(14) D(calc.) = 5.53

Optical Properties: Translucent. *Color:* Dark cochineal-red to dull orange to clove-brown; lemon-yellow in transmitted light. *Streak:* Orange-yellow. *Luster:* Adamantine.
Optical Class: Biaxial negative (–). *Orientation:* $X \sim c$; $Y \sim a$; $Z = b$. *Dispersion:* $r < v$, very strong. $n = \sim 3$ 2V(meas.) = 34° 2V(calc.) = n.d.
R_1–R_2: (400) 33.7–37.1, (420) 32.2–36.1, (440) 30.7–35.1, (460) 29.1–33.7, (480) 27.6–32.1, (500) 26.1–30.1, (520) 24.8–28.5, (540) 23.7–27.3, (560) 22.7–26.3, (580) 22.0–25.5, (600) 21.5–24.8, (620) 21.3–24.4, (640) 21.0–24.1, (660) 20.8–23.8, (680) 20.6–23.6, (700) 20.4–23.5

Cell Data: *Space Group:* $C2/c$. $a = 12.00(1)$ $b = 6.26(1)$ $c = 17.08(1)$ $\beta = 110.0°$ Z = 8

X-ray Powder Pattern: Jáchymov, Czechoslovakia.
3.00 (100), 2.82 (60), 3.14 (30), 2.13 (30), 5.5 (20), 4.02 (20), 3.38 (20)

Chemistry:

	(1)	(2)
Ag	65.15	65.42
As	14.63	15.14
S	19.07	19.44
Total	98.85	100.00

(1) Freiberg, Germany. (2) Ag_3AsS_3.

Polymorphism & Series: Dimorphous with proustite.

Occurrence: Usually associated with other silver sulfosalts in hydrothermal veins.

Association: Proustite, pyrargyrite, acanthite, arsenic, calcite.

Distribution: In small amounts at numerous localities. From Jáchymov (Joachimsthal) and Příbram, Czechoslovakia. In Saxony, Germany, at Freiberg, Annaberg, Marienberg, Johanngeorgenstadt, and Schneeberg; at St. Andreasberg, in the Harz Mountains. From Baia Sprie (Felsőbánya), Romania. At Sainte-Marie-aux-Mines, Haut-Rhin, France. From Chañarcillo, Atacama, Chile. In several mines at Cobalt, Ontario, Canada. At the General Petite mine, Atlanta district, Elmore Co., Idaho, and the Flathead mine, Niarada Co., Montana, USA. From the Batopilas district, Chihuahua, Mexico.

Name: From the Greek for *yellow* and *powder* in allusion to its color.

References: (1) Palache, C., H. Berman, and C. Frondel (1944) Dana's system of mineralogy, (7th edition), v. I, 371–372. (2) Peacock, M.A. (1950) Studies of mineral sulfo-salts: XV. Xanthoconite and pyrostilpnite. Mineral. Mag., 29, 346–358. (3) Engel, P. and W. Nowacki (1968) Die Kristallstruktur von Xanthokon, Ag_3AsS_3. Acta Cryst., B24, 77–81 (in German).

Crystal Data: Monoclinic. *Point Group:* $2/m$, 2, or m. As prismatic crystals, elongated and striated on {010}, to 8 mm. *Twinning:* On (001).

Physical Properties: Hardness = n.d. VHN = 103 average (5 to 200 g load). D(meas.) = 7.08 D(calc.) = 7.07

Optical Properties: Opaque. *Color:* Lead-gray; in reflected light, white with a faint blue tint. *Streak:* Gray. *Luster:* Metallic. *Pleochroism:* Distinct, white to white with blue tint. *Anisotropism:* Distinct, dark gray to gray.

R_1–R_2: (405) 43.2–46.6, (436) 44.5–46.8, (480) 44.3–46.8, (526) 44.2–44.9, (546) 41.8–44.5, (578) 41.4–44.1, (589) 40.9–43.8, (622) 40.5–43.3, (644) 40.0, 43.0, (656) 39.9–43.0, (664) 39.8–43.0, (700) 37.5–41.2

Cell Data: *Space Group:* $C2/m$, $C2$, or Cm. $a = 13.65$ $b = 4.078$ $c = 20.68$ $\beta = 93.0°$ Z = 4

X-ray Powder Pattern: Chaobuleng district, China.
3.386 (100), 2.177 (90), 2.073 (80), 2.051 (70), 1.955 (70), 1.788 (6), 1.396 (5)

Chemistry:

	(1)	(2)	(3)
Pb	52.074	52.06	50.45
Zn	0.653		
Cu	0.16		
Ag	0.75	0.50	
Bi	29.72	29.81	33.93
Sb	0.09		
S	15.09	15.25	15.62
oxides	1.333	1.62	
Total	99.87	99.24	100.00

(1) Chaobuleng district, China; ignoring minor components and oxides, corresponds to $Pb_{3.18}Bi_{1.81}S_{6.00}$. (2) Do.; by electron microprobe. (3) $Pb_3Bi_2S_6$.

Occurrence: In a skarn-type iron deposit.

Association: Magnetite, sphalerite, pyrrhotite, pyrite, arsenopyrite, chalcopyrite, digenite, bornite, molybdenite, galena, bismuth, bismuthinite.

Distribution: In the Chaobuleng district, Xilingola League, Inner Mongolia Autonomous Region, China.

Name: For the Xilingola locality in China.

Type Material: n.d.

References: (1) Hong Huidi, Wang Xiangwen, Shi Nicheng, and Peng Zhizhong (1982) Xilingolite, a new sulfide of lead and bismuth, $Pb_{3+x}Bi_{2-2/3x}S_6$. Acta Petrologica Mineralogica et Analytica, 1, 14–18 (in Chinese with English abs.). (2) (1984) Amer. Mineral., 69, 409 (abs. ref. 1).

Xingzhongite $(Pb, Cu, Fe)(Ir, Pt, Rh)_2S_4$

Crystal Data: Cubic or pseudocubic. *Point Group:* n.d. As a rim around iridosmine.

Physical Properties: Hardness = n.d. VHN = 753 (50 g load). D(meas.) = n.d. D(calc.) = [7.94]

Optical Properties: Opaque. *Color:* Steel-gray; in reflected light, bluish gray. *Luster:* Metallic.
R: (466) 40.5, (544) 38.9, (589) 41.1, (656) 41.0

Cell Data: *Space Group:* n.d. $a = 9.970$ Z = [8]

X-ray Powder Pattern: China.
1.765 (10), 5.80 (8), 3.00 (8), 1.208 (8), 2.49 (6), 2.88 (5), 1.335 (4)

Chemistry:

	(1)
Pb	12.80
Cu	3.81
Fe	1.58
Ir	43.49
Pt	9.67
Rh	7.19
S	21.68
Total	100.22

(1) China; by electron microprobe, giving $(Pb_{0.37}Cu_{0.35}Fe_{0.17})_{\Sigma=0.89}(Ir_{1.33}Rh_{0.41}Pt_{0.29})_{\Sigma=2.03}S_{4.00}$.

Occurrence: In dunite-hosted platinum ores, related to chromium mineralization (China).

Association: Pt–Fe alloy, osmiridium, iridosmine, osmium, iridium, erlichmanite, cooperite, irarsite, osarsite, chromite, magnetite, pyrite, olivine, serpentine, talc (China).

Distribution: From an unstated locality in China. At Fox Gulch, Goodnews Bay, Alaska, USA. From Tiébaghi, New Caledonia.

Name: Presumably for the undefined Chinese type locality.

Type Material: n.d.

References: (1) Yu Tsu-Hsiang, Lin Shu-Jen, Chao Pao, Fang Ching-Sung, and Huang Chi-Shun (1974) A preliminary study of some new minerals of the platinum group and another associated new one in platinum-bearing intrusions in a region in China. Acta Geol. Sinica, 2, 202–218 (in Chinese with English abs.). (2) (1976) Amer. Mineral., 61, 184–185 (abs. ref. 1). (3) Cabri, L.J., Ed. (1981) Platinum group elements: mineralogy, geology, recovery. Can. Inst. Min. & Met., 146–147. (4) Peng Zhiizhong, Chang Chiehung, and Ximen Lovlov (1978) Discussion on published articles in the research of new minerals of the platinum-group discovered in China in recent years. Acta Geol. Sinica, 4, 326–336 (in Chinese with English abs.). [Peng Zhiizhong formerly Pen Chih-Zhong]. (5) (1980) Amer. Mineral., 65, 408 (abs. ref. 4). (6) Institute of Geochemistry, Chinese Academy of Sciences (1981) Platinum deposits in China, geochemistry of the platinum group elements, and platinum group minerals. Science Publishing Agency, Beijing, China, 190 p. (in Chinese). (7) (1984) Amer. Mineral., 69, 412 (abs. ref. 6).

Crystal Data: Hexagonal. *Point Group:* $\overline{3}\ 2/m$, $3m$, or 32. Massive and as stellate aggregates.

Physical Properties: *Cleavage:* {0001}. Hardness = n.d. VHN = 93–98 (15 g load). D(meas.) = n.d. D(calc.) = 4.89

Optical Properties: Opaque. *Color:* Blue; blue with a slight violet tint in oil under reflected light. *Streak:* n.d. *Luster:* Metallic. *Pleochroism:* Blue to bluish white. *Anisotropism:* Strong, orange-red.

R_1–R_2: (400) 19.2–32.9, (420) 18.6–32.3, (440) 17.9–31.4, (460) 16.9–30.5, (480) 15.6–29.5, (500) 14.2–28.2, (520) 12.4–27.0, (540) 10.7–25.6, (560) 9.08–24.3, (580) 7.51–23.1, (600) 6.15–22.2, (620) 5.02–22.2, (640) 4.33–23.0, (660) 4.38–23.6, (680) 5.96–23.4, (700) 9.44–23.0

Cell Data: *Space Group:* $P\overline{3}m1$, $P3m1$ or $P321$. $a = 3.800(1)$ $c = 67.26(4)$ Z = 3

X-ray Powder Pattern: Spionkop Creek, Canada.
1.899 (100), 3.061 (55), 2.767 (35), 5.032 (30), 3.678 (25), 2.849 (25), 5.955 (20)

Chemistry:

	(1)	(2)	(3)
Cu	68.5	68.5	69.03
Fe		0.1	
Ag		0.2	
S	30.4	31.5	30.97
Total	98.9	100.3	100.00

(1) Yarrow and Spionkop Creeks, Canada; by electron microprobe. (2) Cannington Park, England; by electron microprobe. (3) Cu_9S_8.

Occurrence: As weathering-produced lamellar replacements of other copper sulfides in stratabound red-bed copper deposits (Yarrow and Spionkop Creeks, Canada).

Association: Anilite, djurleite, spionkopite, tennantite (Yarrow and Spionkop Creeks, Canada).

Distribution: In the Upper Grinnell Formation, Spionkop Creek and Yarrow Creek areas of southwestern Alberta, Canada. From Cannington Park, Bridgwater, Somerset, England. In the High Rolls district, Otero Co., New Mexico, USA.

Name: For the type locality at Yarrow Creek, Canada.

Type Material: Canadian Geological Survey, Ottawa; Queen's University, Kingston, Ontario, Canada; National Museum of Natural History, Washington, D.C., 149430, 149431; Harvard University, Cambridge, Massachusetts, USA, 122290.

References: (1) Goble, R.J. (1980) Copper sulfides from Alberta: yarrowite, Cu_9S_8, and spionkopite, $Cu_{39}S_{28}$. Can. Mineral., 18, 511–518. (2) (1981) Amer. Mineral., 66, 1279 (abs. ref. 1).

Crystal Data: Hexagonal. *Point Group:* $3m$ (?). Fine flaky, scaly aggregates up to 8 mm, and veinlets to 12 mm.

Physical Properties: *Tenacity:* Plastic. Hardness = < 1 VHN = n.d. D(meas.) = 2.94 D(calc.) = 3.00

Optical Properties: Opaque. *Color:* Pinkish violet. *Anisotropism:* Strong, orange-red with weak lilac tint in isotropic sections, red-lilac to pinkish dark gray in anisotropic sections.
R: (400) — , (420) — , (440) 26.0, (460) 22.4, (480) 19.0, (500) 17.6, (520) 17.8, (540) 18.7, (560) 20.7, (580) 23.3, (600) 26.4, (620) 29.6, (640) 32.8, (660) 35.7, (680) 38.2, (700) 40.1

Cell Data: *Space Group:* $P3m1$ (?). $a = 3.21$ $c = 11.3$ Z = n.d.

X-ray Powder Pattern: Pay-Khoy, USSR.
5.68 (10), 2.76 (6), 1.575 (6), 1.596 (4), 11.4 (3), 1.534 (3), 1.386 (3)

Chemistry:

	(1)
V	32.48
S	32.93
Mg	10.17
Al	5.66
O	19.89
H	1.26
Total	102.39

(1) Pay-Khoy, USSR; by electron microprobe, O and H calculated, average of three analyses, corresponding to $V_{1-x}S \cdot n(Mg, Al)(OH)_2$, n = 0.612, assigned by analogy to valleriite.

Occurrence: In quartz-carbonate veins in carbonate rocks.

Association: Cadmian sphalerite, sulvanite, fluorite.

Distribution: From Pay-Khoy, in the middle stream of the Silova-Yakha River, USSR.

Name: For N.P. Yushkin, Russian mineralogist.

Type Material: Komi Branch of Academy of Sciences; Institute of Geology of Ore Deposits, Petrography, Mineralogy and Geochemistry; A.E. Fersman Mineralogical Museum, Academy of Sciences, Moscow, USSR.

References: (1) Makeev, A.B., T.L. Evstigneeva, N.V. Troneva, L.N. Vyal'sov, A.I. Gorshkov, and N.V. Trubkin (1984) Yushkinite, $V_{1-x}S \cdot [n(Mg, Al)(OH)_2]$ — A new mineral. Mineral. Zhurnal, 6-5, 91–97 (in Russian). (2) (1986) Amer. Mineral., 71, 846 (abs. ref. 1).

Crystal Data: Hexagonal. *Point Group:* $6/m\ 2/m\ 2/m$. As small irregular plates, sometimes striated in multiple directions producing a cross-hatching, and with hummocky surfaces.

Physical Properties: *Cleavage:* Perfect on {0001}. *Tenacity:* Brittle. Hardness = 2 VHN = n.d. D(meas.) = 6.9–7.2 D(calc.) = 7.135

Optical Properties: Opaque. *Color:* White. *Streak:* White and slightly grayish. *Luster:* Metallic.

R_1–R_2: n.d.

Cell Data: *Space Group:* $P6_3/mmc$. $a = 2.665$ $c = 4.947$ $Z = 2$

X-ray Powder Pattern: Synthetic.
2.091 (100), 2.473 (3), 2.308 (40), 1.687 (28), 1.342 (25), 1.1729 (23), 1.332 (21)

Chemistry:

	(1)
Zn	~90
Sn, Pb, Cd	~10
Fe, Mn, B, Si, Cu, Ag, Ca, Ba	trace
Total	

(1) Elsa mine, Canada; combined X-ray fluorescence and spectrographic analyses.

Occurrence: In the oxidized zone of Pb-Zn-Ag deposits, derived from sphalerite by oxidation (Elsa mine, Canada); coatings on fibrous volcanic glass, as a volcanic sublimate (Mount Elbrus, USSR); in platinum concentrates (Aurora deposit, USSR).

Association: Silver, sulfur, oxidized sphalerite, "limonite", manganese oxides, cerussite, anglesite, freibergite, galena (Elsa mine, Canada); copper, Cu–Zn alloy, sphalerite, djurleite, cuprite (Dulcina mine, Chile).

Distribution: In the Elsa mine, Keno Hill-Galena Hill area, Yukon Territory, Canada. From Mount Elbrus, Caucasus Mountains, and the Aurora deposit, locality not otherwise specified, USSR. In the Dulcina de Lampos copper mine, near Copiapó, Chile.

Name: From the German *zink*, of obscure origin.

References: (1) Palache, C., H. Berman, and C. Frondel (1944) Dana's system of mineralogy, (7th edition), v. I, 127. (2) (1953) NBS Circ. 539, 1, 16. (3) Boyle, R.W. (1961) Native zinc at Keno Hill. Can. Mineral., 6, 692–694. (4) Bartikyan, P.M. (1966) Native lead and zinc in the rocks of Armenia. Zap. Vses. Mineral. Obshch., 95, 99–102 (in Russian). (5) (1962) Mineral. Abs., 18, 200. (abs. ref. 4).

Zinkenite $Pb_9Sb_{22}S_{42}$

Crystal Data: Hexagonal. *Point Group:* 6. Thin prismatic with striations along [0001], to 5 cm long, also as columnar or radial fibrous to felted aggregates; massive.

Physical Properties: *Cleavage:* Indistinct on $\{11\bar{2}0\}$. *Fracture:* Uneven. Hardness = 3–3.5 VHN = n.d. D(meas.) = 5.33 D(calc.) = 5.34

Optical Properties: Opaque. *Color:* Steel-gray, occasionally tarnished iridescent; gray-white in reflected light. *Streak:* Steel-gray. *Luster:* Metallic. *Anisotropism:* Distinct.
R_1–R_2: (400) 39.1–44.2, (420) 39.2–44.4, (440) 39.3–44.6, (460) 39.2–44.5, (480) 39.0–44.2, (500) 38.8–43.8, (520) 38.4–43.2, (540) 37.9–42.5, (560) 37.3–41.8, (580) 36.7–41.0, (600) 36.0–40.4, (620) 35.4–39.8, (640) 34.6–39.0, (660) 34.0–38.2, (680) 33.2–37.3, (700) 32.4–36.3

Cell Data: *Space Group:* $P6_3$. $a = 44.15$ $c = 8.62$ Z = 12

X-ray Powder Pattern: Wolfsberg, Germany.
3.45 (100), 2.81 (40), 1.985 (30), 1.828 (30), 3.02 (20), 2.13 (20), 2.06 (20)

Chemistry:

	(1)	(2)	(3)
Pb	34.33	32.77	31.66
Cu	0.70	1.20	
Fe	0.06	0.02	
Sb	42.15	35.00	45.48
As		5.64	
S	22.63	22.50	22.86
rem.		1.58	
Total	99.87	98.71	100.00

(1) Wolfsberg, Germany. (2) Red Mountain district, Colorado, USA. (3) $Pb_9Sb_{22}S_{42}$.

Occurrence: As a constituent of hydrothermal veins, associated with base metal and tin sulfides and sulfosalts.

Association: Stibnite, jamesonite, boulangerite, bournonite, plagionite, fülöppite, cassiterite, stannite, andorite, pyrite, sphalerite, chalcopyrite, arsenopyrite, galena.

Distribution: In Germany, from Wolfsberg, in the Harz Mountains; and at Aldersbach, in the Kinzigtal, Bayern. From Săcărâmb (Nagyág) and Baia Mare (Nagybánya), Romania. At Peschadoire, near Pontgibaud, Puy-de-Dôme, France. Fine crystals from the San José, Itos, and other mines, Oruro, Bolivia. At the Brobdignag prospect, near Silverton, Red Mountain district, San Juan Co., Colorado; from Morey, Nye Co., and Eureka, Eureka Co., Nevada, USA. In Canada, from Bonanza Creek, Bridge River district, British Columbia; and the Yellowknife Bay area, Northwest Territories. From Grainsgill, Carrock Fell, Cumbria, England. Known from a few additional localities.

Name: For J.K.L. Zinken (sometimes Zincken) (1798–1862), German mineralogist and mining geologist.

References: (1) Palache, C., H. Berman, and C. Frondel (1944) Dana's system of mineralogy, (7th edition), v. I, 476–478. (2) Berry, L.G. and R.M. Thompson (1962) X-ray powder data for the ore minerals. Geol. Soc. Amer. Mem. 85, 165–166. (3) Harris, D.C. (1965) Zinckenite. Can. Mineral., 9, 381–382. (4) Portheine, J.C. and W. Nowacki (1975) Refinement of the crystal structure of zinckenite, $Pb_6Sb_{14}S_{27}$. Zeits. Krist., 141, 79–96. (5) Lebas, G. and M.-T. Le Bihan (1976) Étude chimique et structurale d'un sulfure naturel; la zinkénite. Bull. Soc. fr. Minéral., 99, 351–360 (in French). (6) Smith, P.P.K. (1986) Direct imaging of tunnel cations in zinkenite by high-resolution electron microscopy. Amer. Mineral., 71, 194–201.

Crystal Data: Orthorhombic. *Point Group:* n.d. Long, irregular lath-shaped grains to 0.5 mm.

Physical Properties: *Fracture:* Uneven. Hardness = ~3 VHN = 154–170, 166 average (15.2 g load). D(meas.) = n.d. D(calc.) = 5.15

Optical Properties: Opaque. *Color:* Steel-gray; in reflected light, white with a faint yellowish tint to light gray. *Streak:* Black. *Luster:* Metallic. *Anisotropism:* Strong, from light to dark gray with a faint greenish tint.

R_1–R_2: (400) — , (420) 38.6–44.3, (440) 38.5–44.0, (460) 38.4–44.0, (480) 38.0–43.7, (500) 37.9–43.8, (520) 37.9–43.9, (540) 37.6–43.6, (560) 37.4–43.1, (580) 36.9–42.6, (600) 36.5–41.9, (620) 36.2–41.5, (640) 35.6–40.9, (660) 35.4–40.2, (680) 34.9–39.2, (700) 34.2–38.1

Cell Data: *Space Group:* n.d. $a = 18.698(8)$ $b = 6.492(3)$ $c = 4.577(1)$ Z = 1

X-ray Powder Pattern: Příbram, Czechoslovakia.
1.797 (10), 2.222 (9), 1.325 (8), 2.392 (6), 3.070 (5), 1.956 (5), 3.670 (4)

Chemistry:

	(1)	(2)	(3)
Ag	5.75	5.88	6.18
Pb	46.30	47.26	47.52
Zn	0.35	0.22	
Cu	0.27	0.22	
Fe	0.02	0.05	
Sb	27.39	28.53	27.92
S	18.30	18.75	18.38
Total	98.38	100.91	100.00

(1–2) Příbram, Czechoslovakia; by electron microprobe, each an average of 15 analyses on 2 different specimens, corresponding to $(Ag_{0.93}Cu_{0.07})_{\Sigma=1.00}(Pb_{3.92}Zn_{0.10}Fe_{0.01})_{\Sigma=4.03}Sb_{3.94}S_{10.00}$. (3) $AgPb_4Sb_4S_{10}$.

Occurrence: In hydrothermal veins.

Association: Boulangerite, galena, diaphorite, argentian tetrahedrite, sphalerite.

Distribution: From Příbram, Czechoslovakia.

Name: To honor Academician Vladimír Zoubek, former Director of the Geological Survey and Geological Institute of the Czechoslovak Academy of Science, Prague, Czechoslovakia.

Type Material: Mining Museum, Příbram, 432, 220; Department of Mineralogy, National Museum, Prague, Czechoslovakia.

References: (1) Megarskaya, L., D. Rykl, and Z. Táborský (1986) Zoubekite, $AgPb_4Sb_4S_{10}$, a new mineral from Příbram, Czechoslovakia. Neues Jahrb. Mineral., Monatsh., 1–7. (2) (1987) Amer. Mineral., 72, 227 (abs. ref. 1).

Zvyagintsevite $(Pd, Pt, Au)_3(Pb, Sn)$

Crystal Data: Cubic. *Point Group:* $4/m\ \overline{3}\ 2/m$. As irregular grains up to 250 μm; massive in veinlets up to 120 μm long.

Physical Properties: Hardness = n.d. VHN = 241–318, 279 average (15 g load). D(meas.) = 13.32 (synthetic). D(calc.) = 13.42

Optical Properties: Opaque. *Color:* In reflected light, bright white with a cream tint. *Streak:* Black. *Luster:* Metallic.
R: (400) — , (420) — , (440) 62.0, (460) 63.3, (480) 63.0, (500) 63.7, (520) 64.6, (540) 65.5, (560) 66.7, (580) 67.7, (600) 67.4, (620) 67.3, (640) 67.9, (660) 68.0, (680) 68.0, (700) 68.1

Cell Data: *Space Group:* $Pm3m$ (synthetic). $a = 4.02(1)$ Z = 1

X-ray Powder Pattern: Synthetic Pd_3Pb.
2.32 (100), 1.215 (90), 2.01 (80), 1.423 (70), 0.923 (60), 0.900 (50), 1.163 (40)

Chemistry:

	(1)	(2)
Pd	55.0	57.3
Pt	7.5	
Au		3.6
Pb	25.0	38.7
Sn	12.0	
Cu	1.0	
Fe	1.0	
Ni	1.0	
Total	102.5	99.6

(1–2) Noril'sk region, USSR; by electron microprobe, averages.

Occurrence: As small irregular grains and veinlets in copper sulfides, associated with differentiated gabbro-diabase intrusives (Noril'sk region, USSR).

Association: Pt–Fe alloy, polarite, talnakhite, cubanite, pentlandite, magnetite, valleriite, Ag–Au alloy (Noril'sk region, USSR).

Distribution: In the Noril'sk region, western Siberia, USSR. From the Upper and Banded zones of the Stillwater Complex, Montana, USA.

Name: For Professor Orest Evgenevich Zvyagintsev, who did geochemical research on the platinum metals.

Type Material: n.d.

References: (1) Genkin, A.D., I.V. Murav'eve, and N.V. Troneva (1966) Zvyagintsevite, a natural intermetallic compound of palladium, platinum, lead and tin. Geol. Rudn. Mestorozhd., 8, 94–100 (in Russian). (2) (1967) Amer. Mineral., 52, 299 (abs. ref. 1). (3) Cabri, L.J. and R.J. Traill (1966) New palladium minerals from Noril'sk, western Siberia. Can. Mineral., 8, 541–550. (4) (1967) Amer. Mineral., 52, 1587 (abs. ref. 3). (5) Cabri, L.J., Ed. (1981) Platinum group elements: mineralogy, geology, recovery. Can. Inst. Min. & Met., 147.